FERROELECTRICS – MATERIAL ASPECTS

Edited by **Mickaël Lallart**

Ferroelectrics – Material Aspects
Edited by Mickaël Lallart

Published by InTech
Janeza Trdine 9, 51000 Rijeka, Croatia

Publishing Process Manager Silvia Vlase

Technical Editor Teodora Smiljanic

Cover Designer InTech Design Team

Additional hard copies can be obtained from orders@intechopen.com

Ferroelectrics – Material Aspects, Edited by Mickaël Lallart
p. cm.
ISBN 978-953-307-332-3

For EU product safety concerns:
IN TECH d.o.o., Prolaz Marije Krucifikse Kozulić 3, 51000 Rijeka, Croatia,
info@intechopen.com or visit our website at intechopen.com.

Meet the editor

Dr. Mickaël Lallart graduated from Institut National des Sciences Appliquées de Lyon (INSA Lyon), Lyon, France, in electrical engineering in 2006, and received his Ph.D. in electronics, electrotechnics, and automatics from the same university in 2008, where he worked for the Laboratoire de Génie Electrique et Ferroélectricité (LGEF). After working as a post-doctoral fellow in the Center for Intelligent Material Systems and Structures (CIMSS) in Virginia Tech, Blacksburg, VA, USA in 2009, Dr. Lallart has been hired as an Associate Professor in the Laboratoire de Génie Electrique et Ferroélectricité. His current field of interest focuses on electroactive materials and their applications, vibration damping, energy harvesting and Structural Health Monitoring, as well as autonomous, self-powered wireless systems.

Contents

Preface

Ferroelectricity has been one of the most used and studied phenomena in both scientific and industrial communities. Properties of ferroelectrics materials make them particularly suitable for a wide range of applications, ranging from sensors and actuators to optical or memory devices. Since the discovery of ferroelectricity in Rochelle Salt (which used to be used since 1665) in 1921 by J. Valasek, numerous applications using such an effect have been developed. First employed in large majority in sonars in the middle of the 20th century, ferroelectric materials have been able to be adapted to more and more systems in our daily life (ultrasound or thermal imaging, accelerometers, gyroscopes, filters…), and promising breakthrough applications are still under development (non-volatile memory, optical devices…), making ferroelectrics one of tomorrow's most important materials.

The purpose of this collection is to present an up-to-date view of ferroelectricity and its applications, and is divided into four books:

- *Material Aspects,* describing ways to select and process materials to make them ferroelectric.
- *Physical Effects,* aiming at explaining the underlying mechanisms in ferroelectric materials and effects that arise from their particular properties.
- *Characterization and Modeling,* giving an overview of how to quantify the mechanisms of ferroelectric materials (both in microscopic and macroscopic approaches) and to predict their performance.
- *Applications,* showing breakthrough use of ferroelectrics.

Authors of each chapter have been selected according to their scientific work and their contributions to the community, ensuring high-quality contents.

The present volume aims at exposing the material aspects of ferroelectric materials, focusing on synthesis (chapters 1 to 8), emphasizing the importance of adapted methods to obtain high-quality materials; effect of doping and composite design and growth (chapters 9 to 13), showing how the ferroelectric activity may be significantly enhanced by the addition of well-chosen materials; lead-free materials (chapters 14 to 20), addressing the importance of environmentally friendly devices; and ferroelectric thin films (chapters 21 to 23), which show particular effects due to their size and attracted much attention over the last few years.

I sincerely hope you will find this book as enjoyable to read as it was to edit, and that it will help your research and/or give new ideas in the wide field of ferroelectric materials.

Finally, I would like to take the opportunity of writing this preface to thank all the authors for their high quality contributions, as well as the InTech publishing team (and especially the publishing process manager, Ms. Silvia Vlase) for their outstanding support.

June 2011

Dr. Mickaël Lallart
INSA Lyon, Villeurbanne
France

Part 1

Preparation and Synthesis

1

BST and Other Ferroelectric Thin Films by CCVD and Their Properties and Applications

Yongdong Jiang, Yongqiang Wang, Kwang Choi
Deepika Rajamani and Andrew Hunt
nGimat Co.
U.S.A

1. Introduction

Ferroelectric materials, such as $BaTiO_3$ (BTO), $Pb(Zr,Ti)O_3$ (PZT), $SrBi_2Ta_2O_9$ (SBT), and $LiNbO_3$ (LNO), are a category of materials with reorientable spontaneous polarization, a sub-category of pyroelectric materials. Because of their high dielectric constant, large polarization, and high breakdown voltage, ferroelectric materials have a wide range of applications, including infrared (IR) detectors for security systems and navigation, high density capacitors, high-density dynamic random access memory (DRAM), non-volatile ferroelectric random access memory (FRAM), and high frequency devices such as varactors, frequency multipliers, delay lines, filters, oscillators, resonators and tunable microwave devices (Tagantsev, et al., 2003; Cole, et al., 2000; Bao, et al., 2008; Gevorgian, et al., 2001; Dawber, et al., 2005).

Among these ferroelectric materials, BTO based films with Sr dopant, namely $Ba_{1-x}Sr_xTiO_3$ (BST) are the most investigated one for various applications, especially for electric field response (or tunable) components and devices because of its high dielectric constant, reasonable dielectric loss, high tunability, and large breakdown strength. The Curie temperature T_c can be easily adjusted by controlling the Ba to Sr ratio. Studies have revealed that the electrical properties of BST films are influenced by the deposition and post-deposition process, stoichiometry, electrodes, microstructure, thickness, surface roughness, oxygen vacancies in films, and film homogeneity. The composition of the BST film such as the (Ba+Sr)/Ti ratio plays a critical role in determining its electrical properties (Y. H. Xu, 1991; Takeuchi, et al., 1998; Im, et al., 2000). Both the dielectric constant and loss increased with increasing (Ba+Sr)/Ti ratio. The lowest loss tangent (0.0047) and the best figure of merit were achieved with a (Ba+Sr)/Ti ratio of 0.73, but tunability was diminished (Im, et al., 2000). *n*Gimat has also optimized the elemental ratios to achieve some of the highest figures of merit in tunable devices using the enhancements thus optimized.

It has also been reported that dopants influence the electrical properties of BST thin films, but all dopants negatively affect at least one of the desired properties of the solicitation (Copel, et al., 1998 and Chung, et al., 2008). Copel and coworkers (Copel, et al., 1998) investigated the effect of Mn on electrical properties of BST thin films and found that leakage current was improved by introducing Mn. This was attributed to the acceptor Mn doping increasing the depletion width in BST films and the barrier for thermionic emission from a Pt contact into the BST film. Takeuchi and coworkers (Takeuchi, et al., 1998) studied several BST dopants using their combinatorial synthesis technique. The experimental results showed that both W and Mn in small amounts reduced the leakage current dramatically while only slightly decreasing dielectric constant. It was theorized that the W substituted for Ti as a donor and suppressed the formation of oxygen vacancies. *n*Gimat has studied numerous dopants and uses dopants in almost all applications.

Although much success has been made in optimizing physical properties of uniform composition FE materials, especially BST, for various applications, these materials still suffer from decreased performance such as low tunability and high loss in high frequency range. Therefore, compositionally graded and multilayered FE thin films have been attracting much attention in past few years (Zhong, et al., 2007; Misirlioglu, et al., 2007; Katiyar, et al., 2005; Kang, et al., 2006; Pintilie, et al., 2006; Lu, et al., 2008; Liu, et al., 2007; Heindl, et al., 2007). As an example, Zhong (Zhong, et al., 2008) deposited multilayered BST films on Pt/Si substrates. The multiplayer heterostructures consisted of three distinct layers with Ba/Sr ratios of 63/37, 78/22, and 88/12. The first composition is paraelectric while the last two are ferroelectric at room temperature. The film structure has a dielectric constant of 360 with a dielectric loss of 0.012 and a tunability of 65% at 444 kV/cm. These properties exhibited minimal dispersion between -10 and 90°C. As known, while the dielectric loss in BST films can be greatly reduced by various dopants, tunability of monolithic BST is strongly dependent on the temperature. Multilayer and graded FEs display little temperature dependence due to the variations in T_C that results in a diffuse phase transformation. The tunability can be maximized by optimizing the internal electric fields that arise between layers due to the polarization mismatch. *n*Gimat's tunable materials normally consists of at least two compositional layers, with one being <10nm thick.

This chapter covers the following areas: introduction to the CCVD process, depositions and properties of BST, PZT, and $CaCu_3Ti_4O_{12}$ thin films, and fabrication and performance of tunable microwave devices based on BST thin films.

2. Introduction to CCVD

Combustion Chemical Vapor Deposition (CCVD) (Andrew, et al., 1993, 1997, 1999) is an open atmosphere deposition process in which the precursors are dissolved in a solvent, which typically also acts as the combustible fuel. This solution is then atomized to form submicron droplets, which are then conveyed by an oxygen-containing stream to the flame using the Nanomiser® device. In CCVD of thin films, the substrate is coated by simply drawing it over the flame plasma, as shown Figure 1. The flame provides energy required for the precursors to react and to vapor deposit on the substrate. Substrate temperature is an independent process parameter that can be varied to actively control the deposited film's microstructure. Although flame temperatures are usually in excess of 800 °C, the substrate may dwell in the flame zone only briefly, thus remaining cool (<100°C). Alternatively, the substrate can be either allowed to rise in temperature or easily cooled in the open atmosphere. *n*Gimat has utilized its patented CCVD process in depositing over 100 distinct materials compositions for a variety of applications. Due to the inherent compositional flexibility of the NanoSpray Combustion Process, we can fabricate thin films, nanopowders, and composites from a wide range of metals, ceramics, and polymers, as illustrated by the examples in Table 1.

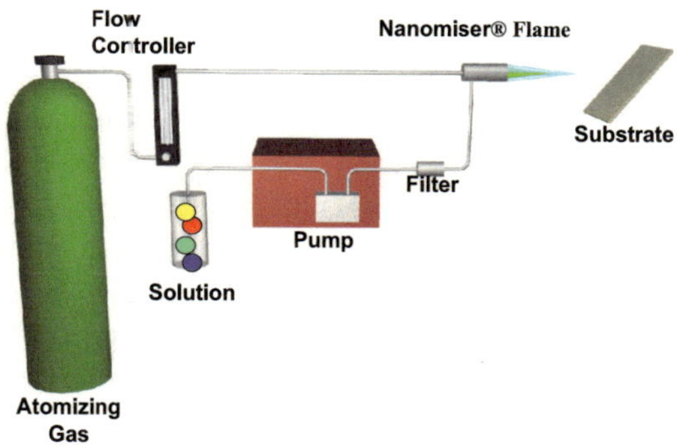

Fig. 1. Schematic of the CCVD system, the thin film NanoSpray combustion process

Metal	Ceramics	Composites
Ag, Au, Cu, Ir, Ni, Rh, Pt, Zn.	Complex oxides: $(Ba,Sr)TiO_3$, $(Pb,La)(Zr,Ti)O_3$, $(La,Sr)CoO_3$, $Pb(Mg,Nb)O_3$, Spinels, $YBa_2Cu_3O_x$, $YbBa_2Cu_3O_x$, $LaAlO_3$, ITO, $Y_3Fe_5O_{12}$, $SrRuO_3$, ZrO_2, Simple oxides: Al_2O_3, SiO_2, Ta_2O_5, In_2O_3, ZnO, ZrO_2, V_2O_5, WO_3, CeO_2, Cr_2O_3, Cu_xO, Fe_2O_3, MgO, Mn_2O_3, MoO_3, Nb_2O_5, NiO, RbO_x, RhO_x, RuO_2, TiO_2	Polymer/metal Polymer/ceramic Ceramic/metal
Substrates Used		
Single crystal ceramics: Si, sapphire, $LaAlO_3$, MgO, $SrTiO_3$, yttrium stabilized ZrO_2, quartz Polycrystalline ceramics: SiC, Si_3N_4, Al_2O_3, silica Metals: platinized Si wafers, Cu, Al, Ag, Pt, Ni, steel, NiCr, superalloys, Ti, TiAl alloy Polymers: Nafion™, Teflon™, polycarbonate		
Applications		
Capacitors, resistors, catalytic applications, corrosion resistance, electronics, engines, ferroelectrics, solar cells, fuel cells, optics, piezoelectrics, buffer layers, superconductors, thermal barrier, thermal control, and wear resistance		

Table 1. Partial list of materials deposited by CCVD

3. Depositions of ferroelectric thin films by CCVD

Many ferroelectric materials, such as BST and PZT, have been deposited successfully by the CCVD technique. These ferroelectric thin films are grown epitaxially on sapphire, single crystal MgO, and single crystal $SrTiO_3$ (STO) substrates.

3.1 Depositions of BST thin films by CCVD and their properties

Compared to polycrystalline or textured thin films, epitaxial dielectric thin films show higher dielectric breakdown and lower dielectric loss. Therefore, epitaxial thin films are preferred for many applications, especially for high frequency microwave applications. Single layer BST and multilayer dielectric thin films have been successfully deposited on sapphire (both *c*- and *r*-orientations). Figure 2 shows typical plan view and cross sectional images on a single layer BST thin film of *c*-sapphire substrate by CCVD. The film is dense and smooth with uniform grains and thickness. Figure 3 shows an area detector XRD pattern and a (110) pole figure of a typical BST thin film on *c*-sapphire. Epitaxy can be determined in about 15 min by area detector XRD. The sample is rotated continuously in ϕ and scanned in ω during signal collection so that all peaks are excited. The (006) plane of sapphire is parallel to the substrate surface and perpendicular to the $\theta/2\theta$ direction. 2θ increases from the right side to the left side. The area detector XRD pattern shows that there are only (111) peak of the BST film and (006) peak of sapphire along the $\theta/2\theta$ direction. The (110) and (111) peaks of the BST film appear as dots and align with (104) and (006) peaks of sapphire, showing the BST film was grown epitaxially on *c*-sapphire substrate. The epitaxiy of the BST film is further confirmed by the (110) pole figure as shown in Figure 3 (b). Pole figure measurement is a powerful method to determine the in-plane alignment between the epitaxial film and its substrate in a relatively large area. BST (110) reflections were selected to perform the pole figure collection and to detect the presence of the in-plane alignment because of its large 2θ separation from the sapphire (104) plane. As shown in Figure 3 (b), six sharp spots of the BST (110) reflections with narrow intensity distribution were observed every 60° along the ϕ direction. These results indicate clearly that the BST thin film was epitaxially grown on *c*-sapphire substrate and has (111) plane parallel to the substrate surface. The orientation relationship between the BST film and *c*-sapphire substrate is BST (111)//sapphire (0001) and BST [110]//sapphire [104]. The pole figure measurements suggest a type (2) or type (3) epitaxial growth of BST film on *c*-sapphire substrate (Baringay & Dey, 1992).

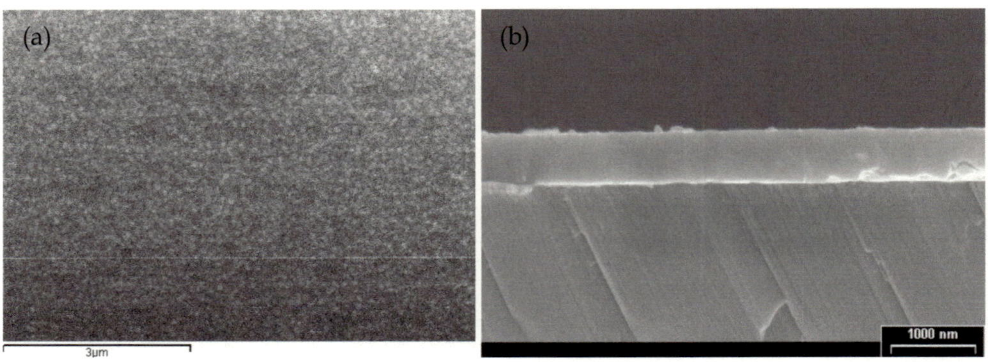

Fig. 2. SEM (a) plan view and (b) cross section images of typical BST thin films by CCVD

Inter-digital capacitors (IDC) with an 8 μm gap between electrodes and co-planar waveguide (CPW) structures were fabricated on the epitaxial BST dielectric thin films by the lift-off process. Dielectric properties were measured on the IDC structures at 1 MHz by a HP 4285A LCR meter. Its tuning and dielectric loss as a function of applied voltage are present in Figure 4. The tuning increases while the dielectric loss decreases with the increase of applied voltage. At an applied voltage of 40 V (which is the limit of the instrument), a tuning of 51% and a dielectric loss of 0.0046 were achieved. The dielectric constant of the film is about 1150.

In addition to single layer BST dielectric thin films, nanostructured multilayer dielectric thin films with alternative ferroelectric and paraelectric phases with a thickness in nanometer range have also been successfully deposited onto various single crystal substrates including *c*-sapphire, single crystal MgO, and single crystal STO, et al. Figure 5 shows the SEM image and area detector XRD pattern of a multilayer dielectric thin film with 36 alternative ferroelectric and paraelectric nano-layers and a total thickness of 500 nm. The film is dense and smooth with uniform fine grains. The XRD pattern shows that the (110) and (111) peaks of the multilayer dielectric film appear as dots, aligning with (104) and (006) peaks of sapphire (the (006) plane is parallel to the substrate surface), showing the multilayer dielectric film was grown epitaxially on the *c*-sapphire substrate as single layer thin films.

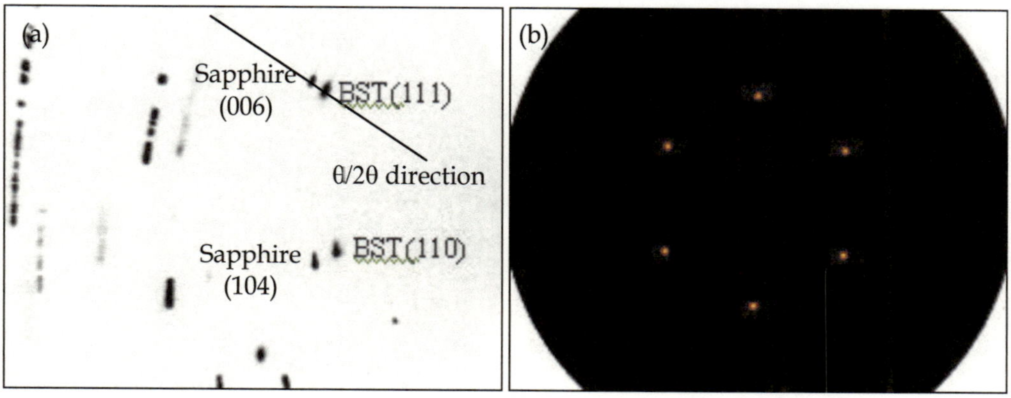

Fig. 3. (a) Area detector XRD pattern and (b) (110) pole figure of a typical single layer BST film on *c*-sapphire substrate

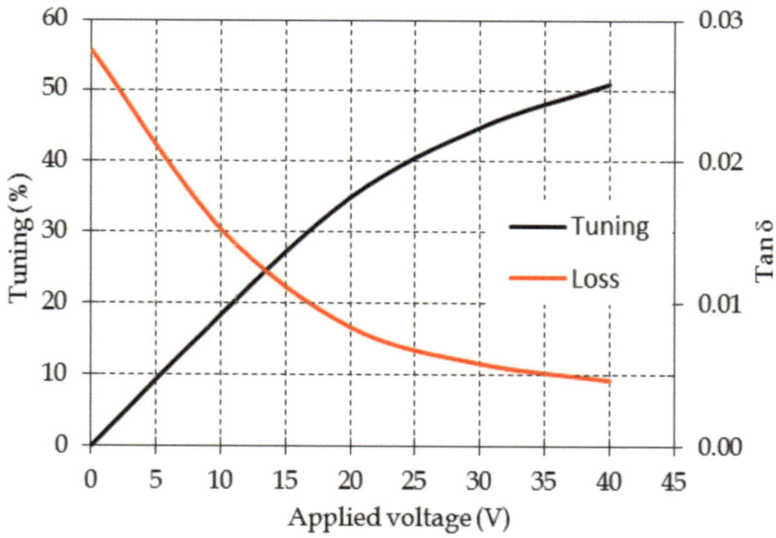

Fig. 4. Tuning and dielectric loss of a single layer BST film on *c*-sapphire substrate as a function of applied voltage

The same IDC and CPW structures were fabricated on the multilayer thin films. The dielectric and microwave properties of a selected multilayer thin film and a standard single layer film are summarized in Table 2. The multilayer thin film has a slightly lower capacitance at 1 MHz compared to the standard single layer film. However, its dielectric loss at 1 MHz and 0 V is about 0.005, which is much lower than that of the single layer thin film (0.028). The figure of merit (FOM), which is defined as (tuning × capacitance)/loss tangent, of the multilayer film is about 3 times as high as that of the standard single layer film. The high FOM and low dielectric loss benefit the applications for high frequency and high power microwave devices.

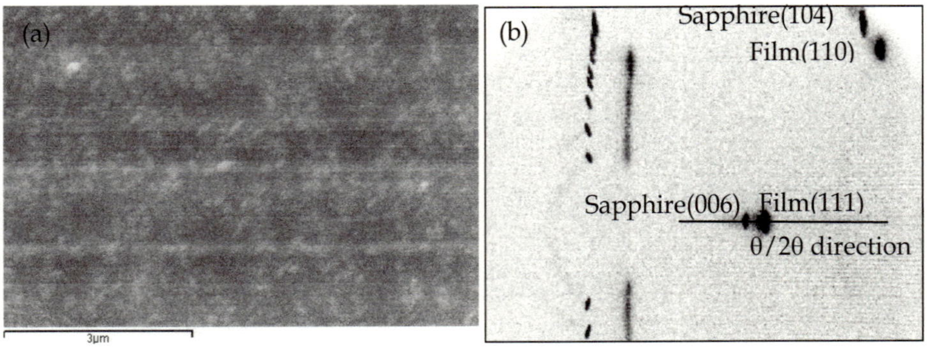

Fig. 5. (a) SEM image and (b) XRD pattern of a multilayer dielectric thin film

Sample ID	Capacitance and loss at 1 MHz				S_{21} at 50 GHz (dB)	Tuning (%)	FOM
	0 V		40 V				
	Cp (pF)	Tanδ	Cp (pF)	Tanδ			
Multilayer	1.14	0.005	0.90	0.003	1.46	21.1	4820
Single layer	1.28	0.028	0.85	0.019	-	33.3	1537

Table 2. Comparison of electrical properties between a multilayer and a single layer film

The tunable BST dielectric thin films, both single layer and multilayer, have been scaled up to 2″ round sapphire wafers. For depositing BST thin films on 2″ wafers, the substrate is maintained at a uniform temperature in a furnace. The substrate rotates on a vacuum chuck and the flame impacts the wafer at a 45° angle through a cutout on the side of the furnace, in which smooth and dense epitaxial thin films are deposited, as shown in Figure 6.

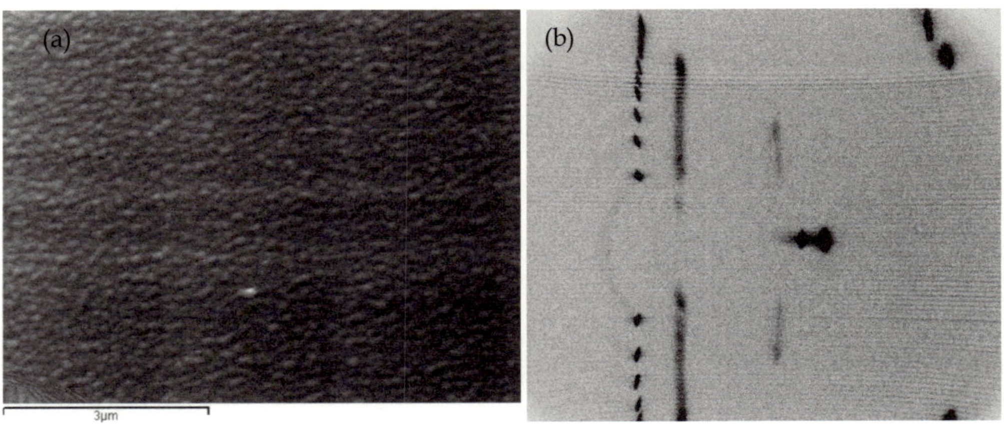

Fig. 6. (a) SEM image and (b) area detector XRD pattern of a BST thin film on 2" *c*-sapphire wafer

3.2 Depositions of PZT thin films by CCVD and their properties

Lead-based ferroelectric materials such as lead zirconate titanate ($Pb(Zr,Ti)O_3$, PZT), a member of the perovskite structure family, is a solid solution of lead titanate ($PbTiO_3$, PTO) and lead zirconate ($PbZrO_3$, PZO) with different Zr/Ti ratios. It is well known that their physical properties can be modified by changing the Zr/Ti ratio and substituting a part of Pb ion by tri-valent ions. Among the tri-valent dopants, lanthanum (La) has been found the most suitable element for increasing the density and other physical properties of the materials (Rukmini et al., 1999; Dimos et al., 1994). PZT and La doped PZT ($Pb_{1-x/100}La_{x/100}(Zr_{y/100}Ti_{z/100})O_3$, PLZT x/y/z) have been extensively investigated for applications, such as DRAM (Hwang et al., 1999; H. H. Kim et al., 1998), FRAM (Ramash et al., 2001; W. S. Kim et al., 1999), sensors and actuators for microelectromechanical systems (MEMS)(B. M. Xu, 1999; Polla and Francis, 1998), infrared detectors (Song et al., 2001; Kobune et al., 2001), due to their excellent dielectric, ferroelectric, piezoelectric, and pyroelectric properties. PLZT is transparent in the visible and near infrared region of the electromagnetic waves and has excellent electro-optical properties. Therefore, it is widely used in electro-optic modulators (Haretling, 1999; Dimos, 1995), and optical displays (Uchino, 1995; Moulson and Herbert, 1997). With the rapid development of optical telecommunications and optical networks, the electro-optical applications of PLZT materials are becoming more and more important. For these applications, it is essential to grow a highly oriented or epitaxial microstructure in order to reduce optical loss, which is mainly caused by light scattering at grain boundaries because of the inhomogeneous refractive indices. Thus the synthesis and processing of epitaxial PLZT thin films have been investigated intensively.

In *n*Gimat, PLZT thin films with various La contents and Zr to Ti ratios have been grown epitaxially on *c*-sapphire substrate with an epitaxial $Pb_{1-x}La_xTiO_3$ (PLT) seed layer. As known, Sapphire has a different crystal structure than $LaAlO_3$ (LAO) and MgO, which have cubic structure and are common substrates for PLZT thin films. The lattice mismatch between PLZT and sapphire is much larger than those between PLZT and LAO or MgO. A PLT seed layer with cubic structure can promote the epitaxial growth of PLZT films on sapphire substrates. Figure 7 shows the XRD patterns, which were created by *Chi* integration along the substrate normal from area detector XRD patterns, of PLZT thin films with various La contents and Zr to Ti ratios on *c*-sapphire substrate. It is clear that the XRD patterns of the PLZT 20/30/70, PLZT 17/40/60, and PLZT 17/50/50 thin films show only (111) peaks, indicating that these films inherited the epitaxy of PLT seed layer and were grown epitaxially on *c*-sapphire substrate with (111) plane parallel to the substrate surface. However, the XRD pattern of the PLZT 15/30/70 film shows small extra peaks of (100) and (110), and those of the PLZT 12/40/60 and PLZT 15/50/50 films show extra (110) peaks, suggesting that these films were grown preferentially with multi-orientations parallel to the substrate surface. Further studies showed that for PLZT films with a Zr to Ti ratio of 30:70, 40:60, and 50:50, when La content is lower than 20, 17, and 16 mol.%, respectively, the PLZT film grew preferentially with multiple out-of-plane orientations with or without randomly oriented grains in plane, which depends on the composition. According to the PLZT phase diagram (Haertling and Land, 1971), for PLZT materials with a Zr to Ti ratio of 30:70, 40:60, and 50:50, when La content is lower than about 20, 17, and 16 mol.%, respectively, the crystal structure is tetragonal. Therefore, the large lattice mismatch between PLZT and *c*-sapphire limits the epitaxial growth of PLZT thin films with these compositions on *c*-sapphire substrate. However, the PLZT films with these compositions can be grown epitaxially on $SrTiO_3$ (100) or *r*-sapphire with (100) or (110) plane parallel to the substrate surface, respectively (Yoon et al., 1994).

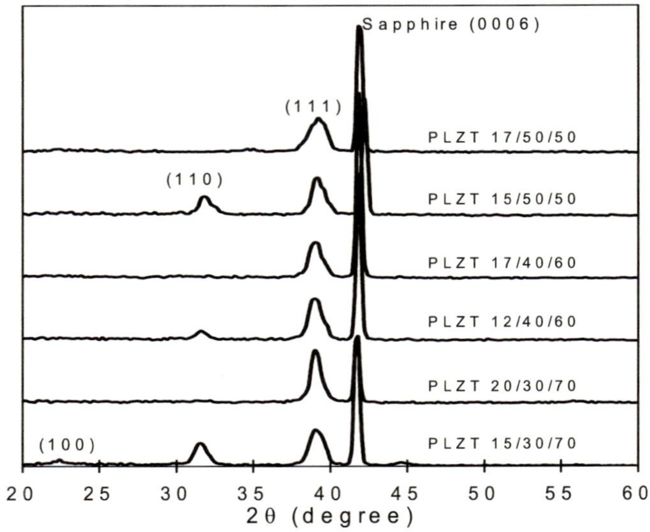

Fig. 7. XRD patterns of the PLZT thin films with various La contents and Zr to Ti ratios by CCVD on *c*-sapphire substrate with a PLT seed layer

Pole figure measurements were performed on the epitaxial films using (110) reflections. The pole figure of a PLZT 20/30/70 thin film is shown in Figure 8. As BST thin films on *c*-sapphire, the PLZT thin film shows six sharp dots of (110) poles with narrow density distributions, which is similar to that of the PLT seed layer. There is no broadening or satellite found from the pole figure, suggesting an excellent crystallinity. The PLZT films grew off the PLT seed layer and keep the crystallographic orientations. The orientation relationship between the PLZT thin film and *c*-sapphire substrate is PLZT (111)//sapphire (001) and PLZT [110]//sapphire [104].

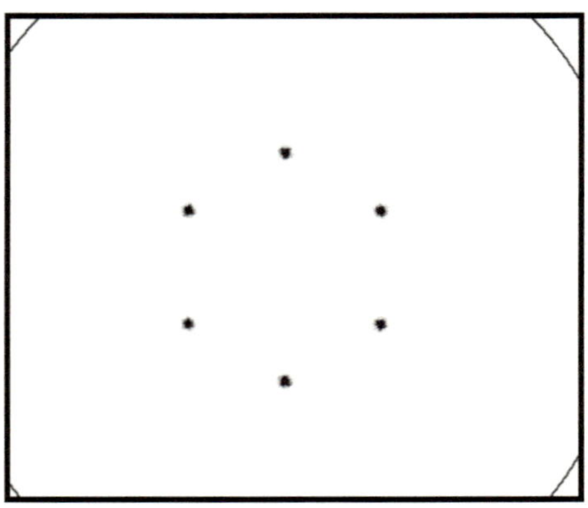

Fig. 8. (110) pole figure of a PLZT 20/30/70 thin film on *c*-sapphire substrate with a PLT seed layer

Figure 9 shows the SEM micrographs of the PLZT thin films with different compositions. These PLZT thin films inherit the microstructure of the PLT seed layer. They contain uniformly distributed fine grains less than 100 nm in size. The film morphology is strongly influenced by the film composition. For the film with a Zr to Ti ratio of 50/50 (not shown in the figure), the grains are not closely packed. Voids and pores were formed in this film. With the increase of Ti content and the decrease of Zr content, the film density increases and the grain size decreases. For the film PLZT 20/30/70, there is no pin hole formed. All of these films are crack free. It is also noticed that particles were formed on these films, which may be attributed to poor atomization or high flame temperature. Further studies show that without the PLT seed layer the PLZT films deposited at the same conditions contain multiple out-of-plane orientations or are random. Pyrochlore phase was also formed at these conditions for variety of compositions without a PLT seed layer. Therefore, the PLT seed layer can markedly enhance the formation kinetics of perovskite phase and improve the crystallization behavior of the PLZT thin films subsequently deposited. The probable cause for the presence of pyrochlore phase is that the lattice mismatch between the sapphire substrate and PLZT thin films hinders the phase transformation process (Kao et al., 2003).

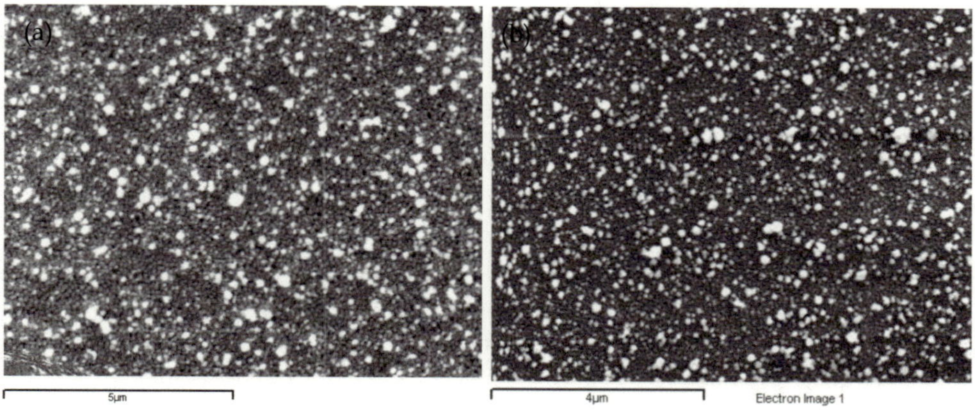

Fig. 9. SEM images of PLZT thin films with different La contents and Zr to Ti ratios on *c*-sapphire substrates with a PLT seed layer, (a) PLZT 17/40/60 and (b) PLZT 20/30/70

The optical properties of these PLZT thin films were measured by a spectrometer in the visible and near infrared regions. For composition, all the transmittance was normalized to the sapphire substrate. As shown in Figure 10, it is found that all the three films have a transmittance of higher than 70% and 90% in the visible region and near infrared region, respectively. High transmittance is necessary for optical applications such as optical modulators and switches. The waveguiding modes and refractive indices of these thin films will be measured by a prism coupler later.

3.3 Depositions of CCT thin films by CCVD and their properties

In recent years, $CaCu_3Ti_4O_{12}$ (CCT) has been attracting much attention due to its extraordinary high dielectric constant of about 10^5 at room temperature and very small temperature dependence of the dielectric constant over a wide temperature range from 100 to 600 K (Subramanian et al., 2000, 2002; Ramirez et al, 2000; Home et al., 2001; Adams, et al., 2002; Sinclair et al., 2002; Maurya, et al. 2008; Prakash, et al., 2008; Zhu, et al., 2008; Kim, et al., 2010). CCT and its family, $ACu_3Ti_4O_{12}$ (A = rare earth or other alkali earth element), were first identified in 1967 (Deschavnres et al., 1967). Since then, this family has been expanded. Its accurate structure was determined in 1979 (Bochu et al., 1979). CCT has a body-centered cubic structure with a centro-symmetric space group *Im3* and two formula units per unit cell. Its cubic structure is related to that of perovskite ($CaTiO_3$), but the TiO_6 octahedra are tilted to produce a square planar environment for Cu^{2+}. Cu atoms are bonded to the four oxygen atoms and the large Ca atoms are without bonds.

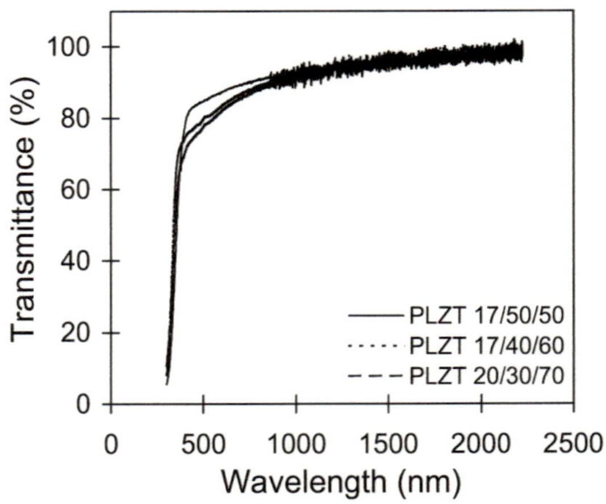

Fig. 10. Optical transmittance spectra of PLZT thin films with various La contents and Zr to Ti ratios grown on *c*-sapphire substrate with a PLT seed layer

Subramanian and coworkers (Subramanian et al., 2000, 2002) prepared CCT based ceramics by sintering related powders. A dielectric constant of higher than 10^5 was achieved at room temperature. The dielectric constant increases rapidly with the increase of temperature and reaches 3×10^5 at 450°C. Based on Subramanian's work, Ramirez and coworkers (Ramirez, et al., 2000) extended the measurement temperature range to cryogenic values and additional measurements were performed on the CCT compound. Dielectric measurements showed that over the temperature range of 100-380 K, ε_r is higher than 10000 and only weakly temperature dependent at 1 kHz.

While its extraordinary dielectric constant has generated huge interest, the origin of the high dielectric constant and its sharp decrease at 100 and 600 K has also attracted intensive studies. Many studies argue against an explanation in terms of ferroelectricity since there has been no phase or structure transition observed. Several other intrinsic physical mechanisms suggested include high tension on Ti-O bonds (Subramanian, et al., 2000), highly polarizable relaxational modes (Ramirez, et al., 2000), and a relaxor-like dynamical slowing down of dipolar fluctuations in nanosize domains (Home, et al., 2001). However, it was also suggested that the giant dielectric constant of this material may be enhanced by its microstructure such as the barrier layer mechanism (Subramanian, et al., 2000; Adams, et al., 2002; Sinclair, et al., 2002; Li, et al., 2009). In their studies, impedance spectroscopy measurements show that CCT ceramics are electrically inhomogeneous, contains semiconducting grains with insulating grain boundaries that is the desired electrical microstructure for internal barrier layer capacitors (BLC).

Although excellent electrical properties have been achieved on CCT bulk ceramics, for microelectronic applications, thin films are preferred since thin films can provide a higher level of integration than can be achieved with discrete components by bulk materials, and hence the devices are faster, lighter, and of lower cost. Furthermore, single-layer, thin-film devices intrinsically have lower inductance than multilayer capacitors because of the high mutual inductance between the internal counter electrodes (Dimos and Mueller, 1998). *n*Gimat has successfully deposited CCT thin films on various single crystal substrates, including single crystal STO, single crystal $LaAlO_3$ (100), *c*-sapphire, and *r*-sapphire, by its proprietary CCVD process. Materials and electrical properties were characterized.

3.3.1 CCT thin films on STO substrate

CCT has a body-centered-cubic structure with a lattice parameter, *a*, of 7.393 Å. STO has a cubic perovskite structure with a lattice constant of 3.905 Å. STO single crystal is one of the most popular substrates for high temperature superconductors and other electronic materials because of its high thermal and chemical stabilities and compatible lattice constant and structure. In this study, stoichiometric CCT thin films with different thickness were first grown on STO (100) substrates between 950 and 1025°C. The XRD spectra integrated along the substrate normal direction from the area detector XRD patterns are shown in Figure 11. It is clear that the main peak of these films is CCT (400) which aligns well with STO (200) diffraction, suggesting a highly (*l*00) preferred growth of the CCT films. There is a small (220) peak for all these films, implying a little portion of random or mis-oriented grains with (220) plane parallel to the substrate surface. Dielectric constant as a function of film thickness is shown in Figure 12. The dielectric constant of the CCT films on STO substrates decreases with the increase of film thickness. At a thickness of 45 nm, which was deposited at 950°C for 10 min, the dielectric constant is about 45,000, while at a thickness of 400 nm, which as deposited at 1025°C for 40 min, the dielectric constant is about 5,900. The dielectric constant is approximately proportional to the reciprocal of film thickness based on the simulation of the data points. The lower dielectric constant of the thicker films deposited at higher temperatures could also be caused by the higher stress between the CCT film and the substrate, which needs to be further understood.

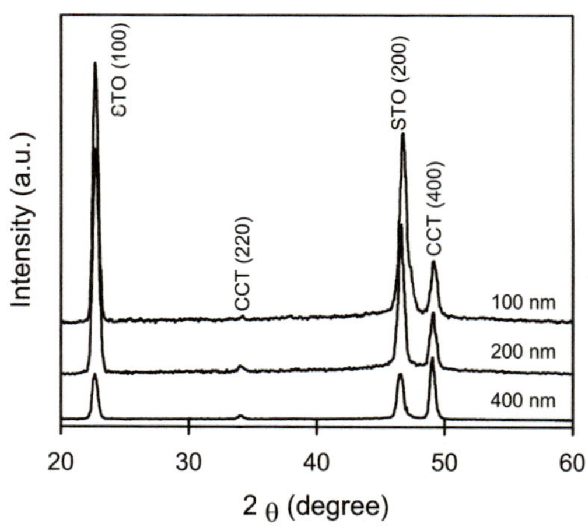

Fig. 11. XRD spectra of CCT thin films deposited at 1025°C on STO substrate with different thickness

The frequency and bias voltage dependence of dielectric constant and quality factor of a 45 nm thick CCT film on STO substrates are shown in Figure 13. In the tested frequency range, dielectric constant decreases slightly while quality factor increases with the increase of bias voltage. For example, at 1 MHz and 0 V, the dielectric constant and quality factor are 45,320 and 730, respectively. When applying a bias voltage of 40 V, at the same frequency, the dielectric constant and quality became 44,950 and 950, respectively. In tested voltage range, when frequency is lower than 1 MHz, dielectric constant decreases gradually with the increasing of frequency. It shows a sharp decrease at the frequency of 10 MHz at all the tested voltages. However, quality factor increases rapidly with the increase of frequency in the tested voltage range.

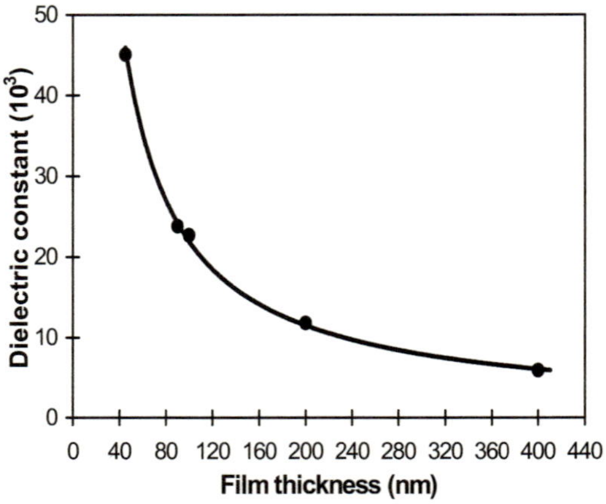

Fig. 12. Dielectric constant of CCT films on STO (100) substrates as a function of film thickness

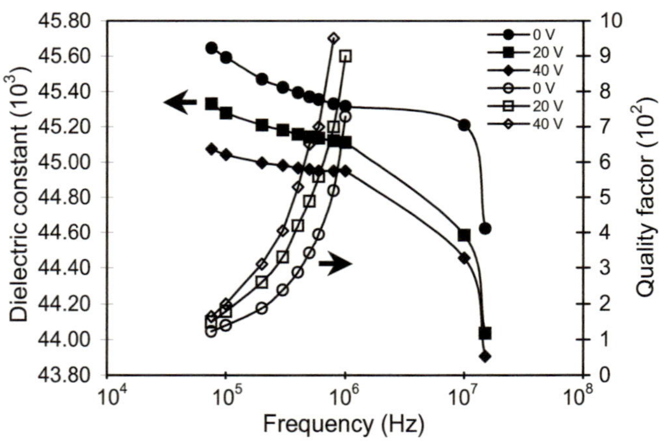

Fig. 13. Dielectric constant and quality factor of a CCT film, deposited at 950°C for 10 min on STO (100) substrate, as a function of frequency at different bias voltages

3.3.2 CCT thin films on *c*-sapphire substrate

Sapphire (α-Al_2O_3) has a hexagonal crystal structure with a = 4.759 Å and c = 12.99 Å. It is widely used to deposit ferroelectric materials for electrooptic applications and radiation hardened electronic components. Sapphire has a different crystal structure than LAO or STO. Also, the lattice mismatch between CCT and sapphire is much larger than those between CCT and LAO or STO. Therefore, the growth of CCT on sapphire is different from that of CCT on LAO or STO. Stoichiometric CCT films with different thickness were deposited onto *c*-sapphire at a temperature of 1025°C. Figure 14 (a) shows the SEM image of a 90 nm thick CCT film. The base layer of the film is dense and smooth with uniform and fine grains. There are a few hillocks on its surface. With the increase of film thickness, the films become rougher and grains become larger. The area detector XRD patterns show preferred growth of these CCT films with multiple orientations, including (220), (222), and (400), which is consistent with the corresponding morphology with square and pyramid shaped grains stacking on each other that can be seen clearly in the thicker films. This is confirmed by the integrated XRD spectra along the substrate normal, shown in Figure 14 (b), which match the characteristics of the CCT compound. There are also some in-plane and out-of-plane random grains formed in these films. It is noticed that (310) and (321) diffractions became stronger with the increase of film thickness as well.

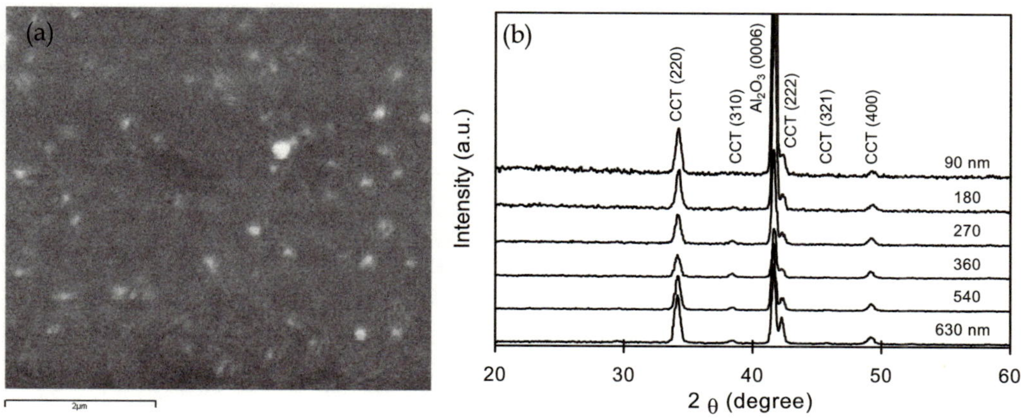

Fig. 14. (a) SEM image of a 90 nm thick CCT film and (b) XRD spectra of CCT films with different thickness

The dielectric constant of CCT films on *c*-sapphire substrates, as shown in Figure 15, decreases with the increase of film thickness, which is similar to those of CCT on STO and LAO, but with lower values. The dielectric constant is about 930 for a 90 nm thick film. It decreases to about 235 when film thickness increases to 630 nm.

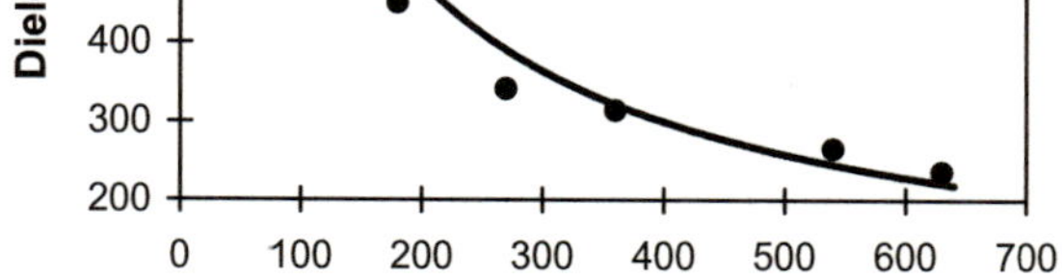

Fig. 15. Dielectric constant of CCT films on *c*-sapphire substrates as a function of film thickness

From the results of CCT films on different substrates, it is clear that CCT films on STO have the highest dielectric constant while the ones on c-sapphire substrates have the lowest. It is worthwhile to point out that the observation of the dielectric constant decreasing with increasing film thickness in this study is contrast to the fact found in other ferroelectric thin films such as $Ba_xSr_{1-x}TiO_3$ (BST) and CCT on platnized Si wafers. Conventionally the dielectric collapse is understood by assuming the existence of "dead layers" with severely depressed dielectric constants at the electrode-dielectric interfaces. These dead layers act as parasitic capacitors in series with the "bulk-like" dielectrics. Hence the decrease in dielectric constant is said to follow the "series capacitor model" (Sinnamon et al., 2002). Various suggestions on the exact nature of the dead layers, although still under debate, have been proposed in these years (Sinnamon et al., 2002; Streiffer et al., 1999), including (1) the interfacial discontinuity affecting the polarization states in the dielectric close to the electrode-dielectric interface (Zhou and Newns, 1997; Natori et al., 1998; as cited in Sinnamon et al., 2002), (2) Schottky barriers formed as a result of mis-match in the band structure between dielectric and electrode (Hwang et al., 1999; Scott, 1999; as cited in Sinnamon et al., 2002), and (3) thickness-related soft-mode hardening (Sirenko et al., 2000, as cited in Sinnamon et al., 2002). The different phenomenon observed in this study could be attributed to different residual stress states, and orientations, etc.

For comparison, Table 3 summarizes the results for all the substrates studied with approximately the same CCT film thickness. For bulk CCT materials, the peak intensity ratio of (220), (222), and (400) is 100:30:80. Therefore, CCT films on STO and LAO substrates are highly (400) preferred while CCT films on sapphire substrates are (220) preferred.

Substrate	Thickness (nm)	Orientation (Intensity ratio)	Dielectric constant
STO (100)	200	(220) + (400) (100:746)	11825
LAO (100)	180	(220) + (400) (100:892)	1188
c-Sapphire	180	(220) + (222) + (400) (100:42:19)	450

Table 3. Comparison between substrates, film thickness, orientation, and dielectric constant

4. Applications of BST based thin films

Telecommunications require the use of filters throughout RF devices. For high frequency receivers, filters must be used to remove signals in unwanted frequency bands to prevent overload of the receiver itself and undesirable interference from signals outside the band of operation. On the transmit side, signal purity must be maintained to minimize interference to other users and the incoming signal, conform with government regulations for radio emissions, and in the case of military applications to minimize detection of the radio source by hostile forces. Current technology in both the commercial and military fields employs ceramic blocks or quartz substrates and complex metallization patterns to define the filter response. The resulting filter structures operate on a narrow set of frequencies, determined at the time of device fabrication. A radar or communications device capable of operating on multiple frequency bands requires complex and expensive banks of filters connected with microwave switches or PIN diodes.

In addition, currently wireless communications can readily be monitored or jammed. Different agencies can't directly contact each other due to assigned specific frequencies. Instead of traditional fixed frequency filters set for specific bands, a new generation of rapidly variable filters will be available to provide tactical capability. These existing issues can be overcome through reconfiguring transmission and reception in a few microseconds using tunable filters. Low cost, low loss, high IIP_3, high speed, robust, and radiation-hard filters can enable wide adaptations. Planar low-voltage capacitor structures with improved power handling capability have been developed using BST films and used in the design and fabrication of tunable filters operated at frequencies of 2, 6-20, and 30-45 GHz. In the past several years, *n*Gimat has worked on materials development, as well as design, fabrication, and testing of tunable ferroelectric thin film based microwave devices.

4.1 RF MEMS filters with tunable bandwidth and tunable center frequency

For this application, a CPW admittance inverter topology was employed to realize a Ka-band tunable filter. Two configurations, wideband (WB) and narrowband (NB), were designed, as shown in Figure 16. The electronic realization of bandwidth control is accomplished by introducing a ground-to-ground connection through the inter-resonator gaps as shown in Figure 16 (b). For proof-of concept, perfect open and short connections were used at first. Figure 17 shows the measured results of wideband and narrowband 3-pole filters without BST capacitors. Their bandwidths are 4% and 8% respectively, with a center frequency of 39.5 GHz.

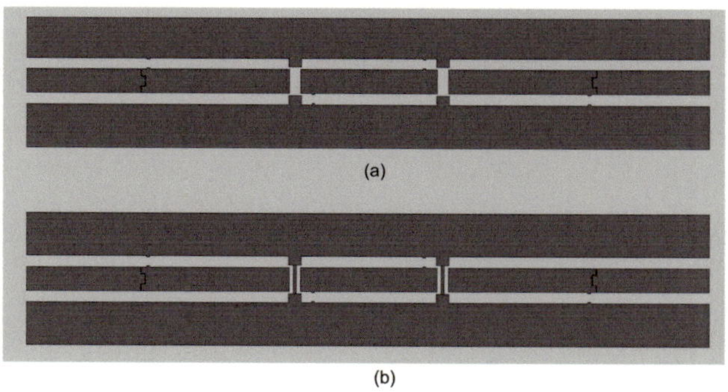

Fig. 16. 3-pole filter topologies: (a) wideband and (b) narrowband configuration

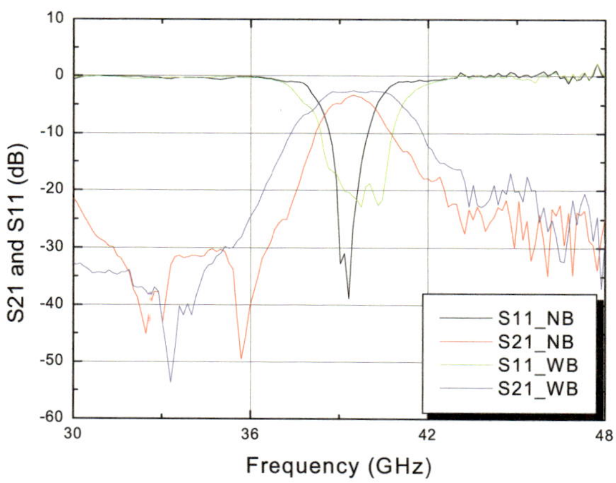

Fig. 17. Measured S parameters of 3-pole wideband and narrowband filters (no BST capacitors)

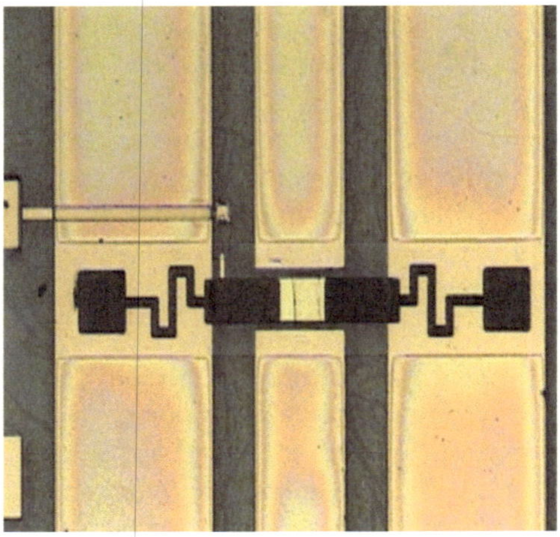

Fig. 18. A 2-pole CPW admittance inverter tunable filter with MEMS switches

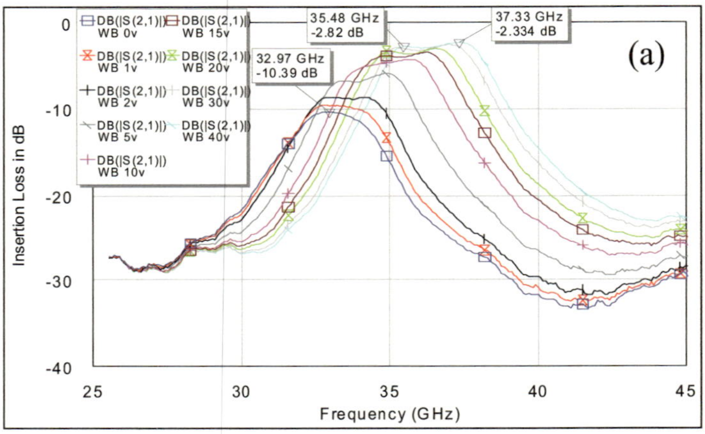

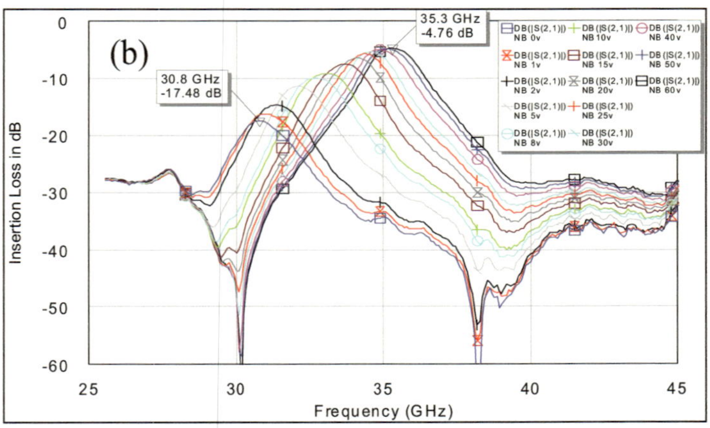

Fig. 19. Measured insertion loss, S_{21}, of a 2-pole CPW admittance inverter filter with MEMS switches (a) at the up and (b) at the down state, each curve representing the insertion loss at one voltage level

RF microelectromechanical systems (MEMS) filters were also designed and fabricated, as shown in Figure 18. Two MEMS switches are used to realize ground-to-ground connection. One bias pad is used to activate the switches. When the MEMS are at the up state, the filter is wideband. A narrowband filter results when the switches are at the down state. The RF MEMS filter was tested up to 45 GHz. Figure 19 (a) presents the insertion loss of the filter. The filter has an insertion loss, S_{21}, of 10.4 dB, a return loss, S_{11}, of ~5 dB, and a bandwidth of 8.3 % at 0 V. The S_{21} and S_{11} are <3 dB and ~10 dB, respectively, when the BST capacitors are biased at 40 V. The center frequency moves >3 GHz (10% tunability). When the MEMS switches are at the down state (narrowband), as shown in Figure 19 (b), the filter has an S_{21} of 17.5 dB, a S_{11} of ~5.3 dB, and a bandwidth of 5 % at 0 V. The S_{21} and S_{11} are <4.8 dB and 17 dB respectively when the BST capacitors are biased to 60 V. The center frequency moves from 30.8 to 35.3 GHz (13.6% tunability). To the best of our knowledge, this is the first time that a BST tunable filter has ever been realized at the mm-wave frequency with such performance.

4.2 CDMA tunable filters operated at the 1.9 GHz band

Code Division Multiple Access (CDMA) is one of the digital cellular technologies deployed worldwide. A prototype 2-pole, 1-zero frequency-agile band-pass CDMA Tx filter was designed and fabricated, as shown in Figure 20 (a). A 2-resonator ceramic block with a BST tunable device fabricated by standard IC technique is used to tune the filter's center frequency. Table 4 shows typical specifications of a US personal communication service (PCS) antenna duplexer. In order to meet these stringent specifications, the fixed-tuned ceramic filters on today's cellular market typically have 6 poles on the receive and 5 on the transmit side. They can be significantly simplified using the tunable filters, which results in smaller physical size and lower insertion loss. Figure 20 (b) shows the CDMA filter response. An insertion loss of 2.0 dB or less and a tuning of required 60 MHz are achieved at a DC bias of 10 V.

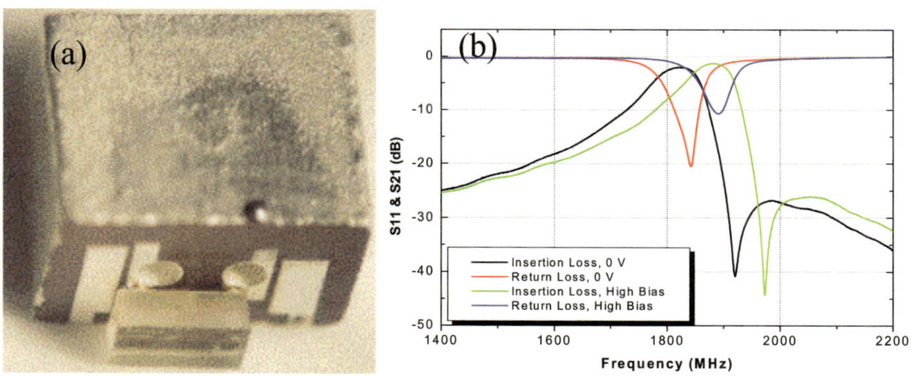

Fig. 20. (a) CDMA filter prototype, (b) CDMA Filter Response at 0V and high DC bias

4.3 X-band to Ku-band tunable filters

Slow wave resonators (SWR) are sections of transmission lines periodically loaded with tunable BST capacitors. By changing simultaneously the capacitance of each BST device, the effective dielectric constant, thus, the electrical length of the resonator is changed so that a different resonance frequency is achieved. In addition, due to the increased effective

Parameter	Values
Tx to antenna frequency fange Insertion loss Return loss Attenuation at Rx	1850 - 1910 MHz 2 dB max. 12 dB 40 dB
Antenna to Rx frequency range Insertion loss Return loss Attenuation at Tx	1930 - 1990 MHz 2.0 dB max 12 dB 45 dB
Tx to Rx isolation 1850-1910 1930-1990	 50 dB 40 dB

Table 4. Typical specifications of a US PCS duplexer

dielectric constant the resulting filters are more compact (smaller size) when compared to the regular resonator (λ/2 or λ/4) filters. The change in the effective dielectric constant also results in a characteristic impedance change.

A prototype of a 3-pole SWR filter with a dimension of 8.30 mm × 2.65 mm is shown in Figure 21. Figure 22 shows the microwave results of a 2-pole and a 3-pole SWR filter. Both filters have the same tuning range, i.e. 4.15 GHz, which corresponds to a tunability of 48%. The 3-pole filter shows higher loss due to its narrower bandwidth and possibly more capacitors, but has much better stop-band rejection than the 2-pole one. Proprietary low voltage electrode (LVE) BST capacitors were also used in the filter design, replacing regular gap capacitors. Figure 23 shows the measured S_{21} and S_{11} of a 2-pole low-voltage SWR filter. The filter shows an S_{21} of 5.4 and 3.3 dB at 0 and 30 V, respectively. The S_{11} is greater than 10 dB. The 3-dB bandwidth is between 14-15% at all bias voltages. The center frequency at 0 V is 11.5 GHz. The tunability is about 2.3 GHz (20% tuning).

Fig. 21. A prototype of a 3-pole SWR filter with a total of 8 DC blocking capacitors

4.4 X-band back-to-back 4-pole band-pass filters

A tunable back-to-back 4-pole filter was built on a flexible organic liquid crystal polymer (LCP) substrate, which consists of two open-loop resonators coupled to the two others through apertures lithographically opened in their common ground plane, resulting in a footprint size reduction of about 50% compared to typical open-loop resonator based filters. The filter frequency is tuned using BST capacitor chips, which are mounted and ribbon-bonded between both ends of each resonator on both sides. An insertion loss of 1.8 - 5.4 dB and a tuning of 12.6% are achieved in frequency range of 9.3 – 10.1 GHz.

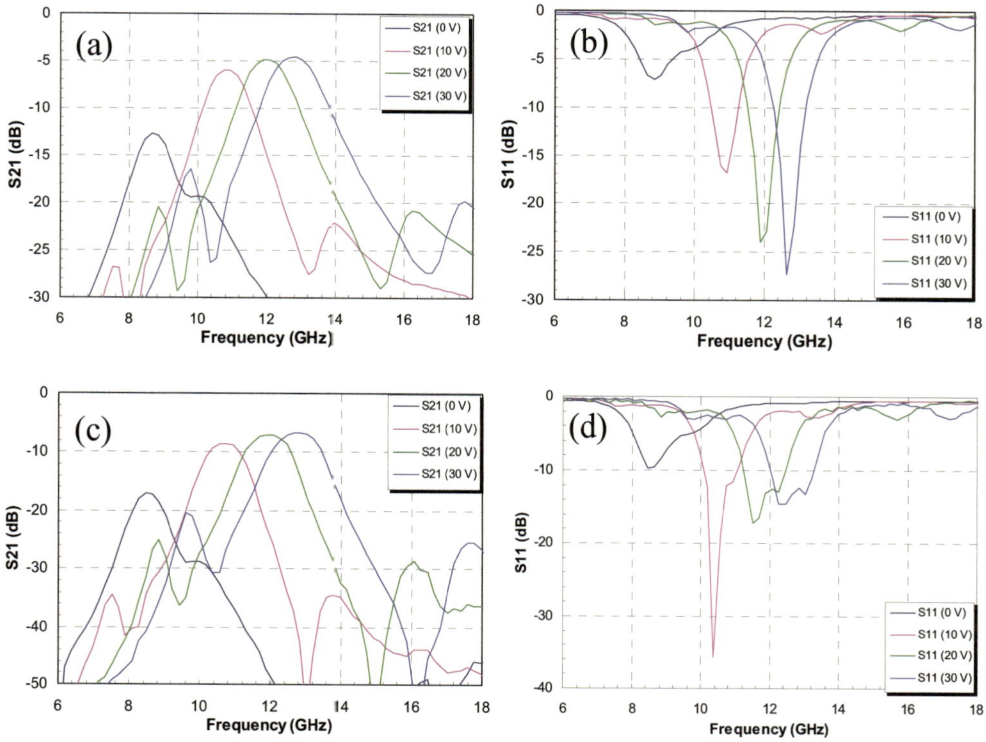

Fig. 22. Measured (a) insertion loss and (b) return loss of a 2-pole and (c) insertion loss and (d) return loss of a 3-pole SWR filter

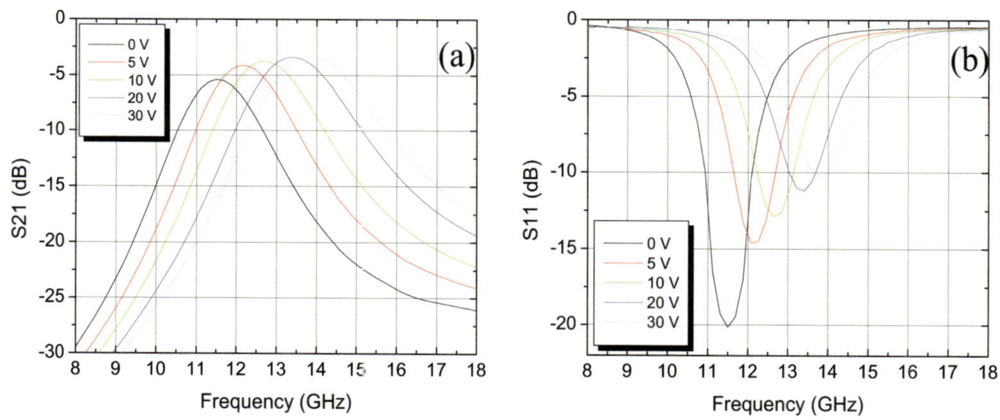

Fig. 23. Measured (a) insertion and (b) return loss (SOLT cal) of a 2-pole low-voltage SWR filter using LVE BST capacitors

The fabrication requires two 200 μm thick LCP layers originally covered by a 9 μm Cu foil. The top resonators (1 and 4) are patterned on one side of the first layer while the coupling apertures are patterned on the other side using conventional lithography techniques. As for the second layer, Cu is completely etched off on one side while bottom resonators (2 and 3) are patterned on the other side. A 25 μm LCP bond ply layer is placed between both layers which are aligned and bonded together with a substrate bonder at a temperature of 280 °C. The BST capacitor chips are then mounted on both sides and ribbon-bonded to the resonators. A 75 µm wide ribbon bond is used in order to limit the induced series inductance. High resistors are also surface mounted on the bottom layer to isolate the RF signal from the bias network. A photograph of both sides of the fabricated filter including the bias network is shown in Figure 24.

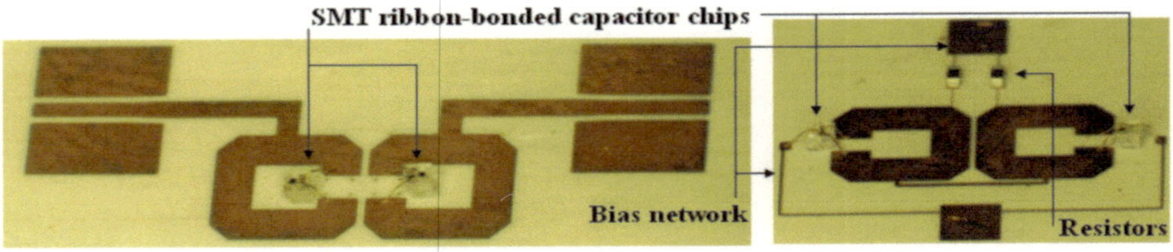

Fig. 24. Prototype of the 4-pole tunable filter with mounted capacitors: oblique views of the (a) top resonators (1 and 4) and (b) bottom resonators (2 and 3)

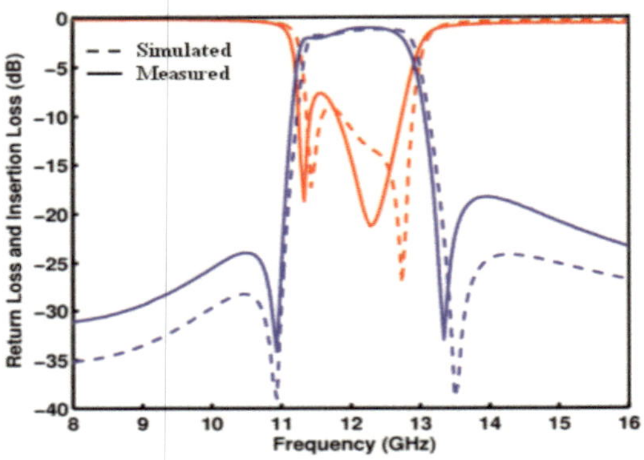

Fig. 25. Measured filter response without capacitors, compared to the simulated response

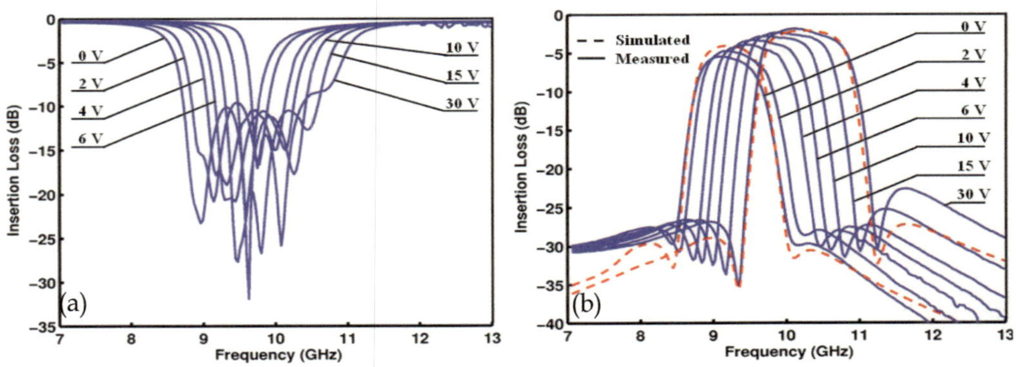

Fig. 26. Measured (a) return and (b) insertion loss of the 4-pole tunable filter

Figure 25 presents measured return and insertion loss of the filter compared to simulated results, showing a good agreement. Measured insertion loss is 1.0 dB at 12.23 GHz with a bandwidth of 11.2% at 1 dB. The measurement includes the via-less coplanar-to-microstrip transition effects which are not de-embedded. BST capacitors are then mounted on the substrate and ribbon-bonded. A bias voltage of 0 to 30 V is applied to the four capacitors to tune the center frequency. Figure 26 presents the measured results. The measured insertion loss is also compared to the simulated ones with a serial resistance of 6 Ω, an inductance of 0.6 nH, and a capacitance of 90 fF at 0 V and 54 fF at 30 V. An insertion loss of 5.4 dB at 9.1 GHz and 1.84 dB at 10.25 GHz is achieved, resulting in an analog tuning of 12.6% with a capacitance ratio of 1.67:1. Return loss is lower than 10 dB over the whole frequency range.

4.5 Ka-band ring tunable filters

Ka-band ring tunable filters with a size of 4.73 × 3.5 mm have also been fabricated on the BST based films, as shown in Figure 27. Each resonator consists of a ring separated by a thin gap, C_{gap}, from a stub protruding from a microstrip. The resonant frequency of each ring is tuned by the BST capacitors between the rings and ground. Radial stubs are used to realize electrical short to the RF ground, simplifying the fabrication process by eliminating via drilling through sapphire. Figure 28 shows the tuning of the fabricated filter. The best

Fig. 27. Photograph of a ring filter

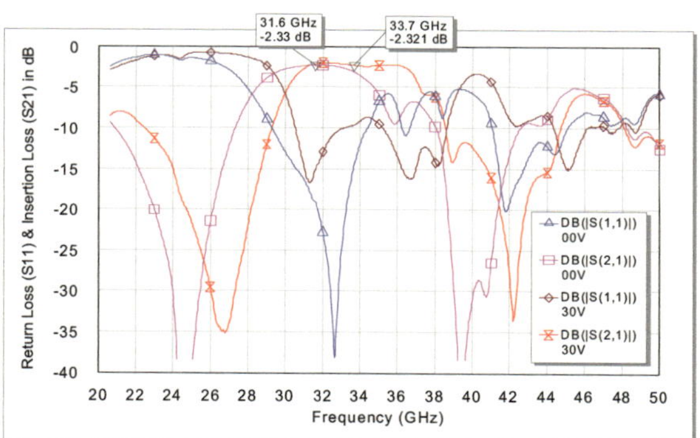

Fig. 28. Measured S-parameters for Ka-band ring filter under different biases (blue=S_{11} and pink=S_{21} at 0V; brown=S_{11} and red=S_{21} at 30 V

insertion loss in pass-band is 2.3 and 2.0 dB at 0 and 30 V, respectively. The 3-dB bandwidth is 20% for both bias states. The filter tunes from 31.6 to 33.7 GHz, a 6.6% tunability.

4.6 Phase shifters

In the meanwhile, phase shifters have also been developed using ferroelectric BST capacitors for frequencies ranging from L- to Ka-band. The all pass network shown in Figure 29 is realized using tunable BST capacitors and planar inductor structures. The resulting phase shifters exhibit low loss and phase shift greater than 360° using a low bias voltage. The main advantages of these phase shifters include small size, full passivation to prevent failure from contaminations, no static power consumption, solder ball termination for direct flip-chip mounting on a carrier, coplanar waveguide input and output to facilitate transition to other circuitry, and lumped element design for low loss. Figure 30 shows the insertion S_{21} and S_{11} of a 3 GHz phase shifter with a dimension of 2.46 × 1.83 × 0.44 mm developed at *n*Gimat. The maximum insertion loss at 3 GHz is 4.3 dB at 15 V. The figure of merit is 89.4°/dB at 0V and 89.2°/dB when the measurements at 0, 5, 10, 15, 20, 25, 30, and 35V are averaged. The S_{11}, shown in Figure 30 (b), is lower than 11.5 dB in all biased states. A phase shift of 376° is measured at 35V as shown in Figure 31. Another example of *n*Gimat's Ka-band phase shifter is discussed by Courrèges, et al (Courrèges et al., 2010). At 36 GHz, the phase shifter has the maximum S_{21} of 6.8 dB at 0V. The S_{11} is <10 dB in all biased states. A phase shift of 361° is measured at 30 V.

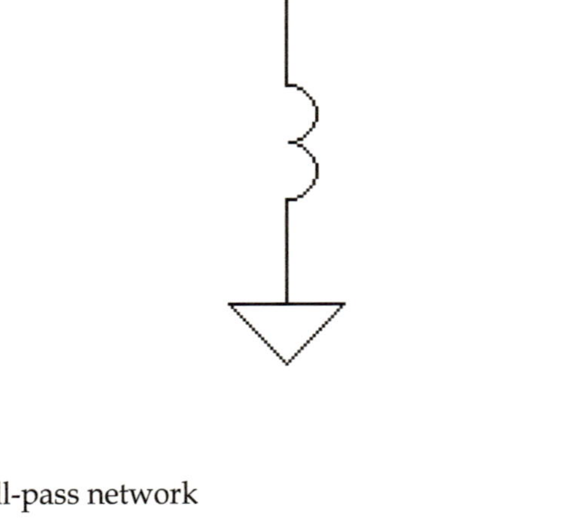

Fig. 29. Schematic of all-pass network

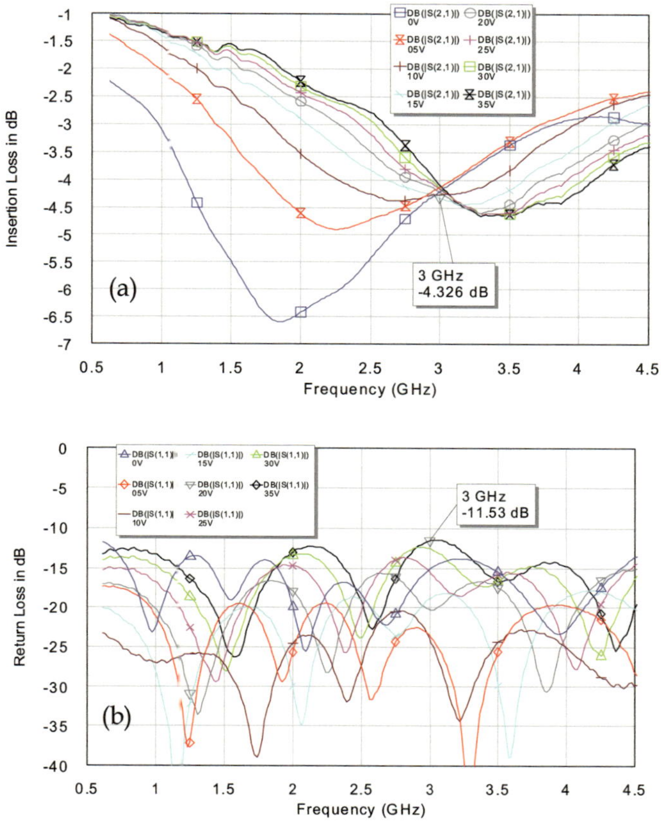

Fig. 30. (a) Insertion loss, S21 and (b) return loss, S11 of 3 GHz phase shifter

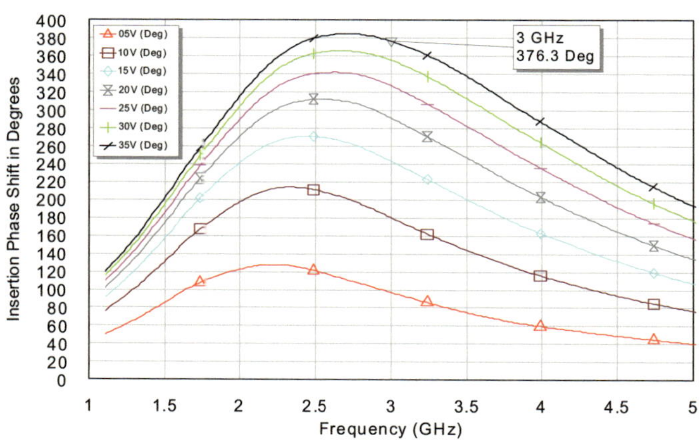

Fig. 31. Phase shift of the 3 GHz phase shifter at different frequencies and bias voltages

5. Conclusions

As a summary, high quality epitaxial or textured ferroelectric and dielectric thin films, including BST (both single layer and nanostructured multilayer), PZT, and CCT, have been successfully deposited by the proprietary CCVD process onto various substrates, including sapphire and single crystal STO, MgO, and LAO etc. Excellent electrical properties have been achieved on these ferroelectric and dielectric thin films. High performance microwave devices that can be used up to Ka band, such as tunable MEMS filters and CDMA filters, have been designed and fabricated on BST based ferroelectric thin films. The performance of these microwave devices are summarized as following:

- MEMS Ka-band tunable bandpass filters (both center frequency and bandwidth are tunable): the best insertion loss of 3 dB when biased, and the bandwidths of 3 and 7.8% for 3-pole narrowband and wideband, respectively;
- CDMA Tx tunable filters: insertion loss <2 dB, VSWR <1.5:1, center frequency shifting from 1.85 to 1.91 GHz, Rx zero (@1.93 GHz) rejection >40 dB, DC bias <10 V;
- X- to Ku-band tunable bandpass filters: insertion loss of ~5 dB @11.5 GHz (0V) to 3 dB @14 GHz (30 V), VSWR <2:1, DC bias <30 V, 6 × 1.5 × 0.5 mm in footprint;
- X-band back-to-back 4-pole bandpass filters: Insertion loss from 5.4 dB at 9.1 GHz to 1.84 dB at 10.25 GHz with an analog tuning of 12.6%; return loss <10 dB over the whole X-band frequency range;
- Ka-band ring filters: insertion loss of 2.3 and 2.0 dB for 0 and 30 V, respectively; 3-dB bandwidth of 20% for both bias states; tuning from 31.6 to 33.7 GHz, a 6.3% tunability;
- 3 GHz phase shifter: The insertion loss at is 4.3 dB at 15 V and 3 GHz. The figure of merit is 89.4°/dB at 0V. A phase shift of 361° is measured at 30V.

6. References

Adams, T. B., Sinclair, D. C., and West, A. R. (2002). Giant Barrier Layer Capacitance Effects in $CaCu_3Ti_4O_{12}$ Ceramics. *Advanced Materials,* Vol.14, No.18, (Sept 2002), pp. 1321-1323, ISSN 1521-4095

Bao, P., Jackson, T. J., Wang, X., and Lancaster, M, J. (2008). Barium Strontium Titanate Thin Film Varactors for Room-Temperature Microwave Device Applications. *J. Phys. D: Appl. Phys.,* Vol.41, No.6, (June 2008), pp. 063001, ISSN 0022-3727

Baringay, C. K. & Dey, S. K. (1992). Observation of sol-gel solid phase epitaxial growth of ferroelectric $Pb(Nb,Zr,Ti)O_3$ thin films on sapphire. *Appl. Phys. Lett.,*Vol. 61, No. 11, (September1992), pp.1278-1280, ISSN 0003-6951

Bochu, B., Deschizeaux, M. N., and Joubert, J. C., (1979). Synthése et Caratérisation d'une Série de Titanate Pérowskite Isotypes de $[CaCu_3](Mn_4)O_{12}$. *J. Solid State Chem.,* Vol.29, No.2, (August 1979), pp. 291-298, ISSN 0022-4596

Chung, U. C., Elissalde, E., Estournes, C., Pate, M., and Ganne, J. P. (2008). Low Loss, Highly Tunable $Ba_{0.6}Sr_{0.4}TiO_3$/MgO Composites. *Appl. Phys. Lett.* Vol.92, No.4, (January 2008), pp. 042902, ISSN 0003-6951

Cole, M. W., Joshi, P. C., Ervin, M. H., and Pfeffer, R. L. (2000). The Influence of Mg Doping on the Materials Properties of BST Thin Films for Tunable Device Applications. *Thin Solid Films,* Vol. 374, No.1, (October 2000), pp. 34-41, ISSN 0040-6090

Copel, M., Baniecki, J. D., Duncombe, P. R., and Shaw, T. M.(1998). Compensation Doping of $Ba_{0.7}r_{0.3}TiO_3$ Thin Films. *Appl. Phys. Lett.* Vol.73, No.13, (September 1998), pp. 1832-1834, ISSN 0003-6951

Courrèges, S., Zhao, Z. Y., Choi, K., Hunt, A. T., and Papapolymerou, J. (2010) Electrically Tunable Ferroelectric Devices for Microwave Applications, In: *Microwave and Millimeter Wave Technologies from Photonic Bandgap Devices to Antenna and Applications*, Igor Minin, pp. 1-20, InTech, ISBN: 978-953-7619-66-4, Vienna, Austria

Dawber, M., Rabe, K. M., and Scott, J. F. (2005). Physics of Thin Film Ferroelectric Oxides. *Reviews of Modern Phys.*, Vol. 77, No.4, (December 2005), pp. 1083-1130, ISSN 0034-6861

Deschanvres, A., Raveau, B., and Tollemer, F. (1967). Replacement de Metal Bivalent par le Cuivre Dans les Titanates de Type Perowskite. *Bull. Soc. Chim. Fr.*, pp. 4077-4078, ISSN 0037-8968

Dimos, D. and Mueller, C. H. (1998). Perovskite Thin Films for High Frequency Capacitor Applications. *Ann. Rev. Mater. Sci.*, Vol.28, No.1, (January 1998). pp. 397-419, ISSN 1531-7331

Dimos, D., Schwartz, R. W., and Lockwook, S. J. (1994). Control of Leakage Resistance in Pb(Zr,Ti)O3 Thin Films by Donor Doping. *J. Am. Ceram. Soc.* Vol. 77, No.11, (November 1994), pp. 3000-3005, ISSN 0002-7820

Gevorgian, S. S. and Kollberg, E. L. (2001). Do We Really Need Ferroelectrics in Paraelectric Phase Only in Electrically Controlled Microwave Devices. *IEEE Trans. Microwave Theory Techn.*, Vol.49, No.11 (November 2001), pp. 2117-2124, ISSN 0018-9480

Heindl, P., Srikanth, H., Weller, T., Tatarenko, A. S., and Srinivasan, G. (2007). Structure, Magnetism, and Tunable Microwave Properties of Pulsed Laser Deposition Growth Barium Ferrite/Barium Strontium Titanate Bilayer Films. *J. Appl. Phys.* Vol. 101, No.9, (July 2007), pp. 09M503, ISSN 0021-8979

Home, C. C., Vogt, T., Shapiro, S. M., Wakimoto, S., and Ramirez, A. P. (2001). Optical Response of High Dielectric Constant Perovskite-Related Oxide. *Science,* Vol.293, No.5530, (July 2001), pp. 673-676, ISSN 0036-8075

Hwang, C. S., Lee. B. T., Kang, C. S., Lee, K. H., Cho, H., Hideki, H., Kim, W. D., Lee, S. I., and Lee, M. Y. (1999), Depletion Layer Thickness and Schorttky Type Carrier Injection at the Interface Between Pt Electrode and $(Ba,Sr)TiO_3$ Thin Films. *J. Appl. Phys.*, Vol.85, No.1, (January 1999), pp. 287-295, ISSN 0021-8979

Hwang, K. S., Manabe, T., Nagahama, T., Yamaguchi, I., Kumagai, T., and Mizuta, S. (1999). Effect of Substrate Material on the Crystallinity and Epitaxy of $Pb(Zr,Ti)O_3$ Thin Films. *Thin Solid Films* Vol. 347, No.1-2, (June 1999), pp. 106-111, ISSN 0040-6090

Hunt, A. T., Carter, W. B., and Cochran, J. K. (1993). Combustion Chemical Vapor Deposition: A Novel Thin Film Deposition Technique. *Appl. Phys. Lett.*, Vol.63, No.2, (July 1993), pp. 266-268, ISSN 0003-6951

Hunt, A. T., Cochran, J. K., and Carter, W. B. (1997). Combustion Chemical Vapor Deposition of Films and Coatings. U.S. Patent No. 5,652,021

Hunt, A. T., Hwang, T. J., and Shao, H. (1999). Combustion Chemical Vapor Deposition of Phosphate Films and Coatings. U.S. Patent No. 5,858,465

Im, J., Auciello, O., Streiffer, S. K., and Krauss, A. R. (2000). Composition Control of Magnetron-Sputter-Deposited $(Ba_xSr_{1-x})Ti_{1+y}O_{3+z}$ Thin Films for Voltage Tunable Devices. *Appl. Phys. Lett.* Vol.76, No.5, (January 2000), pp. 625-627, ISSN 0003-6951

Kang, D. H., Lee, S. Y., Kim, Y. H., Gil, S. K., and Park, D. S. (2006), Control of Dispersion Frequency of $BaTiO_3$-based Ceramics Applicable to Thin Absorber for Millimeter Electromagnetic Wave. *J. Euro. Ceram. Soc.*, Vol. 26, No.10-11, (October 2006), pp. 2169-2173, ISSN 0955-2199

Katiyar, R. S., Jain, M., Miranda, F. A., and Van Keuls, F. W. (2005). Comparative Studies of Ferroelectric Thin Films for High Frequency Phase Shifter Applications. *Integrated Ferroelectrics*, Vol.71, No.1, (January 2005) pp. 11-19, ISSN 1058-4587

Kim, B. K., Lee, H. S., Lee, J. W., Lee, S. E., and Cho, Y. S. (2010). Dielectric and Grain-Boundary Characteristics of Hot Pressed $CaCu_3Ti_4O_{12}$. *J. Am. Ceram. Soc.* Vol.93, No.9, (September 2010), pp. 2419-2422, ISSN 0002-7820

Kim, H. H., Kim, S. T., and Lee, W. J. (1998). The Effects of In-situ Pretreatments of the Substrate Surface on the Properties of PLZT Films Fabricated by a Multi-Target Sputtering Method. *Thin Solid Films* vol. 324, No.1-2, (July 1998), pp.101-106, ISSN 0040-6090

Kim, W. S., Ha, S. M., and Kim, C. E. (1999). The Effects of Cation Substitution on the Ferroelectric Properties of Sol-Gel Derived PZT Thin Films for FRAM Application. *Thin Solid Films* Vol. 355-356, pp. 531-535, No.1, (November 1999), ISSN 0040-6090

Kobune, M., Fuijii, S., Takayama, R., and Tomozawa, A. (1996). Preparation and Pyroelectric Properties of $(Pb,La)(Zr,Ti)O_3$ (PLZT Thin Films. *Jpn. J. Appl. Phys.* Vol. 35, No.9B, (September 1996), pp. 4980-4983, ISSN 0021-4922

Li, M., Shen, Z. J., Nygren, M., Feteira, A., Sinclair, D. C., and West, A. R. (2009). Origin(s) of the Apparent high Permittivity in CaCu3Ti4O12 Ceramics: Clarification on the Contributions from Internal Barrier Layer Capacitor and Sample-Electrode Contact Effects. J. Appl. Phys., Vol.106, No.10, (November 2009), pp. 104106, ISSN 0021-8979

Liu, H., Gong, X. G., Liang, L. E., Li, X. D., Zhu, J. G., and Pu, Z. H. (2007). Enhanced Dielectric and Ferroelectric Properties of $Pb(Zr_{0.8}Ti_{0.2})TiO_3/Pb(Zr_{0.2}Ti_{0.8})TiO_3$ Multilayer Films. *Appl. Phys. Lett.* Vol.91, No.12, (September 2007), pp. 122906, ISSN 0003-6951

Lu, X. Y., Wang, B., Zheng, Y., and Li, C. L. (2008). Adjustable Ferroelectric Properties in Ferroelectric/Paraelectric Trilayers. *J. Phys. D: Appl. Phys.* Vol. 41, No.3, (February 2008), pp. 035303, ISSN 0022-3727

Mauray, D., Singh, D. P., Agrawal, D. C., and Mohapatra, Y. N. (2008). Preparation of High Dielectric Constant Thin Films of $CaCu_3Ti_4O_{12}$ by Sol-Gel. *Bull. Mater. Sci.* Vol.31. No.1, (February 2008), pp.55-59, ISSN 0250-4707

Misirlioglu, I., B., Akcay, G., Zhong, S., and Alpay, S. P. (2007), Interface Effects in Ferroelectric Bilayers and Heterostructures. *J. Appl. Phys.* Vol. 101, No.3, (February 2007), pp. 036107, ISSN 0021-8979

Natori, K., Otani, D., and Sano, N. (1998), Thickness Dependence of the Effective Dielectric Constant in a Thin Film Capacitor. *Appl. Phys. Lett.*, Vol. 73, No.5, (August 1998), pp. 632-634, ISSN 0003-6951

Pintilie, L., Boldyreva, K., Alexe, M., and Hesse, D. (2008). Capacitance Tuning in Antiferroelectric Ferroelectric $PbZrO_3$-$Pb(Zr_{0.8}Ti_{0.2})TiO_3$ Epitaxial Multilayers. *New J. Phys.* Vol. 10, No.1, (January 2008), pp. 013003, ISSN 1367-2630

Polla, D. L. and Francis, L. F. (1998). Processing and Characterization of Piezoelectric Materials and Integration into Microelectromechical Systems. *Ann. Rev. Mater. Sci.* Vol. 28, No.1, (January 1998), pp. 563, ISSN 1531-7331

Prakash, B. S., Varma, K. B., and Maglione, M. (2008). Deposition and Dielectric Properties of CCTO Thin Films Deposited on Pt/Ti/SiO_2/Si Substrates using RF Magnetron Sputtering. Thin Solid Films, Vol.516, No.10, (), pp. 2874-2880, ISSN 0040-6090

Ramesh, R., Aggarwal, S., and Auciello, O. (2001). Science and Technology of Ferroelectric Films and Heterostructures for Non-volatile Ferroelectric Memories. *Mater. Sci. Eng. R* Vol. 32, No.6, (April 2001), pp. 191-236, ISSN 0927-796X

Ramirez, A. P., Subramanian, P. A., Gardel, M., Blumberg, G., Li, D., Vogt, T., and Shapiro, S. M. (2000). Giant Dielectric Constant Response in a Copper Titanate. *Solid State Communications,* Vol.115, No.5, (June 2000), pp. 217-274, ISSN 0038-1098

Rukmini, H. R., Choudhary, R. N. P., and Rao, V. V. (1999). Structural and Dielectric Properties of $Pb_{0.9}(La,K)_{0.09}(Zr_{0.65}Ti_{0.35})_{0.9775}O_3$. *J. Mater. Sci.* Vol.34, No.19, (October 1999), pp. 4815-4819, ISSN 0022-2641

Scott, J. F. (1999). Depletion Width in $SrTiO_3$ and $BaxSr_{1-x}TiO_3$ Films. *Ferroelectrics,* Vol.232, No.1, (January 1999), pp. 25-34, ISSN 0015-0193

Sinclair, D. C., Adams, T. B., Morrison, F. D., and West, A. R. (2002). $CaCu_3Ti_4O_{12}$: One-Step Internal Barrier Capacitor. *Appl. Phys. Lett.,* Vol.80, No.12, (March 2002), pp. 2153-2155, ISSN 0003-6951

Sinnamon, L. J., Saad, M. M., Bowman, R. W., and Gregg, J. M. (2002). Exploring Grain Size as a Cause for "Dead-Layer" Effects in Thin Film Capacitors. *Appl. Phys. Lett.,* Vol.81. No.4, (July 2002), pp.703-705, ISSN 0003-6951

Song, Z. T., Chong, N., Chan, H. L. W., and Choy, C. L. (2001) Electrical and Pyroelectric Properties of In-plane Polarized Lead Lanthanum Titanate Thin Films. *Appl. Phys. Lett.* Vol.79, No.5, (July 2001), pp. 668-670, ISSN 0003-6951

Streiffer, S. K., Basceri, C., Parker, C. B., Lash, S. E., and Kingon, A. I. (1999) Ferroelectricity in Thin Films: The Dielectric Response of Fiber-Textured $(Ba_xSr_{1-x})Ti_{1+y}O_{3+z}$ Thin Films Grown by Chemical Vapor Deposition. *J. Appl. Phys.,* Vol.86, No.8, (October 1999), pp. 4565-4567, ISSN 0003-6951

Subramanian, M. A, Li, D., Duan, N., Reisner, B. A., and Sleight, A. W. (2000). High Dielectric Constant in $ACu_3Ti_4O_{12}$ and $ACu_3Ti_4FeO_{12}$. *J. Solid State Chem.,* Vol.151, No.2, (May 2000), pp.323-325, ISSN 0022-4596

Subramanian M. A. and Sleight, A. W. (2002). $ACuTi_4O_{12}$ and $ACu_3Ru_4O_{12}$ Perovskites: High Dielectric Constants and Valence Degeneracy. *Solid State Sciences,* Vol.4, No.3, (March 2002), pp. 347-351, ISSN 1293-2558

Tagantsev, A. K, Sherman, V. O, Astafiev, K. F. , Venkatesh, J., and Setter, N. (2003). Ferroelectric Materials for Microwave Tunable Applications. *J. Electroceramics,* Vol. 11, No.1, (September 2003), pp. 5-66, ISSN 1385-3449

Takeuchi, I., Chang, H., Gao, C., Xiao, X. D., Downes, M. J., and Venkatesan, T. (1998). Combinatorial Synthesis and Evaluation of Epitaxial Ferroelectric Device Libraries. *Appl. Phys. Lett.*, Vol.73, No.7, (August 1998), pp. 894-896, ISSN 0003-6951

Xu, B., M., Rolcawich, R., G., Trolier-McKinstry, S., Ye, Y., H., and Cross, L. E. (1999) Sensing Characteristics of In-plane Polarized Lead Zirconate Titanate Thin Films. *Appl. Phys. Lett.*, Vol. 75, No.26, (December 1999), pp. 4180-4182, ISSN 0003-6951

Xu, Y. (1991). *Ferroelectric Mater. & Their Appl.*, North Holland, ISBN 0-444-88354-1, Amsterdam, The Netherlands

Zhong, S., Alpay, S. P., Cole, M. W., Ngo, E., Hirsch, S., and Demaree, J. O. (2007). Highly Tunable and Temperature Insensitive Multilayer Barium Strontium Titanate Films. *Appl. Phys. Lett.* Vol. 90, No.9, (February 2007), pp. 092901, ISSN 0003-6951

Zhou, C. and Newns, D. M. (1997). Intrinsic Dead Layer Effect and the Performance of Ferroelectric Thin Film Capacitors. *J. Appl. Phys.*, Vol.82, No.6, (September 1997), pp. 3081-3088, ISSN 0021-8979

Zhu, B. P., Wang, Z. Y., Zhang, Y., Yu, Z. S., Shi, J., and Xiong, R. (2008). Low Temperature Fabrication of the Giant Dielectric Material $CaCu_3Ti_4O_{12}$ by oxalate Coprecipitation Method, *Mater. Chem. Phys.* Vol.113, No.2-3, (February 2008), pp. 746-748, ISSN 0254-0584

2

Synthesis of Ferroelectric $Na_{0.5}Bi_{0.5}TiO_3$ by MSS (Molten Salt Synthesis) Method

Teresa Zaremba
Silesian University of Technology
Poland

1. Introduction

Environmental destruction has been a serious problem worldwide. One problem is the release of harmful materials (e.g., Cd, Hg, Pb) from electrical industries. Thus, the restriction on hazardous substance (RoHS) will be enforced soon to prevent the release of harmful waste (Hiruma et al., 2007; Panda, 2009). Lead - based ferroelectric ceramics represented by Pb $(Zr,Ti)O_3$ (PZT), have been widely used for piezoelectric transducers, sensors and actuators due to their excellent piezoelectric properties. However, the evaporation of harmful lead oxide during preparation causes a crucial environment problem. Therefore, it is necessary to develop environment - friendly lead - free piezoelectric ceramics to replace the PZT - based ceramics, which has become one of the main trends in present development of piezoelectric materials. Sodium bismuth titanate, $Na_{0.5}Bi_{0.5}TiO_3$ (abbreviate as NBT), discovered in 1960 (Smolenskii et al., 1960), is considered to be a promising candidate of lead - free piezoelectric ceramics (Pookmaneea et al., 2001; Isupov, 2005; Panda, 2009; Zhou et al., 2010).

NBT is a relaxor ferroelectric material with the general formula $A'_xA''_{1-x}BO_3$. The ferroelectricity in NBT ceramic is attributed to $(Bi_{1/2}Na_{1/2})^{2+}$ ions, especially Bi^{3+} ions at the „A" site of the perovskite structure (ABO_3) and due to rhombohedral symmetry at room temperature. It has high Curie temperature (T_c = 320°C), and shows diffuse phase transition (Suchanicz et al., 2000; Suchanicz et al., 2001; Raghavender et al., 2006). However, the piezoelectric properties of NBT ceramics are not good enough for most practical uses. In order to enhance the properties and meet the requirements for practical uses, it is necessary to develop new NBT - based ceramics (Raghavender et al., 2006; Zhou et al., 2010). Researches have investigated many dopants into NBT ceramics (Panda, 2009). Also, it is desirable to fabricate ceramics with a textured microstructure in order to improve the properties (Hao et al., 2007).

The ferroelectric ceramic powders are synthesized trough conventional solid - state method which needs high calcination temperature and repeated grindings (Lu et al., 2010). In order to eliminate these defects, the wet chemical synthesis techniques have been developed, for instance hydrothermal method (Cho et al., 2006; Wang et al., 2009), sol - gel method (Xu et al., 2006; Mercadelli et al., 2008), and molten salt method (Zeng et al., 2007; Li et al., 2009). But the hydrothermal and sol - gel synthesis are usually long and complex processes, use hazardous solvents such as 2-methoxyethanol, and result in agglomerated particles (Bortolani & Dorey, 2010). Moreover, in the sol -gel method the cost of starting materials is high (Li et al., 2009).

Molten salt synthesis (MSS) is a process that yields large amounts of ceramic powders in a relatively short period of time. Moreover, it is a suitable method for preparation of complex oxide compounds with anisotropic particle morphologies. In this technique starting materials are mixed together with a salt (usually alkaline chloride and sulphate) and then heat treated at a temperature higher than the melting point of the salt. The melting temperature of the salt system can be reduced by using a eutectic mixture of salts, e.g. the use of NaCl – KCl instead of pure NaCl reduces the melting point from 801 to 657°C. A reaction between the precursors takes place in the molten salt (the flux) and the solid product obtained is separated by washing of the final mixture with hot deionised water. The typical starting materials are oxides, but carbonates, oxalates and nitrates can also be used. There are several requirements for the selection of salt to be used for MSS. First, the melting point of the salt should be relatively low and appropriate for synthesizing of the required phase. Second, the salt should possess sufficient aqueous solubility in order to eliminate it easily after synthesis by washing. Finally, the salt should not react with the starting materials or the product (Bortolani & Dorey, 2010; Hao et al., 2007). MSS has been used to form various ceramic powders such as niobates relaxors (Yoon et al., 1998), $Bi_4Ti_3O_{12}$ (Kan et al., 2003), $ZnTiO_3$ (Xing et al., 2006), $BaTiO_3$ (Zhabrev et al., 2008) and $Pb(Zr, Ti)O_3$ (Cai et al., 2008; Bortolani & Dorey, 2010).

It was found that ternary compound $Na_{0.5}Bi_{0.5}TiO_3$ was formed in the solid - state process through the intermediate binary compound, i.e. bismuth titanate - $Bi_4Ti_3O_{12}$ (Zaremba, 2008). $Bi_4Ti_3O_{12}$ (abbreviate as BIT) belongs to the Aurivillius family with a general formula $(Bi_2O_2)[A_{m-1}(B)_mO_{3m+1}]$, which consists of $(Bi_2O_2)^{2+}$ sheets alternating with $(Bi_2Ti_3O_{10})^{2-}$ perovskite - like - layers (Aurivillius, 1949, as cited in Stojanović et al., 2008). In general formula m represents the number of octahedra stacked along the direction perpendicular to the sheets, and A and B are the 12- and 6- fold coordination sites of perovskite slab, respectively. This kind of structure promotes plate - like morphology (Dorrian et al., 1971, as cited in Stojanović et al., 2008).

In this paper, $Na_{0.5}Bi_{0.5}TiO_3$ powders were prepared by molten salt synthesis in the presence pure NaCl or NaCl - KCl as fluxes. The first stage of the study related to direct synthesis of NBT via MSS from Na_2CO_3, Bi_2O_3 and TiO_2. For comparison, the synthesis of NBT by the conventional method (CMO - conventional mixed oxides) was investigated. The second stage included obtaining intermediate binary compound BIT via MSS from oxide precursors, i.e. Bi_2O_3 and TiO_2, and then synthesis of NBT via MSS using BIT, Na_2CO_3 and TiO_2 as starting materials.

The details pertaining to studies of synthesis of NBT and an Aurivillius - structured BIT precursor are reported in the following sections.

2. Synthesis of ferroelectric $Na_{0.5}Bi_{0.5}TiO_3$

Chemically pure powders of Bi_2O_3, TiO_2 (rutile) and Na_2CO_3 were used as starting materials. The $Na_{0.5}Bi_{0.5}TiO_3$ (NBT) was prepared by the following two routes:

$$Na_2CO_3 + Bi_2O_3 + 4TiO_2 \rightarrow 4Na_{0.5}Bi_{0.5}TiO_3 + CO_2 \quad (1)$$

$$Bi_4Ti_3O_{12} + 5TiO_2 + 2Na_2CO_3 \rightarrow 8Na_{0.5}Bi_{0.5}TiO_3 + 2CO_2 \quad (2)$$

In route (1), the starting materials were weighed in the proportion to yield NBT and mixed in isopropyl alcohol employing an agate mortar and pestle for 1 h. Using MSS method, the dry mixture of the precursors in the stoichiometric composition was mixed with an aqual

weight of salt. Salts used in this experiment were NaCl and eutectic mixture of 0.5NaCl - 0.5KCl, i.e. 43.94% NaCl - 56.06% KCl (by weight). The mixture of the precursors and flux was dried at 120°C for 2 h for complete removal of isopropyl alcohol, placed in a Pt crucible and heated in a sealed alumina crucible (to prevent salt evaporation) at temperatures between 800°C and 1100°C for a different time period. After thermal treatment the chlorides were removed from the products by washing with hot deionized water several times until the filtrates gave no reaction with silver nitrate solution. The powders were finally dried at 100°C for 2 h. NBT powders were also prepared by a conventional mixed oxide method (CMO) for comparison. All the syntheses were carried out in a conventional electric furnace.
Platelike $Bi_4Ti_3O_{12}$ (BIT) particles were obtained by MSS method in 0.5NaCl - 0.5KCl flux (abbreviate as NaCl - KCl) in the same way as described above. Temperature of thermal treatment ranging from 700°C to 1100°C for a different time period.
In route (2), BIT crystals produced earlier were subjected to second molten salt synthesis. Na_2CO_3 and TiO_2 were added to give the total composition of NBT. Again, pure NaCl or NaCl - KCl mixture was added (weight ratio of precursors to flux = 1:1).
Finally, the phase composition of the synthesized samples was analyzed by the powder X-ray diffraction (XRD; model 3003 TT, Seifert) using Ni - filtered Cu K_α radiation. The microstructure was observed by a scanning electron microscope (SEM; model BS 340, Tesla). The samples were coated by a gold layer by using a metal - coating plant under a vacuum. X-ray energy dispersive spectra (EDS) were measured using a Hitachi S-3400 N scanning electron microscope with an EDS system Thermo Noran.

2.1 Synthesis of $Na_{0.5}Bi_{0.5}TiO_3$ from Bi_2O_3, TiO_2 and Na_2CO_3

Fig. 1 represents the XRD patterns of the selected powders synthesized through route (1), i.e. directly from Bi_2O_3, TiO_2 and Na_2CO_3 via MSS (NaCl flux) and CMO. Similar trends were also observed for NBT produced using NaCl- KCl flux. The particle morphology of the starting materials and synthesized powders is compared in Figs 2 - 4.
NBT perovskite phase was observed in all the prepared samples. A comparison of interplanar spacings determined from XRD patterns of the samples prepared by a conventional solid state reaction and via MSS shows that agreement is quite satisfactory. Analysis of XRD patterns of NBT samples obtained via MSS has not shown displacement of maxima of diffraction peaks as the NaCl-KCl flux was used.
Isometric particles are found to exist in the samples of NBT. Typical micrograph of the NBT powder prepared by CMO is shown in Fig. 3a. There is high degree of agglomeration in this powder. The NBT particles prepared directly by CMO and MSS (NaCl flux) are very small (about 1 µm). The size of the particles increased with increasing temperature, especially, as NaCl-KCl flux was used. Probably, this is mainly due to the different melting points for each salt used. NaCl and 0.5 NaCl - 0.5 KCl have melting points of about 800°C and 650°C, respectively.
According to (Cai et al., 2007, as cited in Bortolani & Dorey, 2010) the solubility of the starting materials in the molten salt plays an important role in the synthesis as it has an influence on the final product morphology. For a simple two reactant system, two different cases can be distinguished: either both reactants are equally soluble in the molten salt or one oxide is more soluble than the other (Li et al., 2007, as cited in Bortolani & Dorey, 2010). In the first case (dissolution - precipitation mechanism) both reactants fully dissolve, react in the molten salt and the final product precipitates from the molten salt after formation. The shape of the product has no connection with the shape of the starting materials. In the second case, the more soluble precursor dissolves in the salt and diffuses to the less soluble

one. Here, at the surface, it reacts to final product. This template formation mechanism would result in a product morphology that is similar to that of the less soluble reactant which has acted as a template.

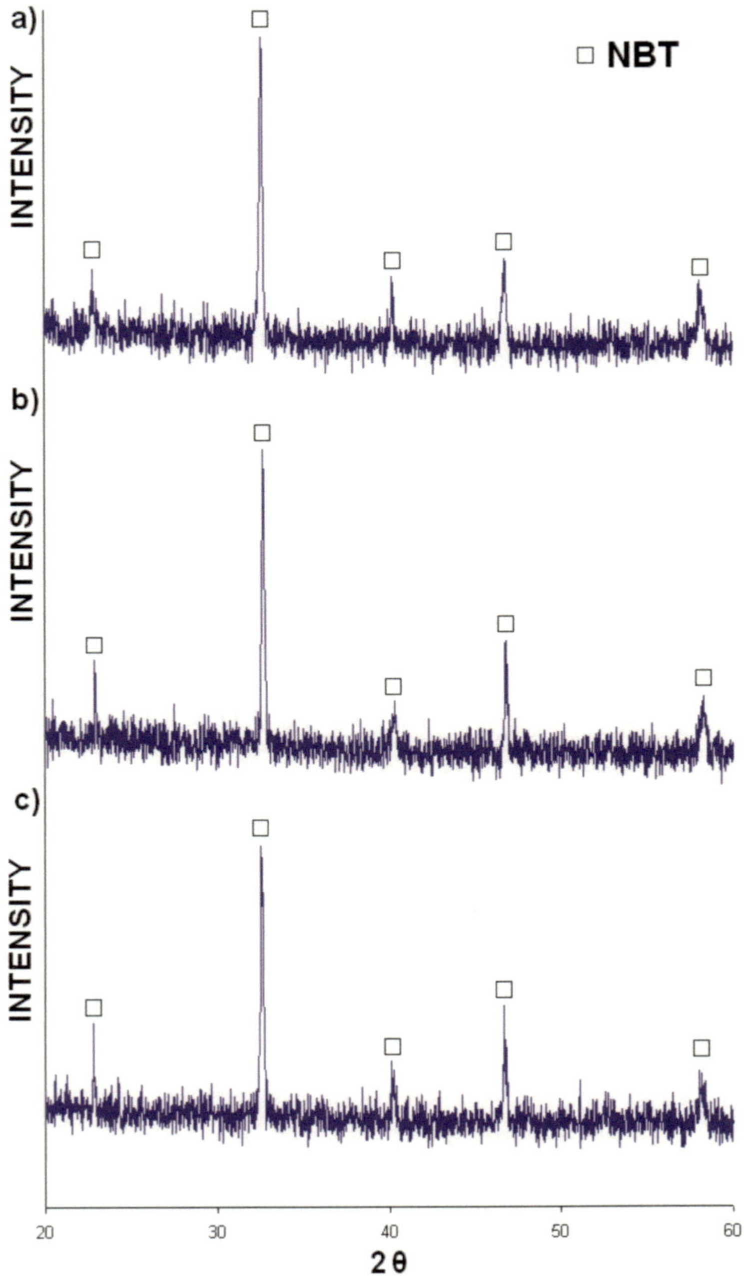

Fig. 1. XRD patterns of $Na_{0.5}Bi_{0.5}TiO_3$ powders fabricated through route (1): (a) via MSS (NaCl flux) at 950°C for 1.5 h; (b) via MSS (NaCl flux) at 1000°C for 4 h; (c) via CMO at 1000°C for 4 h

(a)

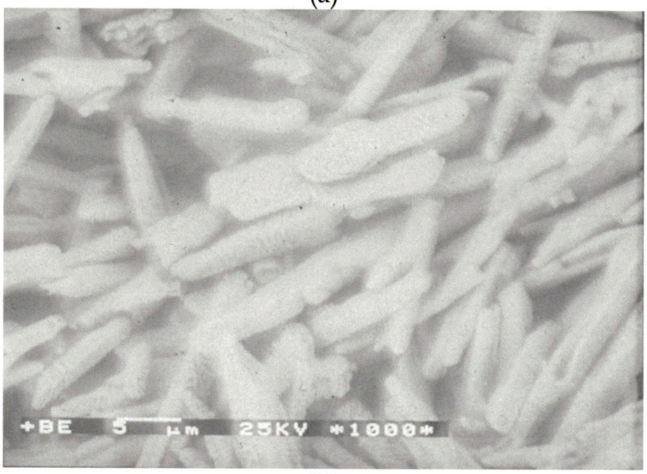

(b)

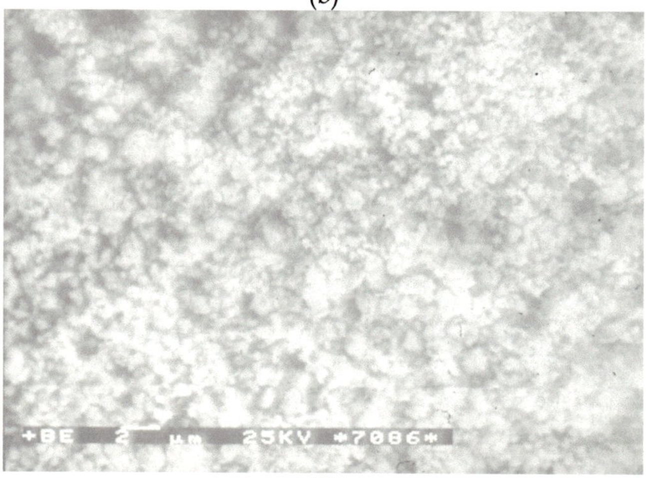

(c)

Fig. 2. SEM micrographs of the starting materials: (a) Na_2CO_3; (b) Bi_2O_3; (c) TiO_2 - rutile

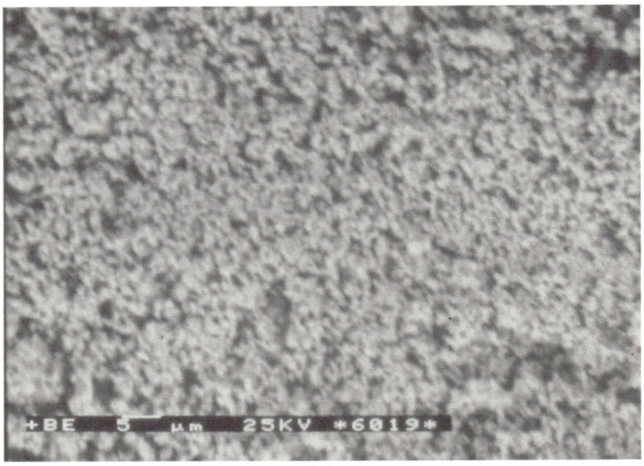

(a)

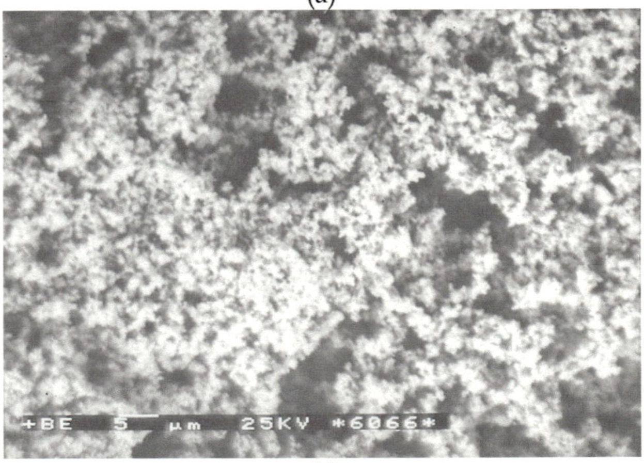

(b)

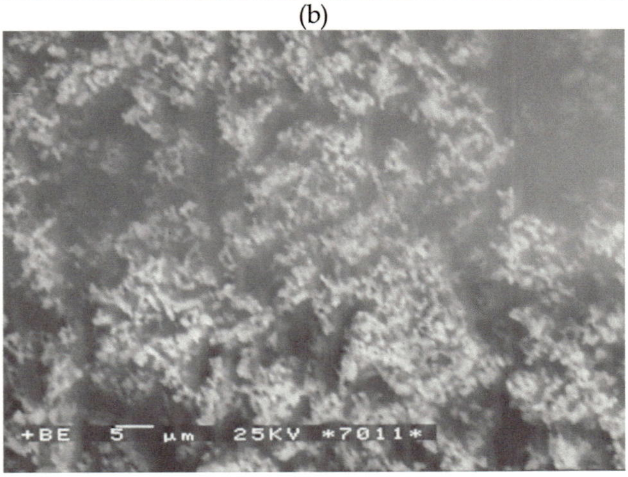

(c)

Fig. 3. SEM micrographs of $Na_{0.5}Bi_{0.5}TiO_3$ powders obtained through route (1): (a) via CMO at 950°C for 1.5 h.; (b) via MSS (NaCl flux) at 950°C for 1.5 h; (c) via MSS (NaCl flux) at 1000°C for 4 h

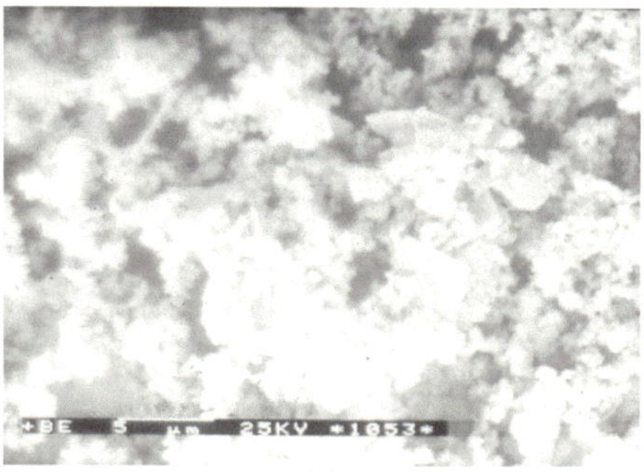

(a)

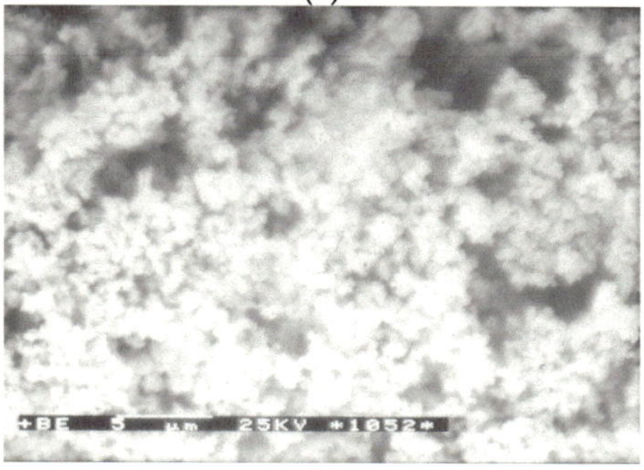

(b)

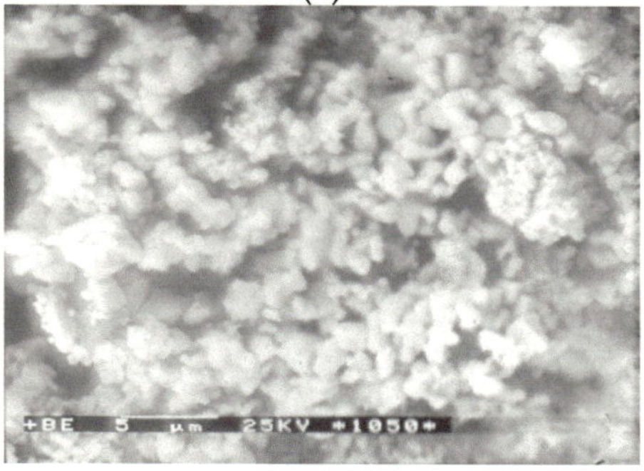

(c)

(d)

Fig. 4. SEM micrographs of $Na_{0.5}Bi_{0.5}TiO_3$ powders obtained through route (1) via MSS (NaCl - KCl flux) at different temperatures for 4 h: (a) 800°C; (b) 900°C; (c) 1000°C; (d) 1100°C

In the case of NBT the mechanism is further complicated due to the presence of (at least) 3 reactants. According to (Cai et al, 2007, as cited in Bortolani & Dorey, 2010) TiO_2 is not soluble in molten alkali chlorides. The final NBT morphology should be similar to the morphology of TiO_2.

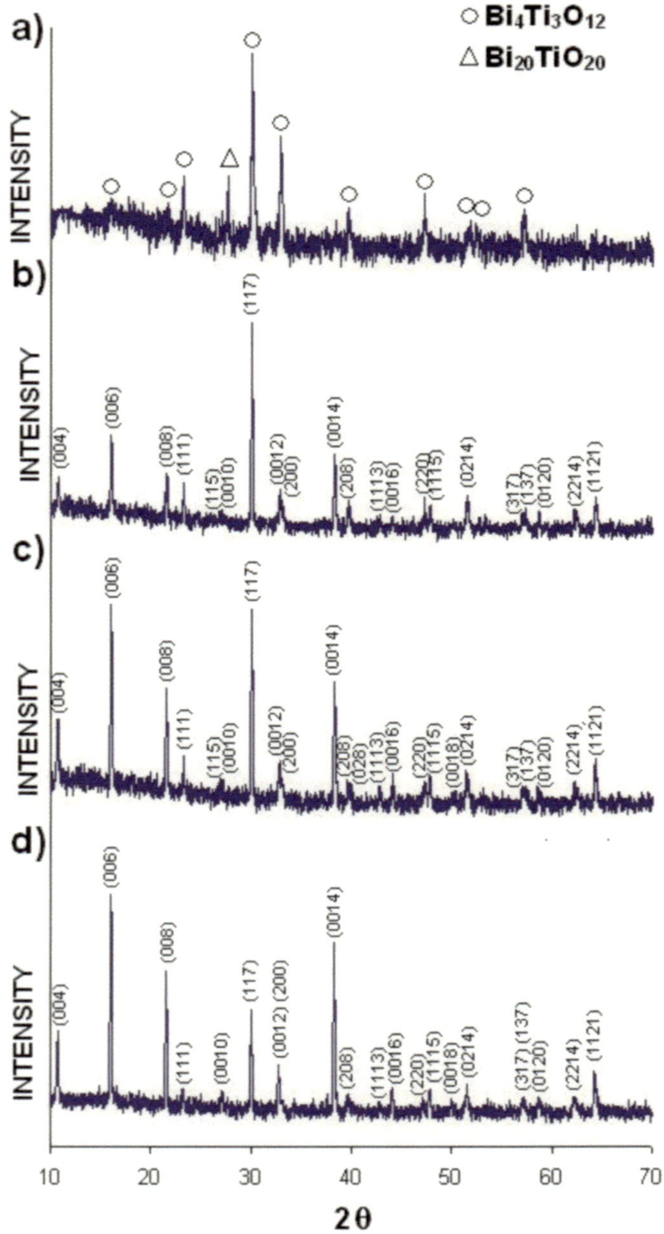

Fig. 5. XRD patterns of $Bi_4Ti_3O_{12}$ powders obtained via MSS at: (a) 700°C for 15 min; (b) 900°C for 30 min; (c) 900°C for 240 min; (d) 1000°C for 15 min (indexed peaks are those of $Bi_4Ti_3O_{12}$)

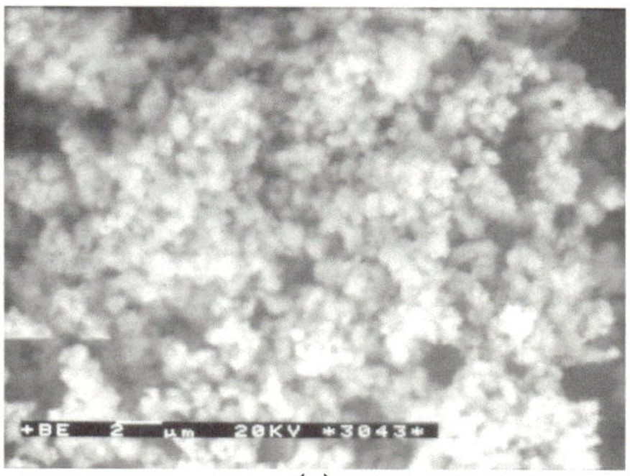

(a)

(b)

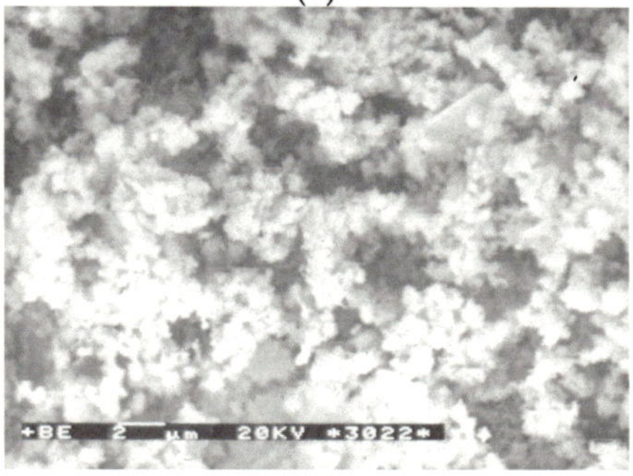

(c)

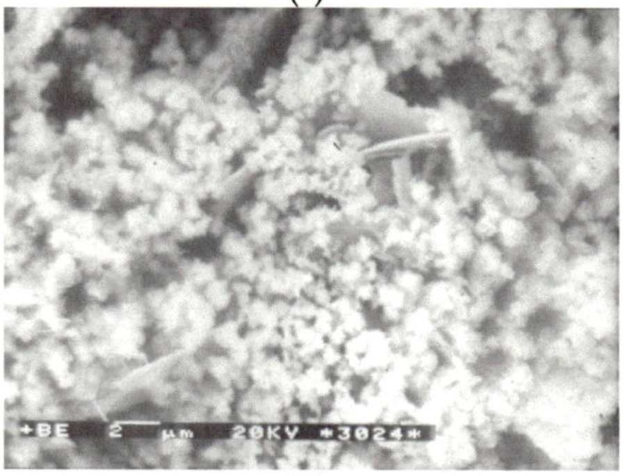

(d)

Fig. 6. SEM micrographs of $Bi_4Ti_3O_{12}$ powders obtained via MSS at 700°C for: (a) 15 min; (b) 30 min; (c) 60 min; (d) 120 min

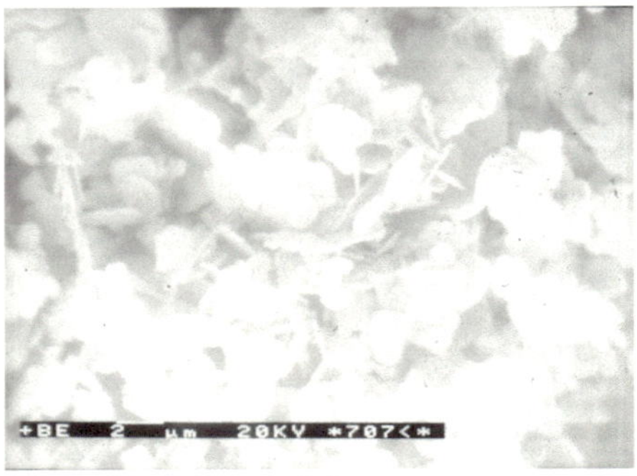

(a)

(b)

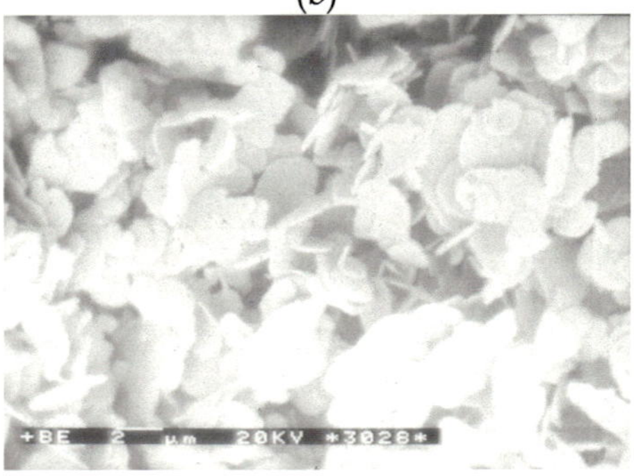

(c)

(d)

Fig. 7. SEM micrographs of $Bi_4Ti_3O_{12}$ powders obtained via MSS at 1000°C for: (a) 15 min; (b) 30 min; (c) 60 min; (d) 120 min

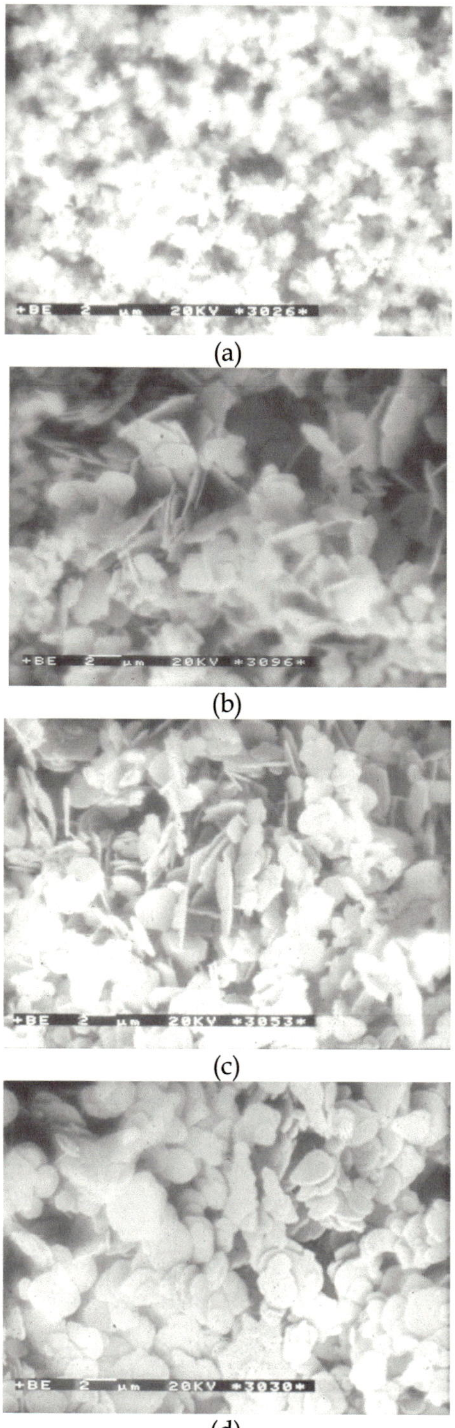

Fig. 8. SEM micrographs of $Bi_4Ti_3O_{12}$ powders obtained via MSS at different temperatures for 4 h: (a) 700°C; (b) 800°C; (c) 900°C; (d) 1000°C

2.2 Synthesis of $Bi_4Ti_3O_{12}$

The XRD patterns of selected samples BIT prepared from the mixture of Bi_2O_3 and TiO_2 via MSS (NaCl-KCl flux) are presented in Fig. 5. At 700°C, the phase $Bi_{12}TiO_{20}$ co-existed with BIT. Pure crystalline BIT was obtained after thermal treatment at 800°C for 15 min.. Increasing the temperature to 1100°C, the intensities of the (00*l*) diffraction lines were increased. These results indicate that during sample preparation for X-ray diffraction characterization, $Bi_4Ti_3O_{12}$ crystals with platelet morphology may align with (00*l*) planes parallel to the flat specimen holder.

Figs 6 – 8 show morphology and size of BIT crystals prepared at different temperatures. The synthesizing temperature could significantly influence the growth rate and crystallization habit of the BIT particles. Between 700°C and 800°C aggregate particles were formed. The powder synthesized at 700°C was composed of fine particles. The size of the primary particles increased and their shape changed from lumpy to plate-like with increasing temperature. Above 800°C discrete plate-like particles with increased particle size were formed. On the other hand, the effect of heating time on morphology and particle size is smaller. The degree of aggregation decreased with increasing prepare temperatures.

According to (Kimura & Yamaguchi, 1987) $Bi_4Ti_3O_{12}$ is formed via MSS by mechanism, when two reactants have comparable dissolution rates in molten salt. If this mechanism dominates during the formation process, the complex oxide powder with a characteristic or lumpy shape is formed, depending on the degree of interaction between the complex oxide and salt.

2.3 Synthesis of $Na_{0.5}Bi_{0.5}TiO_3$ from $Bi_4Ti_3O_{12}$, TiO_2 and Na_2CO_3

Fig. 9 shows the XRD patterns of the selected powders synthesized through route (2), i.e., from $Bi_4Ti_3O_{12}$ (BIT), TiO_2 and Na_2CO_3 via MSS (NaCl-KCl flux) at different temperatures. The diffraction lines were indexed based on the pseudocubic unit cell because of a small rhombohedral distortion of $Na_{0.5}Bi_{0.5}TiO_3$.

BIT has a BLSF (bismuth layer-structured ferroelectric) structure that is highly anisotropic, with the grain growth rate along the *c*-axis much lower than that along the *a (b)*-axis. So it is easy for them to form a plate-like morphology. Similar to the BIT particles, most of the NBT particles laid down with the *c*-axis aligning along the vertical direction during the sample preparation for the XRD analysis. So, they exhibit strong (100) and (200) diffraction peaks, especially, for higher temperatures of synthesis.

Fig. 10 shows the SEM micrographs of the NBT samples prepared from the BIT particles at different temperatures in the presence NaCl-KCl flux. Similar to the BIT particles, most of the NBT particles are large and of plate-like shape. Much larger crystals were grown at 1100°C, but the XRD pattern (Fig. 9d) shows that NBT co-existed with other crystalline phase (phases), probably it was related to the beginning of thermal decomposition of NBT.

It has also been found that the salt has a significant effect on the size of the synthesized particles. Fig. 11 shows the SEM micrographs of the ceramic powders synthesized in different fluxes at 1100°C. The particles for the powder synthesized in NaCl-KCl flux are larger than those synthesized in NaCl flux. NaCl-KCl flux at eutectic has a low melting point (650°C), so ions have a high diffusion rate at the synthesing temperature 1100°C. In order to determine the composition of the prepared samples, energy dispersive X-ray spectroscopy (EDS) data (Fig. 12) performed on a samples synthesized in the presence NaCl-KCl and NaCl flux show that the chemical components of the samples are the elements Na, Bi, Ti and O (without K from KCl for the sample obtained in the presence NaCl-KCl).

As a member of the BLSFs, BIT consist of the $(Bi_2Ti_2O_{10})^{2-}$ (pseudo-) perovskite layers interleaved by $(Bi_2O_2)^{2+}$ fluorite layers. After the reaction with the complementary reactants (Na_2CO_3 and TiO_2), the layer-structured BIT particles were transformed to the perovskite NBT. Although there are works reporting the transformation as a process from a lamellar phase to a perovskite phase (Schaak & Mallouk, 2000), the process involving the $(Bi_2O_2)^{2+}$ layers changing to the perovskite structure is still unclear.

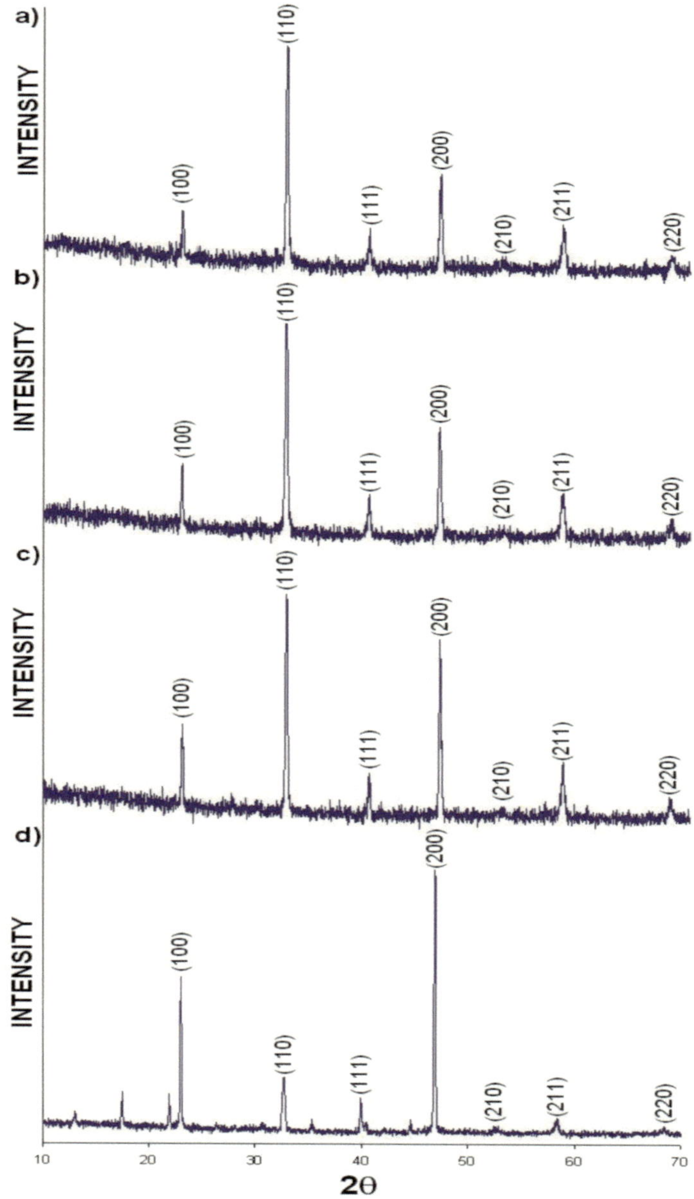

Fig. 9. XRD patterns of $Na_{0.5}Bi_{0.5}TiO_3$ powders fabricated through route (2) via MSS (NaCl - KCl flux) at different temperatures for 4 h: (a) 800°C; (b) 900°C; (c) 1000°C; (d) 1100°C(indexed peaks are those of $Na_{0.5}Bi_{0.5}TiO_3$)

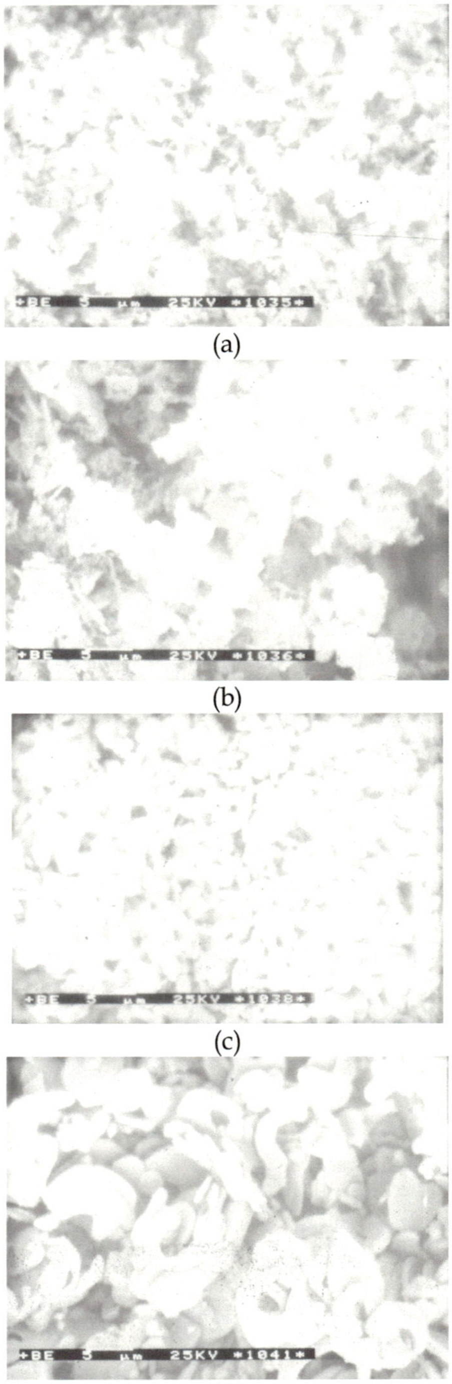

Fig. 10. SEM micrographs of $Na_{0.5}Bi_{0.5}TiO_3$ powders obtained through route (2) via MSS (NaCl - KCl flux) at different temperatures for 4 h: (a) 800°C; (b) 900°C; (c) 1000°C; (d) 1100°C

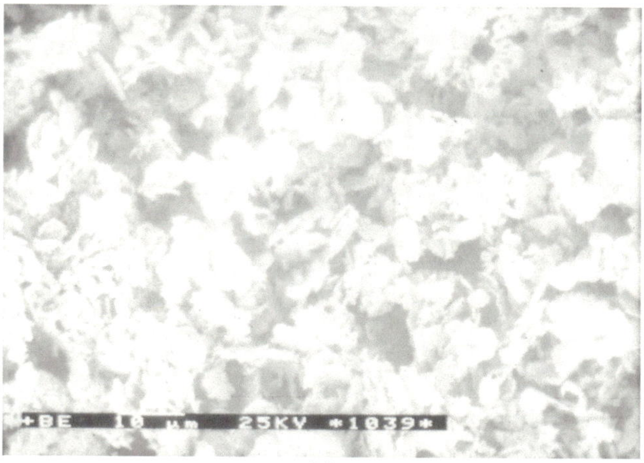

(a)

(b)

(c)

Fig. 11. SEM micrographs of $Na_{0.5}Bi_{0.5}TiO_3$ powders obtained through route (2) via MSS at 1100°C for 4 h: (a) NaCl - KCl flux; (b, c) NaCl flux

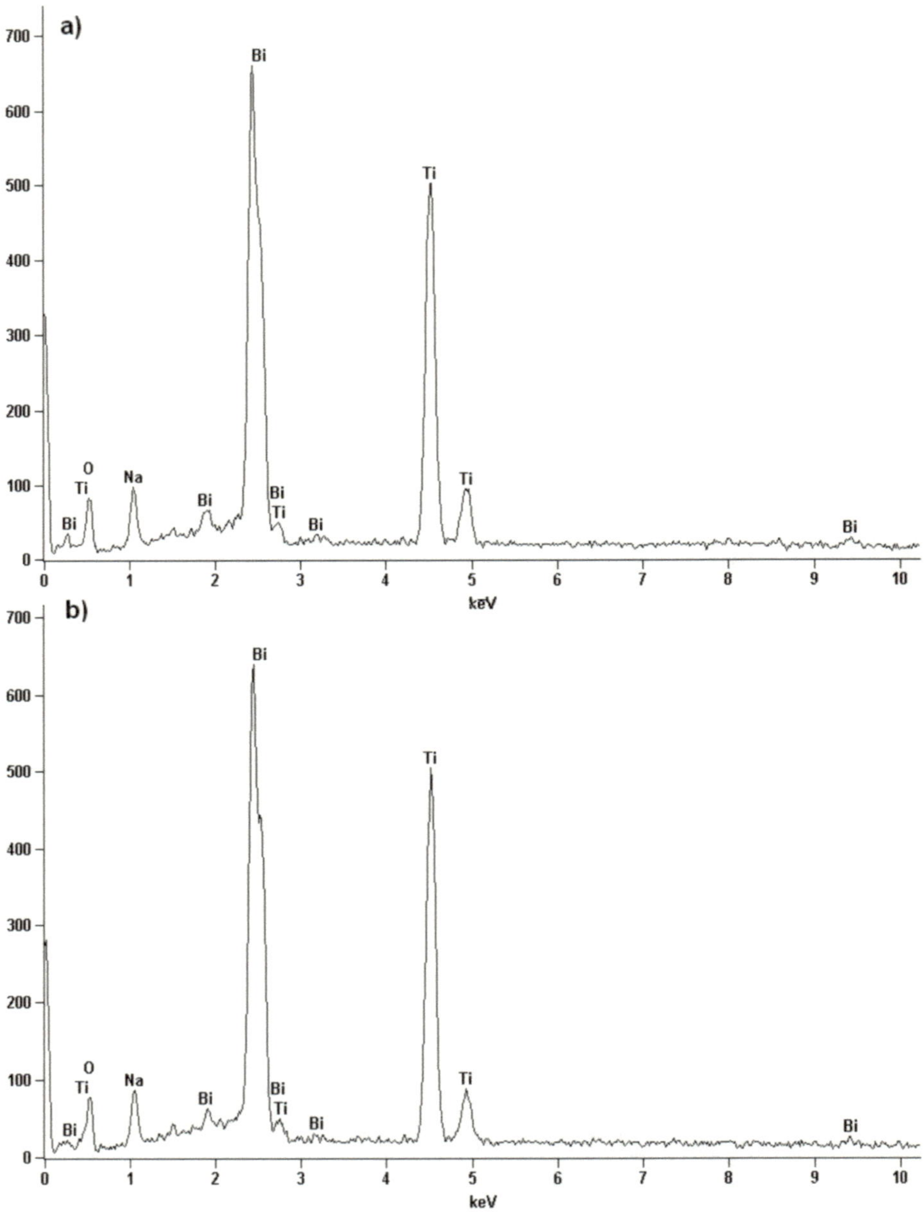

Fig. 12. EDS spectra of of $Na_{0.5}Bi_{0.5}TiO_3$ powders obtained through route (2) via MSS: (a) NaCl- KCl flux; (b) NaCl flux

3. Conclusion

NBT has a perovskite structure with high symmetry, therefore it is difficult to obtain large anisotropic NBT particles by methods as conventional solid-state reaction process or molten salt synthesis. Using Bi_2O_3, TiO_2 and Na_2CO_3 as starting materials, the equiaxed particles NBT were obtained. NBT anisotropic particles with grain orientation were synthesized by conversion of BIT crystals with layered structure. Owing to its highly anisotropic structure, plate-like BIT was firstly fabricated in the NaCl-KCl flux. The plate-like BIT was reacted with the complementary Na_2CO_3 and TiO_2 in the presence of chloride flux, finally transformed to the perovskite NBT and maintained its morphology nearly unchanged. NBT particles show preferred orientation with the (*h*00) plane. The powder synthesized in 0.5 NaCl - 0.5 KCl flux has the larger particles than those synthesized in pure NaCl. The increase of temperature and soaking time of synthesis can make the plate-like grains of NBT more distinct and discrete. The NBT particles prepared in this experiment can be used to prepare ceramics with more uniform grain orientation, i.e., textured ceramics for improving piezoelectric properties.

4. References

Bortolani, F. & Dorey, R.A. (2010). Molten salt synthesis of PZT powder for direct write inks. *Journal of the European Ceramic Society,* Vol.30, (November 2009), pp. 2073-2079, ISSN 0955-2219

Cai, Z.; Xing, X.; Li, L. & Xu, Y. (2008). Molten salt synthesis of lead lanthanum zirconate titanate ceramic powders. *Journal of Alloys and Compounds,* Vol.454, (November 2006), pp. 466-470, ISSN 0925-8388

Cho, J-H.; Ma Y-J.; Lee, Y-H.; Chun, M-P. & Kim, B-I. (2006). Piezoelectric ceramic powder synthesis of bismuth-sodium titanate by a hydrothermal process. *Journal of Ceramic Processing Research,* Vol.7, No.2, pp. 91-94, ISSN 1229-9162

Hao, H.; Liu, H-X.; Liu, Y; Cao, M-h. & Ouyang, S-X. (2007). Lead-Free $SrBi_4Ti_4O_{15}$ and $Bi_4Ti_3O_{12}$ Material Fabrication Using the Microwave-Assisted Molten Salt Synthesis Method. *Journal of the American Ceramic Society,* Vol.90, No.5, (October 2006), pp. 1659-1662, ISSN 0002-7820

Hiruma, Y.; Nagata, H. & Takenaka, T. (2007). Grain Size Effect on Electrical properties of $(Bi_{1/2}K_{1/2})TiO_3$ Ceramics. *Japanese Journal of Applied Physics,* Vol.46, No.3A, (June 2006), pp. 1081-1084, ISSN 0021-4922

Isupov, V.A. (2005). Ferroelectric $Na_{0.5}Bi_{0.5}TiO_3$ and $K_{0.5}Bi_{0.5}TiO_3$ Perovskites and Their Solid Solutions. *Ferroelectrics,* Vol.315, (December 2003), pp. 123-147, ISSN 0015-0193

Kan, Y.; Jin, X.; Wang, P.; Li, Y.; Cheng, Y-B. & Yan, D. (2003). Anisotropic grain growth of $Bi_4Ti_3O_{12}$ in molten salt fluxes. *Materials Research Bulletin,* Vol.38, (June 2002), pp.567-576, ISSN 0025-5408

Kimura, T. & Yamaguchi, T. (1987). Morphology Control of Electronic Ceramic Powders by Molten Salt Synthesis, In: *Advances in Ceramics, Vol.21: Ceramic Powder Science,* G. Messing, K.S., Mazdiyashi, J.W. McCauley & R.A. Haber, (Eds.), 169-177, The American Ceramic Society, Inc., ISBN 978-091-6094-83-6, Westerville, OH, USA

Li, L.; Deng, J.; Chen, J.; Yu, R.; Qiao, L. & Xing, X. (2009). Piezoelectric and ferroelectric properties of $0.96(Na,K)(Nb_{0.9}Ta_{0.1})O_3$-$0.04LiSbO_3$ ceramics synthesized by molten salt method. *Journal of Alloys and Compounds,* Vol.471, (August 2007), pp. 428-431, ISSN 0925-8388

Lu, T.; Dai, J.; Tian, J.; Song, W.; Liu, X.; Lai, L.; Chu, H.; Huang, X. & Liu, X. (2010). Synthesis of $Na_{0.5}Bi_{0.5}TiO_3$ powders through hydrothermal method. *Journal of Alloys and Compounds,* Vol.490, (July 2009), pp. 232-235, ISSN 0925-8388

Mercadelli, E.; Galassi, C.; Costa, A.L.; Albonetti, S. & Sanson, A. (2008). Sol-gel combustion synthesis of BNBT powders. *Journal of Sol-Gel Science and Technology,* Vol.46, (July 2007), pp.39-45, ISSN 0928-0707

Panda, P.K. (2009). Review: environmental friendly lead-free piezoelectric materials. *Journals of Materials Science,* Vol.44, (April 2009), pp. 5049-5062, ISSN 0022-2461

Pookmaneea, P.; Phanichphanta, S. & Heimann, R.B. (2001). Synthesis and Properties of Bismuth Sodium Titanate (BNT). *Berichte der Deutschen Keramischen Gesellschaft,* Vol.78, No.7, (March 2001), pp. 27-30, ISSN 0365-9542

Raghavender, M.; Kumar, G.S. & Prasad, G. (2006). Modification of dielectric relaxations in sodium bismuth titanate with samarium doping. *Journal of Physics and Chemistry of Solids,* Vol.67, (December 2005), pp. 1803-1808, ISSN 0022-3697

Schaak, R.E. & Mallouk, T.E. (2000). Topochemical Synthesis of Three-Dimensional Perovskites from Lammellar Precursors. *Journal of the American Chemical Society,* Vol.122, No.12, (September 1999), pp. 2798-2803, ISSN 0002-7863

Smolenskii, G.A.; Isupov, V.A.; Agranovskaya, A.I. & Krainik, N.N. (1960). New ferroelectrics with complex compounds. IV. *Fizika Tverdogo Tela,* Vol.2, No.11, (June 1960), pp. 2982-2985, ISSN 1063-7834 (in Russian)

Stojanović, B.D.; Paiva-Santos, C.O.; Cilense, M.; Jovalekić, Č. & Lazarević, Z.Ž. (2008). Structure study of $Bi_4Ti_3O_{12}$ produced via mechanochemically assisted synthesis. *Materials Research Bulletin,* Vol.43, (April 2006), pp. 1743-1753, ISSN 0025-5408

Suchanicz, J.; Jeżowski, A.; Poprawski, R. & Dacko, S. (2000). Thermal and Dielectric Behaviour of Pure and Doped $Na_{0.5}Bi_{0.5}TiO_3$ at Low Temperatures. *Physica Status Solidi (b),* Vol.221, (April 2000), pp. 789-795, ISSN 0370-1972

Suchanicz, J.; Mercurio, I.P.; Marchet, P. & Kruzina, T.V. (2001). Axial Pressure Influence on Dielectric and Ferroelectric Properties of $Na_{0.5}Bi_{0.5}TiO_3$ Ceramic. *Physica Status Solidi (b),* Vol.225, (December 2000), pp. 459-466, ISSN 0370-1972

Wang, Y.; Xu, G.; Yang, L.; Ren, Z.; Wei, X.; Weng, W.; Du, P.; Shen, G. & Han, G. (2009). Hydrothermal synthesis and characterization of $Na_{0.5}Bi_{0.5}TiO_3$ microcubes. *Ceramics International,* Vol. 35, (May 2008), pp. 1657-1659, ISSN 0272-8842

Xing, X.; Zhang, C.; Qiao, L.; Liu, G. & Meng, J. (2006). Facile Preparation of $ZnTiO_3$ Ceramic Powders in Sodium/Potassium Chloride Melts. *Journal of the American Ceramic Society,* Vol.89, No.3, (July 2005), pp. 1150-1152, ISSN 0002-7820

Xu, Q.; Chen, S.; Chen, W.; Huang,;D.; Zhou, J.; Sun, H. & Li, Y. (2006). Synthesis of $(Na_{0.5}Bi_{0.5})$ TiO_3 and $(Na_{0.5}Bi_{0.5})_{0.92}Ba_{0.08}TiO_3$ powders by a citrate method. *Journal of Materials Science,* Vol.41, (December 2003), pp. 6146-6149, ISSN 0022-2461

Yoon, K.H.; Cho, Y.S. & Kang, D.H. (1998). Review Molten salt synthesis of lead-based relaxors. *Journal of Materials Science,* Vol.33, (October 1997), pp. 2977-2984, ISSN 0022-2461

Zaremba, T. (2008). Thermoanalytical study of the synthesis of $Na_{0.5}Bi_{0.5}TiO_3$ ferroelectric. *Journal of Thermal Analysis and Calorimetry,* Vol.92, No.2, (June 2007), pp. 583-587, ISSN 1388-6150

Zeng, J.T.; Kwok, K.W. & Chan, H.L.W. (2007). $K_xNa_{1-x}NbO_3$ Powder Synthesized by Molten-Salt Process. *Materials Letters,* Vol.61, (Febuary 2006), pp. 409-411, ISSN 0167-577X

Zhabrev, V.A.; Efimenko, L.P.; Baryshnikov, V.G.; Polyakova, I.G. & Gumennikov, A.V. (2008). Synthesis of $BaTiO_3$ Powders of Different Dispersities by the Exchange Reactions in Molten Salts. *Glass Physics and Chemistry,* Vol.34, No.1, (June 2007), pp. 91-96, ISSN 1087-6596

Zhou, Ch.; Liu, X.; Li, W. & Yuan, Ch. (2010). Dielectric and piezoelectric properties of $YMnO_3$ modified $Bi_{0.5}Na_{0.5}TiO_3$ lead-free piezoelectric ceramics. *Journal of Materials Science: Materials in Electronics,* Vol.21, No.4, (December 2008), pp. 364-367, ISSN 0957-4522

3

Electrical Characterizations of Lead Free Sr and Sn Doped $BaTiO_3$ Ferroelectric Films Deposited by Sol-Gel

Jean-Claude Carru, Manuel Mascot and Didier Fasquelle
Univ. Lille Nord de France, ULCO
France

1. Introduction

Ferroelectric materials (FM) are known for a long time ago but it is only recently that they have been deposited in the form of thin films with a thickness typically inferior to 1 µm. The electrical properties of FM are now used to realize electronic components (Uchino, 2010), thanks to the high quality of ferroelectric thin films. Their main electrical properties are :

- high dielectric permittivity ε'
- tunability of ε' under a DC field E_{DC} : $\varepsilon'(E)$
- non linear polarization P showing an hysteresis cycle : P(E)
- pyroelectric and piezoelectric behaviours

$PbZrTiO_3$ (PZT) is the most used material from a commercial point of view. In fact it has numerous applications such as piezoelectric actuators and transducers, infrared sensors, capacitors, RAM memories, MEMS, etc…However, its main disadvantage is to contain lead with regards to the European directive RoHS (Restriction of Hazardous substances, 2002/95/CE) which prohibits, since july 2006, different polluting materials in electronic components like lead. However lead oxides based FM are still used for their superior piezoelectric properties, but in a near future they will be banned. The search to find an alternative to PZT has begun before the RoHS directive but that concerned mainly the ceramic form. To the best of our knowledge, up to now, no lead free ferroelectric ceramic has superior or even equal piezoelectric performances to PZT.
The chapter contains 5 paragraphs devoted to: material aspects (elaboration and physical characterizations), dielectric, ferroelectric, pyroelectric and piezoelectric, I (V) characterizations of our doped $BaTiO_3$ thin films (either BST or BTS). In the conclusion, their properties will be compared with the ones of PZT films deposited on the same substrate.

2. Materials

The crystallographic structure was characterized by X-Ray $\theta/2\theta$ diffraction using a Rigaku Miniflex+ diffractometer (filtered $CuK_{\alpha 1}$ radiation, $\lambda = 1.5406$ Å).
The morphology of the films was determined using Scanning Electron Microscopy (SEM) with a LEO438 VP apparatus. The thickness of the films was measured both by SEM and

with a TALYSURF INTRA150 profilometer. The typical thickness of our BTS films is about 400 nm.
The electrical properties were determined at room temperature in the frequency range 20 Hz-10^6 Hz as a function of a DC bias (- 40 volts, + 40 volts) with an Agilent HP4284A impedance analyzer coupled to low frequency Cu/Be microprobes from Signatone. The polarization cycles, pyroelectric current and piezoelectric coefficient d_{33} were recorded at room temperature with home-made automated systems.

2.1 Sol-gel deposition

$BaTiO_3$ films doped with Strontium (BST) were prepared from alkaline acetates $Ba(CH_3COO)_2$, $Sr(CH_3COO)_2$ and Ti-isopropoxide $Ti(C_{16}H_{28}O_6)$. Acetic acid ($C_2H_4O_2$) and 2-propanol ($CH_3CHOHCH_3$) were used as solvent agents. Barium acetate and strontium acetate were mixed according to a predetermined ratio of 90/10 and then added in the acetic acid solution. This solution was heated up to 120 °C for 1h. After cooling down to room temperature, Ti isopropoxide was added. Finally 2-propanol was added to control the solution concentration to 0.2 mol%. The final solution was deposited by spin coating at 2500 rpm during 20 s for each layer which was then dried at 300°C for 1 min (Vélu et al, 2003). After deposition of 7 layers, each sample was annealed in air in a tubular oven and these operations were repeated until the wanted final thickness was obtained. BST thin films were deposited by a sol-gel process on commercial (100) silicon wafers from CRYSTAL GmBH with the following layers : 1µm SiO_2 / 40 nm Ti / 200 nm Pt sputtered. Finally gold has been evaporated through a shadow mask to realize the upper electrodes with circular dots ranging from 150 µm to 2 mm in diameter. The Pt layer acted as bottom electrode. The upper electrode diameter was measured by an Olympus BX60 optical microscope.
For $BaTiO_3$ films doped with Tin (BTS), the process was identical to the BST one. The only change concerns dibutyltin oxide $C_8H_{18}OSn$ instead of strontium acetate $Sr(CH_3COO)_2$.
The doping of $BaTiO_3$ with Strontium leads to a substitution of Barium by Strontium (in A site) while the doping by tin entails a substitution of Titanium by Tin (in B site). So, the chemical composition of our BST and BTS films are as follows:

$$BST = Ba_{0.9}Sr_{0.1}TiO_3 \; ; \; BTS = BaTi_{0.98}Sn_{0.02}O_3$$

2.2 Crystallinity

Figure 1 shows the XRD patterns for the BTS thin films annealed at 750 °C for 1 hour (a) and at 950 °C for 15 minutes (b). The films are well crystallized and the perovskite structure is identified. A polycrystalline growth is evidenced for the two films. A change of the orientation from 750 °C to 950 °C is clearly seen with the appearance at 950 °C of the (100) and (111) BTS peaks. Moreover, the large increase of the (110) BTS peak indicates the preferential (110) plane orientation when increasing the annealing temperature to 950 °C. It is a proof of the crystallinity improvement when increasing the annealing temperature. We also observe that the width of the perovskite peaks decreases for the film annealed at 950 °C : this is linked to an increase of the grain size. This fact along with the improvement of the crystallinity will have a favourable effect on the dielectric permittivity and on the ferroelectric properties as will be shown later on (see § 3.1 and 4.1).

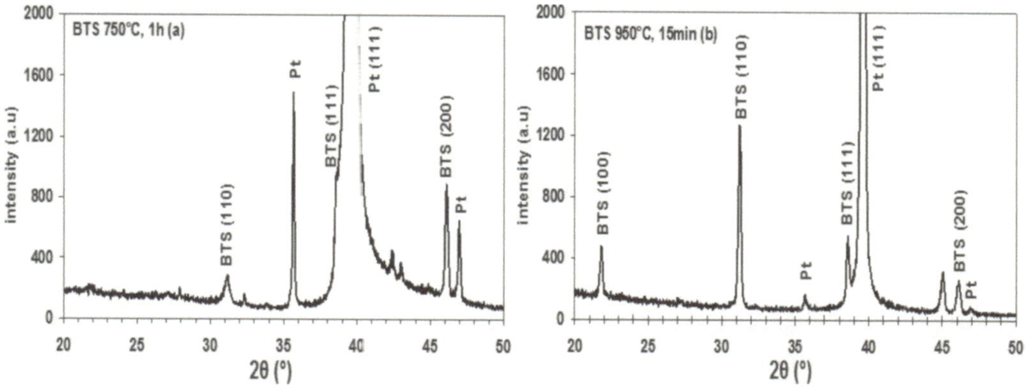

Fig. 1. XRD patterns of BTS thin films annealed at 750 °C during 1 h (a) and at 950 °C during 15 min (b).

2.3 Microstructure

Figure 2 is a SEM micrograph of the films annealed at 750 °C for 1 h (a) and at 950 °C for 15 min (b). We can see a substantial increase of the size of the grains for the film annealed at 950 °C. The mean grain size of this film is about 110 nm against 60 nm for the film annealed at 750 °C. This increase is in agreement with our XRD patterns and with other results on BST films (Malic et al, 2007). Moreover, the surface observation shows a rather good density of grains without cracks.

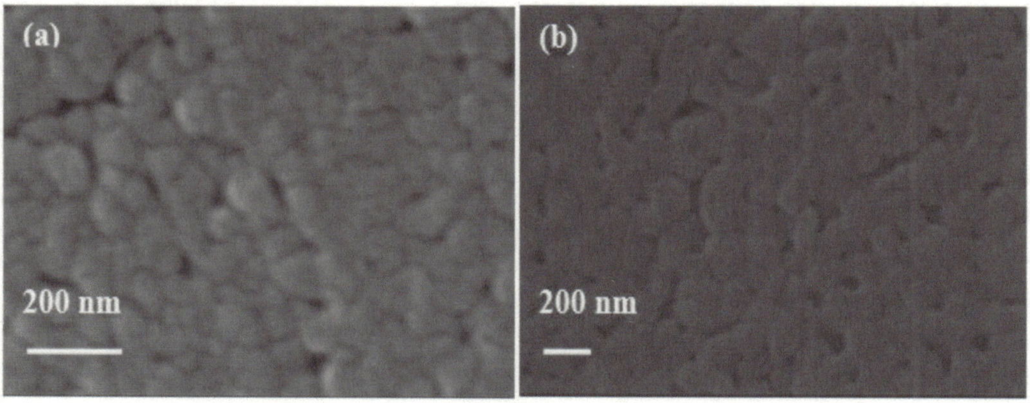

Fig. 2. SEM photographies of BTS thin films annealed at 750 °C during 1 h (a) and at 950 °C during 15 min (b).

3. Dielectric characterizations

The dielectric characterizations consist in determining the complex permittivity ε^* as a function of frequency and temperature :

$$\varepsilon^* (F,T) = \varepsilon'(F,T) - i\varepsilon''(F,T) \tag{1}$$

where ε' is the dielectric permittivity, ε'' is the dielectric losses and the ratio $\varepsilon''/\varepsilon' = tg\delta$ is the loss tangent.

In this section we present the influence : of the annealing conditions of the films, of the thickness of the films and of the substrates. The ferroelectric-paraelectric transition is also studied.

3.1 Influence of the annealing

Annealing of the samples is necessary to crystallize the BST and BTS films in the perovskite structure (Agarwal et al, 2001). Two parameters are important to study : the temperature and the duration of the annealing.

3.1.1 Effect of the temperature

The dielectric permittivity ε′ and loss tangent tgδ responses of the BST thin films annealed at 750 °C, 850 °C and 950 °C during 1 hour are shown as a function of frequency in figures 3a and 3b respectively. It can be noted that for the three samples ε′ decreases in the entire frequency range and in the same manner. For example, the relative decrease ε′(1 kHz) – ε′(1 MHz)/ε′(1 kHz) is equal to about 8 %. This dispersion effect is commonly observed on ferroelectric films and has been attributed to the film-electrode interface effects and domain walls motions. It can be noted that ε′ increases when the annealing temperature increases : indeed, from 750 °C to 950 °C, ε′ for example at 10 kHz has evolved from 330 to 530 which corresponds to a relative increase of 60 %. This significant evolution can be attributed to the increase of the size of the grains and of the enhancement of density. It is in agreement with results obtained on barium titanate ceramics (Arlt et al, 1985; Frey et al, 1998). Previously, we have measured similar values of the dielectric constant for $Ba_{0.8}Sr_{0.2}TiO_3$ films annealed at 750 °C (Mascot et al, 2008).

As concerns the loss tangent presented figure 3.b, for the three samples the curves are very close and merge at 1 MHz towards a value inferior to 0.01. This value is similar to a published one for PZT deposited by a sol-gel route (Sun et al, 2003). We observe for the three samples a decrease of the losses when the frequency increases. So, we could anticipate that the decrease of dielectric losses when increasing the frequency is encouraging for radiofrequencies and microwaves applications.

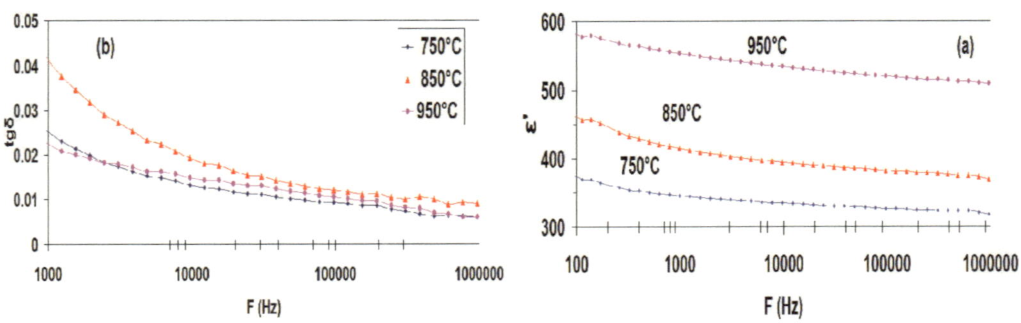

Fig. 3. Dielectric permittivity (a) and loss tangent (b) for BST films annealed at 750 °C, 850 °C and 950 °C during 1 h in air.

3.1.2 Effect of the duration

We present in figure 4 the evolution of the dielectric permittivity for BST films annealed at 950 °C during different durations. It can be seen that above 1 kHz the dielectric permittivity increases when the annealing duration decreases. For example at 1 MHz the values of ε′ are 500, 720 and 830 for duration respectively of 1h, 30min and 15mn. Moreover the film annealed during 15 min has a low value of the loss tangent tgδ = 0.01 at 1MHz.

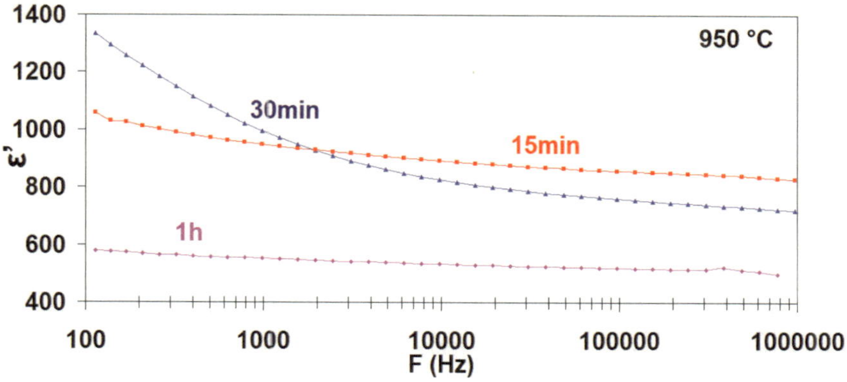

Fig. 4. Dielectric permittivity for BST thin films annealed in air at 950 °C during 15 min, 30 min and 1h.

Then a short time of annealing in air during 15min and at 950°C is most favourable for the dielectric properties.

3.1.3 Microwave annealing

We have made an annealing in a microwave oven (Keyson et al, 2007) at 750 °C during 5 min for a BTS film, 400 nm thick deposited on a Si/Pt substrate. The XRD pattern is presented figure 5.a. It can be compared with one of a BTS film annealed at 750 °C during 1 h in a tubular oven shown figure 1.a. The two BTS films are polycrystalline without preferential orientation whatever the type of annealing. However, with an annealing by microwaves, the amplitude of the peaks is higher and moreover the (100) orientation is observed. So, the crystallisation with a common microwave oven is rather good. It is very interesting as the energy budget is largely inferior to the one with a tubular oven and also as regards the cost, a microwave oven is quite inexpensive.

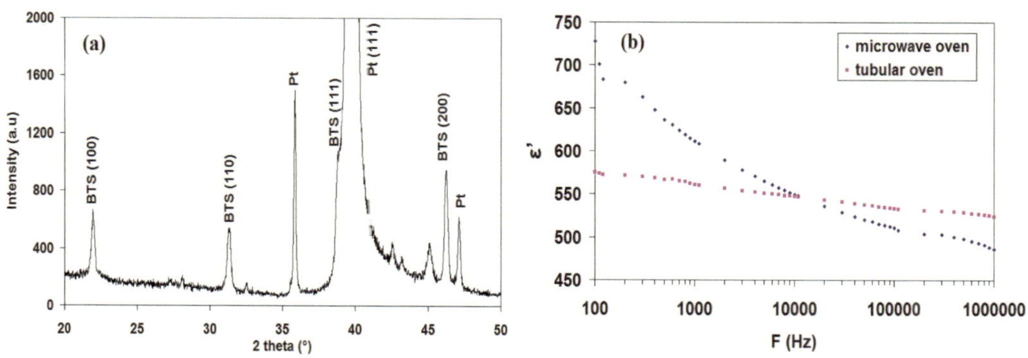

Fig. 5. XRD of a BTS film annealed at 750 °C in a microwave oven (a) and dielectric permittivity of this film and of a BTS film annealed at 750 °C in a tubular oven (b).

The dielectric permittivity measured at room temperature is presented figure 5.b. From 100 Hz to 10 kHz the dielectric permittivity is higher with the microwave annealing whereas it is inferior above 10 kHz. It can be observed that the frequency dispersion is much higher with the microwave annealing than with the conventional one. For example between 100 Hz and 1 MHz it is (728 - 486)/728, i-e 33 % for the microwave annealing and (575 - 524)/575, i-e 9 % for the conventional one annealed at 750 °C, 1 h. As concerns the loss tangent, not shown here, it is higher with a microwave annealing : tgδ = 0.04 at 1 MHz whereas it is only 0.01 with a conventional annealing. Nevertheless, the dielectric properties of a BTS film annealed at 750 °C during 5 min in a microwave oven are very encouraging.

3.2 Influence of the thickness

This study has been made on four BST films with thicknesses varying from 0.4 µm to 1 µm. The dielectric permittivity of these films is presented figure 6.a. We can see that above 1 kHz the dielectric permittivity increases with the thickness. This effect was also observed on a $Ba_{0.7}Sr_{0.3}TiO_3$ film (Parker et al, 2002). In figure 6.b the evolution of ε′ at 1 MHz is presented as a function of the thickness : a linear increase is evidenced from 0.4 to 0.8 µm and then ε′ tends towards saturation at 1 µm.

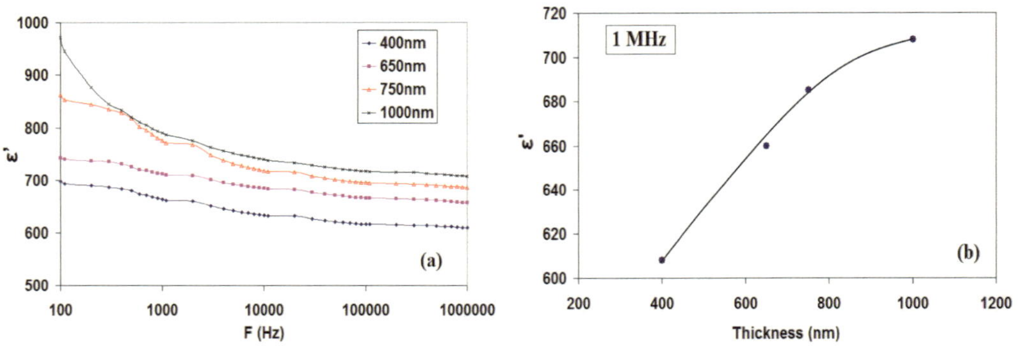

Fig. 6. Dielectric permittivity for four BST films as a function of frequency (a) and at 1 MHz (b) as a function of their thickness.

The increase of the dielectric permittivity can be well explain by the existence of a double layer : a layer with BST material and interface layers (Zhou & Newns, 1997) at the bottom and upper electrodes. These two interface layers are considered in the following as constituting only one interface layer. In fact, the effect of the interface at the bottom Pt electrode is higher than the one at the upper electrode interface due to the deposition conditions. So the measured inverse capacitance $1/C_{meas}$ is the sum of the inverse capacitance interface layer $1/C_i$ and of the BST capacitance layer 1/CBST as follows:

$$1/C_{meas} = 1/C_i + 1/C_{BST} \quad (2)$$

where $1/C_i = h_i/S\varepsilon_0\varepsilon_i$ and $1/C_{BST} = h_{BST}/S\varepsilon_0\varepsilon_{BST}$ with h_i and h_{BST} respectively the thickness of the interface and of the BST layers, ε_i and ε_{BST} the dielectric permittivity respectively of the interface and of the BST layers, S is the area of an electrode dot (250 µm in diameter). The evolution of $1/C_{meas}$ at room temperature and at 100 kHz is given figure 7 as a function of the thickness "h" of the films with:

$$h = h_i + h_{BST} \quad (3)$$

The curve obtained is a straight line, as for other authors (Cho et al, 2005) and whatever the measurement frequency. So, it means that the thickness h_i of the interface is negligible compared with the thickness h_{BST} of the BST layer. Then, if we consider in a first approximation that $h \sim h_{BST}$:

$$1/C_{meas} = 1/C_i + h/S\varepsilon_0\varepsilon_{BST} \quad (4)$$

So, from the slope of the straight line, we can calculate the dielectric permittivity of the BST layer : a value of $\varepsilon_{BST} = 844$ is obtained. Moreover, the extrapolation of the curve for $h = 0$ gives the value of $1/C_i$ which is about $4x10^8$ F^{-1}. With this value it is possible to estimate the thickness of the interface layer in nanometer : h_i (nm) = $0.18\varepsilon_i$. The value of ε_i is not known but surely inferior to ε_{BST} as, from a detailed study of the evolution of C_{meas} with a DC electric field (not discussed here), the interface is non ferroelectric (Chase et al, 2005; Mascot, 2009).
It has to be pointed out that some other effects can have an influence on the dielectric permittivity such as grain boundaries, mechanical strain at the interface with the substrate, internal stress and ferroelectric domains.

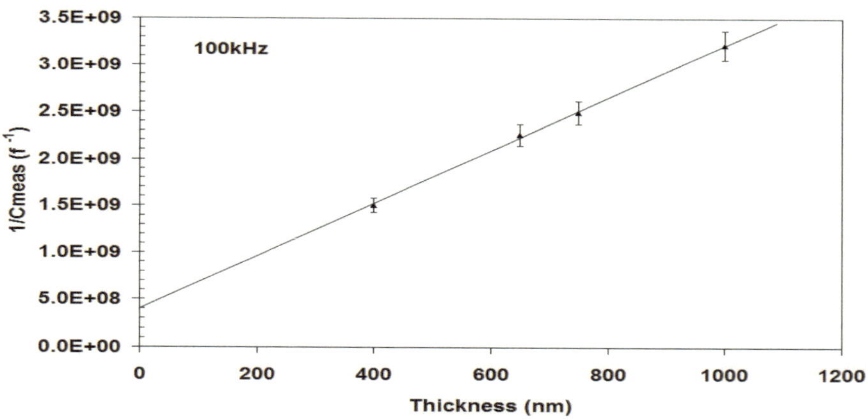

Fig. 7. Inverse of the capacitance at 100kHz for four BST films as a function of their thickness.

3.3 Influence of the substrate

We have deposited BST films, in the same conditions and annealed at 750 °C during 1 hour on stainless steel (type 304) and on platinized silicon substrates with a thickness of 400 nm. In figure 8 we present the evolution of the dielectric permittivity of these two films. It can be seen that the dielectric permittivity of the BST film is much higher with a Si/Pt substrate than with a steel substrate. In particular at 1 MHz, $\varepsilon' = 50$ with a steel substrate whereas it is 350 with a Si/Pt substrate. This can be explained by the existence of a large interface layer, which is not ferroelectric, between the steel substrate and the BST film. Then the capacitance C_i associated to this interface is the dominant term of the measured capacitance C_{meas}. Nevertheless we will show later on (§ 4.2) that the BST film deposited on stainless steel exhibits ferroelectric properties. In order to improve its dielectric properties a buffer layer would be necessary.

We have also deposited BST films on monocrystalline substrates such as sapphire and MgO. The objective was to evaluate the dielectric properties in high frequencies in order to realize tunable microwave components. We give here only some references (Burgnies et al, 2007; Houzet et al, 2008, 2010; Khalfallaoui et al, 2010) concerning our work as it is out of the scope of this chapter which is focused on low frequencies from DC to 1 MHz.

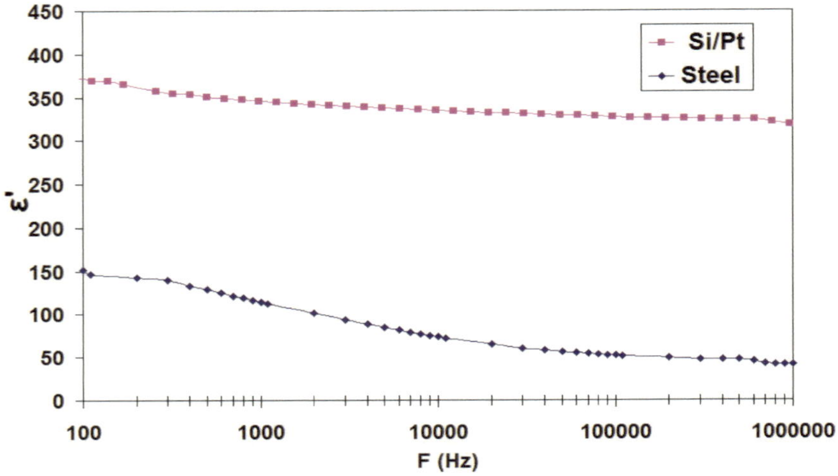

Fig. 8. Dielectric permittivity of BST films deposited on Si/Pt and stainless steel substrates annealed at 750 °C, 1h.

3.4 Ferroelectric-paraelectric transition

It has to be noted that T_C = +70 °C is independent of the duration of the annealing between 15 min and 1 h. The small variation of the dielectric permittivity between 20 °C and 70 °C may be interesting, for example, to realize electronic components such as integrated capacitors.

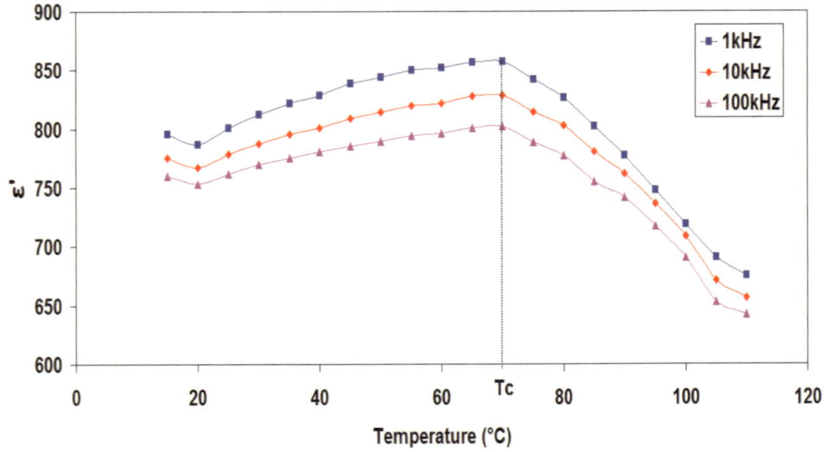

Fig. 9. Dielectric permittivity of a BST film at three frequencies as a function of temperature.

The evolution of the dielectric permittivity as a function of temperature is presented figure 9 for a BST film, 400 nm thick, annealed at 950 °C during 15 min. The plots are given at three frequencies : 1, 10 and 100 kHz. It can be seen that the shape of the three curves is exactly the same whatever the measurement frequency. A maximum is observed at T_C = +70 °C which corresponds to the ferroelectric to paraelectric transition. This transition is not as narrow as with BST in ceramic form : on BST films the transition is diffuse (Lorenz et al, 2003). Moreover, the T_C value is lower than the one found on a BST bulk material (Frayssignes et al, 2005) where Tc = +84 °C. This difference may be related to the small grain size (Frey et al, 1998; Parker et al, 2002) in a film compared with a ceramic and could also be due to a small difference of chemical composition as the transition temperature is very sensitive to Ba/Sr ratio.

4. Ferroelectric characterizations

Ferroelectric characterizations are made with two complementary techniques using a DC electrical field E_{DC} : capacitance measurement C(E) and hysteresis cycle recorded from polarization measurement P(E). These measurements are performed at the same medium frequency such as 1 kHz or 10 kHz.

4.1 Influence of the annealing

Different annealing conditions are studied on BST and BTS films, mainly the temperature and the duration of the annealing. Correlations are evidenced between the dielectric permittivity, the tunability and the size of the grains in the films.

4.1.1 Effect of the temperature

Figure 10.a shows the evolution at 10 kHz and room temperature of the capacitance as a function of a DC electric field for BST films annealed at different temperatures during 1 hour. The corresponding DC voltage variation was in the range [-5 V, +5 V]. The butterfly shape of the curves attests of the ferroelectric behaviour of these films because the polarization and capacitance vary non-linearly with the applied field due to the structure of ferroelectric domains. The shift on left of all these curves is due to the imbalance of space charges at the two different electrodes which may create an internal bias. In our case, BST was deposited on platinized silicon as bottom electrode and gold was used for the top electrode. This difference in metal used can explain the shift to the left of the curves. We also note on each curve a small difference in the height of the two peaks : this effect highlights the existence of a depletion layer capacitance. This depletion layer, also called previously "interface layer", was created between the ferroelectric film and the bottom electrode during the annealing stage. This interface layer, probably non ferroelectric, is due in our case to a diffusion process in our films at high temperature of metals (Pt, Ti) used as bottom electrode (Hu et al, 2006).
The tunability is defined by the following relation :

$$n_r = 1 - C_{min}/C_{max} = 1 - \varepsilon'(E_{max})/\varepsilon'(0) \qquad (5)$$

For the annealing temperatures of 750°C, 850°C and 950°C during 1 hour, the tunability for a DC field of 125 kV/cm is about 30 %, 35 % and 63 % respectively. This property is very interesting for microwaves applications such as phase shifters, tunable filters and electronic antennas arrays.

The polarization cycles were recorded at room temperature also at 10 kHz. The three films showed hysteresis cycles with a quite symmetrical shape as can be seen on figure 10.b. We observe an increase of the remnant polarization when the annealing temperature increases. The remnant polarization P_r has evolved from 1 µC/cm² for an annealing at 750°C to 4.5 µC/cm² for an annealing at 950°C. This evolution is similar to the one of the dielectric permittivity previously observed for the different annealing temperatures (see Fig. 3). The presence of a remnant polarization confirms that our films are in a ferroelectric state. However our results are not as good as those for PZT films, P_r = 22 µC/cm² (Arlt et al, 1985) but they are similar to those for $Ba_{0.8}Sr_{0.2}TiO_3$ films, P_r = 5 µC/cm² (Mascot et al, 2008; Pontes et al, 2001).

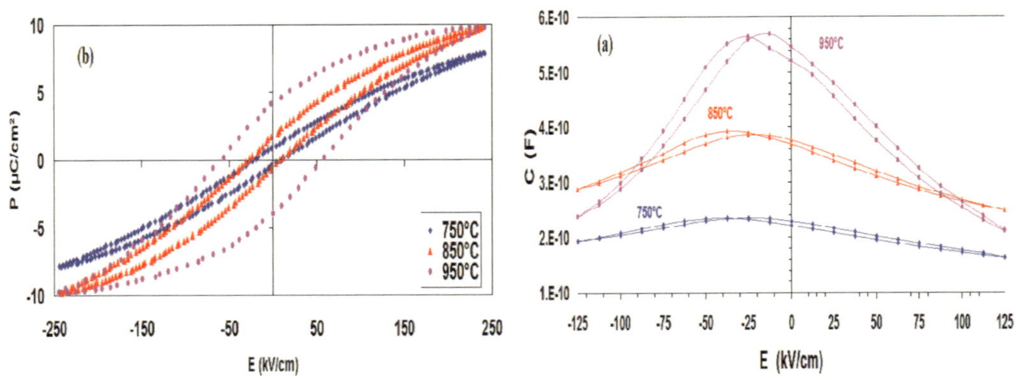

Fig. 10. Evolution at 10kHz as a function of a DC field of the capacitance (a) and of the polarization (b) for BST films annealed at 750 °C, 850 °C and 950 °C during 1h in air.

4.1.2 Effect of the duration

Figure 11 shows the evolution at 1 kHz of respectively the capacitance (a) and polarization (b) as a function of a DC field for 3 annealing duration at the optimum temperature of 950°C. It can be noted that the tunability exhibits a mean value of about 55% whatever the annealing duration. We observe that the C(E) plots of films annealed 15 min and 30 min are centred on the abscissa, while the film annealed 1 hour presents an hysteresis cycle shifted to the left. This shift was observed previously (see Fig. 10.a) on all films annealed during 1 hour. So, we infer that a shorter annealing time induces a lower diffusion process of (Pt, Ti) in the BST film. We think that a shorter annealing time has decreased the interfacial layer thickness because the C(E) curves became centred, showing that the internal bias has disappeared. The interest in a rapid thermal annealing is now well known for thin films and experimented with high power halogen furnaces. As shown previously for the dielectric properties, this minimized effect of the interfacial layer was also evidenced by a significant increased of the dielectric constant value (see Fig. 4). In figure 11.b we observe on the hysteresis cycles that the remnant polarization is quite constant at 5 µC/cm² and also the coercive field at about 50 kV/cm when the annealing time decreases.

We have also studied the influence of the thickness for BST films annealed at 950 °C for 15min. In fact we have observed no effect neither on the tunability, nor on the remnant polarization for film thicknesses ranging from 0.3 µm to 1 µm.

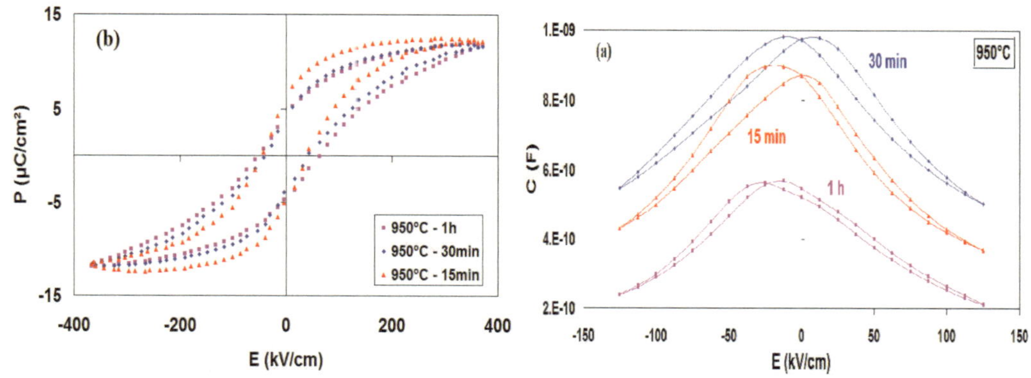

Fig. 11. Evolution at 1 kHz as a function of a DC field of the capacitance (a) and of the polarization (b) for BST films annealed at 950 °C during 1h, 30 min and 15 min in air.

4.1.3 Correlations

We present, figure 12.a , the evolution of the tunability at 10 kHz and of the grain size as a function of the dielectric permittivity for five different annealing temperatures and durations. It is clear that the grain size, the dielectric permittivity and the tunability increase when the annealing temperature increases from 750 °C to 950 °C during the same duration namely 1 hour. A similar evolution was also observed in the case of barium titanate ceramics (Frey et al, 1998). Conversely, at 950 °C, when the annealing duration decreases from 1 hour to 15 minutes, the grain size and the tunability remain constant to about 110 nm and 55 % respectively whereas the dielectric permittivity increases from 530 to 780. So, the annealing duration at 950 °C has no effect on the grain size with a maximum of one hour duration. This is in agreement with other works on BST thin films (Malic et al, 2007).

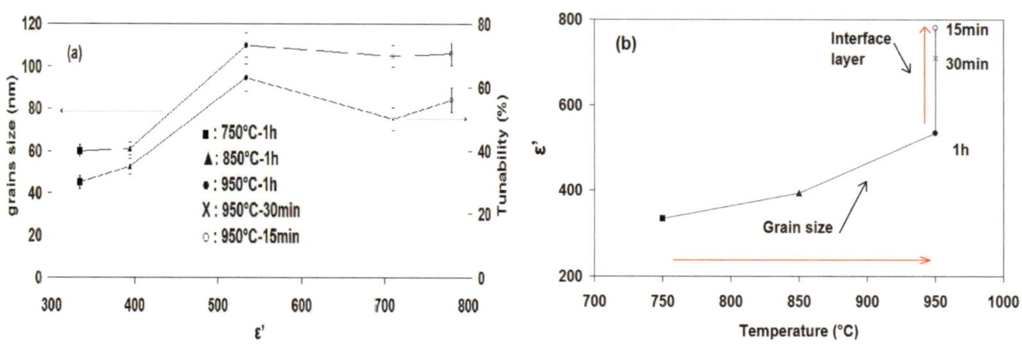

Fig. 12. Correlation of the tunability and of the grain size with the dielectric permittivity for BST films (a) evolution of ε' at 10 kHz as a function of different annealing conditions (b).

From figure 12.a we can see clearly that the tunability is correlated with the grain size. When the grain size increases, the number of grain boundaries decreases. Then, as the tunability increases markedly, this implies that these grain boundaries are non-ferroelectric. In the same way, figure 12.b, the continual increase of the dielectric permittivity can be attributed at first by the grain size increase and then to the interface layer thickness decrease. This is well interpreted by a brick-wall model (Mascot, 2009).

4.1.4 Effect of tin doping : BTS films

The evolution of the capacitance as a function of an electrical field E is given figure 13.a for two BTS thin films annealed at 750 °C during 1 hour and at 950 °C during 15 minutes. The DC electric field variation was in the range [-225 kV/cm; +225 kV/cm] and the measurement frequency was 10 kHz. The butterfly shape of the curves attests to the ferroelectric behaviour of the two films. The tunability for the annealing temperature of 750 °C@1 h is about 40 % and for an annealing 950 °C@15 min it is 76 % under a bias of +225 kV/cm. From figure 13.b it can be seen that for this field the tunability is close to its saturation value. To the best of our knowledge, the value of 76% is the highest reported for a doped $BaTiO_3$ thin film deposited by sol-gel (Mascot et al, 2011). Indeed, for example 45 % have been obtained but with a very high electric field of 400 kV/cm for a $BaSn_{0.05}Ti_{0.95}O_3$ film (Song et al, 2006). Our tunability of 76% corresponds to a variation by a factor 4 of the capacitance and by consequence of the dielectric permittivity. Such a variation can be well compared to the one of a varactor diode and is very interesting, as stated before, for the realization of tunable devices in radiofrequencies and microwaves.

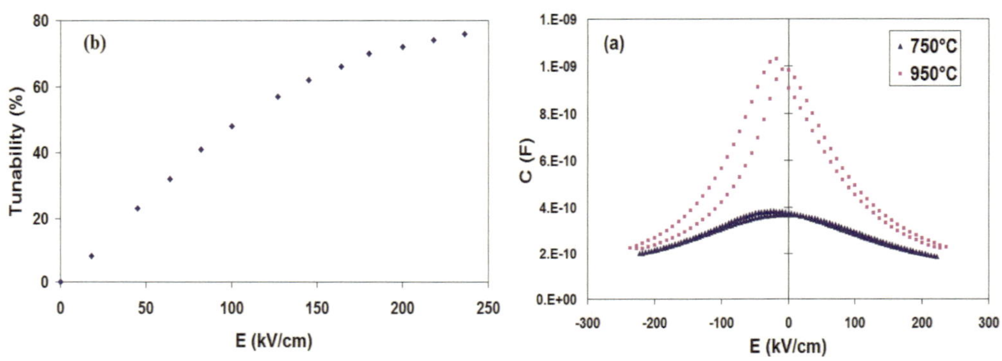

Fig. 13. Evolution at 10kHz of the capacitance as a function of a DC field for BTS films annealed at 750 °C and 950 °C (a) tunability evolution of the BTS film annealed at 950 °C.

4.2 Influence of the substrate

We study here the ferroelectric properties of BST films deposited on a low cost stainless steel substrate compared with the ones on higher cost Si/Pt substrates.

4.2.1 C(E) and hysteresis cycle

We consider the BST films deposited on stainless steel substrates on which we have made dielectric measurements (§ 3.3). In figure 14.a we present the evolution at 1 kHz of the reduced values of the capacitance as a function of a DC field for BST films deposited in the same conditions either on Si/Pt or stainless steel substrates and annealed at 750 °C during 1 hour. We can see that the tunability is largely inferior with the steel substrate than with the Si/Pt one. In fact it is only 9 % under a DC field of 250 kV/cm with a steel substrate whereas it is 30 % at only 125 kV/cm with a Si/Pt substrate. This shows that a thick non-ferroelectric interface layer is present between the BST film and the steel substrate. Consequently, the effective DC field applied to the ferroelectric BST layer is :

$$E_{DC}/(1+ C_{BST}/C_i) \tag{6}$$

with C_{BST} being the capacitance of the BST layer and C_i being the capacitance of the interface layer.

The hysteresis cycle figure14.b confirms the weak ferroelectric nature of the BST film deposited on stainless steel as the cycle is not saturated in comparison with the one of the BST film deposited on Si/Pt substrate.

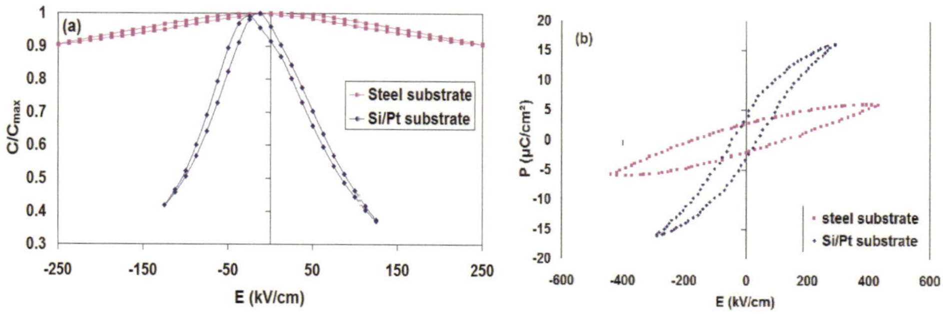

Fig. 14. Evolution at 1kHz as a function of a DC field of the reduced capacitance (a) and of the polarization (b) for BST films deposited on stainless steel and Si/Pt substrates.

4.2.2 Effect of an AC field

We have applied a variable AC field E_{AC} at a frequency of 10 kHz to the BST film deposited on stainless steel. Usually, for the dielectric measurements, the AC field applied is very small, typically 1 kV/cm. We can see, figure 15, that above this field and up to 50kV/cm, the dielectric permittivity has a constant value of $\varepsilon'(E_{AC}) = 50$. This behaviour characterizes a linear dielectric material : in our BST film it is due to the interface layer which is non-ferroelectric. However, above 50kV/cm, the dielectric permittivity increases : it is typical of a non-linear dielectric material. It is due here to the ferroelectric layer of the BST film. It is in relation with the Rayleigh law (Taylor & Damjanovic, 1998) :

$$E'(E_{AC}) = \varepsilon'(0) + \alpha' E_{AC} \quad (7)$$

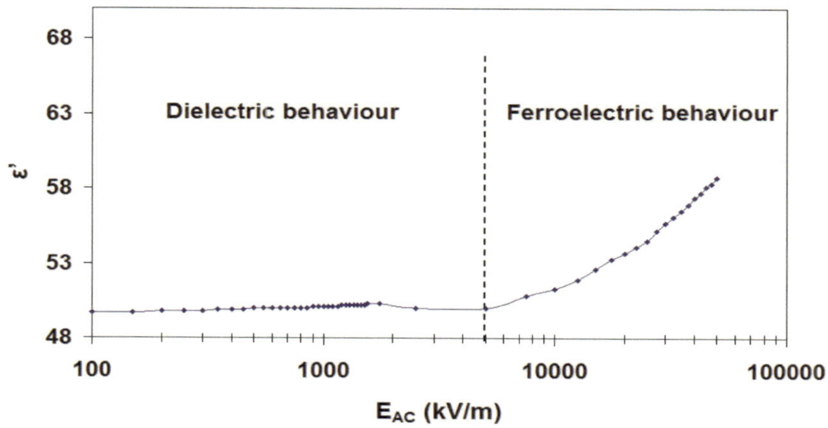

Fig. 15. Evolution at 10kHz of the dielectric permittivity as a function of an AC electrical field for a BST film deposited on a stainless steel substrate.

where ε'(0) is the dielectric permittivity without AC field and α' is a constant linked to the movements of the ferroelectric domain walls under an AC field which increases the size of the domains. This confirms the existence of some ferroelectric domains in our sample. The increase of the dielectric permittivity is from a field of 50kV/cm which is at variance (not presented here) with a BST film deposited on Si/Pt substrate where the Rayleigh law is followed from a zero AC field with $\alpha' = 4.6 \times 10^{-5}$ m/V.

So, this study of the dielectric permittivity as a function of an AC field confirms both the existence of a major non ferroelectric interface layer and of ferroelectric domains in our BST film deposited on a steel substrate.

4.3 Paraelectric state

We have measured the evolution at 10 kHz and at a temperature of 100 °C of the capacitance under a DC field of a BST film annealed at the optimized temperature of 950 °C during 15 min. At this temperature the BST is in the paraelectric state as the curve C(E), figure 16.a, shows no (or a very small) hysteresis effect. We have fitted the corresponding dielectric permittivity following the Landau-Ginzbourg-Devonshire (LGD) model (Johnson, 1962) with the following formula :

$$\varepsilon'(E) = \frac{\varepsilon'(0)}{\left(1 + 12 C_3 \varepsilon_0^{\,3} \varepsilon'(0)^3 E^2\right)^{\frac{1}{3}}} \tag{8}$$

where ε'(0) is the dielectric permittivity without DC field and C_3 is a constant in the LGD model. C_3 was determined from our experimental results (Johnson, 1962; Outzourhit et al, 1995). The agreement is very good between the experimental values of $\varepsilon'(E_{DC})$ and the ones from the LGD model as can be observed figure 16.b. In fact, there is a maximum deviation of 2 %. So, from the LGD model, it can be seen that the dominant parameter for the tunability is the value of $\varepsilon'(0)^3 C_3$. A high value of the dielectric permittivity at zero field ε'(0) is then favourable to obtain a high tunability.

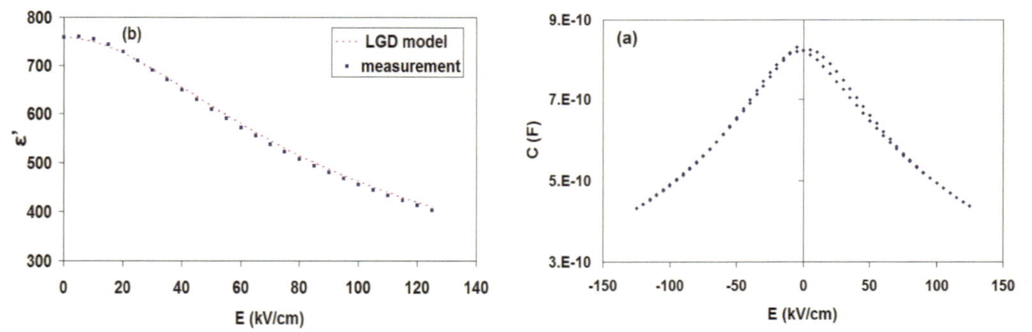

Fig. 16. Evolution at 10 kHz and 100 °C as a function of a DC field of the capacitance (a) and of the dielectric permittivity fitted with LGD model (b) for a BST films annealed at 950 °C.

5. Pyro and piezo-electric characterizations

In the ferroelectric state, contrary to the paraelectric state, it is possible to test the pyroelectric and piezoelectric properties : it is important in view of realizing sensors and actuators.

5.1 Pyroelectric characterizations

The pyroelectric coefficient γ has been determined from the measurement of the pyroelectric current I_p following the Bayer-Roundy method (Bayer & Roundy, 1972) leading to :

$$I_p = \gamma S\, dT/dt \qquad (9)$$

where S is the electrode area, dT/dt is the rate of change of the temperature. This rate is sinusoidal as shown in figure 17a, thanks to a Peltier module. The magnitude of the pyroelectric current, figure 17a, is very low, typically some picoAmps. We have calculated a value $\gamma = 140\ \mu C/m^2K$ at +25 °C for a BTS thin film annealed at 950°C during 15 min.

We present figure 17.b the evolution of the pyroelectric coefficient γ as a function of temperature for two BTS films annealed at 750 °C during 1 h and at 950 °C 15 min. We can see that the pyroelectric coefficient γ varies linearly, in a first approximation, with temperature. For the BTS film annealed at 950 °C@15 min, it increases for example from $140\mu C/m^2K$ at +25 °C to $240\mu C/m^2K$ at +100 °C which is an increase of 70 %. For the BTS film annealed at a lower temperature of 750 °C@1h, the pyroelectric coefficient γ is inferior by a factor of 1.4. So, the pyroelectric properties confirms that the dielectric (see figure 3.a) and the ferroelectric (see figure 10.b) properties are much better with an annealed at 950 °C than at 750 °C. For the two BTS films the pyroelectric coefficient γ reaches a maximum for a temperature of +105 °C which corresponds to the ferroelectric to paraelectric transition.

In order to realize sensors for example, it is necessary to determine the figure of merit defined (Zhang & Ni, 2002) as :

$$FOM = \gamma / d\varepsilon' tg\delta \qquad (10)$$

with d being the thickness of the film. For example, for a BTS film (annealed at 950 °C@15 min) with a thickness of 0.4 µm, we obtain 40 $\mu C/m^3K$ at 10 kHz and +25°C. This value compares well with published data on BST films (Liu et al, 2003).

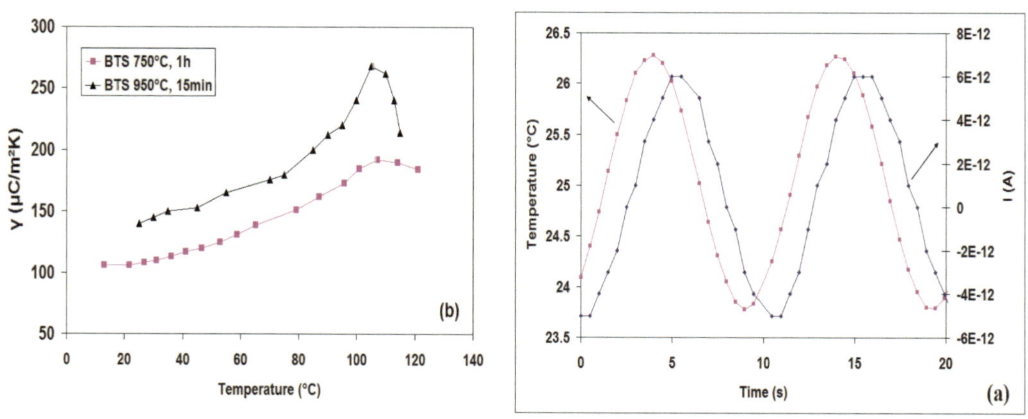

Fig. 17. Temperature and pyroelectric current as a function of time (a) pyroelectric coefficient versus measurement temperature (b) for two BTS films.

5.2 Piezoelectric characterizations

These characterizations are presented on BST films at two different scales : macroscopic one at millimetre level and nanoscopic one at nanometre level.

5.2.1 Macroscopic scale

The piezoelectric properties are determined by tensors with terms d_{ij}. In the case of our doped $BaTiO_3$ films which have a tetragonal crystalline structure, the piezoelectric matrix is mainly characterized by the terms d_{31} and d_{33}. We have developed a set-up to measure d_{33} by modifying a method proposed in the literature (Lefki & Dormans, 1994). It consists in applying a variable force on the film to measure the resulting electrical charge. The d_{33eff} effective coefficient is the derivative of the electrical charge by the force as follows :

$$d_{33eff} = dQ/dF \tag{11}$$

In figure 18 we present, for a BST film of thickness 1μm, the evolution at room temperature of the electrical charge as a function of a force applied from 0 to 10 N. We obtain d_{33eff} = 19 pC/N and 85 pC/N for a 1 μm thick PZT film deposited on the same substrate. This last value is close to 100 pC/N obtained with the same measurement method also for a PZT film (Ren et al, 1997). So the piezoelectric coefficient for BST is about 4.5 less than the PZT one.

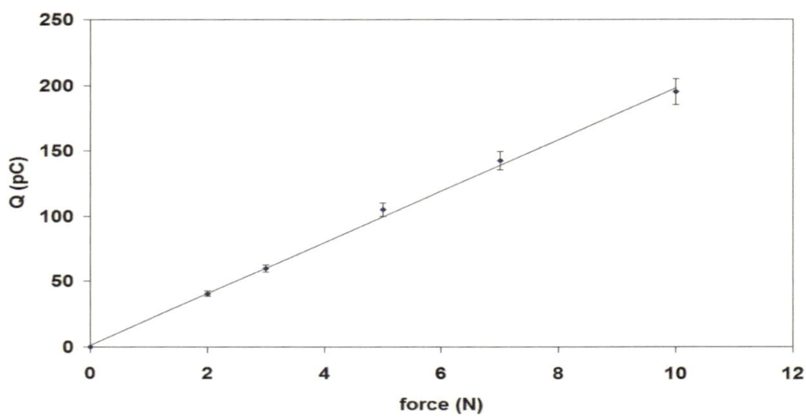

Fig. 18. Direct piezoelectric macroscopic response for a BST film, 1 μm thick.

5.2.2 Nanoscopic scale

Piezoelectric properties at the scale of a grain (size of about 110 nm) are determined by piezo force microscopy (PFM). The piezo response is obtained via the tip of the PFM which operates in contact mode. A DC field is applied between the tip and the bottom electrode of the film. A small AC field is applied to the tip in order to induce the vibration of the grain. We present, figure 19, the inverse piezoelectric response of a BST film deposited on Si/Pt substrate and annealed at 950 °C@15 min.

The magnitude, figure 19.a, represents the mechanical motion of the grain under a variable DC voltage between -10 volts and +10 volts. The butterfly shape of the curve attests of the ferroelectric and of the piezoelectric properties of the grain tested by PFM. The phase, figure 19.b, represents the switching of the electrical polarization of the grain from a negative DC field to a positive DC field and conversely. The hysteresis shape attests of the piezoelectric properties at the nanoscopic level of a grain. We can see that the phase change is about 160 ° which is close to 180 ° for a complete reversal of the polarization of the grain. It can be noted that the field necessary for the switching is 46 kV/cm which is very close to the value of the coercive field, i.e. 50 kV/cm of an hysteresis cycle at a macroscopic scale (see figure 11.b).

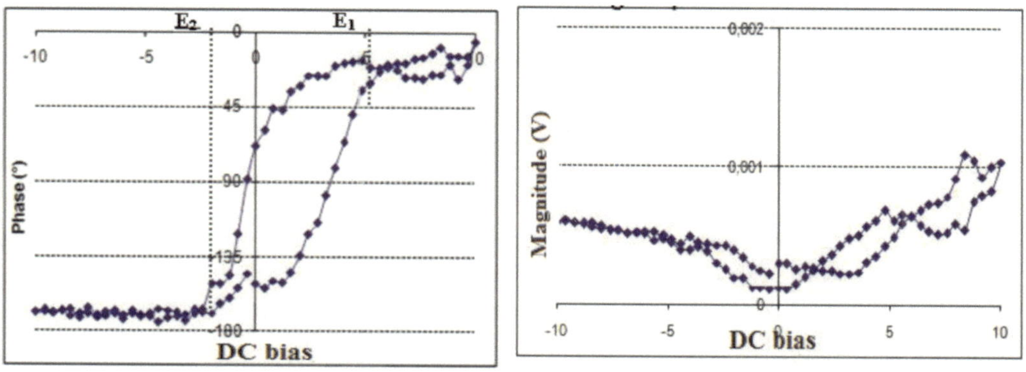

Fig. 19. Inverse piezoelectric nanoscopic response in magnitude (a) and phase (b) for a BST film.

6. J (E) characterizations

We have studied the current density J as a function of a DC field for BST and BTS films annealed at 950 °C during 15min. We present, figure 20, the evolution of J(E) in logarithmic scales for the two films under positive and negative bias. It can be seen that the conduction is very different with the two biases. For example, for the BTS film, J = 0.06 A/m² with E = +25 MV/m whereas J = 12 A/m² with E = -25MV/m. So, there is a ratio of 200 for these two current density which leads to a much higher leakage current with negative DC field bias. As with this bias the electrons come from the gold upper electrode, the electrical conduction is more important than with a positive bias for which the electrons come from the platinum bottom electrode. In order to explain that behaviour we have tested the different physical models of conduction available in the literature. They were mainly formulated for semi-conductor materials and adapted to ferroelectrics (Scott et al, 1991), namely Schottky barrier, Poole-Frenkel, Fowler-Nordheim and space charge mechanisms.

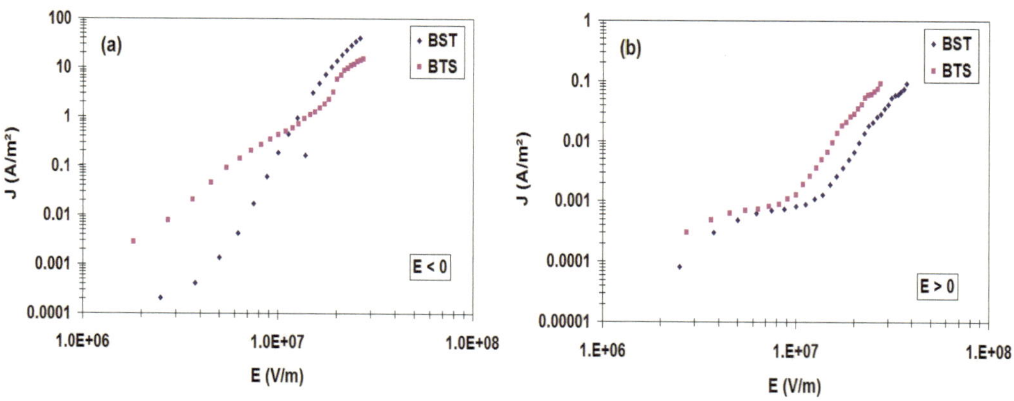

Fig. 20. Evolution of the current density as a function of a negative DC bias (a) and a positive DC bias (b) for BST and BTS films annealed at 950 °C during 15 min.

6.1 Schottky barrier

The Schottky mechanism is the major conduction mechanism at room temperature for ferroelectrics (Scott et al, 1991) from 1MV/m to some ten's of MV/m. The Schottky equation (Hwang et al, 1998) is as follows :

$$J = A^* T^2 \exp\left[\frac{\beta_S\sqrt{E}-\varphi_S}{K_B T}\right] \tag{12}$$

with $A^* = (4\pi e m^* K_B^2/h^3)$ where $e = 1.6x10^{-19}$ C, m* is the effective electron mass, K_B and h are the Boltzmann and the Planck constants, T is the temperature, $\beta_S = (e^3/4\pi\varepsilon_0\varepsilon)^{1/2}$ where $\varepsilon_0 = 1/36\pi x10^9$ F/m and ε is the dielectric permittivity at very high frequency, E is the applied DC field, φ_S is the Schottky barrier height.

It is useful to express log J/T^2 as it varies linearly with $E^{1/2}$:

$$\log\frac{J}{T^2} = \frac{\ln A^*}{\ln 10} - \frac{\varphi_S}{K_B T\ln 10} + \frac{\beta_S}{K_B T\ln 10}E^{1/2} \tag{13}$$

In the following we will consider, as an example, the case of a BTS film.

A linear evolution of log J/T^2 is observed only for a negative DC field as shown figure 21.a. We present also the measurement of the current density at different temperatures from 293 K to 393 K in order to determine φ_S. In this view, from the Schottky equation, we can express $\ln J/T^2$ as a function of 1/T as follows :

$$\ln\frac{J}{T^2} = \ln A^* + \left[\frac{\beta_S\sqrt{E}-\varphi_S}{K_B}\right]\frac{1}{T} \tag{14}$$

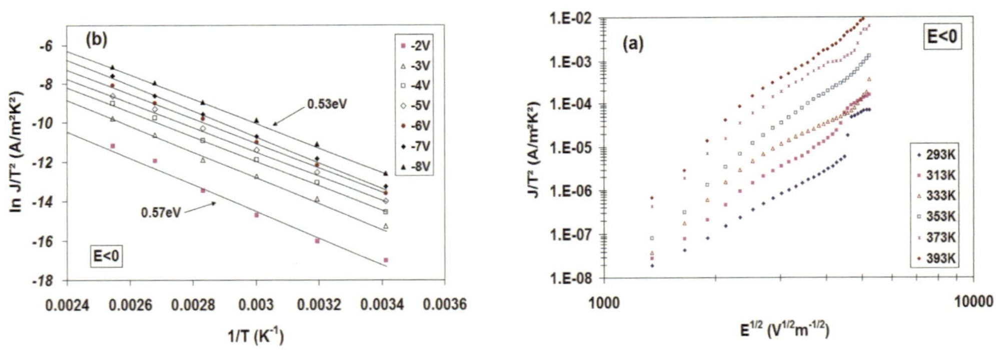

Fig. 21. Evolution of the current density as a function of a negative DC bias (a) and as a function of 1/T (b) for a BTS film.

We have replaced, figure 21.b, the DC electrical field E by V/d where V is the applied voltage and d is the thickness of the film, so $E^{1/2} = (V/d)^{1/2}$

The evolution of the activation energy E_A as a function of the applied voltage V is given by formula (15). The extrapolation, figure 22, at $V^{1/2} = 0$ volt gives the Schottky barrier height. Then we obtain $\varphi_S = 0.62$eV.

$$E_A\ (eV) = (-\varphi_s + \beta_s\sqrt{\frac{V}{d}})\times\frac{1}{e} \tag{15}$$

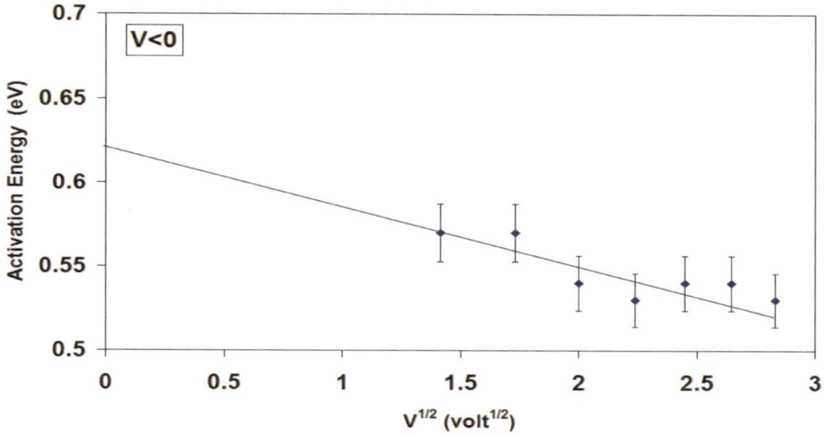

Fig. 22. Evolution of the activation energy as a function of a DC negative bias for a BTS film.

6.2 Space charge mechanism

The formula used to identify a space charge mechanism is a quadratic evolution of the current density as a function of the DC applied field as follows (Scott et al, 1991) :

$$J = aE + bE^2 \tag{16}$$

where « a » is a coefficient of ohmic resistivity in $\Omega^{-1}m^{-1}$ and « b » is a quadratic coefficient of space charge in $\Omega^{-1}V^{-1}$.

We show, figure 23, the evolution of the current density of the BTS film under an applied positive DC field. The experimental values follow, in a first approximation, the quadratic evolution of a space charge mechanism. With this model, the crossing of the tangents at low and high fields gives the threshold V_{TFL} « Trap Filled voltage Limit » beyond which the charges trapped are released. Then it is possible to calculate the number of traps N_t with the following formula (Chang & Lee, 2002) :

$$V_{TFL} = \frac{ed^2 N_t}{2\varepsilon_0 \varepsilon_r} \tag{17}$$

where e, d, ε_0 are defined previously and ε_r ~420 is the dielectric permittivity at low frequency for an applied voltage V_{TFL}. We have obtained N_t = 1.27x10^{18}/cm^3 for V_{TFL} = 8.25V. This value of N_t is comparable to a published one (Chang & Lee, 2002) on a BST film. We have calculated the energy of the traps : $E_t = E_c - 0.21$ eV with the formulae proposed in the literature (Chang & Lee, 2002).

The mechanism of space charge is linked to the existence of free charges in the interfaces. It can be explained by two phenomena (Yang et al, 1998). The first phenomenon is the oxygen vacancies created at the interface of the BTS film and the platinum electrode during the annealing of the BTS film. The second phenomenon occurs when a DC field is applied. In fact oxygen ions can jump from the BTS grains in contact with the Pt electrode into this electrode. So, the lack of oxygen at the interface BTS film/Pt electrode creates a tank of oxygen vacancies. This oxygen vacancies tank contains free electron charges as shown by the following equation (Shen et al, 2002), at the origin of the leakage current:

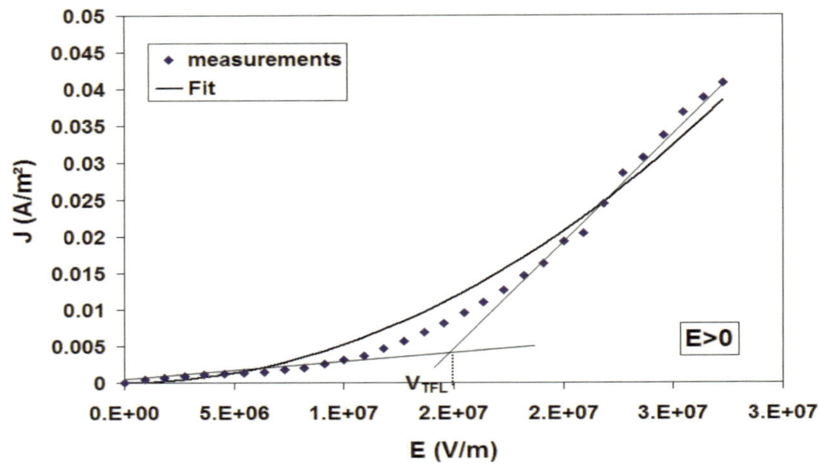

Fig. 23. Evolution of the current density as a function of a positive DC electrical field for a BTS film.

$$O_0 \rightarrow V_0^{2+} + 2e^- + 1/2O_2 \quad (18)$$

This study confirms the existence of an interface at the electrodes levels which was evidenced to be non ferroelectric by our dielectric measurements (§ 3.2).

That allows us to give figure 24 the energy band diagram for the BTS film with Pt and Au electrodes.

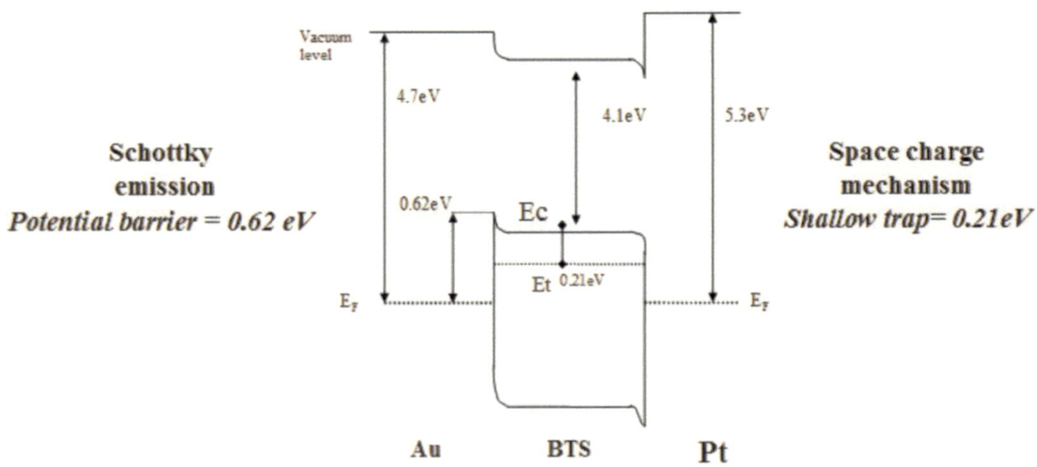

Fig. 24. Energy band diagram of a BTS film deposited on a Si/Pt substrate annealed at 950 °C during 15 min.

7. Conclusion

In conclusion, we have shown that it is possible to deposit by a low cost chemical technique, namely a sol-gel process, good quality ferroelectric films. They were derived from the

$BaTiO_3$ family by substitution with Strontium (BST) or Tin (BTS). We have improved the dielectric and ferroelectric properties by optimizing, first, the annealing conditions in air at 950 °C during only 15 minutes and second, the thickness of the films up to 1 µm. We have also shown that an annealing by microwaves at 750 °C during 5 minutes can crystallize the film in the perovskite structure and gives good dielectric properties. The electrical properties of BST and BTS films are better than the ones of the undoped $BaTiO_3$ films. For radiofrequencies and microwaves applications, the BST or BTS films should be in the paraelectric state to minimize the losses : they are good candidates as the loss tangent of BST may be as low as 6.10^{-3} at 1 MHz and room temperature. For applications in DC or in low frequencies such for infrared or gas sensors, the films should be in the ferroelectric state. A doping with a very low level of tin gives a high yield of usable BTS films and a good reproducibility of their electrical properties. In particular a record tunability of 76 % has been obtained with a BTS film under a DC field of 22.5 V/µm.
Finally, the electrical properties of our optimized doped $BaTiO_3$ (BST and BTS) lead free ferroelectric thin films are comparable to those of PZT films deposited by sol-gel on the same Si/Pt substrates, except for the piezoelectric performances as shown in the following table.

Films	dielectric		ferroelectric			pyroelectric	piezoelectric	
	ε' 1MHz	tg δ 1MHz	Pr (µC/cm²)	Ec (V/µm)	tunability (%)	γ (µC/m²/K)	d_{33} direct (pC/N)	d_{33} inverse (pm/V)
BST	830	0.01	6	4.3	56% @ 12V/µV	350	19	23
PZT	970	0.035	17	5.5	53% @ 25V/µV	200	85	31

Table 1. Comparison of the electrical properties of $Ba_{0.9}Sr_{0.1}TiO_3$ and $PbZr_{0.52}Ti_{0.48}O_3$ films deposited by sol-gel on platinized silicon substrates.

8. Acknowledgment

We would like to thank ADEME and the Conseil Régional Nord-Pas de Calais for the grant given to Manuel Mascot during his Ph.D thesis at the University of Littoral-Côte d'Opale. (Calais, France). We also thank Rachel Desfeux from University of Artois (Lens, France) for the piezoelectric force microscopy measurements and Philippe Belleville from CEA (Le Ripault, France) for the deposition of PZT films by sol-gel on Si/Pt substrates.

9. References

Agarwal, S., Sharma, G.L. & Manchanda, R. (2001). Electrical conduction in (Ba,Sr)TiO_3 thin film MIS capacitor under humid conditions. *Solid State Communications*, Vol. 119, pp. 681-686

Arlt, G., Henning, D. & De With, G. (1985). Dielectric properties of fine-grained barium titanate ceramics. *J. Appl. Phys*, Vol. 58, pp. 1619-1625

Bayer, R.L. & Roundy, C.B. (1972). Pyroelectric coefficient direct measurement technique and application to a nsec response time detector, *Ferroelectrics*, Vol. 3, p. 333-338

Burgnies, L., Vélu, G., Blary, K., Carru, J.-C. & Lippens, D. (2007). Tunability of ferroelectric varactors up to 60 GHz. *Electronics Letters*, Vol. 43, N° 21, pp. 1151-1152

Chang, S.T. & Lee, J.Y. (2002). Electrical conduction mechanism in high-dielectric-constant $Ba_{0.5}Sr_{0.5}TiO_3$ thin films, *App. Phys. Lett.*, Vol. 80, pp. 655-657

Chase, D.R., Chen, L.Y. & York R.A. (2005). Modeling the capactive nonlinearity in thin film BST varactors. *IEEE Trans. Microw. Theory Tech.*, Vol. 53, N° 10, pp. 3215-3220

Cho, Y. W., Choi, S.K. & Venkata Rao, G. (2005). The influence of an extrinsic interfacial layer on the polarization of sputtered $BaTiO_3$ film. *Appl. Phys. Lett.*, Vol. 86, p. 202905

Frayssignes, H., Cheng, B.L., Fantozzi, G. & Button, T.W. (2005). Phase transformation in BST ceramics investigated by internal friction measurements. *Jounal of the European Ceramic Society*, Vol. 25, pp. 3203-3206

Frey, M.H., Xu, Z., Han, P. & Payne, D.A. (1998). The role of interfaces on an apparent grain size effect on the dielectric properties for ferroelectric barium titanate ceramics. *Ferroelectrics*, Vol. 206-207, pp 337-353

Houzet, G., Burgnies, L., Vélu, G., Carru, J.-C. & Lippens, D. (2008). Dispersion and loss of ferroelectric $Ba_{0.5}Sr_{0.5}TiO_3$ thin films up to 110 GHz. *Appl. Phys Lett.*, Vol. 93, p. 053507

Houzet, G., Mélique, X., Lippens, D., Burgnies, L., Vélu, G. & Carru, J.-C. (2010). Microstrip transmission line loaded by split-ring resonators tuned by ferroelectric thin film. *Progress in Electromagnetics Research C*, Vol. 12, pp. 225-236

Hu, W., Yang, C., Zhang, W., Liu, G. & Dong, D. (2006). The diffusion of Pt in BST films on Pt/Ti/SiO2/Si substrate by sol-gel method. *J Sol-Gel Sci. Techn.*, Vol. 39, pp. 293-298

Johnson, K. M. (1962). Variation of dielectric constant with voltage in ferroelectrics and its application to paraelectric devices. *J. Appl. Phys*, Vol. 33, pp 2826-2831

Keyson, D., Volanti, D.P., Cavalcante, L.S., Simoes, A.Z., Souza, I.A., Vasconcelos, J.S., Varela, J.A. & Longo, E. (2007). Domestic microwave oven adapted for fast heat treatment of $Ba_{0.5}Sr_{0.5}(Ti_{0.8}Sn_{0.2})O_3$ powders. *Journal of Materials Processing Technology*, Vol. 189, pp. 316–319

Khalfallaoui, A., Vélu, G., Burgnies, L. & Carru, J.-C. (2010). Characterization of doped BST thin films deposited by sol-gel for tunable microwave devices. *IEEE Trans. Ultrason. Ferroelectr. Freq. Control*, Vol. 57, N° 5, pp. 1029-1033

Lefki, K. & Dormans, G. J. M. (1994). Measurement of piezoelectric coefficients of ferroelectric thin films. *J. Appl. Phys.*, Vol. 76, pp. 1764-1767

Liu, S., Liu, M., Jiang, S., Li, C., Zeng, Y., Huang, Y. & Zhou, D. (2003). Fabrication of SiO_2 doped $Ba_{0.8}Sr_{0.2}TiO_3$ glass ceramic films and the measurement of their pyroelectric coefficient. *Materials Science and Engineering B*, Vol. 99, pp. 511-515

Malic, B., Boerasu, I., Mandeljc, M., Kosec, M., Sherman, V., Yamada, T., Setter, N. & Vakadinovic, M. (2007). Processing and dielectric characterization of $Ba_{0.3}Sr_{0.7}TiO_3$. *Journal of the European Ceramic Society*, Vol. 27, pp. 2945-2948

Mascot, M., Fasquelle, D., Vélu, G., Ferri, A., Desfeux, R., Courcot, L. & Carru, J.-C. (2008). Pyro, ferro and dielectric properties of $Ba_{0.8}Sr_{0.2}TiO_3$ films deposited by sol-gel on platinized silicon substrates. *Ferroelectrics*, Vol. 362, pp. 79-86

Mascot, M. (2009). Ph. D thesis, University of Littoral-Côte d'Opale, Calais, France

Mascot, M., Fasquelle, D. & Carru, J.-C. (2011). Very high tunability of $BaSn_xTi_{1-x}O_3$ ferroelectric thin films deposited by sol-gel. *Functional Materials Letters*, to be published in 2011.

Outzourhit, A., Trefny, J.U., Kito, T. & Yarar, B. (1995). Tunability of dielectric constant of $Ba_{0.1}Sr_{0.9}TiO_3$ ceramics in the paraelectric state. *J. Mater. Res.*, Vol. 10, pp. 1411-1417

Parker, C.B., Maria, J.-P. & Kingon, A.I. (2002). Temperature and thickness dependent permittivity of (Ba, Sr) TiO_3. *Appl. Phys Lett.*, Vol. 81, pp. 340-342.

Pontes, F.M., Longo, E., Leite, E.R. & Varela, J.A. (2001). Study of the dielectric and ferroelectric properties of chemically processed $Ba_xSr_{1-x}TiO_3$ thin films. *Thin Solid Films*, Vol. 386, pp. 91-98

Ren, W., Zhou, H-J., Wu, X.-Q., Zhang, L-Y. & Yao, X. (1997). Measurement of piezoelectric coefficients of lead zirconate titanate thin film by the normal load method using a composite tip. *Materials Letters*, Vol. 31, pp. 185-188

Scott, J.F., Araujo, C.A., Melnick, B.M., Mc Millan, L.D. & Zuleeg, F.L. (1991). Quantitative measurement of space-charge effects in lead zirconate-titanate memories. *J. Appl. Phys.*, Vol. 70, pp. 382-388

Shen, M., Dong, Z., Gan, Z. & Ge, S. (2002). Oxygen-related dielectric relaxation and leakage characteristics of Pt/Ba,Sr.TiO_3/Pt thin-film capacitors, *Appl. Phys. Lett.*, Vol. 80, pp. 2538-2540

Song, S.N., Zhai, J.W. & Yao, X. (2006). The study of the microstructure and tunability of $Ba(Sn_xTi_{1-x})O_3$ thin films. *Integrated Ferroelectrics*, Vol. 78, pp. 337-344

Sun, L.L., Tan, O.K., Liu, W.G., Chen, X.F. & Zhu, W. (2003). Comparaison study on sol-gel $Pb(Zr_{0.3}Ti_{0.7})0_3$ and $Pb(Zr_{0.3}Ti_{0.7})0_3/PbTi0_3$ multilayer thin films for pyroelectric infrared detectors. *Microelectronic Engineering*, Vol. 66, pp. 738-744

Taylor, D. V. & Damjanovic, D. (1998). Domain wall pinning contribution to the nonlinear dielectric permittivity in $PbZrTiO_3$ thin films. *Appl. Phys. Lett*, Vol. 73, pp. 2045-2047

Uchino, K. (2010). *Ferroelectric Devices* (2nd edition), CRC Press, Boca Raton, USA

Vélu, G., Carru, J.-C., Cattan, E., Remiens, D., Mélique, X. & Lippens, D. (2003). Deposition of ferroelectric BST thin films by sol-gel route in view of electronic applications. *Ferroelectrics*, Vol. 288, pp. 59-69

Yang, P., Deng, H. & Chu, J. (1998). Highfield effects of layered perovskite ferroelectric thin films, *Science in China*, Vol. 41, pp. 502-510

Zhang, T. & Ni, H. (2002). Pyroelectric and dielectric properties of sol–gel derived barium–strontium–titanate ($Ba_{0.64}Sr_{0.36}TiO_3$) thin films. *Sensors and Actuators A*, Vol. 100, pp. 252–256

Zhou, C. & Newns, D. M. (1997). Intrinsic dead layer effect and the performance of ferroelectric thin film capacitors. *J. Appl. Phys.*, Vol. 82, pp. 3081-3088

4

Control of Crystallization and Ferroelectric Properties of $BaTiO_3$ Thin Films on Alloy Substrates

Zhiguang Wang,Yaodong Yang, Ravindranath Viswan,
Jie-Fang Li and D. Viehland
Department of Materials Science and Engineering,
Virginia Tech,Blacksburg,
USA

1. Introduction

Growth of complex oxides on different substrates has attracted tremendous interest [1,2] in the past decade due to the possibility that multifunctional devices could be realized by combining various properties of complex oxide phases with that of the functional substrates on which they are deposited. Among a variety of interesting properties that could be explored, the study of ferroelectric and ferromagnetic or magnetoelectric (ME) properties has achieved considerable research progress, driven by the development of thin film deposition technologies, discovery of new magnetoelectric materials and composite structures, and the development of an understanding and measurement methods for the coupling effect. The concept of magnetoelectricity was first proposed by P.Curie in 1872 and studies began with single phase ME materials that exhibited only limited magnetoelectric (ME) coupling. The low ME coupling is principally due to the fact that the magnetization and polarization are favored by different electronic configurations [3].

Accordingly, two phase composite materials became a better choice. Perovskite BTO and PZT materials have excellent piezoelectric properties [4], while iron-based alloys have been proven to have the highest magnetostrictive properties [5]. PZT-Metglas laminated (bulk) composites have shown the highest ME coupling coefficient, and have been used as magnetic sensors that are capable of detecting fields of <10 pTesla [4]. These ME composites consisting of a metallic ferromagnetic magnetostrictive alloy and a perovskite ferroelectric oxide have been the focus of considerable investigations for potential applications in memory and magnetic field sensing. Castel et al. [6,7] have reported preparation of BTO/Ni granular nanocomposite structures by high pressure compacting of BTO and Ni nanoparticles. A large increase of the real and imaginary components of the microwave frequency effective complex permittivity of BTO phase was found that was attributed to the large ME coupling effect between piezoelectric and magnetic phases. Israel et al. [8] used industrially produced multilayer capacitors which consisted of Ni-based magnetostrictive electrodes sandwiching a BTO-based dielectric to study the ME coupling effect. They found that both the magnetization and strain were hysteretic with applied electric field. Geprägs et al. [9] deposited polycrystalline Ni thin films on BTO single crystal substrates by electron

beam evaporation, and studied two different approaches to control the magnetization by electric field. The magnetization could be reversibly changed by more than 20%, due to the combined action of electroelastic strain and inverse magnetostriction. Zou et al. [10] deposited PZT films on Stainless Steel, Titanium and Ni foils using a $LaNiO_3$ buffer layer, by the sol-gel method and studied the dielectric properties. A high dielectric constant, low dielectric loss and symmetric C-V and P-E loops were obtained in all cases. More recently, Ma et al., [11] deposited $Pb_{0.92}La_{0.08}Zr_{0.52}Ti_{0.48}O_3$ (PLZT) thin films on $LaNiO_3$ buffered Ni substrates by chemical solution deposition and studied the dielectric properties. High quality PLZT films with good dielectric properties were obtained. However, there are few reports about the study of magnetic, ferroelectric and the magnetoelectric coupling in ferroelectric thin films deposited on magnetostrictive alloy substrates. This is not only due to the fact that growth of the ferroelectric phase requires higher temperatures which may result in oxidization of the alloy substrate but also, because the ferroelectric films thus deposited may have large leakage currents due to a poor microstructure that may arise from deposition on a metallic substrate with low surface quality: thereby, limiting the possibility of reliable magnetoelectric measurements in such composite structures.
Here, we report the growth of BTO thin films on both amorphous Metglas [12] and $Fe_{81}Ga_{19}$ single crystals. Using this approach, we can realize the combination of ferroelectric and ferromagnetic properties in a semi-monolithic material. In addition to studying the property changes in metal-ceramic multilayers, we have also deposited BTO thin films on $SrTiO_3$ single crystal substrate predeposited with gold particles as seeds. Au layers can buffer BTO on Metglas, as the lattice parameters of Au and BTO closely match. Accordingly, Au can be used to control the growth orientation and grain size in BTO thin films.

2. Experiment details

Pulsed laser deposition (PLD) utilizing a KrF excimer laser (λ= 248 nm) was used for the growth of perovskite piezoelectric thin films. To obtain the necessary high atomic mobility[13] for the crystallization and oriented growth of the BTO layer, a high energy laser density (c.a. 20 J/cm^2) was used to eject high energy particles from the target. Thus we were able to use lower deposition temperatures to prevent oxidation of the metal alloy. The size of the laser spot was 2 mm^2 and the distance between the substrate and target was 8 cm. Deposition was carried out in a 90 mTorr oxygen atmosphere with a base vacuum of 10^{-6} Torr. We used a laser frequency of 10 Hz for the first 10 minutes of deposition onto a substrate held at 300℃. This enabled a layer of BTO to be first deposited, while preventing either melting of the Au layer or the oxidation of the alloy substrate. The substrate temperature was then increased to 600℃, and deposition was carried out under the same conditions for 90 minutes. The surface and cross-sectional morphology of the BTO thin films was studied using a LEO (Zeiss) 1550 Schottky field-emission SEM. The crystal structure was measured using a Philips X'pert high resolution x-ray diffractometer. A FEI Helios 600 NanoLab FIB SEM was used to prepare and lift-out TEM samples. A FEI Titan 300 high-resolution TEM (HRTEM) was used to obtain lattice images. The magnetic properties were measured with a Lakeshore 7300 Series VSM System at room temperature.

3. BTO on Metglas substrate

Metglas has an enormous magnetic permeability, which makes it ideal for the magnetostrictive phase in ME composites. The only problem is how to make the Metglas survive the high temperature process that is necessary for the growth of piezoelectric thin films with good properties. One option is to use a Au buffer layer to protect the Metglas from oxidization, and to use a higher laser energy so that deposition can be done at lower temperature.

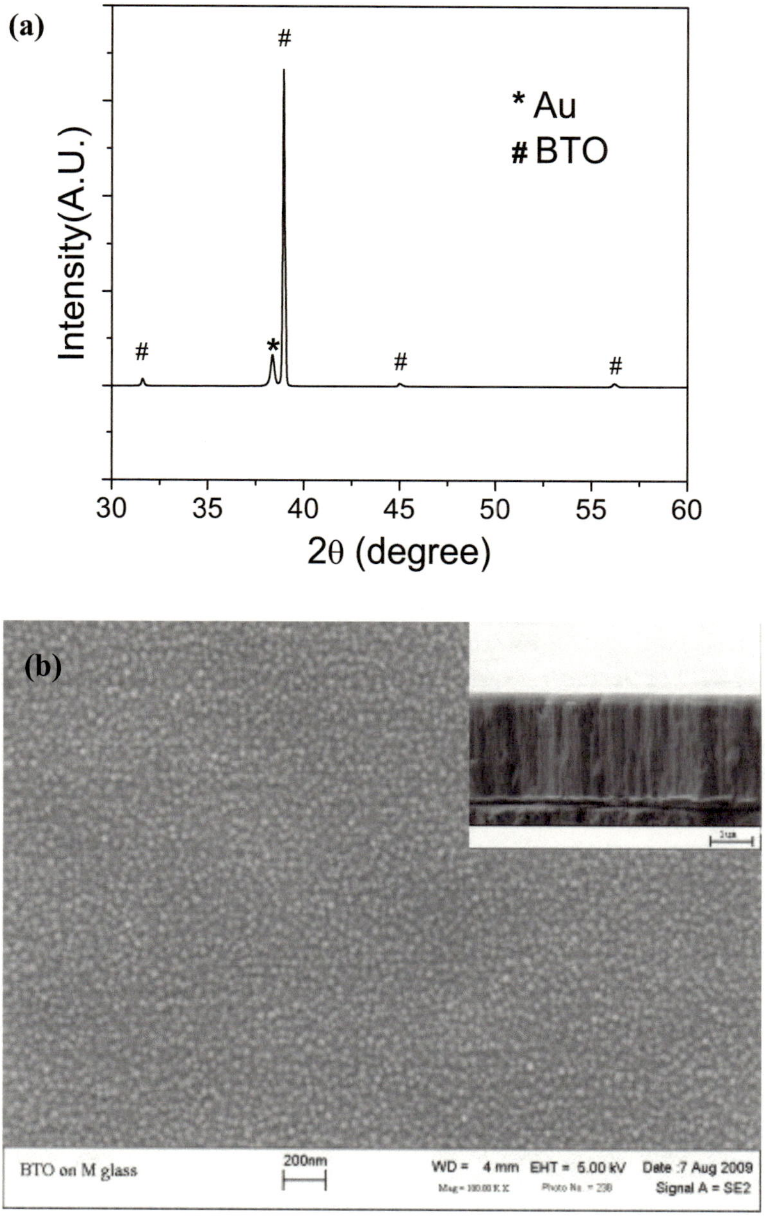

Fig. 1. (a) XRD of BTO/Au/Metglas, and (b) surface SEM result (inset is the cross-section view SEM, scale bar is 1 μm).

Figure1a shows the XRD line scan for a BTO thin film deposited on Au-buffered Metglas foil. The Au-buffer layer reached a thickness of about 60 nm after a deposition time of 12 minutes. In Figure 1, we can observe only a (111) diffraction peak at about $2\theta \approx 38.38°$, which indicates that Au-buffer layer is highly textured. The Au (111) orientation has the lowest surface energy and thus Au thin films grown on amorphous substrates have a (111) texture. The lattice parameter of Au (a=4.08) has a very small crystal mismatch with that of BTO (a=3.994), and thus the subsequent growth of a BTO thin film will also occur along the (111). This is confirmed by the intense BTO (111) peak at $2\theta \approx 38.90°$. Good in-plane alignment and crystalline ordering was confirmed by the small full width at half maximum of FWHM≈0.14° for the BTO (111) peak. There are also several other weaker peaks which are not from the (111) BTO; which may be induced by the relative rough surface of the Au layer, compared with a single crystal substrate.
Figure1.b shows a top-view SEM image for BTO on Au/Metglas. In this image one can see that BTO formed with a uniform columnar morphology with size of about 40 nm in diameter. From the cross-sectional view, we can see that Au could form a dense and uniform interface between Metglas and BTO with a thickness of about 60 nm: which was sufficiently thick to prevent significant oxidation of the Metglas alloy, while preserving the (111) orientation of BTO. The inset shows a higher resolution SEM image of this same structure. The BTO layer thickness reached 2 μm after 2 hours of deposition, while preserving the columnar morphology of diameter about 40 nm which is in accordance with the top-view image.
Figure2a shows a polarization hysteresis measurement of the BTO ferroelectric thin film. A well-defined ferroelectric hysteresis loop can be seen with a saturation polarization of $P_s \approx 25 \mu C/cm^2$ and a remnant polarization of $P_r \approx 7.5\ \mu C/cm^2$. The value of both P_r and P_s are much larger compared with BTO structures deposited on Pt-buffered Si substrates [14, 15]. The coercive field was $E_c \approx 12.5$ kV/cm, which is also much smaller than that of BTO thin films on either STO or Si substrates [16]. After the polarization measurement, the BTO thin films were poled in the out-of-plane direction. Piezoelectric measurement revealed a well-defined butterfly-like shape with a saturation value for the out-of-plane piezoelectric coefficient $D_{33} \approx 11$ pm/V, which is relatively small compared to epitaxial thin films on STO substrates.
Figure3.a shows a typical M-H loop for Metglas after the BTO deposition process, which was measured using a VSM. The saturation field was ~ 400 Oe, and the inset shows that the coercive field was ~60 Oe. Figure3.b shows the magnetostriction of this same Metglas foil after BTO deposition, measured by a strain gauge method. Under a DC magnetic field of 80 Oe, the magnetostriction reached a maximum value of about 14 ppm. Although this magnestriction value is smaller than that of Metglas before deposition, Figure3.c shows that the effective linear piezomagnetic coefficient was about 0.34 ppm/Oe, which is comparable to that of Terfenol-D [5].
These results demonstrate the successful deposition of BTO thin films on Metglas foils with a highly oriented (111) texture. In addition, good ferroelectric and ferromagnetic properties were obtained from the BTO layer and Metglas foil respectively. Thus, the approach offers promise to achieve ME coupling in semi-monolithic heterostructures that have lattice coherency across their interface.

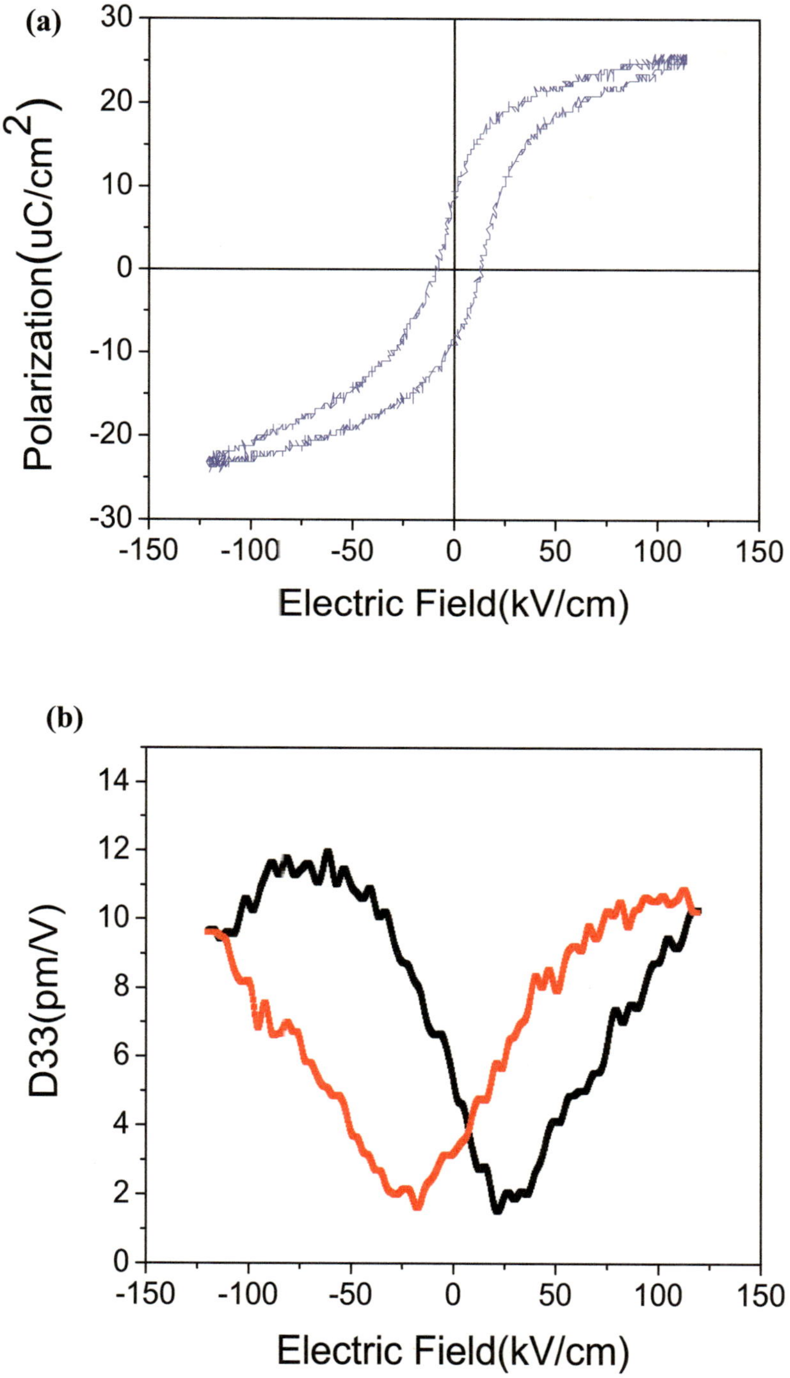

Fig. 2. Ferroelectric hysteresis loop (a) and piezoelectric D_{33} hysteresis loop (b) of BTO on Metglas.

(a)

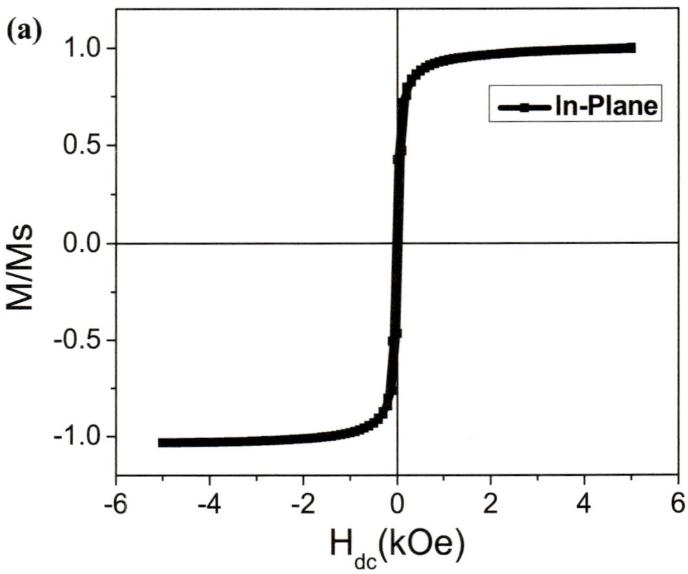
In-Plane
M/Ms
1.0
0.5
0.0
-0.5
-1.0
-6
-4
-2
0
2
4
6
H_dc(kOe)

(b)

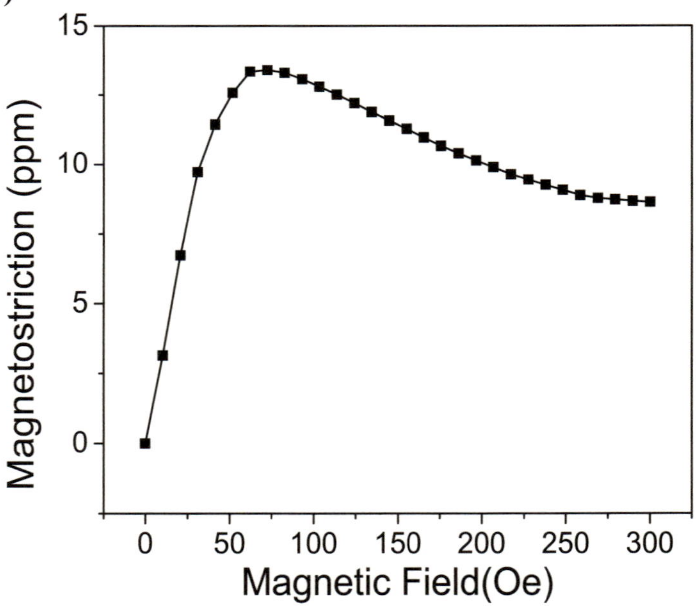
Magnetostriction (ppm)
15
10
5
0
0
50
100
150
200
250
300
Magnetic Field(Oe)

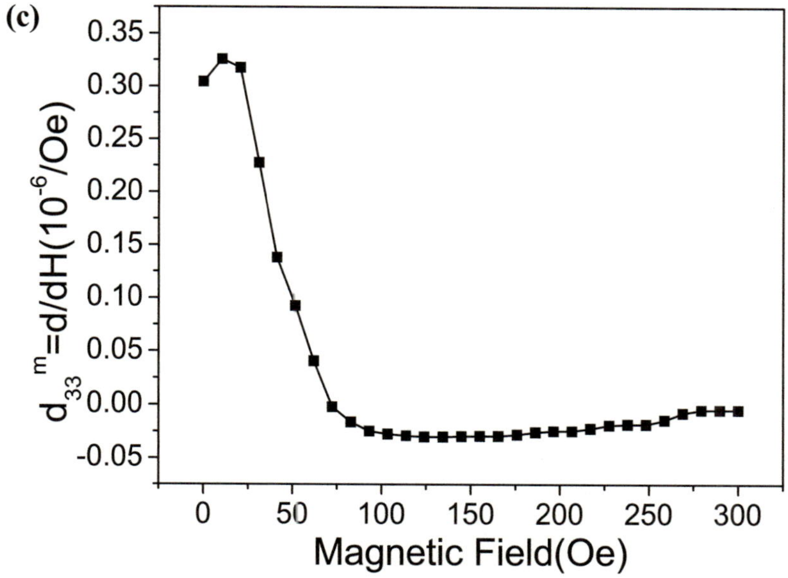

Fig. 3. (a)Normalized magnetization vs applied magnetic field curve of Metglas with BTO on top. (b)Magnetostrictive and (c) piezomagnetic properties of Metglas with BTO on top.

4. BTO on Au-buffered STO

We used a Au-buffer layer to control the orientation of BTO on Metglas foils. "How does Au affect BTO grains?" and "what is the influence of an Au buffer layer on the micro and/or nanostructure of BTO?" are interesting questions that need answers. To obtain a better understanding of the relationship between BTO and the Au layer, a series of experiments were performed. We used (111) oriented single crystal STO substrates that were near atomically flat. Au was deposited via sputtering for different deposition times. Subsequently, we deposited BTO thin films on the different Au buffered layers by PLD.

Figure 4 shows a notable difference between samples with Au layers of different thicknesses. Parts (a) and (b) both exhibit uniform nanostructure although the BTO grains in Figure 4a are smaller than those in Figure 4b. Furthermore, one can see that both areas have pyramidal grain morphologies that are indicative of a preferred (111) texture for BTO. However, in parts (c) and (d), one can see there are numerous upheavals or "hills" of the BTO phase. In part (d) these "hills" are even larger than those in (c). A schematic diagram is given in part (e) that illustrates the BTO growth mechanism for the different cases, which helps explain why the surface consists of numerous "hills" as shown in Figs.4c and 4d. When the buffer layer becomes thicker than about 10nm, the Au layer separates into discontinuous nanoparticles of size about 20nm in diameter. These Au-buffer layers then serve as nucleation sites for the formation of BTO grains. However, on the thicker Au-buffer layers with nanoparticle morphology, BTO grain nucleates and grows out from the Au surface without preserving a unique orientation. This maybe because the Au clusters are too large to serve as a single nucleation site.

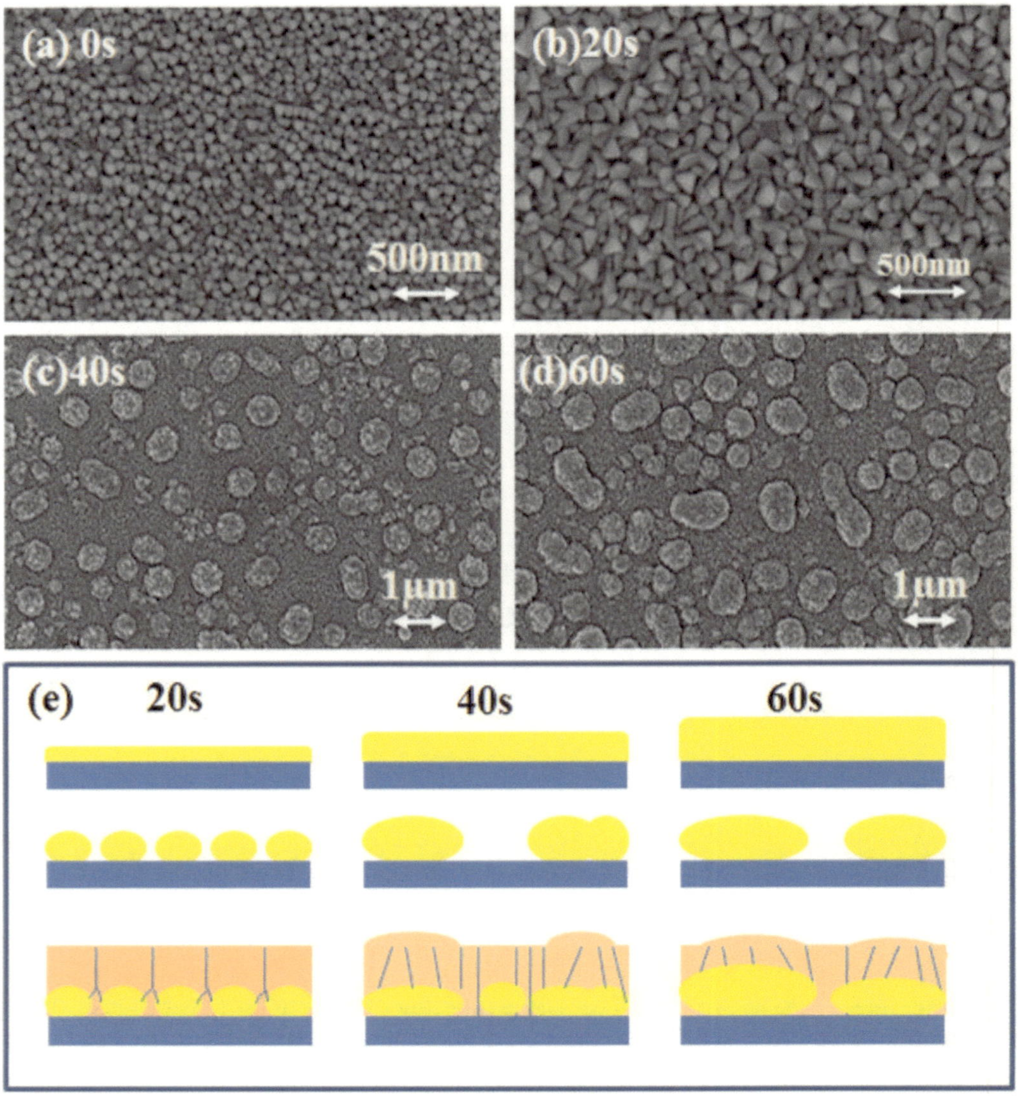

Fig. 4. (a) SEM image of BTO thin film on the Au layers with different thickness. (a) to (d) Au layers were deposited for (a)0s, (b)20s, (c) 40s and (d)60s. (e) is schematic to show a possible growth mechanism.

We then used FIB to lift out a small cross-sectional piece from the same area shown in Fig. 4b, to serve as a TEM sample. HRTEM images are given in Figure 5, which provide more detailed information concerning the interaction between the BTO layers grown on Au. Both the Au and BTO areas can be easily distinguished. Figure 5b shows a boundary between Au and BTO. In this figure, lattice planes can be seen to be well matched to each other. This indicates that it is possible to maintain an epitaxial relationship between the Au nanoparticle and the BTO grain.

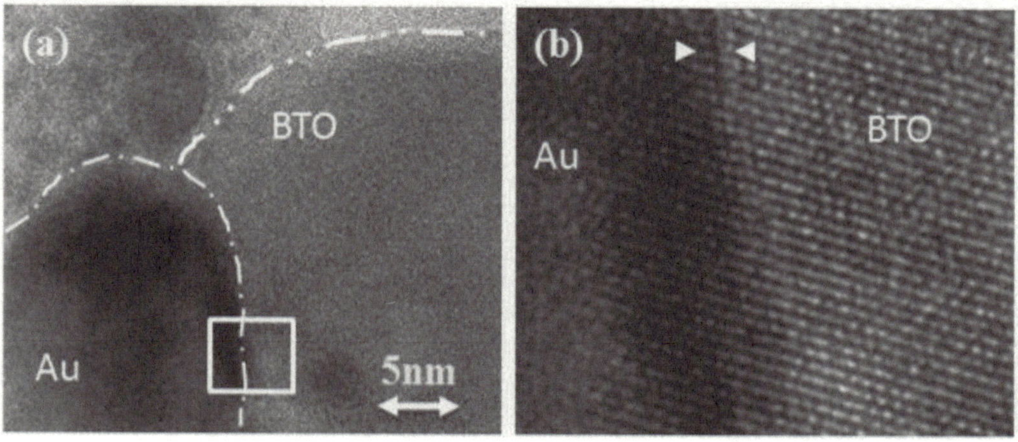

Fig. 5. (a) TEM image of boundary between Au and BTO grain. (b) a high magnification image to show the detail.

5. BTO on $Fe_{81}Ga_{19}$ substrate

The magnetostriction of $Fe_{81}Ga_{19}$ alloys has been proved to be highly dependent on structure and composition [17]. It has been shown that the magnetostriction of $Fe_{81}Ga_{19}$ alloys is highest in a metastable disordered BCC phase (a=2.91Å) [16]. There is a large lattice mismatch between $Fe_{81}Ga_{19}$ and perovskite BTO (a=3.994 Å, c=4.038 Å). However, the difference in atomic spacing along [110] $Fe_{81}Ga_{19}$ and (100) BTO directions is only about 2%. Thus, if growth of $Fe_{81}Ga_{19}$ is carried out at deposition temperatures greater than 575 ^{o}C using a non-equilibrium deposition process, then it may be possible to achieve (101) oriented BTO thin films on (100) oriented $Fe_{81}Ga_{19}$ substrates. In this case, the growth relationship would be BTO(100) | | $Fe_{81}Ga_{19}$(110), and BTO(101) | | $Fe_{81}Ga_{19}$(100).

We deposited BTO thin films on single crystal $Fe_{81}Ga_{19}$ substrates by the PLD technique at 650 ^{o}C. During the first 10 minutes of deposition, a vacuum condition was used and after that an oxygen pressure of 75 mTorr was used for 1 hour. After deposition, the mirror-like surface of the alloy substrate was well-preserved, indicating that oxidation of the substrate has been prevented by the initial deposition in high vacuum conditions.XRD and SEM was used to check the crystallization and the nanostructure of the BTO phase.

Figure6.a shows a XRD line scan for the sample before and after BTO deposition. The black curve indicates the (110) peak of the $Fe_{81}Ga_{19}$ single crystal substrate: please note the lack of other $Fe_{81}Ga_{19}$ peaks in both case. After BTO deposition, the BTO thin film had an intense (111) peak, but much weaker (110) and (100) peaks. This indicates that the BTO thin films develop a (111) texture, but are not highly oriented. Also, a strong $Fe_{81}Ga_{19}$(110) peak, is seen, which indicates that oxidization and phase change could be prevented during the high temperature deposition process. Figure6b shows the SEM top view of BTO thin films. We could clearly see uniform, nano-sized, triangle and polygonal BTO grains aligned densely with each other. The presence of an ordered nanostructure indicates well crystallized BTO nanocrystals which promises good ferroelectric and piezoelectric properties.

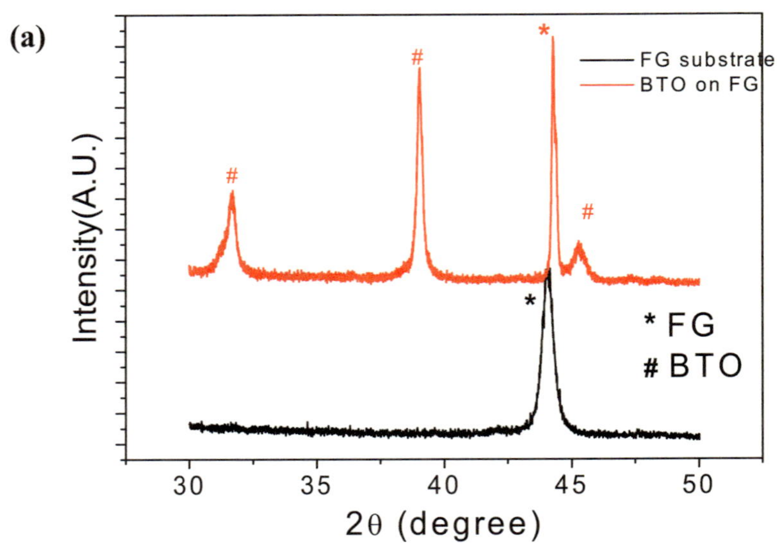

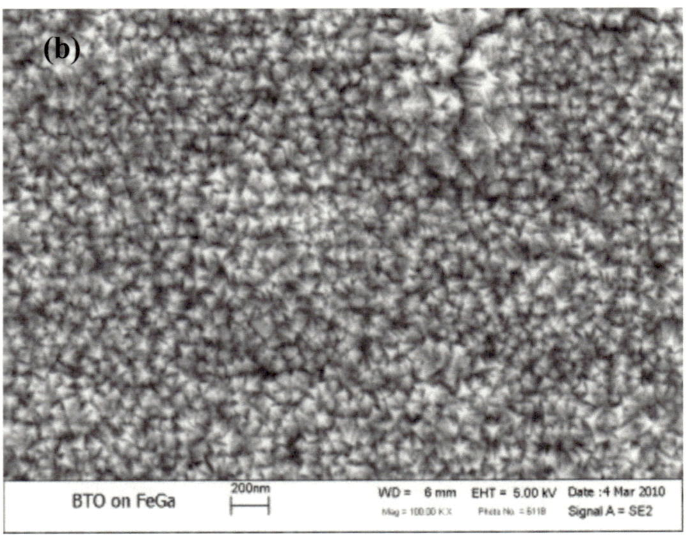

Fig. 6. (a) XRD of $Fe_{81}Ga_{19}$ (FG) substrate and BTO/$Fe_{81}Ga_{19}$, and (b) top-view SEM result.

(a)

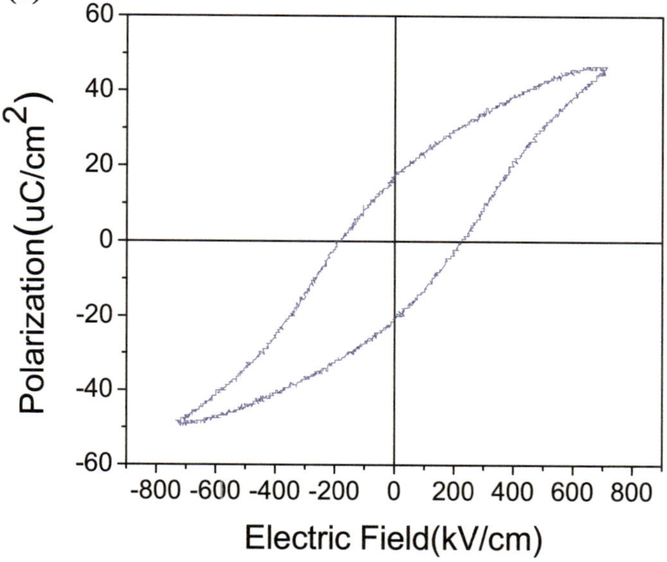

(b)

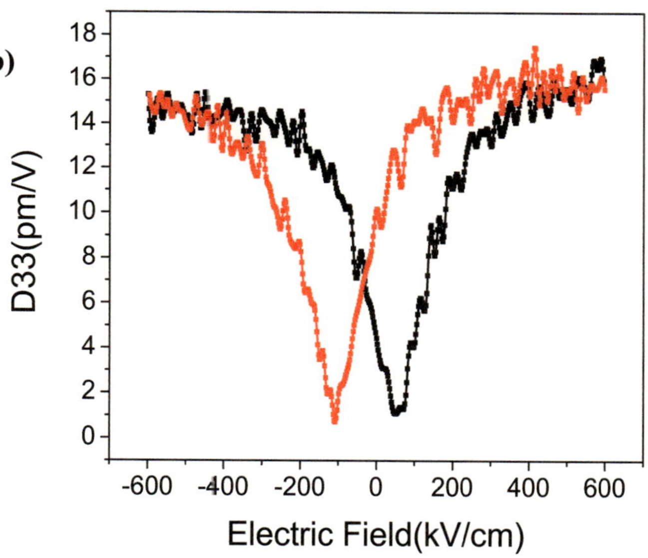

Fig. 7. Ferroelectric hysteresis loop (a) and piezoelectric D_{33} hysteresis loop (b) of BTO on $Fe_{81}Ga_{19}$.

Figure7a is the ferroelectric property measurement of the BTO ferroelectric thin film on $Fe_{81}Ga_{19}$. Well-defined ferroelectric hysteresis loop shows a P_s of ~40μC/cm² and a P_r of about 18 μC/cm². Both P_r and P_s are much larger compared with BTO thin film on Metglas. The coercive field is about 220kV/cm, which was much larger than that for BTO thin films on Metglas. This indicates that the BTO thin films experience much larger strain effects on

$Fe_{81}Ga_{19}$ single crystal substrates than that on amorphous Metglas substrates. Piezoelectric property measurements in Figure7b show well-defined butterfly shape with a saturation out-of-plane piezoelectric coefficient value (D_{33}) of ~ 15 pm/V, which is a little larger than BTO on Metglas. In summary, we have obtained BTO thin films on $Fe_{81}Ga_{19}$ substrate with good ferroelectric and piezoelectric properties.

Also, VSM and strain gauge methods were used to measure the ferromagnetic and magnetostrictive properties of the $Fe_{81}Ga_{19}$ substrate respectively. Figure8a shows a typical M-H loop of $Fe_{81}Ga_{19}$ in the in-plane direction after high temperature deposition process. The saturation magnetic field is about 2000 Oe. Also, since $Fe_{81}Ga_{19}$ is a soft magnetic material, the coercive field is very small. Figure8b shows the magnetostriction measured by a strain gauge method. At a dc magnetic field of about 500 Oe, the magnetostriction reaches a value of about 70 ppm. This is almost the value measured before the PLD process. Also, the calculated piezomagnetic value of 0.32 ppm/Oe is comparable with Metglas as shown in Figure8.c.

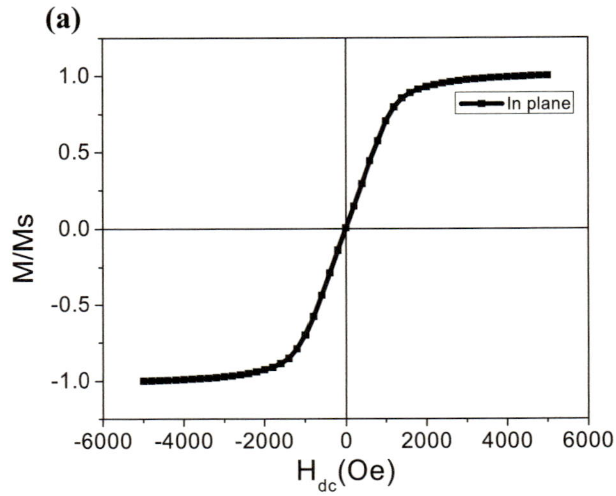

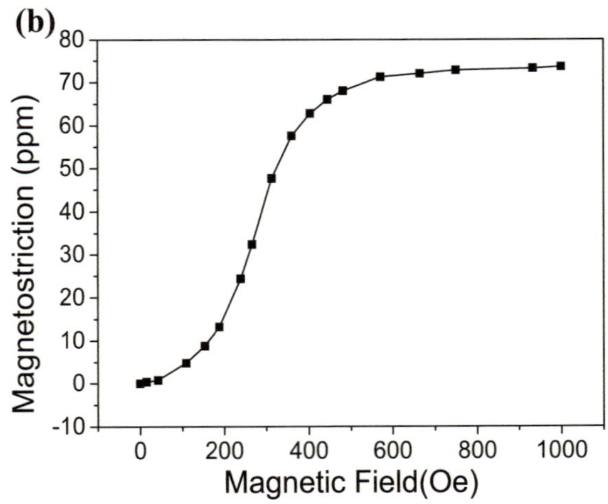

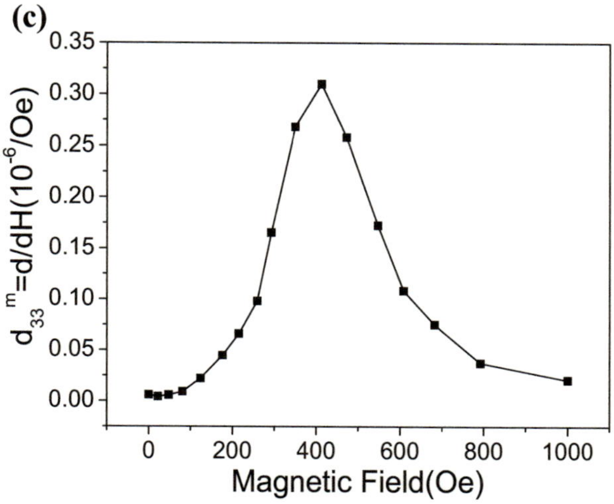

Fig. 8. (a)Normalized magnetization vs applied magnetic field curve of $Fe_{81}Ga_{19}$ with BTO on top. (b)Magnetostrictive and (c) piezomagnetic properties of BTO/$Fe_{81}Ga_{19}$.

6. Conclusion

In summary, we have successfully deposited well-crystallized ferroelectric BTO thin films on both Metglas and $Fe_{81}Ga_{19}$ substrates. The measured values of D_{33} were 11 pm/V and 15 pm/V, respectively. Also, good ferromagnetic and magnetostrictive properties were found after the high temperature BTO deposition process. The relationship between the Au-buffer layer and the BTO thin film was also studied. Ferroelectric thin films on ferromagnetic substrate may enable the coupling of the two different ferroic order parameters, which might in turn be useful for multifunctional devices. The approach is compatible for the growth of various functional oxide thin films on functional amorphous or crystalline alloy materials, and accordingly may allow for different coupling effects.

7. Acknowledgment

Support for this work was provided by the Department of Energy, the Air Force Office of Scientific Research and the National Science Foundation. Authors also give thanks for NCFL in Virginia Tech for all the SEM, FIB and TEM support.

8. References

Ramesh, R. & Spaldin, N. Multiferroics: progress and prospects in thin films. Nature Materials, Vol.6 (January 2007), pp.21, ISSN: 1476-1122

Martin, L. Chu, Y. & Ramesh, R. Advances in the growth and characterization of magnetic, ferroelectric, and multiferroic oxide thin films. Materials Science and Engineering R, Vol.68 (May 2010), pp. 89-133, ISSN: 0927-796X

Ederer, C. & Spaldin, N. Recent progress in first-principles studies of magnetoelectric multiferroics, Curr. Opin. Solid State Mater. Sci. Vol.9, No.3 (June 2005), pp.128, ISSN: 1359-0286

Zhai, J. Xing, Z. Dong, S. Li, J. & Viehland, D. Magnetoelectric Laminate Composites: An Overview. J. Am. Ceram. Soc. Vol.91, No.2 (February 2008), pp.351, ISSN: 1551-2916

Lou, J. Liu, M. Reed, D. Ren, Y. & Sun, N. Giant Electric Field Tuning of Magnetism in Novel Multiferroic FeGaB/Lead Zinc Niobate-Lead Titanate (PZN-PT) Heterostructures. Adv. Mater. Vol.21, No.46 (December 2009), pp.4711, ISSN: 1521-4095

Castel, V. & Brosseau, C. Magnetic field dependence of the effective permittivity in $BaTiO_3$/Ni nanocomposites observed via microwave spectroscopy. Appl. Phys. Lett. Vol.92, No.23 (June 2008) pp.233110, ISSN: 1077-3118

Castel, V. Brosseau, C. & Youssef, J. Magnetoelectric effect in $BaTiO_3$/Ni particulate nanocomposites at microwave frequencies. J. Appl. Phys. Vol.106, No.6 (December 2009), pp.064312, ISSN: 1089-7550

Israel, C. Narayan, S. &Mathur, N. Converse magnetoelectric coupling in multilayer capacitors. Appl. Phys. Lett. Vol.93, No.17 (October 2008), pp.173501, ISSN: 1077-3118

Geprägs, S. Brandlmaier, A. Opel1, M. Gross, R. & Goennenwein, S. Electric field controlled manipulation of the magnetization in Ni/$BaTiO_3$ hybrid structures. Appl. Phys. Lett. Vol.96, No.14 (April 2010). pp.142509, ISSN: 1077-3118

Zou, Q. Ruda, H. & Yacobi, B. Improved dielectric properties of lead zirconate titanate thin films deposited on metal foils with $LaNiO_3$ buffer layers. Appl. Phys. Lett. Vol.78, No.9 (February 2001), pp.1282, ISSN: 1077-3118

Ma, B. Kwon, D. Narayanan, M & Balachandran, U. Fabrication of antiferroelectric PLZT films on metal foils, Mater. Res. Bull., Vol.44, No.1 (January 2009), pp.11, ISSN: 0025-5408

Wang, Z. Yan, L. Yang, Y. et al. Magnetoelectric effect in crystallographically textured $BaTiO_3$ films deposited on ferromagnetic metallic glass foils. J. Appl. Phys. Vol.109, No.3 (February 2011), pp.034102, ISSN: 1089-7550

Schwarzkopf, J. & Fornari, R. Epitaxial growth of ferroelectric oxide films. Progress in Crystal Growth and Characterization of Materials, Vol.52, No.3 (September 2006), pp.159, ISSN: 0960-8974

Cho, Y. Choi, S. & Rao, G. The influence of an extrinsic interfacial layer on the polarization of sputtered $BaTiO_3$ film. Appl. Phys. Lett. Vol.86, No.20 (May 2005), pp.202905, ISSN: 1077-3118

Dkhil, B. Defaÿ, E. & Guillan, J. Strains in $BaTiO_3$ thin film deposited onto Pt-coated Si substrate.Appl. Phys. Lett. Vol.90, No.2 (January 2007), pp.022908, ISSN: 1077-3118

Pompe, W. Gong, X. Suo, Z. & Speck, J. Elastic energy release due to domain formation in the strained epitaxy of ferroelectric and ferroelastic films. J. Appl. Phys. Vol.74, No.10 (August 1993), pp.6012, ISSN: 1089-7550

Srisukhumbowornchai, N. & Guruswamy, S. Influence of ordering on the magnetostriction of Fe-27.5 at.% Ga alloys. J. Appl. Phys. Vol.92, No.9 (July 2002), pp.5371, ISSN: 1089-7550

5

Growth and Characterization of Single Crystals of Potassium Sodium Niobate by Solid State Crystal Growth

Andreja Benčan[1], Elena Tchernychova[1], Hana Uršič[1], Marija Kosec[1] and John Fisher[2]
[1]*Electronic Ceramics Department, Jožef Stefan Institute,*
[2]*School of Materials Science and Engineering, Chonnam National University,*
[1]*Slovenia*
[2]*Korea*

1. Introduction

Due to the toxic nature of PbO in $Pb(Zr,Ti)O_3$ (PZT), the most common type of piezoceramic, many studies on lead-free materials are being conducted worldwide (Röedel et al., 2009). One of the most studied groups of lead-free ferroelectric materials is based on a solid solution of potassium sodium niobate, $K_{0.5}Na_{0.5}NbO_3$ (KNN) (Jaffe et al., 1971; Kosec et al., 2008). However, activity in attempts to find a lead-free replacement for PZT really accelerated with the discovery of textured $(K,Na,Li)(Nb,Ta,Sb)O_3$ ceramics, with properties comparable to those of PZT (d_{33}> 300 pC/N, relative permittivity > 1500, planar coupling coefficient k_p> 60%)(Saito et al., 2004).

One way to enhance the piezoelectric properties of KNN is to grow KNN-based single crystals along certain crystallographic directions, as has been demonstrated for the case of $KNbO_3$ (Wada et al., 2004) and relaxor-based ferroelectric single crystals (Park & Shrout, 1997). The most frequently used methods for growing alkali niobate based single crystals are top seeded solution growth and the flux methods (Chen et al. , 2007; Davis et al., 2007; Kizaki et al., 2007; Lin et al., 2009). However, these methods are not yet fully commercialized due to high costs and poor reproducibility related to compositional inhomogeneity within the crystals. A possible way to improve the homogeneity of crystals with complex composition is to grow them by the low cost Solid State Single Crystal Growth (SSCG) method.

The SSCG method is essentially a form of induced abnormal grain growth, a phenomenon which is very well known in the solid state sintering community. A significant breakthrough in the solid state synthesis of lead-free materials has been achieved in recent years (Kosec at al., 2010; Malič et al., 2008a). However, due to the strongly hygroscopic nature of alkaline carbonates usually used in solid state synthesis, different diffusion rates of involved ionic species during annealing, and the high sublimation rates of the alkaline species at high temperatures (Bomlai et al., 2007; Jenko et al., 2005; Kosec & Kolar, 1975; Malič et al., 2008b), it may be rather difficult to obtain compositionally homogeneous alkali niobate-based single crystals by SSCG.

The chemical homogeneity and porosity, as well as crystal and domain structure of such single crystals strongly affects their ferroelectric behaviour. They depend on many parameters such as the initial growth setup used for SSCG, i.e. the choice of the seed and the presence of a liquid sintering aid, and further growth conditions, such as annealing rate, temperature, pressure etc. In order to be able to understand the composition-property relationship of single crystals their structure and chemical composition has to be accurately determined. This can be achieved using different complementary analytical methods such as optical, scanning and transmission electron microscopes, X-ray diffraction analysis, etc.
The aim of the following chapter is firstly to present basic principles of the SSCG technique and its application to the growth of $K_{0.5}Na_{0.5}NbO_3$ (KNN) and Li,Ta-modified KNN single crystals. Secondly, the use of complementary analytical characterization techniques for obtaining the precise compositional and structural data from the produced crystals and its correlation with the electrical properties will be discussed.

2. Principles of the solid-state crystal growth technique

The SSCG method is a way of growing single crystals from a polycrystalline starting material. In this method, a single crystal (called a seed crystal) is embedded in ceramic powder of the desired crystal composition and pressed. The sample is then sintered and a single crystal with the composition of the ceramic powder grows using the seed crystal as a template (Fig.1). Alternatively, a seed crystal is placed on top of a pre-sintered ceramic sample and annealed (both the faces of the seed crystal and the ceramic that are in contact must be polished). A single crystal then grows downward from the seed crystal into the ceramic.

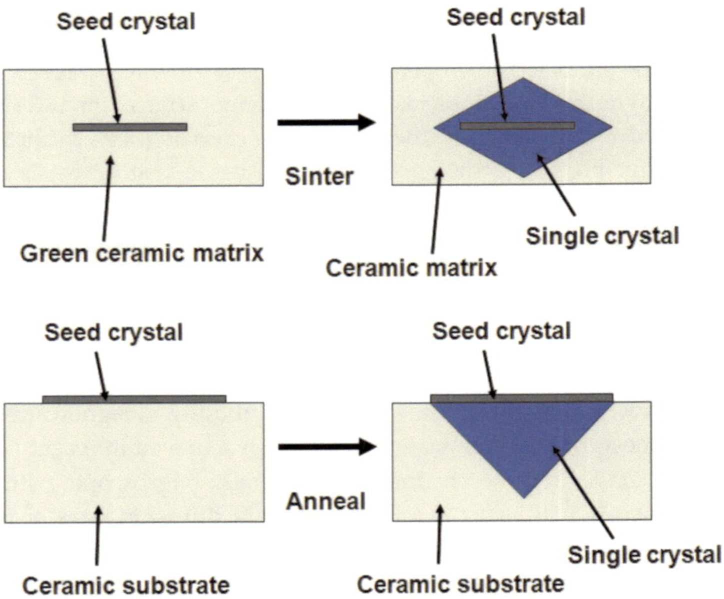

Fig. 1. Schematic of Solid-State Single Crystal Growth

The SSCG method was first used to grow single crystals of $BaTiO_3$ (DeVries, 1964) and has since been used to grow single crystals of manganese zinc ferrite (Kugimiya et al., 1990), $Pb(Mg_{1/3}Nb_{2/3})O_3$–$PbTiO_3$ (Khan et al., 1999), $Pb(Mg_{1/3}Nb_{2/3})O_3$–$PbZrO_3$–$PbTiO_3$ (Zhang et al., 2007), $Ba(Ti,Zr)O_3$ (Rehrig et al., 1999) and $BaTiO_3$ (Yamamoto & Sakuma, 1994). The piezoelectric properties of single crystals grown by the SSCG method are comparable to those of crystals grown by flux-based methods (Lee, 2003).

The SSCG method has several advantages over more traditional flux or melt-based methods of single crystal growth (Lee, 2003). Conventional ceramic processing equipment can be used, avoiding the need for expensive crystal-growing apparatus. The SSCG method is particularly suited to growing single crystals which melt incongruently or which have volatile components. Because the SSCG method does not involve melting and re-solidification of the ceramic powder, problems such as compositional variations in crystals can be avoided. Furthermore, since SSGC is an isothermal crystal growth method, problems of temperature-gradient induced stresses in the crystal are avoided.

However, the SSCG method also has some disadvantages. If pores exist or form in the substrate, they can be included in the growing single crystal (DeVries, 1964; Kim et al., 2006). Once pores are trapped within the crystal they are very difficult to remove, as gases trapped in the pore have to diffuse through the crystal lattice (Choi et al., 2001). Also, if abnormal grain growth takes place in the substrate, these large grains will pin the single crystal boundary, preventing further growth (Kim et al., 2006). So control of the substrate microstructure is vital if good quality single crystals are to be obtained.

The SSGC method utilizes the phenomenon of abnormal grain growth. Abnormal grain growth also frequently occurs in the ceramic matrix/substrate during the sintering/annealing stage. In order to understand and control SSCG, the principles behind abnormal grain growth in ceramics must first be discussed. When a polycrystalline ceramic undergoes abnormal grain growth, a small number of grains (abnormal grains) grow much more rapidly than the other grains (matrix grains). A bimodal microstructure develops, consisting of fine micron-sized matrix grains and a few large abnormal grains, which can be hundreds of microns in diameter. Sometimes, all of the matrix grains are consumed by the abnormal grains, resulting in a coarse microstructure. Examples of systems which display abnormal grain growth include $BaTiO_3$ (Hennings et al., 1987) and $Pb(Mg_{1/3}Nb_{2/3})O_3$–$PbTiO_3$ (Seo & Yoon, 2005).

In a polycrystalline ceramic that is being sintered or annealed, each grain will have its own growth rate $\dot{R}$, given by the equation (Chiang et al., 1997):

$$\dot{R} = 2M_b\Delta\mu \tag{1}$$

where M_b is the interface mobility and $\Delta\mu$ is the driving force for grain growth. The driving force for grain growth of a grain (assuming liquid-phase sintering) is given by (Yoon et al., 2001):

$$\Delta\mu = 2\gamma_{sl}\Omega\left(\frac{1}{\bar{r}} - \frac{1}{r}\right) \tag{2}$$

where γ_{sl} is the solid / liquid interfacial energy, Ω the molar volume of the solid phase, r the grain radius and $\bar{r}$ the radius of a critical grain that neither grows nor shrinks. This is usually considered to be the mean grain radius. For solid-state sintering, γ_{sl} is replaced with the average grain boundary energy γ_{gb}.

Whether a ceramic undergoes abnormal grain growth or not depends upon its interface structure. Interfaces (both grain boundaries and solid/liquid interfaces) can be either rough (disordered) or faceted (ordered) at an atomic level (Hirth & Pound, 1963). If the interfaces are rough, then atom attachment can take place at any point on the growing grain and grain growth is controlled by diffusion of atoms through the liquid phase or across the grain boundary between the shrinking and growing grain. If the interfaces are faceted then attachment of atoms to a grain is energetically unfavourable because a number of broken bonds are formed. After atoms adsorb on the grain surface, they will quickly desorb again unless they can migrate and attach to a low energy site such as the edge of a 2D nucleus. Such a grain can therefore grow by the formation and spread of 2D nuclei on the grain surface. Alternately, if a defect such as a screw dislocation or twin is present, then this can also act as a site for grain growth.

Fig.2 shows the variation of the grain growth rate with driving force for rough and faceted interfaces (Hirth& Pound, 1963). For rough solid/liquid interfaces, interface mobility M_b is constant with $\Delta\mu$ and the growth rate is given by (Greenwood, 1956):

$$\dot{R} = \frac{DC_0\Omega\Delta\mu}{2kTr} \tag{3}$$

where D is the diffusivity of the solid through the liquid phase, C_0 the equilibrium solubility and T the temperature. The growth rate is linearly proportional to the driving force. Grains larger than the mean radius will grow, whilst smaller grains will shrink. For a system with rough solid/liquid interfaces, abnormal grain growth does not occur (Kwon et al., 2000; Lee et al., 2000).

For faceted interfaces, M_b is not constant with $\Delta\mu$. For grain growth caused by 2D nucleation, the growth rate is described by the following equation (van der Eerden, 1993):

$$\dot{R} \cong v_{st}\exp\left(\frac{-\pi\Omega\varepsilon^2}{6h\Delta\mu kT}\right) \tag{4}$$

where v_{st}, ε, and h are the step velocity of the growing nucleus, edge free energy of the nucleus and step height of the nucleus respectively. For low driving forces, the rate at which stable 2D nuclei form is very low. Hence atoms cannot easily attach to the grain and the growth rate is very slow. Above a critical driving force $\Delta\mu_C$, the rate at which stable 2D nuclei form increases exponentially (Hirth& Pound, 1963). Atoms can then easily attach to the grain and grain growth can take place. The grain growth rate therefore varies exponentially with $\Delta\mu$. For $\Delta\mu>\Delta\mu_C$, kinetic roughening takes place: the number of 2D nuclei available for atom attachment on a grain is so large that atoms can easily attach at any point and grain growth is effectively diffusion controlled, as in the case of a rough interface (Kang, 2005). For a system with faceted interfaces, abnormal grain growth can occur. Most of the grains will have $\Delta\mu<\Delta\mu_C$ and will grow very slowly, but a few grains will have $\Delta\mu\geq\Delta\mu_C$ and can grow rapidly to form abnormal grains. This leads to a bimodal microstructure, with large abnormal grains in a matrix of fine grains.

For faceted interfaces, screw dislocations can act as sites for atom attachment and grain growth (Bauser & Strunk, 1982). The growth rate for an interface containing screw dislocations is described by the Burton-Cabrera-Frank theory (Burton et al., 1951). For a crystal face with a growth spiral, the growth rate increases parabolically with $\Delta\mu$ for $\Delta\mu<\Delta\mu_C$, and linearly for $\Delta\mu>\Delta\mu_C$ (Bennema, 1993).

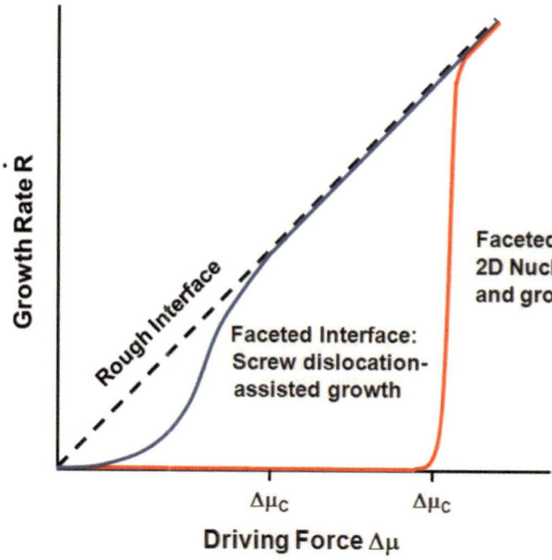

Fig. 2. Variation of growth rate with driving force for rough and faceted interfaces

In the SSCG method, the seed crystal acts as a large grain which can grow abnormally to form a single crystal. Grains in the matrix that are large enough can also undergo abnormal grain growth. For the SSCG method to work, the grains must have faceted interfaces. The seed crystal should have the same crystallographic structure as the desired single crystal, and should have similar unit cell parameters so that epitaxial growth can take place on the seed. In addition, the seed crystal should not react chemically with the matrix material or sintering aid.

3. Characterization methods

Most of the methods that are applicable to the compositional and chemical characterization of bulk ceramics can be employed for the single crystals. One of the most commonly employed methods for exploring ceramic structure is X-ray diffraction (XRD) analysis. Besides crystal structure related analysis, the XRD data gives the information about the bulk phase composition. However, fractions of minor phases below ~3 to ~1 vol%, depending on the XRD configuration, cannot be detected. In case of single crystals, a dedicated XRD setup can be used, if the size of the investigated crystal exceeds 2 mm (Arndt & Willis, 1966).
The microstructure of the material can be explored by the different methods of microscopy, depending on the size of the investigated feature. The optical microscope can be employed for studying larger features and for a general overview of the material. With its specific applications, such as polarization microscopy, the domain structure of the ferroelectric material can be studied with a resolution up to 0.5 μm. Further exploration of the material's microstructure is generally conveyed with a scanning electron microscope (SEM) and a transmission electron microscope (TEM). Here, the local microstructure can be studied at micro- and nano-scales by means of both imaging and electron diffraction. The utilization of the electron back-scattered diffraction (EBSD) technique in SEM allows probing of the

ferroelectric domain structure at the microscale, while the selected area electron diffraction (SAED) patterns obtained in the TEM can be used for studying the nanosized domains.
As well as facillitating the visualization of the microstructure, the electron microscopy techniques make it possible to gain the compositional information from the localized area of the bulk sample. For the bulk material, the complete quantitative elemental composition can be obtained with Inductively Coupled Plasma (ICP) analysis with ppm precision. With energy dispersive X-ray spectroscopy (EDXS) and wavelength dispersive X-ray spectroscopy (WDXS) in SEM it is possible to gain compositional information with micrometer-scale resolution and with typical detection limits of ≈ 0.1wt% and ≈ 0.01 wt.%, respectively (Goldstein et al., 2003). Further improvement of spatial resolution in the compositional analysis can be achieved with TEM-EDXS. Although the detection limits of TEM-EDXS are the same as those in SEM-EDXS, the analyses can be performed with the spatial resolution in the nm range.
It is important to mention that due to the presence of the volatile compounds in alkali based single crystals, the data acquisition and the interpretation of the EDXS analysis have to be performed with caution. Issues such as the spatial resolution, beam-specimen interactions and the use of proper standards should be taken into account to obtain the true chemical composition of KNN single crystals (Benčan et al., 2011). For example, Samardžija et al. (Samardžija et al., 2004) showed that the standardless quantitative EDXS analysis of polycrystalline KNN gave erroneous compositions for the KNN solid solution. Another point which was made by the authors was concerning the proper choice of analytical standards. Glasses containing Na and/or K or natural minerals with Na and/or K (albite, orthoclase) which are usually used as standards in virtual standard packages for standardless SEM-EDXS methods are not reliable enough due to the appearance of compositional variations and the well known beam-induced alkali migration within a single specimen. Additionally, the alkali content in these minerals is rather low, which affects the precision of the analysis. Instead, the use of reliable, homogeneous and stable standard materials such as single crystals of $KNbO_3$ and $NaNbO_3$ is proposed.

4. Growth of $(K_{0.5}Na_{0.5})NbO_3$ single crystals by SSCG.

As was already discussed in Section 2, seed crystals and ceramic powder precursors are needed to grow KNN single crystals by SSCG. For the KNN system, $KTaO_3$ was found to be a suitable material for seed crystals due to the similarity in unit cell parameter between $KTaO_3$ (a=3.9883Å) (ICSD #39673) and $K_{0.5}Na_{0.5}NbO_3$ (a=4.0046Å, b=3.9446Å, c=4.0020Å, and β=90.3327°) (Tellier et al., 2009). Also, $KTaO_3$ does not undergo any phase transitions during cooling which could produce stresses and cracking in the single crystal. For the growth of KNN single crystals, $KTaO_3$ single crystal seeds (FEE GmbH, Germany) oriented in the <001>and <110>directions were used.
The precursor powder with a composition of $(K_{0.5}Na_{0.5})NbO_3$ was prepared by attrition milling K_2CO_3 (Aldrich, 99+%), Na_2CO_3 (Riedel-de Haën, 99.8%) and Nb_2O_5 (H. C. Starck, 99.8%) in acetone with ZrO_2 media, followed by calcination at 700°C for 4 hours. The ceramic powder should have a fine mean particle size and should also have a narrow particle size distribution. The growth rate of the single crystal is inversely proportional to the mean matrix grain size of the ceramic, so a ceramic with a small mean matrix grain size is preferable (Equation 2). To make a precursor ceramic with a small mean matrix grain size, a fine starting powder is necessary. Also, if the starting powder has a wide particle size distribution, the larger particles can grow rapidly to form abnormal grains during sintering

of the ceramic. These abnormal grains will then slow down of even block single crystal growth (Kim et al., 2006).

It may be necessary to add a liquid phase sintering aid to the powder. For the PMN-PT system, addition of PbO as a liquid phase sintering aid increases the growth rate of the single crystal (Gorzkovski et al., 2006). In the experiments described below, $K_4CuNb_8O_{23}$ was used as a liquid phase sintering aid (Matsubara et al., 2005). To make $K_4CuNb_8O_{23}$, K_2CO_3, CuO (Alfa Aesar, 99.7%) and Nb_2O_5 were attrition milled and calcined at 700°C for 4 hours. The effect of $K_4CuNb_8O_{23}$ on single crystal growth of $(K_{0.5}Na_{0.5})NbO_3$ will be discussed in a later Section 4.3.

4.1 Growth of $(K_{0.5}Na_{0.5})NbO_3$ single crystals at atmospheric pressure

$KTaO_3$ seed crystals with the dimensions of 2 mm × 2mm × 0.5 mm were buried in 1.2 g of powder and uniaxially pressed at 100 MPa, followed by cold isostatic pressing at 200 MPa. Samples were sintered in air at 1100°C for 10 h, with heating and cooling rates of 2°C.min^{-1}. A scanning electron microscopy (SEM) micrograph of a vertically-sectioned, polished and thermally etched sample is shown in Fig. 3 (Fisher et al., 2007a). The inset shows the boundary between the single crystal and the matrix. A single-crystal ~160μm thick has grown on the seed crystal. Both the single crystal and the polycrystalline matrix contain large numbers of pores.

Quantitative SEM-EDXS analysis was carried out using single crystals of $KNbO_3$ and $NaNbO_3$ as standards (Samardžija et al., 2004). Analysis shows that both the single crystal and the matrix grains have the composition $(K_{0.5}Na_{0.5})NbO_3$ within the limits of experimental error. Furthermore, the EDXS results show that Ta has not migrated from the seed crystal to the single crystal. High resolution Transmission Electron Microscopy (HRTEM) of the seed/single crystal interface shows that the single crystal grew epitaxially on the seed crystal, as shown by the selected area diffraction pattern of the interface in the inset (Fig. 4).

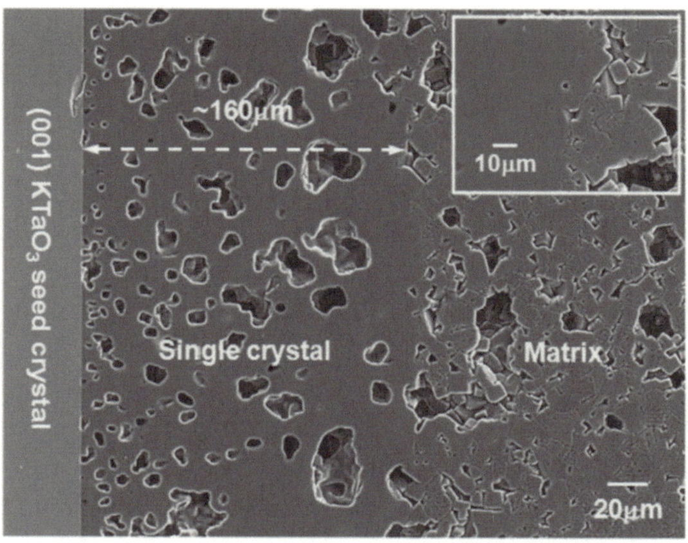

Fig. 3. SEM micrograph of a Single Crystal of $(K_{0.5}Na_{0.5})NbO_3$ grown by SSCG (Fisher et al., 2007a)

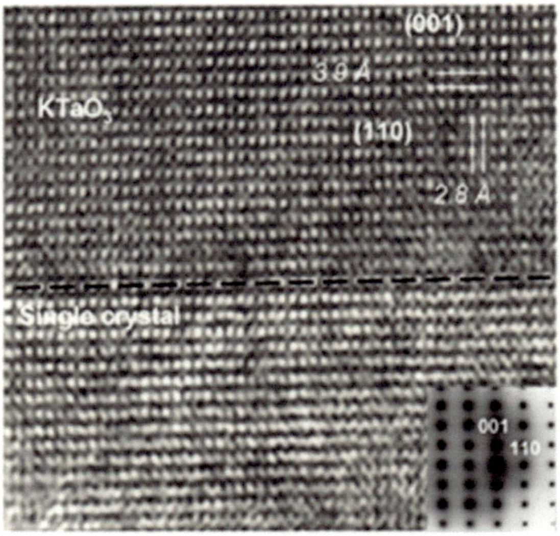

Fig. 4. High resolution TEM (HRTEM) image of the $KTaO_3$ seed crystal / $(K_{0.5}Na_{0.5})NbO_3$ single crystal interface (Fisher et al., 2007a)

4.2 Growth of $(K_{0.5}Na_{0.5})NbO_3$ single crystals in a hot press

As can be seen in Fig. 3 single crystals of $(K_{0.5}Na_{0.5})NbO_3$ grown by the SSCG method are very porous. To solve this problem, single crystals were grown in a hot press (Fisher et al., 2008a). In these experiments, <110>-oriented $KTaO_3$ seed crystals were used.

The sample was pre-densified in the hot press at a temperature of 975°C and a pressure of 50 MPa for 2 h, and then annealed at a temperature of 1100°C and a pressure of 50 MPa for 100 h. In Fig. 5a, the optical micrograph of the resulting single crystal is shown. Compared to a single crystal grown in a conventional box furnace under atmospheric pressure (Fig. 5b), hot pressing made it possible to produce a much denser crystal.

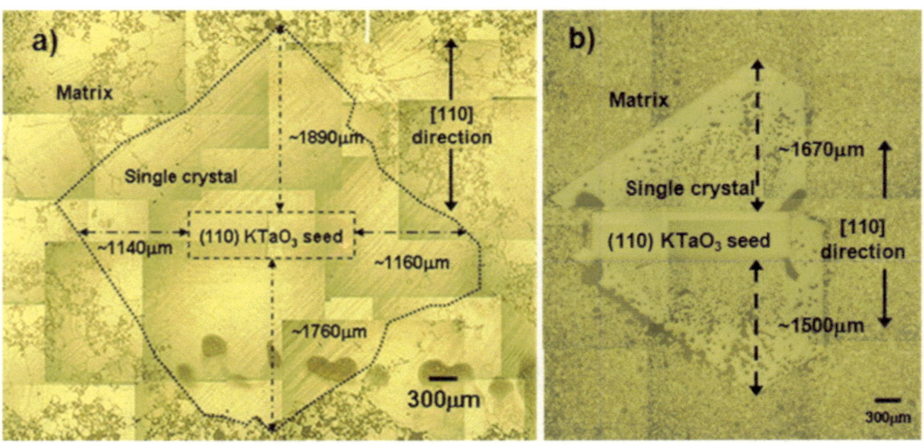

Fig. 5. Comparison of optical micrographs of (a) a $(K_{0.5}Na_{0.5})NbO_3$ single crystal grown in a hot press at 975°C/50 MPa for 2 h, followed by 1100°C/50 MPa for 100 h (b) a $(K_{0.5}Na_{0.5})NbO_3$ single crystal grown in a conventional furnace at 1100°C for 100 h (Fisher et al., 2008a)

During growth of the single crystal in the conventional furnace, single crystal growth, matrix grain growth and matrix densification take place simultaneously. During crystal growth, pores in the matrix can be picked up by the moving single crystal/matrix interface. If the pores then separate from the interface, they will become trapped in the single crystal. The size of the trapped pores increases with crystal growth distance. This is probably due to pore coalescence in the matrix before the crystal/matrix interface reaches them. Application of an external pressure during crystal growth has two benefits. Firstly, during the first stage (975°C/50 MPa for 2 h), the polycrystalline matrix is densified. Application of an external pressure promotes densification without promoting grain growth (Chiang et al., 1997). The sintering temperature can therefore be reduced, allowing the matrix to be densified without much grain growth or single crystal growth. Growing the single crystal in an already dense matrix increases the density of the crystal (Fisher et al., 2007a). Secondly, during the second stage (1100°C/50 MPa for 100 h), the $K_4CuNb_8O_{23}$ liquid phase sintering aid melts and penetrates the grain boundaries, leaving behind pores which must be eliminated. The applied pressure increases the driving force for shrinkage of these pores within the matrix and also of pores that become trapped within the crystal (Kang and Yoon, 1989).

4.3 Effect of sintering aid on crystal growth and composition

The effect of the amount of sintering aid on single crystal growth, matrix grain growth and single crystal composition was investigated (Fisher et al., 2008b). Single crystals were grown from $(K_{0.5}Na_{0.5})NbO_3$ powders with additions of 0, 0.5 and 2 mol % $K_4CuNb_8O_{23}$, using <001>-oriented $KTaO_3$ seed crystals. Before the crystal growth experiments, samples were pre-densified by hot-pressing at 975°C / 50 MPa for 2 h. Crystals were then grown in air under atmospheric pressure at 1100°C for 1-20 h.

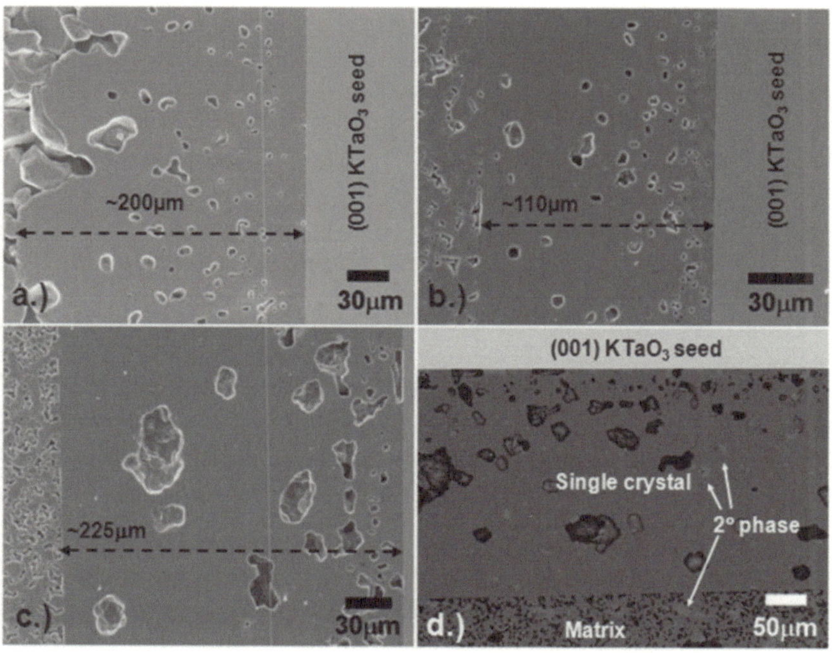

Fig. 6. Single crystals grown from $(K_{0.5}Na_{0.5})NbO_3$ powders with additions of (a) 0, (b) 0.5 and (c) 2 mol % $K_4CuNb_8O_{23}$. Crystals were grown at 1100°C for 10 h. (d) Backscattered electron image of crystal shown in Fig.6 (c) (Fisher et al., 2008b)

Secondary electron SEM images of crystals which had been grown at 1100°C for 10 h are shown in Fig.6 (a) - (c). In the sample with 0 mol % $K_4CuNb_8O_{23}$, the crystal/matrix interface is very irregular. Addition of 0.5 mol % $K_4CuNb_8O_{23}$ causes the interface to become regular but reduces the single crystal growth distance. Addition of 2 mol % $K_4CuNb_8O_{23}$ causes the crystal growth distance to increase again. Fig.6 (d) is a backscattered electron image of the sample with 2 mol % $K_4CuNb_8O_{23}$. It can be seen that there is a second phase trapped within the crystal. EDXS analysis revealed this phase to be the $K_4CuNb_8O_{23}$ sintering aid. This phase was not present within the crystals grown from samples with 0.5 mol % $K_4CuNb_8O_{23}$.

Fig.7 shows the growth distance of the single crystals and mean matrix grain sizes vs. growth time. For the samples with 0 and 0.5 mol % $K_4CuNb_8O_{23}$, crystal growth is initially rapid but tails off with growth time (Fig.7a). Addition of 0.5 mol % $K_4CuNb_8O_{23}$ causes a reduction in single crystal growth distance at all annealing times. For the sample with 2 mol % $K_4CuNb_8O_{23}$, the crystal growth rate decreases after 1 hour and then remains approximately constant up to 20 h. For the samples with 0 and 0.5 mol % $K_4CuNb_8O_{23}$, matrix grain growth is initially rapid but then tails off with annealing time (Fig.7b). For the samples with 2 mol % $K_4CuNb_8O_{23}$, after initial growth for 1 h, the matrix grain size remains almost constant up to 20 h.

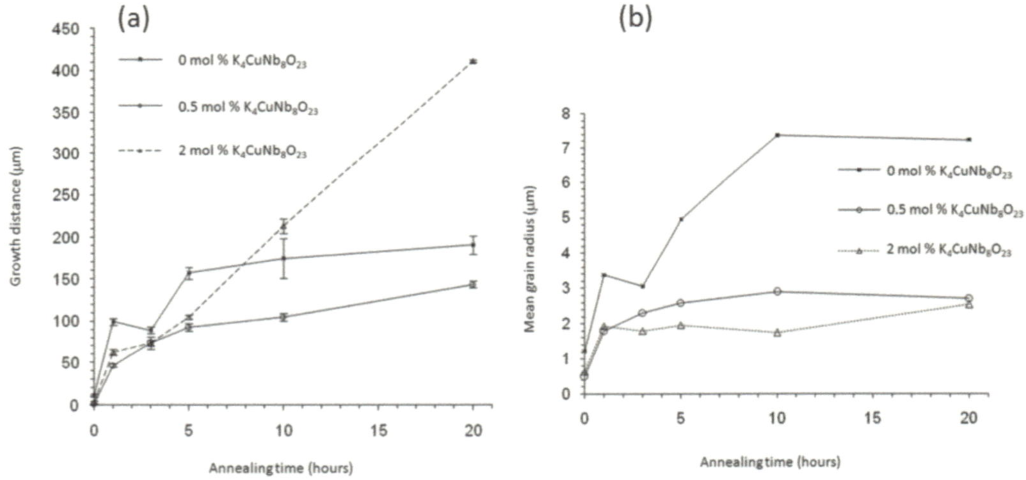

Fig. 7. (a) growth distance of single crystal and (b) mean matrix grain radius vs. growth time at 1100°C (Fisher et al., 2008b)

This behaviour is explained by considering the effect of the liquid phase on both single crystal growth and matrix grain growth. Because the seed crystal acts as a very large grain, for the single crystal equation [2] can be approximated to:

$$\Delta\mu = 2Y_{sl}\Omega\left(\frac{1}{r}\right) \tag{5}$$

Therefore, the single crystal growth rate is inversely proportional to the mean matrix grain size. In the samples with 0 and 0.5 mol % $K_4CuNb_8O_{23}$, matrix grain growth causes the driving force for single crystal growth to decrease with annealing time and the single crystal growth rate to slow down. Addition of 0.5 mol % $K_4CuNb_8O_{23}$ liquid phase sintering aid can further reduce both the crystal and matrix grain growth rates, as the thickness of the solid/liquid interface across which atoms must diffuse increases (Kang, 2005). With addition of 2 mol % $K_4CuNb_8O_{23}$, matrix grain growth effectively ceases after 1 h. This means that the driving for single crystal growth remains constant after 1 h, allowing the crystal to keep growing even for extended annealing times.

Table 1 gives EDXS analyses of single crystals and matrix grains of samples with different amounts of $K_4CuNb_8O_{23}$. Again, single crystals of $KNbO_3$ and $NaNbO_3$ were used as standards. For the samples with 0 and 0.5 mol % $K_4CuNb_8O_{23}$, both the single crystal and matrix grains have compositions close to the nominal composition. For the sample with 2 mol % $K_4CuNb_8O_{23}$, the matrix grains have the nominal composition but the single crystal is Na-rich. According to the $KNbO_3$-$NaNbO_3$ phase diagram, $(K_{0.5}Na_{0.5})NbO_3$ at 1100°C lies just below the solidus line (Jaffe et al., 1971). It is possible that addition of 2 mol % $K_4CuNb_8O_{23}$ lowered the solidus temperature to below 1100°C. This would then cause the equilibrium solid phase to be Na-rich. Indeed, the growing single crystal is Na-rich. The matrix grains retain their original composition as their growth rate is very slow. Therefore, care must be taken when adding a liquid phase sintering aid to promote single crystal growth in this system.

Amount of $K_4CuNb_8O_{23}$ (mol %)	**K (at. %)**	**Na (at. %)**	**K/Na ratio**
0 (single crystal)	10.34 ± 0.58	10.82 ± 0.64	0.96 ± 0.04
0 (matrix)	10.64 ± 0.62	10.53 ± 0.58	1.01 ± 0.05
0.5 (single crystal)	10.41 ± 0.44	10.39 ± 0.41	0.99 ± 0.05
0.5 (matrix)	10.48 ± 0.63	10.42 ± 0.96	1.02 ± 0.13
2 (single crystal)	8.46 ± 0.54	13.35 ± 0.65	0.64 ± 0.06
2 (matrix)	10.58 ± 0.32	10.79 ± 0.99	0.99 ± 0.10
Nominal composition for $(K_{0.5}Na_{0.5})NbO_3$	10	10	1

Table 1. EDXS analyses of single crystals and matrix grains of samples annealed at 1100°C for 10 h (Fisher et al., 2008b).

4.4 Growth of $[(K_{0.5}Na_{0.5})_{0.97}Li_{0.03}](Nb_{0.8}Ta_{0.2})O_3$ single crystals by SSCG.

The SSCG method was successfully applied to the growth of (Li, Ta)-KNN modified single crystals (Fisher et al., 2007b). Powder of a nominal $[(K_{0.5}Na_{0.5})_{0.97}Li_{0.03}](Nb_{0.8}Ta_{0.2})O_3$ composition was prepared in a similar way as before, but with a higher calcination temperature of 900°C. 0.5 mol % of $K_4CuNb_8O_{23}$ was added as a liquid phase sintering aid. A <001>-oriented $KTaO_3$ single crystal was used as a seed. The sample was pre-densified by hot pressing at 975°C / 50 MPa for 2 h. The crystal was grown by annealing in air at 1135°C for 50 hours under atmospheric pressure. A single crystal ~100μm thick grew on the seed (Fig.8). SEM-EDXS analysis showed that the single crystal and the matrix grains have the same composition; however, it was not possible to analyse Li content by means of EDXS.

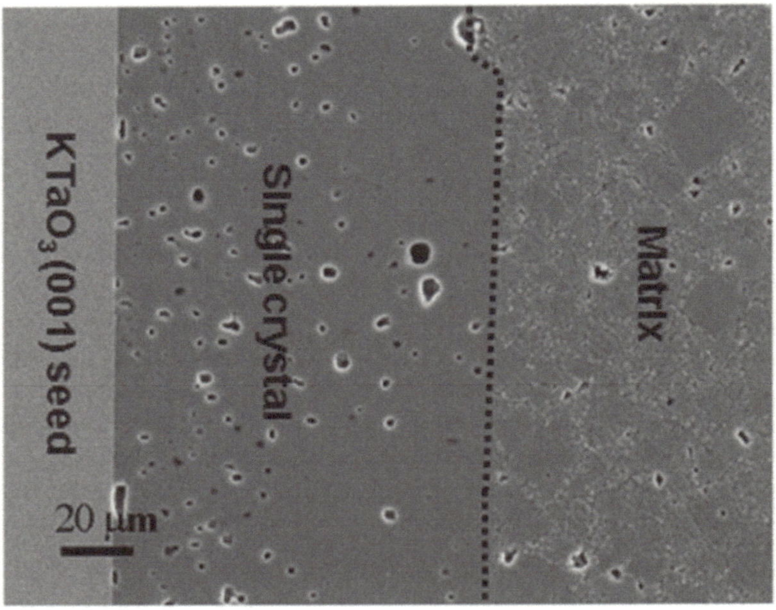

Fig. 8. SEM micrograph of $[(K_{0.5}Na_{0.5})_{0.97}Li_{0.03}](Nb_{0.8}Ta_{0.2})O_3$ Single Crystal grown by SSCG (Fisher et al., 2007b)

5. Dedicated structural and compositional study of a $(K_{0.5}Na_{0.5})NbO_3$ single crystal

The studies of structure and composition were performed on the hot-pressed KNN single crystals (see Fig. 5a). For the single crystal XRD setup, the size of the single crystals after their removal from the matrix was not sufficient. Therefore, the obtained crystals were crushed and a powder XRD setup was used.

In Fig. 9, experimental XRD powder diffraction patterns of the crushed KNN single crystal and a polycrystalline KNN ceramic, as well as calculated a XRD diffraction pattern are shown. The inset in Fig. 9 shows an enlarged view of the 100/001 and 010 diffraction peaks for the KNN single crystal and ceramic. Both the single crystal and ceramic have narrow and well defined peaks. No secondary phases were detected (Benčan et al., 2009). In previous work, different workers have refined KNN unit cell parameters using perovskite unit cells with orthorhombic symmetry (Attia et. al., 2005), monoclinic symmetry (Shiratori et. al., 2005) and also triclinic symmetry (Shiratori et. al., 2005). Our experimental data was fitted using the monoclinic symmetry given by Tellier et al. (Tellier et al. 2009), with unit cell parameters a=4.0046Å, b=3.9446 Å, c=4.0020 Å, and β=90.3327°.

A precise chemical analysis of the KNN single crystal was performed by WDXS and semi-quantitative EDXS analysis in the SEM at twelve selected points across the KNN single crystal. For WDXS analysis, $KNbO_3$ and $NaNbO_3$ single crystals were used as standards. Table 2 shows the determined elemental composition of the KNN single crystal, which is very close to the nominal one. The small variations in the values of standard deviation for both WDXS and EDXS analysis serve as proof of the crystal's homogeneity. The latter makes it possible to use these crystals as reference standards for the quantitative analysis of sodium and potassium in other materials (Benčan et al., 2009).

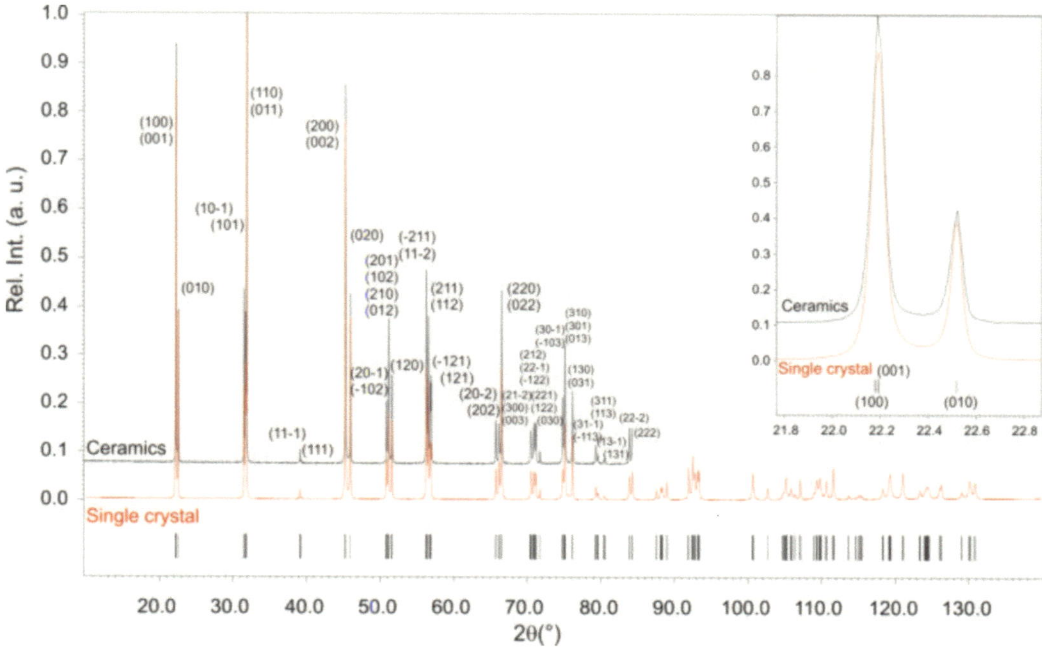

Fig. 9. XRD powder diffraction patterns of the crushed KNN single crystal and polycrystalline KNN ceramic. A calculated XRD pattern using a monoclinic KNN unit cell is added (Benčan et al., 2009)

Element	Nominal composition	WDXS		EDXS	
	at%	at%	STDEV	at%	STDEV
K	10	10.06	0.08	9.5	0.1
Na	10	10.03	0.07	9.8	0.2
Nb	20	19.89	0.10	20.3	0.3
O	60	60.02	0.15	60.4*	0.5

Table 2. Elemental composition in at% of the KNN single crystal, determined by WDXS and EDXS, with standard deviation (STDEV). Nominal composition is shown for comparison. Oxygen (*) is calculated from the stoichiometry (Benčan et al., 2009)

The domain structure of KNN single crystals at micro- and nano-scales was analysed using the techniques of optical, scanning and transmission electron microscopy (Benčan et al., 2009). A polarized light optical micrograph of the KNN single crystal is shown in Fig. 10a. The crystal is still embedded in the KNN ceramic matrix. Large ferroelectric domains from 50 to 100 microns in size are revealed by dark/bright contrast oscillations in the micrograph. These large domains in turn contain smaller domains with dimensions from tens of microns down to hundreds of nm. The smaller domains have a herring bone 90° arrangement, as shown in the inset. in Fig. 10a. The larger domains in the single crystal were also probed by electron backscattered diffraction (EBSD). The EBSD image (Fig. 10b) shows the distribution of the orientations in the crystal and surrounding matrix. Differences in colour inside the single crystal are attributed to the differently oriented ferroelectric domains. From the colour-key inverse-pole-figure it can be seen that the orientation inside the single crystal is changing by 90° and that there are three different orientations rotated to each other by 90° angles.

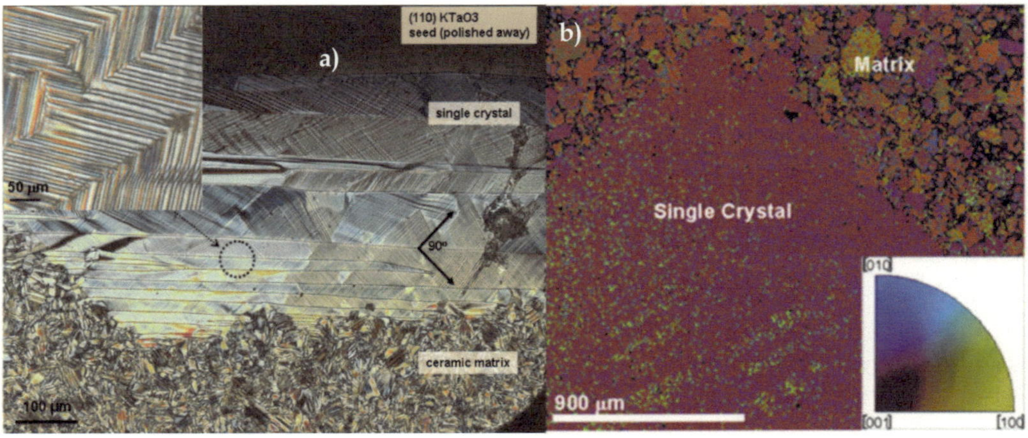

Fig. 10. Optical microscope micrographs of the KNN single crystal and its domain microstructure. The inset shows a herring bone 90° arrangement of smaller domains (a) EBSD orientation map of the KNN single crystal and the corresponding color-key inverse-pole-figure (b) (Benčan et al., 2009)

In order to determine the domain structure at the nanometer scale, the specimen was investigated by TEM (Benčan et al., 2009). Smaller saw-like domains with a size of about 50nm were arranged within the larger ones (Fig.11).

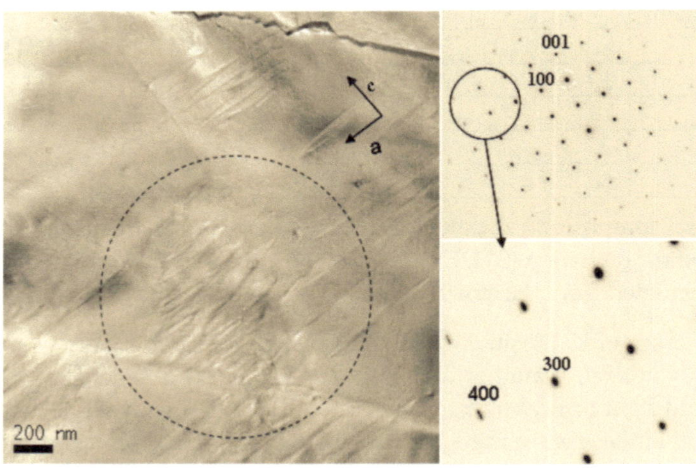

Fig. 11. TEM-BF image of the KNN single crystal with corresponding SAED patterns showing the presence of 180 ° domains. Due to the very small difference in a and c unit cell parameters, a and c axes were chosen arbitrarily (Benčan et al., 2009)

The overlapping of these domains is represented by the selected area diffraction (SAED) pattern in the [010] zone axis, taken from the area of ~1.5 µm. Splitting of the {h00} or {00l} reflections parallel to the <001> or <100> directions is seen. This is due to the β angle (~ 90.3°). Such patterns can be experimentally observed only in the case where the [100] or

[001] direction of one domain is parallel to the [-100] or [00-1] direction of the other one, meaning that these are 180° domains.

6. Dielectric, ferroelectric, piezoelectric and electrostrictive properties of $K_{0.5}Na_{0.5}NbO_3$ single crystals

The dielectric properties of a hot-pressed KNN single crystal (see Fig. 5a for reference) were measured on the as-cut piece of crystal in two perpendicular orientations. These were determined from EBSD analysis and described as $[1\bar{3}1]$ and the $[\overline{323}]$. Fig. 12 shows the temperature dependence of the dielectric constant (ε) and the dielectric losses (tan δ) measured for the KNN single crystal in the above mentioned directions and also for the surrounding polycrystalline KNN matrix. The highest value of ε was obtained for the $[1\bar{3}1]$ direction of the KNN single crystal across whole temperature range. At the same time, two phase transitions from the monoclinic to the tetragonal phase (T_1) at around 193°C, and from the tetragonal to the cubic phase (T_2) at around 410°C were measured (Ursič et al, 2010). The latter are in accordance with the transitions observed in the surrounding polycrystalline KNN ceramic, which is another indication of the obtained crystal compositional homogeneity. Table 3 summarizes the dielectric properties obtained for the KNN single crystal in both directions and for the surrounding polycrystalline KNN matrix, and gives a comparison with the dielectric properties of KNN-based single crystals reported in the literature.

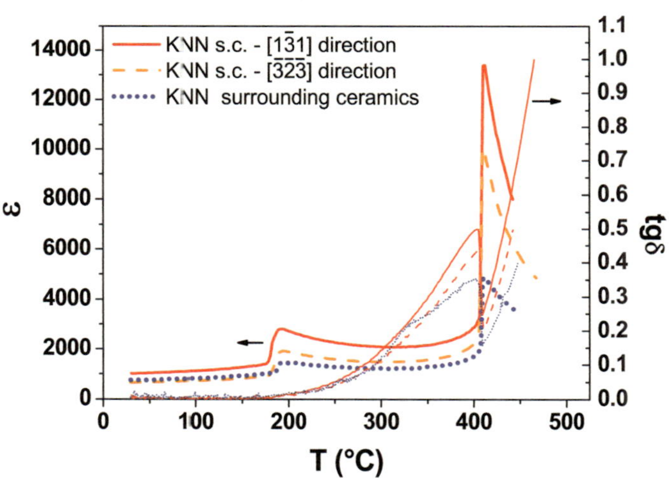

Fig. 12. Comparison of ε (thick lines) and tanδ (thin lines) as a function of the temperature for the KNN single crystal in $[1\bar{3}1]$ and the $[\overline{323}]$ directions and for KNN surrounding ceramics measured at 100 kHz (Uršič et al., 2010).

Due to the high dielectric constant, the $[1\bar{3}1]$ direction of KNN single crystal was chosen for further measurements of the ferroelectric, piezoelectric and electrostrictive properties. The ferroelectric properties, i.e. the remnant polarization (Pr) and coercive field (Ec) measured for the KNN single crystal and surrounding polycrystalline KNN matrix, are compared to the literature and shown in Table 4.

System	Freq. (kHz)	ε Troom	tg δ Troom	T_1 (°C)	T_2 (°C)
KNN s.c. $[1\bar{3}1]$ this study	100	1015	0.01	192	410
KNN s.c. $[\overline{323}]$ this study	100	650	0.01	193	409
KNN ceramics this study	100	750	0.01	193	411
$K_{0.5}Na_{0.5}NbO_3$ s.c. [001] Lin et al., 2009	100	240	0.02	205	393
$K_{0.47}Na_{0.53}NbO_3$ s.c. $[100]_c$ Kizaki et al., 2007	1000	600	below 0.1	190	390
$K_{0.53}Na_{0.47}Mg_{0.004}Nb_{0.996}O_y$ s.c. $[100]_c$ Kizaki et al., 2007	1000	740	below 0.1	160	390
$0.95(K_{0.5}Na_{0.5})NbO_3$-$0.05LiNbO_3$ s.c. [001] Chen et al., 2007	100	185	0.01	192	426
$Li_{0.02}(Na_{0.5}K_{0.5})_{0.98}NbO_3$ s.c. $[001]_c$ Davis et al., 2007	1	205	0.33	177	/

Table 3. The ε, tgδ and monoclinic - tetragonal (T_1) and tetragonal - cubic (T_2) phase transition temperatures for KNN single crystal in the $[1\bar{3}1]$ and the $[\overline{323}]$ directions and for KNN ceramics. For comparison the dielectric properties obtained on KNN based single crystals by different groups are added (Uršič et al., 2010)

System	Freq. (Hz)	P_r (μC/cm²)	E_c (kV/cm)
KNN s.c. $[1\bar{3}1]$ this study	50	17	24
KNN ceramics this study	50	15	24
$K_{0.47}Na_{0.53}NbO_3$ s.c. $[100]_c$ Kizaki et al., 2007	1	/	/
$K_{0.53}Na_{0.47}Mg_{0.004}Nb_{0.996}O_y$ s.c. $[100]_c$ Kizaki et al., 2007	1	40	12
$0.95(K_{0.5}Na_{0.5})NbO_3$-$0.05LiNbO_3$ s.c. [001] Chen et al., 2007	10	9	22

Table 4. Ferroelectric properties of KNN single crystalsin the $[1\bar{3}1]$ direction and for KNN ceramics. For comparison the ferroelectric properties obtained on KNN based single crystals by different groups are added (Uršič et al., 2010)

The displacement signal versus the applied voltage of the KNN single crystal in the $[1\bar{3}1]$ direction and of the surrounding KNN ceramic were measured using an atomic force microscope (AFM). Prior to the analysis, an AFM measurement was performed as a reference on glass under the same experimental conditions as used for the KNN single crystal and ceramics. No strain was observed for the non-piezoelectric glass, confirming that strains observed during AFM analysis of the KNN crystal and ceramics are piezoelectric in nature. The KNN single crystals were not poled before the AFM measurement.

The obtained displacement signal consists of two components. The first component has the same frequency as the applied voltage, i.e., this is the linear piezoelectric component (see Fig. 13). The second component is the pronounced quadratic component with the double frequency (see inset in Fig. 14). The piezoelectric coefficients d_{33}, shown in Fig. 13, were determined from the slopes of the linear fits of the linear component of displacement versus the applied voltage (Uršič et al., 2010).

The d_{33} piezoelectric coefficients for the KNN single crystal and for the surrounding ceramic are approximately 80 pm/V at a measurement frequency of 2 Hz. As frequency increases, the d_{33} value for the KNN single crystal decreases (see Fig.13). Although very small applied electric fields (up to 0.1 kV/mm) were used to measure the piezoelectric coefficient for the KNN single crystal, the obtained d_{33} value (80 pm/V) was in the same range as for the poled KNN ceramic. The explanation of such behaviour can be given by the domain structure of the KNN single crystal. As shown in Section 5 the KNN single crystal consists of large 90° domains with widths of up to 100 microns, and smaller 180° domains with widths ranging between a few tens of nms to 300 nm. Since the contact area of the AFM tip is around 20 nm, it is likely that only the smaller 180° domain walls are moving during the AFM measurements. These small 180° domains probably contribute to the obtained linear response of the KNN single crystal. The inability of the 180° domains to reorientate quickly enough at higher frequencies explains the decrease in d_{33} with increasing measurement frequency. It has been previously demonstrated by McKinstry et al. (McKinstry et al., 2006) that if the mobility of 90° domains is limited, then the 180° domains can contribute to the piezoelectric linear response.

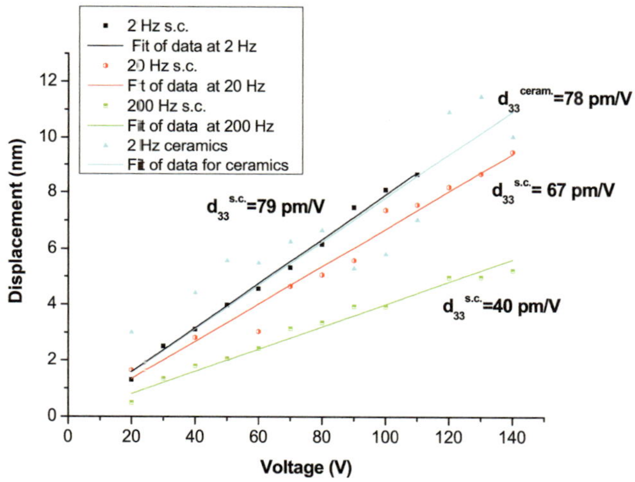

Fig. 13. The linear part of displacement versus voltage amplitude of KNN single crystal in [1 $\bar{3}$ 1] direction measured at 2 Hz, 20 Hz and 200 Hz. The measurement for KNN surrounding ceramics at 2 Hz is added for comparison (Uršič et al., 2010)

The electrostrictive coefficients (M_{33}) were determined from the slope of the linear fit of the relative strain versus the square of the amplitude of the electric field, as shown in Fig. 14. The M_{33} for the surrounding KNN ceramic was lower than that of the KNN single crystal.
The measured values M_{33} for the KNN single crystal are significantly higher than values of M_{33} for PMN-based single crystals. The highest obtained electrostrictive coefficient for a $0.65Pb(Mg_{1/3}Nb_{2/3})O_3$-$0.35PbTiO_3$ single crystal was in the range 1.3 to $4x10^{-15}$ m^2/V^2 at 0.01 Hz; a 90° domain wall contribution to electrostriction was reported (Bokov&Ye, 2002). Such a high M_{33} value for the KNN single crystal can arise from the intrinsic electrostrictive behaviour as well as the extrinsic contribution, i.e., the strain from the domain-wall motion. Most probably in the KNN single crystal, the main contribution to electrostrictive strain arises from the contribution of 180° domain walls. Our results agree with the findings obtained by McKinstry et al. (McKinstry et al., 2006), who showed that 180° domains walls

can contribute to the linear response as well as to electrostrictive strain response in ferroelectric materials. Although the pure electrostrictive response should be frequency independent, they observed in (111) $Pb(Zr_{0.45}Ti_{0.55})O_3$ thin films a decrease of the second-order strain with frequency by 20%, as was also the case for our KNN single crystal.

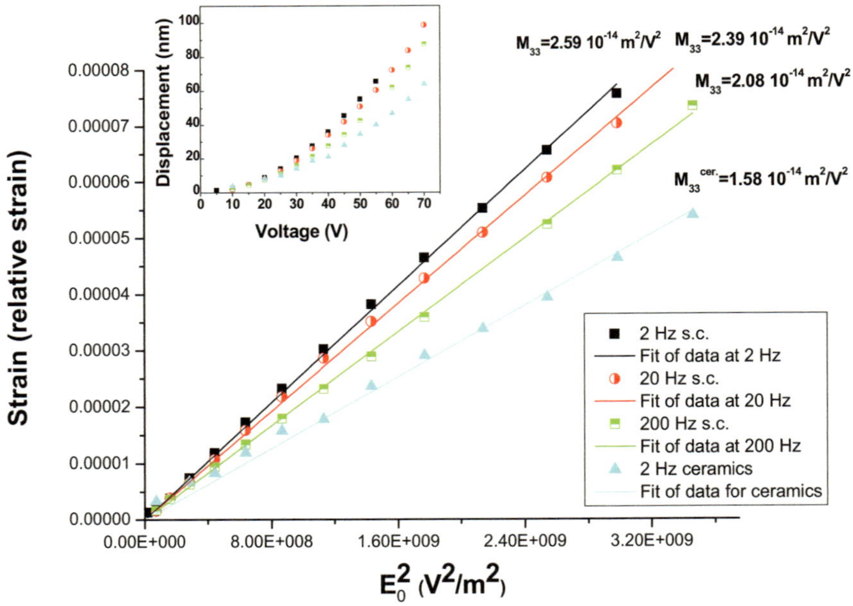

Fig. 14. Relative strain versus square of the amplitude of the electric field of KNN single crystal in the $[1\,\bar{3}\,1]$ direction at 2 Hz, 20 Hz and 200 Hz. Inset shows the quadratic part of displacement versus voltage for KNN single crystal. The measurement for KNN surrounding ceramics at 2 Hz is added for comparison (Uršič et al., 2010)

7. Conclusions

In this chapter the principles of the SSCG technique and its application to the growth of $K_{0.5}Na_{0.5}NbO_3$ (KNN) and Li,Ta-modified KNN single crystals were presented. With the use of the complementary analytical characterization techniques, i.e. XRD, optical microscopy and electron microscopy (SEM, EDXS, WDXS, EBSD, TEM, SAED), the precise compositional and structural analysis of KNN single crystals was performed and the correlation with its electrical properties was given.

There are several possible directions for future work. First, it would be useful to grow larger single crystals. This will enable crystals to be cut in controlled orientations e.g along the [001] or [110] directions and their properties measured and compared with KNN crystals grown by solution-based methods. Furthermore, alternative seed crystals need to be found. Although $KTaO_3$ single crystals make excellent seeds, they are rather expensive and to grow large single crystals, large seed crystals are needed. If cheaper alternatives could be found, this would reduce the cost of growing large KNN single crystals. Work needs to be done in growing single crystals from seeds placed on top of the ceramic substrate. Finally, growth of other compositions such as Li-doped KNN should be carried out.

8. Acknowledgments

Dr. Daniel Rytz, from FEE GmbH, Germany is acknowledged for the preparation of $KTaO_3$ seeds. The authors wish to acknowledge the financial support of the Slovenian Research Agency (P2-105) and the 6FP project IMMEDIATE.

9. References

Arndt, U.W. & Willis, B.T.M. (1966).*Single Crystal Diffractometry* (1st edition), Cambridge University Press, ISBN:978-0-521-04060-0, New York

Attia, J., Bellaiche, M., Gemeiner, P., Dkhil, B. & Malič, B. (2005). Study of potassium-sodium-niobate alloys: A combined experimental and theoretical approach. *Journal de Physique IV (Proceedings),* Vol. 128, No.1, (September 2005), pp. 55–60, ISSN:1155-4339

Bauser, E. & Strunk, H. (1982).Dislocations as Growth Step Sources in Solution Growth and Their Influence on Interface Structure.*Thin Solid Films,* Vol. 93, Nos. 1-2, (July 1982), pp. 185-94,ISSN:0040-6090

Benčan, A., Tchernychova, E., Godec, M., Fisher, J. & Kosec M. (2009).Compositional and Structural Study of a $(K_{0.5}Na_{0.5})NbO_3$ Single Crystal Prepared by Solid State Crystal Growth. *Microscopy and Microanalysis,* Vol.15, No.5, (October 2009), pp. 435-440, ISSN:1431-9276

Benčan, A., Tchernychova, E., Šturm, S., Samardzija, Z., Malič, B. & Kosec, M. (2011). Approches for a reliable compositional analysis of alkaline-based lead-free perovskite ceramics using microanalytical methods. *Journal of Advanced Dielectrics,*Vol.1, No.1, (January 2011), pp. 41-52, ISSN:2010-1368

Bennema P. (1993). Growth and Morphology of Crystals. In:*Handbook of Crystal Growth 1 Fundamentals Part A: Thermodynamics and Kinetics,* D. T. J. Hurle (Ed.), pp. 481-581,ISBN: 0444889086, North-Holland, Amsterdam

Bokov, A. & Ye, Z. G. (2002). Giant electrostriction and stretched exponential electromechanical relaxation in $0.65Pb(Mg_{1/3}Nb_{2/3})O_3$-$0.35PbTiO_3$ crystals, *Journal of Applied Physics,* Vol. 91,No. 10 (May 2002),pp. 6656- 6661, ISSN:0021-8979

Bomlai, P.,Wichianrat, P.,Muensit, S. & Milne, S.J. (2007). Effect of Calcination Conditions and Excess Alkali Carbonate on the Phase Formation and Particle Morphology of $Na_{0.5}K_{0.5}NbO_3$ Powders. *Journal of American Ceramic Society,*Vol. 90, No. 5, (May 2007),pp. 1650-1655, ISSN:0002-7820

Burton, W. K., Cabrera, N. & Frank, F. C. (1951). The Growth of Crystals and the Equilibrium Structure of Their Surfaces. *Philosophical Transactions of the Royal Society of London Series A,* Vol. 243, (June 1951), pp.299-358, ISSN: 0962-8428

Chen, K., Xu, G., Yang, D. ,Wang, X. & Li J. (2007). Dielectric and piezoelectric properties of lead-free $0.95(K_{0.5}Na_{0.5})NbO_3$–$0.05LiNbO_3$ crystals grown by the Bridgman method. *Journal of Applied Physics.*Vol. 101,No. 4, (February 2007), pp. 0441031-4,ISSN:0021-8979

Chiang, Y. M., Birnie III, D. & Kingery, W. D. (1997). *Physical Ceramics: Principles for Ceramic Science and Engineering,* John Wiley & Sons, Inc., ISBN 0-471-59873-9, New York

Choi, J.J., Ryu, J. & Kim, H. E. (2001). Microstructural Evolution of Transparent PLZT Ceramics Sintered in Air and Oxygen Atmospheres. *Journal of the American Ceramic Societ,*.Vol. 84, No. 7 (July 2001), pp. 1465-1469, ISSN:0002-7820

Davis, M., Klein, N.,Damjanovič, D. & Setter, N. (2007). Large and stable thickness coupling coefficients of [001]$_c$ oriented $KNbO_3$ and Li-modified $(K,Na)NbO_3$ single crystals. *Applied Physics Letters.*Vol. 90, No. 6 (February 2007), pp.062904 1-3,ISSN:0003-6951

DeVries.R.C. (1964). Growth of Single Crystals of $BaTiO_3$ by Exaggerated Grain Growth.*Journal of the American Ceramic Society,* Vol. 47, No. 3 (March 1964), pp. 134-136,ISSN:0002-7820

Fisher, J. G., Benčan, A., Holc, J., Kosec, M., Vernay, S. & Rytz, D. (2007a). Growth of Potassium Sodium Niobate Single Crystals by Solid State Crystal Growth. *Journal of Crystal Growth,* Vol. 303, No. 2 (May, 2007), pp. 487-492, ISSN:0022-0248

Fisher, J. G., Benčan, A., Bernard, J., Holc, J., Kosec, M., Vernay, S. & Rytz, D. (2007b). Growth of $(Na,K,Li)(Nb,Ta)O_3$ Single Crystals by Solid State Crystal Growth. *Journal of the European Ceramic Society,* Vol. 27, Nos. 13-15 (2007), pp. 4103-4106,ISSN:0955-2219

Fisher, J. G., Benčan, A., Kosec, M., Vernay, S. & Rytz, D. (2008a).Growth of Dense Single Crystals of Potassium Sodium niobate by a Combination of Solid-State Crystal Growth and Hot Pressing. *Journal of the American Ceramic Society,* Vol. 91, No. 5 (May 2008), pp. 1503-1507,ISSN:0002-7820

Fisher, J. G., Benčan, A., Godnjavec, J. & Kosec, M. (2008b). Growth Behaviour of Potassium Sodium Niobate Single Crystals Grown by Solid-State Crystal Growth Using $K_4CuNb_8O_{23}$ as a Sintering Aid. *Journal of the European Ceramic Society,* Vol. 28, No. 8 (2008), pp. 1657-1663,ISSN:0955-2219

Goldstein, J., Lyman, C.E., Newbury, D.E.,Lifshin, E., Echlin, P., Sawyer, L., Joy D.C. & Michael, J.R. (2003). *Scanning Electron Microscopy and X-Ray Microanalysis,* (3rd edition), Kluwer Academic and Plenum Publishers, ISBN:0-306-47292-9, New York

Gorzkowski, E. P., Chan, H. M. & Harmer, M. P. Effect of PbO on the Kinetics of {001} $Pb(Mg_{1/3}Nb_{2/3})O_3$- 35mol%$PbTiO_3$ Single Crystals Grown into Fully Dense Matrices. *Journal of the American Ceramic Society,* Vol.89, No.3, (March 2006) pp. 856-862,ISSN:0002-7820

Greenwood, G. W. (1956). The Growth of Dispersed Precipitates in Solutions. *ActaMetallurgica,*Vol. 4, No. 3 (May 1956), pp. 243-348, ISSN:0001-6160

Hennings, D. F. K., Janssen, R. & Reynen, P. J. L. (1987). Control of Liquid-Phase- Enhanced Discontinuous Grain Growth in Barium Titanate. *Journal of the American Ceramic Society,* Vol. 70, No. 1 (January 1987), pp.23-27,ISSN:0002-7820

Hirth, J. P. & Pound, G.M. (1963). Condensation and Evaporation. *Progress in Materials Science.*Vol. 11, (1963), pp. 17-192,ISSN:0079-6425

Jaffe, B., Cook Jr., W. R. & Jaffe, H. (1971). Perovskite niobates and tantalates and other ferroelectric and antiferroelectric perovskites, In: *Piezoelectric Ceramics,* Eds. J.P.Roberts & P.Popper, Academic Press, pp. 185-212, ISBN 0123795508, London

Jenko, D., Benčan, A., Malič, B., Holc, J. & Kosec, M. (2005). Electron microscopy studies of potassium sodium niobate ceramics. *Microscopy and microanalysis,* Vol. 11, No.6 (December 2005), pp. 572-580, ISSN:1435-8115

Kang, S. J. L. & Yoon, K. J. (1989). Densification of Ceramics Containing Entrapped Gases. *Journal of the European Ceramic Society,* Vol. 5, No. 2 (1989), pp. 135-139, ISSN:0955-2219

Kang, S. J. L. (2005). Chapter 15, Grain Shape and Grain Growth in a Liquid Matrix, In: *Sintering: Densification, Grain Growth and Microstructure,* pp. 205-26, Elsevier, ISBN:07506 63855, Oxford

Khan, A., Meschke, F. A., Li, T., Scotch, A. M., Chan, H. M. & Harmer, M. P. (1999). Growth of $Pb(Mg_{1/3}Pb_{2/3})O_3$ – 35 mol% $PbTiO_3$ Single Crystals from (111) Substrates by

Seeded Polycrystal Conversion.*Journal of the American Ceramic Society,*Vol. 82, No.11, (November 1999), pp. 2958-62,ISSN:0002-7820

Kim, M. S., Fisher, J. G., Kang, S. J. L. & Lee, H. Y. (2006). Grain Growth Control and Solid-State Crystal Growth by Li_2O/PbO Addition and Dislocation Introduction in the PMN-35 PT System. *Journal of the American Ceramic Society,* Vol. 89, No. 4 (April 2006), pp. 1237–1243,ISSN:0002-7820

Kizaki, Y., Noguchi, Y. & Miyayama, M. (2007). Defect control for Superior Properties in $K_{0.5}Na_{0.5}NbO_3$ Single Crystals. *Key Engineering Materials,* Vol. 350, (October 2007), pp. 85-88, ISSN:1013-9826

Kosec, M., Malič, B., Benčan, A. & Rojac, T. (2008). KNN-based piezoelectric ceramics. In:*Piezoelectric and Acoustic Materials for Transducer Applications,*Eds.: A. Safari and E. K. Akdogan,pp. 81-102, Springer, ISBN: 978-0-387-76538-9., New York

Kosec, M., Malič, B., Benčan, A., Rojac, T. & Tellier, J. (2010). Alkaline niobate-based piezoceramics: crystal structure, synthesis, sintering and microstructure. *Functional materials letters* Vol. 3, No.1, (March 2010), pp. 15-18, ISSN:1793-6047

Kosec, M. &Kolar, D. (1975).On Activated Sintering and Electrical Properties of $NaKNbO_3$. *Material Research Bulletin,* Vol. 10, No. 5, (May 1975), pp. 335–40,ISSN: 0025-5408

Kugimiya, K., Hirota, K. & Matsuyama, K. (1990). Process for Producing Single Crystal Ceramics. US Pat. No. 4900393, 1990.

Kwon, S. K., Hong, S. H., Kim, D. Y. & Hwang, N. M. (2000). Coarsening Behavior of Tricalcium Silicate (C_3S) and Dicalcium Silicate (C_2S) Grains Dispersed in a Clinker Melt. *Journal of the American Ceramic Society,* Vol. 83, No. 5, (May 2000), pp. 1247-1252,ISSN:0002-7820

Lee, B. K., Chung, S. Y. & Kang, S. J. L. (2000). Grain Boundary Faceting and Abnormal Grain Growth in $BaTiO_3$. *Acta Materialia,* Vol. 48, No. 7, (April 2000), pp. 1575-1580ISSN:1359-6454

Lee H. Y. (2003). Solid-State Single Crystal Growth (SSCG) Method: A Cost-effective Way of Growing Piezoelectric Single Crystals. In: *Piezoelectric Single Crystals and their Applications,* S. Trolier-McKinstry, L. E. Cross and Y. Yamashita (Eds.), pp. 160-177, Pennsylvania State University, State College, PA

Lin, D., Li, Z., Zhang, S., Xu Z. & Yao, X. (2009). Dielectric/piezoelectric properties and temperature dependence of domain structure evolution in lead free $K_{0.5}Na_{0.5}/NbO_3$ single crystal. Solid State Communications, Vol. 149, No.39-40, (October 2009), pp. 1646- 1649, ISSN:0038-1098

Malič, B., Benčan, A., Rojac, T. & Kosec, M. (2008a). Lead-free Piezoelectrics Based on Alkaline Niobates: Synthesis, Sintering and Microstructure. *Acta Chimica Slovenica,* Vol. 55, No.4, (December 2008),pp. 709-718, ISSN: 1318-0207

Malič, B., Jenko, D., Holc, J., Hrovat, M. & Kosec, M. (2008b). Synthesis of sodium potassium niobate: a diffusion couples study. *Journal of the American Ceramic Society,* Vol. 91, No.6, pp. 1916-1922, ISSN: 0002-7820

Matsubara, M., Yamaguchi, T., Sakamoto, W., Kikuta K., Yogo, T. & Hirano, S. I. (2005). Processing and Piezoelectric Properties of Lead-Free $(K,Na)(Nb,Ta)O_3$ Ceramics. *Journal of the American Ceramic Society,* Vol. 88, No. 5, (May 2005), pp. 1190-1196,ISSN:0002-7820

McKinstry, T. S., Gharb N. B. & Damjanovic, D. (2006). Piezoelectric nonlinearity due to motion of 180°domain walls at subcoercive fields: A dynamic poling model. *Applied Physics Letters,* Vol. 88, No.20, (May 2006) pp. 202901-3, ISSN:0003-6951

Park, S.E. & Shrout, T.R. (1997). Ultrahigh strain and piezoelectric behavior in relaxor based ferroelectric single crystals. *Journal of Applied Physics,* Vol. 82, No.4, (August 1997) pp. 1804-1811,ISSN:0021-8979

Rehrig, P. W., Park, S. E., Trolier-McKinstry, S., Messing, G. L., Jones, B. & Shrout, T. R.(1999). Piezoelectric Properties of Zirconium-doped Barium Tiotanate Single Crystals Grown by Templated Grain Growth. *Journal of Applied Physics,* Vol. 86, No.3 (August 1999), pp. 1657-1661, ISSN:0021-8979

Rödel, J., Jo, W., Seifert, K., Anton, E. M., Granzow, T. & Damjanovič, D. (2009). Perspective on the Development of Lead-free Piezoceramics, *Journal of the American Ceramic Society*, Vol. 92, No.6, (June 2009), pp. 1153- 1177, ISSN:0002-7820

Saito, Y., Takao, H., Tani, T., Nonoyama, T., Takatori, K., Homma, T., Nagaya, T. & Nakamura, M., (2004). Lead-free piezoceramics, *Nature,* Vol. 432, No. 7013, pp. 84-87 (November 2004), ISSN: 0028-0836

Samardžija, Z., Bernik, S., Marinenko, R. B., Malič, B. & Čeh, M. (2004). An EPMA Study on $KNbO_3$ and $NaNbO_3$ Single Crystals–Potential Reference Materials for Quantitative Microanalysis. *Microchimica Acta,* Vol. 145, Nos. 1-4, (2004), pp. 203–208, ISSN:0026-3672

Seo, C. E. & Yoon, D. Y. (2005).The Effect of MgO Addition on Grain Growth in PMN-35PT.*Journal of the American Ceramic Society,* Vol. 88, No. 4, (April 2005), pp. 963-967, ISSN:0002-7820

Shiratori, Y., Magrez, A. & Pithan C. (2005). Particle Size Effect on the Crystal Structure Symmetry of $K_{0.5}Na_{0.5}NbO_3$. *Journal of the European Ceramic Society,* Vol. 25, No. 12 (2005), pp. 2075-2079, ISSN:0955-2219

Tellier, J., Malič, B., Dkhil, B., Jenko, D., Cilensek, J. & Kosec, M. (2009). Crystal structure and phase transition of sodium potassium niobateperovskite. *Solid State Science,*Vol. 11, No. , (February 2009),pp. 320–324, ISSN:1293-2558

Uršič, H., Benčan, A., Škarabot, M., Godec, M. & Kosec, M. (2010). Dielectric, ferroelectric, piezoelectric, and electrostrictive properties of $K_{0.5}Na_{0.5}NbO_3$ single crystals. *Journal of Applied Physics,* Vol.107, No. 3, (February 2010), pp. 033705-5, ISSN:0021-8979

van der Eerden, J. P. (1993). Crystal Growth Mechanisms, In: *Handbook of Crystal Growth. 1 Fundamentals Part A: Thermodynamics and Kinetics,* D. T. J. Hurle (Ed.), pp. 311-475, North-Holland, ISBN: 2- 88074-246-3, Amsterdam

Wada, S., Muraoka, K., Kakemoto, H., Tsurumi, T. & Kumagai, H. (2004). Enhanced Piezoelectric Properties of Potassium Niobate Single Crystals by Domain Engineering. *Japanese Journal of Applied Physics,* Vol.43, No. 9B, (September 2004) pp.6692-6700, ISSN:0021-4922

Yamamoto, T. & Sakuma, T. (1994). Fabrication of Barium Titanate Single Crystals by Solid-State Grain growth.*Journal of the American Ceramic Society,* Vol. 77, No. 4, (April 1994), pp.1107-1109, ISSN:0002-7820

Yoon, D.Y., Park, C.W.& Koo, J.B. (2001).The Step Growth Hypothesis for Abnormal Grain Growth.In:*Ceramic Interfaces 2,* S.J.L. Kang (Ed.), pp. 2-21, Institute of Materials, ISBN:978 1 861251 18 3, London

Zhang, S., Lee, S. M., Kim, D. H., Lee, H. Y. & Shrout, T. R. (2007). Electromechanical Properties of PMN–PZT Piezoelectric Single Crystals NearMorphotropic Phase Boundary Compositions. *Journal of the American Ceramic Society* Vol. 90, No. 12 (December 2007), pp.3859-3862,ISSN:0002-7820

6

Deposition of $CoFe_2O_4$ Composite Thick Films and Their Magnetic, Electrical Properties Characterizations

W. Chen and W. Zhu
Microelectronics Center, School of Electrical and Electronic Engineering, Technological University Nanyang
Singapore

1. Introduction

In recent years, spinel ferrites have been shown to exhibit interesting electrical conductivity and dielectric properties in their nanocrystalline form compared with that of the micrometer size grains (Ponpandian & Narayanasamy, 2002; Sepelak et al., 2000; Dias & Moreira, 1999). Typical examples of Ni-Zn ferrites and Co-ferrites have been extensively investigated: the former suggests that dielectric constant of nanostructured Ni-Zn ferrite is smaller than that of bulk ceramics (Sivakumar et al., 2008), but the situation is reversed for the Co-ferrites (Sivakumar et al., 2007). Fortunately the dielectric loss of nanostructured ferrites is hence reduced for both of them compared to their bulks. Furthermore, a non-Debye type of dielectric relaxation is observed in these ferrites, which is extensively expressed by electrical modulus (Sivakumar et al., 2008; Sivakumar et al., 2007; Perron et al., 2007). However, the detailed reports on cobalt ferrite, which is one of the potential candidates for magnetic and magneto-optical recording media (Kitamoto et al., 1999; Fontijin et al., 1999), have not drawn enough interests so far. Much attention has been paid on the synthesis of nanostructured cobalt ferrite particles as well as bulk ceramics or thin films (Toksha et al., 2008; Komarneni et al., 1998; Sathaye et al., 2003; Paike et al., 2007; Bhame & Joy, 2008; Gul et al., 2008) and characterizations of their magnetic properties. As for their dielectric properties, which can provide important information on the behavior of localized electric charge carriers, giving rise to a better understanding of the mechanism of dielectric polarization, have attracted little attention except few reports on nanostructured $CoFe_2O_4$ powder (Sivakumar et al., 2007; George et al., 2007). Recently, more attention has been paid to the electric properties of the double-phase multiferroic composites, such as CFO-PZT, and CFO-BTO (Chen et al., 2010; Zhong et al., 2009), or its doping systems (Gul et al., 2007). While pure $CoFe_2O_4$, especially its thick film structure, which is a critical scale range for micro-electro-mechanical systems (MEMS) design, has not been found in the literatures.

In order to explore the processing of cobalt ferrite thick film and its electrical properties for potential MEMS development, the present work has adopted a similar fabrication to typical PZT ferroelectric thick films (Chen et al., 2009). 10 μm of cobalt ferrite composite thick films is successfully prepared via a hybridized sol-gel processing. The influence of annealing temperature on the phase structures, microstructures, Raman shift, magnetic and electrical properties are well characterized. Furthermore, Ac conductivity spectra analysis is employed to investigate the ion motion nature of $CoFe_2O_4$ composite thick films. The detailed electrical investigations were conducted in the frequency range of 100 Hz-1MHz and temperature range between 25 and 200 °C. Real and imaginary parts of impedance (Z′ and Z″) in the above frequency and temperature domain suggested the coexistence of two relaxation regimes: one was induced by electrode polarization; while the other was attributed to the co-effect of grains and grain boundaries, which was totally different from its counterpart of bulks and also not reported in other ferrites. Electrical modulus (M′ and M″) further showed the crossover from grains effect to grain boundaries effect with increasing measured temperature under the suppression of electrode polarization. A non-Debye relaxation behavior and two segments of frequency independent conductivity were observed in dielectric spectra, which was also consistent with the results of ac conductivity spectra. In the conductivity spectra, double power law and single power law were separately applied to the co-effect from grains and grain boundaries and electrode polarization effect. Moreover, the dc conductivity from both effects well obeyed the Arrhenius law and their activation energies were matching to the ones calculated from imaginary impedance peaks, the detailed physical mechanisms on them were finally discussed.

2. Deposition of $CoFe_2O_4$ composite thick films

2.1 Experimental procedural

$CoFe_2O_4$ (abbreviated as CFO) sol-gel solution was prepared by mixing cobalt acetate, ferric nitrate, and polyvinylpyrrolidone together at 60 °C according to the molar ratio of 1: 2: 2 till a clear solution was obtained. Then 40 ml of 2-methoxyethanol was added to get 0.125 M of CFO sol-gel solution. The pH value of resultant dark-red CFO sol-gel solution was 4.2. In addition, modified CFO particles were prepared by a high energy ball milling method as reported previously (Chen et al., 2009), which showed an average particle size of 233 nm. Next, the modified CFO particles were dispersed in the CFO solution with a mass ratio of 2:3, which is similar to the fabrication of hybridized PZT slurry (Chen et al., 2010), to get the uniform CFO slurry via an agate ball milling for 15 hours. The collected CFO slurry showed a black color and was immediately spin coated onto the Pt/Ti/SiO_2/Si substrate alternatively with CFO sol-gel solution to obtain the dense CFO film. After each coating layer, the film was baked at 140 °C for 3 minutes to dry the solvent and then held at 300 °C for another 3 minutes to burn up the organic components. The resulting thick films were annealed in air at various temperatures from 550 °C to 700 °C for 1 hour each, and their thicknesses were measured via a surface profiler to be around 10 μm.

TGA and DTA were performed using a Thermal Analyzer (TA-60WS) with a heating rate of 2 °C/min. Phase structures were evaluated using an X-ray diffractometer (2400, Rigaku, CuK_α radiation). Raman spectroscopic measurements were carried out with a WITEC CRM200 confocal Raman system. The excitation source is 532 nm laser (2.33 eV). Surface and cross-sectional morphologies of the thick films were obtained using a Scanning Electron Microscope (JSM-5600LV). Magnetic properties were detected by a Lakeshore Vibration Sample Magnetometer (7404). After deposition of gold top electrodes with the size of 0.8 mm × 0.8 mm on the surface of thick films using E-beam, impedance spectroscopy was measured by using a Solartron SI1260 impedance/gain-phase analyzer from 0.1 Hz to 1 MHz at room temperature. In addition, the detailed electrical properties of the thick films were measured by an Agilent 4294A precision impedance analyzer over 100 Hz-1MHz and 25-200 °C at the ac oscillation level of 100 mV. Each measured temperature was kept constant with an accuracy of ±1 °C.

2.2 Characterizations

TGA/DTA analysis of the dried CFO slurry, dried at 110 °C for 24 hours, is shown in Fig. 1.

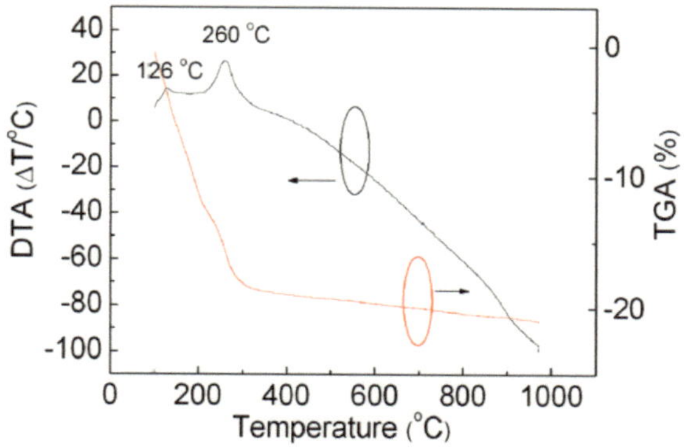

Fig. 1. DTA/TGA curves of the dried CFO slurry.

TGA yields a weight loss of 21 wt% before 300 °C, and then nearly keeps stable until 800 °C. In the DTA curve, two exothermic peaks are observed: one at 126 °C which is due to the organic solvent evaporation; the other at 300 °C symbols the decomposition and combustion of the bound organic species in the CFO slurry. Since that the CFO powder has been presintered at a high temperature of 1200 °C before high energy ball milling, it has almost no effect on TGA/DTA analysis. The observations of weight loss and exothermic peak in DTG can be presumed to be occurring from the sol-gel part of the composite film. That is why 140 °C and 300 °C are selected after each coating processing.

X-ray diffraction patterns of the resultant CFO thick films annealed at different temperatures are exhibited in Fig. 2.

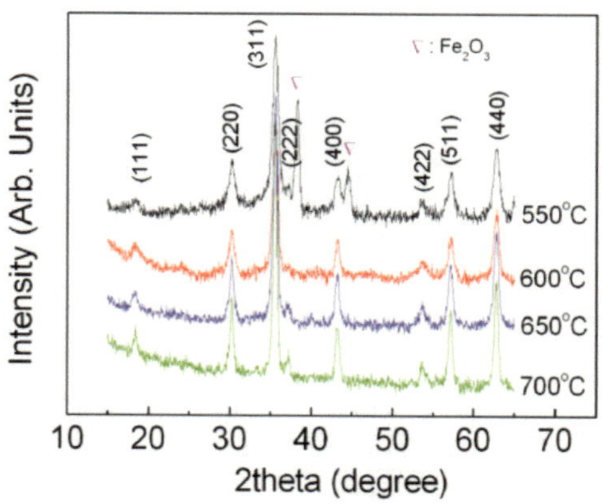

Fig. 2. XRD patterns of CFO composite thick films annealed at different temperatures.

Although major peaks due to $CoFe_2O_4$ are observed for the film annealed at 550 °C, additional peaks (marked) assignable to Fe_2O_3 are also observed indicating the process of

CFO formation is not complete. With the rise of the annealing temperature, complete formation of spinel phase is observed for films annealed above 600 °C. Furthermore, these characteristic peaks of CFO phase become narrow, indicative of an increase of their grain size with increasing annealing temperature.

In order to further verify the chemical impurity in the composite thick films, micro-Raman spectroscopy is performed in Fig. 3.

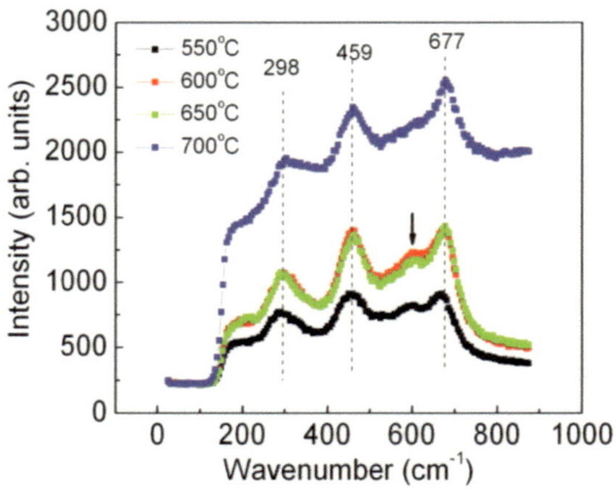

Fig. 3. Micro-Raman spectra of CFO composite thick films annealed at different temperatures.

It can be seen that three main peaks (298 cm^{-1}, 459 cm^{-1}, and 677 cm^{-1}) of the spinel CFO are clearly observed for all the films without any Raman shift (Ortega et al., 2008; Yu et al., 2002). Films annealed below 600 °C show the presence of a peak at 600 cm^{-1}, which can be assigned to CFO, supporting the inference that below 600 °C, formation of CFO does not go to completion. In addition, these mode peaks are gradually becoming sharp with the rise of annealing temperature, suggesting a harden process of CFO modes.

Typical surface morphology and cross-sectional picture of CFO composite thick film annealed at 700 °C are shown in Fig. 4.

Fig. 4. Typical surface morphology (a) and cross-sectional image (b) of CFO composite thick films annealed at 700 °C.

It can be seen from Fig. 4(a) that the thick films have rough, dense microstructure due to agglomeration as is the case for synthesized CFO thin films reported in the literature (Pramanik et al., 2005). The roughness can be attributed to the overlarge thickness, evidenced by its cross-sectional picture in Fig. 4(b), which also indicates a thickness closing to 10 μm. It is far beyond the currently reported ferrite films (Sathaye et al., 2003; Gul & Maqsood, 2008).
In-plane magnetic hysteresis loops are shown in Fig. 5.

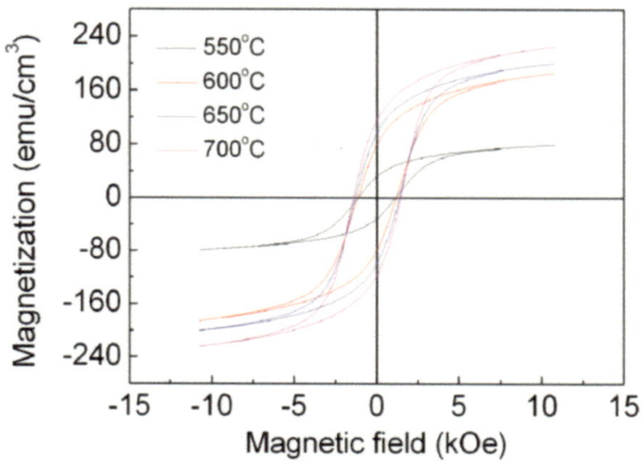

Fig. 5. Magnetic hysteresis loops of CFO composite thick films annealed at different temperatures.

It can be seen that all the films reach saturation below 8 kOe due to the CFO ferrite thick film being in a quasi-free state with negligible shear stress from the substrate compared to chemical synthesized CFO thin film (Sathaye et al., 2003) or pulse laser deposited CFO epitaxial thin film (Lisfi & Williams, 2003). Furthermore, the present composite thick films show an annealing temperature dependent saturation magnetization (*Ms*) and magnetic coercivity (*Hc*). With increasing annealing temperature, both *Ms* and *Hc* values exhibit a monotone enhancement. The enhanced *Ms* values from 79 to 225 emu/cm^3 are due to the enlargement of average cobalt ferrite grains, which has been demonstrated in CFO bulks and thin films (Sathaye et al., 2003; Wang et al., 2008). In the CFO thin films (Sathaye et al., 2003), the *Ms* value was reported as 300 emu/cm^3. Compared with the present composite thick films, the higher *Ms* value in CFO thin film was mainly caused by higher annealing temperature.
The particles used in CFO composite thick films include two parts: one is the sol-gel synthesized particles with a small particle size of dozens of nanometer; the other is high energy ball milling modified CFO particles with a large average particle size of about 233 nm. Since the latter has been presintered at 1200 °C, the growth rates of both kinds of CFO particles under 700 °C of annealing temperature are different, resulting in non-uniform segregation causing the rough surface, which increases the coercivity of CFO composite thick films from 1130 to 1434 Oe. Generally speaking, high coercivity can be obtained in systems with a nanostructure or preferred orientation, such as thin films with preferred crystal texture or nanoparticles with a single domain diameter (Yin et al., 2006; Lee et al., 1998). The single domain diameter of the present CFO is about 40 nm, which is much smaller than the average diameters of our CFO composite thick films (above 100 nm), plus

the polycrystalline state of the present thick films, evidenced by X-ray diffraction. Thus, the lower *Hc* value is mainly attributed to the magnetic multi-domain configuration of the CFO particles in the composites (Lee et al., 1998).
Room temperature impedance spectroscopy for the CFO composite thick films is exhibited in Fig. 6 for frequencies of 0.1 Hz to 1 MHz.

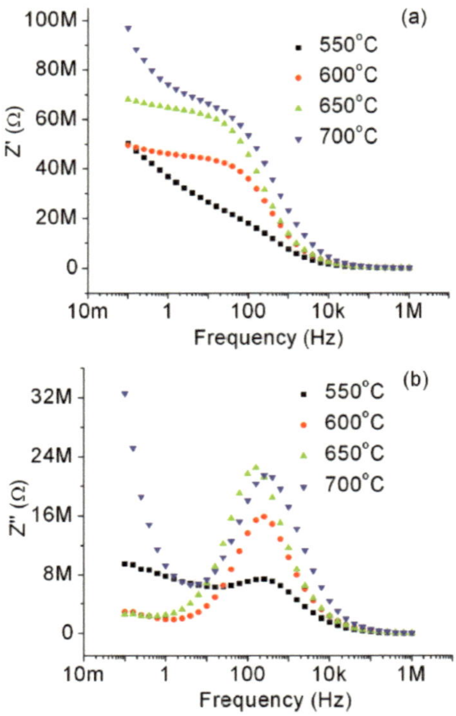

Fig. 6. Frequency dependence of real (a) and imaginary impedance (b) of CFO composite thick films at room temperature from 0.1 Hz to 1 MHz.

Fig. 6(a) shows the frequency dependence of impedance real part (Z′). A step-like decreasing trend is observed in real impedance spectra for all the samples from 10 Hz to 10 kHz, and their specific impedance values are reduced by nearly three orders of magnitude. An apparent imaginary impedance peak appears in all the samples and becomes strong with increasing annealing temperature, as can be seen in Fig. 6(b). It can be seen that the peak frequency is around 100 Hz, which is in the middle point of the step-like decreasing curve in real impedance spectra, indicative of a relaxation behavior. This phenomenon has not been reported in the literatures on CFO ferrite but recent studies on multiferroic $BiFeO_3$ thin films and $BiFeO_3/CoFe_2O_4$ bilayered thin films show a similar behavior (Srivastava et al., 2009; Wu & Wang, 2009). The relaxation peak was initially observed in $BiFeO_3$ thin films at 150 °C of measured temperature (Srivastava et al., 2009), but only 100 °C for $BiFeO_3/CoFe_2O_4$ bilayered thin films (Wu & Wang, 2009), indicating that CFO is beneficial to shift this relaxation peak to low temperature side. This is also why we observe the present relaxation behavior at room temperature. Furthermore, the present composite thick films show a similar characteristic frequency maxima (f_{max}), indicating the relaxation time is independent on annealing temperature. Additionally, above 10 kHz, both real and imaginary curves merge together independent of annealing temperature; while apparent annealing temperature dependent diffusion phenomena is observed below 10 Hz. Normally, grain effects are attributed to the high frequency impedance behavior, while grain boundary effects are responsible for the low frequency phenomena (Nirose & West, 1996). Annealing temperature independent impedance spectroscopy at high frequency side for the present composite thick films reveals that CFO grains are insensitive to the fast switch of applied alternate electric field. However, low frequency diffusion behavior indicates a remarkable grain boundaries effect, which should be attributed to the increased aggregation caused by higher annealing temperature.

In order to further investigate the effect of grains and grain boundaries of CFO composite thick films. Nyquist plots (relation between real and imaginary impedance) at room temperature for all samples are shown in Fig. 7.

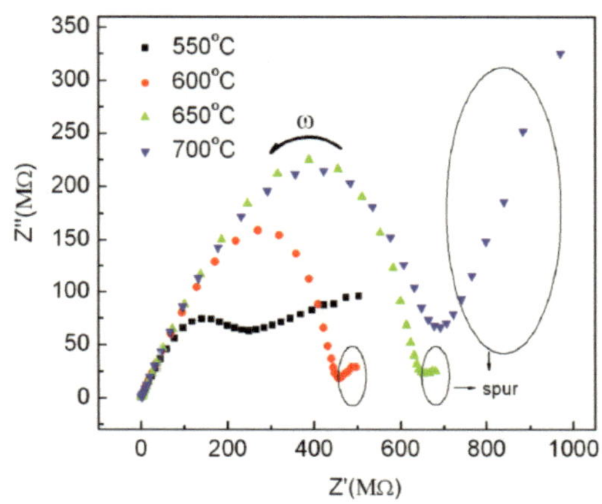

Fig. 7. Nyquist plots of CFO composite thick films annealed at different temperatures, measured in the frequency range from 0.1 Hz to 1 MHz.

The irregular shape of CFO thick film annealed at 550 °C should be attributed to the mixture of the second phase. For the sample above 600 °C of annealing temperature, it can be seen that an approximate semicircle arc is formed at the high frequency side. This semicircle arc is gradually expanded with increasing annealing temperature until 650 °C, where it is almost unchanged any more compared with the one annealed at 700 °C, and the absolute value of impedance also reaches the maximum, indicating that CFO grains effect reach a stable state. On the other hand, the "spur" which appeared at low frequency side is almost unchanged when annealing temperature is increased from 600 to 650 °C, but when the sample is annealed at 700 °C, this "spur" becomes very large, indicating an increased grain boundaries effect caused by more aggregation as mentioned above. From the impedance spectroscopy analysis, we can expect that 650 °C is an optimized temperature for promising electric properties. However, to further learn the ion motion nature of three different regions in Fig. 6, AC conductivity spectra is presented below.

It is known that AC conductivity of a composite thick film can be estimated from its impedance and phase angle through the following relationship,

$$\sigma^* = \frac{d}{AZ^*} = \frac{d}{A(Z'+iZ")} = \frac{d(Z'-iZ")}{A|Z|^2} \tag{1}$$

$$\sigma' = \frac{d\cos\theta}{A|Z|} \tag{2}$$

where d and A are the sample's thickness and its effective area, θ is the impedance phase angle and $|Z|$ is the absolute value of impedance, Z' and Z'' are real and imaginary impedance, and σ^* and σ' are complex conductivity and real conductivity with the latter usually known as the AC conductivity. In terms of equation (2), we can obtain the frequency dependence of AC conductivity in the whole measured frequency.

As can be seen in Fig. 8, three different regions are observed in ac conductivity spectra which is consistent with the three zones mentioned in impedance spectra.

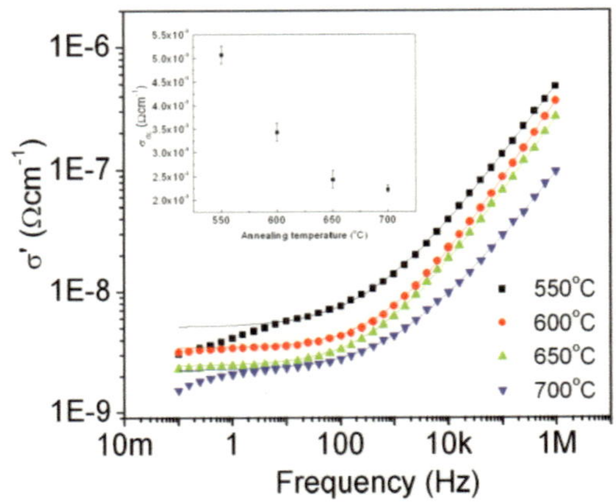

Fig. 8. AC conductivity spectra of CFO composite thick films annealed at different temperatures, the inset is the estimated DC conductivity dependence on annealing temperature.

They are corresponding to the three effects that contribute to the ac conductivity (Jame et al., 2006): (1) low frequency electrode effects; (2) intermediate frequency dc plateau; (3) high frequency ac conductivity effect. It is clearly seen that low frequency electrode effects, represented by the deviation from flat conductivity, are especially remarkable for the thick film annealed at 700 °C, but very faint for the thick films annealed at 600 and 650 °C. In addition, for the ac conductivity spectra at intermediate and high frequency range, the difference in the trend decreases with increasing annealing temperature due to the increased impedance values. This can be attributed to improved crystallization of composite thick films. Furthermore, the dc conductivity estimated from the power law (George et al., 2007) also indicates a decrease trend with increasing annealing temperature, as can be seen the inset picture of Fig. 8. More detailed investigations on ac conductivity spectra are conducted in the following section.

Since there is a lack of detailed impedance spectroscopy analysis of CFO thin films and bulks in the literature, data of $BiFeO_3/CoFe_2O_4$ bilayered thin films is introduced for comparison to our results (Wu & Wang, 2009), where DC plateau and NCL regime are also observed and both of them move to high frequency with increasing measured temperature. This is similar to the present case of CFO composite thick films. However, the decrease in dependence on measured temperature of $BiFeO_3/CoFe_2O_4$ bilayered thin films at high frequency side is attributed to the introducing of low conductive $BiFeO_3$, which can be also confirmed in PZT/CFO multilayered thin films (Ortega et al., 2008) where insulated PZT is combined together with CFO. As for the electrode polarization effect on conductivity spectra, there are no reports in the literature.

Detailed analysis for the CFO composite thick films annealed at 600 °C reveals the complicated ion motion process in this typical ferrite (Chen et al., 2010). In order to further learn the electrical behavior of this magnetic thick film, the film annealed at 600 °C is specifically studied as followed.

3. Electrical properties

3.1 Impedance spectra

Fig. 9(a) and (b) show the variation of real and imaginary parts of impedance (Z′ and Z″, respectively) with frequency from 100 Hz to 1 MHz and temperature between 25 and 200 °C.

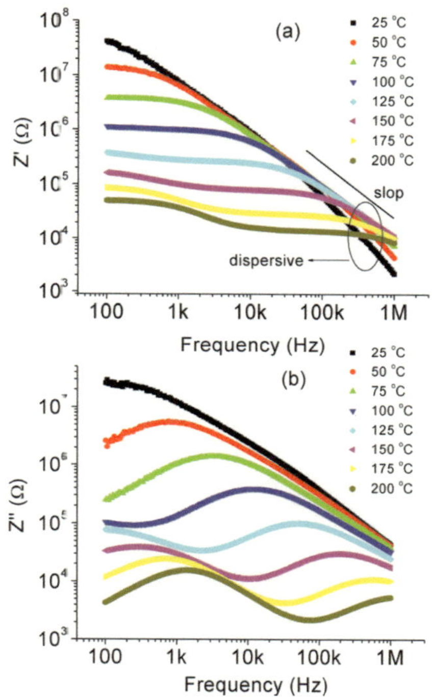

Fig. 9. Frequency dependent of real impedance (a) and imaginary impedance (b) for $CoFe_2O_4$ composite thick film from 100 Hz to 1 MHz and between 25 and 200 °C.

A temperature dependent Z′ plateau is observed initially from low frequency side at 50 °C followed by a nearly negative slope at high frequency side, indicating a crossover from low frequency relaxation behavior to high frequency dispersion phenomenon. Furthermore, this segment of nearly constant real impedance becomes predominated with increasing temperature, suggesting a strengthened relaxation behavior. This is similar to the behavior observed in multiferroic $BiFeO_3$ thin films above 150 °C, where a clear relaxation behavior was smoothing into the frequency window from low frequency side due to the rising temperature (Srivastava et al., 2009). When the measured temperature is above 100 °C, another step-like impedance behavior is smoothing into the frequency window from the low frequency side; in the meanwhile, it pushes the previous high frequency dispersive behavior out of the frequency window, both remarkably relaxations are hence coexisted above 100 °C. This phenomenon has been never reported in ferrites, but an extremely weak impedance relaxation and another strong one were separately observed in different temperature ranges for recent PZT/CFO layered thin films, the strong relaxation found in high temperature was attributed to the thermal activation mechanism (Ortega et al., 2008). Fig. 9(b) shows a broad imaginary impedance peak initially at 50 °C and moves to high frequency side with increasing temperature and finally disappears at 200 °C; meanwhile, another broad peak is also appearing above 100 °C and moves to high frequency side, which corresponds to both plateau relaxations observed in real impedance spectra. The Arrhenius law is hence applied for both relaxations,

$$\tau = \tau_0 \exp\left(-\frac{E_a}{k_B T}\right),\ \tau = 1/2\pi f_p \tag{3}$$

where τ_0 is the characteristic relaxation time, E_a is the activation energy for the relaxation process, k_B is the Boltzmann constant, T is the absolute temperature and f_p is the peak frequency of imaginary impedance. The estimated activation energies from their respective imaginary peaks are 0.675eV and 0.483eV, and the characteristic relaxation times τ_0 are 8.01×10^{-15}s and 4.16×10^{-10}s, respectively.

Nyquist plots of impedance data at different temperatures are exhibited in Fig. 10.

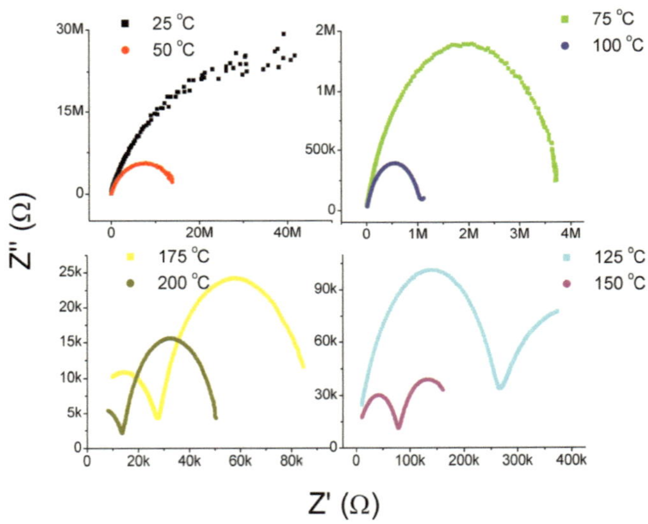

Fig. 10. Nyquist plots of Z' and Z'' for $CoFe_2O_4$ composite thick film at all measured temperatures.

At 25 °C, a semicircle arc is observed and it becomes a whole semicircle till 75 °C, which should be attributed to the grains effect in CFO thick film. Beginning with 100 °C, a slight segment of arc is appeared from low frequency side which is connecting to this semicircle. Furthermore, with further increasing temperature, the second arc is gradually spreading till 150 °C, where the original semicircle is degenerated and this arc continues to strengthen, which is corresponding to the situation of imaginary impedance spectra, where two peaks are coexisted. When the temperature finally reaches 200 °C, it can be seen that the second arc is nearly formed into a whole semicircle at low frequency side and the initial semicircle degenerates into a segment of arc. This suggests a process in which the grains effect is gradually replaced by grain boundaries effect with increasing temperature. Additionally, during this process, the impedance value is decreased by four orders of magnitude, which is due to thermal activation mechanism. The rise of temperature brings with an enhanced conductivity, and hence, decreasing the impedance values.

Observation of two Nyquist semicircles in Fig. 10 naturally leads us to believe that the grains effect and grain boundaries effect contribute to them like other systems (Ortega et al., 2008; Srivastava et al., 2009). However, the activation energies estimated from Arrhenius law for both semicircles suggest a different situation. Normally, the activation energy from grain boundaries was larger than that of grains (Ortega et al., 2008). That was also why the grain boundaries could play a blocking effect in many ionic oxides due to their high barrier. While the present situation is just on the contrary, the so-called grains produced activation energy is higher than that of grain boundaries, which totally cancels the barrier effect of grain boundaries. Therefore, we propose that the grain boundaries and the grains in the thick films co-contribute to the initial semicircle and the second semicircle appeared at high temperature is due to electrode polarization effect.

3.2 Electrical modulus

In order to demonstrate this point, electrical modulus formalism has been introduced due to its special advantage of suppressing the electrode polarization effects (Ponpandian et al., 2002). The electrical modulus is calculated from the following equation:

$$M^* = M' + jM" = j\omega C_0 Z^* = j\omega C_0(Z' - jZ") \tag{4}$$

where ω is the angular frequency and the geometrical capacitance is $C_0=\varepsilon_0 A/d$ (d is the sample thickness, A is the electrode area, and ε_0 is the permittivity of vacuum, $8.854*10^{-14}$ F/cm^{-1}). Through the equation (4), we can calculate the values of $M'$and $M"$using the relationship $M' = \omega C_0 Z"$ and $M" = \omega C_0 Z'$. Frequency dependent M' and $M"$at all temperatures is hence presented in Fig. 11.

As can be seen in Fig. 11(a), unlike the impedance spectroscopy, where two relaxation behaviors were well separated, the real modulus nearly showed a single relaxation behavior, which is mainly featured by a positive slope moving to high frequency side with increasing temperature. Fig. 11(b) shows that the broad peaks are being located from 25 to 100 oC, and beyond 100 °C, the peak is degenerated and finally disappears above 150 °C. Both of them demonstrate that the high temperature electrode polarization reflected on impedance spectroscopy is totally suppressed here (Sivakumar et al., 2007; Srivastava et al., 2009). Additionally, through careful observation we can notice an inflexion in the middle of the increasing real modulus for the samples in between 25 and 100 °C; in the meanwhile, the broad peak mentioned in imaginary modulus shows a depressed behavior in the middle of its peak. This is different from the modulus peak of the kinds of ferrites, including CFO ferrite (Sivakumar et al., 2007), Ni-Zn and Mn-Zn ferrites (Sivakumar, 2007, 2008), in which only a clearly and smoothly relaxation behavior was observed. The present behavior indicated that this broad peak is constituted by two incomplete relaxation peaks, which are almost merging together. It is also why we proposed a co-contribution from grain and grain boundaries to impedance semicircle. In the modulus curves, the relaxation peak at low frequency side is contributed by grain boundaries effect and the one at high frequency side is induced by grains effect.

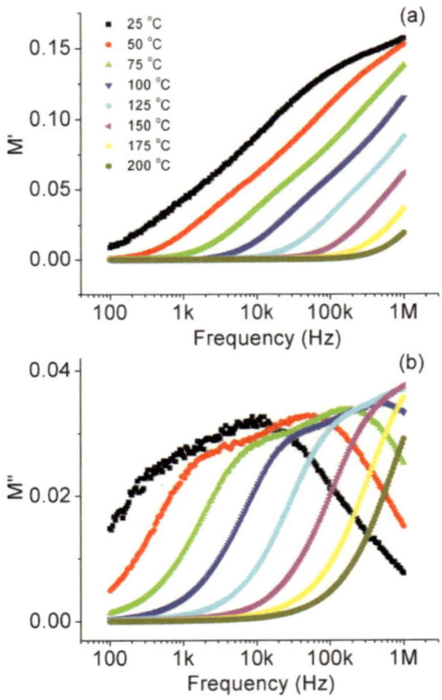

Fig. 11. Frequency dependent of real (a) and imaginary electrical modulus (b) for $CoFe_2O_4$ composite thick film from 25 to 200 °C.

More clearly evidence can be seen in Fig. 12, complex modulus plots of CFO thick films at all temperatures are exhibited.

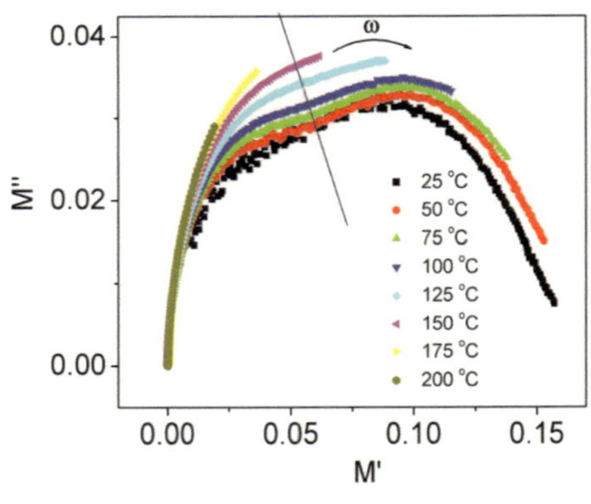

Fig. 12. *M′-M′′* plots of $CoFe_2O_4$ composite thick film at all measured temperatures.

Two incomplete semicircle arcs are forming into a broad semicircle in the temperature range of 25-100 °C. With further increasing temperature, only a segment of arc is left at the low frequency side, which is attributed to the grain boundaries effect induced by higher temperature. In the whole temperature range, we can see that the enhancement of measured temperature strengthens the grain boundaries effect gradually and weakens the grains effect at the same time. This is similar to many materials (Lin et al., 2008; Ahmad et al., 2009), where higher temperature stimulated the ions hopping over their barrier layers instead of hopping within their own sites.

From the results above, we can see that two impedance relaxations are initially observed in the present CFO thick films, especially the coexistence phenomenon above 100 °C, which was never found in literatures. The estimated activation energies from two relaxations indicate a co-effect from grains and grain boundaries to the impedance relaxation appeared at low temperature. Usually, grains effect and grains boundaries effect were being located at individual frequency and temperature range for the homogeneous materials (Srivastava et al., 2009). However, the present films show an abnormal relaxation behavior, remarkably expressed in electric modulus spectra, where the electrode polarization is suppressed. Different from a single modulus peak in other nanostructured ferrites (Sivakumar, 2007a, 2007b, 2008), a temperature dependent crossover between two peaks are observed, which may be induced by the hybridized microstructure consisting of two kinds of different sized CFO particles. The similar phenomenon reported in multiferroic PZT/CFO layered thin films demonstrated it (Ortega et al., 2008), where two electric modulus relaxations were observed, but both relaxations in this multiferroic material were located at different temperature ranges without any overlapping, which should be attributed to the large property difference between insulted PZT and low resistivity CFO phase. While the present CFO thick film is composed by the same CFO phase only with different particle sizes, it is hence expected that this special hybridized microstructure could be the main reason for the double relaxation behavior observed.

3.3 Dielectric permittivity and loss

Fig. 13 shows that real and imaginary dielectric constant and dielectric loss (ε', ε'' and *tanδ*) are plotted against the frequency for all temperatures.

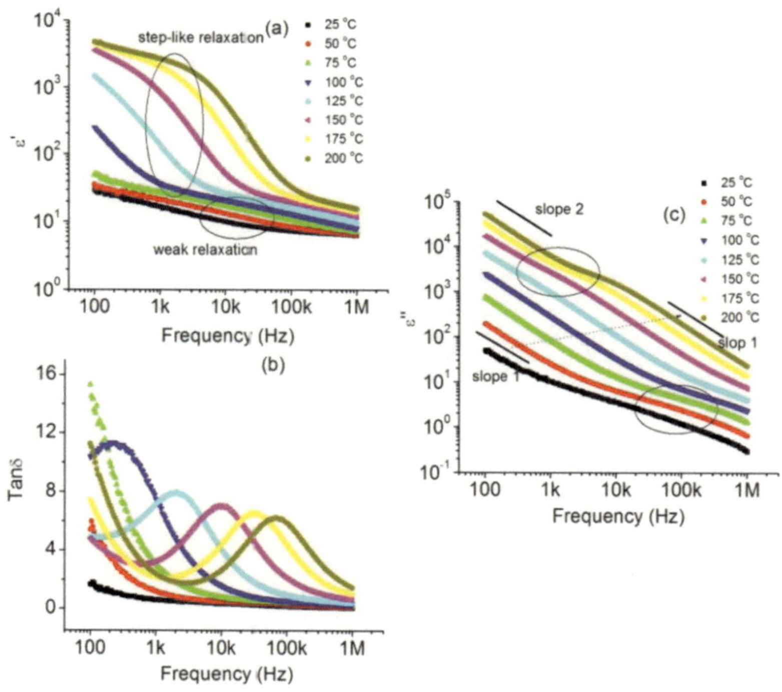

Fig. 13. Frequency dependent of real dielectric constant (a), dielectric loss (b), and imaginary dielectric constant for $CoFe_2O_4$ composite thick film from 25 to 200 °C.

As can be seen in Fig. 13(a), it shows a strong dispersion in the real dielectric constant at low frequencies. In addition, a weak relaxation is initially observed above 1 kHz for the sample below 100 °C and then gradually disappears at high frequency when the temperature reaches 100 °C. With further increasing temperature, a step-like relaxation behavior, which is similar to a Debye relaxation, is smoothing into the frequency window from the left side. In the meanwhile, a corresponding dielectric loss peak is observed in Fig. 13(b) from 100 °C. Furthermore, this peak is moving to high frequency side along with a reduced peak loss value. It is noticed that temperature dependent peak frequency in this Figure is not obeying Arrhenius equation as other materials, suggesting this step-like relaxation a non-Debye type. Imaginary dielectric constant versus frequency is shown in Fig. 13(c), where no peak is observed in the measurable frequency range, but two weak relaxation behaviors are observed in the temperature range of 25-100 °C and 125-200 °C locating at different frequency ranges, respectively. Besides, two negative slopes of the straight lines of log-log plot are observed in Fig. 13(c). Moreover, both of them are moving to high frequency side with increasing temperature. This is a natural result of the frequency independent conduction (Dutta et al., 2004).

3.4 AC conductivity

Real conductivity is usually adopted for studying the ion motion of ionic oxides, glasses or melting. It can be estimated from the equation (2). The real conductivity spectra are hence presented in Fig. 14(a).

A monotonously increasing conductivity curve is observed at 25 °C and a plateau is smoothing into the frequency window from the left side above this temperature, which corresponds to the slopes in Fig. 13(c), demonstrating a constant conductivity at low frequency side. At high frequency range, there is an exponentially increasing conductivity behavior and it moves to the right side of this frequency window along with the conductivity plateau in the all temperature ranges, which was usually called as nearly constant loss regime (NCL) in the literatures (Abbas et al., 2007; Patange et al., 2009), also reflected in Fig. 13(c). According to the jumping relaxation mode (Jonscher, 1977), the frequency independent plateau at a low frequency for higher temperatures is attributed to the long-range translational motion of ions contributing to dc conductivity. According to this model, the conductivity at the low frequency region is associated with the successful hops to its neighborhood vacant site due to the available long time period; such successive jumps result in a long-range transitional motion of ions contributing to dc conductivity. At higher frequency (>10 kHz), two competing relaxation processes may be visualized: one is the jumping ion to jump back to its initial position, i. e., unsuccessful hopping and the other is the neighborhood ions become relaxed with respect to the ion's position, i. e., successful hopping. The increase in the ratio of successful to unsuccessful ion hopping results in a more dispersive conductivity at higher frequency. For the present conductivity plateau, the data have been fitted to a double power law (Jonscher, 1977),

$$\sigma' = \sigma_{dc} + Af^{n_1} + Bf^{n_2} \tag{5}$$

where σ' is the real conductivity, σ_{dc} is the transitional hopping gives the long-range electrical transport in the long time limit, and coefficient A, B and exponent n_1, n_2 are temperature and material intrinsic property dependent constants (Jonscher, 1977; Funke, 1993). The term Af^{n_1} and Bf^{n_2} characterize the contributions from grains and grain boundaries. As can be seen in Fig. 14(a), solid lines are perfectly fitting to the experimental data. Moreover, the estimated σ_{dc} is also obeying the Arrhenius-like law,

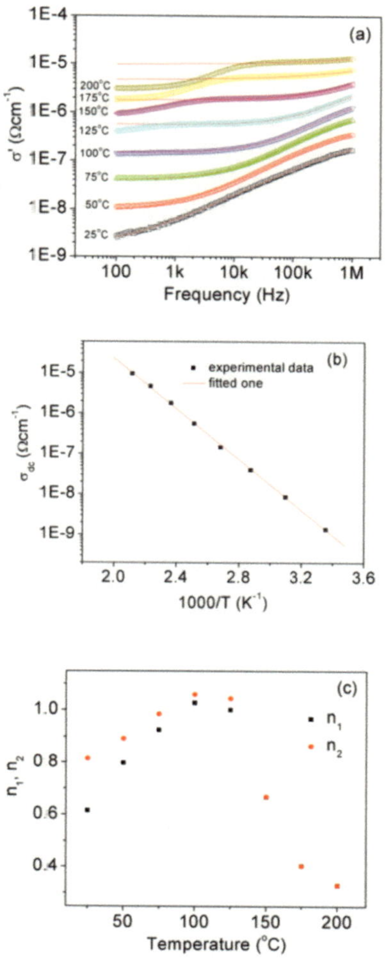

Fig. 14. Frequency dependent of real conductivity and its fitting curves from double power law (a); temperature dependent of dc conductivity and its fitting curve from Arrhenius law (b); double power parameters at all measured temperatures (c).

$$\sigma_{dc} = \sigma_0 \exp\left(-\frac{E_a}{k_B T}\right) \tag{6}$$

where σ_0 is a constant, E_a is the activation energy, k_B is the Boltzmann constant, and T is the absolute temperature, as can be seen in Fig. 14(b). This indicates that ions are more incline to their nearest neighbor hopping. Furthermore, the activation energy calculated here is closing to 0.675eV as mentioned above from imaginary impedance. In addition, parameter n_1 and n_2 mostly locate in the range of (0, 1), which corresponds to a short-range transitional hopping motion. Temperature dependent of n_1 and n_2 values are presented in Fig. 14(c), it is clear that both values are very closing to each other, which are due to the comparative grains and grain boundaries effects. Furthermore, both of them show a peak in the temperature range from 75 to 125 °C, which corresponds to the onset of the crossover from the grain contribution to grain boundaries contribution well supported by frequency and temperature dependent modulus plots. Worthy of noticing is that both parameters are

closing to 1 and even above it at 100 and 125 °C, which is corresponding to the few of localized or reorientational hopping motion (Funke, 1993).
Additional, when the temperature is above 100 °C, a conductivity tail is appearing at low frequency side, such as the situation happened at 125 °C, which is usually believed to be electrode polarization effect (Marczak & Diesinger, 2009; Imre et al., 2009). Followed by this tail, another plateau is observed at low frequencies with further increasing temperature. This behavior is never reported before. It is also why a whole Nyquist semicircle on electrode polarization effect is rarely found in literatures. For the present case, we can see that this electrode polarization effect reflecting on real conductivity is similar to the way of conductivity spectra resulted from the co-effect of grains and grain boundaries. Its dc plateau, which is corresponding to the slope2 in imaginary dielectric constant, is also moving to high frequency as well as the NCL regime with the rise of temperature. Due to this part is mainly contributed by electrode polarization effect, single power law is hence applied,

$$\sigma' = \sigma_{dc} + Af^n \tag{7}$$

It is noticed that the fitted n value from Fig. 14(a) is 1.12, indicating a localized or reorientational hopping motion, which should be attributed to the Au/Ti electrode layer. Furthermore, its dc conductivity is also fitted to the Arrhenius law and the estimated activation energy is nearly same to the value calculated from imaginary impedance, demonstrating again the electrode polarization contribution to the low frequency semicircle in Nyquist plots.

4. Conclusion

Cobalt ferrite composite thick films are prepared on $Pt/Ti/SiO_2/Si$ substrate by a hybrid sol-gel processing. Through annealing at different temperatures, XRD and Raman spectra indicate that pure spinel phase is formed above 600 °C. A 10 μm of thickness is confirmed by cross-sectional SEM imaging. Furthermore, with increasing annealing temperature, saturation magnetization and magnetic coercivity are increased. Room temperature impedance spectroscopy analysis indicates a relaxation behavior from 10 Hz to 10 kHz, and this relaxation behavior is strengthened with increasing annealing temperature. Complex Z′-Z″ plots reveal the main contribution to the relaxation behavior is from grains for all the samples. Additionally, 650 °C of annealing temperature is believed optimal one due to a large grain boundary effect being observed at 700 °C of annealing temperature.
Detailed temperature and frequency dependent impedance spectra were conducted on the CFO thick films annealed at 600 °C for further electrical investigations. Two relaxations were observed in impedance spectra corresponding to both semicircles in Nyquist plots. The high frequency semicircle is induced by co-effect of grains and grain boundaries, whereas the low frequency semicircle is due to the electrode polarization effect. Electrical modulus studies demonstrate that the grain effect is decisive below 100 °C, while the grain boundaries are playing a more important role above this temperature. Non-Debye relaxation is subsequently observed in dielectric spectra, and imaginary dielectric constant spectra further indicates two segments of frequency independent conductivity, which is demonstrated in real conductivity spectra. In the conductivity spectra, on one hand, the dc plateau at high frequency obeys the double power law along with two similar power

parameters, indicative of the comparative contribution from grain and grain boundaries. The peak in temperature dependent power parameters further suggests the crossover from grain effect to grain boundaries effect. Moreover, the dc conductivity well obeys the Arrhenius law and the estimated activation energy is same to the one calculated from high frequency imaginary impedance peaks. On the other hand, the other dc plateau at low frequency obeys the power law, and its power parameter is 1.12, suggesting a localized or reorientational hopping motion probably induced by Au/Ti layer. In addition, the activation energy calculated from its fitted dc conductivity shows the same value to the one from low frequency imaginary peaks, demonstrating the electrode polarization contribution to the low frequency Nyquist semicircle.

5. Reference

Abbas, S. M., Chatterjee, R., Dixit, A. K., Kumar, A. V. R., & Goel, T. C. (2007). Electromagnetic and microwave absorption properties of (Co^{2+}-Si^{4+}) substituted barium hexaferrites and its polymer composite. *Journal of Applied Physics,* Vol.101, No.7, (April 2007), pp. 074105-6, ISSN 0021-8979

Ahmad, M. M., Yamane, Y., & Yamada, K. (2009). Structure, ionic conduction, and giant dielectric properties of mechanochemically synthesized $BaSnF_4$. *Journal of Applied Physics,*Vol.106, No.7, (October 2009), pp. 074106-7, ISSN 0021-8979

Bhame, S. D., & Joy, P. A. (2008). Effect of sintering conditions and microstructure on the magnetostrictive properties of cobalt ferrite. *Journal of the American Ceramic Society,* Vol.91, No.6, (June 2008), pp. 1976-1980, ISSN 0002-7820

Chen, W., Wang, Z. H., Ke, C., Zhu, W., & Tan, O. K. (2009). Preparation and characterization of $Pb(Zr_{0.53}Ti_{0.47})O_3/CoFe_2O_4$ composite thick films by hybrid sol-gel processing. *Materials Science and Engineering B,* Vol.162, No.1 (May 2009) pp. 47-52, ISSN 0921-5107

Chen, W., Zhu, W. G., Chen, X. F., & Wang, Z. H. (2010). Enhanced Ferroelectric and Dielectric Properties of $CoFe_2O_4$-$Pb(Zr_{0.53}Ti_{0.47})O_3$ Multiferroic Composite Thick Films. *Journal of the American Ceramic Society,* Vol.93, No.3, (March 2010), pp. 796-799, ISSN 0002-7820

Chen, W., Zhu, W., Tan, O. K., & Chen, X. F. (2010). Frequency and temperature dependent impedance spectroscopy of cobalt ferrite composite thick films. *Journal of Applied Physics,* Vol.108, No.3, (August 2010), pp. 034101-7, ISSN 0021-8979

Dias, A. & Moreira, R. L. (1999). Chemical, mechanical and dielectric properties after sintering of hydrothermal nickel-zinc ferrites. *Materials Letters,* Vol.39, No.1, (April 1999), pp. 69-76, ISSN 0167-577X

Dutta, S., Choudhary, R. N. P., Sinha, P. K., & Thakur, A. K. (2004). Microstructural studies of $(PbLa)(ZrTi)O_3$ ceramics using complex impedance spectroscopy. *Journal of Applied Physics,* Vol.96, No.3, (August 2004), pp. 1607-1613, ISSN 0021-8979

Fontijin, W. F. J., van der Zaag, P. J., Feiner, L. F., Metselaar, R., & Devillers, M. A. C. (1999). A consistent interpretation of the magneto-optical spectra of spinel type ferrites. *Journal of Applied Physics,* Vol.85, No.8, (April 1999), pp. 5100-5105, ISSN 0021-8979

Funke, K. (1993). JUMP RELAXATION IN SOLID ELECTROLYTES. *Progress in Solid State Chemistry,* Vol.22, No.2, (May 1993), pp. 111-195, ISSN 0079-6786

George, M., Nair, S. S., Malini, K. A., Joy, P. A., & Anantharaman, M. R. (2007). Finite size effects on the electrical properties of sol-gel synthesized $CoFe_2O_4$ powders: deviation from Maxwell-Wagner theory and evidence of surface polarization effects. *Journal of Physics D- Applied Physics,* Vol.40, No.6, (March 2007), pp. 1593-1602, ISSN 0022-3727

Gul, I. H., Abbasi, A. Z., Amin, F., Anis-ur-Rehman, M., & Maqsood, A. (2007). Structural, magnetic and electrical properties of $Co_{1-x}Zn_xFe_2O_4$ synthesized by co-precipitation method. *Journal of Magnetism and Magnetic Materials,* Vol.311, No.2, (April 2007), pp. 494-499, ISSN 0304-8853

Gul, I. H., Ahmed, W., & Maqsood, A. (2008). Electrical and magnetic characterization of nanocrystalline Ni-Zn ferrite synthesis by co-precipitation route. *Journal of Magnetism and Magnetic Materials,* Vol.320, No.3-4, (February 2008), pp. 270-275, ISSN 0304-8853

Gul, I. H., & Maqsood, A. (2008). Structural, magnetic and electrical properties of cobalt ferrites prepared by the sol-gel route. *Journal of Alloys and Compounds,* Vol.465, No.1-2, (October 2008), pp. 227-231, ISSN 0925-8388

Imre, A. W., Schonhoff, M., & Cramer, C. (2009). Unconventional Scaling of Electrical Conductivity Spectra for PSS-PDADMAC Polyelectrolyte Complexes. *Physical Review Letters,* Vol.102, No.25, (June 2009), pp. 255901-4, ISSN 0031-9007

Jame, A. R., Prakash, C., & Prasad, G. (2006). Structural properties and impedance spectroscopy of excimer laser ablated Zr substituted $BaTiO_3$ thin films. *Journal of Physics D-Applied Physics,* Vol.39, No.8, (April 2006), pp. 1635-1641, ISSN 0022-3727

Jonscher, A. K. (1977). Universal dielectric response. *Nature,* Vol.267, No.5613, (June 1977), pp. 673-679, ISSN 0028-0836

Kitamoto, Y., Kantake, S., Shirasaki, F., Abe, M., & Naoe, M. (1999). Co ferrite films with excellent perpendicular magnetic anisotropy and high coercivity deposited at low temperature. *Journal of Applied Physics,* Vol.85, No.8, (April 1999), pp. 4708-4710, ISSN 0021-8979

Komarneni, S., D'Arrigo, M. C., Leonelli, C., Pellacani, G. C., & Katsuki, H. (1998). Microwave-hydrothermal synthesis of nanophase ferrites. *Journal of the American Ceramic Society,* Vol.81, No.11, (November 1998), pp. 3041-3043, ISSN 0002-7820

Lee, J. G., Lee, H. M., Kim, C. S., & Oh, Y. J. (1998). Magnetic properties of $CoFe_2O_4$ powders and thin films grown by a sol-gel method. *Journal of Magnetism and Magnetic Materials,* Vol.177, No.2, (January 1998), pp. 900-902, ISSN 0304-8853

Lin, F. D., Wang, Y. J., & Lonergan, M. (2008). *Journal of Applied Physics,* Vol.104, No.10, (November 2008), pp. 103517-8, ISSN 0021-8979

Lisfi, A., & Williams, C. M. (2003). Magnetic anisotropy and domain structure in epitaxial $CoFe_2O_4$ thin films. *Journal of Applied Physics,* Vol.93, No.10, (May 2003), pp. 8143-8145, ISSN 0021-8979

Marczak, M., & Diesinger, H. (2009). Traveling wave dielectrophoresis micropump based on the dispersion of a capacitive electrode layer. *Journal of Applied Physics,* Vol.105, No.12, (June 2009), pp. 124511-7, ISSN 0021-8979

Nirose, N., & West, A. R. (1996). Impedance spectroscopy of undoped $BaTiO_3$ ceramics. *Journal of the American Ceramic Society*,Vol.79, No.6, (June 1996), pp. 1633-1641, ISSN 0002-7820

Ortega, N., Kumar, A., Bhatacharya, P., Majumder, S. B., & Katiyar, R. S. (2008). Impedance spectroscopy of multiferroic $PbZr_xTi_{1-x}O_3/CoFe_2O_4$ layered thin films. *Physical Review B*, Vol.77, No.1, (January 2008), pp. 014111-10, ISSN 1098-0121

Paike, V. V., Niphadkar, P. S., Bokade, V. V., & Joshiw, P. N. (2007). Synthesis of spinel $CoFe_2O_4$ via the co-precipitation method using tetraalkyl ammonium hydroxides as precipitating agents. *Journal of the American Ceramic Society*, Vol.90, No.9, (September 2007), pp. 3009-3012, ISSN 0002-7820

Patange, S. M., Shirsath, S. E., Toksha, B. G., Jadhav, S. S., & Jadhav, K. M. (2009). Electrical and magnetic properties of Cr^{3+} substituted nanocrystalline nickel ferrite. *Journal of Applied Physics*,Vol.106, No.2, (July 2009), pp. 023914-7, ISSN 0021-8979

Perron, H., Mellier, T., Domain, C., Roque, J., Simoni, E., Drot, R., & Catalette, H. (2007). Structural investigation and electronic properties of the nickel ferrite $NiFe_2O_4$: a periodic density functional theory approach. *Journal of Physics-Condensed Matter*, Vol.19, No.34, (August 2007), pp. 346219-10, ISSN 0953-8984

Ponpandian, N., Balaya, P., & Narayanasamy, A. (2002). Electrical conductivity and dielectric behaviour of nanocrystalline $NiFe_2O_4$ spinel. *Journal of Physics-Condensed Matter*, Vol.14, No.12, (April 2002), pp. 3221-3237, ISSN 0953-8984

Ponpandian, N. & Narayanasamy, A. (2002). Influence of grain size and structural changes on the electrical properties of nanocrystalline zinc ferrite. *Journal of Applied Physics*, Vol.92, No.5, (September 2002), pp. 2770-2778, ISSN 0021-8979

Pramanik, N. C., Fujii, T., Nakanishi, M., & Takada, J. (2005). Preparation and magnetic properties of the $CoFe_2O_4$ thin films on Si substrate by sol-gel technique. *Journal of Materials Science*, Vol.40, No.16, (August 2005), pp. 4169-4172, ISSN 0022-2461

Sathaye, S. D., Patil, K. R., Kulkarni, S. D., Bakre, P. P., Pradhan, S. D., Sarwade, B. D., & Shintre, S. N. (2003). Modification of spin coating method and its application to grow thin films of cobalt ferrite. *Journal of Materials Science*, Vol.38, No.1, (January 2003), pp. 29-33, ISSN 0022-2461

Sepelak, V., Baabe, D., Litterst, F. J., & Becker, K. D. (2000). Structural disorder in the high-energy milled magnesium ferrite. *Journal of Applied Physics*, Vol.88, No.10, (November 2000), pp. 5884-5893, ISSN 0021-8979

Sivakumar, N., Narayanasamy, A., Ponpadian, N., & Govindaraj, G. (2007). Grain size effect on the dielectric behavior of nanostructured $Ni_{0.5}Zn_{0.5}Fe_2O_4$. *Journal of Applied Physics*, Vol.101, No.8, (April 2007), pp. 084116-6, ISSN 0021-8979

Sivakumar, N., Narayanasamy, A., Chinnasamy, C. N., & Jeyadevan, B. (2007). Influence of thermal annealing on the dielectric properties and electrical relaxation behaviour in nanostructured $CoFe_2O_4$ ferrite. *Journal of Physics-Condensed Matter*, Vol.19, No.38, (September 2007), pp. 386201-11, ISSN 0953-8984

Sivakumar, N., Narayanasamy, A., Jeyadevan, B., Joseyphus, R. J., & Venkateswaran, C. (2008). Dielectric relaxation behaviour of nanostructured Mn-Zn ferrite. *Journal of Physcis D-Applied Physics*, Vol.41, No.24, (December 2008), pp. 0245001-5, ISSN 0022-3727

Srivastava, A., Garg, A., & Morrison, F. D. (2009). Impedance spectroscopy studies on polycrystalline $BiFeO_3$ thin films on Pt/Si substrates. *Journal of Applied Physics*, Vol.105, No.5, (March 2009), pp. 054103-6, ISSN 0021-8979

Toksha, B. G., Shirsath, S. E., Patange, S. M., & Jadhav, K. M. (2008). Structural investigations and magnetic properties of cobalt ferrite nanoparticles prepared by sol-gel auto combustion method. *Solid State Communication*, Vol.147, No.11-12, (September 2008), pp. 479-483, ISSN 0038-1098

Wang, Z. L., Liu, X. J., Lv, M. F., Chai, P., Liu, Y., & Meng, J. (2008). Preparation of ferrite MFe_2O_4 (M = Co, Ni) ribbons with nanoporous structure and their magnetic properties. *Journal of Physical Chemistry B*, Vol.112, No.36, (September 2008), pp. 11292-11297, ISSN 1520-6106

Wu, J. G., & Wang, J. (2009). Multiferroic behavior and impedance spectroscopy of bilayered $BiFeO_3/CoFe_2O_4$ thin films. *Journal of Applied Physics*, Vol.105, No.12, (June 2009), pp. 124107-6, ISSN 0021-8979

Yin, J. H., Ding, J., Liu, B. H., Miao, X. S., & Chen, J. S. (2006). Nanocrystalline Co-ferrite films with high perpendicular coercivity. *Applied Physics Letters*, Vol.88, No.16, (April 2006), pp. 162502-3, ISSN 0003-6951

Yu, T., Shen, Z. X., Shi, Y., & Ding, Y. (2002). Cation migration and magnetic ordering in spinel $CoFe_2O_4$ powder: micro-Raman scattering study. *Journal of Physics-Condensed Matter*, Vol.14, No.37, (September 2002), pp. L613-L618, ISSN 0953-8984

Zhong, C. G., Jiang, Q., Fang, J. H., & Jiang, X. F. (2009). Thickness and magnetic field dependence of ferroelectric properties in multiferroic $BaTiO_3$-$CoFe_2O_4$ nanocomposite films. *Journal of Applied Physics*, Vol.105, No.4, (February 2009), pp. 044901-6, ISSN 0021-8979

7

Studies on Electrical Properties and Memory Retention Enhancement of Metal-Ferroelectric-Insulator-Semiconductor with Radical Irradiation Treatments

Le Van Hai[1,2], Takeshi Kanashima[1] and Masanori Okuyama[1]
[1]*Osaka University,*
[2]*Hanoi University of Science and Technology, Viet Nam,*
Japan

1. Introduction

Ferroelectric materials and their applications have been known and developed sine 1920s and nowadays they are one of the most important materials used in the memory, optical device and sensor device production. Although they are possible to utilize ferroelectric characteristics for a wide variety of applications, but a primary object of recent research activities is directed toward the development of nonvolatile memories. In this study we focused on enhancement of electrical and data retention characteristics of memory devices based on metal–ferroelectric–insulator–semiconductor (MFIS) structure by radical irradiation treatments.

Ferroelectric memories has gathered much attention since they show ultimate properties of nonvolatility, low-power consumption, and high-speed operation [Hai, L. V., etc. (2010), Hai, L. V., etc. (2006 a), Tarui Y, ect. (1997), Scott, J. F. (2000), Sakai, S. & Ilangovan, R. (2004), Ishiwara, H. (2001)]. The well-known ferroelectric memory applications of ferroelectric capacitor-type random access memories (FeRAM) have been successfully developed and commercially available at present [Sakai, S. etc. (2010)]. Another type of ferroelectric memory is a ferroelectric field effect transistor (FeFET), in which a ferroelectric layer was inserted between top gate and oxide layer of an FET, as shown in. In recent years, memory cells of the FeFETs were selected as the most suitable candidate to alternate for floating-gate (FG) memory cells in the NAND flash memory, due to the fact that FeFET cells exhibited superior characteristics from viewpoints of downscale ability, long endurance characteristics, and low-voltage operation, in comparison with the FG-cells [Hai, L. V., etc. (2010)]. The new Fe-NAND flash memory based on the FeFETs can open a new generation of the NAND flash memory which meets fully requirements of high speed USB flash drives, MP3 players, flash memory in portable devices and solid-state drive (SSD) applications [Hai, L. V., etc. (2010)]. Although many efforts have been made by many research groups to improve but the FeFET-type memories still need more investigations to satisfy for commercial application as some problems were pointed out. It is extremely difficult to fabrication FeFETs with high-quality ferroelectric and insulator films therefore they always do not show acceptable retention and electrical properties. We attribute the relatively low-

quality films to diffusion of constituent elements. The diffusion between the ferroelectric film and insulator layer has given damages to interface layers, such as the formation of high-density electron or hole surface traps and charge injection into the ferroelectric layer, which seriously degrade device performance because of increases in leakage current and depolarization field [Takahashi, M. (2001)]. In addition, the element diffusions between layers in MFIS stack during fabrication process cause mainly stoichiometric composition change, and lead to quality degradation of insulator and ferroelectric films.
Among potential candidates of gate structure for MFIS-type FET, Pt/SBT/SiO_2/Si stack is the simplest structure, good matching with complementary metal-oxide-semiconductor (CMOS) process [Paz de Araujo, C.A., etc., (1995), Hai, L. V., etc. (2006 b)] and low-cost production. SiO_2 buffer layer was grown simply by thermal oxidation method directly on Si substrate, and did not need a special buffer layer of high-k material which requires a complicate process and unfamiliar with the convenience silicon manufacturing process. It made Pt/SBT/SiO_2/Si stack give advantage in comparison with the other MFIS structures. But the SiO_2 buffer layer has a small dielectric constant and is not good as diffusion barrier layer in comparison with high-k material (Si_3N_4, Al_2O_3, HfO_2, HfAlO, etc.) [Aizawa, K., etc. (2004), Sakai, S. etc. (2004), Youa, I.-K., etc., (2001)]. To overcome these challenges, We have suggested a novel method of using nitrogen radical irradiation to treat the SiO_2 buffer layer in MFIS structure [Hai, L. V., etc. (2008)]. The SiO_2 layer shows enhancements of dielectric constant and thermal stability, and becomes a good buffer layer for suppressing the constituent element diffusion problem. These achievements were demonstrated through our experiment results.
Furthermore, nitrogen and oxygen radical irradiation treatments were employed to modify surfaces of ferroelectric layer for the first time [Hai, L. V., etc. (2006 a)]. We found that ferroelectric interface layers have been formed and demonstrated promising properties of barrier layers. Furthermore, dielectric constant of buffer layer increases, and so depolarization field will be suppressed. It is reported that it could significantly suppress the diffusion of ferroelectric components or chemical reactions with nitrogen treatment [Hai, L. V., etc. (2006 a)]. As a result, the nitrogen radical irradiation treatment is a significant candidate for improving memory retention characteristic of the Pt/SBT/SiO_2/Si MFIS.

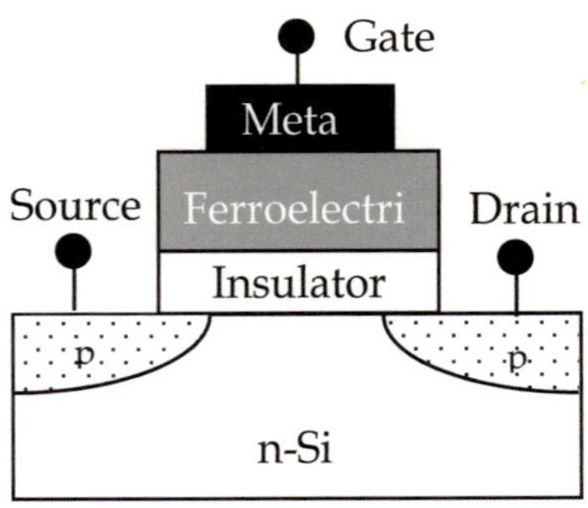

Fig. 1. Schematic of ferroelectric gate FET on n-Si substrate

The goal of this work is to solve the main problems of MFIS structure, namely large leakage current and short retention time, to realize ferroelectric memory applications with the feature of non-destructive readout [Hai, L. V., etc. (2010), Hai, L. V., etc.(2006 a), Tarui Y, ect. (1997), Scott, J. F. (2000), Sakai, S. & Ilangovan, R. (2004), Ishiwara, H. (2001)]. The study results include: demonstrations of the simplest MFIS structure with good characteristics for ferroelectric memory application; using a novel method of radical irradiation to enhance electrical characteristics of MFIS structures such as, decrease of leakage current and improvement of retention property from 3 hours to 23 days.

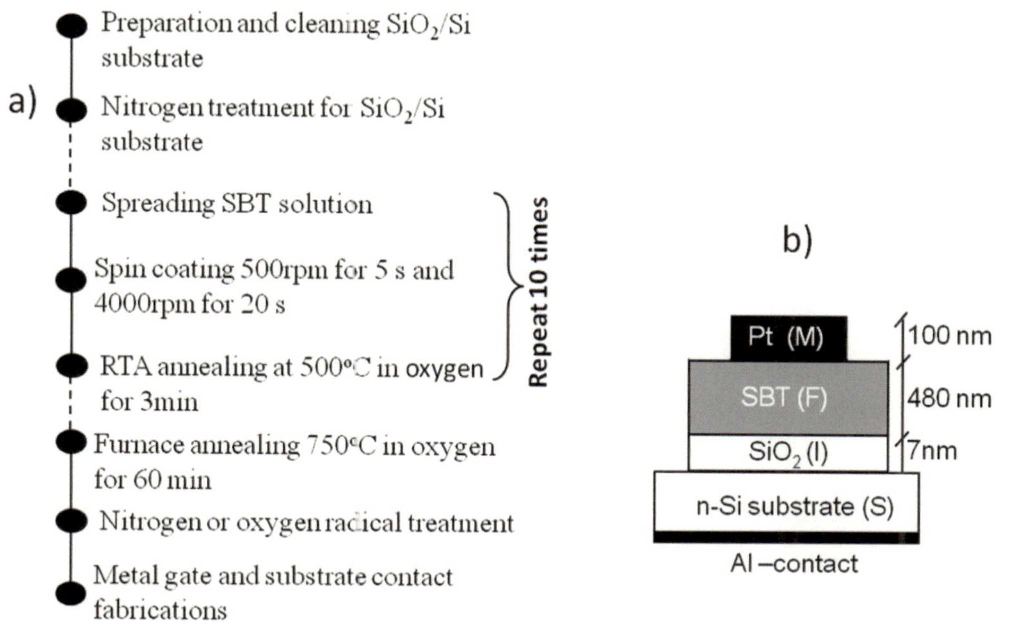

Fig. 2. a) Schematic of fabrication steps for MFIS structure and b) cross-section of and parameters of MFIS stack.

2. Structure and Fabrication processes of Metal-ferroelectric-insulator-semiconductor

2.1 Structure of MFIS devices

The present FeFET structures like the metal-oxide-semiconductor field effect transistor (MOSFET), in which a ferroelectric layer was inserted between top metal gate and an insulator layer, as shown in Fig. 1. The principal structure of the FeFETs are composed from a MFIS stack of metal, ferroelectric, insulator, semiconductor layer, as in Fig. 2b . In a FeFET, polarization direction of the ferroelectric layer depends on application voltages of the gate and drives the drain current between the source and drain regions.

The SiO_2 insulator of thickness 7.5nm was prepared directly from the n-Si semiconductor substrate by thermal oxidization method beforehand. Substrate with SiO_2 layer on surface was cleaned by high purity acetone, propanol and deionized-water in ultra-sonic cleaner before treating by radical irradiation, which will be described in more detail in next section.

2.2 Fabrication of the MFIS stack

First, SBT ferroelectric thin film was prepared on the substrate by metal-organic decomposition method (MOD). The SBT solution used for the MOD was Y-1 type0 (Sr:Bi:Ta = 0.9:2.2:2.0) manufactured by Kojundo Chemical Lab. Co. Ltd. Si substrate with SiO_2 buffer layer was coated with SBT solution by spin-coating method, at 500 rpm for 5 s and subsequently rotated at 4000 rpm for 30 s. Then, the films were dried at 160°C for 3 min by hot plate in atmosphere and subsequently annealed in O_2 by rapid thermal annealing (RTA) for 3 min at 700°C for. This step was repeated 10 times to achieve 480-nm thickness of SBT thin film. Finally, the SBT thin film was atreated at 750°C by furnace annealing in O_2 ambience for 60 min to crystallize SBTs. To enhance basic property of thin film, the SBT were treated in vacuum chamber by nitrogen or oxygen radical irradiation which will be described in more detail in the next section. The Pt circle electrodes were prepared by Ar plasma sputtering method on the SBT thin films with diameter of 150 μm. The Al substrate contact on the back-side of the n-Si substrate was prepared by thermal evaporation.

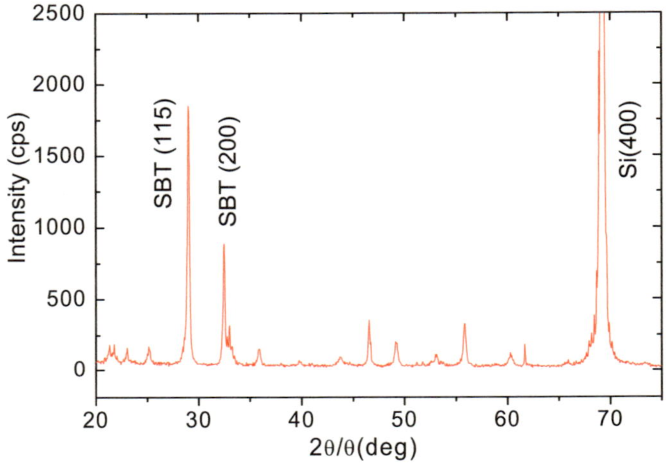

Fig. 3. X-ray diffraction pattern for SBT film grown on SiO_2/n-Si substrate by MOD method and treated by furnace annealing in oxygen ambience at ate 750°C for 60 min.

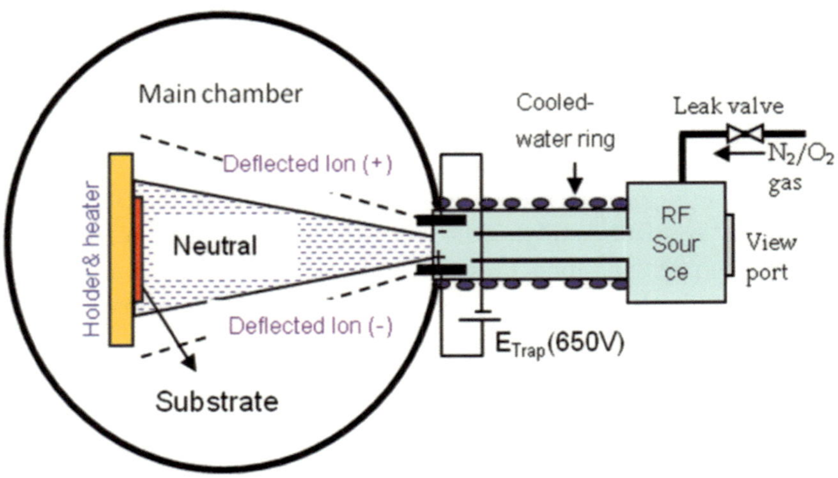

Fig. 4. Schematic diagram of radical irradiation system.

2.3 X-ray diffraction characterization of SBT thin films

X-ray diffraction (XRD) pattern of SBT thin films deposited on SiO_2/n-Si substrates is shown in Fig. 3. The SBT thin film was treated at 750°C by furnace annealing in O_2 ambience for 60 min. The thickness of the SBT is about 480 nm. It can be observed that SBT film deposited on SiO_2 /n-Si shows a highly textured (115) orientation and a minor textured (200) orientation. Some reports of SBT thin films have revealed that typical peak of SBT(115) at 2θ=29.00 is Bi-layered peroskite structure and (222) peak of the pyrochlore SBT is at 2θ=29.45 [J.C.Riviere (1983)]. The figure shows no diffraction peaks from pyrochlore phase.

3. Treatments of nitrogen and oxygen radical irradiation

The nitrogen and oxygen radical irradiation systems employed in this study is shown in Fig. 4. Nitrogen/oxygen radical was generated within a small tube of pyrolytic boron nitride (PBN) by an RF radical gun. When pure nitrogen/oxygen was introduced to the tube with a leak valve into the radical gun, Nitrogen/oxygen plasma was formed and the nitrogen/oxygen radicals were injected into treatment chamber due to the pressure difference between the treatment chamber and radical gun inside. The RF source operates at 13.56 kHz with a typical maximum power of 600 W.

The nitrogen or oxygen radical beam was injected the into the main chamber through an ion trap, which repels ions with a strong voltage of -650V. Ions are almost bent in way to treatment chamber wall when travelling through the ion trap space and never approaching sample. As a result only neutral species of nitrogen or oxygen can go straight and approach at surface of substrate, because they are not Affect by electric field. The substrate was attached on a holder and its surface is perpendicular to the radical beam. Temperature of back-side of substrate was controlled and kept constant during treatment by a heater source.

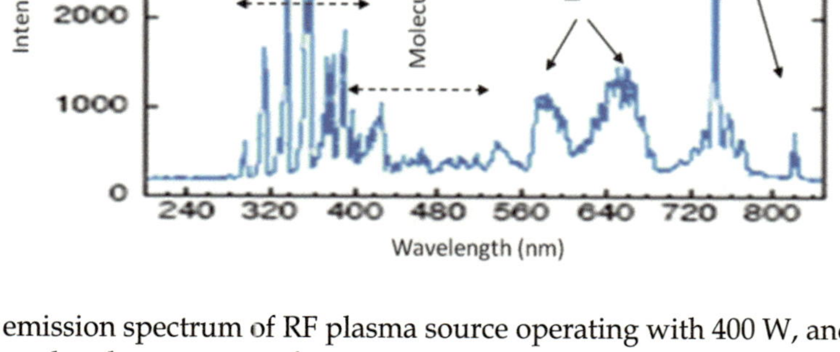

Fig. 5. Optical emission spectrum of RF plasma source operating with 400 W, and using 0.56 Sccm nitrogen at chamber pressure of $7x10^{-3}$ Pa

Fig. 5 shows emission spectrum of the radical source monitored from a quartz window at the end of the radical source. The nitrogen radicals supplied by the radical source are mainly composed of excited molecular neutral (N_2*) and atomic neutral (N*) nitrogen with a small amount of molecular N_2 and atomic N ions. The intensity of N* and N_2* drastically depends on nitrogen flow rate, chamber pressure and the power applied to the radical gun. In this study, we optimized optical emission spectra of nitrogen radical as show in Fig.5. Neutral elements were dominated by optimized parameters in Table 1. Better nitrogen treatment performance can be obtained with high intensity ratios of N* and N_2*.

Parameters	Nitrogen radical irradiation	Oxygen radical irradiation
RF power	400 W	300 W
Reflected power	1 W	3 W
Chamber pressure	7x10^{-3} Pa	8x10^{-3} Pa
Substrate temperature	400°C	400°C
Gas flow	0.56 Sccm	1 Sccm

Table 1.Typical conditions of nitrogen and oxygen radical irradiation treatments

4. Nitrogen radical irradiation treatments for enhancement of property of SiO_2 thin film

4.1 Chemical composition of SiO_2 with nitrogen radical irradiation treatments

After nitrogen treatment, the SiO_2/n-Si substrates were annealed for 30 min at 950°C in nitrogen ambience in furnace to remove fixed charges which were generated during irradiation of SiO_2 surface. Nitrogen incorporated on surface of SiO_2 film were confirmed by surface chemical analysis from x-ray photoelectron spectroscopy (XPS) spectrum.

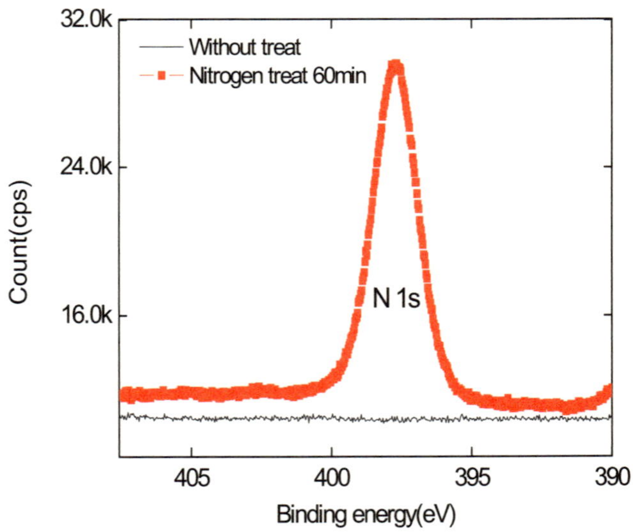

Fig. 6. XPS spectra of N1s state of SiO_2 surface with and without radical treatment for 60min.

Figure 6 shows XPS spectra near N1s state of SiO_2 surface with and without radical treatment. The distribution of the nitrogen concentration near surface of nitrided SiO_2 layer was obviously observed by comparing the intensity of N1s peaks near 398 eV. It was one of evidence to prove incorporation of nitrogen in SiO_2. Nitrogen radicals make bonding with SiO_2 surface and form $SiON_x$ [Hai, L. V. etc., (2006)].

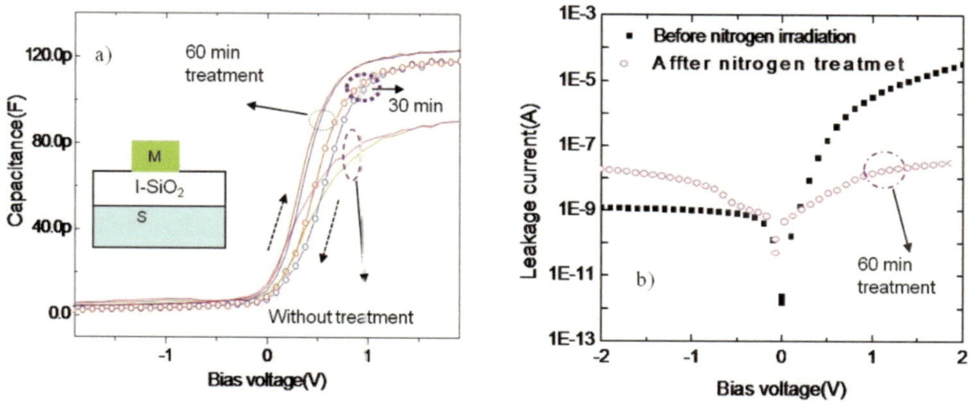

Fig. 7. Electronic properties of $Pt/SiO_2/Si$ MIS diodes with top electrode size of $7x10^4$ μm^2, a) C-V curves of MIS withwith nitrogen treatment 60 min, 30 min and without the treatment, and b) I-V curves of MIS with 60 min and without nitrogen treatment.

4.2. Electrical characteristics of MIS diodes with nitrogen radical treatment

Figure 7 shows C-V and I-V characteristics of MIS diodes which have 7.5-nm SiO_2 insulator layer with and without nitrogen radical. Fig. 7 a) shows capacitance of the MIS structure with different nitrogen treatment period of SiO_2 film. It is confirmed that dielectric constant of insulator layer increases also due to treatment process.

Besides C-V curve improvements, Fig. 7 b) shows the I-V characteristic of sample improved by 60min nitrogen treatment in comparison with sample without treatment. It is believed that neutral nitrogen is incorporated with SiO_2 forming SiON and improves the electrical properties of the insulator layer. All C-V curves of samples with nitrogen treatment show steep transition region and a small hysteresis, while sample without nitrogen treatment has gently sloping and hysteresis in C-V curve which is induced by carrier injection. Furthermore, it was also confirmed that SiO_2 without treatment generates promotion of positive-shift in C-V curve, compared with MIS structures using SiO_2 with nitrogen treatment for 30min or 60min. It is well known that the positive-shift of the flat-band voltage in SiO_2-MOS systems can result from the negative charge trapping in the oxide layer. We believed nitrogen radical treatment is helpful to reduce negative charge trapping in SiO_2 layer. That means the improvements of the Si/SiO_2 interface properties and decrease of negative charge density in the Si/SiO_2 were a primary cause of C-V curve improvements.

5. Nitrogen and oxygen radical irradiation treatment for SBT ferroelectric layer

5.1 Surface morphologies of SBT thin films with nitrogen irradiation treatments

During treatment decrease of oxygen vacancies density in the surface of Si/SiO_2 were primary causes of C-V curve improvements. The decrease of oxygen vacancies density could help to suppress the Bi and other elements from SBT layer in to SiO_2 insulator layer in MFIS structure. Because they react with vacancies in the SiO_2, forming fast-moving complexes [Klee, M. and Macken, U. (1996) ; Tanaka, M. ect. 1998].

Fig. 8 shows SEM micrographs of $SrBi_2Ta_2O_9$ thin films with and without nitrogen treatment. Voids are observed all over the surfaces of the films as there appear different density and size. Surface morphology of as-deposited SBT was not satisfied with many deep voids. However surface morphologies of treated SBT have been remarkably improved by the radical irradiation and the deep voids disappear from the film surfaces, resulting in smooth surfaces. In particular, the film surface morphologies which were investigated by AFM images have confirmed the roughness improvement (Fig. 9). This figure shows the roughness rapidly reduces with the nitrogen radical for 10 min and slowly reduces with increasing irradiation time.

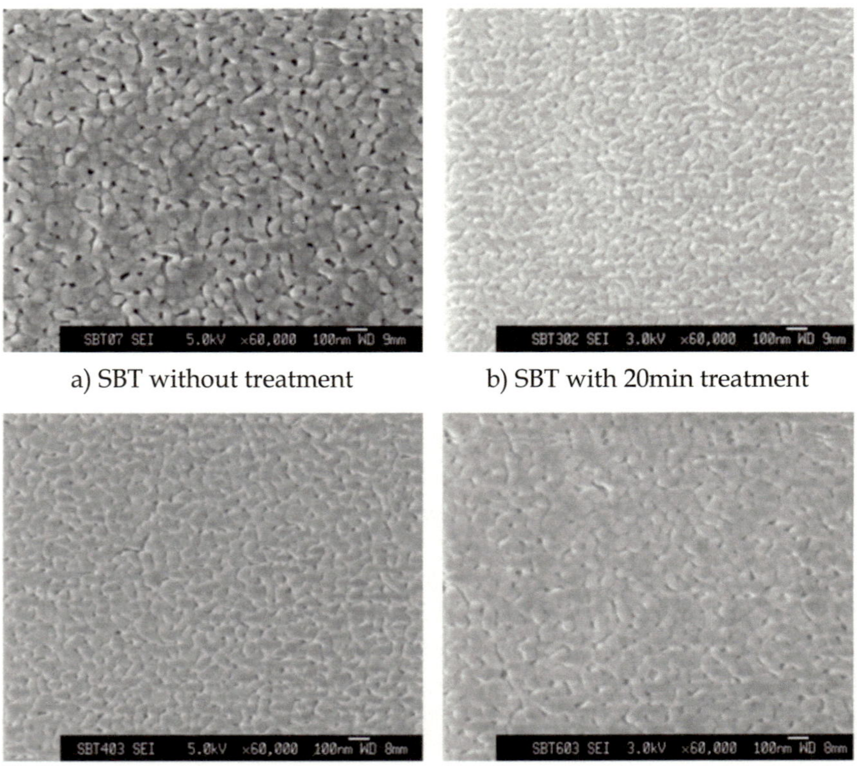

a) SBT without treatment b) SBT with 20min treatment

c) SBT with 40min treatment d) SBT with 60min treatment

Fig. 8. SEM micrographs of surface SBT thin films a) as-deposited , after nitrogen treatment b) for 20min, c) for 40min, and d) for 60min.

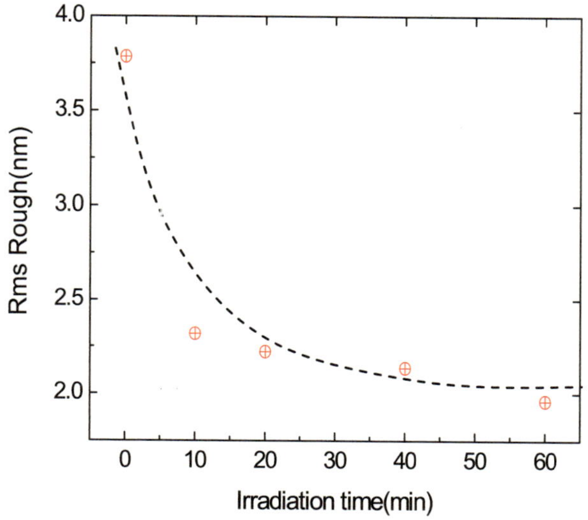

Fig. 9. Surface morphology roughness of SBT thin films and versus treatment time of nitrogen radical irradiation

5.2 Chemical modification of surface SBT thin films induced by nitrogen irradiation

Fig. 10 shows XPS spectra of N1s state of the SBT surfaces with and without radical treatments. As N1s peak intensity which corresponds to nitrogen density of in SBT surface, reaches the maximum value and then reduces with treatment time. The highest nitrogen density can be obtained when irradiation time is around 20 min.

It is suggested that nitrogen is initially incorporated with $(Bi_2O_2)^{2+}$ oxide layer, and replaced oxygen vacancy in defect $(Bi_2O_2)^{2+}$ layer, and even oxygen in Bi-O bonding. If SBT surface was irradiated for long time, it will be damaged by irradiation beam. Appearance of nitrogen on SBT films perhaps modifies energies of Bi-O bonds and N1s state in comparison with general states of them. Binding energy of O and Bi slightly shifts toward lower energy, as XPS spectra of O1s and Bi4f states of the SBT surface shown in Fig. 11. Authors suggested that is due to electro negativity of N-bond (3.04) is smaller than that of O-bond (3.44) and in surface of the SBT layer a small amount of nitrogen atom replace for oxygen atom in Bi-O bond. We found production of oxygen vacancies or free Bi in $(Bi_2O_2)^{+2}$ layer induces a problem in SBT films after thermal crystallization and some interested effects in SBT layer treated by nitrogen radical [Hai, L. V., Kanashima, T., Okuyama, M. (2006 b)]. Work-function and band gap of the SBT surface layer were modified. Barrier energy heights for hole in M-F junction increased, and so the electronic properties of the SBT layer were improved. Composition of SBT surface was changed with decrease of free Bi^0 density. It is considered that oxygen vacancies can be suppressed by nitrogen treatment, because neutral nitrogen radical forms stronger bonding than oxygen and easily reacts with free Bi that remains after crystallization in oxygen. In this study, we found maximum work-function energy of 6.6 eV belongs to SBT film after 20 min nitrogen treatment.

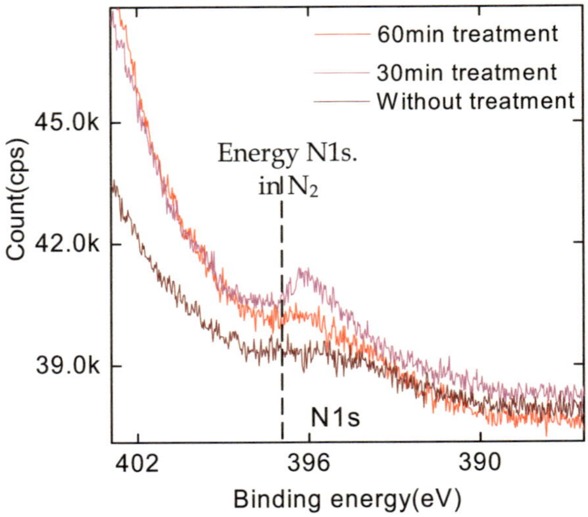

Fig. 10. XPS spectra of N1s state of the SBT surface without and with radical treatment in 30 and 60 min.

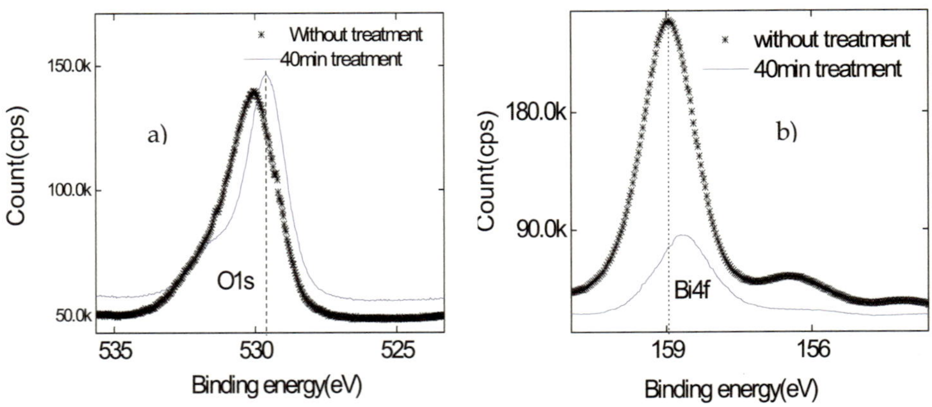

Fig. 11. XPS spectra of SBT before and after irradiation treatment. a) O1s spectra peaks and b) Bi 4f spectra peaks.

5.3 Effect of radical treatments on SBT band gap

X-ray photoelectron spectroscopy (XPS) was used to investigate the binding and composition states of SBT before and after radical treatment. Figure 13 a) shows electron energy levels explaining a typical photo-emission. The binding energies are decided by comparison with carbon peak. The range is concerned with Bi binding, particularly the peaks near 160 eV and 165 eV are attributed to the oxidized Bi^{3+} of -Bi-O binding, 157 and 163 eV are attributed to the metallic Bi^{0} of Bi-metal binding. From the Bi 4*f* XPS spectra of Fig. 12, it is clear that the Bi metallic peaks are affected by nitrogen and oxygen irradiation time, and disappear after 10 min or 20 min treatment, respectively. This behavior indicates that the metallic ion can be reduced by nitrogen or oxygen irradiation. Owning to decrease of metallic Bi atom and Bi defect ion on the surface of thin film, the Bi diffusion - the main reasons of poor metal-ferroelectric interface, should be suppressed. During treatment, a series of chemical reactions took place on the surface of SBT and modified chemical bonding of surface layer.

Fig. 13 b), c) and d) show the O1s XPS peak of both SBT thin films with radical treatment. The O1s XPS signal includes two peaks of oxygen in perovskite structure on the right side with a smaller energy binding and oxygen in the bismuth-deficient (Bi_xO_y) layers on the left side with a larger binding energy. In this figure, the O1s spectra of perovskite structure shifts toward smaller binding energy and O1s spectra of oxygen in bismuth-deficient (Bi_xO_y) become smaller with the treatment. That means the oxygen vacancy in the defected (Bi_xO_y) layers is reduced by nitrogen or oxygen irradiation.

The energy loss spectra of O1s peaks for SBT films have been analyzed to estimate their band gaps between the valence bands and conduction bands. H. Itokawa, et al, discussed on determinations of band gap by analyzing XPS spectral of O 1s core levels for several insulators. The band gap of 4.20 eV is assumed for as-deposited SBT film [Takahashi, M. etc., (2001)]. Fig.s 13 show XPS spectra to estimate band gap of the surface layer of the SBT treated by nitrogen, oxygen irradiation and as-deposited. In results, band gap of 4.20 eV of as-deposited SBT film was confirmed, Fig. 13 b). After 20 min oxygen irradiation band gap energies of SBT of was increased from 4.20 eV to of 4.52 eV, Fig. 13 c).. After 10 min nitrogen irradiation band gap energies of SBT of was increased from 4.20 eV to of 4.72 eV, Fig. 13 d).

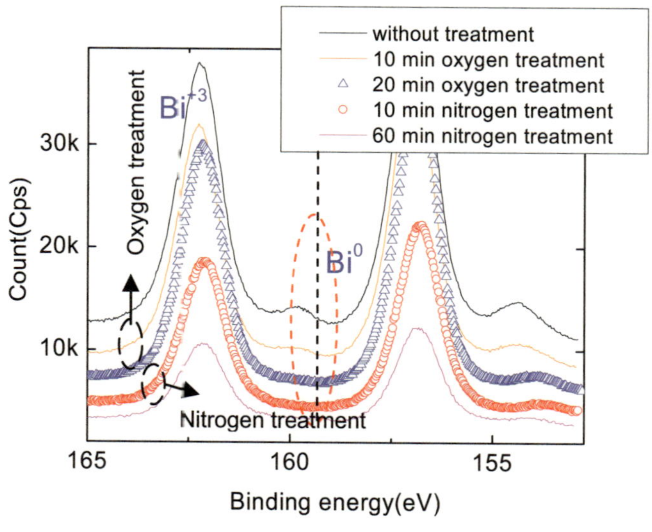

Fig. 12. XPS spectrum near the Bi 4f peaks of SBT film surface with and without radical treatments.

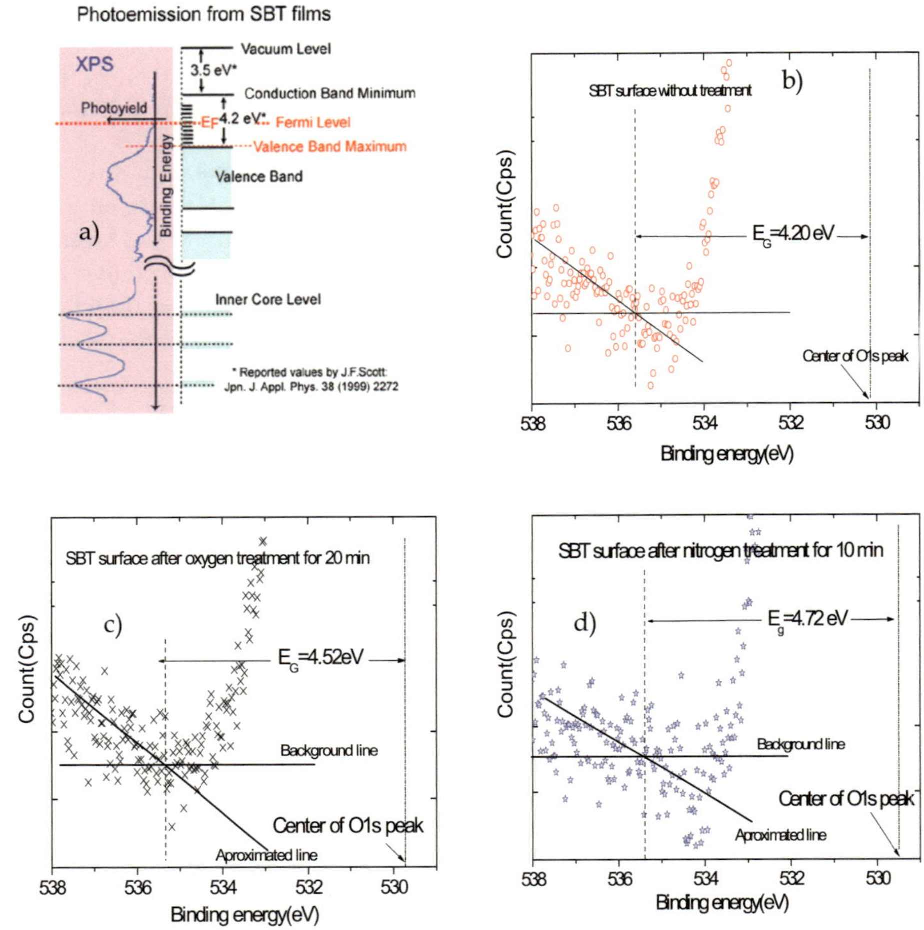

Fig. 13. a) Electron energy levels explaining a typical photo-emission, and XPS spectra of SBT before and after irradiation treatment near O1s spectra peak for b) as-deposited film, c) 20-min oxygen treated and d) 10-min nitrogen radical treated films. Band gap width of SBT were calculated from O 1s core levels

5.4 Effect of radical treatments on Fermi level of SBT thin films

Fermi level energies could be estimated for all nitrogen-treated, oxygen-treated and as-deposited SBT thin films by analyzing their ultraviolet-ray photoyield spectroscopy (UV-PYS) spectra [Takahashi, M., (2003)]. From Fig. 14, Fermi level energy of 5.24 eV was obtained for the as-deposited SBT thin film and it increases due to nitrogen and oxygen irradiation treatments. In estimation, the Fermi level energy of the SBT thin films treated by oxygen and nitrogen radicals are about 5.50 eV and 5.60 eV, respectively. The barrier height of the SBT surface with other layers depends on Fermi level energy of the SBT so absolutely the leakage current through Pt/SBT/SiO_2 will be affected. A detail of this problem will be explained in the nextdiscussion.

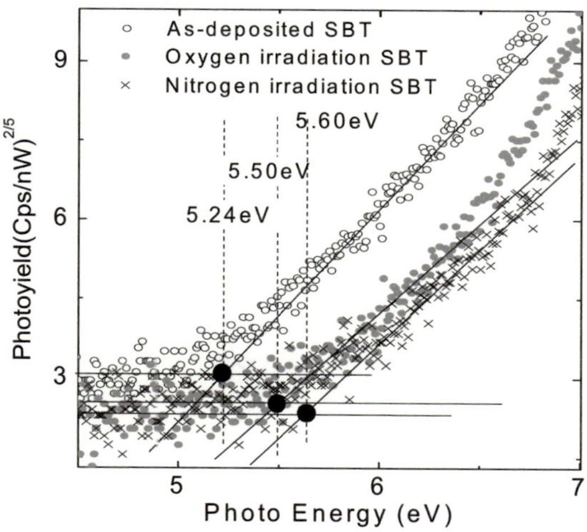

Fig. 14. UV-PYS spectra and estimation of Fermi level in as-deposited and irradiated SBT thin films irradiate by oxygen and nitrogen radicals.

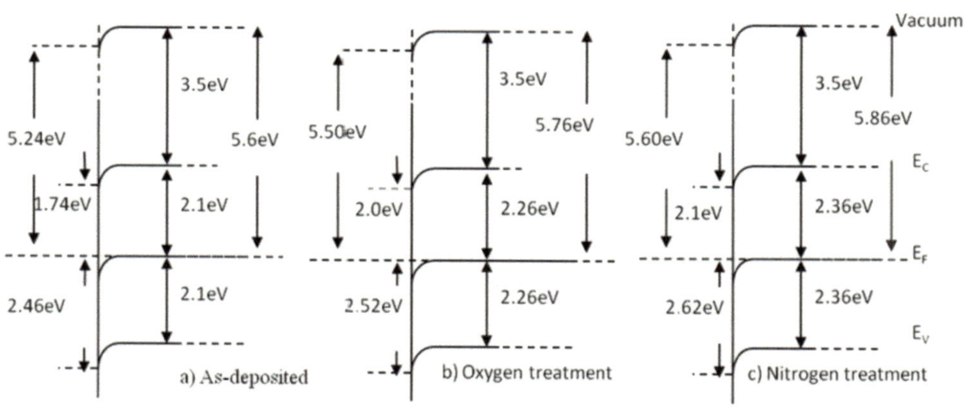

Fig. 15. Band diagrams considered formed-SBT surface before and after irradiation treatment, (a) As-deposited, (b) oxygen radical treatment, and (c) nitrogen radical treatment.

5.5 Calculation of energy diagrams formed-SBT surface with irradiation treatment

From the results of UV-PYS and XPS measurements, we can suggest that a new and very thin layer was formed on surface of the SBT thin film after nitrogen or oxygen irradiation treatment. The composition states of this layer were modified and different from that of the as-deposited SBT thin film. It is considered that both metallic Bi and oxygen vacancy in defected layer (Bi_xO_y) were reduced, that are major causes for modifying the band gap and Fermi level. Band diagrams are considered for SBT surfaces before and after nitrogen irradiation, shown in Fig. 15. If electron affinity of 3.5 eV is assumed for SBT [Klee, M. and Macken, U. (1996).], differences in energy between the Fermi-level and the conduction band minimum, which is considered to be the barrier height for electrons at the metal–ferroelectric interface, are estimated at 1.74 eV for the as-deposited film, 2.00 eV for the oxygen-treated film and 2.10 eV for the nitrogen- treated film. On the other hand, the hole barrier heights are estimated at 2.46 eV for the as-deposited film, 2.52 and 2.62 eV for oxygen and nitrogen irradiation treatment films, respectively.

Fig. 15 suggested barrier versus both electrons and holes that describes the effect of the radical treatment on the SBT surface. Both band offsets for electrons and holes are increased slightly, that means the leakage current will be suppressed.

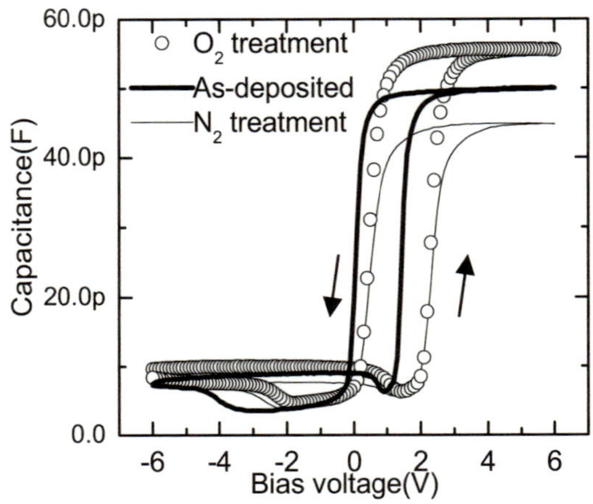

Fig. 16. C-V characteristics of MFIS structures using SBT films with and without irradiation treatments

5.6 Improvements of electrical characteristics of Pt/SBT/SiO_2/n-Si MFIS by application of nitrogen and oxygen radical treatment to SBT layer

Fig. 16 shows the C-V hysteresis characteristics of the Pt/SBT/SiO_2/n-Si structure with as-deposited SBT film, or SBT film after oxygen treatment for 20 min and nitrogen treatment 10 min, which were measured by sweeping the gate voltage from inversion to accumulation region and then sweeping back. The sweeping voltage changes between ±6V with a scan rate of 0.1V/s and frequency of 100 kHz. To separate the effects of the radical treatments on insulator and ferroelectric layer, the SiO_2 used in this experiment was not treated beforehand.

The memory window was slightly were increased about 0.3 V by the nitrogen and oxygen radical treatment. But capacitance of the MFIS in accumulation region was increased with oxygen radical treatment and reduced with nitrogen radical treatment due to the radical treatment processes. It is clear that the good memory window hysteresis are observed, which indicates that the charge injection, the charge trapping, and the ion drift effect are suppressed in the Pt/SBT/SiO_2/n-Si structure with the treated SBT.

The current density through Pt/SBT/SiO_2/n-Si structure, J, using as-deposited and the nitrogen-treated or oxygen-treated SBT thin films were measured as a function of applied voltage V. As shown in Fig. 17, the nitrogen and oxygen treatment succeeded in decreasing the current density. The decreases of currents are considered to be attributed to property of surface of SBT thin films. In after SBT suffering the irradiation treatment, the roughness of surface morphology reduces. The difference barrier height of the SBT surface with SiO_2 and Pt also increase so absolutely the leakage current through Pt/SBT/SiO_2 will be reduced. It is found that the nitrogen radical treatments are more efficient than oxygen radical treatments in term of reduce leakage current. The current density through SBT films were analyzed into two main contributions, from the Schottky and the Frenkel-Poole conduction Fig. 18.

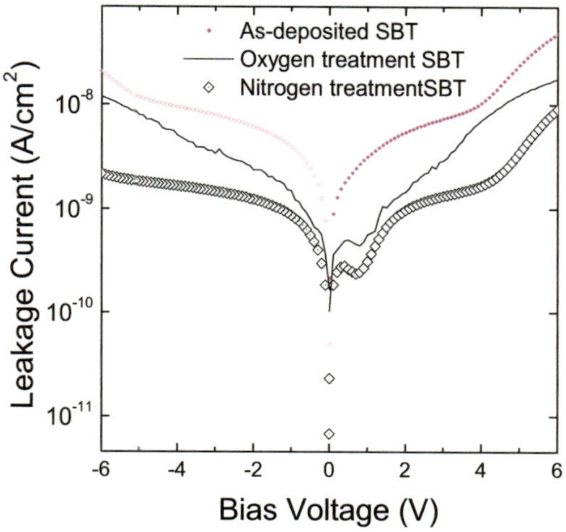

Fig. 17. I-V characteristics of Pt/SBT/SiO2/n-Si with and without radical irradiation treatments

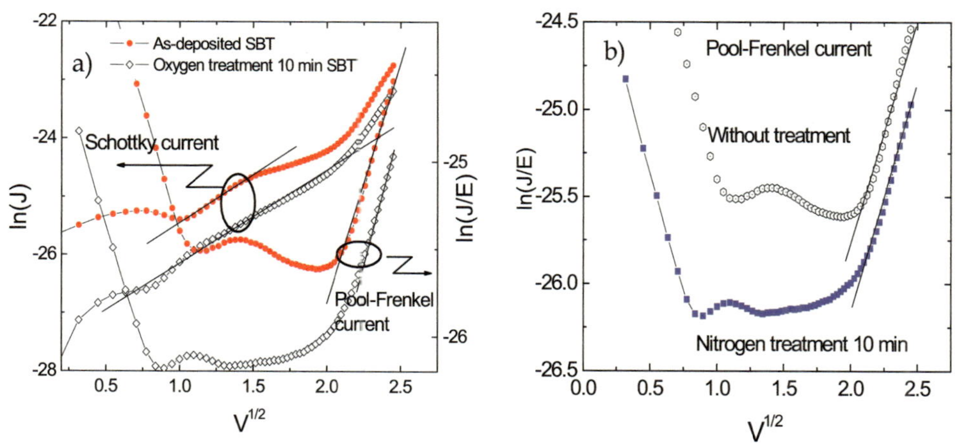

Fig. 18. The leakage current characteristics of MFIS structure, representing Schottky emission at low field and Poole-Frenkel emission at high field, with SBT a) as-deposition and oxygen 10min and b) nitrogen treatment 10 min. E is electric field.

It is found that the Schottky conduction played a key role in total conduction in the nitrogen and oxygen treatment SBT, and consists of carrier transport brought about by thermionic emission across the metal-ferroelectric interface at a low electric field, whereas the Frenkel-

Poole conduction is dominant in the as-deposited SBT, and brought about by field-enhanced thermal excitation of trapped carriers into the band [Takahashi, M. etc., (2001)]. Therefore, reduction of the current density shown in Fig. 17 is attributed to the fact that after the nitrogen and oxygen treatment the SBT have increased the barrier height of the ferroelectric in both accumulate and depletion states. Fig. 18 shows phenomenon of the Frenkel-Poole conduction reduced, became an insignificant minority in modified SBT and so the trap density in the ferroelectric layer may be decreased in the irradiation processes.

6. Improvements of Pt/SBT/SiO_2/n-Si MFIS characteristics with SiO_2 and SBT layers treated by nitrogen radical

To understand more about effects of nitrogen treatments on improve characteristics of Pt/SBT/SiO_2/n-Si MFIS structures, two samples of A and B were investigated by C-V and I-V curves and retention time properties. The first sample was fabricated without using any improvement for reference. The other sample were treated by the nitrogen radical irradiation to the SiO_2 insulator and SBT ferroelectric thin film in fabrication processes to improve MFIS's characteristics. The second sample was treated by nitrogen radical with 60 min for SiO_2 and 20 min for SBT.

6.1 Nitrogen radical treatment to improves MFIS's electrical properties

Fig. 19 a) shows the C-V characteristics of MFIS structures with and without nitrogen treatment, and show counter-clockwise hysteresis loops controlled by polarization of ferroelectric SBT. They were measured at 100 kHz by sweeping the gate voltage from inversion to accumulation region and then sweeping back. The sweeping voltage changes between ±6V with a scan rate of 0.1 V/s. The memory window is about 1.3 V of sample A without nitrogen treatment and 1.8 V of sample B with nitrogen treatment for SiO_2 and SBT. It is found that a larger memory window and flatter depletion capacitance of sample B in comparison with that of sample A. It is believed that suppression of charge injection, charge trapping, and ion drift effect phenomenon are cause of the improvements.

The leakage current density through Pt/SBT/SiO_2/n-Si structures was investigated to verify contribution of the nitrogen treatment to both buffer and ferroelectric layers. As shown in Fig. 19 b), the samples of SiO_2 with the nitrogen treatment for 60 min and SBT with treatment for 20 min succeeded in decreasing the leakage current density in comparison with sample without the treatment. But the measurements exhibits a distinct difference between samples with and without 20 min nitrogen treatment for SBT, and the leakage current reduced one order of magnitude. The currents are considered to be attributed to property of SBT thin films as they are very sensitive to the nitrogen treatment.

In our previous report [Hai, L. V., etc. (2008)], the current density through deposited-SBT films were analyzed into two main contributions, those are the Schottky and the Frenkel-Poole conduction. It is also found that only the Schottky conduction played a key role in total conduction in the nitrogen treatment SBT, and consists of carrier transport brought about by thermionic emission across the metal-ferroelectric interface at a low electric field, whereas the Frenkel-Poole conduction is dominant in the as-deposited SBT, and brought about by field-enhanced thermal excitation of trapped carriers into the band. Therefore, the decreased contributions from the current density shown in Fig. 19 b) suggest that the SBT with the nitrogen treatment have increased the barrier height of the ferroelectric in both accumulation and depletion states.

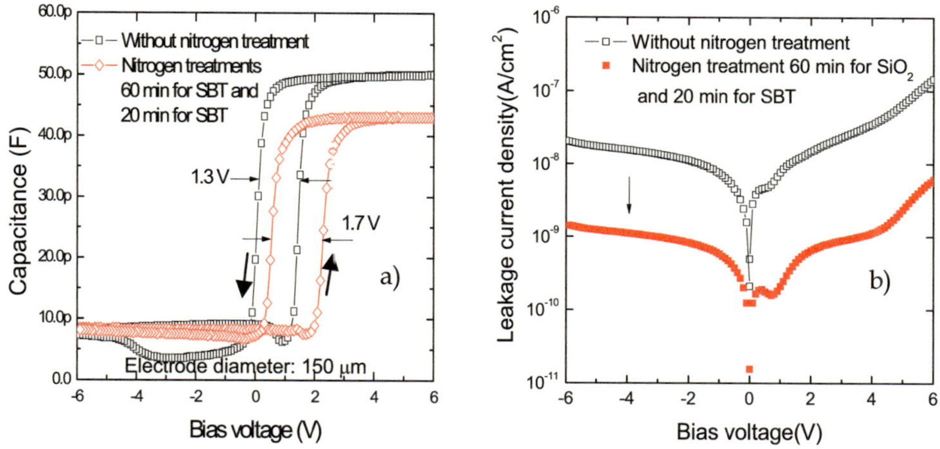

Fig. 19. Pt/SBT/SiO_2/n-Si MFIS structures with and without nitrogen treatment for SiO_2 and SBT thin films, a) C-V characteristics and b) I-V characteristics.

All the results in above part of this study indicated that the chemistry incorporation of nitrogen in the interlayer of both SiO_2 and SBT play a key role in determining electron characteristics of MIS and MFIS structure.

6.3 Memory retention characteristics of capacitance of the MFIS structures

For checking non-volatility of the MFISs, retention characteristics were measured. Fig. 20 shows memory retention characteristics of capacitance of the Pt/SBT/SiO_2/n-Si diodes, which were measured at room temperature. The write pulses of ±6.0V amplitude and 0.1s width were initially applied to the gate, and changes in capacitance versus time were measured. Fig. 20 a) shows retention characteristic of MFIS without treatment, that shows the ON/OFF states can be kept in constant no longer than 3 hours after the write operation in MFIS without treatment. Fig. 20 b) shows retention characteristic of MFIS with using nitrogen treatment SiO_2 and oxygen treatment SBT that shows the ON/OFF states were measured for 7 days after the write operation. Fig. 20 c) shows retention characteristic of MFIS with using nitrogen treatment SiO_2 and SBT that shows the ON/OFF states were measured for 23 days after the write operation.

We believe that the retention is strongly correlated to the magnitude of leakage current density through the stacked gate insulator and ferroelectric layers. The first sample A without the treatment processes exhibited the leakage current larger about 10 times than that of sample B. As we know, nitrogen treatment not only improved surface of SBT but also improved interface layer of SBT and buffer layer. The Ferroelectric SBT films gather many advantage in characteristics over other ferroelectric compounds, for application in ferroelectric memory which include a fatigue-free behavior, good retention properties and low leakage currents [Paz de Araujo, C.A., etc., (1995)]. But they require a high temperature annealing (700°C~800°C) for crystallization that is main cause of constituent- element diffusion from the ferroelectric film into and the insulator layer in Pt/SBT/SiO_2/n-Si MFIS structure [Kim, W. S., (2002), Li, Y.,

(2007)]. Damage of the SiO_2 buffer layer seriously degrades device performance [Kim, W. S., etc., (2002), Li, Y., etc., (2007), Aguilar, G. G., etc., (2006)]. Therefore, Pt/SBT/SiO_2/n-Si MFIS structure could not give good characteristics without any treatment processes and be used for any ferroelectric memory devices.

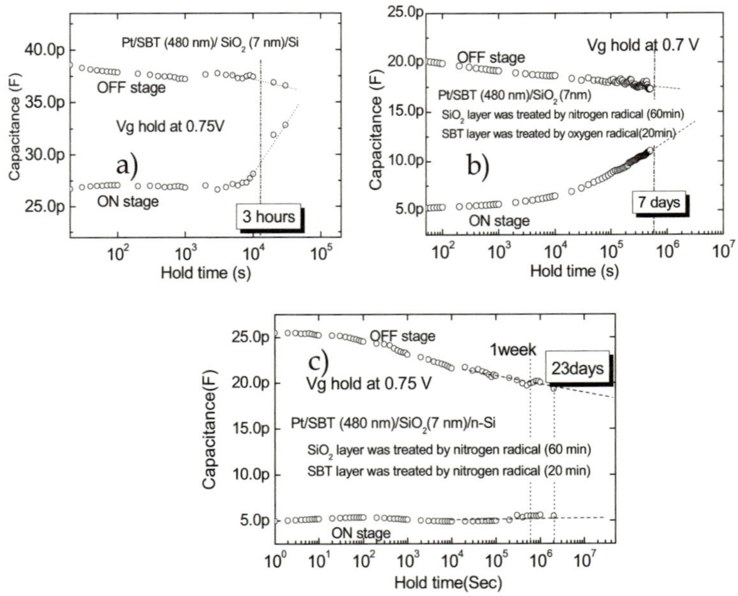

Fig. 20. Retention characteristics of the Pt/SBT/SiO_2/n-Si MFIS structures, a) without using radical irradiation, b) with 60-min nitrogen treatment for SiO_2 and 20-min oxygen treatment for SBT and c) with treatment for SiO_2 and SBT layers by nitrogen radical irradiation for 60 min and 20min, respectively.

With the nitrogen radical treatments the SiO_2 layer was become a stronger barrier to limit the diffusion problem of the Pt/SBT/SiO_2/n-Si structure [Hai, L.V., etc., (2009)]. Damage of SiO_2 by diffusion problems was also suppressed. Both the oxygen radical treatment and nitrogen radical treatment can improve the interface layer of SBT and SiO_2 gate by reducing the interface trap density [Hai, L.V., etc., (2006 a)]. The improvement of C-V hysteresis loop with larger memory window, steep switching were proofs of interface trap density decrease after treatment s.

Although, the oxygen radical irradiation treatment could be effective method to enhance the retention time of Pt/SBT/SiO_2/n-Si MFIS structure, but the nitrogen irradiation treatment is absolutely better. The C-V retention measurement shown in fig. 20 c) revealed that the capacitance ratio of ON/OFF stage does not undergo any significant change after 23 days of measurement. A good retention property of MFIS capacitor indicated 10 year retention times by their extrapolated lines.

7. Conclusions

In summary, we have successfully investigated the characterization of the Pt/SBT/SiO_2/n-Si MFIS structures and demonstrated a novel method to improve their retention properties.

The nitrogen and oxygen radical irradiation treatments applied for the SiO_2 and SBT were analyzed and evaluated about their efficiency and effect on the Pt/SBT/SiO_2/n-Si MFISs' characteristics.

Investigations of treated-SiO_2 and SBT surfaces reveal that nitrogen radicals are incorporated on thin films and modified chemical composition of surface layer. With nitrogen radical treatment $SiON_x$ was formed and is beneficial to the suppression of elemental interdiffusion of the SBT/SiO_2 interface. Chemical bonding of SBT surface layer were by the nitrogen and oxygen radical irradiation treatment modified and become stronger. It is cause of improvement electrical properties of SBT layer.

When nitrogen radical treatment was applied for the SiO_2 and SBT films, the retention and electrical characteristics of Pt/SBT/SiO_2/n-Si was enhanced remarkably. The memory window of C-V hysteresis increased about from 1.3 V to 1.8V when V_g is swept between ± 6V. The leakage current is reduced more than one order of magnitude. The retention characteristic show good behavior for a long retention time, and was measured for 23 day with no significant change.

It is concluded that nitrogen and oxygen radical irradiation treatments contribute to enhance performance of the Pt/SBT/SiO_2/n-Si. Furthermore, These methods can be applied for improving SBT thin films using in various structures of ferroelectric memory devices.

8. References

Paz de Araujo, C.A., Cuchiaro J.D., Mcmillan, L.D., Scott, M.C., Scott, J.F. (1995), Nature 374 (1995) 627.

Riviere, J. C. (1983). Practical Surface Analysis by Auger and X-ray Photoelectron Spectrocopy, John Wiley&Sins, Ltd., chap 2, p.48-53.

Klee, M. and Macken, U. (1996). Integrated Ferroelectric, 12, 11

Park, Y.-B., Lee, J.-K., Jung, H.-J., Park, J.-W.,(1999), Bismuth composition control of $SrBi_2TaNbO_9$ thin films by heat treating Bi_2O_3-inserted heterostructure J. Mater. Res. 14 (1999) 2986.

Tanaka, M., Watanabe, K., Katori, K, Yamoto, H., Yagi, H. (1998). J. Korean Phys. Soc. 32 : S1504.

Hai, L. V., Kanashima, T., Okuyama, M. (2006 a). Improvement of electrical properties and memory retention in MFIS structure by nitrogen radical irradiation of SiO_2 buffer layer, *Proceedings of The 6th Japan-Korea Conference on Ferroelectricity*, Tohoku University, Sendai, Japan.

Hai, L. V., Kanashima, T., Okuyama, M. (2006 b). Improvement of memory retention in metal-ferroelectric-insulator-semiconductor structure by $SrBi_2Ta_2O_9$ surface modification induced by nitrogen and oxygen radical irradiation, Integrated Ferroelectrics, Volume 84, Issue 1, pages 179 –186.

Takahashi, M., Kodama, K., Nakaiso, T., Noda,M. and Okuyama, M.,(2001). Effect of leakage current through ferroelectric and insulator on retention characteristics of metal-ferroelectric-insulator-semiconductor structure, Integrated Ferroelectrics, 40 (2001) 125.

Takahashi, M., Noda, M. and Okuyama, M. (2003). Photoyield and x-ray-photoelectron spectroscopic studies of O_2-annealing effects on $SrBi_2Ta_2O_9$ thin films prepared by pulsed laser deposition, J Appl. Phys. 94, 1912 (2003) .

Hai, L. V., Kanashima, T., Okuyama, M. (2008). Enhancement Of Memory Retention Time Of MFIS With SBT Ferroelectric And SiO2 Buffer Layers Treated By Nitrogen Radical Irradiation, Integrated Ferroelectrics , Vol. 96, Issue 1, 2008, Pages 27-39.

Sakai, S. & Takahashi, M. (2010). Recent Progress of Ferroelectric-Gate Field-Effect Transistors and Applications to Nonvolatile Logic and FeNAND Flash Memory, Materials 2010, 3, 4950-4964.

Ishiwara, H. (2001). Current Status and Prospects of FET-type Ferroelectric Memories, Journal of semiconductor technology and science, vol.1, no. 1, march 2001.

Dawber, M., Rabe, K. M., Scott, J. F.,(2005). Physics of thin-film ferroelectric oxides, reviews of modern physics, volume 77, October 2005.

Hai, L. V., Takahashi, M & Sakai, S. (2010). Fabrication and characterization of sub-0.6-μm ferroelectric-gate field-effect transistors, Semicond. Sci. Technol. 25 (2010) 115013 (5pp).

Tarui Y, Hirai T, Teramoto K, Koike H and Nagashima K (1997).Appl. Surf. Sci. 113 656.

Scott, J. F. (2000). Ferroelectric Memories (Berlin: Springer), chapter 12.

Sakai, S. & Ilangovan, R. (2004). IEEE Electron Device Lett., 25 369 - 71.

Wang. S., Takahashi, M., Li, Q-H, Takeuchi, K. and Sakai, S. (2009). Semicond. Sci. Technol. 24 105029

Hatanaka, T., Yajima, R., Horiuchi, T., Wang, S., Zhang, X., Takahashi, M., Sakai, S. & Takeuchi, K. (2009). *Symp. on VLSI Circuits* (16–18 June 2009, Kyoto, Japan) Dig. of Tech. Papers pp 78 - 79.

Aizawa, K., Park, B. E., Kawashima, Y., Takahashi, K., & Ishiwara, H., (2004). Impact of HfO_2 buffer layers on data retention characteristics of ferroelectric-gate field-effect transistors, Appl. Phys. Lett. 85 (2004) 3199.

Seo, Y. J., Kim, K. C., Kim, H. D., Joo, M. S., An, H. M., Kim, T. G. (2008) Correlation between charge trap distribution and memory characteristics in metal/oxide/nitride /oxide/silicon devices with two different blocking oxides, Al_2O3 and SiO_2, Applied Physics Letters, Aug. 2008,Vol. 93 6,p. 063508 - 063508-3.

Kim, W. S., Yi W., Yu S. G., Heo J., Jeong T., Lee J., Lee, C. S., Kim, J. M.,Jeong, H. J., Shin Y. M., & Lee, Y. H., (2002), Secondary electron emission from magnesium oxide on multiwalled carbon nanotubes, Appl. Phys. Lett. 81, 1098 (2002).

Li, Y., Zhang, S., Fei, W, Sritharan, T., Xu, S., (2007), Nd-substituted $SrBi_2Ta_2O_9$ ferroelectric thin films prepared by radio frequency magnetron sputtering, Thin Solid Films 515 (2007) 8371 - 8375.

Aguilar, G. G., Wu, A., Reis, M. A., Ramos, A. R., Salvado, I. M. M., Alves, E., Costa, M. E. V., (2006), Diffusion processes in seeded and unseeded SBT thin films with varied stoichiometry, Surface Science 600 (2006) 1780–1786

Hai, L. V., Kanashima, T., Okuyama, M. (2009)., Enhancement of Memory Retention Time of Metal/Ferroelectric/Insulator/Semiconductor Structure by Using Fast Annealing and Nitrogen Radical Irradiation, J. of the K. Phys. Soci., Vol. 55, No. 2, August 2009, pp. 884_887.

Youa, I.-K., Leea, W.J., Yanga, II-S.,Yua, B.G., Choa, K.I., Kimb S.H., (2001), Effect of NO(Si_3N_4/SiO_2) layers on the electrical properties of MFIS using SBT ($SrBi_2Ta_2O_9$) materials, Integrated Ferroelectrics: An International Journal, Volume 33, Issue 1 & 4, 2001, Pages 177 - 184.

8

Performance Enhanced Complex Oxide Thin Films for Temperature Stable Tunable Device Applications: A Materials Design and Process Science Prospective

M.W. Cole[1] and S.P. Alpay[2]
[1]*U.S. Army Research Laboratory, Weapons and Materials Research Directorate, Aberdeen Proving Ground,*
[2]*Materials Science and Engineering Program, University of Connecticut, Storrs,*
USA

1. Introduction

The recent growth in the wireless communications area has incurred a large demand for high data rates, broad bandwidth, reliable, and low cost RF tunable microwave devices. Traditional materials, both ferrites and semiconductors, have been exploited for these devices. Unfortunately, semiconductor devices are expensive and have high losses at microwave frequencies while ferrites are not frequency agile and offer slow tuning speeds. Such performance drawbacks have cultivated strong interest in developing new materials, namely perovskite oxide thin films. Recently, barium strontium titanate, $Ba_xSr_{1-x}TiO_3$ (BST), a perovskite oxide solid- solution, has captured the attention of microwave engineers as a candidate material to promote a new generation of passive tunable microwave devices. The intense interest in BST thin films is largely due to the fact that these materials possess the ability to change their dielectric constants as well as their dielectric loss in response to an externally applied field. This feature makes BST thin films ideally suited for electronically tunable microwave devices such as resonators, filters, oscillators and phase shifters (Kalkur et al., 2009). Furthermore BST thin films have several major advantages compared to their conventional ferrite and semiconductor counterparts; they offer fast tuning speeds, low power consumption, and low cost due to affordable fabrication and process science methodologies. Additionally, for microwave applications, the development and implementation of BST materials in thin film form is critical for enabling miniaturization of microwave components and promoting integration with semiconductor microelectronic circuits. Finally, enhanced material performance in concert with scale-up, affordability, integration, and small size are critical metrics for communications systems whether it be a planar phased array antenna system or hand held devices incorporating tunable circuits.

Over the last decade, the growth and process science of uniform composition BST have been investigated intensively for tunable device applications (Weiss et al., 2008, Podpirka et al., 2008); Joshi & Cole, 2000). To achieve optimum device performance, it is critical to fabricate a material with high dielectric tunability, low microwave losses, and a low temperature dependence of the capacitance/ dielectric constants (i.e., good temperature stability) in the device's operational frequency and temperature ranges. Such properties are critical and are required for optimum performance and long-term reliability. The technical literature is laden with experimental and theoretical investigations focused on optimizing uniform composition BST thin film growth, and process science protocols (achieved via doping, thickness variations, buffer layers, stoichiometry, stress modification, annealing procedures, etc.), in order to develop thin films which possess low dielectric loss and high tunability. Although much success has been achieved in optimizing these two material properties, less attention has been devoted to optimizing material temperature stability.

There is significant concern that in practical applications of such tunable devices, in particular, in BST-based phase shifters for electronically scanned antennas (ESAs), the phase shifter performance will be compromised due to the temperature dependence of the device capacitance. The same temperature instability issues are also relevant to tunable filters/preselectos used in small tactical radio systems and handheld communication devices. Specifically, the capacitance of the BST-based device (phase shifter and/or tunable filter) is strongly influenced by temperature changes because the dielectric permittivity (K), or the relative dielectric constant (ε_r)of a single composition paraelectric BST film (e.g., $Ba_{0.5}Sr_{0.5}TiO_3$) follows the Curie-Weiss law,

$$K = C/(T - \theta) \tag{1}$$

where C is the Curie constant, T is the temperature, and θ is the Curie Weiss temperature (Oates et al., 1997). In field applications, communications systems (antenna and/or radio systems) are exposed to a broad range of harsh operational environments, i.e., variable ambient temperatures, and spurious changes in the device capacitance that stem from ambient temperature fluctuations. These will disrupt the phase shifter performance via device-to-device phase shift and/or insertion loss variations, leading to beam pointing errors and ultimately communication disruption and/or failure in the ability to receive and transmit the information. The same is true for BST-based tunable filters where the susceptibility of the capacitance to temperature changes results in the alteration of the band pass window sharpness (window narrows or broadens), or the entire band pass window may shift to higher or lower frequency and/or insertion loss may be degraded. Such poor temperature stability of the capacitance would result in the carrier signal drifting in and out on hot and cold days. Thus, to ensure device performance consistency and reliability, temperature stable devices are essential for the next generation communications systems, ESA's, radios and hand-held communications devices.

This Chapter discusses the temperature stability issues and puts forward a summary of innovative materials designs, and novel process science solutions which serve to help mitigate this temperature sensitivity and render BST thin films more useful and device-relevant for the next-generation RF-microwave devices/systems. Particular emphasis is concentrated on tunable phase shifters and filters to enable phased array antennas, radars and other advanced communications devices. Advances in wireless communications applications are highly dependent upon improvements in microwave materials in concert with achieving balanced property-optimization. Thus, the critical review put forward in this Chapter holds promises to provide the foundation and spawn new materials research solutions to further enable the development of BST thin films for applications in microwave device arena.

2. Temperature instability of BST materials

It is well known that for single crystal and polycrystalline bulk BST ceramics, the dielectric permittivities are strongly temperature dependent, with a sharp dielectric anomaly at the ferroelectric to paraelectric phase transition, *Tc* (Lemanov et al., 1996). Moreover, due to the functional dependence of the dielectric permittivity with temperature, the resonant frequency of a fabricated microwave device also becomes strongly temperature dependent resulting in carrier signal drift in an ambient surrounding (Cava, 2001). The temperature dependent drift of resonant frequency poses serious problems in using bulk ceramic BST for practical device applications, that must be addressed. Initially, BST in thin film form was considered as a first order solution towards realizing temperature stability. Compared to bulk ceramic BST, thin film BST (of the same composition) does not possess such a pronounced dielectric anomaly (Fig 1). For example, Fig. 1 compares the temperature

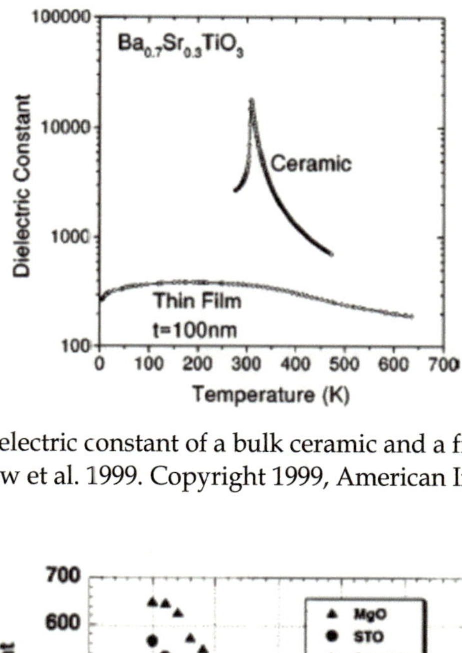

Fig. 1. Variation of the dielectric constant of a bulk ceramic and a film as a function of temperature. [From: Shaw et al. 1999. Copyright 1999, American Institute of Physics.]

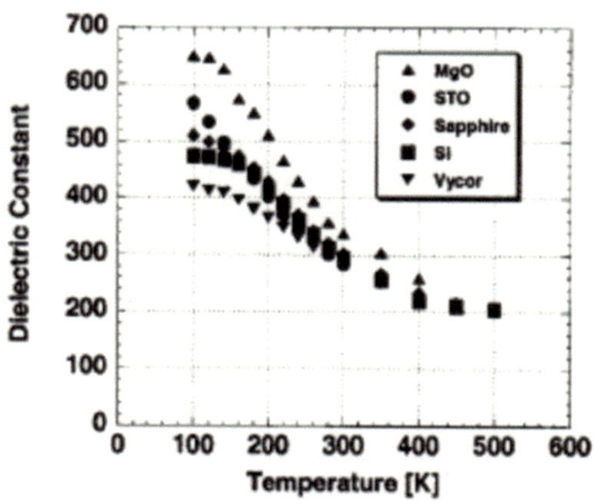

Fig. 2. Dielectric constant as a function of temperature for BST/Pt/substrate structures. From: Taylor et al., 2002. Copyright 2002, American Institute of Physics.]

dependent dielectric constant of a bulk ceramic $Ba_{0.7}Sr_{0.3}TiO_3$/BST70/30 to that of a thin film of the same composition (Shaw et al., 1999). For the BST thin film, not only is the dielectric constant much lower, it also does not have a sharp peak as a function of temperature. This broad dielectric anomaly, indicative of a diffuse phase transition has been attributed to the finer grain sizes, residual strains, composition heterogeneities inherent to synthesis (Kim et al., 2000, Zhang et al., 2010, Mantese et al., 1995). This observed flattening of the dielectric-temperature peak in thin film BST with respect to that of bulk ceramic BST has led many to incorrectly conclude that BST in thin film form is temperature stable. Unfortunately, this is not the case. When compared to bulk ceramics, thin film BST exhibits less temperature sensitivity, (i.e., it has a smaller temperature coefficient of capacitance/TCC). However, it is not temperature stable. Fig. 2 illustrates this temperature instability whereby Taylor and co-workers (Taylor et al.,2002) experimentally explored the dielectric response as a function of temperature for five BST75/25 thin films on a variety of substrates with different thermal expansion coefficients (TECs). This work confirms that there is indeed a temperature dependent dielectric response for thin film BST. However, BST films grown on substrates with smaller TECs (i.e., larger tensile in-plane thermal strain) display a reduced dielectric permittivity and a smaller (although still quite pronounced), temperature dependence of the dielectric response. Therefore, the capacitance of any device based on such a film would be highly temperature dependent, making its use difficult to accommodate in circuit design. Thus, it is important to compensate for the temperature coefficient of the dielectric constant (TCK) and the commensurate TCC. The challenge here is to accomplish this without degrading the other device critical properties, i.e., without decreasing the tunability or increasing the dielectric loss. This notable temperature dependence of the dielectric response is a potential point of concern for the utilization of BST thin film in microwave devices. As such, solutions, whether via engineering or material design must be critically reviewed and considered.

3. Traditional temperature stability solutions

Traditional approaches to address the issue of device (phase shifter and/or tunable filter) temperature instability have focused on employing hermetic or robust packaging, where the package serves to protect the tunable device from the harsh environmental extremes. Although this approach is successful, hermetic/robust packaging would add significant cost, size, and weight to both ESA and radio systems. Other concepts to achieve temperature stability compliance involve the use of system heat sinks and/or cooling apparatuses such as mini-fans, temperature compensation circuits, and/or mini-ovens. Such thermal management solutions may be utilized with ESAs or radios; however, they will add extra weight, size, and cost to the overall system and, as such, are deemed unacceptable. Temperature compensation can also be achieved using either a curve fit or a look-up table approach. The curve fit methodology centers on the formulation of a temperature dependent mathematical expression, which represents the drift of each BST tunable device. A microprocessor utilizes this equation and the ambient temperature data (obtained from a thermocouple mounted on the printed circuit board) to calculate the tuning voltage. The look-up table approach, as its name implies, involves using a look-up table. In order to obtain the relevant coefficients, the phase shifter/filter characteristics must be measured at discrete temperatures. Then the BST bias voltage is manually adjusted to maintain the phase shifter/filter specifications. In the worst-case scenario, one would have to obtain a set of points for each temperature (i.e., 23°C, 24°C etc). Typically, one would expect to have a small subset of temperature/bias points for each bias line. The exact number of points is, of course, dependent on the BST devices, the other phase shifter/filter components, and the phase/filter topology. Unfortunately, both the curve fit and look-up table approaches are quite complex as there is usually not a one-size-fits-all solution. The calibrations are also labor and time intensive and are useful if only a limited number of ESAs, radio, and communication devices are to be fielded.

In contrast to the above described engineering methodologies, there are also viable novel materials science approaches. Conventional materials science methodology for reducing the temperature dependence of an active material involve the selection of the temperature interval of operation well above the temperature corresponding to the active material's permittivity maximum. Unfortunately, this approach results in reduced material tunability and the TCC is still too high for practical military/commercial communication system applications. More useful materials science methods for achieving material/device temperature stability are based on utilization of artificial structures which generally involve the synthesis of BST multilayers or compositionally graded BST structures. Such BST heterostructures were shown to possess unique and desirable dielectric properties, i.e., a low dependence of capacitance on temperature, high permittivities, and high tunabilities (Zhu et al., 2003, Lu et al., 2003, Tian et al., 2003, Zhang et al., 2006). Although these experimental and theoretical studies have produced very promising results, most of the work focused on compositionally graded/multilayer BST films fabricated by techniques that are non-industry standard such as pulsed laser deposition (PLD). Additionally, many of these graded films were deposited on ceramic small-area expensive substrates, utilized "designer" nonstandard electrodes or asymmetric electrodes, and employed high annealing temperatures which are not compliant with conventional silicon integrated circuit (IC) processing protocols. Specifically, the use of small-area ceramic substrates and designer electrodes is not practical from a scale-up, manufacture, and affordability point of view and high annealing temperatures would deteriorate the quality of the films due to the strong diffusion between films and substrates. In the case of the metal-insulator-metal (MIM) design, heating the film above 800 °C would damage the structure of the bottom electrode which will degrade the dielectric loss, leakage characteristics, and tunability of the device. Furthermore, most of the published results in the relevant literature are incomplete in that there is a lack of systematic experimental data which determine and compare the dielectric properties (loss, tunability, and permittivity) to those of uniform composition BST prepared using the same fabrication technique and post-deposition anneal process protocol. Additionally, there are relatively few investigations which evaluate the temperature dependence of dielectric response at microwave (MW) frequencies. Nonetheless, these studies contain important ideas and methodologies for temperature compensation and it is important to summarize these results and populate a materials data base so that future work can benefit from this knowledge and perhaps spawn innovative industry standard and frequency relevant materials solutions to resolve the temperature stability dilemma.

4. Temperature stability via materials solutions

Since the concept of compositionally graded materials was originally proposed for reducing the thermal stresses associated with dissimilar materials research in this area has been greatly expanded from structural materials to functional materials and ultimately to thin film electronic materials. Since the late 1990's extensive interest in frequency agile materials for electronics has been cultivated within the materials science community. This interest has encouraged materials research to address performance issues, via hybrid material designs including multilayer and compositionally graded perovskite oxide thin film materials (e.g., BST and other ferroelectrics). It should be noted that compositionally graded or multilayer ferroelectric thin films exhibit properties not previously observed in conventional monolithic ferroelectric materials. There have been many noteworthy publications which have provided experimental data concentrated on the dielectric performance of tunable devices using graded/multilayer BST thin films. The common thread of this body of literature is focused on the thesis that it is necessary to grow BST thin films with a low temperature dependence of dielectric constant to realize temperature stable/reliable tunable device applications. Furthermore, such results have introduced a variety of functionally graded/multilayer material designs which serve to demonstrate that the low temperature coefficient of the dielectric response can be achieved by producing BST films with graded compositions through the deposition of layers of BST with different Ba/Sr ratios. In order to critically evaluate this technical area and to hone in on a practical materials solution to mitigate the temperature sensitivity of BST in frequency agile communications devices, we summarize here the relevant work, highlighting the technical concepts, results, and solution shortfalls.

4.1 Summary of the relevant literature: non-standard process science

It is well established that the capacitance of any ferroelectric/BST-based device will be highly temperature dependent, making its use difficult to accommodate in circuit design. It is therefore important to compensate for the temperature coefficient of the dielectric constant without decreasing the dielectric tunability and/or increasing the dielectric loss. To maintain such a trade-off, a simple and effective way is to prepare compositionally graded BST films by depositing successive layers with different Ba/Sr ratios. Thus, the graded films are essentially composed of multiphase compositions with different phase transition temperatures, where a high dielectric constant and a relatively smaller TCC and/or TCK values are maintained, making such graded materials highly preferable in applications where the capacitance of a device should be only weakly dependent on temperature.

For microwave device applications, epitaxially grown films are often utilized because of their high dielectric constants and large tunabilities (Lin et al., 2008). Since the pulsed laser deposition (PLD) method of film fabrication is particularly adapted to the growth of stoichiometric epitaxial films with single and multiphase compositions many studies utilize this method for growth of BST graded films. While PLD is a wonderful research tool to study and demonstrate complex film compositions (e.g., compositionally graded/multilayer BST films) it must be kept in mind that this growth technique is not scalable or industry standard for foundry friendly device manufacture (Cole et al., 2002b, Cole et al., 2005). Nevertheless, for research demonstrations, the PLD method of preparing compositionally graded/multilayer BST films has provided some very useful results. In particular Zhu et al., (2003) have grown epitaxial compositionally graded BST with $0.0 \leq x \leq 0.25$ thin films on (100) $LaAlO_3$ (LAO) substrates using $La_{0.5}Sr_{0.5}CoO_3$ (LSCO) bottom electrodes. This study was focused on 2 distinct compositionally graded structures: (1) Down-Graded Films: films with with Ba/Sr ratio varied from $BaTiO_3$ (BTO) at the film/substrate interface to BST 75/25 at the film surface and (2) Up-Graded Films: films with Ba/Sr ratio varied from BST75/25 at the film/substrate interface to BTO at the film surface. The temperature dependence of the dielectric properties measured between -20°C and +130°C as a function of frequency between 1 kHz and 1 MHz is displayed in Fig. 3 which shows that that the dielectric constants and the dissipation factors vary only slightly with frequency. The values of ε_r and the loss tangent (tan δ) at 100 kHz were 380 and 0.013 for the up-graded films, and 650 and 0.010 for the down-graded films, respectively. It can be seen that dielectric behavior was enhanced in the down-graded BST film. The enhanced dielectric behavior observed in the down-graded films was attributed the fact that since the bottom layer (BTO) is under compressive strain in the $b-c$ plane and elongated along the a-axis; such compressive strain would increase the ionic displacement and promote the polarization of electric dipoles along the electric field parallel to the a-axis. Therefore, the enhanced dielectric behavior (larger dielectric response) observed in the down-graded films could be ascribed to the pure BTO layer in the down-graded BST films not only serving as a bottom layer but also acting as an excellent seeding layer to enhance the subsequent film growth, resulting in higher film crystallinity and larger grain sizes in the down-graded films.

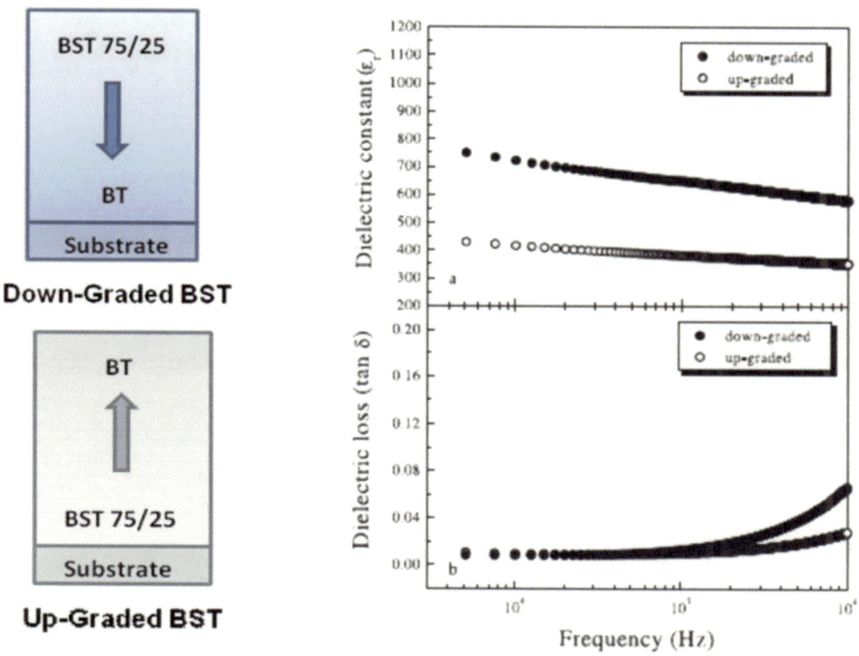

Fig. 3. Schematic drawings of the up-graded and down-graded materials designs (left side). Dielectric constant and dissipation factor of the up-graded and down-graded BST films as a function of measured frequency (right-side). [From Zhu et al. 2003. Copyright 2000.]

The temperature dependence of the dielectric properties of the down-graded BST films is shown in Fig. 4. Fig. 4(a) shows the broadened and flattened transition peak (tetragonal-cubic phase transition) to be at about 90°C. The authors suggest that the flat profile of the dielectric constant versus temperature for the down-graded BST thin films is explained by the presence of compositional and/or residual strain gradients in the epitaxial graded film. The local variation of the composition and/or the strain leads to local changes in the permittivity, which broadens the peak in the dielectric constant as a function of temperature.

Furthermore, similar to relaxor ferroelectrics, the Curie maximum temperature *Tc* shifts towards higher temperatures with increasing frequency. Fig. 4(b) compares TCC of down-graded BST with homogeneous BTO. This plot demonstrates that T_C of the monolithic BTO is about 120°C, and that such an isotropic thin film exhibits a narrow plateau region around the dielectric constant maximum compared with the down-graded films, whereas the graded capacitor has a low temperature coefficient for the capacitance and high dielectric permittivity for a wide range of temperature. This work demonstrates that the broad-flat profile of the dielectric constant versus temperature relationship (low TCC) observed in the graded films is paramount and is useful for building dielectric thin-film capacitors having a capacitance which has a low temperature dependence over a broad temperature regime.

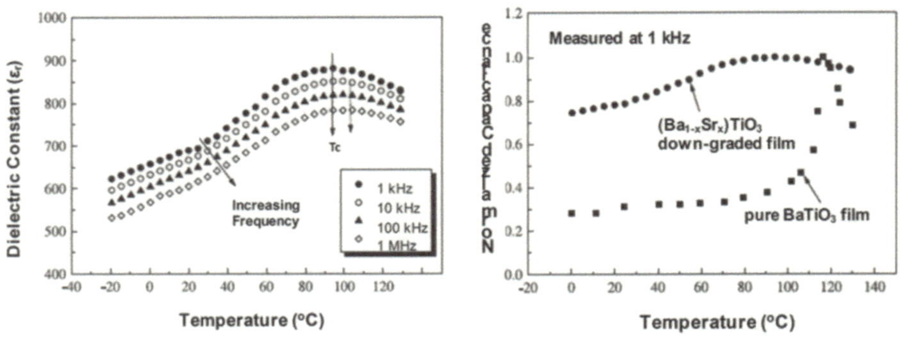

(a) (b)

Fig. 4. Temperature dependence of the dielectric response for the down-graded BST films; (a) The dielectric constant vs. temperature as a function of frequency. A comparison of the temperature dependence of the capacitance between the down-graded BST film and pure $BaTiO_3$ film deposited in the same system is shown in (b). [From: Zhu et al. 2003. Copyright, 2000.]

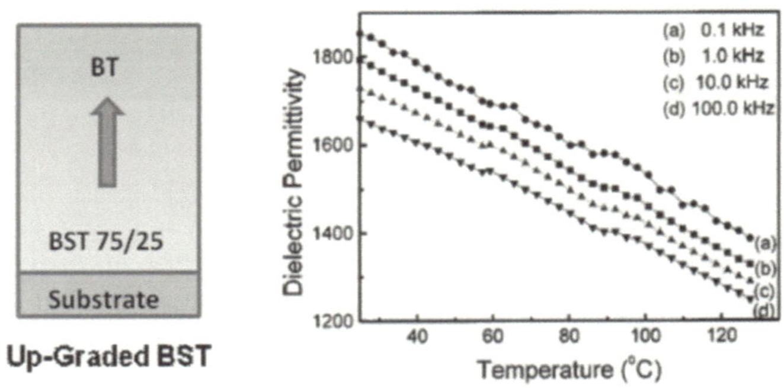

Fig. 5. Temperature relationship of dielectric permittivity for PLD fabricated upgraded BST films on $LaAlO_3$ (001) substrates with $(La_{0.7}Sr_{0.3})MnO_3$ (LSMO) bottom electrodes. [From Lu et al., 2003. Copyright 2003.American Institute of Physics.]

Similar work to promote temperature stable BST films was performed by Lu et al., (2003). However, in this study the authors focused on the up-graded BST thin film structure i.e.,

LAO substrate – $(La_{0.7}Sr_{0.3})MnO_3$ bottom electrode – BST 75/25 – BST 80/20 – BST 90/10 – BTO In this case, considering that the gradient in the film is in an ideal situation, and neglecting the effect of interlayer interactions at the interfaces between the layers, the whole film can be regarded as an ensemble containing a graded distribution of Ba^{2+} and Sr^{2+} ions normal to the growth surface. For this up-graded structure, when the temperature decreases from a temperature higher than 120°C, the phase transition will occur from the top layer to the bottom layer one by one. Because of the induction of polarization, the "layer" with a higher permittivity will dominate the permittivity–temperature characteristics (Jeon et al., 2001).

The temperature dependence of permittivity over the range of 25-130°C and for frequencies of 0.1, 1, 10 and 100 kHz is displayed in Fig. 5. Similar to the work of Zhu et al., (2003) (down- graded BST films), the films exhibit a broad and flat variation profile of the dielectric constant over a wide range of temperature, where there was no Curie–Weiss law behavior observed. However, two broad anomalous peaks were seen, indicating some change associated with the broadened phase transition process. These two peaks occurred at about 59°C and 92°C, near the phase transition temperatures of BST 80/20 (57°C) and BST 90/10 (89°C), respectively. It was suggested that these two broadened peaks might be the result of incomplete diffusion of Ba^{2+} and Sr^{2+} ions, resulting in the trace of BST 90/10 and BST 80/20 compositions within the graded film structure. In addition, over the same temperature intervals (25 - 130°C) Lu et al., (2003) up-graded films of possessed enhanced dielectric constant (~1650 - 1250/upgraded films at 100 kHz/negative slope) with respect to Zhu and co-workers (Zhu et al., (2003) down-graded films (690-850/down-grade at 100 kHz/positive slope). While the absolute values of the permittivities differed between these two studies (Zhu et al., 2003, and Lu et al., 2003) the per cent change in permittivity as the temperature was elevated from 20 to 120°C was similar, i.e. the permittivity of Lu et al., (2003) up-graded film decreased by 25% and Zhu et al., (2003) down-graded films' permittivity increased by 23 % over this temperature interval. The loss tangents of the two films displayed similar trends, i.e., broadened peaks over the entire temperature range from 20-130°C; however, the room temperature dielectric loss was much higher for Lu et al., (2003) 0.05 vs. that of Zhu et al., (2003) 0.010-0.013. Furthermore, the high dielectric permittivity and broad temperature characteristics were obtained in conjunction with a very high tunability of over 70% (at 1 MHz,) which is higher than that of monolithic BST 75/25 with the same compositions and applied fields (~60%). Such results suggest that compositionally graded BST films sustain temperature stability with enhanced performance of high tunability and low loss.

Xia et al., (2006) investigated doping BST to modulate T_C and achieved temperature independent tunability over a broad temperature range. A method to vary the T_C is by chemical substitution in such BST films. Sun et al., (2004) reported that the doping of PbO to BST 70/30 thin films derived by sol-gel method enhanced T_C. Moreover, it is also known that adding more kinds of cations to the *A* site in the ABO_3 perovskite structure usually results in a diffuse phase transition, which can broaden the transition peak and induce low temperature dependence. In order to investigate the temperature stability of the enhanced tunable properties of BST, Xia and co-authors (Xia et al., 2006) leveraged these concepts and fabricated PLD BST films with Ba partially substituted by Pb, i.e., $(Ba_{0.25}Pb_{0.25})Sr_{0.5}TiO_3$ (BPST) on Pt/Si substrates. For comparison BST50/50 films with the same thickness and processing protocols as the BPST films were produced. This study shows that the dielectric constant of BPST exhibits a maximum at the temperature of 7°C, whereas, in the case of BST, the maximum occurs at −79°C. The temperature corresponding to this peak of dielectric constant is considered to be T_C. As is known, BST is a continuous solid solution between BTO and $SrTiO_3$ (STO). $PbTiO_3$ (PTO) has a higher T_C (485°C) than that of BTO (120°C). That is why with the addition of PTO into the BST system, the T_C of the solid solution BPST could be adjusted to shift to higher temperatures.

It is well known that varying the Ba/Sr ratio can also result in a T_C around room temperature (RT) such as in BST 75/25, but it also produces a sharp dielectric peak and a commensurate strong temperature dependence. For BPST films, the addition of Pb cations to the *A* site in perovskite structure, gives rise to an enhanced relaxor behavior and a diffuse phase transition. This study observed that there was a significant broadening of the transition peak with respect to that of the monolithic BST50/50, and that there was a notable increase of peak temperature with increasing frequency from 100 kHz to 1 MHz. This is indicative of a typical relaxor behavior that is not present in the undoped BST50/50 films. It is well known that the relaxor behavior is related to the heterogeneity of the nanoscale composition. Thus, adding Pb ions into the *A* site of perovskite structure of BST system could be responsible for the relaxor behavior of BPST films. Specifically, the regions with *A* site rich in Pb^{2+} would display some PTO-like behavior and become polar at RT, whereas other regions could have a tendency to remain non-polarized since BST with $0.25<x<1$ is indeed paraelectric at RT. The existence of nanopolar regions in a paraelectric environment results in the distribution of T_C in BPST films, giving rise to the broadening of the transition peak and low temperature-dependent dielectric constant. This result is similar to the performance of the graded multilayer BST films reported by Lu et al., (2003) and Zhu and co-workers (Zhu et al., 2003, Zhu et al., 2002a). Evaluation of the 1 MHz ε -V characteristics determined that the tunability of BPST is about 60% at the applied voltage of 5 V (160 kV/cm), which is much larger than the value of 34% for monolithic BST films. Furthermore, such enhanced tunability may also be related to the existence of nanopolar regions in BPST films, since such regions may induce ferroelectricity in the paraelectric matrix due to the internal electrostatic fields that are formed because of the polarization mismatch across the interfaces. The analysis of the temperature dependence of the tunability at 5 V indicates that in the temperature ranging from 25 to 100°C, the BPST films still exhibits relatively high tunability from 60% to 54%. In addition, it was speculated that this wide variation of tunability with regard to temperature could be attributed to the broadened dielectric constant peak caused by the diffuse phase transition. Since the dielectric constant decreases slowly with the increasing temperature, a relatively stable tunability can be obtained. Thus, this study demonstrated that the addition of Pb to BST resulted in a relaxor behavior and a diffuse phase transition, which broadened the dielectric peak. Hence, the tunability of BPST films displayed good stability against the temperature. Although this temperature stability of high tunability bodes well for tunable devices made of BPST films which would be operated at environment with different temperatures, the authors did not investigate the behavior of dielectric loss with respect to the Pb addition and/or temperature stability. In addition, no microwave (MW) frequency analyses were presented. However, as a first approximation, these results show promise toward providing a feasible and simple approach for realizing simultaneously high tunability and good temperature stability of BST based thin films.

4.2 Summary of the relevant literature: industry standard process science

Although these experimental studies highlighted above show very promising results, most of the studies focused on compositional graded BST films fabricated by techniques that are non-industry standard such as pulsed laser deposition (PLD). Additionally, many of these graded films were deposited on ceramic small-area expensive substrates (LAO), utilized "designer" non-standard electrodes or asymmetric electrodes, and employed high annealing temperatures which are not compliant with conventional silicon IC processing protocols.

There has been only a limited quantity of published studies focused on graded BST films whereby the film fabrication method employed was either a sol-gel or metal organic solution deposition (MOSD) method, both of which are industry standard growth techniques. As such, these are scalable and thus promote affordable manufacture/scale-up. The work of Tian et al., (2003) focused on fabricating BST thin films on Si substrates with artificial gradients in composition normal to the growth surface via sol-gel deposition. Specifically, the graded BST thin films were composed of BST 70/30 - BST 80/20 - BST 90/10 layers which were annealed via a special 4-step heating procedure to form the compositional gradient from BST 70/30 to BST 90/10, similar to the up-graded structure devised by Lu et al., (2003) and Zhu et al., (2003) for PLD films, differing slightly in the strength of the composition gradient (a slightly lower strength). In addition, Tian et al., (2003) up-graded structure was unintentionally capped (artifact of film processing) at the top and bottom by narrow surface layers which were off-stoichiometric compositions (Fig. 6). Auger electron spectroscopy (AES) depth profiling determined that composition gradients of Ba and Sr were gradual and uniform without obvious interfaces between the different layers, i.e. no step variation in Ba and Sr concentration. For comparison, the authors also prepared BST thin films with uniform composition, e.g., BST 70/30, employing the same processing method. The low frequency (10-100 kHz) dielectric properties of the graded BST thin films and the uniform composition BST films (BST 70/30, BST 80/20, BST 90/10) were measured in the metal-insulator-metal (MIM) capacitor configuration (Fig. 7). The maximum of the capacitance was 271 pF (corresponding to a dielectric constant about 300) and loss factor around 0.02 for the graded BST thin film. The tunability of the dielectric constant was calculated in terms of dielectric constant (capacitance) variation rate: $\Delta C/C_0$, where ΔC is the change in capacitance relative to zero bias capacitance C_0. The tunability of the graded BST thin film was ~ 35% while the tunabilities of the uniform composition BST films were around 23–25% obtained at an applied electric field of 240 kV cm^{-1} (Fig. 7). This tunability value of 35 % is significantly lower than that achieved for up-graded BST films by Lu et al., (2003) (~70 %). Although both films were in an up-graded material design, the broader compositional strength (BST75/25 - BTO vs. BST 70/30 - BST90/10; i.e., 25% broader compositional strength) of Lu and co-workers study (Lu et al., 2003) may be the reason for the higher tunability with respect to that of Tian et al., (2003). The enhanced tunabilities for the graded BST films suggest that these films are promising candidates as tunable microwave elements and integrated capacitors. Unfortunately this work did not report results related to the temperature dependence of the dielectric response, hence there no conclusions regarding temperature stability of these films can be inferred.

Zhang and co-workers (Zhang et al., 2006) also investigated compositionally up-graded BST thin films fabricated on Pt/Si substrates by the sol-gel technique. However, in this work the compositional gradient was broadened compared to the studies of Tian et al., (2003), Lu et al. (2003), and Zhu et al., (2003) to span from BST 60/40 to BTO. The results demonstrated the dielectric constant and dielectric loss measured at 200 kHz to be 335 and 0.045,

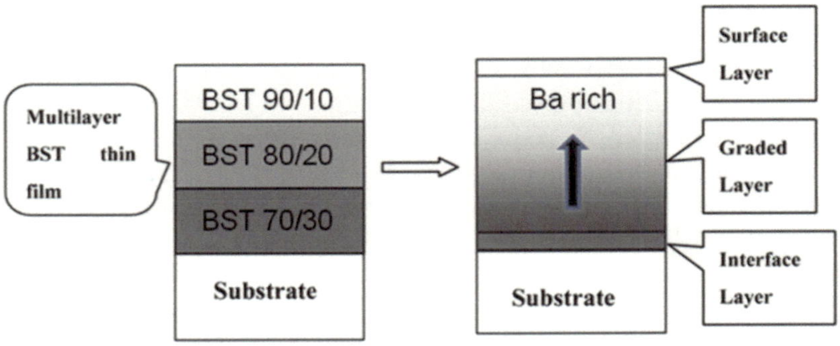

Fig. 6. Schematic of the BST graded thin film. The as-deposited thin film structure (left) and configuration model after heat treatment (right). [Modified Diagram from Tian et al., 2003. Copyright 2003, Elsevier.]

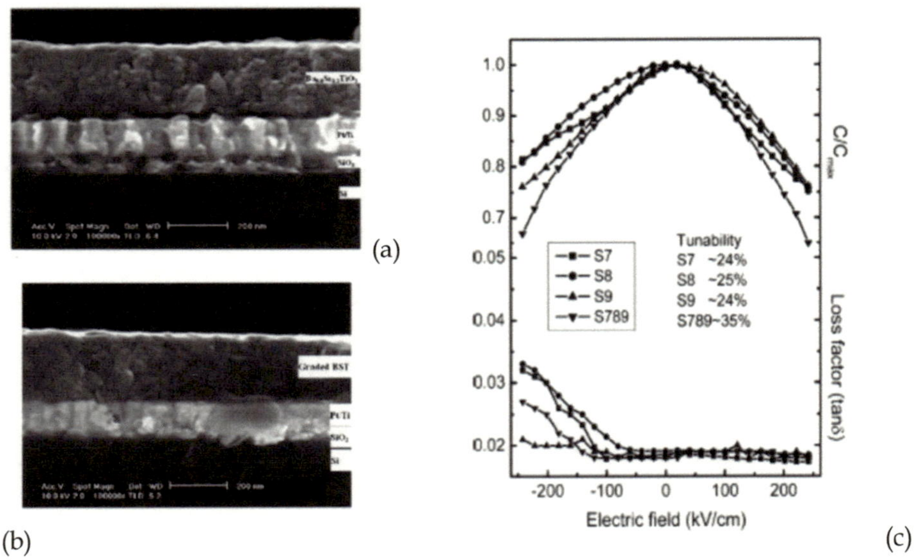

Fig. 7. SEM images of the (a) uniform composition and (b) graded BST films. The field dependence of capacitance (dielectric constant) and dielectric loss at room temperature (100 kHz) is shown in (c). [From Tian et al., 2003. Copyright 2003, Elsevier.]

respectively, and a tunability of 42.3% at an applied field of 250 kV/cm. Unfortunately, this study like that of Tian et al., (2003) did not report the temperature dependence of dielectric response data. The research group of Wang et al., (2008) published an interesting compositionally graded BST study whereby they combined a broad compositional gradient (BST 60/40 to BTO) with a yttria stabilized zirconia (YSZ) buffer layer sandwiched between the film and the Pt bottom electrode on Si substrates. The compositionally graded films were prepared on Pt/Si substrates with and without a buffer layer of YSZ via the sol-gel technique. The results showed that the YSZ buffer layer served to improve the dielectric properties of the graded films, i.e., lower losses (tan δ = 0.026 vs. 0.065), enhanced permittivity (ε_r=339 vs. 318) and improved leakage current characteristics (3.1 x 10^{-9} A/cm^2 vs. 1.7 x10^{-6} A/cm^2 at 4V). The improved materials properties of the graded films with the YSZ buffer layer was attributed to the fact that the buffer layer serves as an excellent seeding layer which enhanced BST growth and mitigated the interfacial dead layer, i.e., the parasitic, in the MIM capacitor structure. As with the studies discussed above there was no data reporting the temperature dependence of the dielectric response.

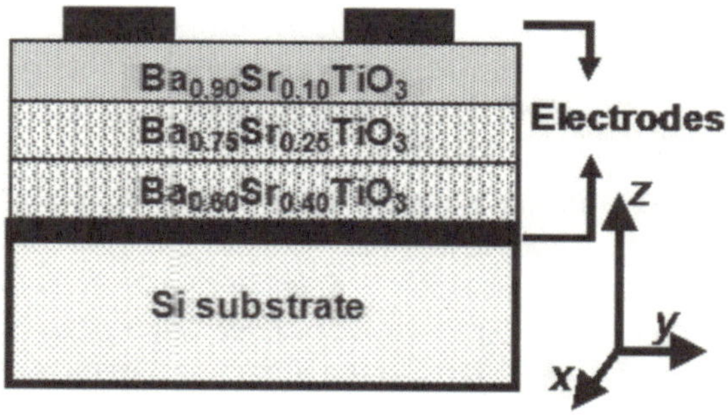

Fig. 8. Schematic diagram of the thin film multilayered material design.[From Cole et al., 2007, Copyright 2007. American Institute of Physics.]

Recently, the collaborative research activities between the U.S. Army Research Laboratory (ARL) and the University of Connecticut (UConn) has resulted in a series of publications focused on the dielectric response and temperature stability of MOSD fabricated compositionally stratified/multilayered BST (BST60/40 - BST75/25 - BST90/10) thin films on Pt-Si substrates (Cole et al., 2007, Zhong et al., 2007, Cole et al., 2008a, Cole et al., 2008b). This series of papers evaluated the dielectric response for both RT material/device performance as well as temperature stability at 100 kHz and microwave (10 GHz) frequencies. The results of this research determined that the effect of the of BST compositional layering, i.e., the "*quasi-up-graded*" film design (Fig. 8) was significant in terms of materials performance with respect to that of uniform composition BST 60/40 prepared by the same MOSD film fabrication method. Specifically, at 100 kHz the multilayer film exhibited a higher permittivity (ε_r=360) and lower dissipation factor (tan δ=0.012) with respect to that of the uniform composition BST film (ε_r=176, tan δ=0.024). The higher permittivity is expected in the multilayered BST films due to the presence of the ferroelectric BST 75/25 and BST 90/10 layers, which possess higher permittivities (and higher tunabilities) with respect to uniform composition paraelectric BST60/40. The dielectric response can be further improved by electrostatic interactions between layers (Roytburd et al., 2005). It has been reported and experimentally demonstrated that permittivity, hence tunability, for single layer uniform composition BST increases with increasing Ba/Sr content (Lu et al., 2003). As discussed earlier in this section, others have reported elevated dielectric constants, 426 to 1650, for compositionally graded BST films prepared via PLD (Lu et al., 2003, Zhu et al., 2003, Zhu et al., 2002a, Zhu et al., 2002b, Zhu et al., 2002c) and 300–320 for graded films prepared by sol-gel methods (Jain et al., 2003, Tian et al., 2003, Zhang et al., 2006). Unfortunately, the literature reports for dielectric loss of compositionally modified (graded and/or layered) films are not as consistent as the reported values for permittivity. Low frequency (100 kHz - 1 MHz) RT dielectric losses have been reported to be as low as 0.007 (Zhu et al., 2002a) for PLD compositionally graded BST (BTO - BST 90/10 - BST 80/20

– BST 75/25) films on MgO substrates to extremely high values of 0.05 (Zhang et al., 2006) for compositionally graded BST sol-gel films (BST 60/40 – BST 70/30 – BST 80/20 – BST 90/10 – $BaTiO_3$) on Pt/Si substrates. However, several research groups have reported dielectric losses which cluster between values of ~ 0.02–0.03 (Lu et al., 2003, Tain et al., 2003, Zhu et al., 2002c). The dielectric loss value, tan δ =0.012, which the ARL-UConn group obtained for their compositionally multilayered BST thin film, was quite low with respect to the reported literature values for graded BST films. Additionally, the fact that these multilayered films also possessed lower loss than the monolithic BST60/40 films prepared by the same method on the same substrate is quite surprising. The reasons for this are not well understood, however, the authors speculated that the lower loss of the compositionally multilayered BST films with respect to both uniform and continuously graded composition BST films may have resulted from the fact that the defects within the film were no longer mobile and were trapped at the compositionally distinct interfaces to compensate for the polarization difference between the layers (Roytburd et al., 2005, Zhong et al., 2006). Such defect trapping at the compositional interfaces may have immobilized the defects such that they were not able to reach the electrodes, thereby allowing the film to possess an improved dielectric loss over films without this interface trapping mechanism, i.e., films which are compositionally uniform and/or continuously graded. Thus, interlayer interfaces appear to promote enhanced material dielectric properties.

The ARL-UConn's researchers also presented experimental data focused on the temperature dependence of the dielectric response for both their quasi-up-graded multilayered BST film and for a homogenous uniform composition BST60/40 film fabricated via the same processing method (Fig. 9). Figure 9(a) demonstrates that the permittivity and dissipation factor of the multilayered film exhibited minimal dispersion as a function of temperature ranging from 90 to −10°C. Lu et al., (2003) also reported a flat variation of dielectric permittivity for compositionally upgraded (BST 75/25 – BST 80/20 – BST 90/10) PLD deposited BST films over a wide temperature range (20−130°C), although there are two peaks at 59 and 92°C, near the phase transition temperatures of BST 80/20 (59°C) and BST 90/10 (89°C), respectively. However, the ARL-UConn results did not show did not display peaks corresponding to the (bulk) phase transition temperatures of the constituents of the multilayered film. Compared to the monolithic paraelectric BST60/40 films (Fig. 9b), the multilayered film possesses a broader, more flat, diffuse dielectric response as a function of temperature. Specifically, the effective change in TCC with respect to uniform BST is significant; i.e., from 20 to 90°C, TCC was lowered by 70% with respect to the uniform composition BST corresponding to an ~3:1 change. Similarly from 20 to −10 °C the TCC value of the multilayered BST film was lowered by 15% with respect to the uniform composition BST film. From Fig. 9 it is also important to note that the dielectric loss exhibits negligible dispersion over the measured temperature range. This is important, since in phase shifter devices the loss must be consistent or predictable to ensure antenna performance consistency and reliability with respect to variable temperature.

In order fully appreciate the BST material property temperature stability enhancements achieved by the ARL-UConn group and that of other researchers, it is important to compare the literature values for both uniform composition BST and compositionally stratified BST thin films. Table I summarizes literature derived dielectric properties, evaluated at 20°C, for monolithic BST films fabricated via sputtering (Taylor et al., 2002), PLD (Lookman et al., 2004), and sol-gel techniques (Jain et al., 2003). In addition, reported literature results for proposed temperature stable material designs, namely, compositionally graded BST of Lu et al., (2003), BST/MgO layered film structures Jain et al., 2003), are also listed in Table I. The TCC was calculated for the literature studies using the published plots of the capacitance C_F (or ε_r) versus temperature from −10 to 20°C via the equation:

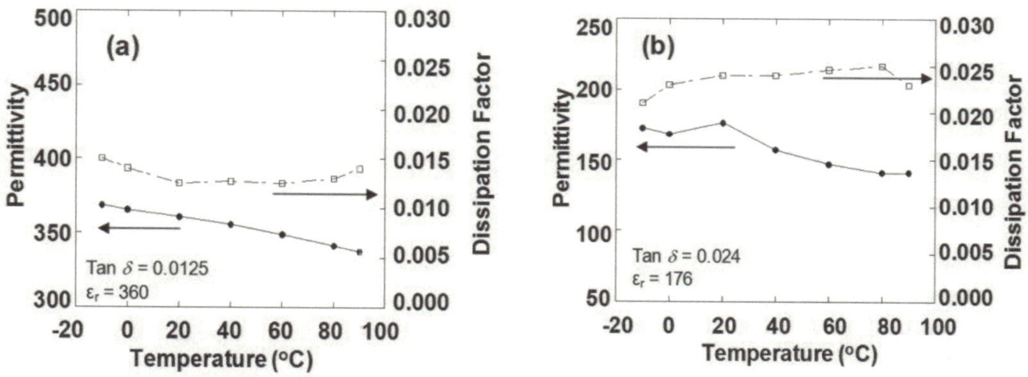

Fig. 9. The temperature dependence of the dielectric response for (a) the post deposition annealed multilayered BST film and (b) homogenous uniform composition paraelectric (BST60/40) BST film measured at 1 MHz. [From Cole et al., 2007. Copyright 2007, American Institute of Physics.]

$$TCC = \Delta C/(C_0 \Delta T), \tag{2}$$

where ΔC is the change in capacitance with respect to reference capacitance C_0 at 20°C and ΔT is the change in temperature relative to 20°C. TCC values reported in Table 1 show that the magnitude of the TCC for homogenous uniform composition BST thin films is quite high, which is indicative of the large temperature dependence of capacitance/permittivity for these films. Specifically, the decrease in the permittivity as the temperature is raised from 20 to 90°C for paraelectric uniform BST films ranged from 20% to 60% (Taylor et al., 2002). For these same films, the permittivity increase ranged from 6.2% (Taylor et al., 2002) to 35% (Jain et al., 2003) as the temperature was lowered from 20 to −10°C. Thus, the largest change in C_F/ε_r occurred for the high end of the temperature spectrum. The uniform composition BST film, fabricated in the ARL-UConn study, possessed a much lower TCC value than the other literature results (Lu et al., 2003, Lookman et al., 2004, Jain et al., 2003). Table 1 also displays the TCC results from three heterogeneous material designs: a thin film multilayered BST/MgO heterostructure (Jain et al., 2003) and two compositionally graded BST thin film material designs (Lu et al., 2003, Zhu et al., 2002a) that attempt to lower the TCC, i.e., to broaden the dielectric anomaly over a wide temperature range. Compared to uniform composition BST, the heterogeneous material designs were quite successful in lowering the TCC for over the temperature range of −10 to 90°C. The dielectric response of such compositionally graded ferroelectrics as a function of temperature exhibits characteristics of a diffuse phase transformation, which is reflected by a low TCC value, and is inherently linked, with the distribution of the phase transformation temperature resulting from the composition gradient across the ferroelectric (Ban et al., 2003a, Ban et al., 2003b). It is important to note that the TCC results, of the ARL-UConn's BST multilayered compositional design, was significantly improved over the three heterogeneous material designs represented in Table I. Specifically, this heterostructure exhibited a 6.4% decrease in permittivity as the temperature was elevated from 20 to 90°C and only a 2.1 increase in permittivity as the temperature was lowered from 20 to −10°C. This very small change in permittivity and low TCC value suggest that the compositional multilayered BST design is temperature insensitive over the temperature range of −10 to 90°C.

Composition Substrate/deposition Method	ε_r (20° C)	Tanδ (20°C)	$TCC_{20\text{-}90}$ (ppt/°C)	$\Delta\ \varepsilon_r$ (20-90 %)	$TCC_{20(-10)}$ (ppt/°C)	$\Delta\ \varepsilon_r$ (20-(-10)%)
$Ba_{0.50}Sr_{0.50}TiO_3$ LSCO/MgO(PLD-210 nm) (Taylor et al., 2003)	410	0.021	-7.32	51 dec	-4.1	12.2 inc
Ba0.24Sr0.76TiO3 PtSi (Sputer-100 nm) (Pervez et al. 2004)	320	--	-4.0	20 dec	-2.1	6.2 inc
Ba0.50Sr0.50TiO LAO (Sol-gel-600 nm) (Lookman et al., 2004)	2934	0.01	-8.52	60 dec	-11.52	35 inc
$Ba_{0.60}Sr_{0.40}TiO_3$ PtSi (MOSD-240nm) (Cole et al., 2007)	176	0.024	-2.9	20 dec	0.83	3 dec
$Ba_{0.50}Sr_{0.50}TiO_3$/MgO-layer LAO (Sol-gel-600 nm) (Lookman et al., 2004)	1932	0.005	-6.89	52 dec	-14.3	43 inc
UG (75/25-80/20-90/10-BT)LSMO/LAO (PLD-800 nm) (Lu et al., 2003)	1650	0.225	-2.16	15 dec	--	--
DG (BT-90/10-80/20-75/25) MgO (PLD-450nm) (Zhu et al., 2002a)	475	--	6.66	44.7 inc	2.25	25 inc
UG (60/40-75/25-90/10) PtSi (MOSD-270 nm) (Cole et al., 2007)	360	0.122	-0921	6.4 dec	-0.716	2.1 inc
UG (60/40-75/25-90/10)+Mg 5 mol%, PtSi (MOSD-270 nm) (Cole et al., 2008a)	316	0.018	-0.94	6.6 dec	-1.14	3.4 inc

Table 1. A comparison of the dielectric properties, TCC, and the percent change of permittivity with respect to 20 °C for heterogenous and uniform composition BST thin films. (Note: UG, up-graded; DG, down-graded; dec, decrease; inc,, increase.) [Modified table from Cole et al., 2007. Copyright 2007, American Institute. of Physics.]

In order to understand why the data for the ARL-UConn group was improved over that of Lu et al,. (2003) and Zhu et al., (2002a) is explained by theoretical models based on a thermodynamic analysis of graded ferroelectric materials and multilayered heterostructures

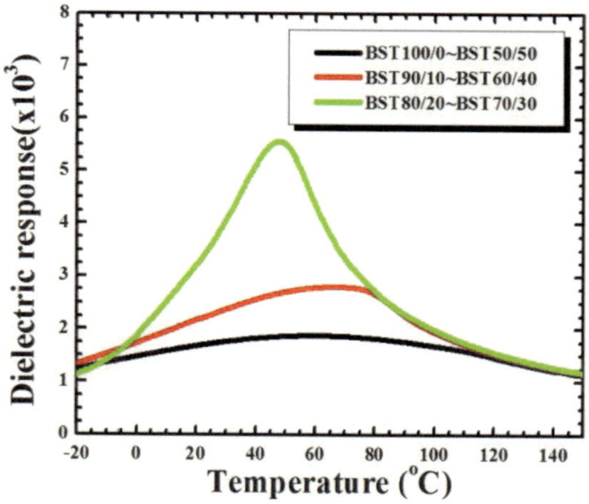

Fig. 10. The theoretical average dielectric response as a function of temperature for three compositionally graded $Ba_xSr_{1-x}TiO_3$ systems with the same nominal average composition. [From Cole et al., 2007. Copyright 2007, American Institute of Physics.]

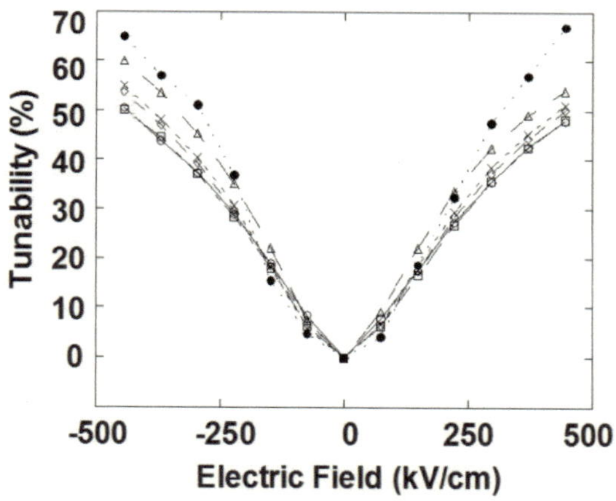

Fig. 11. The temperature dependence of the dielectric tunability for the multilayered BST film from 90 to −10 °C. The symbols on the plot represent the following temperatures: 90 °C (open circles), 80 °C (open squares), 60 °C (open diamonds), 40 °C (crosses), 20 °C (filled circles), and −10 °C (open triangles). [From Cole et al., 2007. Copyright 2007, American Institute of Physics].

(Ban et al., 2003a, Ban et al., 2003b). Very briefly, this formalism considers a single-crystal compositionally graded ferroelectric bar. It basically integrates free energies of individual layers, taking into consideration the energy due to the polarization (spontaneous and induced), electrostatic coupling between layers due to the polarization difference, and the elastic interaction between layers that make up the graded heterostructure. The mechanical interaction arises from the electrostrictive coupling between the polarization and the self-strain and consists of two components: the biaxial elastic energy due to the variation of the self-strain along the thickness and the energy associated with the bending of the ferroelectric due to the inhomogeneous elastic deformation. Based on this approach, the temperature dependence of average dielectric response of compositional BST with the same nominal composition (BST75/25) can be calculated using average thermodynamic expansion coefficients and elastic constants available in the literature (Mitsui et al., 1981), as shown in Fig. 10. In comparison with a sharp peak of the dielectric permittivity at T_C for bulk homogenous ferroelectrics, a diffused dielectric response with the temperature can be expected for compositionally graded ferroelectrics as a result of the polarization grading and interlayer interactions. It should be noted that this model is developed for bulk compositionally graded ferroelectrics. However, it is possible to extend it to thin films by incorporating the internal stresses due to thermal strains as well as the clamping effect of the substrate (Roytburd et al., 2005). While these factors tend to decrease the overall dielectric response compared to bulk graded structures, the temperature dependence of the dielectric permittivity displays the same trend (Ban & Alpay, 2003). The maximum in the dielectric permittivity is broadened over a wide range of temperature depending on the strength of the composition gradient, as shown in Fig.10. A steeper composition gradient will give rise to a broader maximum. Thus, since the ARL-UConn multilayered compositional design BST (BST 60/40 -BST 75/25 - BST 90/10) has a steeper compositional gradient compared to that of Lu et al., 2003 (BST 75/25 - BST 80/20 - BST 90/10) and (Zhu et al., 2002a, Zhu et al., 2003) (BST 90/10 - BST 80/20 -BST 75/25) based on these theoretical results, one would expect the ARL-UConn multilayered film to possess a flatter/broader dielectric anomaly, hence a lower TCC, with respect to that of Lu et al., (2003) and Zhu et al., (2002a).

The ARL-UConn researchers also evaluated the temperature dependence of the dielectric tunability for their multilayered BST film (Fig. 11). From Fig. 11 it is clear that over the temperature range of −10 to 90°C, the tunability was not significantly degraded. The bias tunability trends are temperature independent; however, the absolute value of tunability is slightly modified. Thus, this multilayered BST design will allow the antenna phase shift to be temperature stable over the ambient temperature range of −10 to 90°C. This result is significant, as microwave voltage tunable phase shifter devices are expected to be operated in environments with different ambient temperatures with excellent reliability and accuracy. The fact that this multilayered BST material design possesses outstanding dielectric properties and that both tunability and dielectric loss are stable over a broad temperature range bodes well for its utilization in the next generation temperature stable microwave telecommunication devices.

Although excellent temperature stability results have been achieved via compositional grading of BST, there is still need to further reduce the dielectric loss of these new materials. It is well known that acceptor doping of BST is an excellent method to reduce dielectric loss. It has been shown that losses in BST can be reduced via acceptor doping. Dopants (such as Ni^{2+}, Al^{2+},Ga^{3+}, $Mn^{2+,3+}$, $Fe^{2+,3+}$, Mg^{2+}, etc.) typically occupy the *B* site of the ABO_3 perovskite structure, substituting for Ti^{4+} ions. The charge difference between the dopant and Ti^{4+} can effectively compensate for oxygen vacancies and thereby have been shown to decrease dielectric losses (Cole et al., 2001). Thus, it is well known that doping of BST with Mg is an excellent avenue to reduce dielectric losses in monolithic BST films especially with low Sr content, although the addition of MgO causes a reduction in dielectric response and its tunability (Cole et al., 2003). Dielectric constant, loss tangent, and tunability (at 237 kV/cm) of BST 60/40 and 5 mol % MgO doped BST 60/40 thin films were reported as 720, 0.1, and 28% and 334, 0.007, and 17.2%, respectively. Thus, acceptor doping combined with compositional grading of BST presents an intriguing opportunity to develop new materials for tunable device applications with stringent demands focused on low dielectric losses and temperature insensitivity, while still maintaining moderate/good tunabilities. The ARL-UConn researchers extended this idea to a proof-of-concept study. Specifically, they leveraged prior work on Mg-doped BST(Cole et al., 2000, Cole et al., 2002a) to reduce the dielectric loss and blended this acceptor doping approach with their MOSD fabricated quasi-upgraded compositional multilayer design (BST60/40 - BST 70/30 - BST 90/10 - Pt/Si). Both Mg-doped (5 mol%) and undoped quasi-up graded BST films were fabricated via MOSD technique on Pt/Si substrates (Cole et al., 2008a).

The temperature dependence of dielectric constant and loss tangent of thin films are shown in Fig. 12. At a constant temperature, a higher dielectric constant was measured for both doped and undoped multilayered thin films than the uniform BST60/40 film. This can be attributed mostly to the BST 75/25 layer for which T_C is close to RT and, thus, has a significantly higher dielectric response in monolithic form. It is evident that the dielectric constant was somewhat lowered upon MgO doping which also resulted in a decrease in T_C (Cole et al., 2007). These findings, together with the volumetric expansion with the addition of MgO to BST in the ARL-Uconn films, seem to suggest an effective suppression of ferroelectricity with increased MgO doping due to the substitution of Ti (the displacement of which results in a permanent dipole and, thus, ferroelectric behavior) in the perovskite lattice with Mg cations. TCC was evaluated as the variation of capacitance with temperature relative to the capacitance value at 20°C. Both MgO-doped and undoped multilayered BST films exhibited a lower dielectric dispersion in the range of −10 to 90°C than monolithic BST 60/40 thin films. As the temperature was elevated from 20 to 90°C, 6.6% (TCC=−0.94 ppt/°C), 6.4% (TCC=−0.92 ppt/°C), and 13% (TCC=−1.8 ppt/ °C) decrease in permittivity was observed for doped multilayered, undoped multilayered, and monolithic BST films, respectively. In the case of lowering temperature from 20 to −10°C, 3.4% (TCC=1.14 ppt/°C), 2% (TCC =0.67 ppt/°C), and 4.5% (TCC=1.5 ppt/°C) increase in permittivity were noticed for doped multilayered, undoped multilayered, and monolithic BST films, respectively. Additionally, dielectric loss tangent of MgO-doped films was the lowest one among the samples produced in Cole et al. (2008a). From Fig. 12, it can be seen that, on the average, dielectric loss tangents were 0.009, 0.013, and 0.024 for doped multilayered, undoped multilayered, and uniform BST 60/40 thin films, respectively. This fairly "flat," i.e., temperature insensitive, and low loss tangent makes it feasible for such MgO doped multilayered BST films to be employed in tunable devices operating over a broad temperature range.

The variation of tunability in MgO-doped BST films at various temperatures is given in Fig. 13. A slight increase in tunability was observed with increasing temperature. At low electric field strengths (~250 kV/cm), dispersion in tunability with temperature was quite negligible. However, the tunability of doped multilayered films was lower than both undoped and uniform BST thin films reported earlier by Cole et al., (2007). For example at RT and at an electric field strength of 444 kV/cm, tunability was measured as 65.5%, 42%, and 29% for undoped upgraded, uniform, and doped upgraded BST films, respectively. The results achieved in this body of work are important, as the tailoring of BST material design and composition (grading and Mg-doping) is a promising tool to achieve desired material properties. However, it is important to marry this materials performance with the proper/specific tunable device applications. In other words, actual selection and implementation of a specific materials design (Mg-doped vs. undoped graded or uniform composition BST) must be considered in terms of system requirements. For example, for phase shifters, one would require a large tunability. In this case, undoped multilayered or compositionally graded BST films would be an appropriate choice. On the other hand, for frequency-agile filters operating in the microwave regime, low dielectric losses are a premium. Therefore, acceptor doping combined with compositional grading would yield significantly better loss properties with a reasonable dielectric tunability. It is important to state that such a materials design, Mg-doped quasi-up-graded multilayer BST films, are promising materials for tunable device applications which advocate stringent demands of reduced dielectric loss and temperature stability while maintaining moderate tunability.

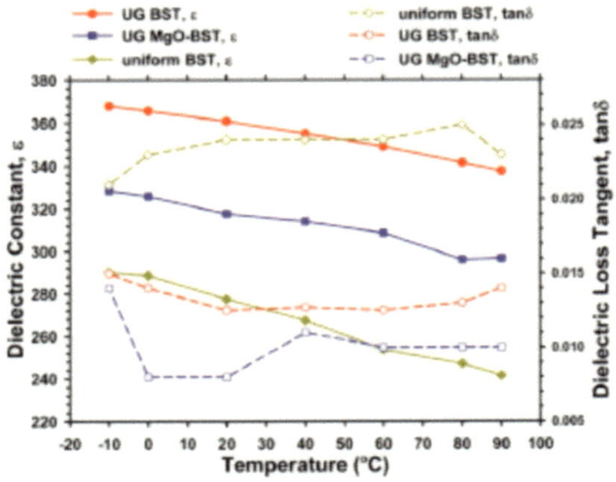

Fig. 12. Temperature dependence of dielectric constant and dielectric loss tangent of MgO-doped multilayered, undoped multilayered, and uniform BST films. [From Cole et al., 2008a. Copyright 2008. American Institute of Physics.]

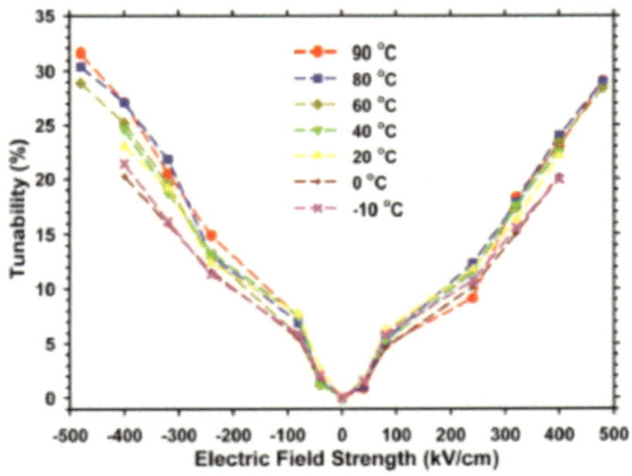

Fig. 13. Variation of tunability of MgO-doped multilayered BST thin film at various temperatures. [From Cole et al., 2008a. Copyright 2008, American Institute of Physics.]

4.3 Summary of the relevant literature: microwave frequency studies

The research summary presented above has discussed the dielectric response/temperature dependence results only within the low frequency (<300 MHz) domain. Since tunable devices for telecommunications are operated in the microwave range (300 MHz to 300 GHz), it is important to evaluate the dielectric properties of these compositionally stratified BST thin films materials at higher frequencies. Unfortunately, there are relatively few published results that have considered microwave characterization of these complex BST thin film materials designs. One of the more comprehensive studies that focuses on the microwave performance of up- and down-graded BST films is that of Lee et al., (2003). Similar to the low frequency studies of Zhu et al., (2003) and Lu et al. (2003) the films were fabricated via PLD; however, the support substrate was MgO (not LAO) and the strength of the compositional gradient was extremely steep. Specifically, compositionally graded BST (Ba_xSr_{1-x})TiO_3 (x=0, 0.2, 0.4, 0.6, 0.8, and 1.0) films were deposited in both the up-graded (STO - BTO) and down-graded (BTO - STO) configurations. The microwave performance (8 to 12 GHz) of the graded BST thin films were investigated with coplanar waveguide (CPW) meander-line phase shifters as a function of the direction of the composition gradient at RT.

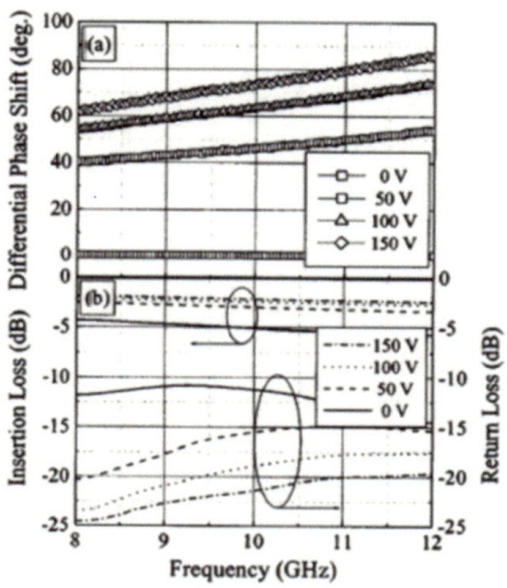

Fig. 14. (a) Differential phase shift and (b) s-parameters of the phase shifter using the graded BTO/STO film. [From Lee et al., 2003. Copyright 2003, American Inst. of Physics.] 2003, American Institute of Physics.]

Fig. 14 shows the measured microwave properties of the CPW meander-line phase shifter based on the down-graded (BTO - STO) thin film. The results in Fig. 14(a) show that as the frequency increased from 8 to 12 GHz, the differential phase shift (at all dc bias values evaluated) also increased. A phase shift of 73° was obtained at 10 GHz with a dc bias of 150 V. Fig. 14 (b) shows the insertion loss (S_{21}) and return loss (S_{22}) as a function of frequency and applied bias voltages. The insertion loss (S_{21}) decreased with an increasing frequency and improved with bias voltage, which is a typical trend of ferroelectric CPW phase shifters. The measured insertion loss at 10 GHz ranged from 5.0 to -2.1 dB with 0 and 150 V, respectively. The return loss (S_{22}) was less than -11 dB over all phase states. The figure of merit of a phase shifter is defined by the differential phase shift divided by the maximum insertion loss for a zero voltage state, which was 14.6 °/dB at 10 GHz. Similar microwave characterization was performed on the up-graded (STO to BTO) BST film (Fig. 15). In this case the differential phase shift was much lower than that of the down-grade BST film, i.e., 22° at 10 GHz with a dc bias of 150 V. The insertion loss (S_{21}) measured at 10 GHz ranged from -2.2 to -1.7 with 0 and 150 V, respectively. The return loss (S_{22}) of the phase shifter was less than -21 dB with good impedance matching over all phase states. The figure of merit at 10 GHz and 150 V was about 10 °/dB.

The differential phase shift for the graded films was analysed as a function of applied dc bias voltages up to 150 V at 10 GHz. The measured differential phase shifts were 73° and 22° for the down-graded and upgraded BST film, respectively. The down-graded BTO - STO film showed a larger phase tuning and insertion loss than the up-graded STO - BTO thin film. Thus the microwave response is strongly related to the direction of the composition gradient of the graded BST thin films. The down-graded materials design has larger phase tuning and higher insertion loss with respect to the up-graded film. However, in terms of figure of merit (FOM =phase shift/insertion loss) at 10 GHz, the up-graded film has the best overall microwave performance (14.6°/dB down-graded vs. 10°/dB up-graded).

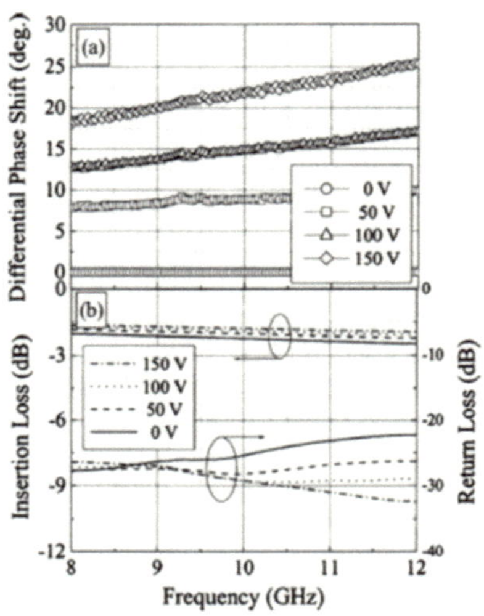

Fig. 15. (a) Differential phase shift and (b) s parameters of the phase shifter with graded BTO/STO film. [From Lee et al., 2003. Copyright 2003, American Inst. of Physics.]

The ARL-UConn group has also contributed to the body of knowledge focused on microwave performance of compositionally graded BST films. Specifically, the dielectric properties of their Mg-doped and undoped quasi-up-graded multilayer BST heterostructures at GHz frequencies were reported whereby they achieved high dielectric tunability (15%–25% at 1778 kV/cm) and low losses (0.04–0.08) (Cole et al., 2008b). The microwave characterization of both BST materials designs were carried out at frequencies ranging from 0.5 to 10 GHz using a coplanar inter-digitated capacitor (IDC) device configuration.

Fig. 16 displays a plot showing the microwave dielectric loss as a function of applied electric field at 0.5, 5, and 10 GHz for the up-graded and the Mg-doped up-graded BST films. As expected, the frequency increases, the loss increases. It should be noted that at each frequency the loss is lower at each frequency for the Mg doped up-graded vs. the undoped up-graded film. For example, the loss at 10 GHz in the undoped film is 0.078 compared to 0.039 in the Mg-doped heterostructure at the same frequency. While both values are significantly larger than the loss at 100 kHz (0.008) (Cole et al., 2008a), these still are within acceptable tolerances for tunable devices. The increase in the dielectric losses in the microwave frequency range can be due to a number of reasons, both of intrinsic (due to the interaction of the ac field phonons, including quasi-Debye losses) and of extrinsic (e.g., mobile charged defects, such as oxygen vacancies) nature.

Fig. 17 shows the dielectric tunability as a function of the applied electric field at 0.5, 5, and 10 GHz for the same two samples. In the undoped up-graded BST, the tunability displays little frequency dependence and is ~25% at 1778 kV/cm for all the test frequencies. In the Mg-doped films, the tunability at 1778 kV/cm decreases from 23% at 0.5 GHz to 15% at 10 GHz. This was also observed at 100 kHz in identical samples; 65% vs. 29% at 444 kV/cm for graded and Mg doped graded films, respectively (Cole et al., 2008a). This reduction in tunability for the Mg-doped up-graded BST film was accompanied by a significant reduction in the dielectric response, (i.e., permittivity). For example, at 10 GHz, the dielectric response of up-graded BST was 261, whereas it was 189 in Mg-doped BST. This is expected as Mg additions are known to lower the ferroelectric transformation temperature, as discussed above. Furthermore, a smaller grain size might also lower the dielectric response (Potrepka et al., 2006).

A comparison of the MW tunability results to that of the 100 kHz performance shows that there is a notable decline in dielectric tunability at the GHz frequencies (Cole et al., 2008a). This behavior may not be entirely intrinsic. It is well known that one can expect a precipitous fall in the dielectric response (and hence its tunability) at higher frequencies for materials where the significant portion of the polarization is due to ionic displacements and/or molecular rearrangement in the presence of an external field. However, the decrease in the tunability noted in comparing the GHz and 100 kHz ARL-UConn studies may also be related to completely different device geometries. The low frequency measurements were acquired in the MIM device configuration, while the GHz measurements were obtained in a co-planar IDC device configuration. Since the device geometry is coplanar, the tunability that is reported for GHz frequencies is actually the lower limit since only a portion (typically less than 50%) of the field is confined within the film (Acikel, 2002). In other words, MIM/parallel plate varactor structures offer higher tunability compared to the coplanar IDC structures since the electric fields are fully confined within the film, as compared to IDCs where there is a large fringing field in the air.

A practical approach to obtaining temperature stabilization of BST varactors was proposed by Gevorgian et al., (2001). The fundamental concept centers on a capacitor which is composed of two ferroelectrics with different Curie temperatures. One of the ferroelectrics is in a paraelectric phase, while and the other is in the ferroelectric state in the temperature interval between T_1 and T_2 (Fig. 18). In the temperature interval between the peaks, the permittivity of the ferroelectric phase increases with increasing temperature, while the permittivity of the paraelectric phase decreases. In a capacitor, the two thin film materials are "connected in parallel;" hence, the decreased permittivity of the paraelectric phase is compensated by the increased permittivity of ferroelectric phase. This concept was experimentally validated using a co-planar capacitor/varactor composed of PLD fabricated epitaxial BST 25/75 and BST 70/25 thin films inter-layered with a MgO seed and a MgO barrier layer (Fig. 19). Here, the lower "seed" layer serves as a strain mitigator and the middle MgO layer serves as a diffusion barrier to ensure that the two ferroelectric layers do not form intermediate phases via diffusion during synthesis.

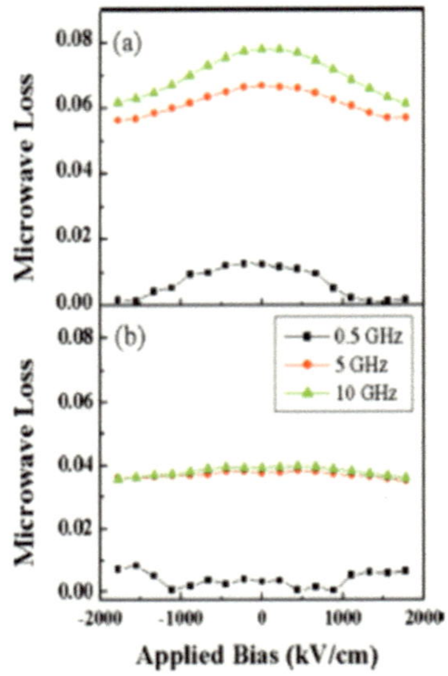

Fig. 16. Microwave loss as a function of the applied bias at 0.5, 5, and 10 GHz for (a) undoped UG-BST and (b) Mg-doped UG-BST. [From Cole et al., 2008b. Copyright 2008 American Institute of Physics.]

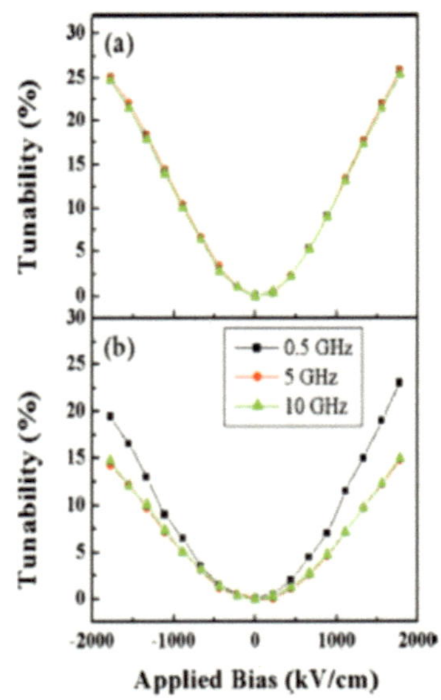

Fig. 17. High frequency tunability as a function of applied bias at 0.5, 5, and 10 GHz for (a) undoped UG-BST and (b) Mg-doped UG-BST. [From Cole et al., 2008b. Copyright 2008, American Institute of Physics.]

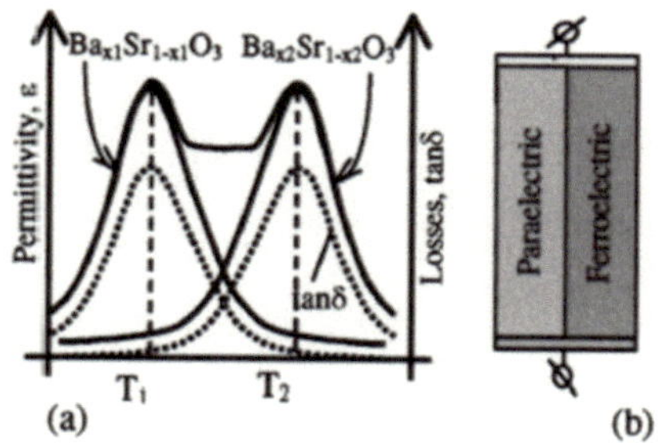

Fig. 18. Temperature dependencies of permittivity and loss (a) and a capacitor (b) with ferroelectrics connected "in parallel". [From Gevorgian et al., 2001. Copyright 2001, Ameer. Inst. of Physics.]

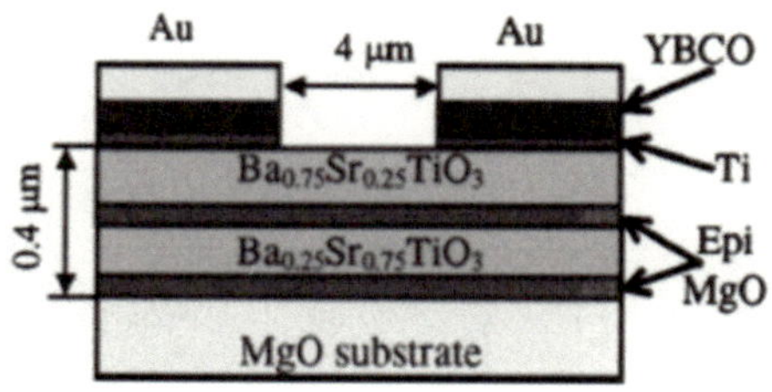

Fig. 19. Cross section of the varactor. Top layer $Ba_{0.75}Sr_{0.25}TiO_3$:0.2mm, bottom layer Ba0.25Sr0.75TiO3 :0.2mm, middle MgO. [From Gevorgian et al., 2001. Copyright 2001, American Institute of Physics.]

The RT frequency dependence of the loss tangent and capacitance was evaluated and the results are displayed in Fig 21. Two relaxation frequencies were observed at 2.15 and 4.61 GHz. The authors suggested that the mechanism for the relaxation may be associated with the interfaces (f_r <1.0 GHz) of BST 25/75 and BST 75/25 films, including electrodes (Sayer et al., 1992). Aside from these relaxation anomalies; it should be noted that tan δ is quite high (~0.1 at 10 GHz). The temperature dependencies of the capacitance and the Q factor ($Q=1/(r\dot{\omega}C=1/$ tan δ) of the varactor at 1 MHz is shown in Fig. 21. The capacitance is almost independent of temperature in a rather wide temperature interval (120 -300 K). The TCC is less than 2 x10^{-4} in the temperature range 150–250 K, which is comparable with the TCC of commercial non-tunable capacitors. However, it is noteworthy to mention that at temperatures above 300K the capacitance is no longer temperature independent and increases dramatically which is a major drawback of this material design. On a positive note, due to the overlapping "tails" of the temperature dependencies of the permittivities of the top and bottom ferroelectric films, the tunability of such a varactor is expected to be larger than if the varactor was composed of only uniform composition BST 25/75 or BST 75/25 films. Although the quality factor of the varactor is highest over same temperature interval where the capacitance is temperature stable, the 1MHz Q-value is somewhat low, Q ~36,/ tan δ~0.028, (Q~ 10/ tan δ~0.1 at 10 GHz) with respect to the that obtained for the graded films and multilayer quasi graded films (Cole et al., 2008b).

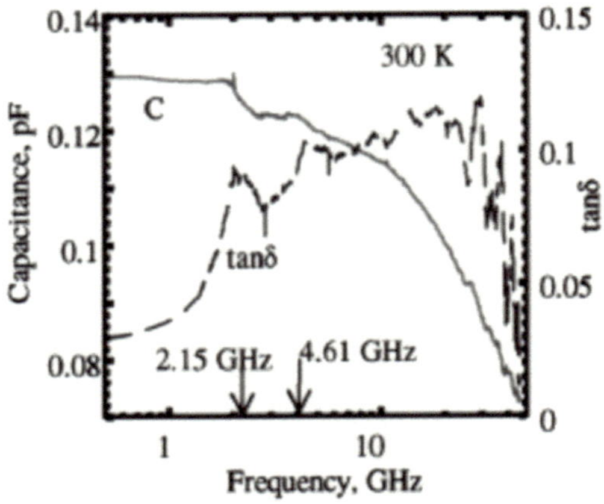

Fig. 20. Frequency dependence of the capacitance and losses at zero dc bias. [From Gevorgian et al., 2001. Copyright 2001, American Institute of Physics.]

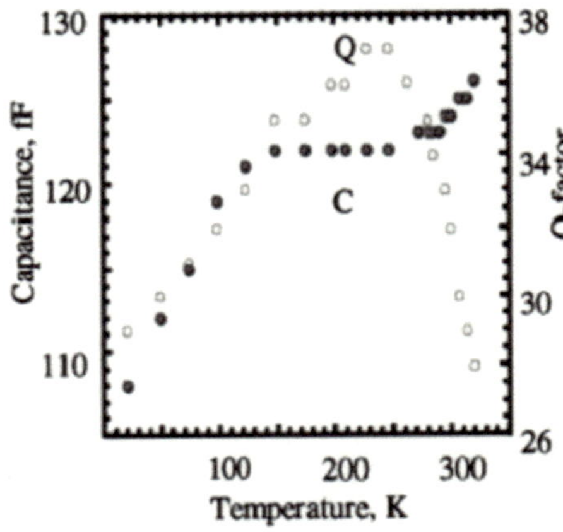

Fig. 21. Experimental dependencies of capacitance and quality factor at 1.0 MHz. [From Gevorgian et al., 2001. Copyright 2001, American Institute of Physics.]

5. Conclusions

This Chapter presented a critical review of the relevant literature pertaining to optimization of BST-based thin film for temperature stable tunable RF-devices. Although, traditional engineering solutions serve to promote material/device temperature stability these add significant cost, size, and weight and/or violate the affordability criteria associated with the systems requirements. Hence, as an alternative to the engineering solutions the material design, approach for eliminating temperature sensitivity was summarized and discussed. Novel material designs, via compositional layering and grading were shown to be effective for achieving dielectric properties with low temperature dependence over a broad temperature regime. While the proof-of-concept of such material designs appears promising, further optimizations are still required to effectively insert these novel designs into practical device applications. In particular there is a need to continue material research solutions with emphasis on systematic studies employing industry standard film growth techniques and process science protocols, large area low cost device relevant substrates, and industry standard electrode metallizations. Additionally, if such novel material designs are to be useful and device-relevant for the next-generation RF-microwave devices/systems, it is critical to evaluate these material designs at microwave frequencies and operational environments. Finally, the temperature stability criteria must be attained considering the trade-offs of material property balance. In other words, materials temperature stability must be accomplished in concert with achieving balanced property-optimization, i.e., high tunability, low dielectric loss and reduced leakage characteristics.

6. Acknowledgements

SPA gratefully acknowledges financial support from the U.S. Army Research Office through Grants W911NF-05-1-0528 and W911NF-08-C-0124.

7. References

Acikel, B. High Performance Barium Strontium Titanate VaractorTechnology for Low Cost Circuit Applications. (2002). *Ph.D. thesis,* pp. 177 University of California, Santa Barbara.

Ban, Z.G. and Alpay S.P. (2003). Optimization of the tunability of barium strontium titanate films via epitaxial stresses. *J. Appl. Phys.* Vol. 93,No. 1 ,(JAN 2003), pp.504-511.

Ban, Z.G., Alpay, S.P., Mantese, JV. (2003a). Hysteresis offset and dielectric response of compositionally graded ferroelectric materials. *Integr. Ferroelectr.* Vol. 58, pp. 1281-1291.

Ban, Z.G., Alpay, S.P., Mantese, J.V . (2003b). Fundamentals of graded ferroic materials and devices. *Phys. Rev. B* Vol. 67, No. 18, (MAY 2003), pp.184104.

Cava,R.J. (2001). Dielectric materials for applications in microwave communications *J. Mater. Chem.* Vol. 11,No. 1., pp. 54 -62.

Cole, M.W., Joshi, P.C., Ervin, M.H., Wood, M.C., Pfeffer, R. L. (2000). The influence of Mg doping on the materials properties of $Ba_{1-x}Sr_xTiO_3$ thin films for tunable device applications . *Thin Solid Films* . Vol. 374,No. 1, (OCT 2000), pp. 34-41.

Cole, M.W., Joshi, P.C., Ervin, M..H. (2001). La doped $Ba_{1-x}Sr_xTiO_3$ thin films for tunable device applications. *J. Appl. Phys.* Vol. 89, No. 8, (APR 2001), pp.6336-6340.

Cole, M.W., Hubbard, C. , Ngo, E. , Ervin, M. , Wood, M., Geyer, R. G.(2002a) . Structure-property relationships in pure and acceptor-doped $Ba_{1-x}Sr_xTiO_3$ thin films for tunable microwave device applications. *J. Appl. Phys.* Vol. 92, No. 1, (JUL 2002), pp. 475-483.

Cole, M.W., Joshi, P.C., Ervin, M. , Wood, M. Pfeffer, R.L. (2002b) . Evaluation of Ta_2O_5 as a buffer layer film for integration of microwave tunable Ba1-xSrxTiO3 based thin films with silicon substrates . *J. Appl. Phys.* Vol. 92, No. 7, (OCT 2002), pp. 3967-3973.

Cole, M.W., Nothwang, W.D., Hubbard, C, Ngo,E., Ervin, M. (2003). Low dielectric loss and enhanced tunability of $Ba_{0.6}Sr_{0.4}TiO_3$ based thin films via material compositional design and optimized film processing methods. *J. Appl. Phys.* Vol. 93, No. 11, (JUN 2003), pp. 9218-9225.

Cole, M.W., Nothwang, W.D., Demaree, J.D., Hirsch, S. (2005). Integration of $Ba_{1-x}Sr_xTiO_3$-based active thin films with silicon-compatible materials and process science protocols to enable affordable on-the-move communications technologies *J. Appl. Phys.* Vol. 98,No. 2, (JUL 2005) pp. 024507.

Cole, M.W., Ngo, E, Hirsch, S., Demaree, J.D., Zhong, S. , Alpay,.SP. (2007). The fabrication and material properties of compositionally multilayered $Ba_{1-x}Sr_xTiO_3$ thin films for realization of temperature insensitive tunable phase shifter devices *J. Appl. Phys.* Vol. 102,No. 3, (AUG 2007), pp. 034104..

Cole, M.W., Ngo, E., Hirsch, S., Okatan, M. B., Alpay, S.P. (2008a). Dielectric properties of MgO-doped compositionally graded multilayer barium strontium titanate films *Appl. Phys. Lett.* Vol. 92,No. 7, (JUN 2008a), pp. 072906.

Cole, M.W., Weiss, C.V., Ngo, E., Hirsch,S., Coryell, L.A. , Alpay, S.P. (2008b). Microwave dielectric properties of graded barium strontium titanate films. *Appl. Phys. Lett.* Vol. 92,No. 18, (JUL 2008), pp. 182906.

Gevorgian, S., Petrov, P.K., Ivanov, Z., Wikborg, E. (2001). Tailoring the temperature coefficient of capacitance in ferroelectric varactors. *Appl. Phys. Lett.* Vol. 79,No. 12, (SEP 2001), pp. 1861-1863.

Jain, M., Majumder, S.B., Katiyar, R.S.,Miranda, F.A.,Van Keuls, F. W. (2003). Improvement in electrical characteristics of graded manganese doped barium strontium titanate thin films. *Appl. Phys. Lett.* Vol. 82,No. 12 ,(Mar 2003), 1911-1913.

Jeon, J.H., Hahn, Y.D., Kim, H.D. (2001). Microstructure and dielectric properties of barium-strontium titanate with a functionally graded structure. *J. Eur. Ceram. Soc.* Vol. 21,No. 10-11, pp. 1653 -1656.

Joshi, P.C., and Cole, MW. (2000). Mg-doped $Ba_{0.6}Sr_{0.4}TiO_3$ thin films for tunable microwave applications. *Appl. Phys. Lett.* Vol. 77,No. 2, (JUL 2000), pp. 289-291.

Kalkur, T.S., Sbrockey, N.M., Tompa, G.S., Alpay,S.P., Spanier , J.E., Galow , E.M. Cole , M.W. (2009). Low Voltage Tunable Band Pass Filters Using Barium Strontium Titanate Parallel Plate Capacitors. *Integrated Ferroelectrics* Vol 112, pp. 1-7.

Kim, W.J., Wu, H.D., Chang, W., Qadri, S. B. , Pond, J. M., Kirchoefer, S. W., Chrisey, D. B. and Horwitz, J. S. (2000). Microwave dielectric properties of strained $(Ba_{0.4}Sr_{0.6})TiO_3$ thin films. *J. Appl. Phys.* Vol. 88,No. 9, (Nov 2000), pp. 5448-5451.

Lemanov, V.V, Smirnova, E.P., Syrnikov, P.P., Tarakanov, E. A. (1996). Phase transitions and glasslike behavior in $Sr_{1-x}Ba_xTiO_3$. *Phys.Rev. B* Vol. 54,No. 5, (AUG 1996), pp. 3151-3157.

Lee, S.J., Moon, S.E., Ryu, H.C., Kwak,M.H., Kim,Y.T., Han, S.K. (2003). Microwave properties of compositionally graded (Ba, Sr)TiO3 thin films according to the direction of the composition gradient for tunable microwave applications. *Appl. Phys. Lett.* Vol. 82, No. 13, (MAR 2003), pp. 2133-2135.

Lin, Y; Chen, X; Liu, J, Collins , G. , Chen ,CL, Jiang, J.C. , Meletis, E.I. , Chen, C.L.P, Bhalla, A. Cole, M.W. (2008). Highly Epitaxial Ferroelectric Lead Strontium Titanate ($(Pb,Sr)TiO_3$) thin films with Extra Large Dielectric Microwave Elements Tunability: A Good Candidate for Room Temperature Tunable Microwave Elements. (2008). *Integrated Ferroelectrics* Vol. 100,pp. 1-15.

Lookman, A., Bowman, R.M., Gregg, J.M., Kut, J. , Rios, S. , Dawber, M. , Ruediger, A. , Scott, J. F. (2004). Thickness independence of true phase transition temperatures in barium strontium titanate films. *J. Appl. Phys.*Vol. 96,No. 1, (JUL 2004), pp. 555-562.

Lu, S.G., Zhu, X.H., Mak, C.L.,Wong, K. H. , Chan, H. L. W., Choy, C. L. (2003). High tunability in compositionally graded epitaxial barium strontium titanate thin films by pulsed-laser deposition. *Appl. Phys. Lett.* Vol. 82,No. 17, (ARP 2003), pp. 2877-2879.

Mantese, J. V. , Schubring, N. W. , Micheli, A. L. and Catalan, A. B. (1195). Ferroelectric Thin Films with Polarization Gradients Normal to the Growth Surface. *Appl. Phys. Lett.* Vol. 67, No. 5 , (JUL 1995), pp. 721-723.

Mitsui,T, Nakamura,E, and Gesi,K. In *Ferroelectrics and Related Substances*, Landolt-Börnstein, New Series, Group III, Vol. 16, Pt. A, edited by K.-H. Hellwege and A. M. Hellwege (Springler, Berlin, 1981).

Oates, D.E., Dionne, G.F., Temme, D.H., Weiss, J. A. (1997). Superconductor ferrite phase shifters and circulators. *IEEE Trans. Appl. Supercond.* Vol. 7,No. 2, pp. 2347-2350.

Pervez, N. K. , Hansen, P. J. , and York, R. A. (2004). High tunability barium strontium titanate thin films for rf circuit applications. *Appl. Phys. Lett.* Vol. 85,No. 19, (NOV. 2004) pp. 4451-4453.

Podpirka, A., Cole, M.W., Ramanathan, S, (2008). Effect of photon irradiation on structural, dielectric, and insulating properties of Ba0.60Sr0.40TiO3 thin films.*Appl. Phys. Lett.* Vol. 92, No. 21, pp. 212906 (2008).

Potrepka, D.M., Hirsch, S., Cole, M.W., Nothwang, W. D., Zhong, S., and Alpay, S. P. (2006). Effect of strain on tunability in $Ba_{0.60}Sr_{0.40}TiO_3$ thin films on Pt-Si substrates. *J. Appl. Phys.* Vol. 99, No. 1, (JAN 2006), pp. 014108.

Roytburd, A.L., Zhong, S., Alpay, S.P., (2005). Dielectric anomaly due to electrostatic coupling in ferroelectric-paraelectric bilayers and multilayers. *Appl. Phys. Lett.* Vol. 87,No. 9 (AUG 2005) pp. 092902.

Sayer, M., Mansingh, A., Arora, A.K., Lo, A. (1992). Dielectric response of ferroelectric thin films on non-metallic electrodes. *Integr. Ferroelectr.* Vol. 1,No. 1, pp. 129-146.

Shaw, T.M., Suo, Z., Huang, M, Liniger, E. , Labibowits, R.B. .(1999). The effect of stress on the dielectric properties of barium strontium titanate thin films. *Appl. Phys. Lett.* Vol. 75,No. 14 (OCT 1999) pp. 2129-2131.

Sun, P., Matsuura, N., Ruda, H.E. (2004) Crystallization and properties of PbO-doped $Ba_{0.7}Sr_{0.3}TiO_3$ films . *J. Appl. Phys.* Vol. 96, No. 6, (SEP 2004), pp. 3417-3423.

Taylor, T.R, Hansen, P.J., Acikel, B., Pervez, N. , York, R. A. , Streiffer, S. K. , Speck, J. S. . (2002). Impact of thermal strain on the dielectric constant of sputtered barium strontium titanate thin films. *Appl. Phys. Lett.* Vol. 80, No. 11, (MAR 2002)pp. 1978 - 1980.

Tian, H.Y, Chan, H.L.W., Chung, L.C., No, K. (2003). The effects of composition gradients of $Ba_xSr_{1-x}TiO_3$ thin films on their microstructures, dielectric and optical properties. *Materials Science and Engineering B Vol.* 103, No. 3 (DEC 2003) pp. 246 -252.

Wang, J., Zhang, T.J., Xia, H.Y., Xiang, J., Li, W., Duo, S. (2008). Influence of YSZ buffer layer on the electric properties of compositionally graded (Ba, Sr)TiO_3 thin film. *J Sol-Gel Sci Technol.* Vol. 47, No. 1, pp. 102-106.

Weiss, C.V., Cole, M.W., Alpay, S.P., Ngo, E., Toonen, R.C. , Hirsch, S.G., Demaree, J.D., Hubbard C. (2008). Dielectric Response of Variable Thickness $Ba_{0.6}Sr_{0.4}TiO_3$ Films for Property –Specific Device Applications. *Integrated Ferroelectrics* Vol. 100, 36-47.

Xia, Y.D., Cai, C., Zhi, X.Y., Pan, B., Wu, D., Meng, X, Liu, Z (2006). Effects of the substitution of Pb for Ba in (Ba,Sr)TiO_3 films on the temperature stability of the tunable properties. *Appl. Phys. Lett.* Vol. 88,No. 18, (MAY 2006), pp. 182909.

Zhang, J., Cole, M.W., Alpay, S.P. (2010). Pyroelectric properties of barium strontium titanate films: Effect of thermal stresses *J. Appl. Phys.* Vol. 108, No. 5, pp. 054103 .

Zhang, T.J., Wang, J, Zhang, B.S., Wang,J.Z., Wan,N., Hu,L. (2006). Preparation and dielectric properties of compositionally graded (Ba,Sr)TiO_3 thin film by sol-gel technique. *Trans. Nonferrous Met. Soc. China* Vol. 16, pp. 119-122.

Zhong, S., Alpay, S.P., Cole, M.W., Ngo,E., Hirsch,S., and Demaree, J. D. (2007). Highly tunable and temperature insensitive multilayer barium strontium titanate films. *Appl. Phys. Lett.* Vol. 90, No. 9,, (FEB 2007) pp. 092901.

Zhong, S., Alpay, S.P., Roytburd, A.L., and Mantese, J.V. (2006). Interlayer coupling in ferroelectric bilayer and superlattice heterostructures. *IEEE Trans. Ultrason. Ferroelectr. Freq. Control.* Vol. 53,No. 12, (DEC 2006), pp. 2349-2356.

Zhu, X.H., Chong, N., Chan, H.L.W., Choy, C.-L. , Wong, K.-H. ,. Liu, Z, Ming, N. . (2002a). Epitaxial growth and planar dielectric properties of compositionally graded ($Ba_{1-x}Sr_x$)TiO_3 thin films prepared by pulsed-laser deposition. *Appl. Phys. Lett.* Vol. 80,No. 18, (MAY 2002), pp. 3376-3378.

Zhu, X.H., Chan, H.L.W., Choy, C.L, Wong, K. H. (2002b). Microstructural and morphological evolutions in compositionally-graded ($Ba_{1-x}Sr_x$)TiO_3 thin films and related dielectric properties. *Integr. Ferroelectr.* Vol. 45,No. 45, pp. 131-140.

Zhu, X., Chan, H.L.W., Choy, C.L., Wong, K.-H. (2002c). Epitaxial growth and dielectric properties of functionally graded ($Ba_{1-x}Sr_x$)TiO_3 thin films with stoichimetric variation. *J. Vac. Sci. Technol. A* Vol. 20,No. 5, pp. 1796-1801.

Zhu, X., Lu, S., Chan, H.L.W., Wong, , K.H. , Xu, J. , Shi, S. . (2003). Preparation of epitaxial compositionally graded ($Ba_{1-x}Sr_x$)TiO_3 thin films with enhanced dielectric properties. *Appl. Phys. A- Materials Science & Processing,* Vol. 77,No. 3-4, (AUG 2003), pp. 499-505.

Part 2

Doping and Composites

9

The Effect of Mn Doping on the Dielectric Properties of Lead Strontium Titanate (PST)

Arne Lüker[2], Qi Zhang[1] and Paul B. Kirby[1]
[1]*Materials Department, Cranfield University, Bedford*
[2]*Instituto Superior Técnico - Departamento de Física, Av. Rovisco Pais,Lisboa,*
[1]*UK*
[2]*Portugal*

1. Introduction

Lead Strontium Titanate (PST) is a ferroelectric perovskite similar to the well known Barium Strontium Titanate (BST). Its transition point between the cubic paraelectric and tetragonal ferroelectric state (Curie temperature) can be shifted linearly by varying the Pb/Sr ratio and is just below room temperature with a ratio of 40/60. For voltage tunable applications the paraelectric state is favored because it offers low dielectric losses (tanδ) due to the absence of the permanent electric dipole, which implies that on removal of an electric field the polarization in the material returns to zero, viz. P/E measurements show no hysteresis at room temperature. The maximum tunability, defined as $n(\%) = \frac{C_{max} - C_{min}}{C_{max}} \cdot 100$, where C_{max} is the capacitance value at zero bias and C_{min} the value under bias, of ferroelectrics is observed in the paraelectric state close to the Curie temperature.

At present, the research on PST thin films focuses on the diffused phase transition and relaxor behavior, the dielectric response of dc bias, and the growth kinetics on different kind of substrates and oxide buffer layers. It is commonly known that proper element doping at A or B site in ABO_3 perovskite-type ferroelectrics is an effective way to alter the ferroelectric and dielectric properties. It has been shown for BST that some dopants including Mg^{2+}, Ni^{2+}, Fe^{3+}, Mn^{2+}, Mn^{3+}, Co^{2+}, Co^{3+}, Al^{3+}, Cr^{3+} and Bi^{3+}, which can occupy the B site of the ABO_3 perovskite structure, behave as electron acceptors and can lower the dielectric loss, enhance the dielectric constant and thus the tunability and figure of merit ($FOM = \frac{n(\%)}{\tan\delta(\%)}$) [1 - 12].

For PST, being a relatively new system under increasing investigation for voltage tunable devices, only a few papers on B site doping can be found in the literature [e.g. 13 - 15] and the presented results are not very clear and satisfying. The aim of this study is to improve the material performance of PST 40/60 by doping of the B site with manganese (Mn^{2+}), which has an ionic radius of 0.67 Å, comparable to that of Ti^{4+} (0.68 Å), and was used in our lab to improve the ferroelectric and pyroelectric properties of sol-gel derived PZT tin films [16 - 18], to gain a deeper understanding of the effect of doping.

A big advantage of the sol-gel route is the ease of doping. By varying only milligrams of the dopant in the initial steps of the solution preparation, one can study the implications on the final properties of the film time and cost effective. Starting with pure, undoped PST 40/60 we sequently increased the Mn^{2+} doping level from 1 to 5 mol% and studied the effects on morphologies, dielectric properties with and without dc bias and the ferroelectricity of the resulting thin films. It is found and explained that a doping level of 2 mol% Mn^{2+} results in optimal properties in terms of tunability and loss.

2. Experimental procedures

To prepare a 40/60 composition of (Pb,Sr) TiO_3 with, for example, 3 mol% of manganese, the stoichiometric amounts of lead acetate and strontium acetate were dissolved with slightly warming in a mixed solvent of propanediol and acetic acid. Meanwhile the stoichiometric amounts of titanium butoxide and manganese acetate were mixed in a N_2 glove box in acetic acid. Both solutions were mixed and stirred at room temperature overnight. The final solution was diluted with 2-methoxyethanol to adjust the concentration of the solution to 0.4 M.
The thus prepared precursor solution was deposited via spin coating onto silicon substrates with a Ti/Pt bottom electrode at 3000 rpm for 30 seconds. In each trial the sample was placed on a hotplate at 350°C for 10 min to evaporate the solvents and annealed on a second hotplate at 650°C for 15 min. A single layer thickness was approximately 50 nm. To obtain thicker films (~300 nm) the process was repeated six times.
An Edwards E480 thermal evaporator was used to deposit Cr/Au top electrodes onto the ferroelectric thin films through a shadow mask with hole-areas of 0.48 mm^2. A Siemens D5005 diffractometer was used for all the XRD measurements. Scanning electron microscopy (S-FEG SEM) was performed on a Philips XL30. To analyse the surface topography of the samples a Digital Instruments Dimension 3000 AFM was used. Contact mode was used to make high-resolution topographical images.
An RT66A Standardised Ferroelectric Test System was used to analyse the ferroelectric hysteresis properties of the thin films produced during this study.

3. Results and discussion

Fig. 1 shows X-ray diffractograms of PST 40/60 doped with 0, 1, 3 and 5 mol% Mn. All the films show a major (110) orientation at $2\theta = 32.2°$ and are well crystallised as confirmed by SEM (Fig. 2) and AFM (Fig. 3). The SEM photographs show clearly that an increasing Mn content has a remarkable influence on the microstructure of (Pb,Sr)TiO_3. The clearly visible grains in Fig 2(a) have an average grain size of 150 nm. With increasing Mn content the grains become smaller, ~ 80 nm in 2(b), and indistinct in 2(c), and finally in 2(d), the microstructure has a sponge-like appearance.
AFM surface scans of an area of 2 x 2 µm are depicted in Fig. 3. It indicates a decrease of the RMS-roughness with increasing Mn content from 3.37 nm for pure PST down to 1.69 nm for PST with 5 mol% Mn.
The dielectric constant and loss of these films at zero bias are depicted in Fig. 4. At first the dielectric constant increases with increasing Mn content or with decreasing roughness/grain size. But there is a sharp decrease in the dielectric constant after 3 mol% Mn while the roughness/grain size is further decreasing. The dielectric constant of PST with 5 mol% Mn is even lower (~300) than that of pure PST. Sun *at al.* [19] have doped the A site of BST with K^+ and showed a clear relationship between grain size/surface roughness and

dielectricproperties of BST thin films. Both, the roughness/grain size and dielectric constant increased with the doping level up to 7.5 mol% K^+ and decreased afterwards. However, no such relationship could be found for B site Mn doped PST in this study.

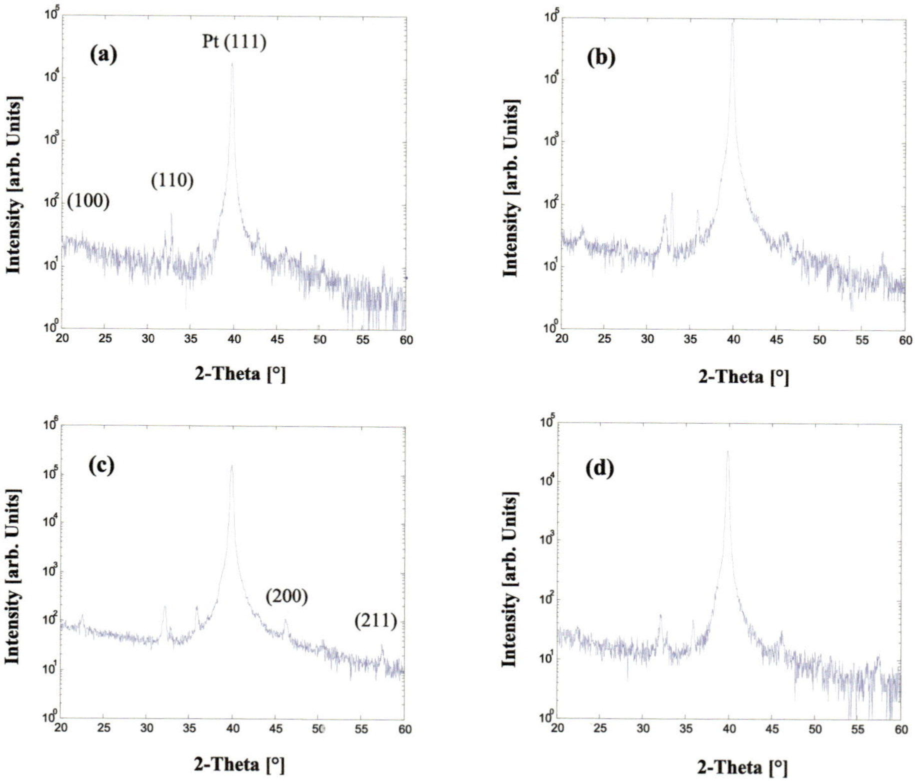

Fig. 1. X-ray diffractograms of $(Pb_{0.4},Sr_{0.6})(Ti_{1-x},Mn_x)O_3$. (a) x=0, (b) x=0.01, (c) x=0,03 and (d) x=0.05.

Li *et al.* have doped the B site of PST 40/60 with Mg^{2+} [14]. They found that the dielectric constant increases with increasing doping level up to 3 mol% Mg^{2+} with a sharp decrease afterwards. They attributed this behaviour to the generation of oxygen vacancies which induce a positive charged defect in the crystal.

Generally, oxygen vacancies are generated by heat treatment under non-oxidising atmosphere [14]. In thin films that are annealed in ambient atmosphere, they form a so-called dead-layer at the interface between the bottom electrode and the ferroelectric thin film. In the case of Mg doped PST, Mg^{2+} ions replace Ti^{4+} ions in the PST lattice due to a similar ionic radii of Mg^{2+} (r = 0.72 Å) and Ti^{4+} (r = 0.68 Å), hence B site doping. According to Li *et al.* proper Mg addition ($0 < x < 0.03$) in $(Pb_{0.4},Sr_{0.6})(Mg_x,Ti1_{-x})O_3$ thin films could be used as acceptors in the ABO_3 perovskite structure. It induces a negative charge and thus balances the positive of the oxygen vacancies. Then with the charge being compensated, the lattice distortion ratio in the system decreases, viz. the lattice structure of $Pb_{0.}4Sr0_{.6}Mg_xTi_{1-x}O_3$ becomes more "perfect" (cubic). According to the thermodynamic theory, the phase formation ability of the crystal therefore increases with increasing Mg content up to x = 0.03.

At the same time, more polarization paths may be provided when the lattice structure becomes more "perfect". This results in an increase of the dielectric constant.

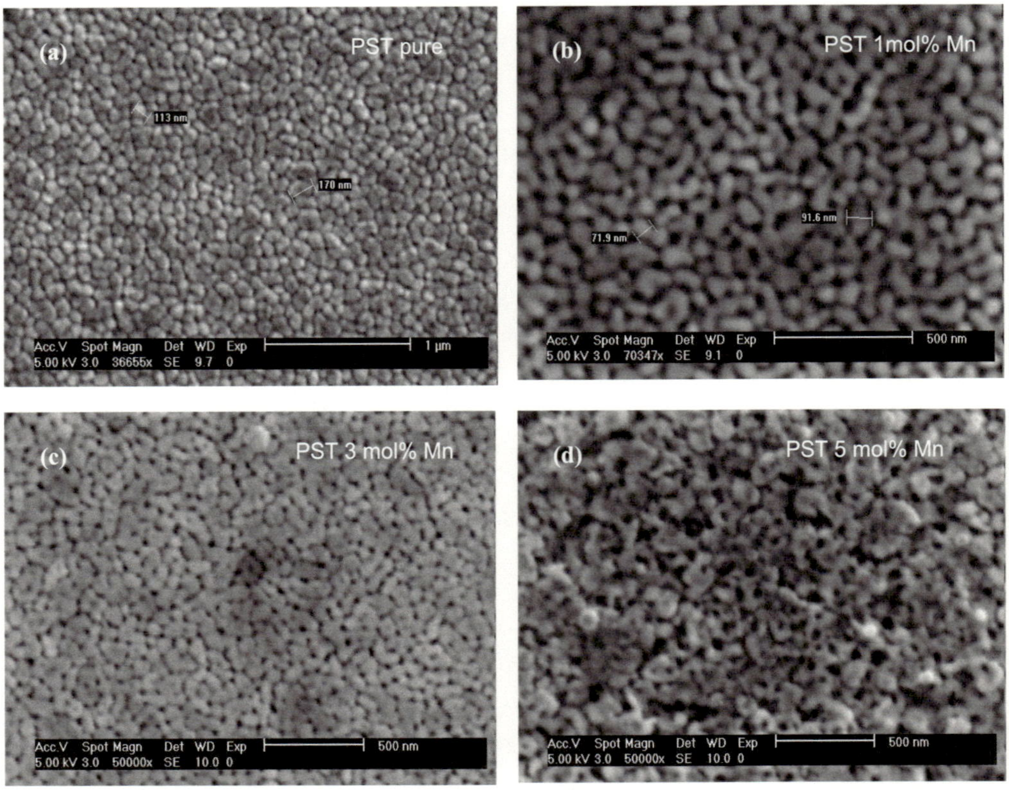

Fig. 2. SEM images of $(Pb_{0.4},Sr_{0.6})(Ti_{1-x},Mn_x)O_3$. (a) x=0, (b) x=0.01, (c) x=0,03 and (d) x=0.05

However, with further Mg doping, excessive oxygen vacancies would be created in the system. The lattice distortion ratio of the perovskite phase structure of the film would increase and then the phase formation ability decreases. The dielectric constant degrades with increasing Mg doping.

At first glance the explanation from Li *et al.* for the behaviour of the dielectric constant in Mg doped PST may as well fit for Mn doped PST. However, Mn is a multivalence ion - it can appear as Mn^{2+}, Mn^{3+} and Mn^{4+}, all having different impacts on the charge balance of the crystals. Mn^{2+} was used to improve the ferroelectric and pyroelectric properties of PZT [16 – 18] in our lab and it has been suggested that the formation of oxygen vacancies is facilitated by the conversion of Ti^{4+} to Ti^{3+} during heat treatment:

$$2Ti^{4+} \rightarrow 2Ti^{3+} + V_O^{\bullet\bullet} \quad [1] \tag{1}$$

[1] Kröger-Vink nomenclature [20]: $V_O^{\bullet\bullet}$: main character: chemical species, V=vacancy; subscript: site (for example, O=regular oxygen site); superscript: charge relative to perfect lattice; •, ′, x correspond to singly positive, singly negative, and neutral effective charge.

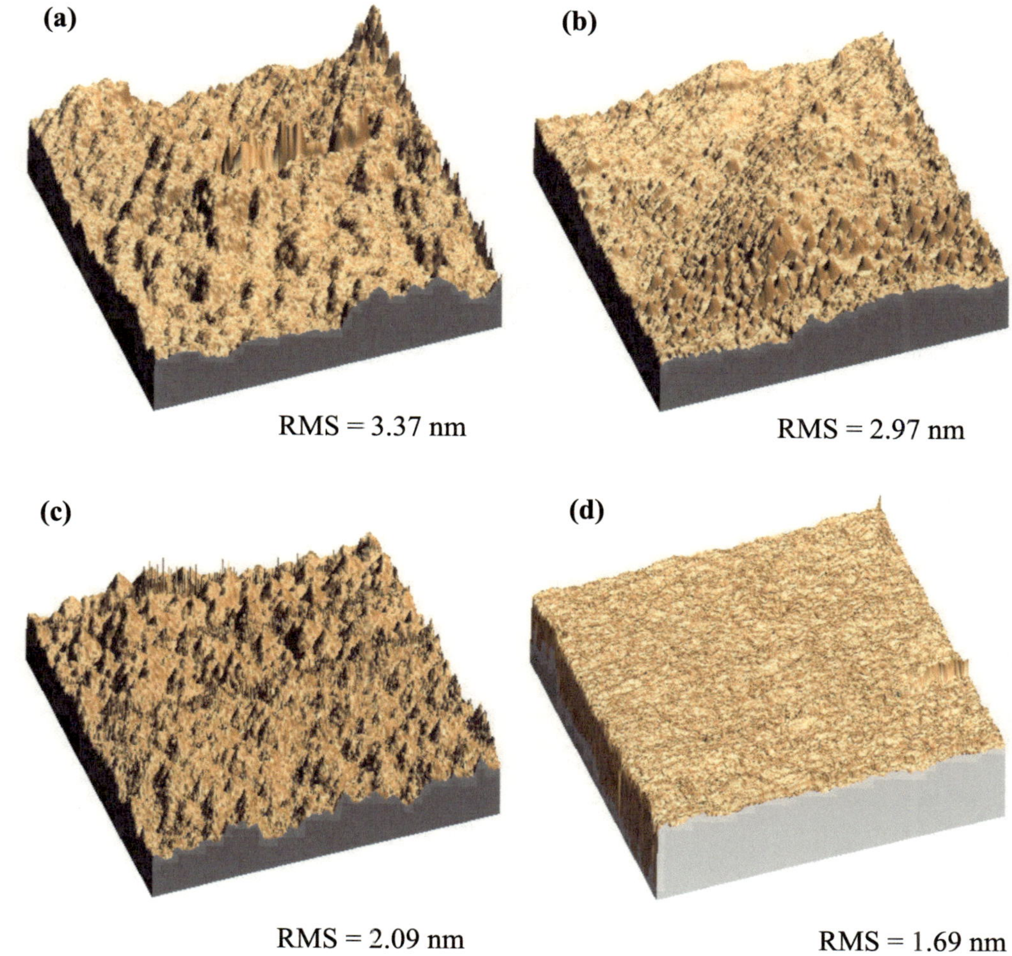

Fig. 3. AFM images of $(Pb_{0.4},Sr_{0.6})(Ti_{1-x},Mn_x)O_3$. (a) x=0, (b) x=0.01, (c) x=0,03 and (d) x=0.05. Scan size: 2*2 μm

By adding Mn^{2+} ions (and some oxygen from e.g. the atmosphere) we may find

$$\frac{1}{2}O_2 + Mn^{2+} + V_O^{\bullet\bullet} \rightarrow Mn^{4+} + O_O^X \tag{2}$$

or

$$\frac{1}{4}O_2 - Mn^{2+} + \frac{1}{2}V_O^{\bullet\bullet} \rightarrow Mn^{3+} + \frac{1}{2}O_O^X \tag{3}$$

Combining these two reactions leads to two possible total reactions [21]:

$$2Ti^{4+} + Mn^{2+} + \frac{1}{2}O_2 \leftrightarrow 2Ti^{3+} + Mn^{4+} + O_O^X \tag{4}$$

or

$$2Ti^{4+} + Mn^{2+} + \frac{1}{4}O_2 \leftrightarrow 2Ti^{3+} + Mn^{3+} + \frac{1}{2}O_O^X \qquad (5)$$

In other words: Mn^{2+} doping actually consumes oxygen vacancies to get incorporated as Mn^{3+}/Mn^{4+} at the Ti^{4+} site of the $(Pb_{0.4},Sr_{0.6})(Mn_x,Ti_{1-x})O_3$ perovskite crystal structure. It is easy to imagine that, at some point, no oxygen vacancies are available anymore. Yang *et al.* [15] doped PST 50/50 thin films with up to 5 mol% Mn and investigated the dc conductivity and the dielectric response at various temperatures. Their results reveal that there are obvious hopping conductions and dielectric relaxations at low frequencies, which can be only ascribed to the thermal-activated short range hopping of localised charge carriers through trap sites separated by potential barriers with different heights, namely, localised electron hopping between Mn^{2+}, Mn^{3+} and Mn^{4+}. They specified the activation energy for the dc-conductivity and charge carrier hopping to be 0.42 and 0.47 eV, respectively - far to low for oxygen vacancies hopping, very common in perovskite-type oxides, whose activation energy is reported to be ~1 eV [22, 23].

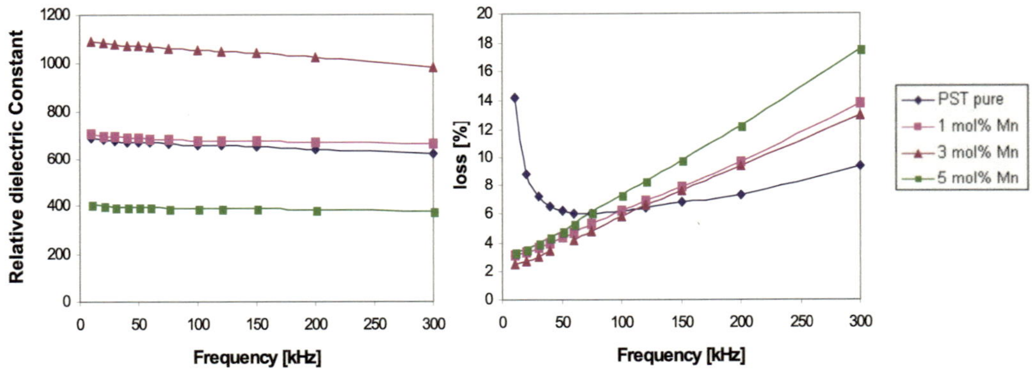

Fig. 4. Dielectric Constant and loss vs. frequency for PST 40/60 with different Mn content.

The hopping conduction due to the hopping of the charge carriers between Mn sites begins to occur in 2 mol% Mn doped PST, and then becomes distinct in 5 mol% doped films. This type of hopping process is therefore associated with a certain amount of Mn dopant and more Mn amount can provide more pathways for the total hopping process. It has been reported elsewhere that the activation energies of carrier hopping between mixed valence Mn sites are about 0.4 - 0.5 eV in Mn doped perovskite-type oxides like $LaGaO_3$ and $Bi_3PbSr_3Ca_3Cu_4O_3$ glasses [24, 25], which is in agreement with the value given by Yang *et al.*. The evidence that there are multivalence Mn ions (namely Mn^{2+}, Mn^{3+} and Mn^{4+}) in 0.2 - 1 mol% Mn doped $BaTiO_3$ has been extensively observed by electron-spin-resonance spectrum [26]. Moreover Tkach *et al.* have studied Mn-doped $SrTiO_3$ using electron paramagnetic resonance (EPR) measurements [27] and Raman spectroscopy [28]. They found that Mn^{4+} substitutes for Ti^{4+} and Mn^{2+} for Sr^{2+}. In addition they sintered $SrTi_{0.95}Mn_{0.05}TiO_3$ ceramics in O_2 and N_2 atmosphere. Sintering in O_2 favors the formation of

Mn^{4+}, whereas sintering in N_2 promotes $Mn^{2+/3+}$ on the B site, typically compensated by $V_O^{\bullet\bullet}$. However, they assumed an ionic radius of ~1.27 Å for Mn^{2+}, a value which is nearly twice as much as reported elsewhere.

For the purpose of this paper the theoretical work presented by J.Yang *et al.* is adopted and it is believed that the hopping conduction due to the hopping of the charge carriers between Mn sites lowers in the end the dielectric constant and increases the loss in PST thin films

3.1 Tunability and ferroelectricity of Mn doped PST

Fig. 5 confirms the rule of thumb for tunable ferroelectrics "a high dielectric constant gives a high tunability". It shows the tunability and loss vs. electric field at 150 kHz of $(Pb_{0.4},Sr_{0.6})(Mn_x,Ti_{1-x})O_3$ thin films with different Mn contents. The overall loss remains under 7.5 % for x = 0, 0.01, and 0.03 and increases for x = 0.05. The increase is attributed to the hopping conduction due to the hopping of the charge carriers between Mn sites, which begins to occur at x = 0.02. The tunability seems to reach a maximum of 78 % for x = 0.03 and an applied field of 350 kV/cm (10.5 V)

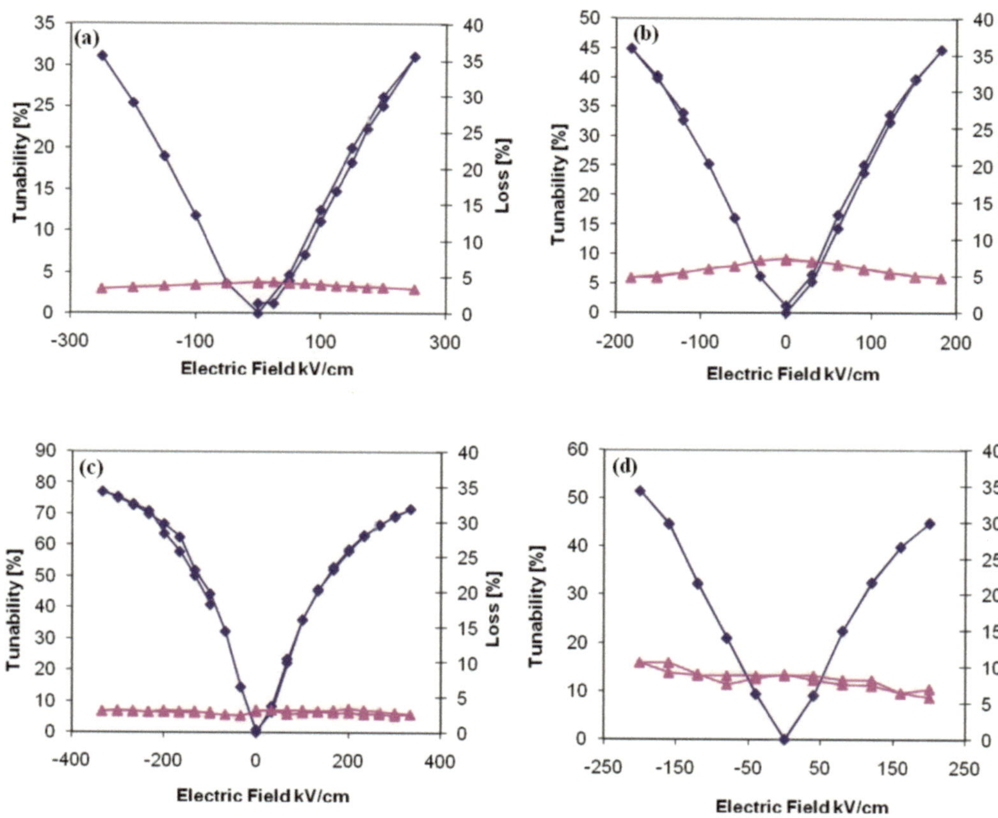

Fig. 5. Tunability and loss vs electric field at 150 kHz of $(Pb_{0.4},Sr_{0.6})(Ti_{1-x},Mn_x)O_3$. (a) x=0, (b) x=0.01, (c) x=0,03 and (d) x=0.05.

Fig. 6 compares the measured tunability with theoretical results using the expression

$$C(V) = \frac{C_{max}}{\cosh\left[\frac{2}{3}\sinh^{-1}\left(\frac{2V}{V_2}\right)\right]-1} \tag{6}$$

for the voltage controlled capacitance [29]. C_{max} is the measured capacitance under zero bias and V_2 is the "2:1" voltage at which $C(V_2)=C_{max}/2$, normally an easily measured quantity. The figure shows that the tunability of $(Pb_{0.4},Sr_{0.6})(Mn_{0.01},Ti_{0.99})O_3$ is slightly higher than the tunability of $(Pb_{0.4},Sr_{0.6})(Mn_{0.03},Ti_{0.97})O_3$. Actually no surprise as we discovered that the hopping of the charge carriers between the Mn sites begins to occur in 2 mol% Mn doped PST. As a consequence the perovskite PST crystal is already slightly degraded at 3 mol% of Mn, as it can already be seen in Fig. 2.

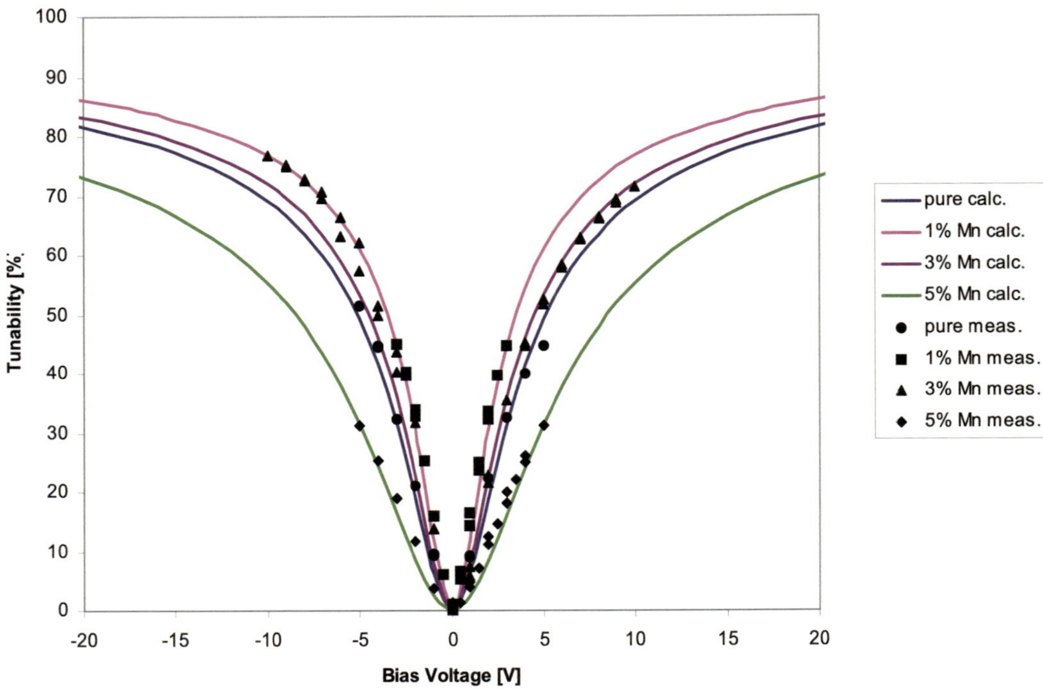

Fig. 6. Comparison of the measured and calculated tunability of $(Pb_{0.4},Sr_{0.6})(Ti_{1-x},Mn_x)O_3$.The fitting parameters C_{max}[nF]/V_2[V] are: pure 9.5/5.1; 1% 15/3.5;3% 13.8/4.5; 5% 5.7/8.5

It should be noted here that V_2 could not be measured directly for PST with 1 and 5 mol% Mn due to the early blistering of the top electrode in same cases. In these cases V_2 was calculated using the formulation

$$V_2 = \frac{4V_n}{(n+2)\sqrt{n-1}} \tag{7}$$

where V_n is the voltage at which the capacitance is reduced by n viz. $n = C_{max}/C_{min}$ [29]. The calculated curves were fitted to the values measured with positive bias because a clear relationship between Mn content and tunability can be seen within these values, whereas the measured values for PST doped with 1 and 3 mol% are not really distinguished under negative bias.

The decrease in the oxygen vacancy concentration due to the generation of higher valance Mn ions leads as well to a restraint domain pinning, and in turn to an improvement of ferroelectric properties because oxygen vacancies are always considered as the major pinning cause of ferroelectric domain wall motions [30]. The enhancement of ferroelectric properties in $(Pb_{0.4},Sr_{0.6})(Mn_x,Ti_{1-x})O_3$ with increasing Mn content is shown in Fig. 7.

The polarisation-voltage dependence of pure PST 40/60 shows a typical paraelectric behaviour - a straight line at room temperature. With increasing Mn content both the remnant polarisation and the coercive field increase, indicating an enhancement of ferroelectricity. It should be noted here that only the first hysteresis was measured on each sample, therefore the loops are not closed.

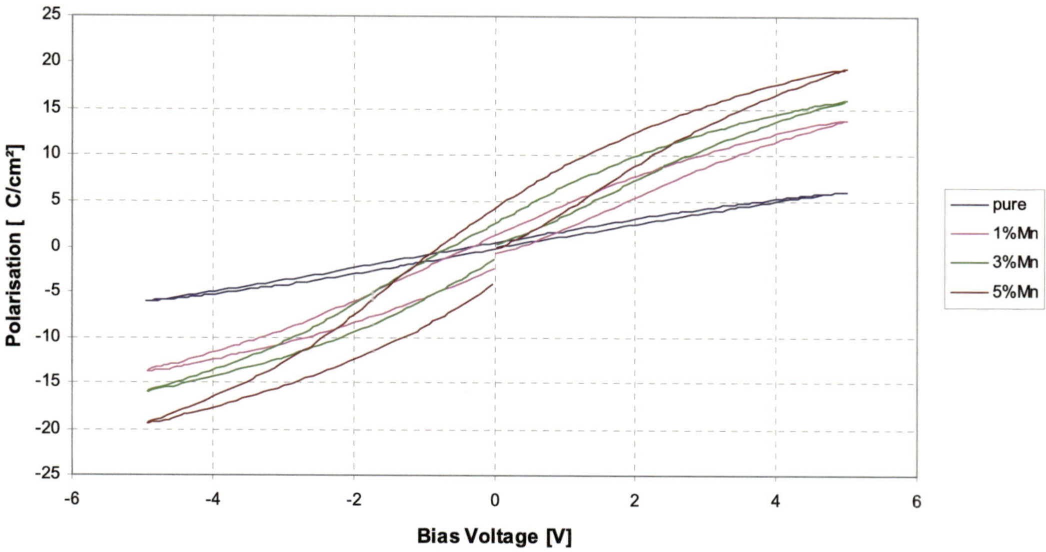

Fig. 7. Hysteresis loops of $(Pb_{0.4},Sr_{0.6})(Ti_{1-x},Mn_x)O_3$. The film is paraelectric with x=0 and the ferroelectricity improves with x.

The hysteresis loops are very slim compared to those of real ferroelectric materials like PZT and are similar to those observed in relaxor ferroelectrics in which the dielectric constant maximum does not correspond to a transition from a non-polar phase to a ferroelectric polar phase, such as observed in Lead Magnesium Niobate (PMN) [31].

On the fundamental science side it is still a challenge to develop an understanding of the many interesting and peculiar features by this kind of materials, because the interactions responsible for the relaxor ferroelectric phenomena are on the macroscopic scale. On the application side, this class of materials offers a high dielectric constant and high electrostriction, which are attractive for a broad range of devices [32].

That PST shows a relaxor behaviour was demonstrated by Hua Xu *et al.* [33] for instance. $Pb_{0.5}Sr_{0.5}TiO_3$ ferroelectric films were deposited onto $Pt/Ti/SiO_2/Si$ substrates by pulsed laser deposition. The state of the films was described as a mixed state, with both ferroelectric and relaxor-like features. The films exhibited high dielectric constant and tunability at room temperature. At 10 kHz, the dielectric constants of the 200-nm- and 400-nm-thick films were

771 and 971, with the tunability of 60.2% and 70.9%, respectively. The temperature-dependent dielectric properties were studied and the relaxor-like behavior was observed in both the thinner and thicker PST films, which could be established in terms of diffuse phase transition characteristics and Vögel-Fulcher relationship. In addition, the contribution of the film-electrode interface layer to the dielectric properties was evaluated and the true dielectric properties of the films were reconstructed. Consequently, the relaxor-like character of the PST films was mainly ascribed to the effect of the film-electrode interfaces.

4. Conclusion

Fig. 8 summarizes the main findings of this study. The dielectric constant reaches a maximum of 1100 with 3 mol% Mn; the maximum value of the tunability with 10 V is 76.72% with 1 mol% Mn and the figure of merit (FOM) reaches 23.96 with 3 mol% Mn. This compares well with results from Du *et al.* [13], who reported a tunability of 80% and a FOM of 14.17 in pure PST 40/60 and 70% and 7 in La doped PST, or Sun *et al.* [34], who achieved a tunability of 69.4% and a FOM of 28.9 in $(Pb_{0.25}Ba_{0.05}Sr_{0.7})TiO_3$.

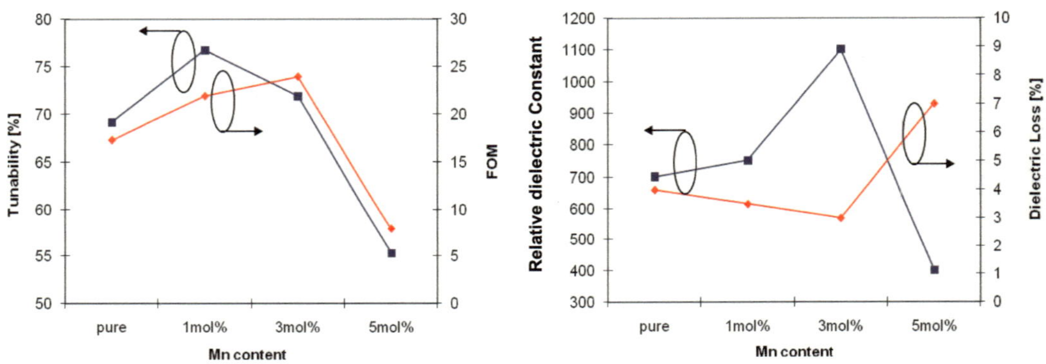

Fig. 8. Tunability and figure of merit at 10V and dielectric constant and loss at zero bias of PST 40/60 with different Mn contents

All these values drop significantly when the Mn doping level exceeds 3 mol% and we identified two possible reasons for this behaviour. First, Mn^{2+} doping consumes oxygen vacancies to get incorporated as Mn^{3+} and/or Mn^{4+} at the Ti^{4+} site of the $(Pb,Sr)TiO_3$ perovskite crystal structure. The negative charged Mn ions balance the positive induced charge of the oxygen vacancies leading to a more "perfect" (cubic) and electronically saturated perovskite. At the same time, more polarisation paths may be provided when the lattice structure becomes more "perfect". This results in an increase of the dielectric constant, tunability and FOM.

At a doping level of 2 mol% Mn, the crystal is totally saturated. With further doping a hopping conduction due to the hopping of the charge carriers between Mn sites begins to occur and becomes distinct in 5 mol% doped films. This type of hopping process is therefore associated with a certain amount of Mn dopant and more Mn amount can provide more pathways for the total hopping process. The dielectric constant, tunability and FOM decreases.

Hysteresis measurements show the effect of an enhanced ferroelectric characteristic in Mn doped PST and give rise to the question whether a relaxor like behaviour is also observable or not.

At the end it is worth pointing out that localised electron hopping between mixed-valence Mn ions provides a possibility to induce double exchange effects of Mn^{2+} and Mn^{3+} or Mn^{4+} and thus brings about magnetic properties [15]. This may be the mechanism behind the magnetic effect in Mn doped $PbTiO_3$ observed by Kumar et al. [35]. The coexisting of ferroelectric and ferromagnetic properties in a single PST thin film would provide a fresh method to obtain multiferroics.

5. Acknowledgements

The authors would like to thank Mr. Andrew Stallard and Mr. Matthew Taunt for their never ending effort to keep our labs running and Benjamin Jacquet and Cédric Fourn, summer students from the University of Rennes/France and helping hands in this project. This research was supported by EPSRC (EP/C520297/1).

6. References

M.W. Cole, C. Hubbard, E. Ngo, M. Ervin, M. Wood, and R.G. Geyer; J. Appl. Phys. 92, 475 (2002)
P.C. Joshi and M.W. Cole; Appl. Phys. Lett. 77, 289 (2000)
M.W. Cole, W.D. Nothwang, C. Hubbard, E. Ngo, and M. Ervin; J. Appl. Phys. 93, 9218 (2003)
Y.A. Jeon, T.S. Seo, and S.G. Yoon; Jpn. J. Appl. Phys., Part 1, 40, 6496 (2001)
L. Radhapiyari, A.R. James, O.P. Thakur, and C. Prakash, Mater. Sci. Eng., B 117, 5 (2005)
M. Jain, S.B. Majumder, R.S. Katiyar, F.A. Miranda, and F.W. Vam Keuls, Appl. Phys. Lett. 82, 1911 (2003)
S.Y. Wang, B.L. Cheng, C. Wang, S.Y. Dai, H.B. Lu, Y.L. Zhou, Z.H. Chen, and G.Z. Yang, Appl. Phys. Lett. 84, 4116 (2004)
S.Y. Wang, B.L. Cheng, C. Wang, H.B. Lu, Y.L Zhou, Z.H. Chen, and G.Z. Yang, J. Cryst Growth 259, 137 (2003)
K.B. Chong, L.B. Kong, L. Chen, L. Yan, C.Y. Tan, T. Yang, C.K. Ong, and T. Osipowicz, J. Appl. Phys. 95, 1416 (2004)
K.T. Kim and C.I. Kim, Thin Solid Films 472, 26 (2005)
A. Lüker, Q. Zhang, P. B. Kirby; *Thin Solid Films,* 518, 14, 3763 (2010)
D. Bhattacharyya, A. Lüker, Q. Zhang, P.B. Kirby; *Thin Solid Films,* 518, 12, 3382 (2010)
P. Du, X. Li, Y. Liu, G. Han, W. Weng; J. Europ. Ceram. Soc. 26, 2147 (2006)
X.T. Li, W.L. Huo, C.L. Mak, S. Sui, W.J. Weng, G.R. Han, G. Shen, and P.Y. Du; Mater. Chem, and Phys 108 (2008) 417 - *420*
J. Yang, X.J. Meng, M.R. Shen, L. Fang, J.L. Wang, T. Lin, J.L. Sun, and J.H. Chu, J. Appl. Phys. 104, 104113 (2008)
Q. Zhang and R.W. Whatmore, J. Appl. Phys. 94, 5228 (2003)
Q. Zhang and R.W. Whatmore, Mater. Sci. and Eng. B 109, 136 (2004)
Q. Zhang, J. Phys. D: Appl. Phys. 37, 98 (2004)
X. Sun, B. Zhu, T. Liu, M. Li and X.Z. Zhao, J. Appl. Phys. 99, 084103 (2006)
F.A. Kröger, *Chemistry of Imperfect Crystals,* North-Holland, Amsterdam, 1964
F .W.Poulsen, Solid State Ionics 129 (2000) 145- 162
N.S. Almodovar, J. Portelles, O. Raymond, J. Heiras, and J.M Siqueirosa, J. Appl. Phys. 102, 124105 (2007)

Z.H. Zhou, J.M Xue, W.Z. Li, J. Wang, H. Zhu, and J.M. Miao, J. Phys. D 38, 642 (2005)
N. Noginova, G.B. Loutts, E.S. Gillman, V.A. Atsarkin, and A.A. Verevkin, Phys. Rev. B 63, 174414 (2001)
S. Bhattacharya, D.K. Modaka, P.K. Pal, and B.K. Chaudhuri, Mater. Chem. Phys. 68, 239 (2001)
X. Wang, M. Gu,, B. Yang, S.N. Zhu, and W.W. Cao, Microelectron. Eng. 66, 855 (2003)
V. V. Laguta, I. V. Kondakova, I. P. Bykov, M. D. Glinchuk, A. Tkatch, P. M. Vilarinho; Phys. Rev. B: Condens. Matter Mater. Phys. 2007, 76, 054104
A. Tkach, P. M. Vilarinho, A. L. Kholkin, I. M. Reany, J. Pokorny, J. Petzelt; Chem. Mater. 2007, 19, 6471
A. Lüker; Sol-*Gel derived Ferroelectric Thin Films for Voltage Tunable Applications*, ISBN 978-3-639-31446-5, VDM Publishing House Ltd. (2010)
X. Wang and H. Ishiwara; Appl. Phys. Lett. 82, 2479 (2003)
G. A. Smolenskii and A. I. Agronovskaya. Sov. Phys. Tech. Phys., 3, 1380 (1958)
L. E. Cross. Ferroelectrics, 76, 29 (1987)
H. Xu, M. Shen, L. Fang, D. Yao and Z. Gan; Thin Solid Films; 493 (2005) 197
X. Sun, H. Huang, S. Wang, M. Li, and X. Zhao; Thin Solid Films 516 (2008) 1308
M. Kumar and K.L. Yadav, J. Phys.: Condens. Matter. 19, 242202 (2007)

10

Enhanced Electro-Optical Properties of Liquid Crystals Devices by Doping with Ferroelectric Nanoparticles

Hao-Hsun Liang and Jiunn-Yih Lee
Department of Materials Science and Engineering, National Taiwan University of Science and Technology
The Republic of China at Taiwan

1. Introduction

Over the past few decades, liquid crystal (LC) displays (LCDs) have been at the leading edge of their field. Many scientists and manufacturers have devoted studies to improve the performance of LCD characteristics such as a fast response, high contrast ratio, and wide viewing angle. Therefore the in-plane switching (IPS) mode (Oh-e et al., 1995) was developed for wide viewing angle; vertical alignment (VA) mode for a higher contrast ratio; and ferroelectric (Clark et al., 1980; Meyer et al., 1975) and antiferroelectric LC (Chandani et al., 1988, 1989) for faster responses, instead of the common nematic LC and twist nematic (TN) modes. Recent studies of liquid crystals doped with nanoparticles have given rise to a number of novel practical applications and pointed the way toward innovative improvement of the physical and electro-optical properties of liquid crystal by means of chemical synthesis (Kobayashi & Toshima, 2007).

Enhancement of the electro-optical properties of liquid crystal is dependent on the size, type, concentration, and intrinsic characteristics of the nanoparticles used for doping. The nanoparticles should share similar attributes to the liquid crystal molecules and be of a size that would not significantly disrupt the order of the liquid crystal. Low doping concentrations (<3% by weight) are usually chosen to yield a more stable and even distribution in the liquid crystal, which lowers the interaction forces between particles.

Commonly used doping nanoparticles include ferromagnetic nanoparticles, metallic nanoparticles, inorganic nanoparticles, and ferroelectric nanoparticles. In the case of ferromagnetic nanoparticles, the large permanent magnetic moments couple with the LC direction, leading to improvements in their magnetic properties. This is known as ferronematics (Brochard & Gennes, 1970). In the case of metallic nanoparticles, due to the surface plasmon resonance and depolarization of the electric field, the metallic nanoparticles can enhance the memory effect of the ferroelectric liquid crystal (FLC) (Kaur et al., 2007) and dielectric properties of nematic liquid crystal (NLC) (Miyama et al., 2004; Shiraki et al.,2004). In the case of inorganic nanoparticles, their intrinsic structures can affect the vertical alignment without the need for an alignment layer (Jeng et al., 2007). Due to the large permanent dipole moments, ferroelectric nanoparticles induce realignment of neighboring liquid crystal molecules, thereby increasing the order parameter and lowering the threshold voltage (Reznikov et al., 2003).

Schurian and Bärner discovered that doping ultra-fine (less than 1 µm) dielectric particles into an isotropic liquid enhances its sensitivity to electric field (Schurian & Bärner, 1996); this heralded studies into the doping of ferroelectric nanoparticles ($BaTiO_3$, $Sn_2P_2S_6$) into NLC (Cheon et al., 2005; Kaczmarek et al., 2008; Li et al., 2006, 2006). Doping of nanoparticles was initially restricted to NLCs due to their widespread applications, technological maturity, and relatively simple liquid crystal structure (which allows the doping of nanoparticles to be less likely to destroy the alignment of the liquid crystal molecules). After the successful enhancement of the electro-optical properties of NLC, attention has turned to other candidates, such as cholesteric liquid crystal and smectic liquid crystal. In the case of cholesteric liquid crystals, there were significant improvements in the contrast of reflectance of planar-focal conic state, and driving voltage of the cholesteric liquid crystal after doping with ferroelectric nanoparticles. In the case of SmA liquid crystals, there were also significant improvements in the driving voltage (Glushchenko et al., 2006).

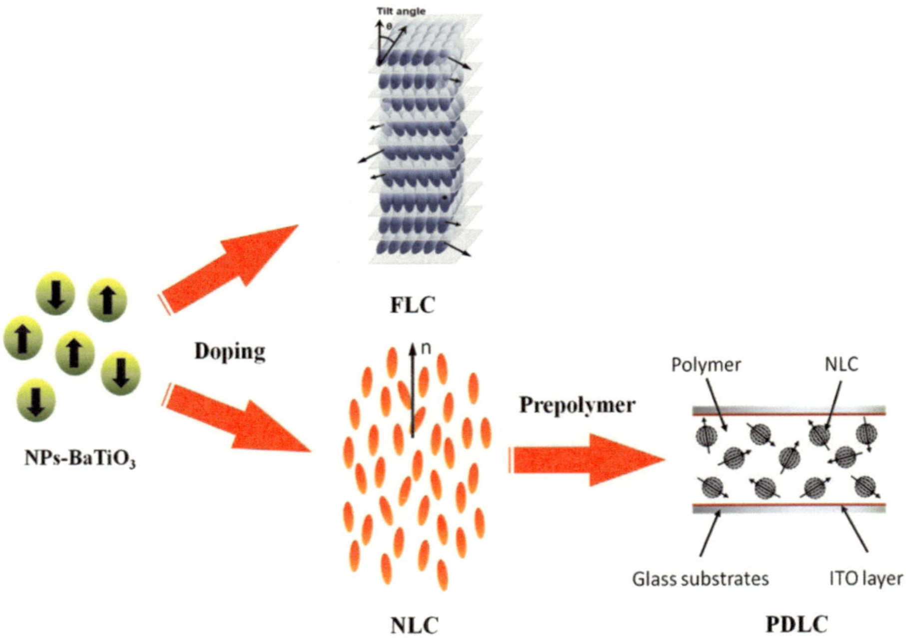

Fig. 1. Schematic representation of the proposed LC devices doped with ferroelectric $BaTiO_3$ nanoparticles.

Among all LC devices, the surface stabilised ferroelectric liquid crystal (SSFLC) mode, prepared using FLC, and the polymer dispersed liquid crystals (PDLC) mode, prepared using NLC, have received the most attention. By using an alignment layer and sufficiently thin cell gap (d <P, the helix pitch) to inhibit the formation of helical structures, the SSFLC structured device has many advantages including an enhanced memory effect, microsecond response time, and wide viewing angle (Lagerwall, 1999). On the other hand, PDLC light shutters combine the superior characteristics of both LCs and polymers, which provide several advantages: they require no polarizer, have fast response times, and can easily be fabricated into a large, flexible display (Crawford & Žumer, 1996; Drzaic, 1998). These are currently the most commonly used in switchable windows in modern buildings. However,

these two modes of light shutters still possess minor issues. In the case of SSFLC mode, the spontaneous polarization (P_s) value is an important factor when considering the response time of the light shutter, and often improvement can only be made via chemical synthesis; the PDLC mode has shortcomings including a high driving voltage and off-axis haze.

The content of this chapter focuses on the improvement of the electro-optical performance of two types of LC materials (FLC, NLC) by doping with ferroelectric $BaTiO_3$ nanoparticles (NPs-$BaTiO_3$). As shown in figure 1, the ferroelectric nanoparticles were well dispersed in the FLC and NLC respectively via appropriate wet grindings. By filling into the homogeneous cells, the former became a SSFLC mode; the latter formed an electrically controlled birefringence (ECB) mode. Subsequently, the addition of UV-curable prepolymer to fill the untreated substrate surface of the cell meant that the PDLC mode consisted of NLC, NPs-$BaTiO_3$ and the polymer. We introduced NPs-$BaTiO_3$ with a large electric dipole moment into the LC in order to enhance the physical properties (liquid crystal phase performance, dielectric properties, spontaneous polarization) and the electro-optical performance (V-T characteristics, and response time). Moreover, in the case of the PDLC mode, part of the NPs-$BaTiO_3$, coated by the polymer, can modify the refractive index of the polymer during phase separation, in order to increase the viewing angles of the 'field-on state' and reduce the off-axis haze effect.

2. Experiments

The characterization of materials (NPs-$BaTiO_3$, FLC, NLC, and prepolymer), sample preperation of NPs-$BaTiO_3$/ LC/polymer composites, and the measurements of the physical and electro-optical properties are described in this section.

2.1 Materials characterization

We made use of commercially available $BaTiO_3$ nanoparticles with an average size of 30-50 nm, polyhedron particle shapes (99+%, from Aldrich), tetrahedral crystal structure, [001] polar axis and a spontaneous polarization of 26 $\mu C/cm^2$ at room temperature. The dielectric constant of the $BaTiO_3$ single crystal is 168 in the direction parallel to the polar axis and 2,920 in the direction perpendicular to the polar axis. In contrast to other studies, which used oleic acid as the surfactant and heptane as the solvent, we made use of a solution of polymeric dispersant as a surfactant, taking advantage of the steric effect where the polymer chains stick to the ferroelectric nanoparticles enhance the dispersion. For the solvent, we used tetrahydrofuran (THF). The phase transition temperatures of the FLC CS1024 (fromChisso) we used are as follows: Cr $\xrightarrow{-12°C}$ SmC* $\xrightarrow{62°C}$ SmA $\xrightarrow{83°C}$ N* $\xrightarrow{93°C}$ Iso. At 25 °C, the spontaneous polarization was -46.9nC/cm^2 while the tilt angle was 25 degrees. The NLC DN113245 (from Daily Polymer) with positive dielectric anisotropic of $\Delta\varepsilon$=13.03 and birefringence Δn=0.24 at 25 °C exhibits a thermotropic transition sequence of Cr $\xrightarrow{-32°C}$ N $\xrightarrow{83°C}$ Iso. The UV-curable prepolymer NOA65 (from Norland Optical Adhesive) with a refractive index of n_p=1.52 (the same as the NLC) was employed for polymerization induced phase separation.

2.2 Sample preparation

We used wet grinding dispersion equipment and yttria-stabilized zirconia (YSZ) as the grinding media. The commercially available ferroelectric $BaTiO_3$ nanoparticles, polymeric surfactant and THF were evenly mixed according to the weight ratio of 1: 0.15: 11. YSZ beads of the appropriate size were then chosen for 2 hour wet grinding. After ultrasonic dispersion, the $BaTiO_3$ suspension was added into FLC and NLC, and a vacuum was employed to evaporate the THF. After ultrasonic dispersion for an hour, we successfully prepared samples with different doping concentrations: Pure FLC, FLC+0.1wt% $BaTiO_3$. FLC+1wt% $BaTiO_3$, pure NLC, NLC+0.1wt% $BaTiO_3$, and NLC+0.5wt% $BaTiO_3$. The homogenous cells consisted of two indium tin oxide (ITO) coated glass substrates with a polyimide (PI) layer rubbed in anti-parallel directions to obtain the homogeneous alignment. The 2 μm and 4 μm homogeneous cells, which were controlled by dispersion of cylindrical glass spacers, were filled with $BaTiO_3$ doped FLC and $BaTiO_3$ doped NLC at above both clearing temperatures T=95°C > Tc, to produce the surface stabilized ferroelectric liquid crystal (SSFLC) and electrically controlled birefringence (ECB) modes. For fabrication of PDLC, the NOA65, equivalent to NLC, was added to the $BaTiO_3$ doped NLC. The NLC- $BaTiO_3$/prepolymer mixture was injected into an empty cell whose inner surfaces were coated with a thin indium-tin-oxide (ITO) electrode. The cell gap was measured to be d=16 μm. The filled cell was then exposed to 3 minutes of UV light (20 mW/cm^2) at room temperature for polymerization induced phase separation (PIPS) to obtain the PDLC devices.

2.3 Instruments

The dispersion was implemented by a wet grinding dispersion equipment (Just Nanotech, JBM-B035). Particle size analyzers (PSA, Brookhaven 90Plus/BI-MAS) and a transmission electron microscope (TEM, JEOL 2000FXII) were used for the measurement of the size distribution of the $BaTiO_3$ nanoparticles. Polarizing optical microscopy (POM, OLYMPUS Optical Co., Ltd., Models BHSP-2, BX-51) was used to observe the liquid crystal phases, differential scanning calorimeter (DSC, PerkinElmer PYRIS 1) for the identification of the liquid crystal phase transition temperatures, scanning electron microscopy (SEM, Leica LEO 420) for PDLC morphology analysis, and low frequency impendence analysis (LFIA, Hewlett Packard 4192A) for measuring the dielectric properties. A Fourier transform infrared (FTIR) spectrometer, mass spectrometer (Finnigan TSQ- 700), and an elemental analyzer (HERAEUS Vario EL-III) were used for qualitative. To determine the value of the spontaneous polarization, we used triangular wave measurements (Miyasato et al, 1983).

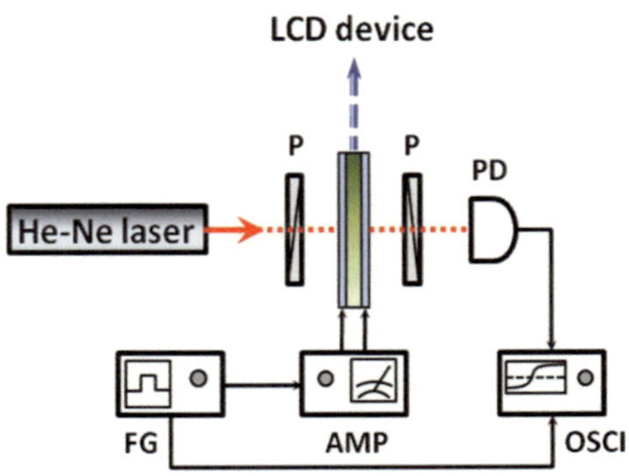

Fig. 2. Experimental setup for measuring the electro-optical properties of the LCD devices with an applied AC voltage. P, PD, FG, AMP, and OSCI represent polarizer, photodiode detector, function generator, high-voltage amplifier, and oscilloscope, respectively.

The electro-optical properties, shown in Figure 2, were obtained using a He-Ne laser (632.8nm, 10mW) as a light source, with orthogonal polarizers either side of the PDLC samples, and the light transmitted through the sample was collected by a photodiode (PD). The electrical signal from the PD was fed to an oscilloscope (OSCI). The AC electric field was supplied by a function generator (FG) in conjunction with a high-voltage amplifier (AMP).

3. Results and discussions

3.1 Identification of the host LC materials and NPs-$BaTiO_3$ after sample preparation

Vacuum treatment was used to remove the solvent (heptane), which could possibly result in the loss of lower molecular weight components in the nematic liquid crystal (Li et al, 2007). To ensure the experiment isolated the effects of the addition of nanoparticles, we performed qualitative and quantitative analysis to ascertain that the composition of the FLC and NLC mixture before and after vacuum treatment, and found that it did not change. The former involved the use of a FTIR spectrometer and mass spectrometer whereas the latter utilized elemental analyzer measurements. The qualitative and quantitative results confirmed that the composition of the FLC CS1024 and NLC DN113245 mixtures was unchanged after vacuum treatment. Combining both the qualitative and quantitative analyses, we can conclude that the changes in the electro-optical properties were entirely the result of the contribution from the $BaTiO_3$ suspension and not due to changes in the composition of the host LC materials during the preparation process.

Before doping, the PSA and TEM were used to determine the particle size distribution of the $BaTiO_3$ nanoparticles. After wet grinding to create the $BaTiO_3$ suspension, we added THF to yield a diluted 0.15 wt% $BaTiO_3$ suspension and performed particle size analysis using the dynamic light scattering technique. Figure 3 (a) shows the results of the PSA measurement, with 99% of the $BaTiO_3$ particles having a diameter of less than 100 nm. The mean diameter was 39.9 nm. Figures 3 (b) and (c) show the TEM measurements before and after grinding. In Figure 3 (b), the average $BaTiO_3$ particle size before grinding was about 90 nm; whereas in Figure 3 (c), the average $BaTiO_3$ particle size after grinding was about 31 nm. This indicates that the wet grinding and dispersion had successfully disrupted the aggregation of the nanoparticles, yielding a optimal nanoscale for ease of doping.

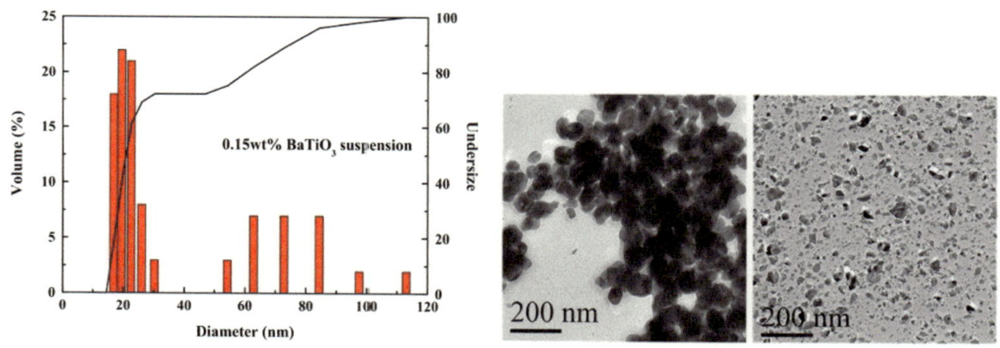

(a) (b) (c)

Fig. 3. (a) The results of the PSA measurement of 0.15wt% $BaTiO_3$ suspension. Transmission electron micrographs of $BaTiO_3$ nanoparticles (a) before and (b) after wet grinding.

3.2 Effect of doping NPs-$BaTiO_3$ on the physical and electro-optical properties of SSFLC mode

The range of liquid crystal phase transition temperatures after doping were measured by POM and DSC. Since the FLC material, CS1024, is a compound whose phase transition temperature is unlike a single component liquid crystal material, it is particularly hard to discern, especially for the SmA and SmC* phase transitions. Table 1 shows the phase transition temperatures of FLC with different doping concentrations. We can see that the pure FLC and FLC+0.1wt% $BaTiO_3$ had almost identical phase transition temperatures at $T_{N^*\text{-}I}$, $T_{N^*\text{-}SmA}$, $T_{SmA\text{-}SmC^*}$, whereas there were slight decreases in the various phase transition temperatures of the FLC+1.0 wt% $BaTiO_3$. In particular, there was a decrease of about 4 °C for the phase transition temperature of $T_{SmA\text{-}SmC^*}$. This result can alternatively be verified by measurements of the dielectric properties and spontaneous polarization.

Sample	Phase transition temperature
pure FLC	$SmC^* \xleftrightarrow{61.15°C} SmA \xleftrightarrow{82.58°C} N^* \xleftrightarrow{90.35°C} Iso$
FLC+0.1wt%$BaTiO_3$	$SmC^* \xleftrightarrow{61.01°C} SmA \xleftrightarrow{82.35°C} N^* \xleftrightarrow{90.12°C} Iso$
FLC+1.0wt%$BaTiO_3$	$SmC^* \xleftrightarrow{56.78°C} SmA \xleftrightarrow{81.89°C} N^* \xleftrightarrow{89.68°C} Iso$

Table 1. The phase transition temperature of FLC with different doping concentrations of $BaTiO_3$ suspensions.

After infusing the pure FLC and the FLC suspension into a 2 μm liquid crystal cell to create a SSFLC mode, one can observe the texture of the higher doping concentration FLC+1.0wt% $BaTiO_3$ under 200 times POM magnification (Figure 4). One can see clearly that there exists almost no characteristic texture of SmC*, indicating that we had successfully created a SSFLC mode. Other samples with different doping concentrations exhibited similar textures.

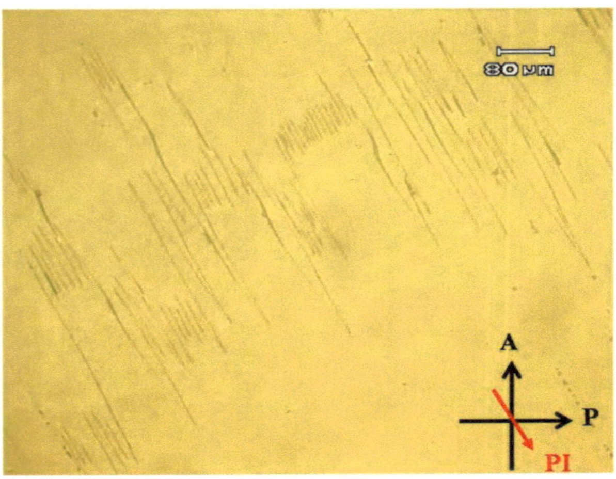

Fig. 4. The SSFLC texture of the FLC+1wt% $BaTiO_3$ at 30°C, under 200 times POM magnification. (The red arrow indicates the PI alignment rubbing direction)

To investigate the effect of doping concentration on the spontaneous polarization, we adopted a two-pronged approach: varying the applied voltage at a constant temperature and varying the temperature at the saturation voltage, using the triangular wave measurements for samples with different doping concentrations. Figure 5 shows the relationship between the temperature and spontaneous polarization for different doping concentrations of $BaTiO_3$ suspensions at f=10 Hz and V_{p-p}=20 V. The horizontal axis is the phase transition temperature from SmA to SmC*, for samples with different doping concentrations. We can see that the spontaneous polarization increased rapidly without reaching a maximum as the temperature was cooled to the SmC* phase. In order to understand the sensitivity of the applied voltage by the doping concentration, we attempted to fix the temperature at 35 °C in the SmC* liquid crystal phase, apply triangular waves with identical frequency (f=10 Hz) but different voltages and observe the relationship between the voltage and spontaneous polarization for different doping concentrations of $BaTiO_3$ suspensions (Figure 6). We can see a significant increase in the absolute values of the spontaneous polarization. The P_s of the FLC+0.1 wt% $BaTiO_3$ (slope=14.47099) and FLC +1.0wt% $BaTiO_3$ (slope=31.21684) also increased with the applied voltage. The slopes for the P_s values were also greater than those of pure FLC (slope=10.10697). This further proves that the doping had significantly improved the sensitivity of the FLC under an applied electric field. In particular, the FLC+0.1 wt% $BaTiO_3$ exhibited the largest spontaneous polarization, even reaching a value that is twice that of pure FLC (80 nC/cm^2).

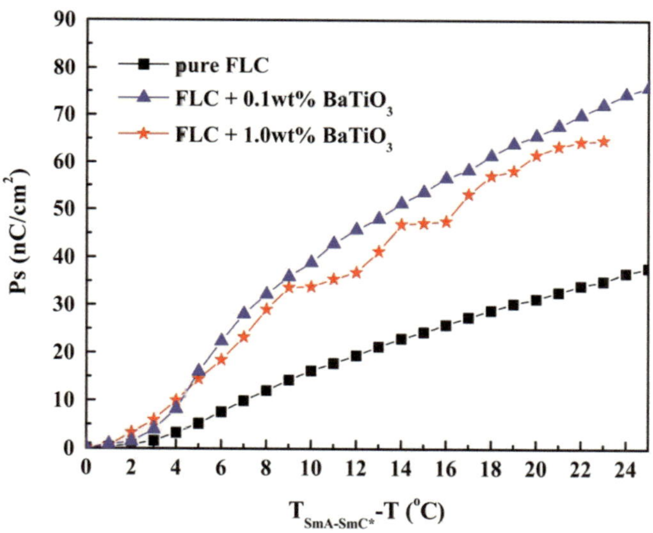

Fig. 5. The dependence of the spontaneous polarization for FLC with different doping concentrations of $BaTiO_3$ suspensions on the temperature at f=10Hz and V_{p-p}=20V.

Assuming that the P_s resulted from the contributions of the $BaTiO_3$ and FLC, we can apply the zeroth order approximation to estimate the P_s value after doping:

$$P_s^{susp.} = (1 - f_w)P_s^{LC} + f_w P_s^{particles} \tag{1}$$

where f_w is the weight ratio of the $BaTiO_3$ suspension. Using the above formula, one obtains a P_s value of about 68 nC/cm^2 for the lower doping concentration FLC+0.1 wt% $BaTiO_3$ and about 300 nC/cm² for the higher doping concentration FLC+1.0wt% $BaTiO_3$ These calculated values are much higher than the experimental value of 65 nC/cm². One possible explanation for the discrepancy is that when the interactions between particles were ignored, the larger size of the nanoparticles than the liquid crystal molecules led to disruption of the FLC stacking, resulting in a smaller P_s value than expected.

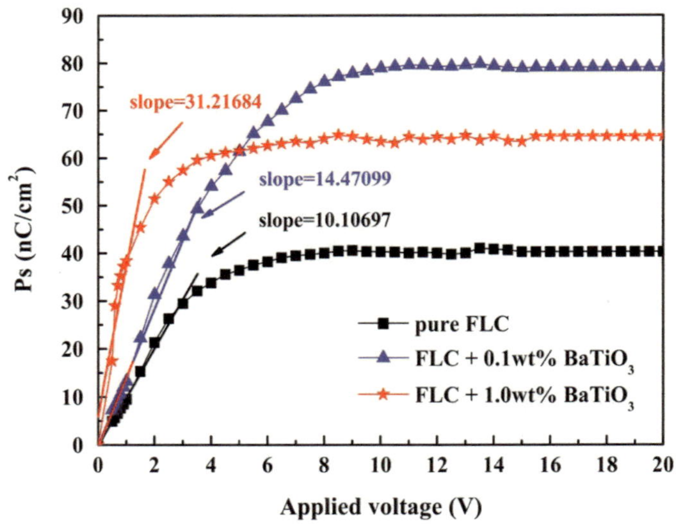

Fig. 6. The dependence of the spontaneous polarization for FLC with different doping concentrations of $BaTiO_3$ suspensions on the applied voltage at 35°C

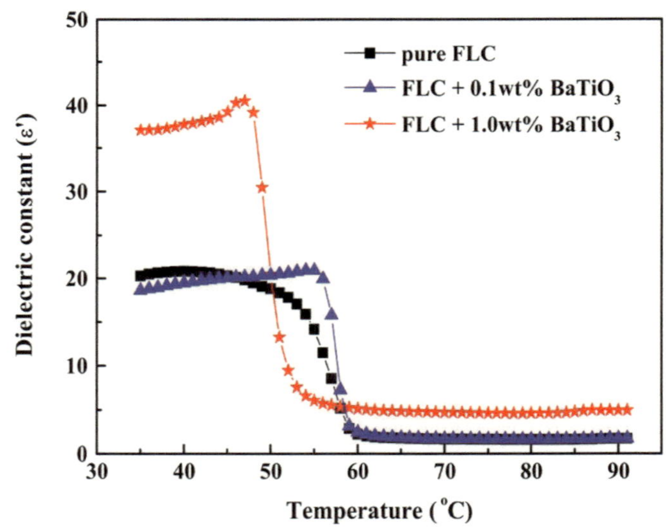

Fig. 7. The dependence of the dielectric constant (ε') for FLC with different doping concentrations of $BaTiO_3$ suspensions on the temperature at 1 kHz.

Figure 7 shows the relationship between the permittivity and the temperature of the pure FLC, FLC+0.1wt% $BaTiO_3$ and FLC+1.0wt% $BaTiO_3$ at a frequency of 1 kHz. The permittivity increased drastically when cooling to 60 °C in the SmC* phase. We can see from Figure 7 that there was very little difference in the permittivity for the different liquid crystal phases of pure FLC and FLC+0.1 wt% $BaTiO_3$ whereas the permittivity of the various liquid crystal phases of FLC+1.0 wt% $BaTiO_3$ were twice those of the others. In particular, the maximum permittivity, 42.9, occurred at 49 °C while the average permittivity of its SmC* phase was approximately 1.5 times those of the pure FLC and FLC+0.1wt% $BaTiO_3$ Therefore, one can see that the doping of $BaTiO_3$ effectively enhanced the permittivity of the liquid crystal material with its large electric dipole moment. In addition, one can also observe the significant differences in the slopes of the permittivity curves when the pure FLC, FLC+0.1wt% $BaTiO_3$ and FLC+1.0wt% $BaTiO_3$ samples entered the SmC* phase. A comparison of the pure FLC and FLC+0.1wt% $BaTiO_3$ revealed that while there was little difference between the permittivity, there was a very significant increase in the slope of the permittivity curve. The effect was especially prominent in the FLC+1.0 wt% $BaTiO_3$, thereby further affirming the observations regarding spontaneous polarization. The doping of NPs-$BaTiO_3$ into the liquid crystal material had enhanced their sensitivity to applied electric fields, and the permittivity curve exhibited a rapid increase upon entering the SmC* phase, before rising to the maximum value.

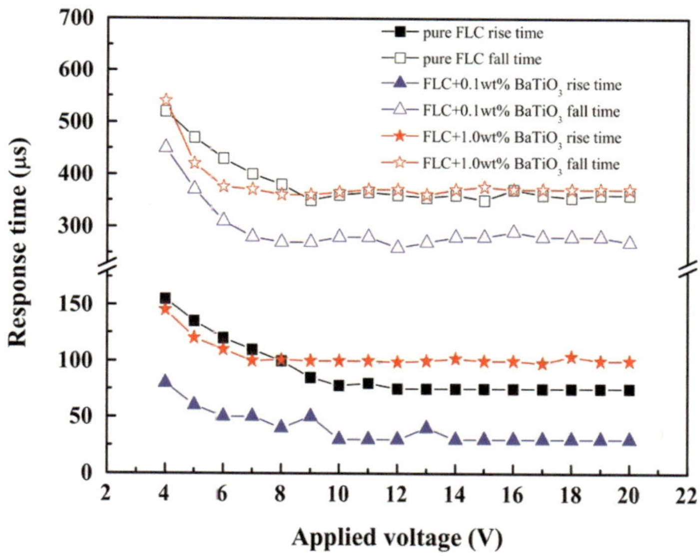

Fig. 8. The dependence of response time for FLC with different doping concentrations of $BaTiO_3$ suspensions on the applied electric field.

To investigate the response time of the SSFLC mode, $V_{p\text{-}p}$ = 20 V, f =10 Hz was applied at a constant temperature of 35 °C. Figure 8 shows the relationship between the applied voltage and the response time. One can see that the response time for the pure FLC, FLC+0.1 wt% $BaTiO_3$ and FLC+1.0wt% $BaTiO_3$ decreased rapidly before saturating with increased applied voltage. This is evidence that the response time will saturate regardless of the applied voltage once the saturation voltage had been exceeded. The response times for all three are tabulated in Table 2. The FLC+1.0 wt% $BaTiO_3$ had the minimum rise and fall times. The rise and fall time values in descending order were found in FLC+1.0 wt% $BaTiO_3$, pure FLC and FLC+0.1 wt% $BaTiO_3$. The response time is the sum of the rise and fall times, and took values of 435 μs, 310 μs and 470 μs, with increased doping concentrations. In addition, from Equation (Kimura et al, 1987):

$$\tau^{-1} = \frac{\tau P_s E}{1.76\gamma_{\varphi}} \tag{2}$$

where τ is the response time, γ_{φ} is the intrinsic viscosity, P_s is the spontaneous polarization, E is the electric field strength, we can infer that the FLC+0.1wt% $BaTiO_3$ with low doping concentration and the largest spontaneous polarization under the same electric field, will have a shorter response time. On the other hand, while the spontaneous polarization of the FLC+1.0 wt% $BaTiO_3$ was greater than that of pure FLC, the larger molecular weight of the polymeric surfactant in the suspension resulted in an overall increase in viscosity. The interplay of the two led to an increase in the response time. Taking into consideration the rise and fall time performances of the different doping concentrations, we can conclude that the FLC+0.1 wt% $BaTiO_3$ is optimal.

Sample	Rise time (μs)	Fall time (μs)	Response time (μs)
pure FLC	75	360	435
FLC+0.1wt% $BaTiO_3$	30	280	310
FLC+1.0wt% $BaTiO_3$	100	370	470

Table 2. The response time of the SSFLC mode of FLC with different doping concentrations of $BaTiO_3$ suspensions.

The V-shaped switching of the SSFLC mode is shown in Figure 9, and we compared two scenarios: identical concentration but different frequencies as well as identical frequency but different concentrations. First of all, applying triangular waves of different frequencies at identical doping concentration, one can see that the hysteresis phenomenon became more pronounced with increasing frequency of the applied electric field (5 Hz to 10 Hz), resulting in a pseudo W-shaped switching. Therefore, when the curve passed through zero electric field, it was not possible to obtain a relatively dark state. There was also a phase shift in the relatively dark state, due to the fact that as the frequency of the applied electric field was increased; the liquid crystal molecules became unable to catch up with the switching frequency. On the other hand, for the V-shaped switching of triangular waves with identical frequency but different doping concentrations, the FLC+0.1 wt% $BaTiO_3$ exhibited the best V-shaped switching at 5 Hz, with no hysteresis phenomenon observable in the figure. The V-shaped switching properties of the FLC+0.1 wt% $BaTiO_3$ were superior to the pure FLC at different frequencies, proving that doping $BaTiO_3$ resulted in an enhancement of the V-shaped switching.

In particular, we examined in detail the case with an electric field applied at a frequency of 5 Hz and high doping concentration (FLC+1.0wt% $BaTiO_3$). We found that the gray scale performance were inferior to the pure FLC and FLC+0.1wt% $BaTiO_3$, but it was worth noting that the voltage required for switching between the two ferroelectric states (the region demarcated by the red dashed line in the figure) were smaller than those for the pure FLC and FLC+0.1wt% $BaTiO_3$. From this phenomenon, we can indirectly infer that doping $BaTiO_3$ in the liquid crystal materials enhances sensitivity to applied electric fields.

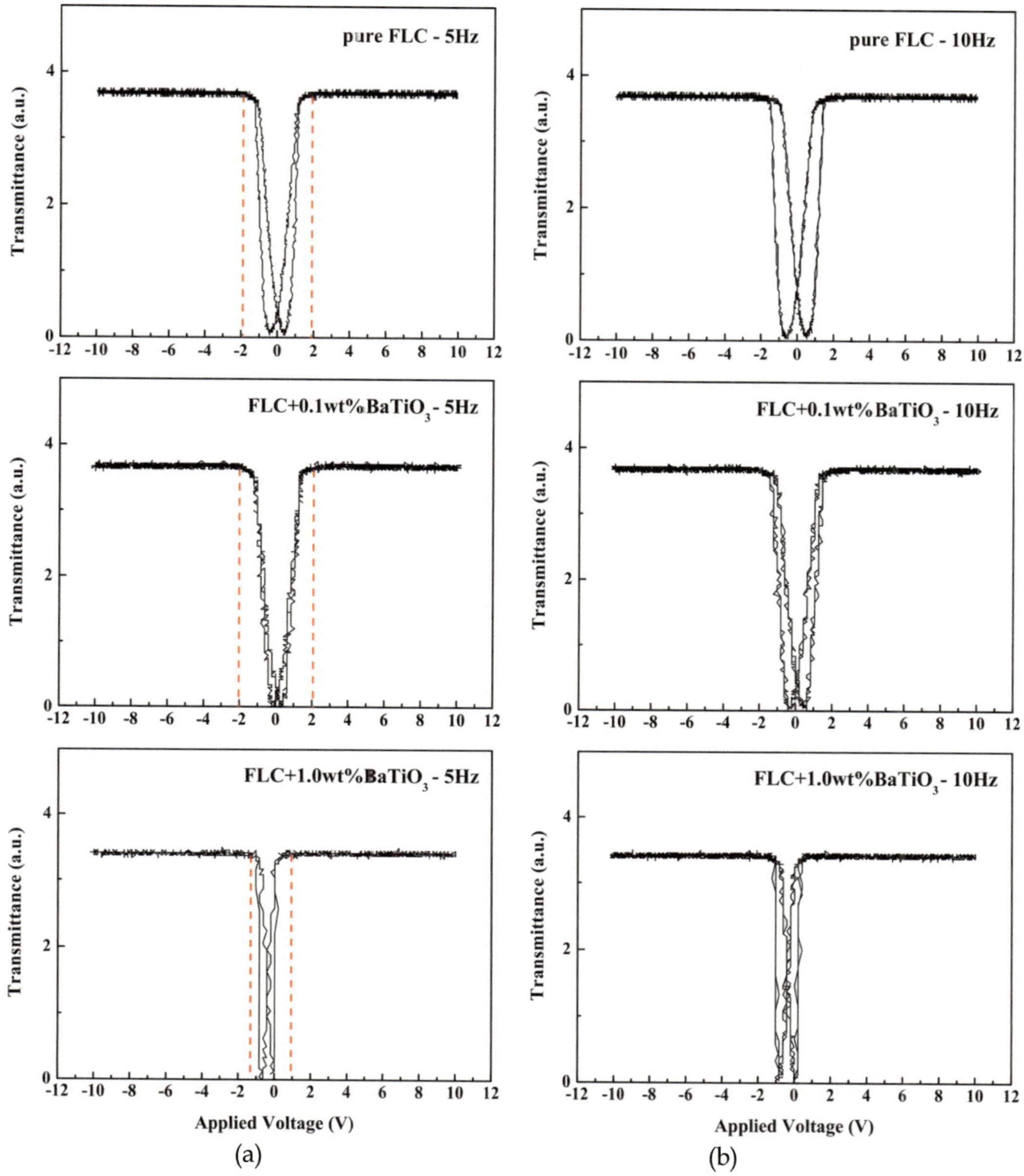

Fig. 9. The dependence of the transmittance of FLC with different doping concentrations of $BaTiO_3$ suspensions on the applied triangular waveform voltage at (a) 5 Hz and (b) 10 Hz.

(The red dashed line represents the switching between the two ferroelectric states)

3.3 Effect of doping NPs-$BaTiO_3$ on the physical and electro-optical properties of PDLC mode

The PDLC device is controlled by the micro nematic droplets, coated by the polymer matrix. To understand the effect of doping NPs-$BaTiO_3$ in the PDLC, one must first determine the changes in the physical and electro-optical properties of the NLC after doping NPs-$BaTiO_3$.

By POM and DSC measurements, we found that the nematic-isotropic transition temperatures (T_{NI}) for the samples with various concentrations were almost identical, at $T_{NI\text{-}pure}$= 83.1°C, $T_{NI\text{-}0.1wt\%}$= 81.9°C and $T_{NI\text{-}0.5wt\%}$= 79.9°C respectively. On the other hand, texture observation results revealed that a small increase in the concentration of defects occurred with increasing concentrations of NPs-$BaTiO_3$ in the NLC.

The anisotropic dielectric constants were measured using a single-cell method and were obtained from the characteristic relationship between capacitance and voltage (Wu et al., 1991). When the voltage was lower than the threshold voltage, the electric field direction was perpendicular to the liquid crystal director. The measured capacitance value is represented as $C_\perp$ and $\varepsilon_\perp$ was calculated; whereas $C_{/\!/}$ and $\varepsilon_{/\!/}$ were obtained by extrapolating the relationship of the capacitance and V_{th}/V, where V_{th} is the threshold voltage. From Figure 10, it can be observed that when the temperature is reduced into the range of the liquid crystal phase, $\varepsilon_\perp$ decreases according to the decreasing temperature, but $\varepsilon_{/\!/}$ increases inversely. After doping NPs-$BaTiO_3$, which has a large electric dipole moment, $\varepsilon_\perp$ and especially $\varepsilon_{/\!/}$ also become significantly larger. Comparing these two results, anisotropic dielectric constants increase according to the increasing concentration of the dopant, which are respectively, $\Delta\varepsilon_{\text{-}pure}$ = 13.03, $\Delta\varepsilon_{\text{-}0.1wt\%}$ = 13.88, $\Delta\varepsilon_{\text{-}0.5wt\%}$ = 14.74.

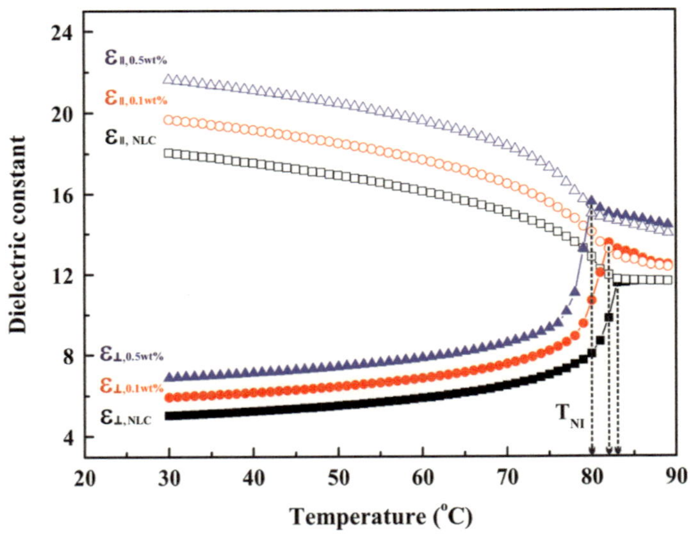

Fig. 10. The dependence of the dielectric constants ($\varepsilon_\parallel$ and $\varepsilon_\perp$) for NLC with different doping concentrations of $BaTiO_3$ suspensions on the temperature.

Next, we studied the V-T characteristics of NLC doped NPs-$BaTiO_3$. We made use of the homogeneous ECB mode to measure the threshold voltage before and after doping. The transmittance under the homogeneous ECB mode is given by the following formula (Chigrinov, 1999):

$$T = \frac{1}{2}\left[\cos^2(\varphi_1 - \varphi_2) - \sin 2\varphi_1 \sin 2\varphi_2 \sin^2\left(\frac{\delta}{2}\right)\right] \tag{3}$$

where φ_1 and φ_2 are the angles between the orientation direction and the two polarizers, and δ is the phase retardation. We set the polarizers such that φ_1=φ_2=45°, and in the absence of an applied electric field, the transmittance reached its maximum and varied periodically with variations in the electric field, as shown in figure 11(a). The derived V_{th} in figure 11(b) is consistent with the V_{th} obtained from the relationship of voltage and capacitance. The results showed that, after doping, V_{th} would be 10% and 26% lower respectively compared with before doping. It was reduced from $V_{th\text{-}pure} = 0.95$ V to $V_{th\text{-}0.1wt\%} = 0.85$ V and finally to $V_{th\text{-}0.5wt\%} = 0.70$ V. The threshold voltage (V_{th}) relationship for homogeneous ECB equation was used:

$$V_{th} = \pi \sqrt{\frac{K_{11}}{\varepsilon_0 |\Delta\varepsilon|}} \qquad (4)$$

where K_{11} is the splay elastic constant of the NLC, ε_0 is the vacuum permittivity and $\Delta\varepsilon$ is the anisotropic dielectric constant. Using the equation to substitute for $\Delta\varepsilon$ and V_{th}, we calculated the splay elastic constants to be $K_{11\text{-}pure}$=10.55 pN, $K_{11\text{-}0.1wt\%}$=8.99 pN, and $K_{11\text{-}0.5wt\%}$=6.48 pN respectively. Note that the splay elastic constants changed significantly after doping.

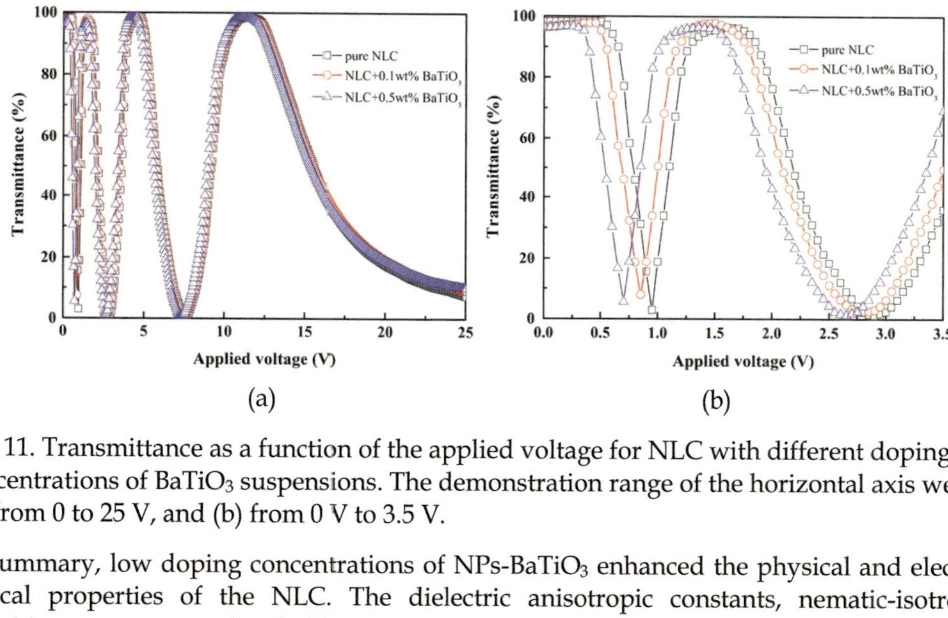

(a) (b)

Fig. 11. Transmittance as a function of the applied voltage for NLC with different doping concentrations of $BaTiO_3$ suspensions. The demonstration range of the horizontal axis were (a) from 0 to 25 V, and (b) from 0 V to 3.5 V.

In summary, low doping concentrations of NPs-$BaTiO_3$ enhanced the physical and electro-optical properties of the NLC. The dielectric anisotropic constants, nematic-isotropic transition temperature, thershold voltage, and splay elastic constant are shown in Table 3.

Sample	$\Delta\varepsilon$	V_{th} (V)	T_{NI} (°C)
pure NLC	13.03	0.95	83.2
NLC+0.1wt%$BaTiO_3$	13.88	0.85	81.9
NLC+0.5wt%$BaTiO_3$	14.74	0.70	79.9

Table 3. Comparison of dielectric anisotropy, threshold voltage, and phase transition temperatures for pure liquid crystals and liquid crystals with ferroelectric nanoparticles.

After preparation of the PDLC films, a square wave electric field (1 kHz) was applied to measure the V-T characteristics of the three PDLC films with different doping concentrations, as shown in Figure 12. All three PDLC films exhibited typical V-T characteristics of PDLC. As the applied electric field was increased, the transmittance increased unit a saturation threshold was reached (saturated transmittance, T_s). $T_{s\text{-}pure}$ = 99.7%, $T_{s\text{-}0.1wt\%}$ = 98.9% and $T_{s\text{-}0.5wt\%}$ = 96.6%, displaying a tendency to decrease as doping concentration increases. On the other hand, the decrease in the driving voltage (V_d) was observed from the Figure 12. Although the degree of voltage decline was incomparable to the voltages during the ECB mode, doped with NPs-$BaTiO_3$, here the 0.1wt% doping and 0.5wt% doping were respectively 4% and 15% lower compared to the pure PDLC.

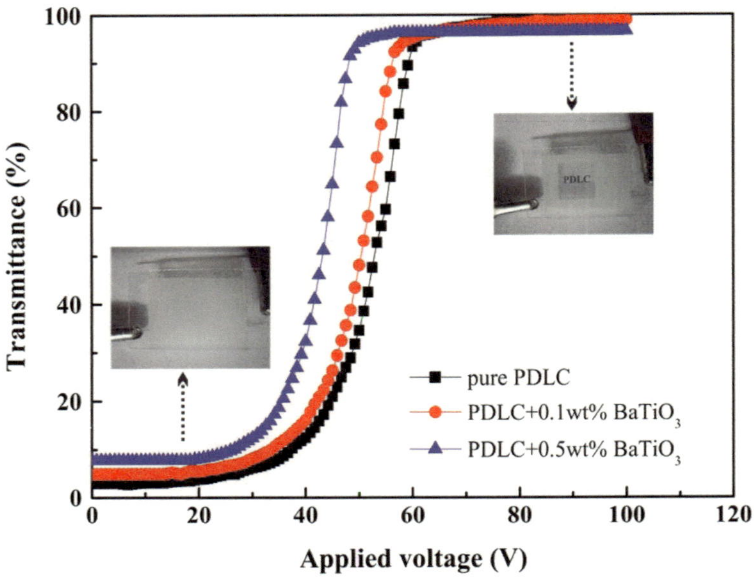

Fig. 12. The dependence of the transmittance for PDLC with different doping concentrations of $BaTiO_3$ suspensions on the applied voltage at 1 kHz. The photographs of PDLC+0.5wt% $BaTiO_3$ film are shown in the inset.

In order to understand the T_s and V_d after doping, we assessed the PIPS method. During the UV polymerization process, the increase in the polymer molecular weight led to a decrease in the immiscibility of the polymer and LC. When the immiscibility is sufficiently low, phase separation will begin. The decline in T_s with respect to increasing doping concentration indirectly confirmed that NPs-BaTiO3 was in the polymer phase during the phase separation. As the refractive index of NPs-$BaTiO_3$ (n_{NPs} = 2.42) is higher than the refractive index of the polymer (n_p = 1.52), the refractive index of the polymer would increase after doping, compared to the initial value while matching with the liquid crystal (n_p and $n^o{}_{LC}$). When the NPs-$BaTiO_3$ dopant was introduced, there was a refractive index mismatch. When the incoming light was incident in the same direction as the applied electric field (perpendicular to the cell surface), a small portion of the light was scattered, resulting in a slight decline in the value of T_s (Yaroshchuk & Dolgov, 2007). On the other hand, the driving voltage relationship for the PDLC is given as (Drzaic, 1998):

$$V_d = \frac{d}{3a}\left(\frac{\rho_p + 2}{\rho_{LC}}\right)\left[\frac{K(l^2-1)}{\Delta\varepsilon\varepsilon_0}\right]^{\frac{1}{2}} \tag{5}$$

where d is the layer thickness; $l=a/b$ is the ratio of a, the length of the semi-major axis, and b, the length of the semi-minor axis; K is the average elastic constant; ρ_p is the resistivity of the polymer and ρ_{LC} is the resistivity of the liquid crystal. Comparing with the results from the ECB mode, under the assumption that the NPs-$BaTiO_3$ dopant does not affect the size and shape of nematic droplet (which we will confirm in the next section), we can reasonably infer that only a portion of NPs-$BaTiO_3$ remains in the droplet after phase separation. This limits the alteration on the inversely proportional relationship of V_d to the anisotropic dielectric constant and the directly proportional relationship of V_d to the elastic constant. Summarizing the measured results of T_s and V_d, the NPs-$BaTiO_3$ dopant was in polymer phase and altered the n_p. Some part remained in the droplet and altered the physical properties of LC. The insert of Figure 12 illustrates the vertical view of the PDLC+0.5wt% $BaTiO_3$ film. When the applied voltage was below V_d, the PDLC light shutter was scattering and could block the characters behind. When the saturation voltage was applied, The shutter was transparent and the images with the characters "PDLC", which were placed at 2 cm behind the cells, were clearly visible.
The LC droplet size in PDLC is a critical factor in determining the electro-optical properties of these devices. To confirm the hypothesis of the size and shape of the droplets, the sections of the PDLC films were carried out through SEM. The SEM results indicated that the LC droplet shape was spherical and almost the same both before and after the doping. An SEM photograph of a cross section of the PDLC+0.5wt% $BaTiO_3$ film is shown in figure 13(a). The droplet sizes of different doping concentration were precisely measured and the respective number distributions N(D) are summarized in figure 13(b). The results showed that all droplet sizes have a peak distribution with average values of $D_{pure\ PDLC} = 2.15 \pm 0.05\mu m$, $D_{PDLC+0.1\ wt\%} = 2.15 \pm 0.06\mu m$, $D_{PDLC+0.5\ wt\%} = 2.16 \pm 0.06\mu m$. In conclusion, the effect of NPs-$BaTiO_3$ on the size and shape of droplets were not significant, which is also consistent with the inference above.

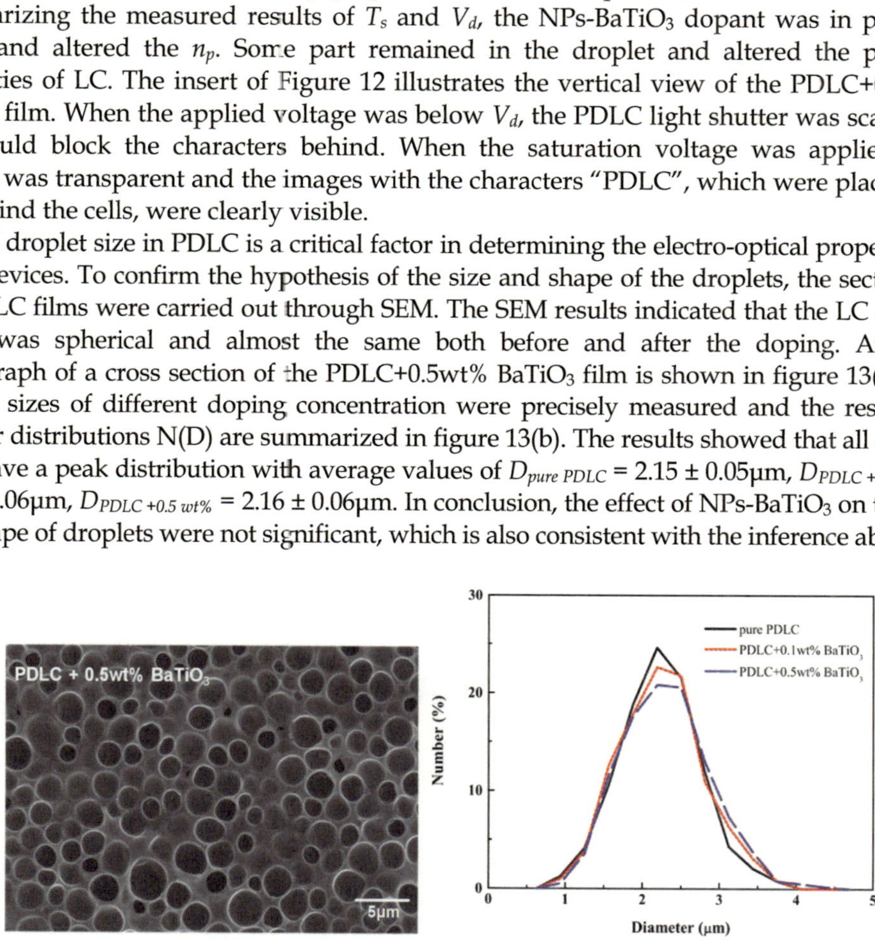

(a) (b)

Fig. 13. (a)Scanning electron microscope photograph of a cross section of the PDLC+0.5wt% BaTiO3 film. (b)The number-weighted distributions for PDLC films with different doping concentrations of $BaTiO_3$ suspensions.

Figure 14 shows the dependences of T_s on the incidence angle of the laser beam for PDLC doped with different concentrations of NPs-$BaTiO_3$. It can be clearly observed that T_s has a tendency to stay in the center of the peaks, $a = 0°$. As the angle between the incidence light and electric field increased, T_s decreased. When $a = 90°$, $T_s \approx 0\%$. For pure PDLC, as a result of material selection, the ordinary refractive index, n^o_{LC}, of the selected LC is almost identical to the refractive index of the polymer, n_p. However, it is less than the extraordinary refractive index of LC, therefore

$$n_{LC}^{o} \approx n_p < n_{LC}^{e} \tag{6}$$

This equation creates two phenomena. The first is a high saturated transmittance that is due to the electric field effect of cell substrates in the vertical direction. Under this effect, the NLC droplets with random orientation were gradually aligned to be parallel with the electric field and n^{o}_{LC} became nearer to n_p, allowing PDLC to have high transmittance under normal light incidence. The second effect is the enhanced scattering of oblique light due to refractive index mismatches. This scattering effect becomes more obvious with increasing angles, which is recognized as an off-axis haze effect. In summary, when the equation above is met, the T_s of PDLC is more sensitive toward the changes in the angle of incident light.

As the doping concentration increased, the amount of NPs-$BaTiO_3$ in the polymer increased. Further, the refractive index of NPs-$BaTiO_3$, n_{NP}, is larger than n_p, so n_p would gradually become larger than n^{o}_{LC} after doping, giving

$$n_{LC}^{o} < n_{p-NP} < n_{LC}^{e} \tag{7}$$

Although this result gradually reduced T_s, which is similar to the V-T characteristic results in the previous section, the peak of the viewing angle becomes wider, as shown in figure. Among the experiment samples, PDLC+0.5wt% $BaTiO_3$ reduced the off-axis haze effect, and provided the best viewing performance. While PDLC+0.1wt% $BaTiO_3$ also performed better than pure PDLC, despite the lower doping concentration. Although using the modified refractive index of polymer matrix to reduce the off-axis haze effect and widen the on-state view results in a decline in T_s, it had little effect on the contrast ratio of the transparent-scattering state. The competition of these two phenomenons, introducing NPs-$BaTiO_3$ into PDLC still needs to be studied further.

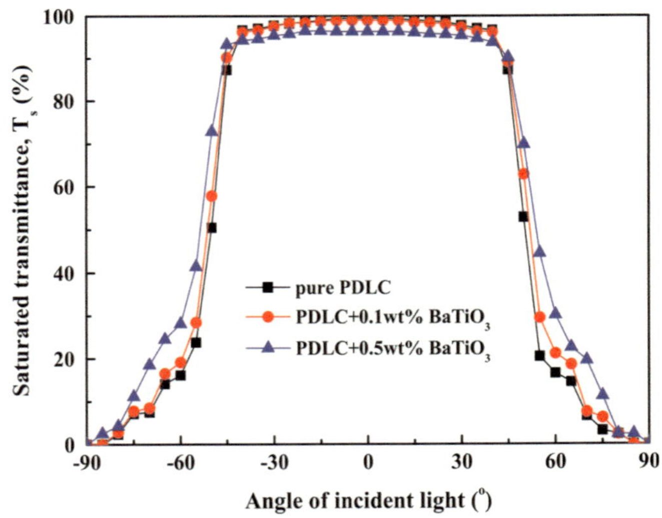

Fig. 14. The saturated transmittance T_s for PDLC films with different doping concentrations of $BaTiO_3$ suspensions as a function of the angle of incident light.

4. Conclusions

Low concentration NPs-$BaTiO_3$-doped LCD demonstrated very promising results. Without disrupting the structure and composition of the host LC, NPs-$BaTiO_3$ shares its intrinsic features with the host LC by enhancing the dielectric properties, spontaneous polarization and other vital physical properties of the host LC. This further improves the electro-optical properties of the LC device. For the case of FLC, the spontaneous polarization of FLC+ 0.1 wt% $BaTiO_3$ was about twice that of pure FLC. This also means that we can adjust the spontaneous polarization of FLC by doping with NPs-$BaTiO_3$ and eliminate the need for time-consuming molecular design and chemical synthesis. After completing the SSFLC light shutter, the V-shaped switching, response time and other electro-optical performance also have been significantly improved. Considering NLC, after doping increases in the anisotropic dielectric constants resulted in decreases in the threshold voltage in both ECB and PDLC modes. It is worth noting that part of the NPs-$BaTiO_3$ in the polymer altered the refractive index of the polymer, resulting in a wider viewing angle and improved the off-axis haze. When compared to previous methods of improve the viewing angle by placing an additional polarizer in front of the PDLC light shutter (West et al., 1992) NPs-$BaTiO_3$ doping provides better light transmittance and is more practical. In summary, using a simple doping technique to modify material properties not only provides non-chemical synthesis methods to improve the applicability of LC devices with shorter means, but also means that the drive modules for LC devices are cheaper.

5. References

Brochard, F. & de Gennes, P. G. (1970). Theory of magnetic suspensions in liquid crystals. *J. Phys. (France)*, Vol. 31, No. 7, pp. 691-708.

Chandani, A. D. L.; Hagiwara, T.; Suzuki, Y.; Ouchi, Y.; Takezoe, H. & Fukuda, A. (1988). Tristable Switching in Surface Stabilized Ferroelectric Liquid Crystals with a Large Spontaneous Polarization . *Jpn. J. Appl. Phys.*, Vol. 27, No. 5, pp. L729-L732.

Chandani, A. D. L.; Gorecka, E.; Ouchi, Y.; Takezoe, H. & Fukuda, A. (1989). Antiferroelectric Chiral Smectic Phases Responsible for the Trislable Switching in MHPOBC. *Jpn. J. Appl. Phys.*, Vol. 28, No. 7, pp. L1265-L1268.

Cheon, C. I.; Li, L.; Glushchenko, A.; West, J. L.; Reznikov, Y.; Kim, J. S. & Kim, D. H. (2005). Electro-Optics of Liquid Crystals Doped with Ferroelectric Nano-Powder, *Society for Information Display Digest of Technical Papers*, Vol. 36, pp. 1471-1473.

Chigrinov, V. G. (1999). *Liquid Crystal Devices: Physics and Applications*, Artech House Publishers, ISBN 0-89006-898-4, Boston, London.

Clark, N. A. & Lagerwall, S. T. (1980). Submicrosecond bistable electro-optic switching in liquid crystals. *Appl. Phys. Lett.*, Vol. 36, No. 11, pp. 899-891.

Crawford, G.P. & Žumer, S. (1996). *Liquid Crystals in Complex Geometries Formed by Polymer and Porous Networks*, Taylor & Francis, ISBN 0-7484-0464-3, London.

Drzaic, P. S. (1998). *Liquid Crystal Dispersions*, World Scientific, ISBN 981-02-1745-5, Singapore.

Glushchenko, A.; Cheon, C.; West, J.; Li, F.; Büyüktanir, E.; Reznikov, Y. & Buchnev, A. (2006). Ferroelectric Particles in Liquid Crystals: Recent Frontiers. *Mol. Cryst. Liq. Cryst.*, Vol. 453, pp. 227-237.

Jeng, S.-C.; Kuo, C.-W.; Wang, H.-L. & Liao, C.-C. (2007). Nanoparticles-induced vertical alignment in liquid crystal cell. *Appl. Phys. Lett.*, Vol. 91, No. 6, pp. 061112-1-3.

Kaczmarek, M.; Buchnev, O. & Nandhakumar, I. (2008). Ferroelectric nanoparticles in low refractive index liquid crystals for strong electro-optic response. *Appl. Phys. Lett.*, Vol. 92, No. 10, pp. 103307-1-3.

Kaur, S.; Singh, S. P.; Biradar, A. M.; Choudhary, A. & Sreenivas, K. (2007). Enhanced electro-optical properties in gold nanoparticles doped ferroelectric liquid crystals. *Appl. Phys. Lett.*, Vol. 91, No. 2, pp. 023120-1-3.

Kimura, S.; Kimura, S.; Nishiyama, S.; Ouchi, Y.; Takezoe, H. & Fukuda, A. (1987). Viscosity Measurement in Ferroelectric Liquid Crystals Using a Polarization Switching Current. *Jpn. J. Appl. Phys.*, Vol. 26, No. 4, pp. L255- L257.

Kobayashi, S. & Toshima, N. (2007). Nanoparticles and LCDs: It's a Surprising World. *SID.*, Vol. 9, No. 7, pp. 26-32.

Lagerwall, S. T. (1999). *Ferroelectric and Antiferroelectric Liquid Crystals*, Wiley-VCH, ISBN 3-527-29831-2, Weinheim.

Li, F.; Buchnev, O.; Cheon, C. I.; Glushchenko, A.; Reshetnyak, V.; Reznikov, Y.; Sluckin, T. J. & West, J. (2006). Orientational Coupling Amplication in Ferroelectric Nematic Colloids. *Phys. Rev. Lett.*, Vol. 97, No. 14, pp. 147801-1-4.

Li, F.; West, J.; Glushchenko, A.; Cheon, C. I. & Reznikov, Y. (2006). Ferroelectric nanoparticle/liquid-crystal colloids for display applications. *Journal of the SID.*, Vol. 14, No. 6, pp. 523-527.

Li, F.; Buchnev, O.; Cheon, C. I.; Glushchenko, A.; Reshetnyak, V.; Reznikov, Y.; Sluckin, T. J. & West, J. (2007). Erratum: Orientational Coupling Amplication in Ferroelectric Nematic Colloids. *Phys. Rev. Lett.*, Vol. 99, No. 21, pp. 219901.

Meyer, R. B.; Lievert, L.; Strzelecki, L. & Keller, P. (1975). Ferroelectric liquid crystals. *J. Physique Lett.*, Vol. 36, No. 3, pp. 69-71.

Miyama, T.; Thisayukta, J.; Shiraki, H.; Sakai, Y.; Shiraishi, Y.; Toshima, N. & Kobayashi, S. (2004). Fast Switching of Frequency Modulation Twisted Nematic Liquid Crystal Display Fabricated by Doping Nanoparticles and Its Mechanism. *Jpn. J. Appl. Phys.*, Vol. 43, No. 5A, pp. 2580-2584.

Oh-e, M. & Kondo, K. (1995). Electro-optical characteristics and switching behavior of the in-plane switching mode. *Appl. Phys. Lett.*, Vol. 67, No. 26, pp. 3895-3897.

Reznikov, Yu.; Buchnev, O.; Tereshchenko, O.; Reshetnyak, V. & Glushchenko, A. (2003). Ferroelectric nematic suspension. *Appl. Phys. Lett.*, Vol. 82, No. 12, pp. 1917-1919.

Schurian, A. & Bärner, K. (1996). Stable suspensions of ferroelectric nm-$LiNbO_3$—and nm-$PbTiO_3$—particles in hydrocarbon carrier liquids. *Ferroelectrics*, Vol. 20, No. 5. pp. 169-176.

Shiraki, H.; Kundu, S.; Sakai, Y.; Masumi, T.; Shiraishi, Y.; Toshima, N. & Kobayashi. (2004). Dielectric Properties of Frequency Modulation Twisted Nematic LCDs Doped with Palladium (Pd) Nanoparticles. *Jpn. J. Appl. Phys.*, Vol. 43, No. 8A, pp. 5425-5429.

West, J.; Fredley, S. & Carrell, J. (1992). Haze-free polymer dispersed liquid crystals utilizing linear polarizers. *Appl. Phys. Lett.*, Vol. 61, No. 17, pp. 2004-2005.

Wu, S.-T.; Coates, D. & Bartmann, E. (1991). Physical properties of chlorinated liquid crystals. *Mol. Cryst. Liq. Cryst.*, Vol. 10, No. 5, pp. 635-646.

Yaroshchuk, O. V. & Dolgov, L. O. (2007). Electro-optics and structure of polymer dispersed liquid crystals doped with nanoparticles of inorganic materials. *Opt. Mater.*, Vol. 29, pp. 1097-1102.

11

Ferroelectric-Dielectric Solid Solution and Composites for Tunable Microwave Application

Yebin Xu and Yanyan He
Huazhong University of Science and Technology
China

1. Introduction

Electric field tunable ferroelectric materials have attracted extensive attention in recent years due to their potential applications for tunable microwave device such as tunable filters, phased array antennas, delay lines and phase shifters (Maiti et al. 2007a; Rao et al. 1999; Romanofsky et al. 2000; Varadan et al 1992.; Zhi et al. 2002). $Ba_{1-x}Sr_xTiO_3$ and $BaZr_xTi_{1-x}O_3$ have received the most attention due to their intrinsic high dielectric tunability. However, the high inherent materials loss and high dielectric constant has restricted its application in tunable microwave device. Various methods have been investigated to lower the dielectric constant and loss tangent of pure ferroelectrics.

Forming ferroelectric-dielectric composite is an efficient method to reduce material dielectric constant, loss tangent and maintain tunability at a sufficiently high level. For binary ferroelectric-dielectric composite (such as BST+MgO) (Chang & Sengupta 2002; Sengupta & Sengupta 1999), with the increase of dielectrics content, the dielectric constant and tunability of composites decrease. In order to decrease the dielectric constant of binary composite, it is necessary to increase the content of linear dielectric, and the tunability will decrease inevitably due to ferroelectric dilution. Replacing one dielectric by the combination of dielectrics with different dielectric constants and forming ternary ferroelectric-dielectric composite can decrease the dielectric constant of composite and maintain or even increase the tunability. This is beneficial for tunable application. The $Ba_{0.6}Sr_{0.4}TiO_3$-Mg_2SiO_4-MgO and $BaZr_{0.2}Ti_{0.8}O_3$-Mg_2SiO_4-MgO composites exhibited relatively high tunability in combination with reduced dielectric permittivity and reduced loss tangent (He et al. 2010, 2011). With the increase of Mg_2SiO_4 content and the decrease of MgO content in $Ba_{0.6}Sr_{0.4}TiO_3$-Mg_2SiO_4-MgO composite, the dielectric constant decrease and the tunability remain almost unchanged. For $BaZr_{0.2}Ti_{0.8}O_3$-Mg_2SiO_4-MgO composite, an anomalous relation between dielectric constant and tunability was observed: with the increase of Mg_2SiO_4 content (>30 wt%), the dielectric constant of composite decreases and the tunability increases. The anomalous increased tunability can be attributed to redistribution of the electric field. $Ba_{1-x}Sr_xTiO_3$-Mg_2TiO_4-MgO can also form ferroelectric ($Ba_{1-x}Sr_xTiO_3$)-dielectric (Mg_2TiO_4-MgO) ternary composite and the dielectric constant can be decreased. With the increase of Mg_2TiO_4 content and the decrease of MgO content, the tunability of $Ba_{1-x}Sr_xTiO_3$-Mg_2TiO_4-MgO composite increase. The multiple-phase composites might complicate method to effectively deposit films, particularly if the dielectrics and ferroelectric are not compatible for simultaneous deposition or simultaneous adhesion with a substrate or with each other. But ferroelectric-dielectric composite bulk ceramics show promising application, especially in accelerator: bulk ferroelectrics composites can be used as active elements of electrically controlled switches and phase shifters in pulse compressors or power distribution circuits of future linear colliders as well as tuning layers for the dielectric based accelerating structures (Kanareykin et al. 2006, 2009a, 2009b).

Forming ferroelectric-dielectric solid solution is another method to reduce material dielectric constant and loss tangent. Ferroelectric $Ba_{0.6}Sr_{0.4}TiO_3$ can form solid solution with dielectrics $Sr(Ga_{0.5}Ta_{0.5})O_3$, $La(Mg_{0.5}Ti_{0.5})O_3$, $La(Zn_{0.5}Ti_{0.5})O_3$, and $Nd(Mg_{0.5}Ti_{0.5})O_3$ that have the same perovskite structure as the ferroelectrics (Xu et al. 2008, 2009). With the increase of the dielectrics content, the dielectric constant, loss tangent and tunability of solid solution decrease. $Ba_{0.6}Sr_{0.4}TiO_3$-$La(Mg_{0.5}Ti_{0.5})O_3$ shows better dielectric properties than other solid solutions. Compared with ferroelectric-dielectric composite, forming solid solution can decrease the dielectric constant more rapidly when the doping content is nearly the same, and can also improve the loss tangent more effectively. On the other hand, ferroelectric-dielectric solid solution shows lower tunability than composites. The advantage of ferroelectric-dielectric solid solution is that single phase materials is favorable for the thin film deposition. The high dielectric field strength can be obtained easily in thin film to get high tunability.

In this chapter, we summarize the microstructures, dielectric tunable properties of ferroelectric-dielectric solid solution and composites, focusing mainly on our recent works.

2. Ferroelectric-dielectric composite

2.1 $Ba_{1-x}Sr_xTiO_3$ based composites

Various non-ferroelectric oxides, such as MgO, Al_2O_3, ZrO_2, Mg_2SiO_4 and $MgTiO_3$, were added to $Ba_{1-x}Sr_xTiO_3$ to reduce the dielectric constant and loss tangent and maintain the tunability at sufficient high level (Chang & Sengupta 2002; Sengupta & Sengupta 1997, 1999). It is better that non-ferroelectric oxide doesn't react with ferroelectric $Ba_{1-x}Sr_xTiO_3$. MgO has low dielectric constant and loss tangent, can form ferroelectric ($Ba_{1-x}Sr_xTiO_3$)-dielectric (MgO) composite. BST-MgO composite shows better dielectric properties. Mg_2SiO_4 is also a linear dielectrics with low dielectric constant, but it can react with $Ba_{1-x}Sr_xTiO_3$ to form $Ba_2(TiO)(Si_2O_7)$, as shown in Fig. 1. For 10 mol% Mg_2SiO_4 mixed $Ba_{0.6}Sr_{0.4}TiO_3$, the major phase is $Ba_{0.6}Sr_{0.4}TiO_3$, and no Mg_2SiO_4 phase can be found except for two unidentified peaks at 27.6^o and 29.7^o (relative intensity: ~1%). As the content of Mg_2SiO_4 increases from 20 to 60 mol%, the impurities phase of $Ba_2(TiO)(Si_2O_7)$ is observed obviously and the relative content is increased with respect to the content of Mg_2SiO_4. For 60 mol% Mg_2SiO_4 mixed $Ba_{0.6}Sr_{0.4}TiO_3$ ceramics sintered at 1220°C, the strongest diffraction peak is the (211) face of $Ba_2(TiO)(Si_2O_7)$ (not shown in Fig. 1). Therefore, for Mg_2SiO_4 added $Ba_{0.6}Sr_{0.4}TiO_3$, it is not as we expected that the ferroelectric ($Ba_{0.6}Sr_{0.4}TiO_3$)-dielectric (Mg_2SiO_4) composite formed. The dielectric constants and unloaded Q values at microwave frequency were measured in the $TE_{01\delta}$ dielectric resonator mode using the Hakki and Coleman method by the network analyzer. Table 1 summarizes ε_r and the quality factor ($Q\times f=f_0/\tan\delta$, where f_0 is the resonant frequency) at microwave frequencies for some $Ba_{0.6}Sr_{0.4}TiO_3$-Mg_2SiO_4 ceramics. Increasing the Mg_2SiO_4 content results in a decrease of dielectric constant but has no obvious effect on the Q×f value. The low Q×f of $Ba_{0.6}Sr_{0.4}TiO_3$-Mg_2SiO_4 ceramics restricts their microwave application, and so the tunability has not been measured furthermore. The low Q×f is due to $Ba_2(TiO)(Si_2O_7)$ which is a ferroelectrics with promising piezoelectric uses.

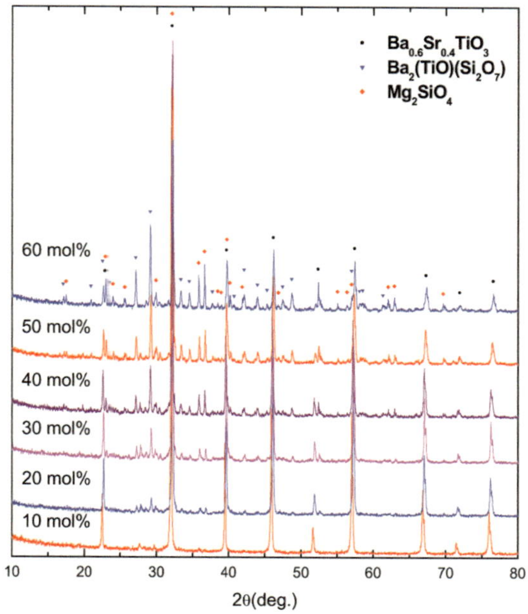

Fig. 1. The XRD patterns of $Ba_{0.6}Sr_{0.4}TiO_3$-Mg_2SiO_4 ceramics. The Mg_2SiO_4 content is 10-60mol%.

Mg_2SiO_4 content (mol%)	Sintering temperature (°C)	f_0(GHz)	ε	tanδ	Q×f(GHz)
20	1260	1.79	683.7	0.016	112
40	1240	2.98	169.2	0.024	124

Table 1. Microwave dielectric properties of $Ba_{0.6}Sr_{0.4}TiO_3$-Mg_2SiO_4 ceramics

For Mg_2SiO_4-MgO added $Ba_{0.6}Sr_{0.4}TiO_3$, ferroelectric ($Ba_{0.6}Sr_{0.4}TiO_3$)-dielectric (Mg_2SiO_4-MgO) composite is formed, as shown in Fig. 2 (He et al., 2010). With the decrease of MgO content and the increase of Mg_2SiO_4 content, the diffraction peaks from MgO decrease gradually and the diffraction peaks from Mg_2SiO_4 increase. Therefore, Mg_2SiO_4-MgO combination can prohibit the formation of $Ba_2(TiO)(Si_2O_7)$ phase.
Fig. 3 shows the FESEM images of $Ba_{0.6}Sr_{0.4}TiO_3$-Mg_2SiO_4-MgO composites sintered at 1350°C for 3h. The FESEM image and element mapping of 40$Ba_{0.6}Sr_{0.4}TiO_3$-12$Ba_{0.6}Sr_{0.4}TiO_3$-48MgO as determined by energy dispersive spectroscopy (EDS) are shown in Fig. 4. Three kind of different grains can be found clearly: light grains with average grain size of about 2μm, nearly round larger grains and dark grains with sharp corners. The element mapping of Si Kα1 and Ti Kα1 in Fig. 4 can show the distribution of Mg_2SiO_4 and $Ba_{0.6}Sr_{0.4}TiO_3$ grains clearly. Therefore, we can identify that light grains are $Ba_{0.6}Sr_{0.4}TiO_3$, the dark, larger grains are MgO, and dark grains with sharp corners are Mg_2SiO_4. With the decrease of MgO content and the increase of Mg_2SiO_4 content, more and more Mg_2SiO_4 grains with different size can be found (Fig. 4). It is consistent with the XRD results. We can conclude that Mg_2SiO_4 and MgO were randomly dispersed relative to ferroelectric $Ba_{0.6}Sr_{0.4}TiO_3$ phase.

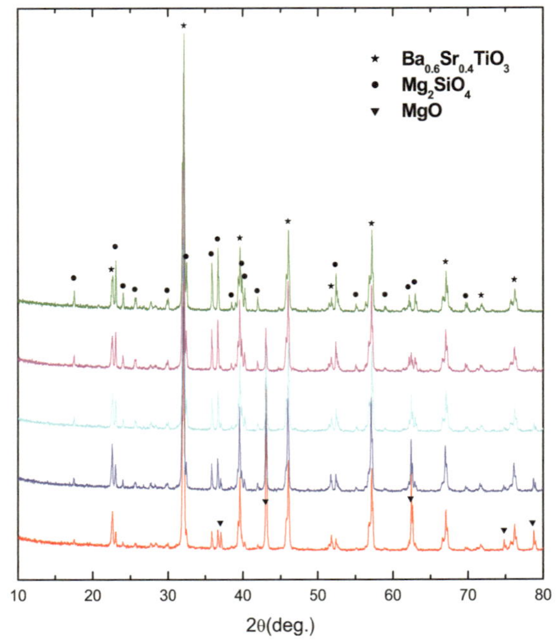

Fig. 2. The XRD patterns of $40Ba_{0.6}Sr_{0.4}TiO_3$-$60(Mg_2SiO_4$-$MgO)$ composite ceramics sintered at 1350°C for 3h. From bottom to top, the MgO content is 48 wt%, 36 wt%, 30 wt%, 24 wt% and 12 wt%, respectively.

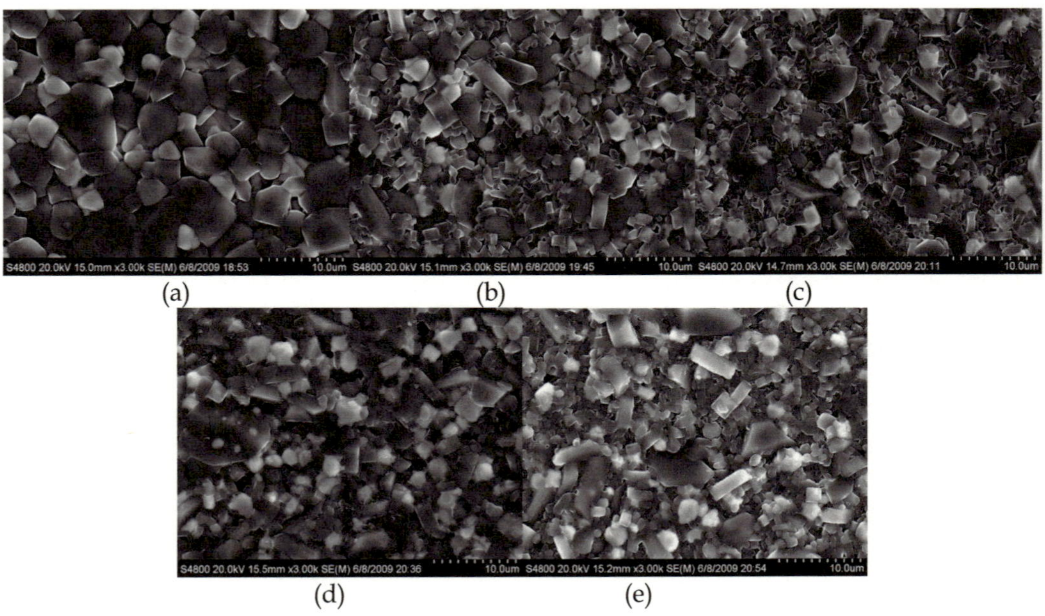

Fig. 3. FESEM images of $40Ba_{0.6}Sr_{0.4}TiO_3$-$60(Mg_2SiO_4$-$MgO)$ composites ceramics sintered at 1350°C for 3h. From (a) to (e), the MgO content is 48 wt%, 36 wt%, 30 wt%, 24 wt% and 12 wt%, respectively.

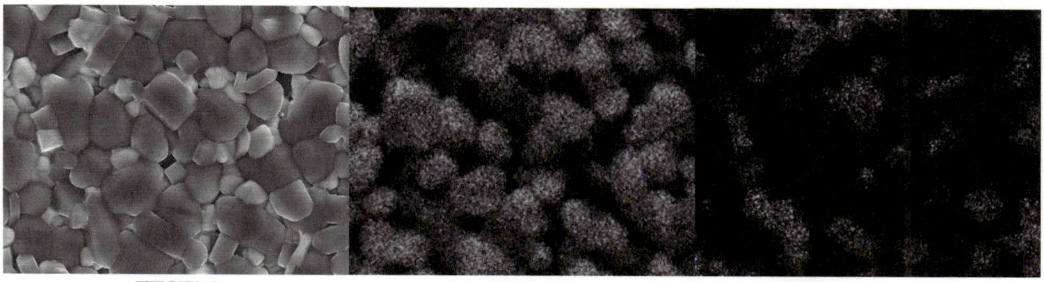

FESEM Mg Kα1_2 Si K α1

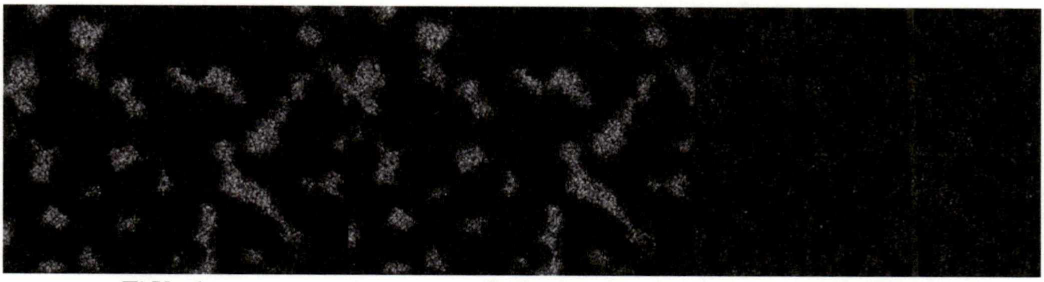

Ti K α1 Ba L α1 Sr K α1

Fig. 4. FESEM image and element mapping of $40Ba_{0.6}Sr_{0.4}TiO_3$-$12Mg_2SiO_4$-$48MgO$ as determined by energy dispersive spectroscopy (EDS).

Because of the relatively low dielectric constant and loss tangent of Mg_2SiO_4 and MgO, it is expected that $Ba_{0.6}Sr_{0.4}TiO_3$-Mg_2SiO_4-MgO composites have lower dielectric constant and loss tangent. Fig. 5 shows the dielectric constant and loss tangent of $Ba_{0.6}Sr_{0.4}TiO_3$-Mg_2SiO_4-MgO composite ceramics at 1MHz. The dielectric constant of composites is much smaller than that of $Ba_{0.6}Sr_{0.4}TiO_3$ (ε~5160 at 1MHz) (Chang & Sengupta, 2002; Sengptal & Sengupta 1999;). The loss tangent of $Ba_{0.6}Sr_{0.4}TiO_3$-Mg_2SiO_4-MgO composites sintered at 1350°C is ~0.0003-0.0006, but the loss tangent of $Ba_{0.6}Sr_{0.4}TiO_3$ is ~0.0096 (Sengptal et al. 1999). Therefore, the composites have much smaller loss tangent than $Ba_{0.6}Sr_{0.4}TiO_3$.

The temperature dependence of dielectric properties for various $Ba_{0.6}Sr_{0.4}TiO_3$-Mg_2SiO_4-MgO composites (sintering temperature: 1350°C) measured at 100kHz is illustrated in Fig. 6. Broadened and suppressed dielectric peaks and shifts of Curie temperature T_C are observed. For $40Ba_{0.6}Sr_{0.4}TiO_3$-$12Mg_2SiO_4$-$48MgO$ ceramics, its ε_{max} is ~ 176.5 at T_c ~224K. As the relative content of Mg_2SiO_4 increase, T_c is shifted slightly to lower temperatures, thus resulting in a decrease in dielectric constant at a given temperature; at the meantime, ε_{max} decreases also. For $40Ba_{0.6}Sr_{0.4}TiO_3$-$30Mg_2SiO_4$-$30MgO$, ε_{max} is ~140.1 at ~216K and for $40Ba_{0.6}Sr_{0.4}TiO_3$-$48Mg_2SiO_4$-$12MgO$, ε_{max} is ~126.8 at ~214K. With the decrease of temperature, the loss tangent increase.

Fig. 6 shows the effect of applied field on the tunability of the $Ba_{0.6}Sr_{0.4}TiO_3$-Mg_2SiO_4-MgO composites at 100kHz. The tunability of $40Ba_{0.6}Sr_{0.4}TiO_3$-$12Mg_2SiO_4$-$48MgO$ at 100kHz under at 2kV/mm is 10.5%. With the increase of Mg_2SiO_4 content, the tunability of $40Ba_{0.6}Sr_{0.4}TiO_3$-$24Mg_2SiO_4$-$36MgO$ decreases slightly to 9.2%. Further increasing Mg_2SiO_4 content results in a slight increase of tunability: $40Ba_{0.6}Sr_{0.4}TiO_3$-$48Mg_2SiO_4$-$12MgO$ composite has tunability of 10.2%.

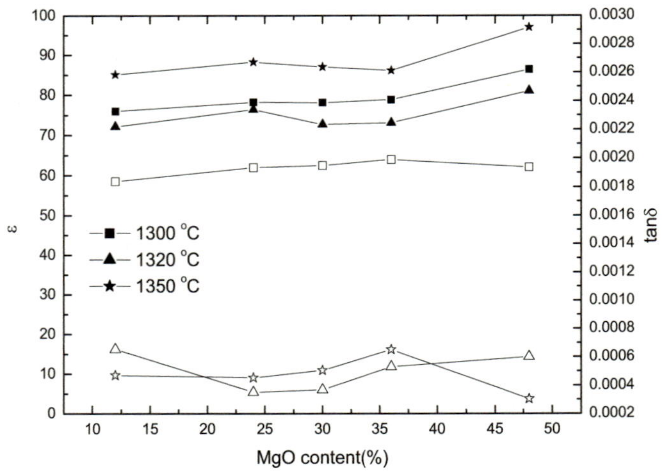

Fig. 5. Dielectric constant (solid) and loss tangent (open) of $40Ba_{0.6}Sr_{0.4}TiO_3$-$60(Mg_2SiO_4$-$MgO)$ composite ceramics sintered at different temperature (measure frequency: 1MHz).

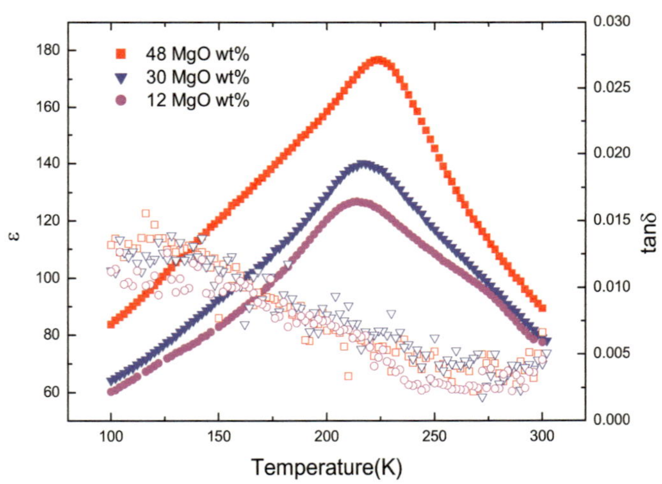

Fig. 6. Variation of dielectric constant (solid) and loss tangent (open) with temperature for $40Ba_{0.6}Sr_{0.4}TiO_3$-$60(Mg_2SiO_4$-$MgO)$ ceramics measured at 100kHz.

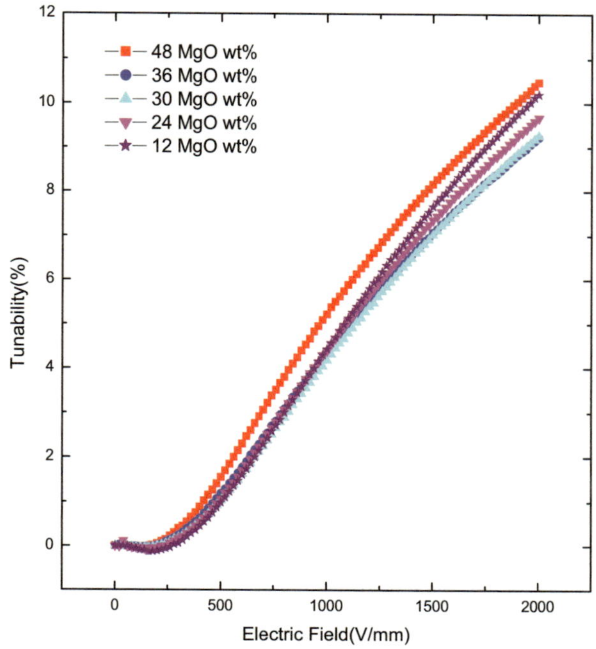

Fig. 7. The tunability of $40Ba_{0.6}Sr_{0.4}TiO_3$-$60(Mg_2SiO_4$-$MgO)$ composites at 100kHz (sintering temperature: 1350°C).

MgO content (wt.%)	f_0(GHz)	ε	tanδ	Q×f(GHz)
12	5.74	74.59	0.023	250
24	5.74	77.72	0.019	302
30	5.80	77.12	0.021	276
36	5.96	74.39	0.017	351
48	5.33	93.86	0.014	381

Table 2. Microwave Dielectric Properties of $40Ba_{0.6}Sr_{0.4}TiO_3$-$60(Mg_2SiO_4$-$MgO)$ ceramics

The room temperature microwave dielectric properties of $40Ba_{0.6}Sr_{0.4}TiO_3$-$60(Mg_2SiO_4$-$MgO)$ composites were summarized in Table 2. With the increase of Mg_2SiO_4 content, the dielectric constant remain almost the same and the Q×f value decrease.

Mg_2TiO_4 is a low loss tangent linear dielectrics and Mg_2TiO_4 added $Ba_{1-x}Sr_xTiO_3$ shows better tuanble dielectric properties (Chou et al. 2007; Nenasheva et al. 2010). The XRD patterns of $40Ba_{0.6}Sr_{0.4}TiO_3$-$xMgO$-$(60-x)Mg_2TiO_4$ (Fig. 8) show that ferroelectric ($Ba_{0.6}Sr_{0.4}TiO_3$)-dielectric (MgO-Mg_2TiO_4) composite is formed. On the other hand, impurity phase $BaMg_6Ti_6O_{19}$ is found in Mg_2TiO_4 doped $Ba_{0.6}Sr_{0.4}TiO_3$. The fomation of $BaMg_6Ti_6O_{19}$ depends on Ba/Sr ratio. $BaMg_6Ti_6O_{19}$ forms in Mg_2TiO_4 doped $Ba_{0.6}Sr_{0.4}TiO_3$ and $Ba_{0.55}Sr_{0.45}TiO_3$ but not $Ba_{0.5}Sr_{0.5}TiO_3$. Mg_2TiO_4-MgO combination can prohibit the formation of $BaMg_6Ti_6O_{19}$ phase. The FESEM images (Fig. 9) show clearly three kind of grains: $Ba_{0.6}Sr_{0.4}TiO_3$, Mg_2TiO_4 and MgO.

Table 3 shows the microwave dielectric properties of $40Ba_{0.6}Sr_{0.4}TiO_3$-$xMgO$-$(60-x)Mg_2TiO_4$ ceramics. With the increase of MgO content, the dielectric constant decrease due to lower dielectric constant of MgO. For x=0-36 wt%, the Q×f value remain unchanged. As a whole, the loss tangent is too high to be used for tunable microwave application.

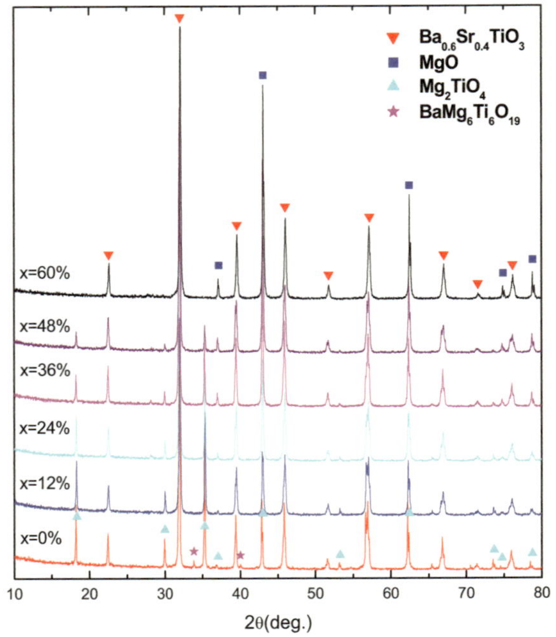

Fig. 8. The XRD patterns of $40Ba_{0.6}Sr_{0.4}TiO_3$-xMgO-(60-x)Mg_2TiO_4 ceramics

Fig. 9. FESEM images of $40Ba_{0.6}Sr_{0.4}TiO_3$-xMgO-(60-x)Mg_2TiO_4 composites ceramics sintered at 1400°C for 3h.

MgO content (wt.%)	f_0(GHz)	ε	tanδ	Q×f(GHz)
0	2.83	193.40	0.034	83
12	2.67	226.76	0.034	79
24	2.96	220.25	0.035	85
36	3.00	207.66	0.036	83
48	3.53	199.71	0.034	104
60	4.80	109.63	0.013	369

Table 3. Microwave dielectric properties of $40Ba_{0.6}Sr_{0.4}TiO_3$-xMgO-(60-x)Mg_2TiO_4 ceramics

Increasing Sr/Ba ratio can decrease the dielectric constant and loss tangent of $Ba_{1-x}Sr_xTiO_3$. $40Ba_{0.5}Sr_{0.5}TiO_3$-xMgO-(60-x)Mg_2TiO_4 will has lower dielectric constant and loss tangent than $40Ba_{0.6}Sr_{0.4}TiO_3$-xMgO-(60-x)Mg_2TiO_4. We prepared $40Ba_{0.5}Sr_{0.5}TiO_3$-xMgO-(60-x)Mg_2TiO_4 ceramics and measured the tunability (Fig. 10). With the increase of Mg_2TiO_4 content, the tunabity of composite increases. The tunability of $40Ba_{0.5}Sr_{0.5}TiO_3$-12MgO-48Mg_2TiO_4 is 16.6% at 2kV/mm and 28.5% at 3.9kV/mm, respectively. The corresponding value of $40Ba_{0.5}Sr_{0.5}TiO_3$-60Mg_2TiO_4 is 13.6% and 24.0% respectively. The higher tunability of $40Ba_{0.5}Sr_{0.5}TiO_3$-12MgO-48Mg_2TiO_4 is due to its higer dielectric constant (ε=150.2) than $40Ba_{0.5}Sr_{0.5}TiO_3$-60Mg_2TiO_4 (ε=127.8).

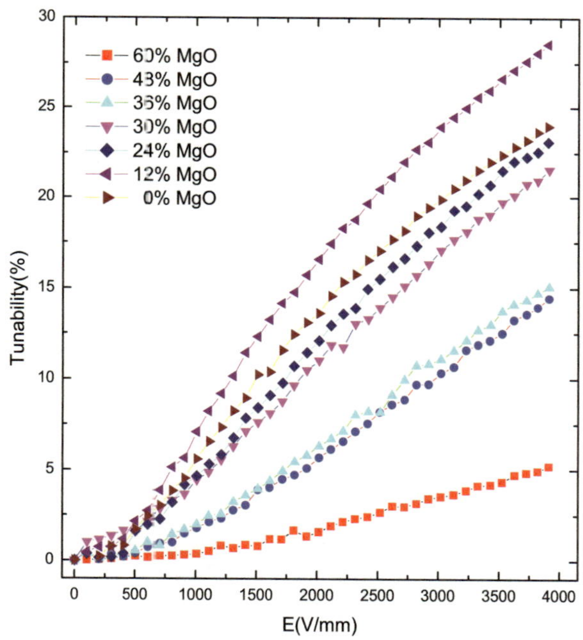

Fig. 10. The tunability of $40Ba_{0.5}Sr_{0.5}TiO_3$-xMgO-(60-x)Mg_2TiO_4 composites at 10kHz.

2.2 $BaZr_xTi_{1-x}O_3$ based composites

$BaZr_xTi_{1-x}O_3$ can form ferroelectric-dielectric composite with MgO (Maiti et al. 2007b, 2007c, 2008). High tunability and low loss tangent of the $BaZr_xTi_{1-x}O_3$: MgO composites are

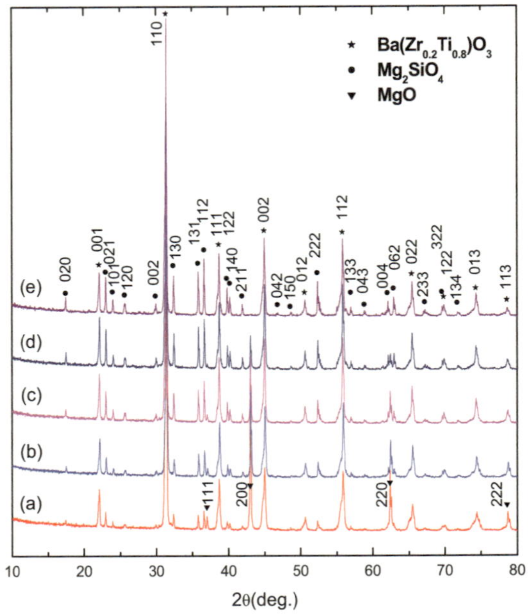

Fig. 11. The XRD patterns of $40BaZr_{0.2}Ti_{0.8}O_3$-(60-x)Mg_2SiO_4-xMgO composites ceramics sintered at 1350°C for 3h. (a) x=48wt%, (b) x=36wt%, (c) x=30wt%, (d) x=24wt%, (e) x=12wt%.

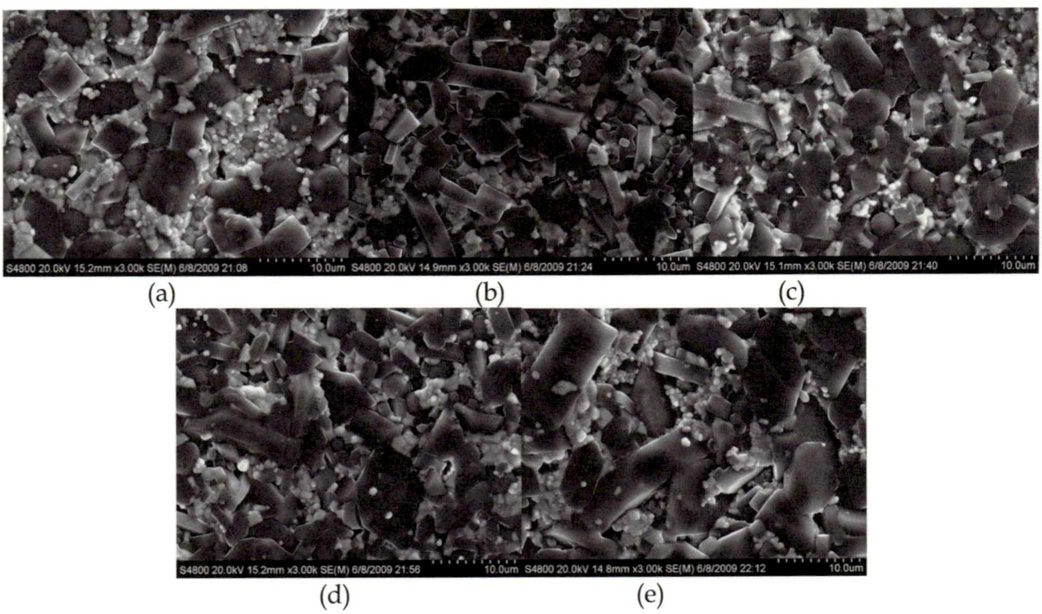

Fig. 12. FESEM images of $40BaZr_{0.2}Ti_{0.8}O_3$-(60-x)Mg_2SiO_4-xMgO composites ceramics sintered at 1350°C for 3h. From (a) to (e), x=48 wt%, 36 wt%, 30 wt%, 24 wt% and 12 wt%, respectively.

reported, but the sintering temperature is as high as 1500°C. We prepared $BaZr_{0.2}Ti_{0.8}O_3$-Mg_2SiO_4-MgO composite ceramics at 1350°C (He et al. 2011). The formation of ferroelectric ($BaZr_{0.2}Ti_{0.8}O_3$)-dielectric (Mg_2SiO_4-MgO) composite was proved by XRD patterns (Fig. 11). Similar to $Ba_{0.6}Sr_{0.4}TiO_3$-Mg_2SiO_4-MgO composites, three kind of grains: $BaZr_{0.2}Ti_{0.8}O_3$, Mg_2SiO_4 and MgO, can be identified (Fig. 12 and Fig. 13).

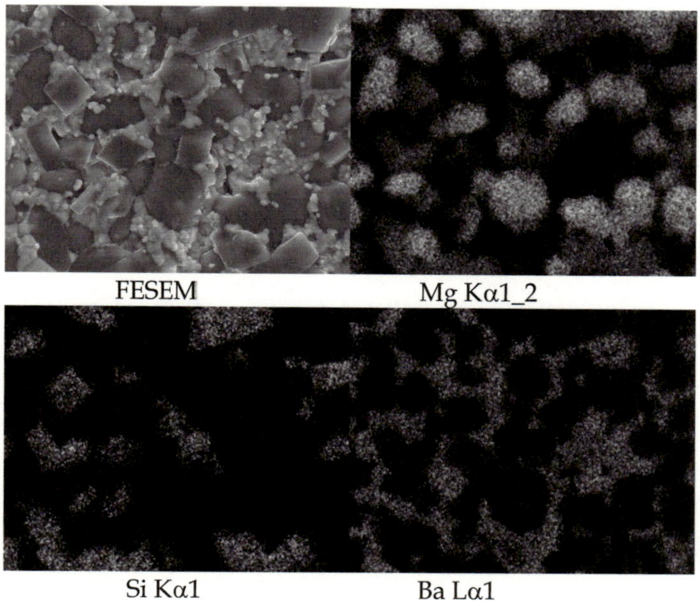

Fig. 13. FESEM image and element mapping of $40BaZr_{0.2}Ti_{0.8}O_3$-$12Mg_2SiO4$-48MgO as determined by energy dispersive spectroscopy (EDS).

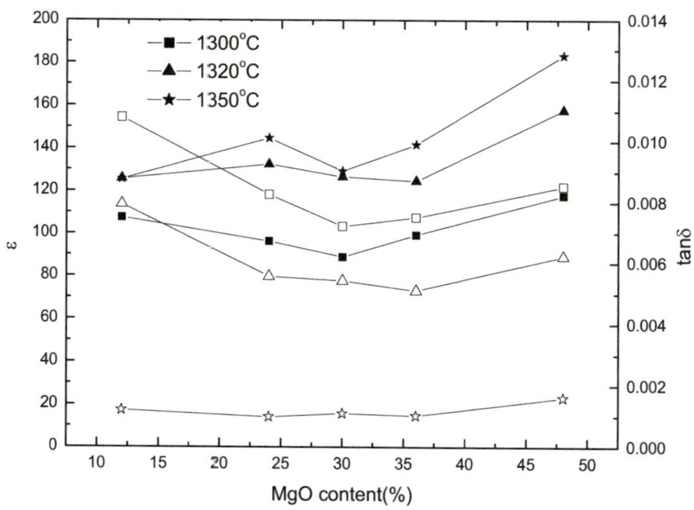

Fig. 14. Dielectric constant (solid) and loss tangent (open) of $40BaZr_{0.2}Ti_{0.8}O_3$-$(60-x)Mg_2SiO_4$-$xMgO$ composites ceramics sintered at various temperature (measure frequency: 1MHz).

Fig. 14 shows the dielectric constant and loss tangent of $BaZr_{0.2}Ti_{0.8}O_3$-Mg_2SiO_4-MgO composite ceramics at 1MHz. With the increase of sintering temperature from 1300°C to 1350°C, the dielectric constant of the composites increase and the loss tangent decrease.

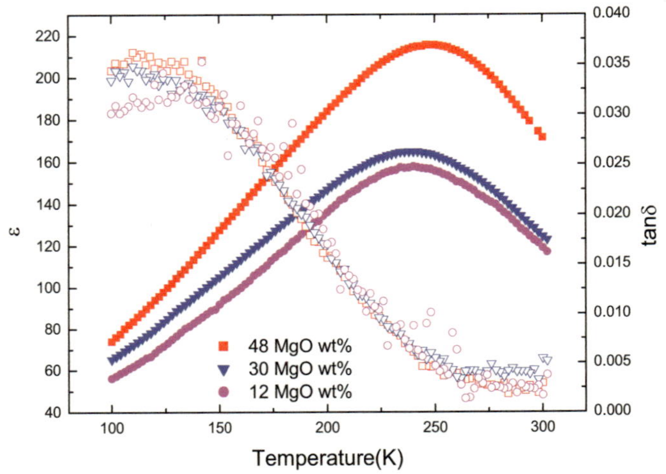

Fig. 15. Variation of dielectric constant (solid) and loss tangent (open) with temperature for $40BaZr_{0.2}Ti_{0.8}O_3$-$(60-x)Mg_2SiO_4$-$xMgO$ ceramics (sintering temperature: 1350°C) measured at 100kHz.

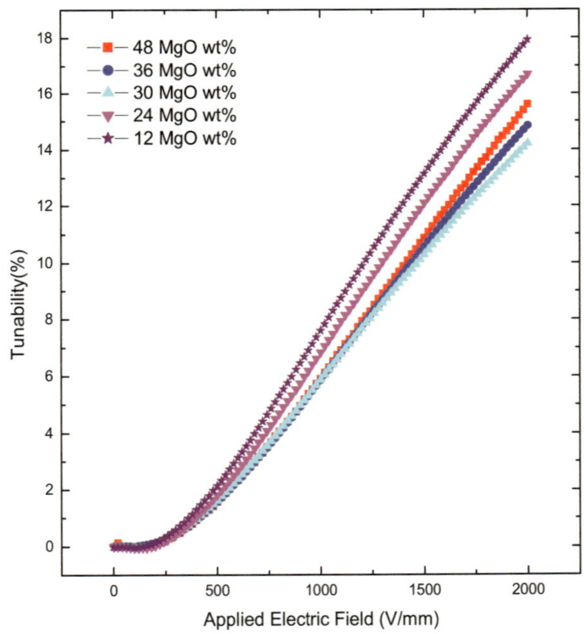

Fig. 16. The tunability of $40BaZr_{0.2}Ti_{0.8}O_3$-$(60-x)Mg_2SiO_4$-$xMgO$ composite ceramics at 100kHz at room temperature (sintering temperature: 1350°C).

Increasing Mg_2SiO_4 content tends to decrease the dielectric constant of composites. The dielectric constant and loss tangent of composite sintered at 1350°C is ~125-183 and ~0.0010-0.0016, respectively, which is smaller than that of $BaZr_{0.2}Ti_{0.8}O_3$ (Maiti et al. 2007b).
The temperature dependence of dielectric properties for $BaZr_{0.2}Ti_{0.8}O_3$-Mg_2SiO_4-MgO composites (sintering temperature: 1350°C) measured at 100kHz is illustrated in Fig. 15. Compared with pure $BaZr_{0.2}Ti_{0.8}O_3$ bulk ceramic (Maiti et al. 2007b), broadened and suppressed dielectric peaks and shifts of Curie temperature T_C are observed with the addition of Mg_2SiO_4 and MgO. The results are similar to that of $Ba_{0.6}Sr_{0.4}TiO_3$-Mg_2SiO_4-MgO. For $40BaZr_{0.2}Ti_{0.8}O_3$-$12Mg_2SiO_4$-48MgO ceramics, its ε_{max} decreases to ~ 215.5 and T_c shifts to lower temperature ~246K. For $40BaZr_{0.2}Ti_{0.8}O_3$-$48Mg_2SiO_4$-12MgO, ε_{max} is ~157.7 at ~240K.
Fig. 16. shows the tunability of the $BaZr_{0.2}Ti_{0.8}O_3$-Mg_2SiO_4-MgO composites at 100kHz at room temperature. The tunability of $40BaZr_{0.2}Ti_{0.8}O_3$-$12Mg_2SiO_4$-48MgO under 2kV/mm is 15.6%. With the increase of Mg_2SiO_4 content, the tunability of $40BaZr_{0.2}Ti_{0.8}O_3$-$30Mg_2SiO_4$-30MgO decreases slightly to 14.2%. Further increasing Mg_2SiO_4 content results in an anomalous increase of tunability: $40BaZr_{0.2}Ti_{0.8}O_3$-$48Mg_2SiO_4$-12MgO composite has lower dielectric constant than $40BaZr_{0.2}Ti_{0.8}O_3$-$12Mg_2SiO_4$-48MgO but slightly higher tunability (17.9%).

3. Ferroelectric-dielectric solid solution

Forming ferroelectric-dielectric solid solution is another method to reduce material dielectric constant and loss tangent. Some non-ferroelectric complex oxides with perovskite structures have relatively low dielectric constant and low loss tangent. It is expected that they can be combined with barium strontium titanate to reduce material dielectric constant and loss tangent. Furthermore, it is possible for them to form solid solutions with barium strontium titanate because they have the same perovskite structure as barium strontium titanate. Single phase material is favorable for the thin film deposition. On the other hand, some perovskite oxide has positive temperature coefficient of dielectric constant and it can decrease the temperature coefficient of dielectric constant of barium strontium titanate above Curie temperature.

3.1 $Ba_{0.6}Sr_{0.4}TiO_3$-$Sr(Ga_{0.5}Ti_{0.5})O_3$ solid solution

$Sr(Ga_{0.5}Ta_{0.5})O_3$ has a comparatively small dielectric constant (27 at 1MHz), a positive temperature coefficient of dielectric constant (120ppmK^{-1}) and a low dielectric loss (Q=8600 at 10.6 GHz) (Takahashi et al. 1997). The lattice constant (a=0.3949nm) of cubic perovskite structure $Sr(Ga_{0.5}Ta_{0.5})O_3$ is very close to that of $Ba_{0.6}Sr_{0.4}TiO_3$ (a=0.3965nm). Therefore, $Sr(Ga_{0.5}Ta_{0.5})O_3$ will be possible to form solid solution with $Ba_{0.6}Sr_{0.4}TiO_3$ and reduce the dielectric constant of $Ba_{0.6}Sr_{0.4}TiO_3$. The XRD results (Fig. 17.) prove that solid solution can be formed between $Ba_{0.6}Sr_{0.4}TiO_3$ and $Sr(Ga_{0.5}Ta_{0.5})O_3$ under the preparative conditions (Xu et al. 2008).
Fig.18 shows the FESEM images of $Ba_{0.6}Sr_{0.4}TiO_3$-$Sr(Ga_{0.5}Ta_{0.5})O_3$ ceramics sintered at 1600°C for 3h. The effect of $Sr(Ga_{0.5}Ta_{0.5})O_3$ content on the average grain size in not very obvious. We can also see that $0.9Ba_{0.6}Sr_{0.4}TiO_3$-$0.1Sr(Ga_{0.5}Ta_{0.5})O_3$ has higher porosity than other compositions. The morphology of $0.5Ba_{0.6}Sr_{0.4}TiO_3$-$0.5Sr(Ga_{0.5}Ta_{0.5})O_3$ shows difference from that of other three compositions.
The temperature dependence of dielectric properties for various $Ba_{0.6}Sr_{0.4}TiO_3$-

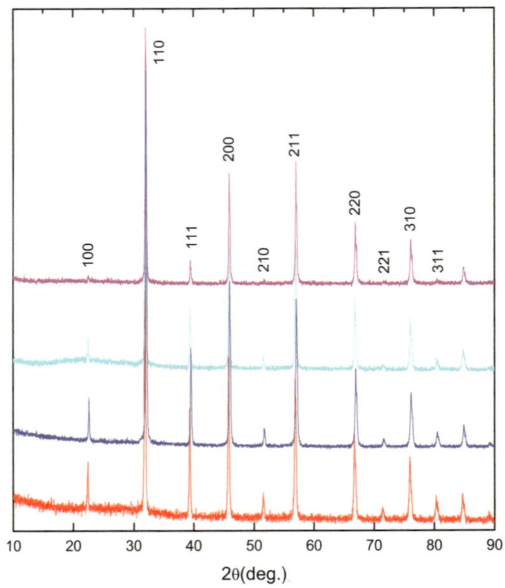

Fig. 17. The XRD patterns of $Ba_{0.6}Sr_{0.4}TiO_3$-$Sr(Ga_{0.5}Ta_{0.5})O_3$ ceramics sintered at 1600°C for 3h. From bottom to top, the $Sr(Ga_{0.5}Ta_{0.5})O_3$ content is 10, 20, 30 and 50mol%, respectively. The intensity is plotted on a log scale.

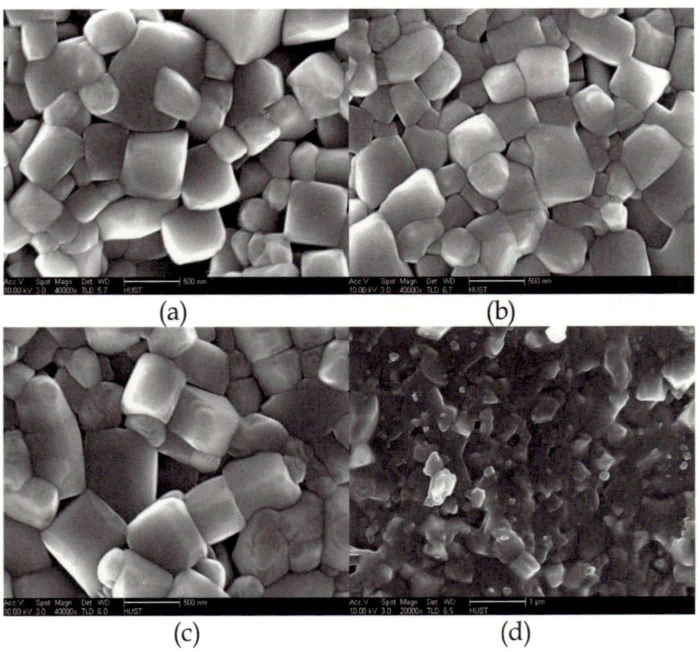

Fig. 18. FESEM images of $Ba_{0.6}Sr_{0.4}TiO_3$-$Sr(Ga_{0.5}Ta_{0.5})O_3$ ceramics sintered at 1600°C for 3h. From (a) to (d), the $Sr(Ga_{0.5}Ta_{0.5})O_3$ content is 10, 20, 30 and 50mol%, respectively.

$Sr(Ga_{0.5}Ta_{0.5})O_3$ ceramics (sintering temperature: 1600°C) measured at 100kHz is illustrated in Fig. 19. Broadened and suppressed dielectric peaks and shifts of Curie temperature T_C are observed with the addition of $Sr(Ga_{0.5}Ta_{0.5})O_3$. For $0.9Ba_{0.6}Sr_{0.4}TiO_3$-$0.1Sr(Ga_{0.5}Ta_{0.5})O_3$ ceramics, its ε_{max} decreases to ~ 686 and T_c shifts to lower temperature ~250K. As more $Sr(Ga_{0.5}Ta_{0.5})O_3$ is added to $Ba_{0.6}Sr_{0.4}TiO_3$, T_c shifts to lower temperatures, thus resulting in a decrease in dielectric constant at a given temperature and ε_{max}. For $0.8Ba_{0.6}Sr_{0.4}TiO_3$-$0.2Sr(Ga_{0.5}Ta_{0.5})O_3$, ε_{max} is ~335 at ~200K and for $0.5Ba_{0.6}Sr_{0.4}TiO_3$-$0.5Sr(Ga_{0.5}Ta_{0.5})O_3$, ε_{max} is ~95 at ~100K. On the other hand, loss tangent increases on cooling. For $0.9Ba_{0.6}Sr_{0.4}TiO_3$-$0.1Sr(Ga_{0.5}Ta_{0.5})O_3$ ceramics, there is small peak around ~250K. The loss tangent of $0.5Ba_{0.6}Sr_{0.4}TiO_3$-$0.5Sr(Ga_{0.5}Ta_{0.5})O_3$ ceramics (not shown) is almost independent on temperature and fluctuates around 0.004 at the temperature range of 60K-300K.

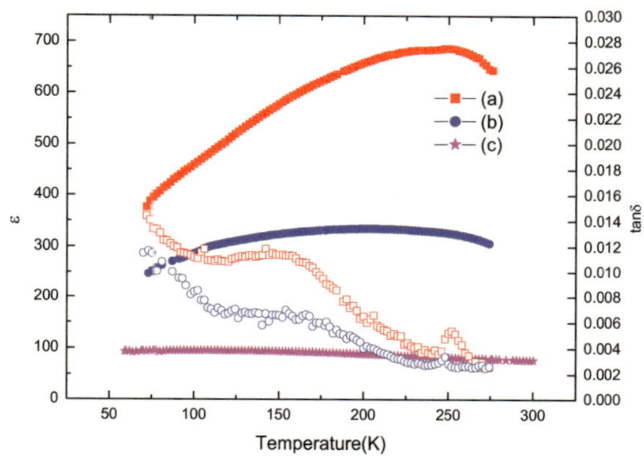

Fig. 19. Variation of dielectric constant (solid) and loss tangent (open) with temperature for $Ba_{0.6}Sr_{0.4}TiO_3$-$Sr(Ga_{0.5}Ta_{0.5})O_3$ ceramics (sintering temperature: 1600°C) measured at 100kHz: From (a) to (c), the $Sr(Ga_{0.5}Ta_{0.5})O_3$ content is 10, 20, and 50 mol%, respectively.

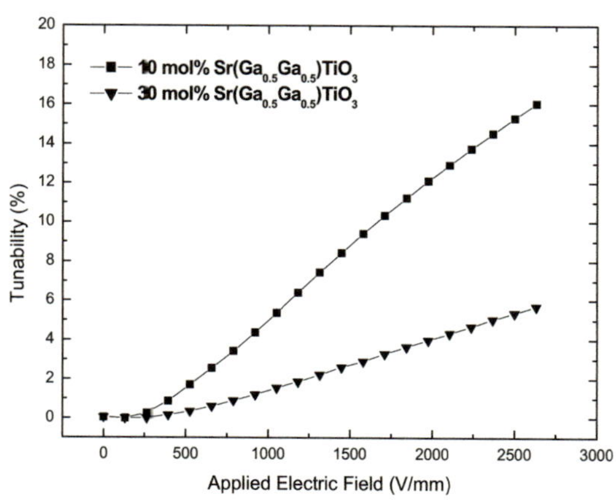

Fig. 20. The tunability of $0.9Ba_{0.6}Sr_{0.4}TiO_3$-$0.1Sr(Ga_{0.5}Ta_{0.5})O_3$ and $0.7Ba_{0.6}Sr_{0.4}TiO_3$-$0.3Sr(Ga_{0.5}Ta_{0.5})O_3$ at 100 kHz (sintering temperature: 1600°C).

Fig. 20 shows the tunability of the $Ba_{0.6}Sr_{0.4}TiO_3$-$Sr(Ga_{0.5}Ta_{0.5})O_3$ solid solutions at 100kHz, showing that the tunability decreases as the dielectric $Sr(Ga_{0.5}Ta_{0.5})O_3$ content increases. The decrease in the dielectric constant and tunability of $0.9Ba_{0.6}Sr_{0.4}TiO_3$-$0.1Sr(Ga_{0.5}Ta_{0.5})O_3$ results from the Ga and Ta substitution into B-site Ti and Sr substitution into A-site Ba in barium strontium titanate. $0.9Ba_{0.6}Sr_{0.4}TiO_3$-$0.1Sr(Ga_{0.5}Ta_{0.5})O_3$ has a dielectric tunability 16% under 2.63kV/mm versus a dielectric constant ε=534. The tunability of 0.7 $Ba_{0.6}Sr_{0.4}TiO_3$-$0.3Sr(Ga_{0.5}Ta_{0.5})O_3$ drops to be 5.7% under 2.63 kV/mm.

The microwave dielectric properties of $Ba_{0.6}Sr_{0.4}TiO_3$-$Sr(Ga_{0.5}Ta_{0.5})O_3$ solid solutions were listed in Table 4. With the increase of $Sr(Ga_{0.5}Ta_{0.5})O_3$ content, the dielectric constant decrease and the Q×f value increase. The Q×f value of the solid solution is not high except $0.5Ba_{0.6}Sr_{0.4}TiO_3$-$0.5Sr(Ga_{0.5}Ta_{0.5})O_3$. The low relative density maybe is the main reason: the relative density of $0.9Ba_{0.6}Sr_{0.4}TiO_3$-$0.1Sr(Ga_{0.5}Ta_{0.5})O_3$, $0.8Ba_{0.6}Sr_{0.4}TiO_3$-$0.2Sr(Ga_{0.5}Ta_{0.5})O_3$ and $0.7Ba_{0.6}Sr_{0.4}TiO_3$-$0.3Sr(Ga_{0.5}Ta_{0.5})O_3$ ceramics sintered at 1600°C for 3h is 82%, 89% and 88%, respectively.

$Sr(Ga_{0.5}Ta_{0.5})O_3$ content(mol%)	f_0(GHz)	ε	tanδ	Q×f(GHz)
10	1.73	592.4	0.015	115
20	2.05	375.3	0.013	158
30	2.49	236.9	0.012	208
50	4.68	79.6	0.0039	1200

Table 4. Microwave dielectric properties of $Ba_{0.6}Sr_{0.4}TiO_3$-$Sr(Ga_{0.5}Ta_{0.5})O_3$ solid solutions

3.2 $Ba_{0.6}Sr_{0.4}TiO_3$-$La(Mg_{0.5}Ti_{0.5})O_3$ solid solution

$La(Mg_{0.5}Ti_{0.5})O_3$ with low dielectric constant and loss tangent can form solid solutions with $BaTiO_3$ or $SrTiO_3$ in the whole compositional range (Avdeev 2002; Lee 2000). As shown in Fig. 21., $La(Mg_{0.5}Ti_{0.5})O_3$ form solid solution with $Ba_{0.6}Sr_{0.4}TiO_3$ (Xu et al. 2009).

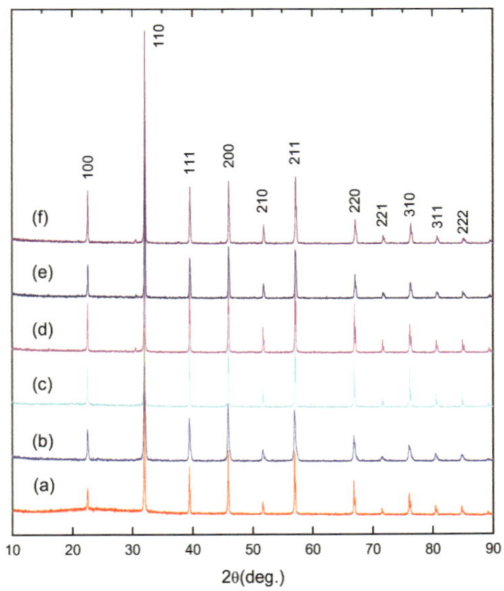

Fig. 21. The XRD patterns of (a) 10, (b) 20, (c) 30, (d) 40, (e) 50 and (f) 60 mol% $La(Mg_{0.5}Ti_{0.5})O_3$ mixed $Ba_{0.6}Sr_{0.4}TiO_3$ ceramics.

The microwave dielectric properties of $Ba_{0.6}Sr_{0.4}TiO_3$-$La(Zn_{0.5}Ti_{0.5})O_3$ ceramics were investigated. For different composition, the optimal sintering temperature is different. If the sintering temperature exceeds the corresponding value, the sample's rim and then interior became dark-blue in color, due to partial reduction of Ti^{4+} (d^0) to Ti^{3+} (d^1) associated with the oxygen loss from the lattice. Fig. 22 show the dielectric constant and Qf of $Ba_{0.6}Sr_{0.4}TiO_3$–$La(Mg_{0.5}Ti_{0.5})O_3$ ceramics sintered at optimal temperature. ε decreases with the increase of $La(Mg_{0.5}Ti_{0.5})O_3$ content, from ε_r=338.2 for $0.9Ba_{0.6}Sr_{0.4}TiO_3$-$0.1La(Mg_{0.5}Ti_{0.5})O_3$ to ε_r=47 for $0.4Ba_{0.6}Sr_{0.4}TiO_3$-$0.6La(Mg_{0.5}Ti_{0.5})O_3$. Qf value increases with increasing amounts of $La(Mg_{0.5}Ti_{0.5})O_3$. High Qf value of 9509 GHz with dielectric constant of 46.7 was obtained for $0.4Ba_{0.6}Sr_{0.4}TiO_3$-$0.6La(Mg_{0.5}Ti_{0.5})O_3$ at 5.69 GHz.

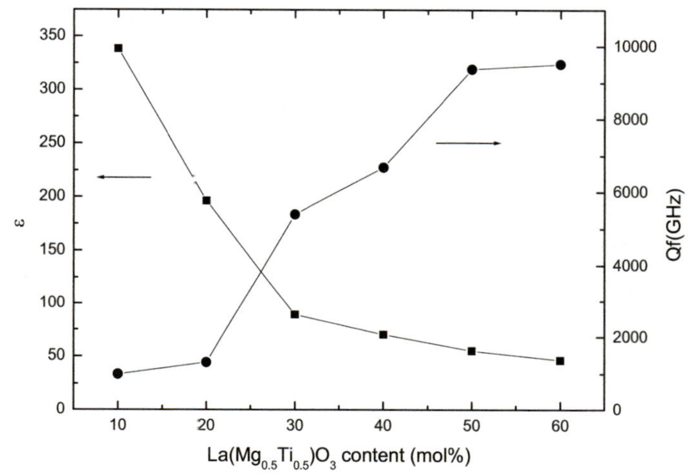

Fig. 22. Dielectric constant and quality factor of $Ba_{0.6}Sr_{0.4}TiO_3$–$La(Mg_{0.5}Ti_{0.5})O_3$ compositions as a function of $La(Mg_{0.5}Ti_{0.5})O_3$ content.

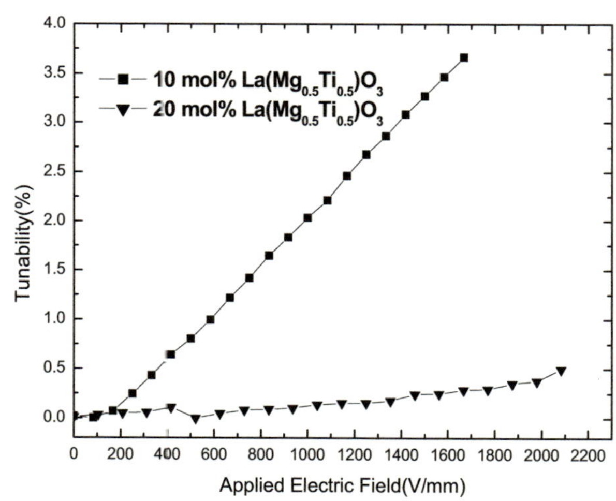

Fig. 23. The tunability of $Ba_{0.6}Sr_{0.4}TiO_3$–$La(Mg_{0.5}Ti_{0.5})O_3$ compositions measured at 100kHz and room temperature.

The tunability of $Ba_{0.6}Sr_{0.4}TiO_3$-$La(Mg_{0.5}Ti_{0.5})O_3$ ceramics is shown in Fig. 23. $La(Mg_{0.5}Ti_{0.5})O_3$ decreases the tunability of $Ba_{0.6}Sr_{0.4}TiO_3$ abruptly. The tunability of $0.9Ba_{0.6}Sr_{0.4}TiO_3$-$0.1La(Mg_{0.5}Ti_{0.5})O_3$ is only 3.7% under 1.67 kV/mm, although its Qf reaches 979GHz. Increasing $La(Mg_{0.5}Ti_{0.5})O_3$ content decreases the tunability further: the tunability of $0.8Ba_{0.6}Sr_{0.4}TiO_3$-$0.2La(Mg_{0.5}Ti_{0.5})O_3$ is 0.5% under 2.08 kV/mm.

The typical FESEM images of $Ba_{0.6}Sr_{0.4}TiO_3$-$La(Mg_{0.5}Ti_{0.5})O_3$ ceramics sintered at optimal temperature and the energy dispersive spectroscopy of $0.9Ba_{0.6}Sr_{0.4}TiO_3$-$0.1La(Mg_{0.5}Ti_{0.5})O_3$ were shown in Fig. 24. For $0.4Ba_{0.6}Sr_{0.4}TiO_3$-$0.6La(Mg_{0.5}Ti_{0.5})O_3$, dense ceramics were obtained, but higher porosity can be observed for the other three compositions. The high Qf value of $0.4Ba_{0.6}Sr_{0.4}TiO_3$-$0.6La(Mg_{0.5}Ti_{0.5})O_3$ can be related to its higher relative density. The chemical composition calculated from energy dispersive spectroscopy were listed in Table 5. We can see that the measured At% is consistent with the theoretical value within the error range. The result also proves the formation of solid solution further.

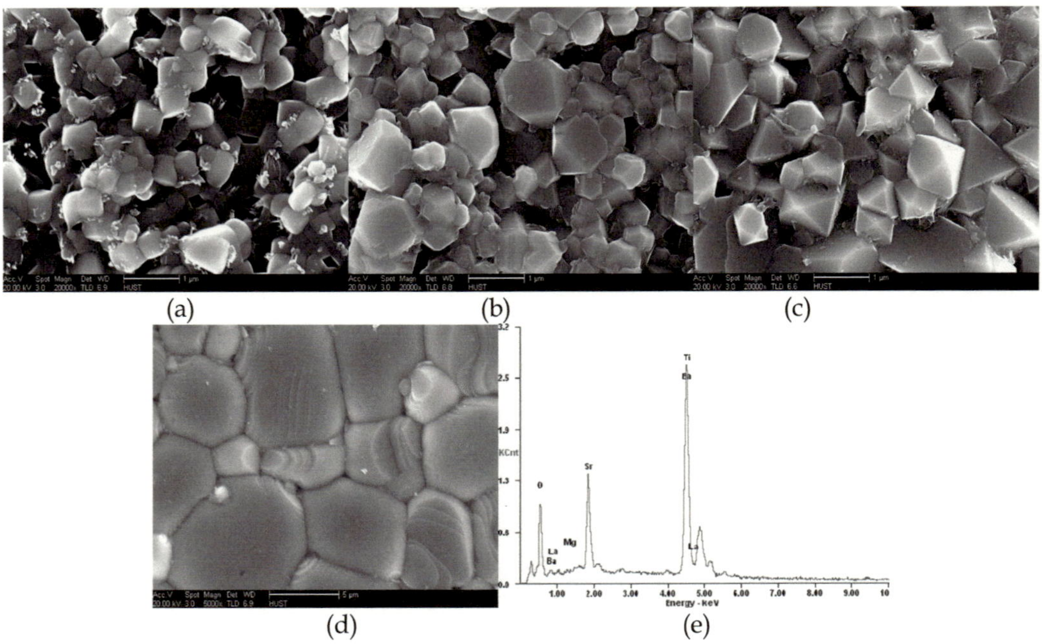

Fig. 24. FESEM images of $Ba_{0.6}Sr_{0.4}TiO_3$-$La(Mg_{0.5}Ti_{0.5})O_3$ ceramics and the energy dispersive spectroscopy of $0.9Ba_{0.6}Sr_{0.4}TiO_3$-$0.1La(Mg_{0.5}Ti_{0.5})O_3$ (f). From (a) to (d), $La(Mg_{0.5}Ti_{0.5})O_3$ content is 10, 20, 30 and 60 mol%, respectively.

Element	Wt%	At%	Theoretical At%
OK	21.15	56.99	60.61
MgK	00.53	0.95	1.01
SrL	18.65	9.18	7.27
BaL	28.76	9.03	10.91
TiK	24.19	21.77	19.19
LaL	06.71	2.08	2.02

Table 5. The chemical composition of $0.9Ba_{0.6}Sr_{0.4}TiO_3$-$0.1La(Mg_{0.5}Ti_{0.5})O_3$

3.3 $Ba_{0.6}Sr_{0.4}TiO_3$-$La(Zn_{0.5}Ti_{0.5})O_3$ and $Ba_{0.6}Sr_{0.4}TiO_3$-$Nd(Mg_{0.5}Ti_{0.5})O_3$ solid solution

$La(Zn_{0.5}Ti_{0.5})O_3$ have a comparatively small dielectric constant (ε=34 at 10GHz), a negative temperature coefficient of the resonance frequency (τ_f=-52ppmK^{-1}) and a low dielectric loss

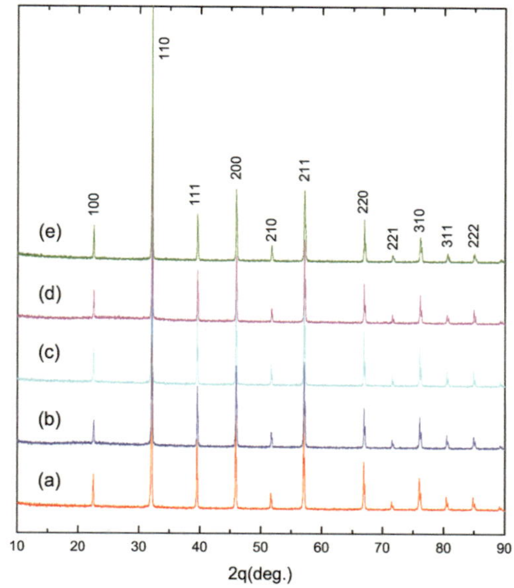

Fig. 25. The XRD patterns of (a) 10, (b) 20, (c) 30, (d) 40, and (e) 50 mol% $La(Zn_{0.5}Ti_{0.5})O_3$ mixed $Ba_{0.6}Sr_{0.4}TiO_3$ ceramics.

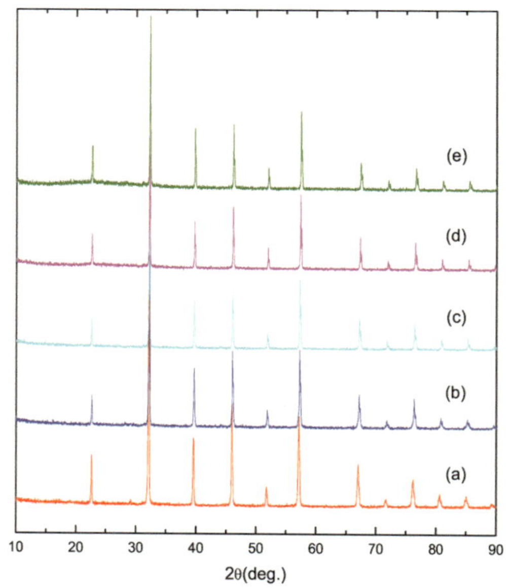

Fig. 26. The XRD patterns of (a) 10, (b) 20, (c) 30, (d) 40, and (e) 50 mol% $Nd(Mg_{0.5}Ti_{0.5})O_3$ mixed $Ba_{0.6}Sr_{0.4}TiO_3$ ceramics.

(Qf=59000GHz) (Cho et al. 1997). For $Nd(Mg_{0.5}Ti_{0.5})O_3$, the corresponding value is 26, -49 ppmK^{-1} and 36900GHz, respectively (Cho et al. 1999). XRD analysis showed that they can form solid solution with $Ba_{0.6}Sr_{0.4}TiO_3$ (Fig. 25 and 26), but their microwave dielectric properties is inferior to that of $Ba_{0.6}Sr_{0.4}TiO_3$-$La(Mg_{0.5}Ti_{0.5})O_3$.

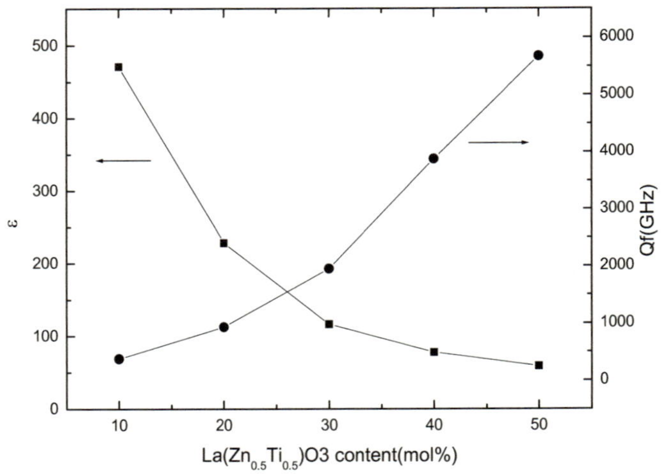

Fig. 27. Dielectric constant and quality factor of $Ba_{0.6}Sr_{0.4}TiO_3$–$La(Zn_{0.5}Ti_{0.5})O_3$ compositions as a function of $La(Zn_{0.5}Ti_{0.5})O_3$ content.

Fig. 27 show the dielectric constant and Qf of $Ba_{0.6}Sr_{0.4}TiO_3$–$La(Zn_{0.5}Ti_{0.5})O_3$ ceramics. The dielectric constant of $Ba_{0.6}Sr_{0.4}TiO_3$-$La(Zn_{0.5}Ti_{0.5})O_3$ solid solution decrease as the $La(Zn_{0.5}Ti_{0.5})O_3$ content increases. The Qf values of $Ba_{0.6}Sr_{0.4}TiO_3$-$La(Zn_{0.5}Ti_{0.5})O_3$ increase monotonously with increasing $La(Zn_{0.5}Ti_{0.5})O_3$ content. The highest Qf value of 5674 GHz was achieved in $0.5Ba_{0.6}Sr_{0.4}TiO_3$-$0.5La(Zn_{0.5}Ti_{0.5})O_3$ but reduced to 377GHz for $0.9Ba_{0.6}Sr_{0.4}TiO_3$-$0.1La(Zn_{0.5}Ti_{0.5})O_3$. The effect of $La(Zn_{0.5}Ti_{0.5})O_3$ on the microwave dielectric properties of $Ba_{0.6}Sr_{0.4}TiO_3$ solid solution is similar to that of $La(Mg_{0.5}Ti_{0.5})O_3$. The Qf value of $Ba_{0.6}Sr_{0.4}TiO_3$-$La(Zn_{0.5}Ti_{0.5})O_3$ is lower obviously than that of $Ba_{0.6}Sr_{0.4}TiO_3$-$La(Mg_{0.5}Ti_{0.5})O_3$ system although their relative density is higher than that of the corresponding $Ba_{0.6}Sr_{0.4}TiO_3$-$La(Mg_{0.5}Ti_{0.5})O_3$.

$Nd(Mg_{0.5}Ti_{0.5})O_3$ content (mol%)	sintering temperature (°C)	f_0(GHz)	ε	Qf(GHz)
20	1500	2.68	198.3	535
20	1550	2.83	193.0	615
30	1500	4.05	93.0	880
30	1550	4.30	94.7	1137

Table 6. Microwave dielectric properties of $Ba_{0.6}Sr_{0.4}TiO_3$-$Nd(Mg_{0.5}Ti_{0.5})O_3$ solid solutions

Table 6 lists the microwave dielectric properties of some $Ba_{0.6}Sr_{0.4}TiO_3$-$Nd(Mg_{0.5}Ti_{0.5})O_3$ ceramics. Although increasing $Nd(Mg_{0.5}Ti_{0.5})O_3$ content can increase the Qf value, the Qf value is not ideal: they are even lower than that of $Ba_{0.6}Sr_{0.4}TiO_3$-$La(Zn_{0.5}Ti_{0.5})O_3$ system.

The tunability of $Ba_{0.6}Sr_{0.4}TiO_3$-$La(Zn_{0.5}Ti_{0.5})O_3$ ceramics is shown in Fig. 28. The tunability of $0.9Ba_{0.6}Sr_{0.4}TiO_3$-$0.1La(Zn_{0.5}Ti_{0.5})O_3$ is only 2.7% under 1.67 kV/mm. It is even smaller than that of $0.9Ba_{0.6}Sr_{0.4}TiO_3$-$0.1La(Mg_{0.5}Ti_{0.5})O_3$ although $0.9Ba_{0.6}Sr_{0.4}TiO_3$-$0.1La(Zn_{0.5}Ti_{0.5})O_3$ has higher dielectric constant and loss tangent than that of $0.9Ba_{0.6}Sr_{0.4}TiO_3$-$0.1La(Mg_{0.5}Ti_{0.5})O_3$. Similarly, increasing $La(Zn_{0.5}Ti_{0.5})O_3$ content decreases the tunability of $Ba_{0.6}Sr_{0.4}TiO_3$-$La(Zn_{0.5}Ti_{0.5})O_3$ further. We can see that the dielectric properties of $Ba_{0.6}Sr_{0.4}TiO_3$-$La(Mg_{0.5}Ti_{0.5})O_3$ is better than that of $Ba_{0.6}Sr_{0.4}TiO_3$-$La(Zn_{0.5}Ti_{0.5})O_3$.

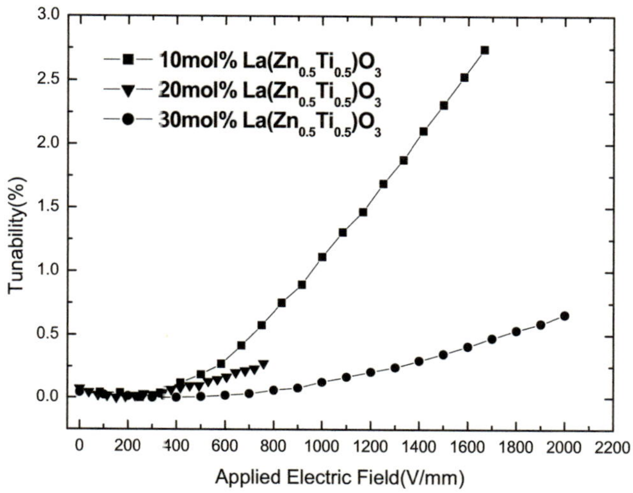

Fig. 28. The tunability of $Ba_{0.6}Sr_{0.4}TiO_3$–$La(Zn_{0.5}Ti_{0.5})O_3$ compositions measured at 100kHz and room temperature.

4. Discussion

Forming ferroelectric-dielectric solid solution and composite both can reduce the dielectric constant of ferroelectrics efficiently, but has different effect on the dielectric properties of ferroelectrics. (1). Forming solid solution can decrease the dielectric constant of ferroelectrics more rapidly when the doping content is nearly the same. The dielectric constant of $0.5Ba_{0.6}Sr_{0.4}TiO_3$-$0.5La(Mg_{0.5}Ti_{0.5})O_3$ is 55, which is far lower than that of 60 wt% MgO-mixed $Ba_{0.6}Sr_{0.4}TiO_3$ (ε=118) (Chang & Sengupta 2002; Sengupta & Sengupta 1999) although the doping content in 60 wt% MgO-mixed $Ba_{0.6}Sr_{0.4}TiO_3$ is higher and MgO has lower dielectric constant than $La(Mg_{0.5}Ti_{0.5})O_3$. (2). Forming solid solution can improve the loss tangent of ferroelectrics more effectively. The Qf value of $0.5Ba_{0.6}Sr_{0.4}TiO_3$-$0.5La(Mg_{0.5}Ti_{0.5})O_3$ and 60 wt% MgO-mixed $Ba_{0.6}Sr_{0.4}TiO_3$ is 9367GHz and 750GHz (Chang & Sengupta 2002; Sengupta & Sengupta 1999), respectively. Even for loose $0.9Ba_{0.6}Sr_{0.4}TiO_3$-$0.1La(Mg_{0.5}Ti_{0.5})O_3$ ceramics, its Qf value (979GHz) is much higher than that of $Ba_{0.6}Sr_{0.4}TiO_3$-Mg_2TiO_4-MgO. In the preparation process of microwave dielectric ceramics, the formation of secondary phase should be prevented. (3). Forming multiphase composite can maintain sufficiently high tunability. $0.5Ba_{0.6}Sr_{0.4}TiO_3$-$0.5La(Mg_{0.5}Ti_{0.5})O_3$ has lost tunability completely, but the tunability of 60 wt% MgO-mixed $Ba_{0.6}Sr_{0.4}TiO_3$ at 2kV/mm and 8kV/mm is 10% and 38%, respectively (Chang & Sengupta 2002; Sengupta & Sengupta 1999).

Some authors addressed the dielectric response of ferroelectric-dielectric composites theoretically and various composite models were used to evaluate the dielectric constant, tunability, and loss tangent (Astafiev 2003; Sherman et al. 2006; Tagantsev et al. 2006). As Tagantsev stated (Tagantsev et al. 2006), mixing a tunable ferroelectric with a linear dielectric may modify the electrical properties of the material due to mainly two effects: (i) "doping effect",–effect of doping of the ferroelectric lattice via the substitution of the ions of the host material and (ii) "composite effect"–effects of redistribution of the electric field in the material due to the precipitation of the non-ferroelectric phase at the grain boundaries or in the bulk of the material. The first effect results primarily in a shift and smearing of the temperature anomaly of the permittivity. The second effect leads to a redistribution of the electric field in the material. For ferroelectric-dielectric solid solutions, "composite effect" can be excluded. We can deduce that the addition of low loss perovskite dielectric influenced the chemistry and microstructure of the material, which resulted in the change of dielectric properties of materials. In ferroelectric-dielectric solid solution, a high degree of structural disorder due to random cation arrangement in both A- and B-sites is present, addition of pervoskite dielectrics apparently destroys the ferroelectric state, leading to the sharp decrease of tunability. For ferroelectric-dielectric composites, "doping effect" can be ignored. The effect of the dilution-driven field redistribution in the material is the main manifestation of addition of the dielectric into ferroelectrics in two-phase or multiphase composite. The reduction of the volume of ferroelectric, which is responsible for tuning, causes suppression of the tunability of the material. In ferroelectric-dielectric composites, ferroelectrics host lattice remains unchanged and the decrease of tunability is mainly due to ferroelectric dilution. "Destruction" in solid solution and "dilution" in composite has different effect on the tunability. Therefore, forming ferroelectric-dielectric solid solution can cause the decrease of tunability more sharply.

In ferroelectric-dielectric composite, the big contrast in the values of dielectric constants of linear dielectrics and the ferroelectric affects the redistribution of the electric field around the dielectrics. The dielectric constant of the ferroelectric under applied electric field becomes in-homogeneously distributed over the volume of the ferroelectric. The overall tunability of the composite, thus changes. Two competitive phenomena affect the tunable properties of the ferroelectric when it is diluted with a dielectric (Sherman et al. 2006). First, the reduction of the volume of ferroelectric, which is responsible for tuning, will cause suppression of the tunability of the material. Second, the redistribution of the electric field surrounding the linear dielectrics will affect the local tuning of the ferroelectric. Depending on the shape of the linear dielectrics and on the dielectric constants of the components, the impact of each of these two effects on the composite tunability is different and the second effect may be stronger (Sherman et al. 2006). In ferroelectric-dielectric composite $BaZr_{0.2}Ti_{0.8}O_3$-Mg_2SiO_4-MgO, with the increase of Mg_2SiO_4 content and the decrease of MgO content, the volume of ferroelectric $BaZr_{0.2}Ti_{0.8}O_3$ decrease due to smaller density of Mg_2SiO_4 than that of MgO, the tunability of composite will be suppressed. It is the fact as MgO content decreases from 48 wt% to 30 wt%. The anomalous increased tunability in $BaZr_{0.2}Ti_{0.8}O_3$-Mg_2SiO_4-MgO composite with MgO content < 30wt% can be attributed to redistribution of the electric field. Mg_2SiO_4 and MgO have different dielectric constants, they will have different effects on the redistribution of the electric field. The combination of linear dielectrics with different dielectric constants can result in the change of dielectric constant and loss tangent and even increase the tunabilty by affecting the redistribution of the electric field in the composite. As MgO content decreases from 30 wt% to 12 wt%, the increase of the tunability due to redistribution of the electric field exceeds the decrease of the tunability due to ferroelectric dilution, so the tunability of composite ceramics increases anomalously. The almost unchanged tunability in $Ba_{0.6}Sr_{0.4}TiO_3$-Mg_2SiO_4-MgO composite can also be explained. No similar result was observed in BST-Mg_2TiO_4-MgO composite: with the increase of MgO content and the decrease of Mg_2TiO_4 content, the dielectric constant and tunability decrease monotonously. The tunability of 40$Ba_{0.5}Sr_{0.5}TiO_3$-48Mg_2TiO_4-12MgO ceramics at 2kV/mm is 16.6%, but the corresponding tunability of 40$Ba_{0.5}Sr_{0.5}TiO_3$-12Mg_2TiO_4-48MgO is only 5.7%.

We can also see that ternary compositions ferroelectric-dielectric composite shows some advantages over binary compositions. (1). We can decrease the dielectric constant of ternary composites and remain the tunability almost unchanged (in $Ba_{0.6}Sr_{0.4}TiO_3$-Mg_2SiO_4-MgO), even increase the tunability (in $BaZr_{0.2}Ti_{0.8}O_3$-Mg_2SiO_4-MgO), without increasing the content of linear dielectrics. In order to decrease the dielectric constant of $Ba_{0.6}Sr_{0.4}TiO_3$-MgO, it is necessary to increase the content of linear dielectrics MgO, and the tunability will decrease inevitably. The tunability of $Ba_{0.5}Sr_{0.4}TiO_3$-MgO decreases with the increase of MgO content (Chang & Sengupta 2002). For $Ba_{0.6}Sr_{0.4}TiO_3$-Mg_2SiO_4-MgO composite, their dielectric constant decrease to 85-97, the tunability at 2kV/mm can be kept at around 10%. (2). The sintering temperature of ternary compositions $BaZr_{0.2}Ti_{0.8}O_3$-Mg_2SiO_4-MgO and $Ba_{0.6}Sr_{0.4}TiO_3$-Mg_2SiO_4-MgO can be reduced to 1350°C, which is 100°C and 150°C lower than the normal sintering temperature of BST-MgO (~1450°C) and BZT-MgO (~1500°C), respectively. The sintering temperature of BST-Mg_2TiO_4-MgO composite is also lower than that of BST-MgO.

On the other hand, ternary composition is helpful for the formation of ferroelectric-dielectric composite and can prevent from the formation of undesired phase. Forming ferroelectric-dielectric composite is an effective method to reduce the dielectric constant and loss tangent of ferroelectric and maintain higher tunability. The key is that the linear dielectrics with low dielectric constant and loss tangent should not react with ferroelectrics. For binary composition $Ba_{0.6}Sr_{0.4}TiO_3$-Mg_2SiO_4, it is expected to form ferroelectric ($Ba_{0.6}Sr_{0.4}TiO_3$)-dielectric (Mg_2SiO_4) composite, but undesired impurity phase $Ba_2(TiO)(Si_2O_7)$ is formed among $Ba_{0.6}Sr_{0.4}TiO_3$-Mg_2SiO_4 composite. $Ba_2(TiO)(Si_2O_7)$ deteriorate the properties of composites. MgO and Mg_2SiO_4 combination can prevent from the formation of $Ba_2(TiO)(Si_2O_7)$ and ferroelectric ($Ba_{0.6}Sr_{0.4}TiO_3$) and dielectric (Mg_2SiO_4 and MgO) composite is obtained. Similarly, Mg_2TiO_4 can react with $Ba_{1-x}Sr_xTiO_3$ (x=0.5 and 0.6) to form $BaMg_6Ti_6O_{19}$, but no $BaMg_6Ti_6O_{19}$ formed in BST-Mg_2TiO_4-MgO composite. Therefore, maybe ternary compositions can open a new route to decrease the dielectric constant and loss tangent of ferroelectrics and remain higher tunability. In future work, it is necessary to search new combination of linear dielectrics. Even if one linear dielectric may react with ferroelectrics, some linear dielectrics combination is possible to form ferroelectric-dielectric composite with ferroelectrics. This will expand the select range of linear dielectrics.

The multiple-phase ferroelectric-dielectric composites are useful for tunable microwave applications requiring low dielectric constant and make the impedance match more easily. The ferroelectric-dielectric composite bulk ceramics show promising application, especially in accelerator, as active elements of electrically controlled switches and phase shifters in pulse compressors or power distribution circuits of future linear colliders as well as tuning layers for the dielectric based accelerating structures (Kanareykin et al. 2006, 2009a, 2009b). The ferroelectric bulk ceramics can also be used in ferroelectric lens (Rao et al. 1999).

The ferroelectric-dielectric composites might complicate method to effectively deposit films. Therefore, the advantage of ferroelectric-dielectric solid solution over composite is that single phase materials is favorable for the thin film deposition. At the meantime, the tunability of solid solution can be increase to relatively high level by increasing applied electric field. Generally, linear dielectric with perovskite structure can form solid solution with ferroelectric BST. Different linear dielectrics has different effects on the dielectric properties. Among studied solid solution, $Ba_{0.6}Sr_{0.4}TiO_3$-$La(Mg_{0.5}Ti_{0.5})O_3$ shows better properties. It is necessary to increase the density of the solid solution, meanwhile, prevent the reduction of Ti^{4+}.

5. Conclusion

Forming ferroelectric-dielectric composite and solid solution can reduce the dielectric constant of ferroelectrics efficiently, but the mechanisms affecting dielectric properties differ in composites and solid solutions. Forming ferroelectric-dielectric solid solution can improve the loss tangent of ferroelectrics more effectively and is beneficial to film deposition. Forming ferroelectric-dielectric composite is more efficient to decrease the dielectric constant of ferroelectrics to a low value and maintain tunability at a sufficiently high level.

6. Acknowledgment

This work is supported by the Natural Science Foundation of China under grant number 10975055 and 60771021.

7. References

Astafiev K. F., Sherman V. O., Tagantsev A. K. & Setter N. (2003). Can the Addition of a Dielectric Improve the Figure of Merit of a Tunable Material? *J. Eur. Ceram. Soc.*, Vol. 23, No. 14, (2003), pp. 2381-2386, ISSN 0955-2219

Avdeev M., Seabra M. P. & V. M. Ferreira (2002). Crystal Structure of Dielectric Ceramics in the $La(Mg_{0.5}Ti_{0.5})O_3$–$BaTiO_3$ System. *J. Mater. Res.*, Vol.17, No. 5, (May 2002.), pp. 1112-1117, ISSN 0884-2914

Chang W. & Sengupta L. (2002). MgO-mixed $Ba_{0.6}Sr_{0.4}TiO_3$ Bulk Ceramics and Thin Films for Tunable Microwave Applications. *J. Appl. Phys.*, Vol.92 No.7, (October 2002), pp. 3941-46, ISSN 0021-8979

Cho S. Y., Seo M. K., Hong K. S. & Park S. J. (1997). Influence of ZnO Evaporation on the Microwave Dielectric Properties of $La(Zn_{1/2}Ti_{1/2})O_3$. *Mater. Res. Bull.*, Vol.32, No.6, (June 1997), pp. 725-735, ISSN 0025-5408

Cho S. Y., Kim C. H., Kim D. W., Hong K. S. & Kim J. H. (1999). Dielectric Properties of $Ln(Mg_{1/2}Ti_{1/2})O_3$ as Substrates for High-Tc Superconductor Thin Films. *J. Mater. Res.*, Vol.14, No.6, (June 1999), pp. 2484-87, ISSN 0884-2914

Chou X. J., Zhai J. W. & X. Yao. (2007). Dielectric tunable Properties of Low Dielectric Constant $Ba_{0.5}Sr_{0.5}TiO_3$–Mg_2TiO_4 Microwave Composite Ceramics. *Appl. Phys. Lett.*, Vol.91, No.10, (September 2007), pp. 122908, ISSN 0003-6951

He Y. Y., Xu Y. B., Liu T., Zeng C. L., Chen W. P. (2010). Microstructure and Dielectric Tunable Properties of $Ba_{0.6}Sr_{0.4}TiO_3$-Mg_2SiO_4-MgO Composite. *IEEE Trans. Ultrason. Ferroelectr. Freq. Control*, Vol.57, No.7, (July 2010), pp. 1505-1512, ISSN 0885-3010

He Y. Y., Xu Y. B., Liu T., Zeng C. L., Chen W. P. (2011). Tunable Dielectric Properties of $BaZr_{0.2}Ti_{0.8}O_3$-Mg_2SiO_4-MgO composite ceramics. *J. Alloy. Compd.*, Vol.509, No.3, (January 2011), pp. 904-908, ISSN 0925-8388

Kanareykin A., Nenasheva E., Yakovlev V., Dedyk A., Karmanenko S., Kozyrev A., Osadchy V., Kosmin D., Schoessow P. & Semenov A. (2006). Fast switching ferroelectric materials for accelerator applications. *AIP Conf. Proc.*, Vol. 877, (2006), pp. 311-319, ISSN 0094-243X

Kanareykin A., Nenasheva E., Kazakov S., Kozyrev A., Tagantsev A., Yakovlev V. & Jing C. (2009a). Ferroelectric based technologies for accelerators. *AIP Conf. Proc.*, Vol.1086, (2009), pp.380-385, ISSN 0094-243X

Kanareykin A., Jing C., Nenasheva E., Schoessow P., Power J. G. & Gai W. (2009b). Development of a Ferroelectric Based Tunable DLA Structure. *AIP Conf. Proc.*, Vol. 1086, (2009), pp.386-391, ISSN 0094-243X

Lee D. Y., Yoon S. J., Yeo J. H., Nahm S., Paik J. H., Whang K. C. & B. G. Ahn (2000). Crystal Structure and Microwave Dielectric Properties of $La(Mg_{1/2}Ti_{1/2})O_3$ Ceramics. *J. Mat. Sci. Lett.*, Vol.19, No.2 , (January 2000), pp. 131-134, ISSN 0261-8028

Maiti T., Guo R. & Bhalla A. S. (2007a). Enhanced Electric Field Tunable Dielectric Properties of $BaZr_xTi_{1-x}O_3$ Relaxor Ferroelectrics. *Appl. Phys. Lett.*, Vol.90, No.18, (April 2007), pp. 182901, ISSN 0003-6951

Maiti T., Guo R., Bhalla A. S. (2007b). Ferroelectric relaxor behaviour in $Ba(Zr_xTi_{1-x})O_3$: MgO composites. *J. Phys. D-Appl. Phys.*,Vol.40, No.14, (July 2007), pp. 4355-4359, ISSN 0022-3727

Maiti T., Guo R., Bhalla A. S. (2007c). Tailored dielectric properties and tunability of lead free relaxor $Ba(Zr_xTi_{1-x})O_3$: MgO composites. *Ferroelectr.*, Vol.361, No.1, (2007), pp. 84-91, ISSN 0015-0193

Maiti T., Guo R., Bhalla A. S. (2008). Structure-Property Phase Diagram of $Ba(Zr_xTi_{1-x})O_3$ System. *J. Am. Ceram. Soc.*, Vol.91, No. 6, (June 2008), pp. 1769–1780, ISSN 0002-7820

Nenasheva, E. A., Kartenko, N. F., Gaidamaka, I. M., Trubitsyna, O. N., Redozubov, S. S., Dedyk, A. I.& Kanareykin, A. D. (2010). Low loss microwave ferroelectric ceramics for high power tunable devices. *J. Eur. Ceram. Soc.*, Vol.30, No.2, (January 2010), pp. 395-400, ISSN 0955-2219

Rao J. B. L., Patel D. P. & Krichevsky V. (1999). Voltage-controlled Ferroelectric Lens Phased Arrays. *IEEE Trans. Antennas Propaga.*, Vol.47, No.3, (March 1999), pp. 458-468, ISSN 0018-926X

Romanofsky R. R., Bernhard J. T., Van Keuls F. W., Miranda F. A. & Canedy C. (2000). K-Band Phased Array Antennas based on $Ba_{0.60}Sr_{0.40}TiO_3$ Thin-Film Phase Shifters. *IEEE Trans. Microwave Theory Tech.*, Vol.48, No.12, (December 2000), pp. 2504-2510, ISSN 0018-9480

Sengupta L. C. & Sengupta S. (1997). Novel Ferroelectric Materials for Phased Array Antennas. *IEEE Trans. Ultrason. Ferroelectr. Freq. Control*, Vol. 44, No.7, (July 1997), pp. 792-797, ISSN 0885-3010

Sengupta L. C. & Sengupta S. (1999). Breakthrough Advances in Low Loss, Tunable Dielectric Materials. *Mat. Res. Innovat.*, Vol.2, No.5, (March 1999), pp. 278-282, ISSN 1432-8917

Sherman V. O., Tagantsev A. K. & N. Setter. (2006). Ferroelectric-Dielectric Tunable Composites. *J. Appl. Phys.*, Vol.99, No.7, (April 2006), pp. 074104, ISSN 0021-8979

Tagantsev A. K., Sherman V. O., Astafiev K. F., Venkatesh J. & N. Setter. (2003). Ferroelectric Materials for Microwave Tunable Applications. J. Electroceram., Vol.11, No.1-2, (Sept./Nov. 2003) pp. 5-66, ISSN 1385-3449

Takahashi J., Kageyama K., Fujii T., Yamada T. & Kodaira K. (1997). Formation and Microwave Dielectric Properties of $Sr(Ga_{0.5}Ta_{0.5})O_3$-Based Complex Perovskites. *J. Mater. Sci., Mater. in Electron.*, Vol. 8, No. 2, (April 1997), pp. 79-84, ISSN 0957-4522

Varadan V. K., Varadan V. V., J. F. Kelly & Glikerdas P. (1992). Ceramic Phase Shifters for Electronically Steerable Antenna Systems. *Microwave J.*, Vol.35, No.1, (January 1992), pp. 116-127, ISSN 0192-6225

Xu Y. B., Liu T., He Y. Y. & Yuan X. (2008). Dielectric Properties of $Ba_{0.6}Sr_{0.4}TiO_3$-$Sr(Ga_{0.5}Ta_{0.5})O$ Solid Solutions. *IEEE Trans. Ultrason. Ferroelectr. Freq. Control,* Vol.56, No.11, (November 2008), pp. 2369-2376, ISSN 0885-3010

Xu Y. B., Liu T., He Y. Y. & Yuan X. (2009). Dielectric Properties of $Ba_{0.6}Sr_{0.4}TiO_3$-$La(B_{0.5}Ti_{0.5})O_3$ (B=Mg, Zn) Ceramics. *IEEE Trans. Ultrason. Ferroelectr. Freq. Control,* Vol.56, No.11, (November 2009), pp. 2343-2350, ISSN 0885-3010

Zhi Y., Chen A., Guo R. & Bhalla A. S. (2002). Dielectric Properties and High Tunability of $Ba(Ti_{0.7}Zr_{0.3})O_3$ Ceramics under dc Electric Field. *Appl. Phys. Lett..*, Vol.81, No.7, (August 2002), pp.1285–87, ISSN 0003-6951

12

New Multiferroic Materials: Bi_2FeMnO_6

Hongyang Zhao[1], Hideo Kimura[1], Qiwen Yao[1], Yi Du[2], Zhenxiang Cheng[2] and Xiaolin Wang[2]
[1]National Institute for Materials Science,
[2]Institute for Superconducting and Electronics Materials, University of Wollongong,
[1]Japan
[2]Australia

1. Introduction

The term "ferroic" was introduced by Aizu in 1970, and presented a unified treatment of certain symmetry-dictated aspects of ferroelectric, ferroelastic, and ferromagnetic materials. Ferroelectric materials possess a spontaneous polarization that is stable and can be switched hysteretically by an applied electric field; antiferroelectric materials possess ordered dipole moments that cancel each other completely within each crystallographic unit cell. Ferromagnetic materials possess a spontaneous magnetization that is stable and can be swithched hysteretically by an applied magnetic field; antiferromagnetic materials possess ordered magnetic moments that cancel each other completely within each magnetic unit cell. By the original definition, a single-phase multiferroic material is one that possesses more than one 'ferroic' properties: ferroelectricity, ferromagnetism or ferroelasticity. But the classification of multiferroics has been broadened to include antiferroic order. Multiferroic materials, in which ferroelectricity and magnetism coexist, the control of magnetic properties by an applied electric field or, in contrast, the switching of electrical polarization by a magnetic field, have attracted a great deal of interest. Now we can classify multiferroic materials into two parts: one is single-phase materials; the other is layered or composite heterostructures. The most desirable situation would be to discover an intrinsic single-phase multiferroic material at room temperature. However, $BiFeO_3$ is the only known perovskite oxides that exhibits both antiferromagnetism and ferroelectricity above room temperature. Thus, it is essential to broaden the searching field for new candidates, which resulted in considerable interest on designed novel single phase materials and layered or composite heterostructures.

2. Material designation and characterization

For ABO_3 perovskite structured ferroelectric materials, they usually show antiferromagnetic order because the same B site magnetic element. While for the $A_2BB'O_6$ double perovskite oxides, the combination between B and B′ give rise to a ferromagnetic coupling. They are also expected to be multiferroic materials. The ferroelectric polarization is induced by the distortion which usually causes a lower symmetry. For device application, a large magnetoelectric effect is expected in the $BiFeO_3$ and bismuth-based double perovskite oxides ($BiBB'O_6$), many of which have aroused great interest like Bi_2NiMnO_6, $BiFeO_3$-$BiCrO_3$. But far as we know, few researches were focused on Bi_2FeMnO_6.

Multiferroic material is an important type of lead-free ferroelectrics. While they usually showed leaky properties and not well-shaped *P-E* loops. Dielectricity includes piezoelectricity, and piezoelectricity includes ferroelectricity. Therefore, it is essential to characterize the dielectric, piezoelectric and ferroelectric properties together. Firstly we have designed several multiferroic materials, and then we studied their properties using efficient techniques which include *P-E* loop measurement, positive-up-negative-down (PUND) test and piezoresponse force microscopy (PFM). All the fabricated materials were found to be multiferroics, so the magnetic properties were also characterized.

2.1 Material designation

Magnetism and ferroelectricity exclude each other in single phase multiferroics. It is difficult for designing multiferroics with good magnetic and ferroelectric properties. Our interest is to design new candidate multiferroics based on $BiFeO_3$. According to the Goodenough-Kanamori (GK) rules, many ferromagnets have been designed in double perovskite system ($A_2BB'O_6$) through the coupling of two B site ions with and without e_g electrons. Because the complication of the double perovskite system, there are still some questions about the violation of the GK rules in some cases and the origin of the ferromagnetism or antiferromagnetism. Nevertheless, it is believed that the B site superexchange interaction, the oxygen defects and the mixed cation valences are the important factors in determining the magnetic properties of the double perovskites. Therefore, the preparation methods and conditions will show a large influence on the magnetic properties of the fabricated double peroskites. In order to modify the antiferromagnetic properties of $BiFeO_3$, novel single-phase Bi_2FeMnO_6 series materials were designed. We have obtained very interesting results and firstly succeeded in proofing that the designed Bi_2FeMnO_6 is another promising single-phase room temperature multiferroic material. Then we designed Nd: $BiFeO_3/YMnO_3$, Nd: $BiFeO_3/Bi_2FeMnO_6$ to further study the B site superexchange interaction between Fe and Mn. Surprisingly, they also showed room temperature multiferroic properties. These exciting results provided us with more confidence in designing devices based on multiferroic materials. Different preparation methods also show large influence to their properties. The comparison between the samples of bulk, nano-powder and films is essential for the understanding of the underlying physics and the development of ferroelectric concepts.

2.1.1 Bi_2FeMnO_6 (BFM) and $(La_xBi_{1-x})_2FeMnO_6$ (LBFM)

$BiFeO_3$ is a well-known multiferroic material with antiferromagntic with a Neel temperature of 643 K, which can be synthesized in a moderate condition. In contrast, $BiMnO_3$ is ferromagnetic with $T_c = 110$ K and it needs high-pressure synthesis. Single phase Bi_2FeMnO_6 (BFM) ceramics could be synthesized by conventional solid state method as the target. The starting materials of Bi_2O_3, Fe_3O_4, $MnCO_3$ were weighed according to the molecular mole ratio with 10 mol% extra Bi_2O_3. They were mixed, pressed into pellets and sintered at 800 °C for 3 h. Then the ceramics were crushed, ground, pressed into pellets and sintered again at 880 °C for 1 h. BFM films were deposited on (100) $SrTiO_3$ substrate by pulsed laser deposition (PLD) method at 650°C with 500 ~ 600 mTorr dynamic oxygen.

The stabilization of the single-phase Bi-based perovskites are difficult because of their tendency of multiphase formation and the high volatility of bismuth. Stabilization can be facilitated by a partial replacement of Bi^{3+} cations by La cations. In addition, $LaMn_{1-x}Fe_xO_3$ including La_2FeMnO_6 has been also reported to be an interesting mixed-valence manganite with perovskite structure. Therefore, La was chosen to partially substitute Bi in Bi_2FeMnO_6 to stabilization the phase. Polycrystalline 20 mol% La doped Bi_2FeMnO_6 (LBFM) ceramic and film were also obtained using the similar preparation methods mentioned above.

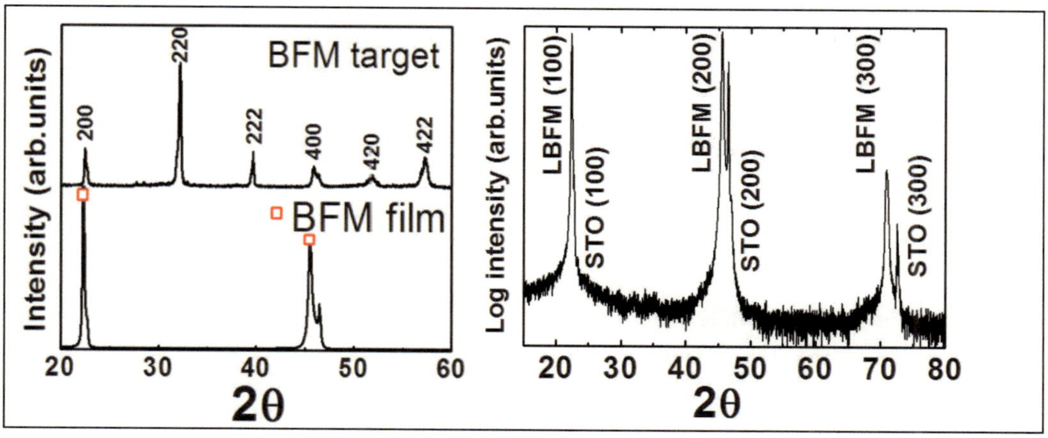

Fig. 1. XRD spectra for BFM target and film fabricated on (100) STO (left); XRD for LBFM film (right).

Figure 1 (left) shows the XRD patterns of the BFM target and the film. Because $BiFeO_3$ has a rhombohedral $R3c$ structure whereas $BiMnO_3$ has a monoclinic structure, it is natural that the BFM will show a different structure due to the coexist of two transition metal octahedral with different_distortions. Bi et al has calculated three structures of BFM with the space group of $Pm\bar{3}m$, $R3$ and $C2$. In this work, the bulk BFM target shows a cubic $Pm3m$ structure and it was indexed using the data from Bi et al. The second phase ($Bi_2Fe_4O_9$) was observed in the BFM ceramics, which often appears in the $BiFeO_3$ ceramics. While the thin film on the (100) STO substrate fabricated in high oxygen pressure condition shows a single phase with a bulk-like structure with no traceable impurity. In this study we focused mainly on the single phase film, because the impurities will have large influences on magnetic properties and blind the observation of the intrinsic property. As shown in Figure 1 (right), the LBFM diffraction peaks of (100), (200) and (300) were observed in the XRD pattern. It indicates the epitaxial growth of LBFM film on the (100) STO substrate. There is no traceable impurity in the film which is believed to have a bulk-like cubic structure. But there are unavoidable impurities of bismuth oxides in the LBFM ceramics, which reduces further the crystalline quality of the ceramic compared with the LBFM film..

The Scanning electron microscopy (SEM) was used for the film morphology characterization. The SEM images of the BFM films were shown in Figure 2. The film on Si shows fiber shaped morphology with different orientations, as marked as parallel fibers and inclined fibers. In the contrast, the film on STO substrate shows fibers with almost the same orientation. It is essential to understand the orientation and anisotropy properties to optimize and design functional devices. In the previous work, it is proved that BFM on (100)

STO shows large magnetic anisotropy and out-of-plane is the easy magnetization direction. In this work, we focus mainly on the BFM film fabricated on STO substrates.

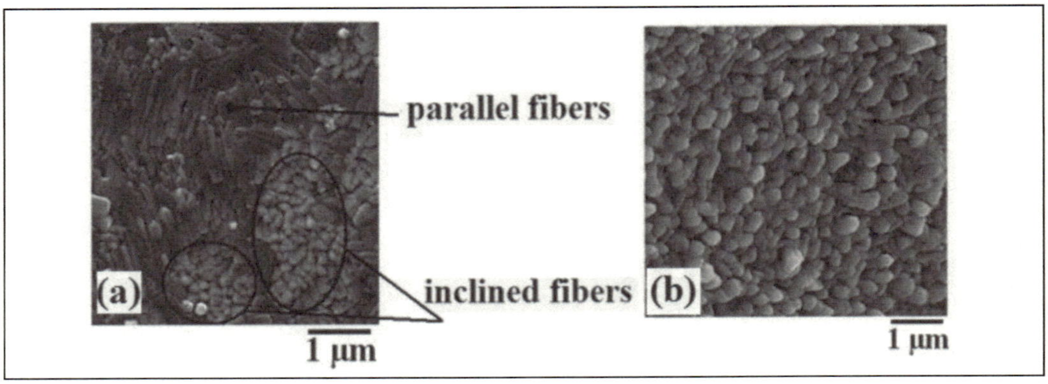

Fig. 2. SEM images of BFM film on (a) Si and (b) STO substrate.

2.1.2 Nd: $BiFeO_3$/ Bi_2FeMnO_6 (BFO/BFM)

In our former works, the doping of Nd into $BiFeO_3$ was found to further improve the ferroelectric properties. The Bilayered $Nd_{0.1}Bi_{0.9}FeO_3$ (Nd: $BiFeO_3$)/ BFM films on $Pt/Ti/SiO_2/Si$ substrate were fabricated using a PLD system. Nd: $BiFeO_3$ films were fabricated at 550 ~ 580 °C with 200 mTorr dynamic oxygen pressure, and the BFM films were fabricated at 550 ~ 580 °C with ~10^{-5} Torr.

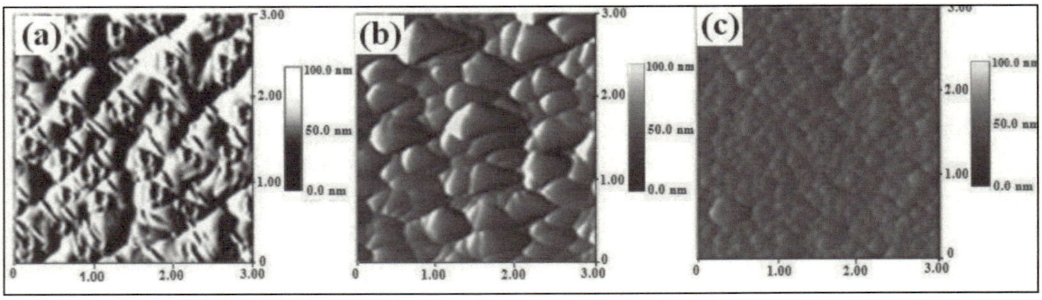

Fig. 3. Surface morphology of (a) Nd: $BiFeO_3/Bi_2FeMnO_6$, (b) Bi_2FeMnO_6 and (c) Nd: $BiFeO_3$.

The surface morphology of the Nd: $BiFeO_3/Bi_2FeMnO_6$ and Nd: $BiFeO_3$ films were studied using an atomic force microscope (AFM), as shown in Fig. 3. It can be found that the corresponding root-mean-square roughness (R_{rms}) and the grain size (S) are different: R_{rms} (Nd: $BiFeO_3$) < R_{rms} (Nd: $BiFeO_3/Bi_2FeMnO_6$) < R_{rms} (Bi_2FeMnO_6), and S (Nd: $BiFeO_3$) < S (Nd: $BiFeO_3/Bi_2FeMnO_6$) < S (Bi_2FeMnO_6). Fig. 3 (a) revealed the morphology of the Nd: $BiFeO_3$ film on the $Bi_2FeMnO_6/Pt/Ti/SiO_2/Si$, which indicated that Nd: $BiFeO_3$ had a larger growth rate on Bi_2FeMnO_6 than on $Pt/Ti/SiO_2/Si$ substrate.

2.1.3 Nd: $BiFeO_3$/$YMnO_3$ (BFO/YMO)

Another well-studied muliferroic material $YMnO_3$ was chosen to from the Nd: $BiFeO_3/YMnO_3$ (BFO/YMO) heterostructure. The hexagonal manganite $YMnO_3$, which shows an antiferromagnetic transition at T_N=75 K and a ferroelectric transition at T_C=913

K, is one of the rare existing single phase multiferroics. The hexagonal $YMnO_3$ is ferroelectric, but the orthorhombic YMnO3 is not ferroelectric. The (111) planes are special for $BiFeO_3$, the Fe spins are coupled ferromagnetically in the pseudocubid (111) planes and antiferromagnetically between neighbouring (111) planes. In this study, the BFO/YMO film was fabricated on (111) Nb: $SrTiO_3$ (STO) substrate the Nd: $BiFeO_3$ and $YMnO_3$ ceramics were synthesized by conventional solid state method as the targets. The Nd: $BiFeO_3/YMnO_3$ (BFO/YMO) film were deposited on (111) STO substrate using a pulsed laser deposition (PLD) system at 530-700°C with 10^{-3} ~10^{-1} Torr dynamic oxygen. The two separate targets were alternately switched and the films were obtained through a layer-by-layer growth mode. After deposition, the film was annealed at the same condition for 15 minutes and then cooled to room temperature. In this report, the film comprised of four layers: (1) Nd: $BiFeO_3$ (2) $YMnO_3$ (3) Nd: $BiFeO_3$ and (4) $YMnO_3$. The deposition time of each layer is 10 min.

2.2 Ferroelectric characterization

The methods and special techniques for materials with weak ferroelectric properties will be explained and summarized in detail. For typical ferroelectric materials, it is easy to identify their ferroelectricity because we could obtain well-shaped ferroelectric polarization hysteresis loops (*P-E* loop). However, as the definition of ferroelectricity is strict, it is difficult to characterize weak ferroelectricity and to check whether it has ferroelectric property or not. Here we will introduce our experience for characterization and identification of such materials.

2.2.1 P-E loop measurement

For the P-E loop measurment, Pt upper electrode with an area of 0.0314 mm^2 were deposited by magnetron sputtering through a metal shadow mask. The ferroelectric properties were measured at room temperature by an aixACCT EASY CHECK 300 ferroelectric tester. Figure 4 shows the ferroelectric hysteresis loops of the Nd: $BiFeO_3/Bi_2FeMnO_6$ film, the upper inset shows the polarization fatigue as a function of switching cycles up to 10^8 and the lower inset shows frequency dependence of the real part of dielectric permittivity. The remnant polarization P_r is 54 $\mu C/cm^2$ and E_c is 237 kV/cm. Some anomalies were observed in the P-E loop: the loop is asymmetry and the polarization decreased as the increasing of the electric field. It can be caused by many effects but some of them can be neglected like the macroscopic electrode influence and nonuniform polarization on the surface of the film. We consider there are two main reasons. The film is insulating so there is no movable carriers to balance the bound charge. Therefore, the polarization gradient will be arisen in the film and induced the depolarization field. In addition, there are inhomogeneous domains with different coercivity in the film, some of which are difficult to switch with applied field. Evidence can also be seen in the fatigue results which showed that the polarization increased with the increasing of the switching cycles. The fatigue can be caused by domain nucleation, domain wall pinning due to space charges or oxygen vacancies, interface between electrode and film, thermodynamic history of the sample and so on. For the unusual profile of fatigue (polarization increased with that the increasing of switching cycles), we consider the different domain wall played important roles during the polarization reversal. The dielectric properties were measured using a HP4248 LCR meter. Frequency dependence of the real part of the permittivity was measured at room temperature. There is a notable increase at low frequencies (as shown in the lower inset of Fig. 4). In such bilayered films, it is believed that there are space charges at the interface between the two layers of the Nd: $BiFeO_3$ and Bi_2FeMnO_6 which will affect the ferroelectric properties.

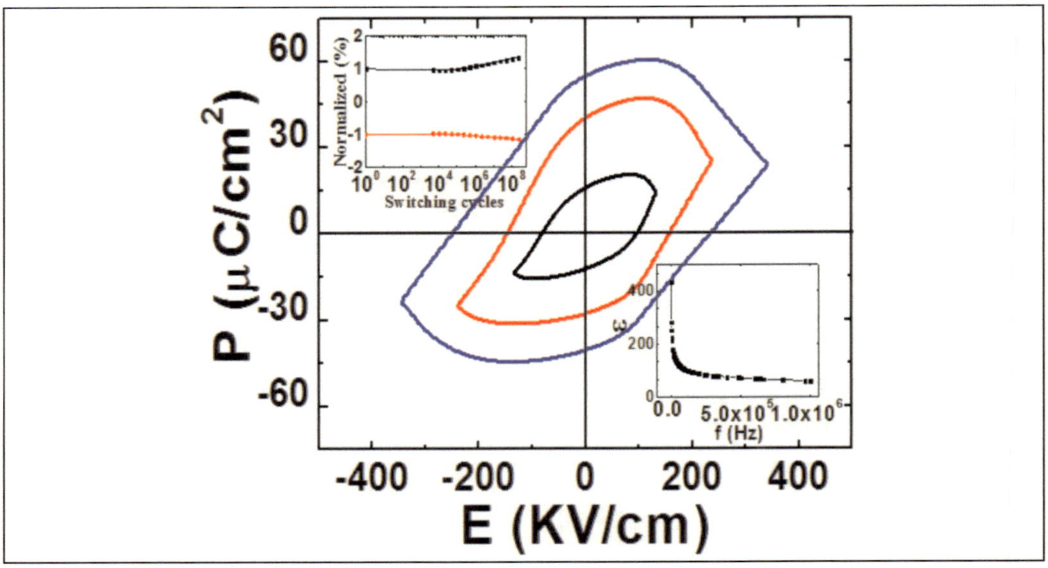

Fig. 4. Ferroelectric hysteresis loops of Nd: $BiFeO_3$/BFM film, the polarization fatigue as a function of switching cycles (upper inset) and the frequency dependence of the real part of dielectric permittivity (lower inset).

2.2.2 PUND: positive-up-negative-down test

As the definition of ferroelectricity is strict, a not-well-saturated loop might not be a proof of ferroelectricity, we have also measured the so-called positive-up-negative-down (PUND) test for Nd: $BiFeO_3$/ BFM film. The applied voltage waveform is shown in Fig. 5. The switching polarization was observed using the triangle waveform as a function of time as shown in Fig. 5.

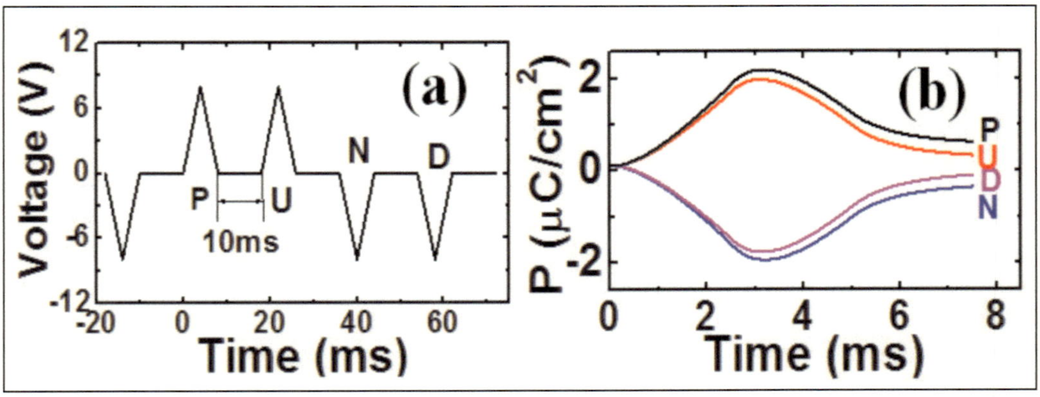

Fig. 5. (a) PUND waveform and (b) corresponding switching polarization.

2.2.3 PFM characterization for BFM and LBFM film

Until now there is no report about the ferroelectric properties of BFM because the difficulty of obtaining well-shaped polarization hysteresis loops. Thus, it is important to study and understand the ferroelectric properties and leakage mechanisms in the BFM system. The emerging technique of piezoresponse force microscopy (PFM) is proved to be a powerful tool to study piezoelectric and ferroelectric materials in such cases and extensive contributions have been published. In PFM, the tip contacts with the sample surface and the deformation (expansion or contraction of the sample) is detected as a tip deflection. The local piezoresponse hysteresis loop and information on local ferroelectric behavior can be obtained because the strong coupling between polarization and electromechanical response in ferroelectric materials. In the present study, we attempts to use PFM to study the ferroelectric/piezoelectric properties in BFM and LBFM thin films. PFM response was measured with a conducting tip (Rh-coated Si cantilever, k~1.6 N m^{-1}) by an SII Nanotechnology E-sweep AFM. PFM responses were measured as a function of applied DC bias (V_{dc}) with a small ac voltage applied to the bottom electrode (substrate) in the contact mode, and the resulting piezoelectric deformations transmitted to the cantilever were detected from the global deflection signal using a lock-in amplifier.

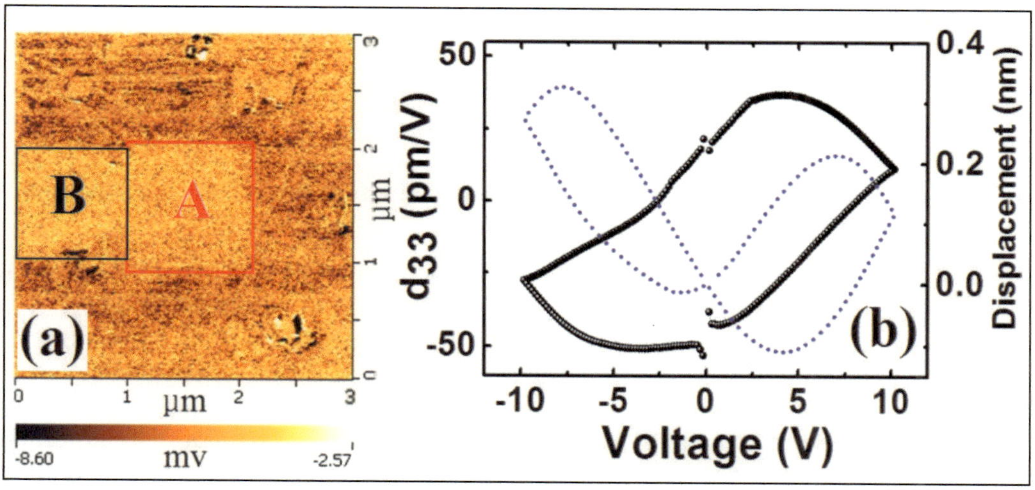

Fig. 6. (a) OP PFM image polarized by ±10 V and (b) which curve is associated with the left y-axis and which one is with the right y-axis as well as Fig.7 (c)local piezoresponse hysteresis loop of BFM film.

In Figure 6 (a), the smaller part A marked in red square was firstly poled with -10 V DC bias, and the total area of 3×3 μm^2 was subsequently poled with +10 V DC bias. The domain switching in red square area was observed, while another similar area beside 'A' was also observed and marked as B in black square. It may be because the expansion of ferroelectric domain under the DC bias. To further understand its ferroelectric nature, the local piezoelectric response was measured with a DC voltage from -10 V to 10 V applied to the sample. The typical "butterfly" loop was observed but it is not symmetrical, and it is not well-shaped due to the asymmetry of the upper and bottom electrodes. According to the equation $d_{33}=\Delta l/V$, where Δl is the displacement, the effective d_{33} could be calculated. At the voltage of -10 V, the sample has the maximum effective d_{33} of about -28 pm/V.

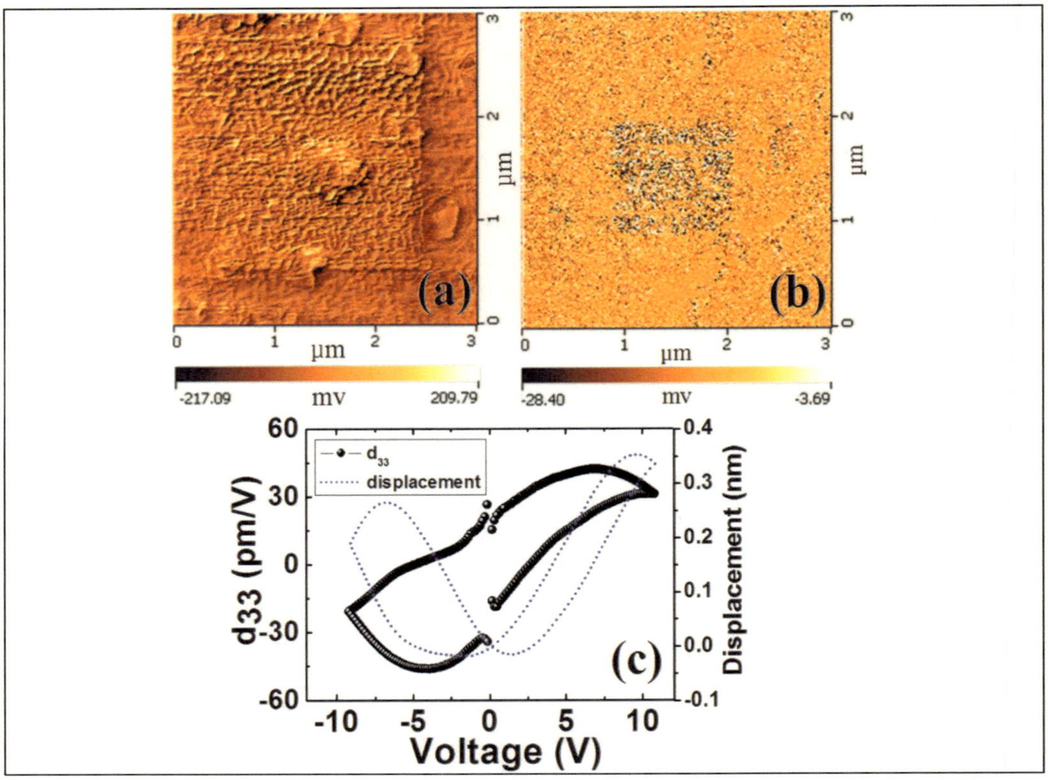

Fig. 7. (a) OP PFM image, (b) IP PFM image polarized by ±10 V and (c) local piezoresponse hysteresis loop of LBFM film.

Figure 7 shows the OP (a) and IP (b) PFM images of the LBFM film which was also fabricated on (100) STO substrate. Under ±10 V DC bias, PFM images were observed in the scans of the LBFM film, demonstrating that polarization reversal is indeed possible and proving that the LBFM film is ferroelectric at room temperature. At the voltage of 10 V, the sample has a maximum effective d_{33} of about 32 pm/V. The LBFM film shows improved piezoelectric and ferroelectric properties compared to the BFM film, indicating that through the doping or changing of other conditions, the ferroelectric property of BFM system could be improved as in the $BiFeO_3$. The domain boundary is very clear and regular in LBFM, while in BFM it is obscure and expanded over the poled area. The propagation of domain wall is strongly influenced by local inhomogeneities (e.g. grain boundaries) and stress in polycrystalline ferroelectrics, which results in strong irregularity of the domain boundary. After the La substitution, it is assumed that the crystallization is better both in ceramics and films.

2.3 Magnetic characterization for BFM film

BFM is considered to be a new multiferroic material, it is important to study their magnetic properties. Magnetic properties were measured using the commercial Quantum Design SQUID magnetometer (MPMS). In the following, we will discuss the XPS measurements, the magnetization hysteresis loops, and the ZFC and FC courves for the BFM film fabricated on the (100) STO susbtrate.

2.3.1 XPS measurements

The valance states of Fe and Mn in the BFM film were carried out using PHI Quantera SXM x-ray photoelectron spectrometer (XPS). Figure 8 shows the Fe 2p and Mn 2p photoelectron spectra of BFM film. It was reported that Fe 2p photoelectron peaks from oxidized iron are associated with satellite peaks, which is important for identifying the chemical states. The Fe^{2+} and Fe^{3+} $2p_{3/2}$ peaks always show the satellite peaks at 6 eV and 8 eV above the principal peaks at 709.5 eV and 711.2 eV, respectively. In Figure 8 (a), the satellite peaks were found just 8 eV above the $2p_{3/2}$ principal peak. It indicates that in this system Fe is mainly in the Fe^{3+} state. Figure 8 (b) shows typical XP spectra of Mn 2p. There are two main peaks corresponding to the $2p_{1/2}$ and $2p_{3/2}$ peaks, respectively. The peaks with higher binding energy above the main peaks as well as the splitting of the main peaks were observed in the film. It indicates the existence of Mn^{2+}. Such shake-up satellite peaks were considered to be a typical behavior in Mn^{2+} systems.

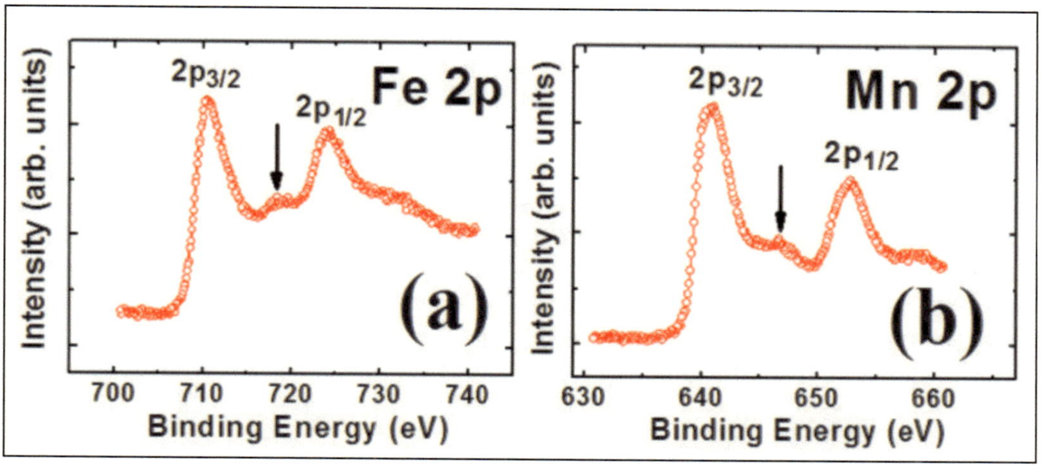

Fig. 8. (a) Fe 2p and (b) Mn 2p XP spectra for BFM film on (100) STO

2.3.2 Magnetic hysteresis loops

For the BFM thin films, different substrates of $Pt/Ti/SiO_2/Si$ and STO were used and different fabrication conditions were attempted. Some unavoidable impurities and different structures were observed for the films on $Pt/Ti/SiO_2/Si$ substrates. In order to discuss the origin of the ferromagnetic properties in BFM film, films on (100) STO were used for the study of magnetic properties. Figure 9 (a) shows the hysteresis loops measured at different temperatures. There is no significant change in the loop width from 5 K to 300 K. Figure 9 (b) shows the in-plane and out-of-plane magnetic field dependence of magnetization measured at 5 K. The film shows stress induced anisotropy from film/substrate mismatch which is an evidence of a Jahn-Teller effect and the out-of-plane is the easy magnetization direction. However, we observed experimentally that Mn shows multiple valence states despite the higher stability of the compound only containing Mn^{3+} ions in the film. It is possibly because the Mn^{2+} and Mn^{4+} cations could decrease the Jahn-Teller effect from Mn^{3+} in the film, which may result in less lattice distortion caused by Mn^{3+}.

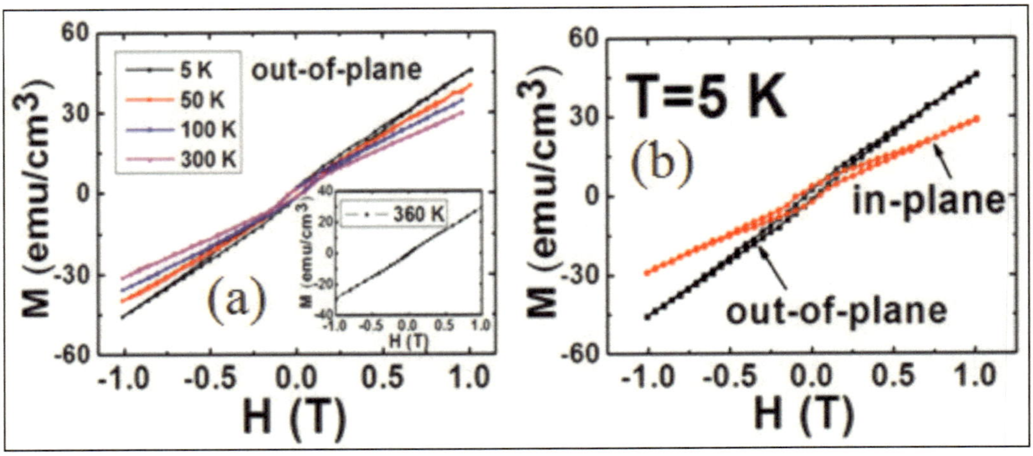

Fig. 9. Magnetic hysteresis loops of BFM film. (a) at different temperatures and (b) with magnetic field applied parallel and perpendicular to the sample plane

2.3.3 ZFC and FC measurements

Figure 10 shows temperature dependence of out-of-plane magnetization measured under zero-field-cooling (ZFC) and field-cooling (FC) conditions and in different magnetic fields. Similar to $BiFeO_3$ (with a cusp at around 50 K) a spin-glass-like behavior below 100 K was observed with the cusp at about 25 K. As shown in Figure 10 (a), the irreversibility below 100 K between FC curve and ZFC curve is clear with applied field of 500 Oe and 1000 Oe, but it was suppressed in higher field above 5 kOe and shift to much lower temperature, which is a typical behavior of spin glass ordering. Above the temperature of 100 K for spin-glass-like behavior appearing, another magnetic transition at about 360 K was observed in Figure 10 (b). Hysteresis behavior disappears above this temperature as shown in Figure 9 (a), which indicated an antiferromagnetic transition happens at this temperature.

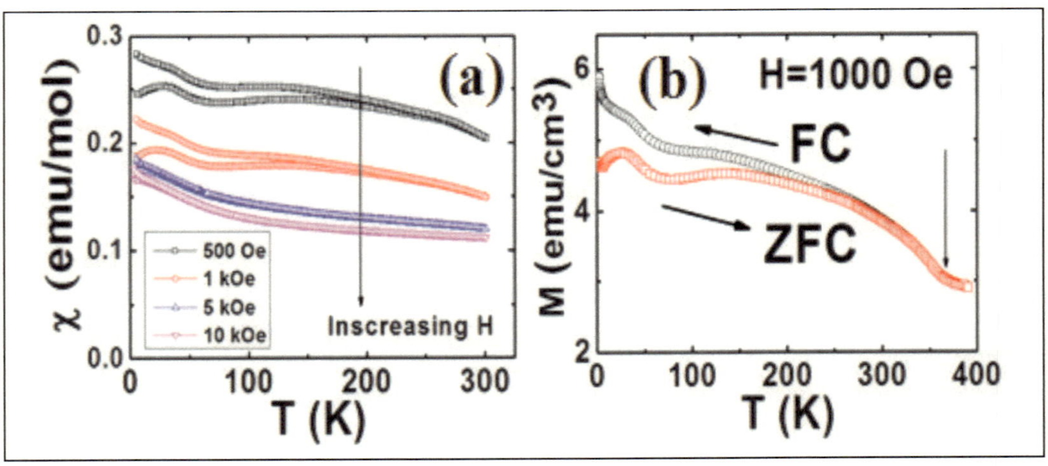

Fig. 10. ZFC and FC results of BFM film.

The film on STO was fabricated at higher temperature and higher oxygen pressure resulted in a good crystalline quality, less oxygen vacancies and no traceable impurity. BFM film on

(100) STO made under these improved fabrication conditions will display enhanced magnetic properties. The magnetizations of BFM film at 1 T are estimated from *M-H* loops as 0.30 μ_B, 0.26 μ_B, 0.23 μ_B, 0.21 μ_B and 0.19 μ_B per B site ion at temperatures of 5 K, 50 K, 100 K, 300 K and 360 K, respectively. These values are smaller than 0.5 μ_B for antiferromagnetic ordering of Fe^{3+} and Mn^{3+}, which is probably due to the local inhomogeneities in the film and some antisite disorders in the B-site. Actually the magnetic moment should be much larger than 0.5 μ_B at per B site if Mn and Fe are homogenously distributed, because both Fe^{3+}-O^{2-}-Mn^{2+} and Fe^{3+}-O^{2-}-Mn^{3+} in 180-degree bonds will produce orthoferrite, i.e. canted antiferromagnet and result in a larger moment based on the Goodenough-Kanamori rules. Therefore, in our films the arrangements of Mn^{3+}-O-Mn^{3+} and Fe^{3+}-O-Fe^{3+} with both resulting strong antiferromagnetism will have significant contribution to the observed magnetic properties. Due to Mn^{3+} ($3d^4$) is a Jahn-Teller ion, a strong Jahn-Teller effect will cause significant structure distortion in BFM film and produce the anisotropy effects. An external stress originating from BFM/STO lattice mismatches can greatly enhance the resulting cooperative strain and enhance the magnetic anisotropy. However, the multiple valence states of Mn ions in the film, the Mn^{2+} and Mn^{4+} can decrease the lattice distortion caused by Mn^{3+} and result in better lattice matching between film/substrate and decrease the anisotropic property. All of the curves shown here are corrected from the diamagnetic background of the STO substrate. The *M-H* and *M-T* data of the (100) STO substrate were obtained using the same system.

3. Conclusion

The piezoelectric/ferroelectric and magnetic properties of BFM series materials, which include BFM film and ceramic, LBFM film and ceramic, Nd: $BiFeO_3$/ BFM film and Nd: $BiFeO_3$/$YMnO_3$ film, were studied in detail. In this chapter, we mainly focus on the BFM film. It was proved that stabilization can be facilitated by a partial replacement of Bi^{3+} cations by La cations. The film and ceramic showed different properties and after La doping, both ferroelectric and magnetic properties were improved.

The piezoelectric/ferroelectric properties of BFM series materials have been studied using different methods, including *P-E* loop measurement, positive-up-negative-down (PUND) test and piezoresponse force microscopy (PFM). PFM was used to investigate the domain configurations and local piezoresponse hysteresis loops for BFM and LBFM films. The PFM images confirmed that the domain could be poled and switched in both films. The clearer domain boundary in the LBFM film indicated better crystallization and ferroelectric properties compared to the BFM film. The local butterfly-type piezoresponse hysteresis loops were obtained. All the observations suggest that BFM and LBFM films are room temperature ferroelectric materials. Improved ferroelectric properties are expected in the BFM system through the adjustment of doping ions and fabrication conditions to obtain promising multiferroic candidates.

The magnetic hysteresis loops and temperature dependent magnetization were also studied. BFM film with good crystalline quality and with enhanced magnetic properties was obtained on (100) $SrTiO_3$ substrate through the optimization of the fabrication conditions. Similar to $BiFeO_3$, the spin-glass-like behavior is observed below 100 K with the cusp at 25 K. The ZFC and FC curves measured from 2 K to 400 K show a kink at around 360 K and hysteresis disappears at 360 K, revealing a antiferromagnetic transition at this temperature. The observed anisotropy effects were caused by Jahn-Teller ions of Mn^{3+}. Mn tends to form

multiple valence states as in the film it is possibly because the Mn^{2+} and Mn^{4+} cations decrease the Jahn-Teller effect caused by Mn^{3+}.
Several questions in weak ferroelectric materials still remained to be anwsered. We wish to share these questions and have more discussion based on the as-designed materials for further development of such ferroelectrics.

4. Acknowledgment

The authors gratefully acknowledge Dr. Shigeki Nimori, Dr. Hideaki Kitazawa, Dr. Minora Osada, Dr. Baowen Li of NIMS, Prof. Huarong Zeng of Shanghai Institute of Ceramics for the valuable discussions and Dr. Hideo Iwai of NIMS for the XPS measurement. This work was supported in part by grants from JSPS and ARC under the Japan-Australia Research Cooperative Program, and Grant-in-Aid for JSPS Fellows (21-09608).

5. References

Lines, M.E. & A.M. Glass, (1977), *Principles and applications of ferroelectrics and related materials,* Oxford University press, ISBN 0198512864

Eerenstein,W.; Mathur, N.D. & Scott, J.F. (2006). Multiferroic and magnetoelectric materials. *Nature,*Vol. 442, No. 17, (August 2006), pp. 759-765, ISSN 0028-0836

Dass, R. I.; Yan, J. Q. & Goodenough, J. B. (2003). Oxygen stoichiometry, ferromagnetism and transport properties of $La_{2-x}NiMnO_{6+\delta}$, *Phys. Rev. B* Vol. 68, pp. 064415-064427, ISSN 1098-0121

Dass, R.I. & Goodenough, J.B. (2003). Multiple magnetic phases of $La_2CoMnO_{6-\delta}$ ($0<\sim\delta<\sim0.05$), *Phys. Rev. B* Vol.67, pp. 014401.1-014401.9, ISSN 1098-0121

Blasse, G. (1965). Ferromagnetic interactions in non-metallic perovskites, *J. Phys. Chem. Solids,* Vol. 26, No. 12, pp. 1969-1971, ISSN 0022-3697

Azuma, M.; Takata, K.; Saito, T.; Ishiwata, S.; Shimakawa, Y. & Takano, M. (2005). A Designed New Ferromagnetic Ferroelectric Bi_2NiMnO_6, *J. Am. Chem. Soc.* Vol. 1, pp. 8889-8892, ISSN 0002-7863

Rogado, N.S.; Li, J.; Sleight, A.W. & Subramanian, M.A. (2005). Magnetocapacitance and Magnetoresistance Near Room Temperature in a Ferromagnetic Semiconductor: La_2NiMnO_6, *Adv. Mater.* Vol. 17, pp. 2225-2227, ISSN 1501-4095

Wang, J.; Neaton, J. B.; Zheng, H.; Nagarajan, V.; Ogale, S. B.; Liu, B.; Viehland, D.; Vaithyanathan, V.; Schlom, D. G.; Wuttig, M. & Ramesh, R. (2003). Epitaxial BiFeO3 Multiferroic Thin Film Heterostructures, *Science* Vol. 299, (March 2003), pp. 1719-1722, ISSN 0036-8075

Atou, T.; Chiba, H.; Ohoyama, K.; Yamaguchi, Y. & Syono, Y. (1999). Structure Determination of Ferromagnetic Perovskite $BiMnO_3$, *J. Solid State Chem.* Vol. 145, No. 2, pp. 639-642, ISSN 0022-4596

Kimura, T.; Kawamoto, S.; Yamada, I.; Azuma, M.; Takano, M. & Tokura, Y. (2003). Magnetocapacitance effect in multiferroic $BiMnO_3$, *Phys. Rev. B* Vol. 67 (R), (May 2003), pp. 180401-180404, ISSN 1098-0121

Fujino, S.; Murakami, M.; Lim, S. H.; Salamanca-Riba, L.G.; Wuttig, M. & Takeuchi, I. (2007). Multiphase growth in Bi-Mn-O thin films, *J. Appl. Phys.* Vol. 101, No. 1, pp. 013903, ISSN 0021-8979

De, K.; Ray, R.; Panda, R.N.; Giri. S.; Nakamura, H. & Kohara, T. (2005). The effect of Fe substitution on magnetic and transport properties of $LaMnO_3$, *J. Magn. Magn. Mater.* Vol. 288, pp. 339-346, ISSN 0304-8853

De, K.; Thakur, M.; Manna, A. & Giri, S. (2006). Unusual glassy states in $LaMn_{0.5}Fe_{0.5}O_3$: Evidence of two distinct dynamical freezing processes, *J. Appl. Phys.* Vol. 99, No. 1, pp. 013908, ISSN 0021-8979

Tong, W.; Zhang, B.; Tan, S. & Zhang, Y. (2004). Probability of double exchange between Mn and Fe in $LaMn_{1-x}Fe_xO_3$, *Phys. Rev. B* Vol. 70, (July 2004), pp. 014422, ISSN 1098-0121

Gajek, M.; Bibes, M.; Fusil, S.; Bouzehouane, K.; Fontcuberta, J.; Barthelemy, A. & Fert, A. (2007). Tunnel junctions with multiferroic barriers, *Nat. Mater.* Vol. 6, pp. 296-302, ISSN 1476-1122

Gajek, M.; Bibes, M.; Wyczisk, F.; Varela, M.; Fontcuberta, J. & Barthelemy, A. (2007). Growth and magnetic properties of multiferroic $La_xBi_{1-x}MnO_3$ thin films, *Phys. Rev. B* Vol. 75, (May 2007), pp. 174417, ISSN 1098-0121

Langenberg, E.; Varela, M.; Garcia-Cuenca, M.V.; Ferrater, C.; Polo, M.C.; Fina, I.; Fabrega, L.; Sanchez, F. & Fontcuberta, J. (2009). Epitaxial thin films of $(Bi_{0.9}La_{0.1})_2NiMnO_6$ obtained by pulsed laser deposition, *J. Mag. Mag. Mat.* Vol. 321, No. 11, pp. 1748-1753, ISSN 0304-8853

Marin, L.W.; Crane, S.P.; Chu, Y-H.; Holcomb, M.B.; Gajek, M.; Huijben, M.; Yang, C-H.; Balke, N. & Ramesh, R. (2008). Multiferroics and magnetoelectrics: thin films and nanostructures, *J. Phys.: Condens. Matter* Vol. 20, No. 43, pp. 434220, ISSN 1742-6588

Yakel, H.L.; Koehler, W.C.; Bertaut, E.F.& Forrat, E.F. (1963). On the crystal structure of the manganese(III) trioxides of the heavy lanthanides and yttrium, *Acta Crystallogr.* Vol. 16 pp. 957-962, ISSN 0365-110X

Dho, J.; Leung, C.W.; MacManus-Driscoll, J.L. & Blamire, M.G. (2004). Epitaxial and oriented $YMnO_3$ film growth by pulsed laser deposition, *J. Crystal Growth* Vol. 267, No. 3-4, pp. 548-553, ISSN 0022-0248

Ruette, B.; Zvyagin, S.; Pyatakov, A.P.; Bush, A.; Li, J.F.; Belotelov, V.I.; Zvezdin, A.K. & Viehland, D. (2004). Magnetic-field-induced phase transition in $BiFeO_3$ Cycloidal to homogeneous spin observed by high-field electron spin resonance: order, *Phys. Rev. B* Vol. 69, (February 2004), pp. 064114, ISSN 1098-0121

Jang, H.W.; Ortiz, D.; Baek, S.; Foliman, C.M.; Das, R.R.; Shafer, P.; Chen, Y.B.; Nelson, C.T.; Pan, X.Q.; Ramesh, R. & Eom, C. (2009). Domain Engineering for Enhanced Ferroelectric Properties of Epitaxial (001) BiFeO Thin Films, *Adv. Mater.* Vol. 21, No. 7, pp. 817-823, ISSN 1501-4095

Wu, J.G.; Kang, G.Q.; Liu H. J. &Wang, J. (2009). Ferromagnetic, ferroelectric, and fatigue behavior of (111)-oriented $BiFeO_3/(Bi_{1/2}Na_{1/2})TiO_3$ lead-free bilayered thin films, *Appl. Phys. Lett.* Vol. 94, No. 17, pp. 172906, ISSN 0003-6951

Scott, J.F. (2008). Ferroelectrics go bananas, *J. Phys: Condens. Matter,* Vol. 20, No. 2, pp. 021001, ISSN 1742-6588

Scott, J. F. (2000). *Ferroelectric memories,* Springer, ISBN 978-3-540-66387-4, Berlin

Balke, N.; Bdikin, I.; Kalinin, S.V. & Kholkin, A.L. (2009). Electromechanical Imaging and Spectroscopy of Ferroelectric and Piezoelectric Materials: State of the Art and Prospects for the Future, *J. Am. Ceram. Soc.,* Vol. 92, No. 8, pp. 1629-1647, ISSN 1551-2916

Kalinin, S.V. & Bonnell, D.A. (2001). Local potential and polarization screening on ferroelectric surfaces, *Phys. Rev. B* Vol. 63, (March 2001), pp. 125411, ISSN 1098-0121

Kalinin, S.V.; Rodriguez, B.J.; Borisevich, A.Y.; Baddorf, A.P.; Balke, N.; Chang, H.J.; Chen, L.Q.; Choudhury, S.; Jesse, S.; Maksymovych, P.;. Nikiforov, M.P & Pennycook, S.J.

(2010). Defect-Mediated Polarization Switching in Ferroelectrics and Related Materials: From Mesoscopic Mechanisms to Atomistic Control *Adv. Mater.* Vol. 22, No. 3, pp. 314-322, ISSN 1501-4095

Shafer, P.; Zavaliche, F.; Chu, Y.H.; Yang, P.L.; Cruz, M.P. & Ramesh, R. (2007). Planar electrode piezoelectric force microscopy to study electric polarization switching in $BiFeO_3$, *Appl. Phys. Lett.* Vol. 90, pp. 202909, ISSN 0003-6951

Catalan, G.; Bea, H.; Fusil, S.; Bibes, M.; Paruch, P.; Barthélémy, A. & Scott, J.F. (2008). Fractal Dimension and Size Scaling of Domains in Thin Films of Multiferroic $BiFeO_3$, *Phy. Rev. Lett.* Vol. 100, No. 2, pp. 027602, ISSN 0031-9007

Keeney, L.; Zhang, P.F.; Groth, C.; Pemble, M.E. & Whatmore, R.W. (2010). Piezoresponse force microscopy investigations of Aurivillius phase thin films, *J. Appl. Phys.* Vol. 108, pp. 042004, ISSN 0021-8979

Cheng, Z.X.; Wang, X.L.; Dou, S. X.; Kimura, H. & Ozawa, K. (2008). Improved ferroelectric properties in multiferroic $BiFeO_3$ thin films through La and Nb codoping, *Phys. Rev. B* Vol. 77, pp. 092101, ISSN 1098-0121

Cheng, Z.X.; Li, A.H.; Wang, X.L.; Dou, S.X.; Ozawa, K.; Kimura, H.; Zhang, S.J. & Shrout, T.R. (2008). Structure, ferroelectric properties, and magnetic properties of the La-doped bismuth ferrite, *J. Appl. Phys.* Vol. 103, pp. 07E507, ISSN 0021-8979

Zhao, H.Y.; Kimura, H.; Cheng, Z.X.; Wang, X.L. &Nishida, T. (2009). Room temperature multiferroic properties of Nd:$BiFeO_3$/Bi_2FeMnO_6 bilayered films, *Appl. Phys. Lett.* Vol. 95, No. 23, pp. 232904, ISSN 0003-6951

Zhao, H.Y.; Kimura, H.; Cheng, Z.X.; Wang, X.L.; Ozawa, K. & Nishida, T. (2010). Magnetic characterization of Bi_2FeMnO_6 film grown on (100) $SrTiO_3$ substrate *Phys. Status Solidi RRL* Vol. 4, No. 11, pp. 314, ISSN 1862-6270

Zhao, H.Y.; Kimura, H.; Cheng, Z.X.; Wang, X.L.; Ozawa, K. & Nishida, T. (2010). Magnetic properties of La doped Bi_2FeMnO_6 ceramic and film, *J. App. Phys.* Vol. 108, pp. 093903, ISSN 0021-8979

Du, Y.; Cheng, Z.X.; Dou, S.X.; Wang, X.L.; Zhao, H.Y. & Kimura, H. (2010). Magnetic properties of Bi_2FeMnO_6: A multiferroic material with double-perovskite structure, *Appl. Phys. Lett.* Vol. 97, pp. 122502, ISSN 0003-6951

Wandelt, C. (1982). Photoemission studies of adsorbed oxygen and oxide layers, *Surf. Sci. Pep.* Vol. 2, No. 1, pp. 1-121, ISSN 0127-5729

Beyreuther, E.; Grafstrom, S.; Thiele, L.M. & Dorr, K. (2006). XPS investigation of Mn valence in lanthanum manganite thin films under variation of oxygen content, *Phys. Rev. B* Vol. 73, No. 15, pp. 155425, ISSN 1098-0121

Wang L. & Gao, J. (2009). Electronic structures and Hall effect in low-doped $La_{0.9}Hf_{0.1}MnO_3$ epitaxial films, *J. Appl. Phys.* Vol. 105, pp. 07E514, ISSN 0021-8979

Dionne, G.F. (1979). Origin of the magnetostriction effects from Mn_{3+}, Co_{2+}, and Fe_{2+} ions in ferrimagnetic spinels and garnets, *J. Appl. Phys.* Vol. 50, pp. 4263, ISSN 0021-8979

Dionne, G.F. (2007). Evidence of magnetoelastic spin ordering in dilute magnetic oxides, *J. Appl. Phys.* Vol. 101, pp. 09C509, ISSN 0021-8979

Dionne G.F. & Kim, H-S. (2008). *J. Appl. Phys.* Vol. 103, pp. 07B333, ISSN 0021-8979

Bi, L.; Taussig, A.R.; Kim, H-S.; Wang, L.; Dionne, G.F.; Bono, D.; Persson, K.; Ceder, G.& Ross, C.A. (2008). Structural, magnetic, and optical properties of $BiFeO_3$ and Bi_2FeMnO_6 epitaxial thin films: An experimental and first-principles study, *Phys. Rev. B* Vol. 78, No. 10, pp. 104106, ISSN 1098-0121

Singh, M.K.; Prelier, W.; Singh, M.P.; Katiyar, R. S. &. Scott, J. F (2008). Spin-glass transition in single-crystal $BiFeO_3$,*Phys. Rev. B* Vol. 77, No. 14, pp. 144403, ISSN 1098-0121

13

Lead Titanate-Based Nanocomposite: Fabrication, Characterization and Application and Energy Conversion Evaluation

Walter Katsumi Sakamoto[1], Gilberto de Campos Fuzari Jr[1],
Maria Aparecida Zaghete[2] and Ricardo Luiz Barros de Freitas[3]
[1]Faculdade de Engenharia, UNESP - Univ. Estadual Paulista - Campus de Ilha Solteira,
Depto de Física e Química, Grupo de Polímeros,
[2]Instituto de Química, Universidade Estadual Paulista - UNESP,
[3]Faculdade de Engenharia, UNESP - Univ. Estadual Paulista - Campus de Ilha Solteira,
Departamento de Engenharia Elétrica
Brazil

1. Introduction

Within the past 5 decades the use of ferroelectric composite made of ferroelectric ceramic immersed in polymer matrix has expanded significantly. One of the goals to embedding ceramic grains within a polymer matrix to form a 0-3 composite film is to combine the better properties of each phase, such as high piezoelectric activity of the ceramic and the mechanical resistance, formability and flexibility of the polymer, also because the 0-3 composite is the easier and cheaper way to fabricate this alternative electro-active material.

Some desirable properties for composite materials are: high piezoelectric charge and voltage coefficients for passive piezoelectric devices; large piezoelectric charge coefficient and low relative dielectric constant for active devices. Furthermore, the poling process of the composite film should be effective, which impose the composite must be homogeneously fabricated, i.e., the composite film should have uniformly dispersed ceramic grain.

Piezoelectric devices have some specific advantages over electromagnetic such as suitability to be miniaturized; there is no need of magnetic shielding; it is more efficient at least in the lower power range. There is a very large range of applications of piezoelectric materials, either as sensing element or as actuators. One of the most recent interests on piezoelectric materials is energy harvesting, converting mechanical to electrical energy. The aim of this research area is to provide clean energy attending the needs of the world in the fight against pollution.

Conventional ferroelectric materials such as lead zirconate titanate (PZT) [Jaffe 1969, Ren 2003, Klee 2010], modified lead titanate [Wang 2000, Chu 2004, Pontes 2001] and ferroelectric polymers [Lovinger 1983, Kawai 1969, Bauer 2000] have been used in applications which use either piezo or pyroelectric properties. Concerned to the piezoelectric applications, they have been employed in a large range of transducers such as for hydrophone [Lau 2002, Boumchedda 2007], dynamic strain measurements [Soman 2011], medical ultrasound [Zhang 2006, Muralt 2004] and non-destructive evaluation of structures [Brown 1996, Ciang 2008, Edwards 2006].

On the pyroelectric applications, ferroelectric materials can be used as infrared detectors [Sosnin 2000, Huang 2002, Guggilla 2006] and X-ray intensity measurements [De Paula 2005, Pontes 2010, De Carvalho 1997]. Using the pyroelectric property of the sensing element, KUBE Electronics AG (Switzerland) has developed a flame detection and gas analysis device [www.kube.ch].

Ferroelectric ceramics have high piezo and pyroelectric properties but, for some applications, their poor mechanical properties and the mismatch of the acoustic impedance with water and human tissue restrict their usage. On the other hand, ferroelectric polymers have mechanical flexibility and formability but their piezo and pyroelectric activity are low. To overcome these problems composite materials made of ferroelectric ceramic and polymer have been investigated as an alternative material which combine the better properties of ceramic and polymer [Furukawa 1976, Dias 1996, Sakamoto 2006, Wong 2006, Kumar 2005, Estevam 2011, Feng 2010].

According to Newnham and co-workers [Newnham 1978] there are ten connectivity patterns in which a two phase composite system can be fabricated, ranging from unconnected 0-0 pattern to a 3-3 pattern in which both phases are three dimensionally self-connected. The easier and cheaper composite to obtain is the 0-3 pattern, that means the ceramic grains are dispersed (unconnected) into a polymer matrix (self-connected three dimensionally). The main goal of embedding ferroelectric ceramic grains within a polymer matrix is to obtain a material which displays the combined better properties of each single phase. However it is very difficult to obtain a 0-3 composite with high ceramic content. There are basically two problems: high ceramic content will provide a mixed connectivity due to the percolation of the grains; high ceramic concentration means low flexibility of the composite material.

But these problems have not drawn the interest of researchers in using the 0-3 composite, on the contrary, it intensified the search for optimum results and many studies on the polarization of the composites have been conducted [Furukawa 1986, Lau 2007, Ploss 2001, Wong 2002]. Still seeking a more effective polarization of the composite material, studies with the inclusion of a semiconductor phase were carried out [Sa-Gong 1986, Sakamoto 2002, Renxin 2006, Ploss 2006, Chau 2007]. These efforts were not in vain and 0-3 composites are being used as sensors and transducers, and is now a well-established alternative to conventional ferroelectric materials for many applications. New methods of preparing ferroelectric ceramics have also been studied and the latest is the hydrothermal method for obtaining ceramic powder [Shimomura 1991, Morita 2010]. The grain size and structure are also objects of study.

This work presents the preparation and characterization of PZT ceramic obtained by different methods. The influence of the synthesis method on the grain size and the morphology are also object of study. The fabrication and characterization of composite films with 0-3 connectivity, immersing nanoparticles of PZT into the non-polar poly(vinylidene fluoride) – PVDF as the polymer matrix were presented. For comparison there are some results obtained with composite samples made of ceramic particles recovered with a conducting polymer and also using the conducting polymer as a third phase. Moreover it presents the results obtained with the new material that includes a semiconductor phase, polyaniline - PAni as a sensing material and as a piezoelectric material for energy harvesting. In this sample the PZT grain was partially covered by PAni, which allowed better distribution of grains in the polymer matrix in comparison with the inclusion of the 3rd phase separately, avoiding a continuous electrical flux path which does not allow the polarization process of the composite sample. The use of this composite as sensor and power converter is an indicative that it is a good alternative for technological applications.

2. Experimental

2.1 Ceramic

The control of some parameters is important to achieve the desired properties in lead zirconate titanate (PZT) materials. These parameters include the absence of intermediate crystalline phases, a defined and fixed stoichiometry, as well as a homogeneous distribution of lead in the material. Lead zirconate titanate (PZT) is a very interesting ceramic that has good piezoelectric properties used to making ultrasonic transducers, filters and pyroelectric detectors [Haertling 1999]. This material can be prepared using different ways but the most important is using low temperature to obtain the crystalline phase. This condition promotes the homogeneous lead distribution and consequently occurs the formation of pure phase PZT. The presence of secondary phases reduces the dielectric and piezoelectric constants [Zaghete 1999, Zaghete 1992]. To minimize theses problems some chemicals processes has been proposed as the optional procedure. Methods such as sol-gel [Ishikawa 1994], hydrothermal synthesis [Pan 2007, Abothu 1999] and Pechini's method [Zaghete 1992] can be used.

It is known that ceramic materials prepared from chemical solutions routes are transformed via a nucleation and growth process, often requiring high temperatures to surmount the large energy barriers of the nucleation and growth of the stable phase. Consequently, these energy barriers frequently determine the calcinations conditions and therefore the characteristics such as, particle size, morphology and degree of aggregation of the precursor powder. The most significant advance in this field, however, consists of the ability to control phase development at low calcinations temperatures to avoid lead evaporation.

The expected phase equilibrium (perovskite) may grow starting from a gel matrix under a nucleus along a certain crystallographic orientation. This type of heterogeneous nucleation eliminates the need for the system to exceed the activation energy required to form the nucleus, as in the case of systems with homogeneous nucleation. As a result, the perovskite phase may crystallize at lower temperatures.

The present study show the influence of synthesis method on size and morphologic distribution of particle and the amount of perovskits phase synthesized at different temperatures. The procedure of PZT synthesis, based on Pechini's method [Zaghete 1992] makes use of the capability that certain α-hydroxycarboxylic organic acids possess of forming polybasic acid chelates with several cations. When mixing with a polyhydroxylic alcohol and heating, the chelate transforms into a polymer, maintaining the cations homogeneously distributed.

The organic part is eliminated at low temperatures forming reactive oxides with well-controlled stoichiometry. Pure PZT with composition $Pb(Zr_{0.48}Ti_{0.52})O_3$ can be prepared from the metal-citrate complex polyesterified in ethylene glycol. Appropriate quantities of Zr, Ti and Pb solutions were mixed and homogenized by stirring at 90°C for 3 h. Next, the temperature was increased to 130-140°C, yielding a high viscous polyester resin. The powder was calcined at 600, 700 and 800°C for 2h and ball milled for 2h in isopropylic alcohol medium.

Recently, the hydrothermal synthesis has been widely used in the study of these materials for the production of particles with nanometric sizes, high purity and crystallinity, good stoichiometric control and good yield. There are few reports relating to the study on the PZT synthesized by the hydrothermal method in literature. In one of these works, PZT powder was obtained with cubic morphology and crystalline phase by hydrothermal synthesis [PAN 2007]. By controlling the variables process of the synthesis, it becomes possible to change the morphology, the particle size, as well as the hydrothermal synthesis assisted by microwave method that has the advantage of producing rapid heating, thereby promoting homogeneous nucleation of particles [Moreira 2009, Rao 1999].

Hydrothermal media provide an effective reaction environment for the synthesis of numerous ceramic materials because of the combined effects of solvent, temperature, and pressure on ionic reaction equilibrium. The conventional hydrothermal method has become an effective synthetic route in Materials science by dramatically increasing the control of the micro/nanometric morphology and orientation [LUO 2008]. In addition, this method is environmentally friendly and depends on the solubility of the chemical salts in water under temperature and pressure conditions.

The key factors in this method are the vapor pressure and solubility of the chemical salts in water [Lencka 1995]. In contrast to the conventional hydrothermal method which requires a long time typically several days and high electric power (over thousand watts) [Dutta 1994], microwave-assisted heating is a greener approach to synthesize materials within a shorter time typically several minutes to some few hours less than the duration of the conventional method and with lower energy consumption (hundreds of watts).

The desired PZT product can be synthesized using $Pb(NO_3)_2$, $ZrOCl_2.8H_2O$, TiO_2 , KOH . At first, a suspension containing $ZrOC_2.8H_2O$, $Pb(NO_3)_2$ and TiO_2 was prepared in aqueous medium. After that, KOH aqueous solution containing 3.31g of KOH (1.84 $mol.L^{-1}$, pH=14). was add to the precursor suspension and then kept at room temperature under stirring for approximately 20 minutes. It was further placed containing all the reagents in Teflon jars, sealed and taken to the microwave for the synthesis of PZT powder.

The synthesis temperature was 180°C and the lower time used to obtain PZT was 0.5 hour, under constant pressure of approximately 10 Bar. The PZT powders were synthesized using microwave-assisted hydrothermal digester (MARS CEM, USA). The precursor was further again loaded into a 90 mL Tefflon autoclave reaching 30% of its volume. The autoclave was sealed and placed into a microwave-assisted hydrothermal system using 2.45 GHz microwave radiation with a maximum output power of 800 W. Then the solid product was washed with distilled water until a neutral pH was obtained and was further dried at room temperature.

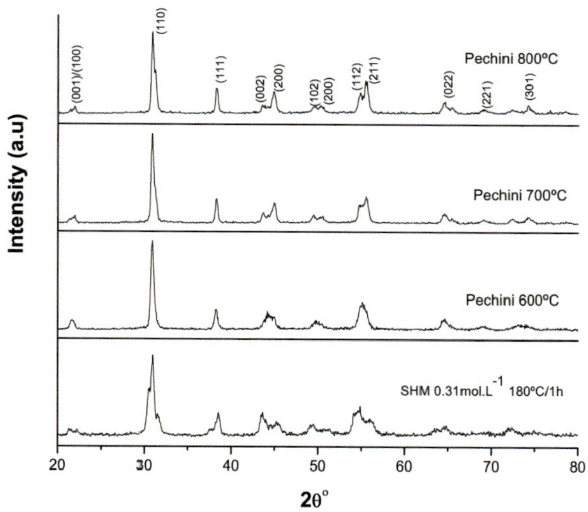

Fig. 1. X-ray diffraction patterns for PZT powders synthesized by Pechini method at different temperatures of calcination and hydrothermal synthesis assisted by microwaves.

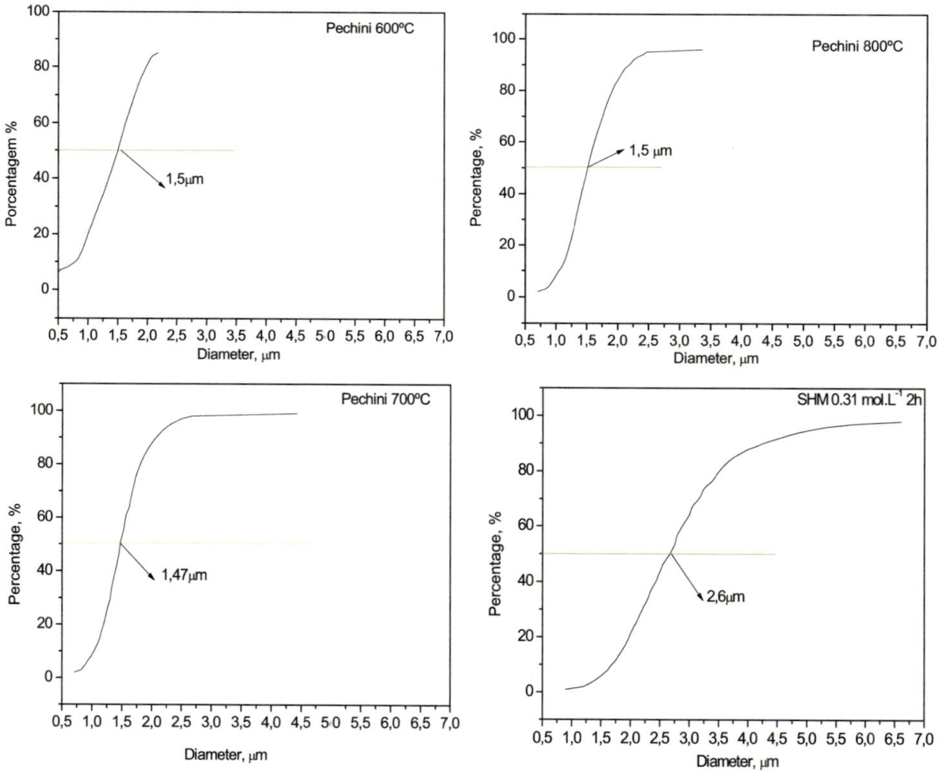

Fig. 2. Diagrams of particles size distribution of PZT synthesized by the Pechini's method calcimine at (a) 600°C; (b) 700°C ; (c) 800°C and (d) synthesized by hydrothermal microwave method at 180°C/1h.

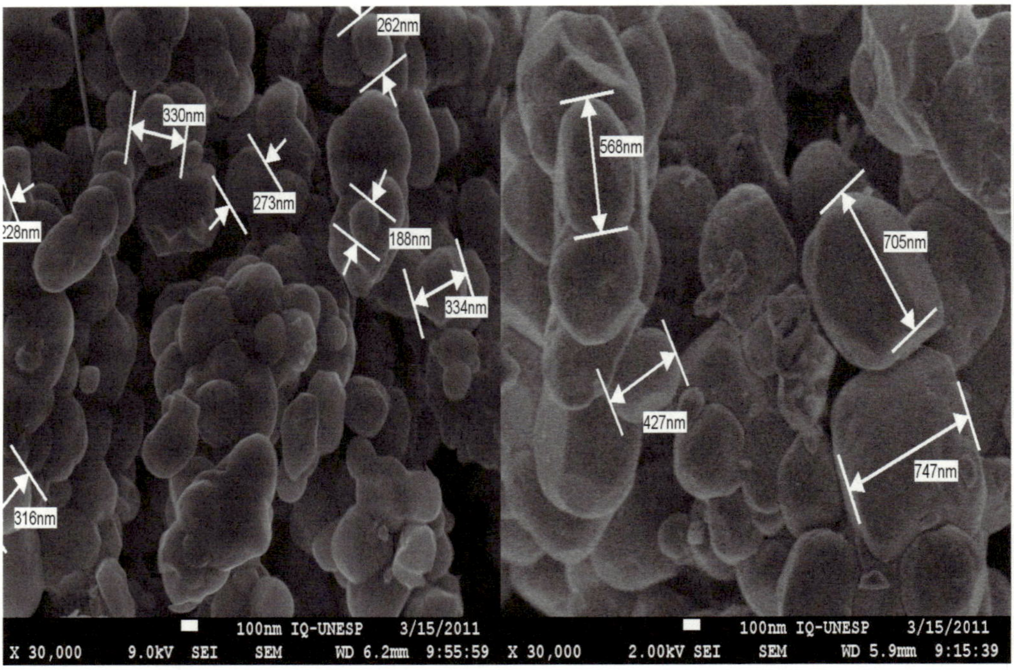

(A) (B)

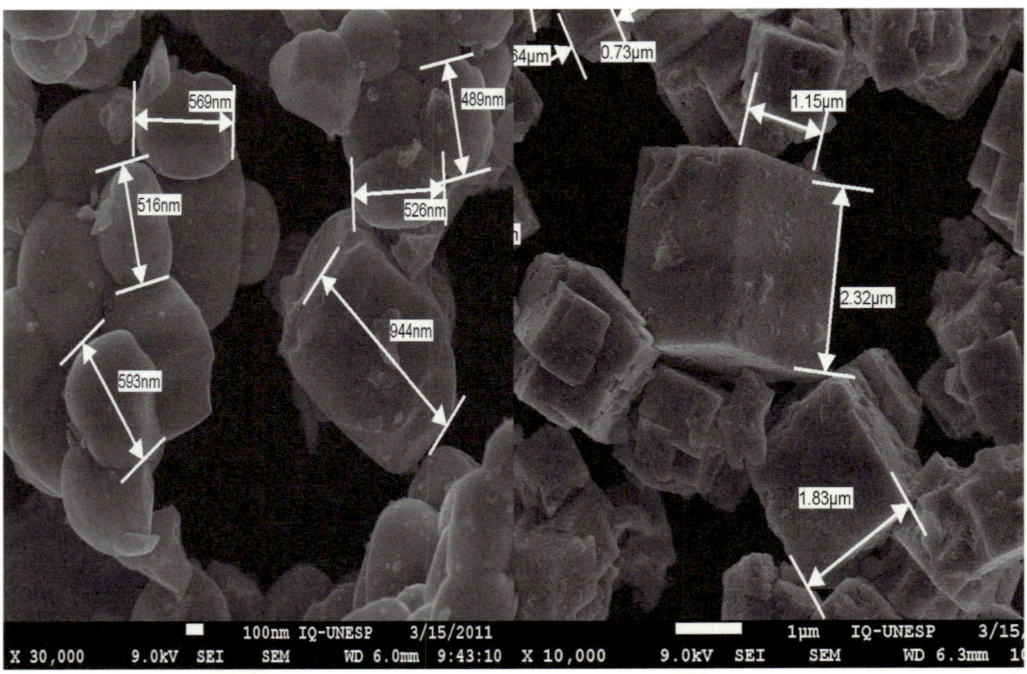

(C) (D)

Fig. 3. FEG-SEM images of PZT nanostructures synthesized by Pechini's method: (a) 600°C/3h; (b) 700°C/3h, (c) 800°C/3h and (d) synthesized by hydrothermal microwave method at 180°C/1h.

The obtained powders were characterized by X-ray powder diffraction using a Rigaku, DMax 2500PC with rotator anode at 50 kV and 150 mA, Cu Kα radiation in the 2θ range from 20° to 80° with 0.02°min-1. A field emission gun scanning electron microscope (SEM-FEG)-ZEISS SUPRA 35 microscope was used to analyze the shape and size of particles; energy dispersive X-ray microanalysis spectroscopy (EDS) was used for compositional determination. All measurements were taken at room temperature.
Today it is well known the effectiveness of the synthesis rote on the perovskits phase formation but it isn't well understood the influence of the particle size on the composite properties. Some of results obtained for PZT prepared by pechini´s method and synthesis hydrothermal assisted by microwave were presented to show the different characteristic as function of the synthesis way.
The analysis of the crystal structure of the material indicated mixture of the tetragonal and rhombohedral phase that is characteristics of the morphotropic phase transition region (YU 2007) Figure 1. When prepared by the Pechini's method has been the formation of pure crystalline phase from 600°C, the same result can be observed with hydrothermal treatment at 180°C for 1 hour. Also is possible to observe that the distribution of particle size and average particle size are directly affected by the temperature of thermal treatment as well as by the synthesis process, Figures 2 and 3. The purity of the composition was analyzed by EDS and found a homogeneous distribution of Pb, Ti, Zr on the surface of the entire sample as showed in Figure 4.

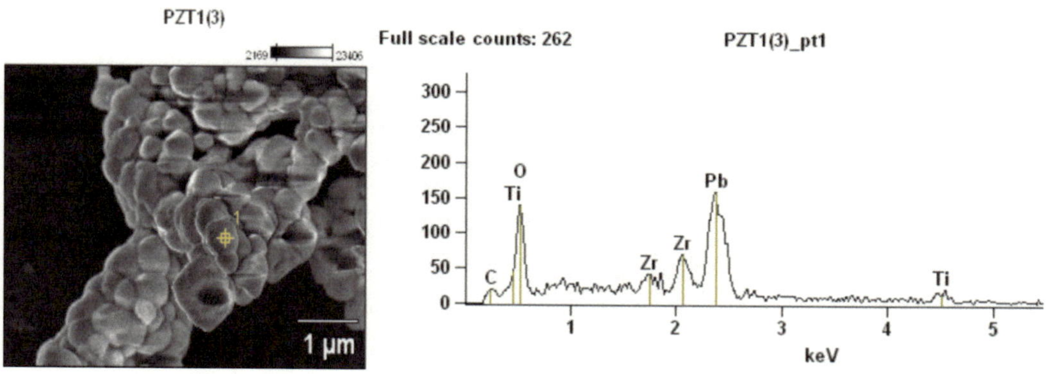

(a)

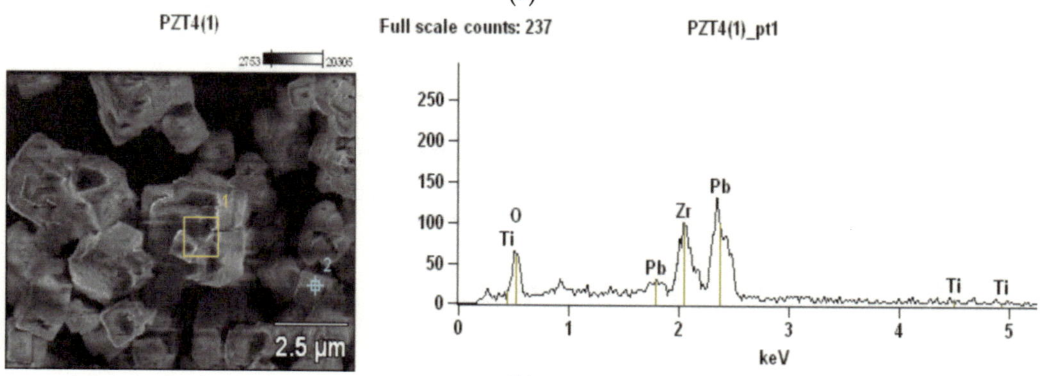

(b)

Fig. 4. Energy dispersive scanning, EDS, results of PZT prepared by Pechini's Method at (a) 700°C/3h and (b) synthesized by the hydrothermal microwave method. At 180°C/1h.

2.2 Polymer matrix

Poly(vinylidene fluoride) (PVDF) is a thermoplastic with excellent mechanical, optical and thermal properties, and showing resistance to attack of various chemicals [Lovinger 1982]. Formed by repeated units of $(-H_2C-CF_2-)_n$, has a molecular weight around 105 g / mol. Depending on the means of acquiring or thermal history, PVDF can possess the degree of crystallinity from 45 to 60%, melting temperature (Tm) in the range from 165 to 179°C and glass transition temperature (Tg) of about -34°C. Its crystal structure is spherulitic (composed of lamellar crystalline radial). The range of relatively low melting temperature and some properties of the polymer described above ensure easy processing by melting and blending, which means great advantage in large scale production. The PVDF can also be processed by casting that may result in thin films.

Relative to the molecular structure, PVDF is a linear polymer that has permanent electric dipoles approximately perpendicular to the direction of their chains. These dipoles are formed by the electronegativity difference between atoms of hydrogen and fluorine. PVDF can be found in four distinct structural phases α, β, γ and δ. α phase is the most common, this being non-polar usually obtained by cooling molten. The β phase (polar) is very attractive technical-scientific because of its piezoelectric and pyroelectric activity.

2.3 Getting PZT grains coated with PAni

The monomer aniline ($C_6H_5NH_2$) was purchased from Sigma-Aldrich and used in the synthesis after vacuum distillation. For the polymerization of aniline was employed oxidant ammonium persulfate from MERCK. To obtain the PZT particles partially coated with polyaniline, the PZT powder was incorporated into the solution of aniline and 1M cloridric acid under stirring at a temperature around 2°C for approximately 2 h. The solution was filtered and washed with 0.1M hydrochloric acid and the product was dried in an oven at 50°C for 3h.

Figure 5 shows the FEG-SEM micrograph of (a) the PZT and (b) PAni-coated PZT particles. It can be seen the lack of smooth of the coated-particle surface.

2.4 Composite

The PVDF in the form of powder was mixed with pure PZT, the PZT particles coated with PAni and PAni and PZT placed separately. The mixtures were then placed between sheets of Kapton and pressed close to the melting temperature of PVDF. To find the optimum condition for preparation of the composite film the effect of pressing temperature, time and pressure to be taken by the mixtures were studied.The optimum conditions were found to be: temperature of 185°C for about 1 minute at a pressure of about 7.6 MPa. The thickness of the films was in the range from 100 to 420 μm depending on the ceramic content. The composite films were obtained with different volume fractions of ceramic, which was calculated using the equation below [Marin-Franch 2002]:

$$m_c = \frac{m_p \rho_c}{\rho_p} \frac{\Phi_c}{1-\Phi_c} \tag{1}$$

where m is the mass and ρ is the density. The subscript c and p are related to ceramic and polymer, respectively. Φc is the volume fraction of ceramic. Figure 6 shows FEG images of the composite sample. It can be seen the homogeneous distribution of the ceramic nanoparticles recovered with PAni.

(a)

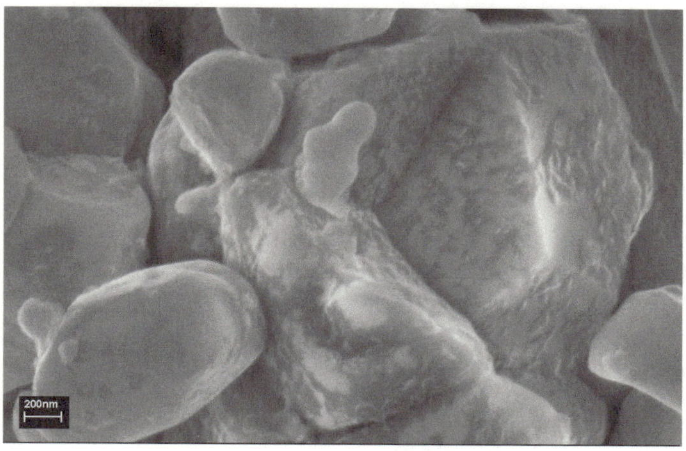

(b)

Fig. 5. FEG-SEM images: (a) PZT, (b) PZT recovered with PAni.

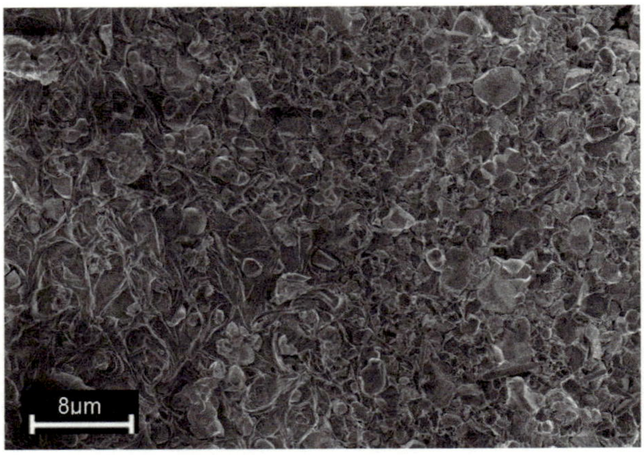

(a)

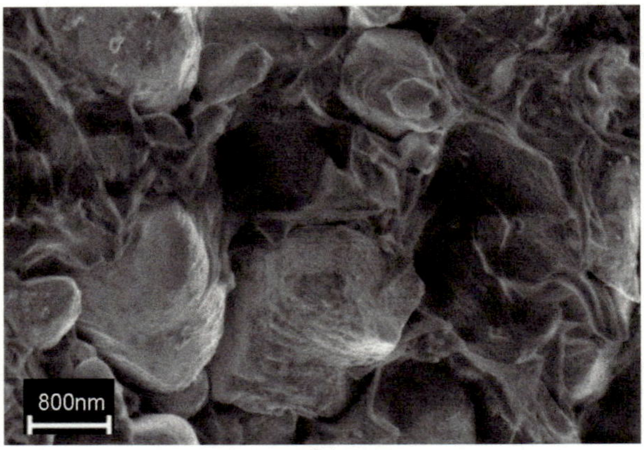

(b)

Fig. 6. FEG-SEM images (profile). (a) particle distribution into matrix e (b) particle recovered with PAni composite with 30 vol% of PZT-PAni.

The protonation degree of polyaniline can be controlled so that the conductivity of the composite, i. e, their permittivity can be changed and facilitate the polarization process [Wei 2007, Wong 2005, Or 2003, Zhou 2005]. Pani can have its conductivity controlled by the pH [Blinova 2008], which means that in more acidic - pH below 4.0 - it begins to undergo the process of doping and its conductivity gradually increased with decreasing pH.

3. Characterization of the composite

Aluminum electrodes with 1.0 cm of diameter were vacuum evaporated onto both sides of the sample for electrical contact. The composite films were poled with several electric field strengths and times in silicone oil bath with controlled temperature. A TREK high voltage power supply was used for the dc poling process. The dielectric data were taken using an impedance analyzer HP 4192 A in circular samples with 1.5 cm diameter.
An important aspect to note in materials is its electrical response when subjected to an alternating electric field. So it can be observed by measuring the impedance spectroscopy the behavior of dielectric constant versus frequency. Figure 7 shows the behavior of the real

relative permittivity ε' and loss factor ε″ which have their highest values with the preparation of PZT-PAni at lower pH values, and in ε" this increase is more significant. This is also an indication that the conductivity σ of the material increases, since it is directly related to the permittivity ε of the material by the relationship [Poon 2004]:

$$\sigma = \omega\varepsilon'' - i\omega(\varepsilon' - \varepsilon_0) \qquad (2)$$

where ω is the angular frequency of an external electric field applied and ε_0 is the electric permittivity in vacuum. In systems which contain grains covered with PAni doped partially or completely, ε' and especially ε" reach high values at lower frequencies, and for these cases the magnitude of ε" is bigger than ε', indicating high charge mobility, characteristics of the polymeric phase.

The piezoelectric activity of the composite films was studied by measuring the longitudinal piezoelectric coefficient d_{33}. The samples were poled in different poling conditions, changing parameters such as poling field, poling temperature and poling time. The best conditions of poling the sample were found as it can be seen in Figure 8 (a), (b) and (c).

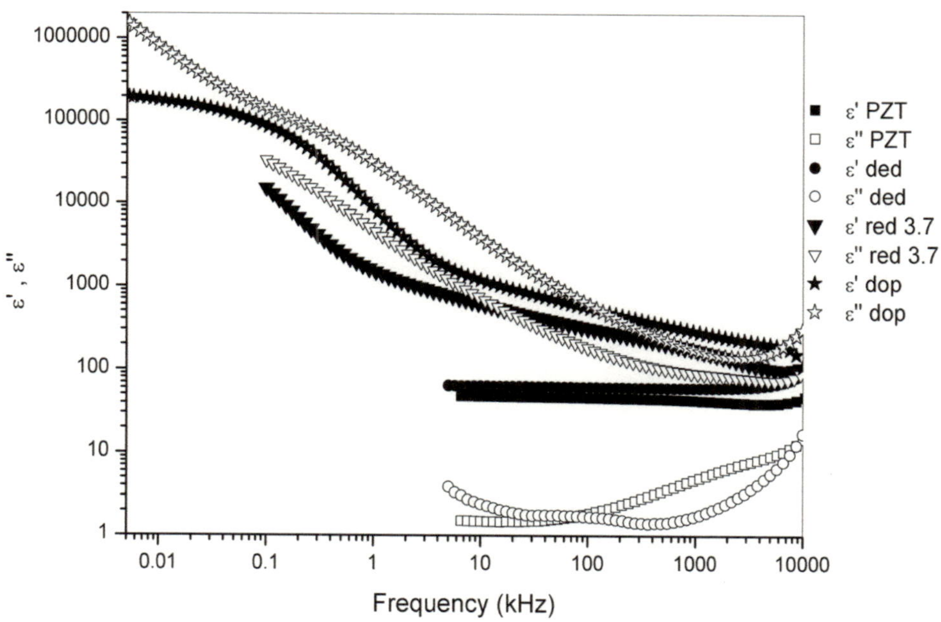

Fig. 7. Dielectric Constant as a function of frequency for PZT e PZT-PAni with different protonation degree.

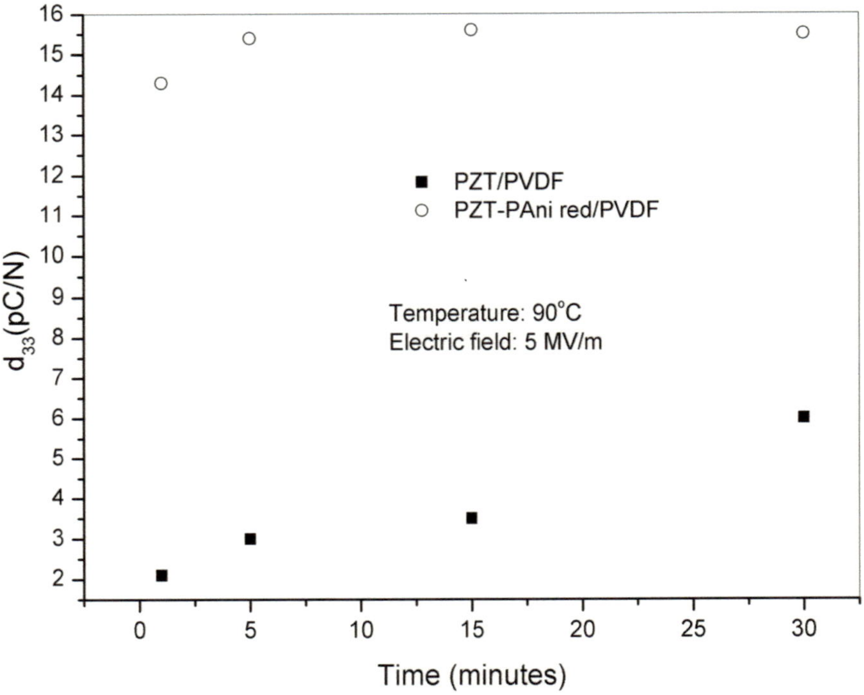

PZT/PVDF
PZT-PAni red/PVDF
Temperature: 90°C
Electric field: 5 MV/m
d33(pC/N)
Time (minutes)

(a)

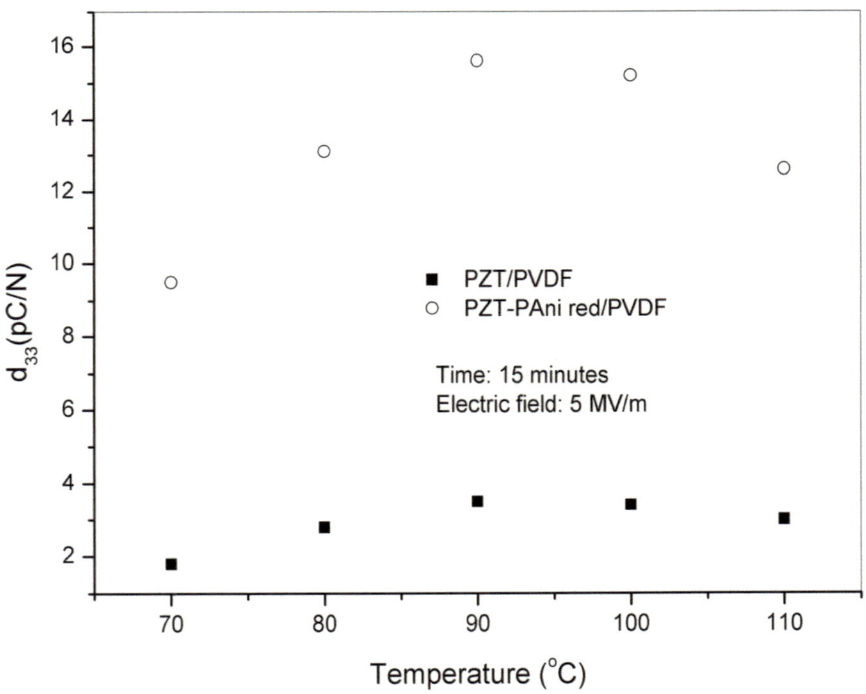

PZT/PVDF
PZT-PAni red/PVDF
Time: 15 minutes
Electric field: 5 MV/m
d33(pC/N)
Temperature (°C)

(b)

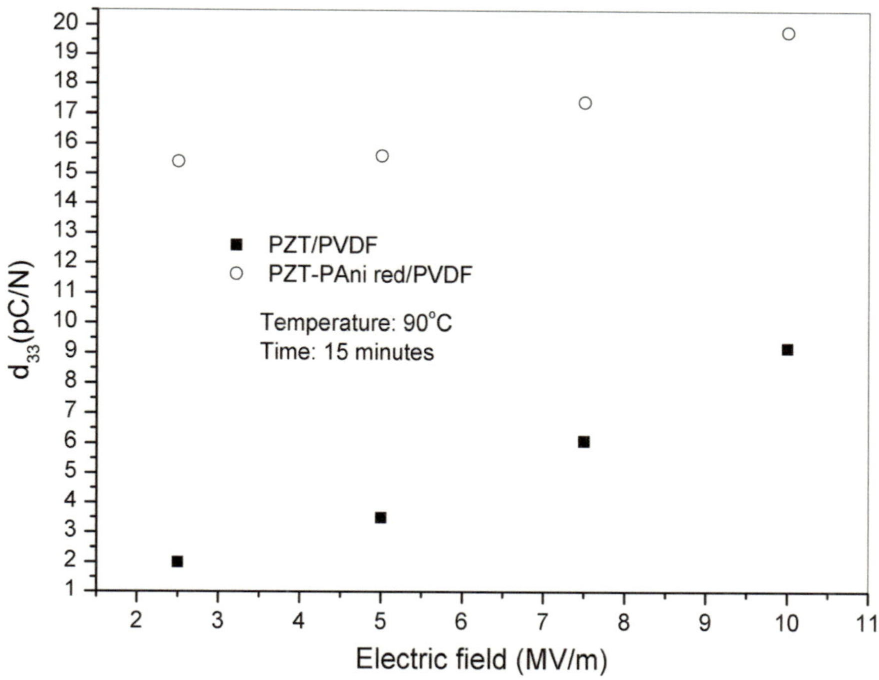

(c)

Fig. 8. Longitudinal piezoelectric coefficient d_{33} :(a) As a function of poling time; (b) as a function of poling temperature and (c) as a function of applied electric field.

The longitudinal piezoelectric coefficient d_{33} was measured with a Pennebaker Model 8000 Piezo d33 Tester, (American Piezo Ceramics Inc) coupled with a multimeter 34401A, (Hewlett Packard). To avoid problems related to non-uniformity of the composites, the measurement is made at least in 10 different points for each sample and the average of these points is taken as the coefficient d_{33}. Table 1 shows the values of d_{33} piezo constant for some composite materials with respective volume fraction of ceramic. There is a clear indication that even for lower ceramic content, the piezoelectric activity of the PZT-PAni/PVDF composite, i.e., composite with ceramic particle covered with a conducting polymer phase is comparable with other composites with higher ceramic phase and much better than the PZT/PVC composite with the same ceramic volume fraction.

By means of thermally stimulated depolarization current (TSDC) is possible to obtain the pyroelectric coefficient p(T). However it is necessary to clean up this curve, or leave it free from unwanted effects, such as the fluctuation of space charges injected during the polarization, since the interest is only the dipolar contribution, because it will remain even after heating the sample, if the provided temperature does not exceed the bias.

Figure 9 shows that for the first heating the depolarization current is greater than for the next heat, in which current tends to stabilize. This stabilized curve is the pyroelectric current, i.e., that due to dipolar relaxation.

Composites	Volumetric fraction of ceramic (%)	d_{33} (pC/N)
PSTM/PEKK [PELAIZ-BARRANCO 2005]	50	21
PTCa/PEKK [PELAIZ-BARRANCO 2005]	50	28
PZT/PU with 1% of graphite (SAKAMOTO 2001)	49	13
PZT/PVC [LIU 2006]	30	6
PZT/PVC [LIU 2006]	50	17
PZT/PVDF with 10% of PAni [RENXIN 2006]	50	28
PZT-PAni/PVDF redoped in pH=3.7	30	15-20*

Table 1. D_{33} piezo constant for some composite materials.

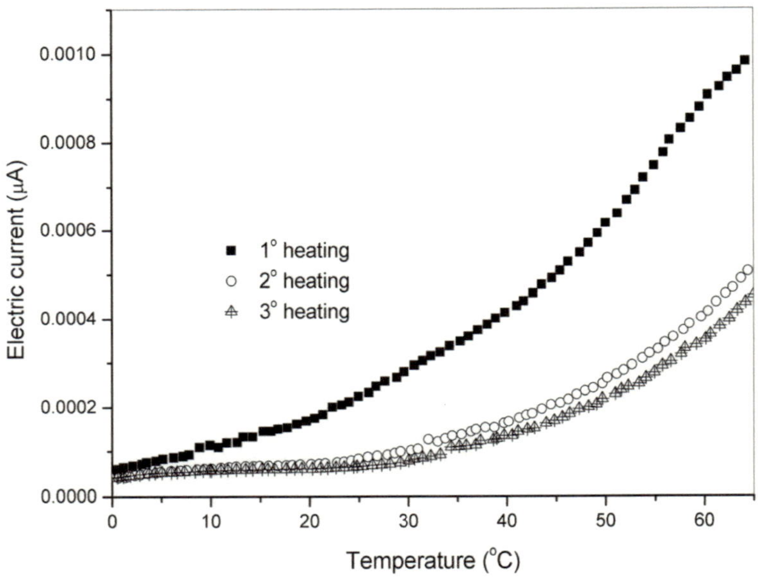

Fig. 9. Depolarization current for PZT-PAni/PVDF 30/70 vol% with PAni redoped in pH=3.7.

The pyroelectric coefficient p(T) can be obtained using the equation below:

$$p(T) = \frac{dP}{dT} = \frac{d(Q/A)}{dT} = \frac{1}{A}\frac{dQ}{dT} = \frac{1}{A}\frac{dQ/dt}{dT/dt} = \frac{1}{A}\frac{I_p}{dT/dt} \quad (3)$$

where P is the polarization, A the area of the electrode, Ip the pyroelectric current and dT/dt the heating rate. Figure 10 shows the pyroelectric coefficient as a function of the temperature for different composites with 30 vol.% of PZT.

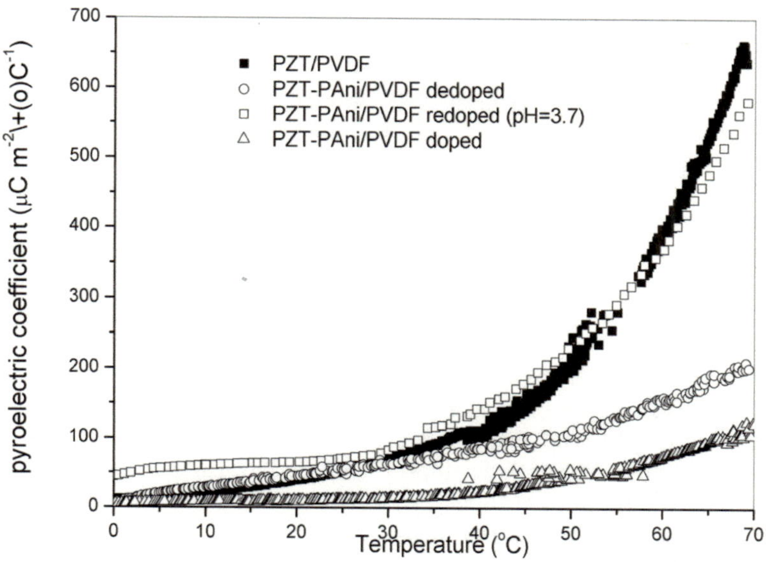

Fig. 10. Pyroelectric coefficient of composite with 30 vol.% of PZT-PAni.

It can be observed low values of pyroelectric coefficient for composites PZT-PAni/PVDF fully doped (Δ). This effect is expected, since its polarization is hindered by the high conductivity of the films. For the composites PZT/PVDF and PZT-PAni/PVDF dedoped in the range of 0 to 40°C the behavior of the pyroelectric coefficients are similar, while the composite PZT-PAni/PVDF redoped (pH 3.7) has larger values. At temperatures above 40°C, the composite with PAni dedoped has little increase in pyroelectric coefficient with increasing temperature, while the composites PZT/PVDF and PZT-PAni/PVDF redoped having similar behavior, suffer a sharp increase. Table 2 shows the value of pyroelectric coefficient at room temperature for some composite materials. The composite with ceramic particle partially covered with PAni after suitable protonation degree display pyroelectric coefficient comparable with composite samples described in literature, which uses high ceramic volume fraction.

Composite	Volumetric fraction of ceramic (%)	p(T) ($\mu C.m^{-2}.^{o}C^{-1}$)
PZT/Epoxy [DAS-GUPTA 1999]	40	40
BaTiO3/rubber [DAS-GUPTA 1999]	30	60
TGS/PVDF [DAS-GUPTA 1999]	80	90
PT/PVDF [DAS-GUPTA 1999]	62	130
PZT/PVDF [DAS-GUPTA 1999]	50	10
PTCa/P(VDF-TrFE) [DAS-GUPTA 1999]	65	130
PZT-PAni/PVDF redoped in pH=3.7	30	70

Table 2. Pyroelectric coefficient for some composite at room temperature

4. On the energy conversion

For material to be used as a sensor, the relationship between output voltage and input power should be linear, i.e., the responsivity should remain constant. And regarding the magnitude of responsivity, it may indicate the quality of the sensor, i.e. the larger the magnitude, more input signal is being converted into output signal.

Four samples: two of PZT/PVDF 30/70 vol%, one previously polarized with a electric field of 10 MV/m for 1 hour and another with a field of 5 MV/m for 15 minutes, a PZT/PVDF 50/50 vol% polarized with 10 MV/m for 1 hour, and one of PAni with 30/70 vol% PZT-PAni/PVDF redoped at pH 3.7 and polarized with 5 MV/m for 15 minutes were compared. Figure 11 illustrates the behavior of the output voltage as a function of frequency and input power.

A close analysis of the results allows observing that the composite with polyaniline was superior in all respects in comparison to the other composites. Although the PZT/PVDF 50/50 vol% has close values of output voltage (Figure 11), the electric field and the time spent to polarize it were high. Furthermore, the ferroelectric ceramics content is higher. Without conducting phase in the composite the poling conditions have to be actually higher for the best properties of electroactivity. It can easily be observed when comparing the two composites PZT/PVDF 30/70 in volume. The composite polarized with higher electric field and spending more time showed higher output voltage. Yet comparing the composites with 30% load PZT or PZT-PAni, the latest presents better results.

In photopyroelectricity the upper face of the sample is painted with black ink to optimize the energy absorption. Since the sample works as a thermal transducer, absorbed modulated radiation increases the temperature of the sample and the heat is transformed into an electrical signal generated by the potential difference between the two faces of the sample. According to Mandelis and Zver [Mandelis 1985] the photopyroelectric voltage ($V(\omega)$) can be written, for the optically opaque and thermally thick pyroelectric sample, as:

$$V(\omega) = \frac{A_0}{\omega} \exp\left[-\left(\frac{\omega}{2\alpha_s} \right)^{1/2} L_s - i\varphi \right] \tag{4}$$

where :

$$A_o = \frac{p}{\varepsilon_0 \kappa} \eta I_o F \tag{5}$$

is the amplitude, η is quantum efficiency, I_0 is the light intensity, F is the thermal coefficient which depends on the thermal parameters of the sample, ω is the angular frequency of light modulation, α_s is the thermal diffusivity of the sample, L_s the thickness and φ is the phase of the photopyroelectric signal. In the amplitude p is the pyroelectric coefficient, ε_0 the vacuum permittivity and κ the relative permittivity.

The concept of energy harvesting must be related to capture the ambient energy and convert it into usable electrical energy without environment attack i.e., a clean electric energy. Although there are a number of sources of harvestable ambient energy, such as solar energy and energy from wind [Schwede 2010, Chang 2002], piezoelectric materials are very interesting due to their ability to convert applied strain energy into usable electric energy. Some countries are working hard on it and Israel, for example, using piezoelectric plates under the track, can obtain power enough to provide electricity to a medium house.

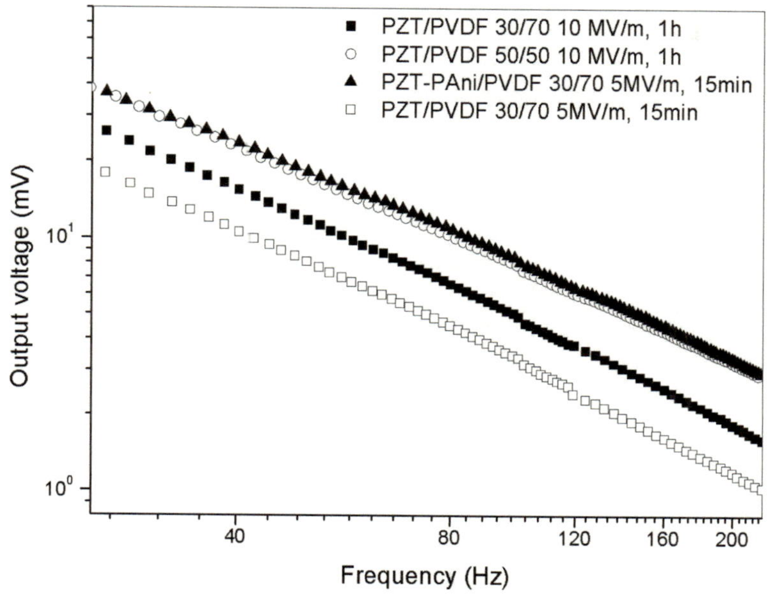

(a)

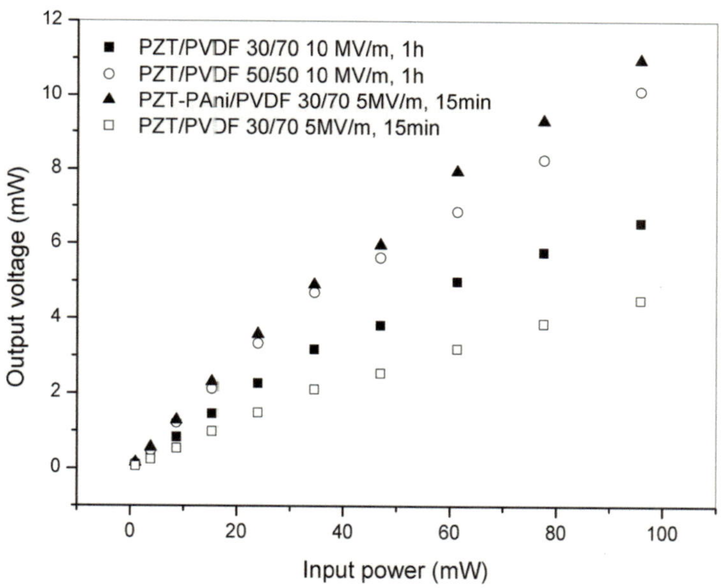

(b)

Fig. 11. (a) output voltage x frequency; (b) output voltage as a function of incident Power for different composites.

The energy recovery from the wasted energy used or not was a topic of discussion for a long time. Unused energy exists in various forms, such as vibration, water, wind, sun, heat, cold, human and vehicle movement, and shock waves. In today's world, there is a strong technological breakthrough in the way of life. More and more people are carrying portable

electronic devices. These devices have and enable incredible power and versatility in communication and problem-solving. But as the technologies of portable computers and microcontrollers have grown tremendously, the battery energy and the storage technology did not follow them. New technology allows these portable devices become ever smaller, but the sizes of cells or batteries are still the same and the limited operating life is a great problem.
An alternative to the batteries and cells is the implementation of a method to obtain energy surrounding devices that could power supply them. Piezoelectric materials can be used to convert mechanical energy into electrical energy that can be used to power other devices. Energy harvesting using piezoelectric materials have attracted many attention of the researchers around the world. Many works have been published in this area [Koyama 2009, Umeda 1997, Sodano 2004, Zheng 2009, Wang 2010, Anton 2007] and the focus is to find a material which gives power enough to allow its use commercially.

Composite	Thickness (μm)	Piezoelectric Coefficient d_{33} (pC/N)
A	800	28
B	600	28
C	600	20
D	700	10

Table 3. Characteristic of the composite samples

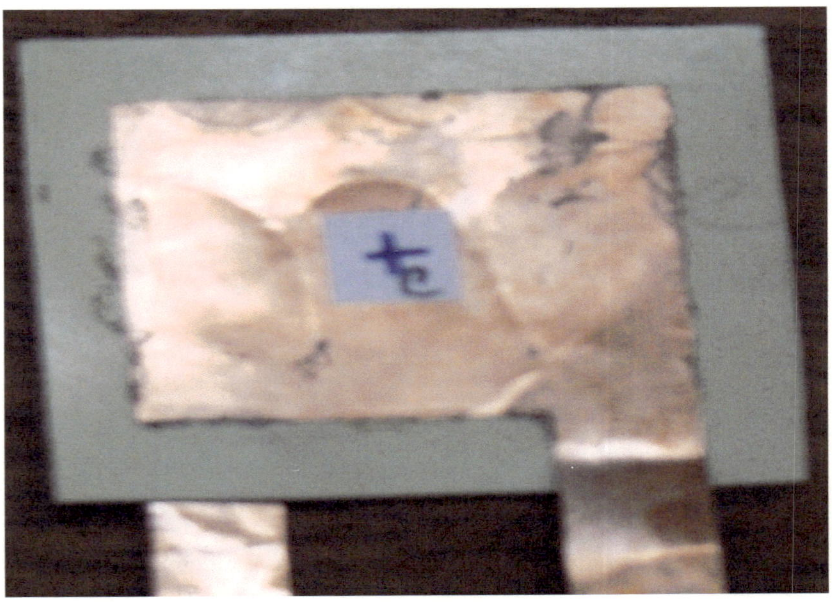

Fig. 12. Composite sample 50 vol% (grey) with cooper foil for electric contact.

The cantilever beam structure is one commum setup for energy harvesting. It uses PZT thin film to transforme the mechanical vibration into electrical energy. In the present work composite films were put under a track which will simulates car traffic or people movement.

Within this context, four square composite samples with 4.5 cm^2 were poled with suitable electric field and copper foil (1 mm thick) was glued for electrical contact as show in Figure 12. Table 3 shows the longitudinal d_{33} piezoelectric coefficient for each one.
To evaluate the power generated by these samples, they were pressed by the blue car continuosly as shown in Figure 13. The weight and the frequency of the blue car which will impact the composite samples can be controlled and fixed during the experiment. The output voltage provided by the piezoelectric composite can be measured with an oscilloscope.

Fig. 13. System used to simulate the vehicle traffic or people walking.

A track is project and constructed with two parts. A bottom steel base with electrical tape on its top, fixed to a press device plane. A top made of aluminum with the bottom with duct tape, and attached to external screws that make this part of mobile resource, since the composite is between bottoms and top part of the track it receives the impact of the track above it. The composites were used as transducer individually, in series and in parallel. Then they were connected directly (open circuit) to acquire the waveforms from the digital oscilloscope. Further, the composites were connected in circuit (closed circuit) with the oscilloscope at the entrance acquiring waveforms again. Finally, voltages were measured at the capacitors for every minute during 10 minutes. Acquisition board was used to get the electrical signal provided by the composite. This board consists of a retifier circuit AC/DC and a output capacitor.
The experiments starts using a force of 200 kgf, to stroke the composites with a frequency of 3.0 Hz, and a capacitor of 3300 µF. The open circuit (directly on the composites) and the

closed circuit measurements for each composite, and combined composites in parallel (//) which are showed in Figure 14.

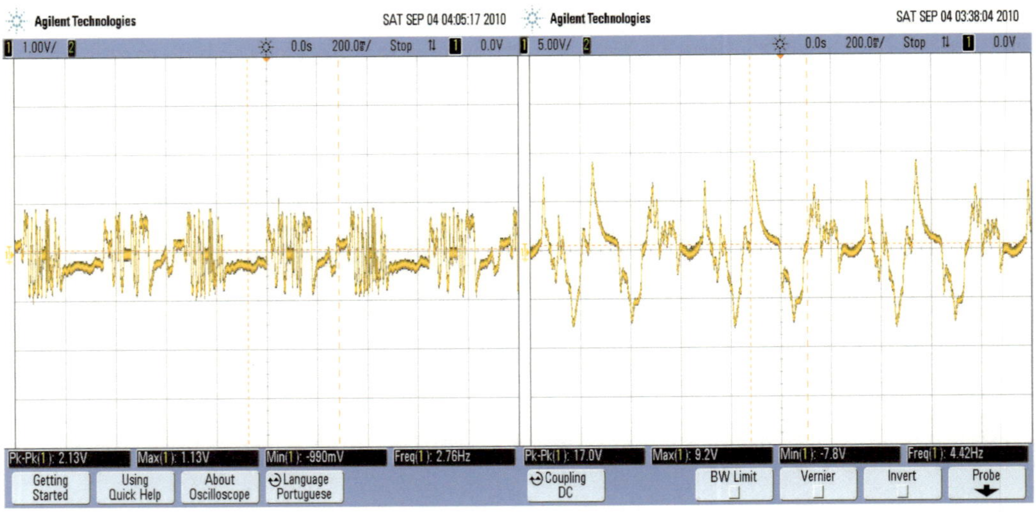

Fig. 14. Composites A///B//C//D voltage measurement with open circuit (right)and closed circuit (left).

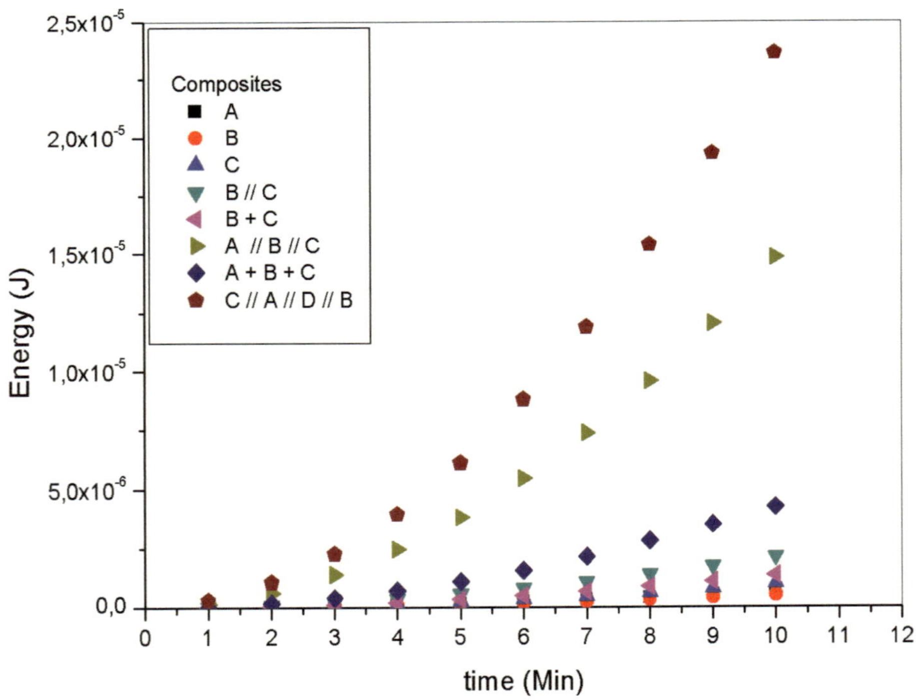

Fig. 15. Energy harvesting analysis. // means parallel connection; + represents a series connection.

Experimental results show that an open circuit output voltage of 17.0 V_{pp} are generated while in closed circuit the peak to peak voltage generated is 2.13 V because of the impedance of the capacitor.

Figure 15 shows the energy analysis of the experiments for different configurations of the composite films. It can be seen the increasing energy supplied when the composite films are connected in parallel. The useful energy, after 10 min, by four composite films is about ten times higher than the energy generated by one composite film. The values of energy in Figure 15 were calculated from the measurement of the output voltage against time, using the following relation:

$$U = \frac{1}{2} C V^2 \tag{6}$$

where U is the available energy and V is the voltage measured on the capacitor. The voltage was measured during the charge of the capacitor due to the deformation of the composite films by the applied stress.

5. Conclusions

Composite films made of PZT ceramic immersed in PVDF polymer matrix were obtained with 0-3 connectivity. The method of synthesis can provide different structure of the ceramic and also can provide ceramic particles with different size distribution which are important parameters for the electroactive properties of the sample. The inclusion of a semiconductor phase, separately or coating the ceramic particles improve the poling process of the composite, avoiding timing consuming and high applied electric field to polarize the ferroelectric ceramic particles immerse into the polymer matrix. The advantages of recovered particles is the better control of the homogeneity of the particle distribution avoiding percolation of conductive particles that may form a continuous path which not allow the poling process.

Using small amount of ceramic (30 vol%) the composite was used as infrared detector, indicating the possibility of its use as intruder detector or fire alarm. Using the right protonation (doping) degree of the PAni, the composite display piezo and pyroelectric coefficients high as many composite materials with higher ceramic content even when poled with lower electric field and shorter poling time. The study of energy harvesting simulating people walking or vehicle traffic showed low power generated by each small composite sample (4.5 cm^2 area) but the association of four samples enhanced the converted electrical energy from the energy wasted during vehicle traffic. These preliminary results show that the composite material deserves to be deeply studied as alternative material to obtain clean energy.

6. Acknowledgment

This work has financial support from the Brazilian Agencies: Fundação de Amparo à Pesquisa do Estado de São Paulo - FAPESP and Conselho Nacional de Desenvolvimento Científico e Tecnológico - CNPq through the Instituto Nacional de Ciência e Tecnologia de Materiais em Nanotecnologia - INCTMN.

7. References

Abothu, I. R.; et. al. (1999) Processing of Pb($Zr_{0.52}Ti_{0.48}$)O_3 (PZT) ceramics from microwave and conventional hydrothermal powders. *Materials Research Bulletin*, 34, (9), 1411-1419.

Anton, S. R. & Sodano, H. A. (2007) A review of power harvesting using piezoelectric materials (2003 - 2006), *Smart Mater. Struct.*, 16, R1 - R21.

Bauer, F. (2000) PVDF shock sensors: applications to polar materials and high explosives, *IEEE Trans. Ultrason. Ferroelectr. Freq. Control*, 47 (6), 1448 -1454.

Blinova, N. V., et. al. (2008) Control of polyaniline conductivity and contact angles by partial protonation, *Polym. Int.* 57, 66 - 69.

Boumchedda, K., Hamadi & M., Fantozzi, G. (2007) Properties of a hydrophone produced with porous PZT ceramic, *J. European Ceram. Soc.*, 27, 4169 - 4171.

Brown L. F.& Mason, J. L. (1996) Disposable PVDF Ultrasonic Transducers for Nondestructive Testing Applications, *IEEE Trans. Ultrason. Ferroelectr. Freq. Control*, 43 (4), 560 - 567.

Chang, L. (2002) Wind energy conversion systems, *IEEE Canadian Review –Spring/Printemps*, 12 - 16.

Chau, K. H., Wong, Y. W., & Shin, F. G. (2007) Enhancement of piezoelectric and pyroelectric properties of composite films using polymer electrolyte matrix, *Appl. Phys. Lett.*, 91, 252910.

Chu, S-Y.& Chen, T-Y. (2004) Fabrication of modified lead titanate piezoceramics with zero temperature coefficient and its application on SAW devices, *IEEE Trans Ultras. Ferroelctr. Freq. Contr.*, 51 (6), 663 - 667.

Ciang, C. C., Lee, J-R.& Bang, H-J. (2008) Structural health monitoring for a wind turbine system: a review of damage detection methods, *Meas. Sci. Technol.* 19, 122001.

Das-Gupta, D. K. (1999) Ferroelectric composite sensor materials, *Mater. Eng.*, 10 (2), 97 - 125.

De Carvalho, A. A. & Alter, A. J. (1997) Measurement of X-ray intensity in the medical diagnostic range by a ferroelectric detector, *Ultrason. Ferroelectr. Freq. Control*, 44 (6), 1198 - 1203.

De Paula, M. H., et al. (2005) Microcontrolled pyroelectric instrument for measuring X-ray intensity in mammography, *Med, Biol. Eng. Comput.*, 43, 751 - 755.

Dias, C. J.& Das-Gupta, D. K. (1996) Inorganic ceramic/polymer ferroelectric composite electrets, *IEEE Trans. Dielectr. Electr. Insul.*, 3 (5), 706 - 734.

Dutta, P.K., et. al. (1994) Hydrothermal Synthesis and Dielectric Properties of Tetragonal BaTiO3, *Chem Mater.*, 6 (9):1542-1548.

Edwards, G., et. al. (2006) PMN-PT single-crystal transducer for non-destructive evaluation, *Sensors and Actuators A*, 132, 434 - 440.

Estevam, G. P., De Melo, W. L. B. & Sakamoto, W. K. (2011) Photopyroelectric response of PTCa/PEEK composite, *Review of Scientific Instruments*, 82, 023903.

Feng Y., et. al. (2010) A model for 0-3 piezoelectric composites with an interlayer, *Polymer Composites*, 1922 - 1927.

Furukawa, T., Fujino, K. & Fukada, E. (1976) Electromechanical properties in the composites of epoxy resin and PZT ceramics, *Jpn. J. Appl. Phys.* 15 (11), 2119 - 2129.

Furukawa, T., Suzuki, K. & Date, M. (1986) Switching process in composite systems of PZT ceramics and polymers, *Ferroelectrics*, 68, 33 - 44.

Guggilla, P. et. al. (2006) Pyroelectric ceramics for infrared detection applications, *Materials Letters*, 60, 1937 - 1942.

Haertling, G. H. (1999) Ferroelectric ceramics: history and technology, *Journal of the American Ceramic Society*, 82, (4), 797–818.

Huang, Z., et. al. (2002) Infrared optical properties of $PbTiO_3$ ferroelectric thin films, J. *Phys. D: Appl. Phys.*, 35, 246 - 248.

Ishikawa, K., et. al. (1994) Crystallization and growth process of lead titanate fine particles from alkoxide-prepared powders, *Jpn. J. Appl. Phys.* 33, 3495.

Jaffe, H., Piezoelectric applications of ferroelectrics, IEEE Trans. Electr. Devices, 16 (6), 544 - 554 (1969).

Kawai, H. (1969) The piezoelectricity of poly(vinylidene fluoride), *Jpn. J. Appl. Phys.*, 8, 975 - 976.

Klee, M. et. al. (2010) Ferroelectric and piezoelectric thin films and their applications for integrated capacitors, piezoelectric ultrasound transducers and piezoelectric switches, IOP Conf. Series: *Mater. Sci. and Eng.* 8, 012008.

Koyama, D. & Nakamura, K. (2009) Array configurations for higher power generation in piezoelectric energy harvesting, *IEEE International Ultrasonics Symposium Proceedings*, 1973 - 1976.

Kumar N. & Nath, R. (2005) Ferroelectric phase stability studies in potassium nitrate: Polyvinylidene fluoride composite films, *J Appl. Phys.*, 97, 024105.

Lau, S. T., et. al. (2002) Piezoelectric composite hydrophone array, *Sensors and Actuators A*, 96, 14 - 20.

Lau, S. T., et. al. (2007) A poling study of lead zirconate titanate/polyurethane 0-3 composites, *J. Appl. Phys.*, 102, 044104.

Lencka, M.M. & Riman, R. E. (1995) Thermodynamics of the Hydrothermal Synthesis of Calcium Titanate with Reference to Other Alkaline-Earth Titanates, *Chem. Mater.*, 7: 18-25.

Liu, X-F., et. al. (2006) Piezoelectric and dielectric properties of PZT/PVC and graphite doped with PZT/PVC composites, *Mater. Sci. Eng. B*, 127, 261 - 266.

Lovinger, A. (1982) *Developments in Crystalline Polymers-1*, D. C. Basset Ed., Applied Science Publishers, 195 - 273 London.

Lovinger, A. (1983) Ferroelectric Polymers, *Science*, 220, 1115 - 1121.

Luo, Z., et. al. (2008) Self-assembly of BaMoO4 single-crystalline nanosheets into microspheres, *Materials Chemistry and Physics*, 110, 17- 20.

Mandelis, A. & Zver, M. M. (1985) Theory of photopyroelectric spectroscopy of solids, *J. Appl. Phys.*, 57 (9), 4421 - 4430.

Marin-Franch, P., et. al. (2002) PTCa/PEKK piezo-composites for acoustic emission detection, *Sensors and Actuators A*, 99 236 - 243.

Moreira, M. L., et. al. (2009) Synthesis of Fine Micro-sized BaZrO3 Powders Based on a Decaoctahedron Shape by the Microwave-Assisted Hydrothermal Method, *Crystal Growth & Design*, 833-839.

Morita, T. (2010) Piezoelectric Materials Synthesized by the Hydrothermal Method and Their Applications, *Materials*, 3, 5236 - 5245.

Muralt, P.& Baborowski, J. (2004) Micromachined Ultrasonic Transducers and Acoustic Sensors Based on Piezoelectric Thin Films, *Journal of Electroceramics*, 12, 101 - 108.

Newnham, R. E., Skinner, D. P. & Cross, L. E. (1978) Connectivity and piezoelectric-pyroelectric composites, *Mat. Res. Bull*, 13, 525 - 536.

Or, Y., et. al. (2003) Modeling of poling, piezoelectric, and pyroelectric properties of ferroelectric 0-3 composites, *J. Appl. Phys.* 94 (5).

Pan, Q., et. al. (2007) Intense blue-light emission from hydrothermally synthesized lead zirconate titanate platelets. *Materials Letters*, 61, (4-5), 1210–1213.

Pelaiz-Barranco, A. & Marin-Franch, P. (2005) Piezo- pyro, ferro-, and dielectric properties of ceramic/polymer composites obtained from two modifications of lead titanate, *J. Appl. Phys.*, 97, 03411 - 034414.

Ploss, B., & Kopf, S. (2006) Improving the pyroelectric coefficient of ceramic/ polymer composite by doping the polymer matrix, *Ferroelectrics*, 338, 145 - 151.

Ploss, B., et. al. (2001) Poling study of PZT/P(VDF-TrFE) composites, *Composites Science and Technology*, 61, 957 - 962.

Pontes, D. S. L., et. al. (2001) Microstructural, dielectric and ferroelectric properties of calcium-modified lead titanate thin films derived by chemical processes, *J. European Ceram. Soc.*, 21, 1107 - 1114.

Pontes, W., et. al. (2010) PZT, for measuring energy fluence rate of X-ray used in superficial câncer therapy, *Instr. Sci. Technol.*, 38, 210 - 219.

Poon, Y. M. & Shin, F. G. (2004) A simple explicit formula for the effective dielectric constant of binary 0-3 composites, *J. Mater. Sci.*, 39, 1277 - 1281.

Rao, K. J., et. al. (1999) Synthesis of Inorganic Solids Using Microwaves, Chem. Mater., 11, 882-895.

Ren, T-L., et.al. (2003) Piezoelectric and ferroelectric films for microelectronic applications, *Mater. Sci.and Eng.* B99, 159 - 163.

Renxin X., et. al. (2006) Dielectric and piezoelectric properties of 0-3 PZT/PVDF composite doped with polyaniline, *Journal of Wuhan University of Technology - Mater. Sci. Ed.*, 21 (1), 84 - 87.

Sa-Gong, et. al. (1986) Poling flexible piezoelectric composites, *Ferroelectrics Letters*, 5, 13' - 142.

Sakamoto, W. K., De Souza, E. & Das-Gupta, D. K. (2001) Electroactive properties of flexible piezoelectric composites, *Materials Research*, 4 (3), 201 - 204.

Sakamoto, W. K., et. al. (2006) PTCa/PEEK composite acoustic emission sensors, *IEEE Trans. Dielectr. Electr. Insul.*, 13 (5), 1177 - 1182.

Sakamoto, W. K., Marin-Franch, P. & Das-Gupta, D. K. (2002) Characterization and application of PZT/PU and graphite doped PZT/PU composite, *Sensors and Actuators A*, 100, 165 - 174.

Schwede, J. W., et. al. (2010) Photon-enhanced thermionic emisión for solar concentrador Systems, *Nature Materials*, 9, 762 - 767.

Shimomura, K., et. al. (1991) Preparation of lead zirconate titanate thin film by hydrothermal method, *Jpn. J. Appl. Phys.*, 30, 2174 - 2177.

Sodano, H. A., Park, G. & Inman, D. J. (2004) Estimation of electric charge output for piezoelectric energy harvesting, *Strain*, 40, 49 - 58.

Soman J.& O'Neal, C. B. (2011) Fabrication and Testing of a PZT Strain Sensor for Soil Applications, *IEEE Sensors Journal*, 11 (1), 78 - 85.

Sosnin, A. (2000) Image infrared converters based on ferroelectric-semiconductor thin-layer systems, *Semiconductor Physics, Quantum Electronics & Optoelectronics*, 3 (4), 489 - 495.

Umeda, M., Nakamura, K. & Ueha, S. (1997) Energy storage characteristics of a piezo-generator using impact induced vibration, *Jpn. J. Appl. Phys.*, 36, 3146 - 3151.

Wang, C.M.; et. al. (2000) Calcium modified lead titanate thin films for pyroelectric applications, *ISAF 2000. Proceedings of the 12th IEEE International Symposium on Applications of Ferroelectrics - ISAF*, vol. 2, 771 - 774.

Wang, D-A. & Ko, H-H. (2010) Piezoelectric energy harvesting from flow-induced vibration, *J. Micromech. Microeng.* 20, 025019.

Wei, N., et. al. (2007) Effect of electrical conductivity on the polarization behaviour and pyroelectric, piezoelectric property prediction of 0-3 ferroelectric composites, *J. Phys. D: Appl. Phys.*, 40, 2716 - 2722.

Wong, C. K. & Shin, F. G. (2006) Effect of electrical conductivity on poling and the dielectric, pyroelectric and piezoelectric properties of ferroelectric 0-3 composites, *J. Mater. Sci.*, 41, 229 - 249.

Wong, C. K. & Shin, F. G. (2005) Electrical conductivity enhanced dielectric and piezoelectric properties of ferroelectric 0-3 composites, *J. Appl. Phys.* 97, 064111.

Wong, C. K., Wong, Y. W. & Shin, F. G. (2002) Effect of interfacial charge on polarization switching of lead zirconate titanate particles in lead zirconate titanate/polyurethane composites, *J. Appl. Phys.*, 92 (7), 3974 - 3978.

Yu. N. Zakharov, A. V. Borodin, & V. Z. Borodin (2007) Pyroelectric properties of PZT -type ferroelectric ceramics in the morphotropic phase-transition region , *Bulletin of the Russian Academy of Sciences: Physics* , 71, (5), 709-710.

Zaghete, M. A., et. al. (1999) The Effect of Isostructural Seeding on the Microstructure and Piezoelectric Properties of PZT Ceramics, *Ceramics International*, 25, .239 - 244.

Zaghete, M. A., et. al. (1992) Phases Characterization In PZT Obtained From Organic Solutions Of Citrates. *Journal of the American Ceramic Society*, 75, 2088 - 2093.

Zhang, Q. Q., et. al. (2006) High frequency broadband PZT thick film ultrasonic transducer for medical imaging applications, *Ultrasonics*, 44 , e711 - e715.

Zheng B., Chang, C-J. & Gea, H. C. (2009) Topology optimization of energy harvesting devices using piezoelectric materials, *Struct. Multidisc. Optim.*, 38, 17 - 23.

Zhou, Y., et. al. (2005) Effects of polarization and permittivity gradients and other parameters on the anomalous vertical shift behavior of graded ferroelectric thin films, *J. Appl. Phys.*, 98, 034105.

Part 3

Lead-Free Materials

14

Barium Titanate-Based Materials – a Window of Application Opportunities

Daniel Popovici[1], Masanori Okuyama[2] and Jun Akedo[1]
[1]National Institute of Advanced Industrial Science and Technology,
[2]Osaka University
Japan

1. Introduction

Since it was discovered in 1945, barium titanate (BT) attracted much attention to researchers, becoming one of the most investigated ferroelectric materials due to good electrical properties at room temperature, mechanical and chemical stability and the easiness in its preparation. It is known that above the Curie temperature, the crystalline barium titanate has a cubic, perovskite-like structure as shown in figure 1.

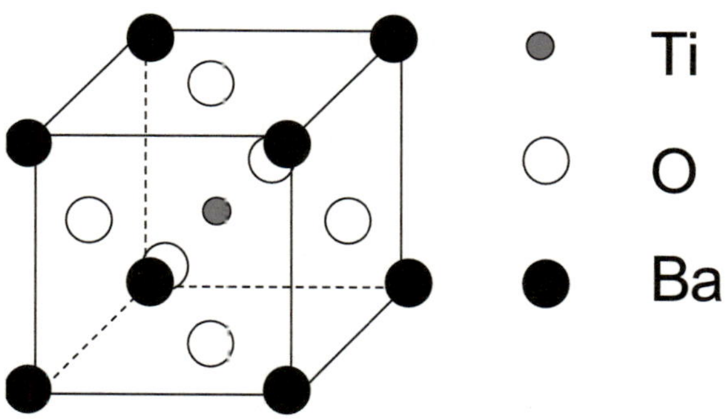

Fig. 1. Crystalline structure of BT material above Curie temperature

Below the Curie temperature, the crystalline cell is suffering a series of changes: to tetragonal (at 120^{0}C), from tetragonal to orthorhombic (at 0^{0}C) and from orthorhombic to rhombohedral (at -90^{0}C) in which the material has ferroelectric properties.
Theories concerning the ferroelectric behavior of crystalline materials that have a perovskite structure pinpoint the important role played by the spatial oxygen arrangement having an ion in its center, to the ferroelectrical properties. Taking this into consideration it is easy to predict that a change in spatial alignment of the oxygen octahedra or a substitution of the central ion (B-site substitution) can modify the ferroelectric behavior of the material. Change in spatial alignment of the oxygen octahedra can be also made by (so called) A-site substitution, when an A-site ion is substituted with another ion. In the case of barium titanate, it has been found that substitutions can make the temperature of paraelectric to ferroelectric transition to shift towards lower or higher values and, in some conditions, the

temperature of dielectric constant maxima will be affected by the frequency of the applied field (relaxor behavior). An A-site substitution, for example substituting Ba^{2+} with Sr^{2+} or Pb^{2+}, is responsible for shifting the temperature region in which the ferroelectric properties are present while the values of permittivity remain relatively large. This is good from the applications viewpoint because the possibility of shifting Curie temperature and the selection of the sector for the temperature dependence for dielectric constant and dielectric loss broadens the application area of these BT-based materials. The representative material in this class is $(Ba_{1-x},Sr_x)TiO_3$ (BST), one of the most studied solid solution due to its stability and the wide range of possible applications that can use its electrical properties.

A B-site substitution is also responsible of changing the degree of ordering in the solid solution resulting in a shift of Curie temperature and the appearance of the relaxor behavior when the local ordering of B-sites will make it favorable. In this category there is no widely studied BT-based material because their properties were comparable to other ferroelectric materials such as lead zirconate titanate (PZT), pure barium titanate, lead titanate or even barium strontium titanate. However, it has been found that small amounts of $BaZrO_3$ or $BaHfO_3$ included in BT can make it a candidate material for pyroelectric sensor, having electrical characteristics superior of those of lead lanthanum zirconate titanate (PLZT) or BST, materials that were commonly used for such applications.

As mentioned earlier, $(Ba,Sr)TiO_3$ (BST) solid solutions are one of the most investigated ceramic materials because the shift of ferroelectric phase transition towards lower temperatures can easily be controlled by adjusting the Ba/Sr ratio while maintaining acceptable high dielectric constants coupled with good thermal stability. $Ba(Ti,Sn)O_3$ (BTS) solid solutions are another subclass of materials that can be used for specific application. For a given application, to achieve the desired properties in the BST or BTS system, compositional control should be considered along with the preparation method and/or deposition method in the final device structure.

From many applications that can incorporate BT-based materials, here only optimization for two applications will be discussed in detail: dielectric bolometer mode of infrared sensor and embedded multilayered capacitor structures. Since the requirements for ferroelectric materials suitable for dielectric bolometer mode of infrared sensor and embedded multilayered capacitor structures are different, a good selection of ferroelectric material and fabrication method is necessary to ensure high quality ceramic layers for these applications. As a result, BTS thin films have been fabricated using metal-organic decomposition method as a suitable process to ensure good quality films for dielectric bolometer mode of infrared sensing applications. In the case of films for embedded multilayered capacitor applications, since the target require a low temperature fabrication technique, BST thick films have been fabricated using a relatively new deposition technique called aerosol deposition method, developed at National Institute of Advanced Industrial Science and Technology by Dr. Akedo, one of the coauthors of this paper, a fabrication method that allows fabrication of thick and dense ceramic layers at room temperature.

2. Preparation and characterization of BTS thin films for dielectric bolometer mode of infraredd sensor applications

One important characteristic for a material to be suitable for dielectric bolometer (DB) mode of infrared sensor applications is to have a large *Temperature Coefficient of Dielectric constant* (TCD).

$$TCD = \frac{(\varepsilon_r(T_2) - \varepsilon_r(T_1))}{\frac{(\varepsilon_r(T_2) + (\varepsilon_r(T_1))}{2}} \cdot \frac{100}{T_2 - T_1} \quad (1)$$

From 1990, $Ba(Ti_{1-x},Sn_x)O_3$ solid solution captured the attention of the researchers because of his stable ferroelectric properties in the vicinity of the Curie point that makes it a good candidate for specific applications. Because it belongs to a class of ferroelectric materials that show a diffuse phase transition (DPT) who have promising properties behavior that can be used for various applications such as sensors, actuators or high permittivity dielectric devices, this solid solution captured the attention of many research groups as a suitable active material. Investigation made with bulk $Ba(Ti_{1-x},Sn_x)O_3$ samples revealed that, if $BaSnO_3$ content is 30% or more, the solid solutions of $Ba(Ti_{1-x},Sn_x)O_3$ have relaxor behavior (Mueller et al., 2004; Lu et al., 2004; Yasuda et al., 1996; Xiaoyong et al., 2003). Moreover, Yasuda and al. observed a deviation of the dielectric constant from the Curie-Weiss law (that is specific for relaxor ferroelectrics) even when $BaSnO_3$ content is between 10 and 20%, but only in a narrow temperature region above Curie point, and a relaxor behavior for samples in witch the $BaSnO_3$ content is above 20%.
More recently, some authors see in $Ba(Ti_{1-x},Sn_x)O_3$ a candidate to replace $(Ba,Sr)TiO_3$ in microwave applications (Lu et al., 2004; Jiwei et al., 2004). Jiwei et al. showed that, in some conditions, tunability of a metal-ferroelectric-metal (MFM) structure could be as high as 54% at an applied field of 200 kV/cm and a frequency of 1 MHz.
A more important indirect result has shown by Tsukada et al. where, from the dielectric constant versus temperature for a $Ba(Ti_{1-x},Sn_x)O_3$ ($BaSnO_3$ content of 15%) thin film with a thickness of 400 nm deposited by PLD on $Pt/Ti/SiO_2/Si$, a value close to 11% at 25^{0}C can be calculated.

2.1 Fabrication of BTS thin films by metal-organic decomposition process

In the processing of the thin films, the goal is not only to reduce the cost and time in fabrication process but, more important, is to optimize the film properties for specific applications. Metal-organic decomposition process (MOD) has some advantages in comparison with other widely used deposition techniques: precise control of stoichiometry, high homogeneity, large area of deposition and simple equipment and process flow. However, one of the biggest problems implying this technique is that is not possible to fabricate crystalline thin films with epitaxial or columnar structure and that the density of the material is lower than the one obtained by other technique. High quality films can still be obtained by this process comparing with other techniques and, along with the advantages offered by MOD convinced many researchers to use it in their investigations.
Liquid solution of BTS was prepared by mixing barium isopropoxide $[Ba[OCH(CH_3)_2]_2]$, titanium butoxide $[Ti[O(CH_2)_3CH_3]_4]$ tin isopropoxide $[Sn[OCH(CH_3)_2]_4]$ and 1-methoxy-2-propanol supplied by Toshima MGF. CO.LTD.
The $Ba(Ti_{0.85},Sn_{0.15})O_3$ (BTS) solution was deposited on Pt(240nm)/Ti(60nm) $/SiO_2$(600nm)/Si substrates by spin-coating at 500 rpm for 5 seconds followed by another 20 seconds at 2200 rpm. This step was performed in enriched N_2 atmosphere (1-5 l/min flow) to avoid moisture, because the solution is highly hygroscopic. After spin coating, the film was moved quickly on a hot plate and dried at 250^{0}C for one minute followed by 10 minutes drying into an oven at the same temperature in air. After drying, the BTS films were pyrolyzed at 450^{0}C for 10 minutes into an oven in enriched O_2 atmosphere (1 l/min

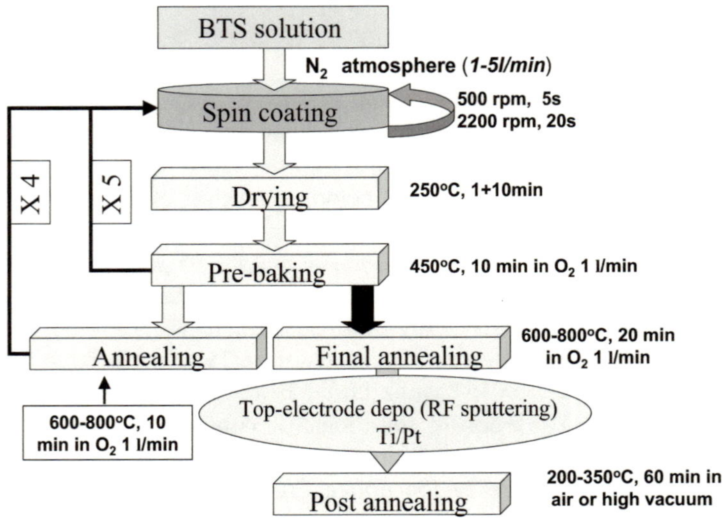

Fig. 2. Process flow of the BTS thin films prepared by MOD

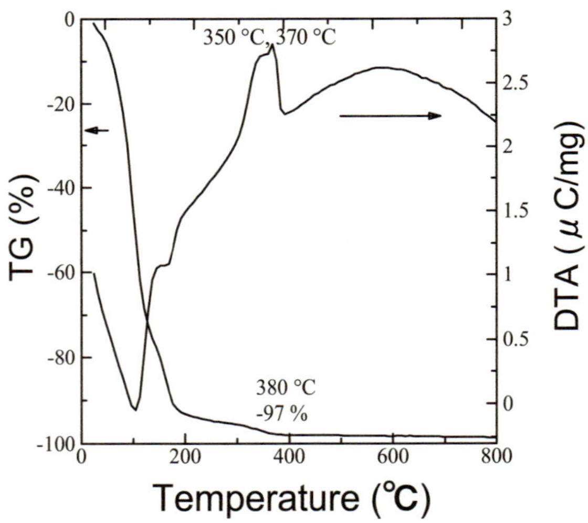

Fig. 3. TG-DTA analysis results of the BTS MOD-solution

flow). Spin-coating / drying / pyrolyzing sequence was repeated another 4 times before annealing in enriched O_2 atmosphere (1 l/min flow) for 10 minutes was performed. The BST15 thin films were annealed at 600°C, 700°C, 750°C or 800°C. The deposition and heat treatment were repeated 20 times before a final annealing was performed for 20 minutes in O_2 enriched atmosphere. The schematic representation of the deposition steps is shown in Figure 2. Differential thermal analysis (DTA) and thermo-gravimetric analysis (TG) (Figure 3) were used to determine the thermal decomposition behavior of the BTS solution and to select the appropriate temperatures for drying and baking. DTA curve shows an endothermic peak at 103°C corresponding to solvent evaporation point and two exothermic peaks at 350 and 370⁰C, temperatures that correspond to precursor decomposition and formation of BTS compound. The TG curve showed that the total mass of the investigated liquid decreases rapidly at the beginning, the solution loosing almost 94% of its mass at 180⁰C and slowly loosing more, reaching -97% at 380⁰C. The weight loss is insignificant above 380⁰C.

According to TG-DTA results, a drying temperature over 180⁰C and a baking temperature over 370⁰C are necessary. A drying temperature of 250⁰C and baking temperature of 450⁰C were selected to ensure full solvent evaporation in short time and to minimize as much as possible the stress and defects caused by a further weight reduction during annealing and a rapid complete precursor decomposition and BTS formation.

The thickness of the BTS15 films obtained by this process was about 360nm.

After BTS thin films preparation was completed, Pt/Ti electrodes were formed on the film by RF sputtering to make BTS capacitors. After completion of BTS capacitor fabrication, for films annealed at 700⁰C, a post electrode-forming annealing was performed at temperature varying from 200 to 350⁰C in air and at 300⁰C in high vacuum for 60 minutes.

In order to obtain high quality films suitable for DB-mode of infrared sensing applications (high values of TCD), the BTS thin film properties have been studied for different fabrication conditions and the results were used to optimize the deposition conditions for improved BTS thin films. The influence of annealing temperature and postannealing treatment on physical and electrical properties of the fabricated BTS thin films was investigated aiming an increase in TCD values near room temperature. The temperature of maximum permittivity for the fabricated BTS thin films was found to be near 13⁰C.

2.2 Annealing and postannealing treatment effect on BTS thin film properties

The annealing effect on the properties of the fabricated BTS thin films has been checked first in order to optimize the fabrication conditions.

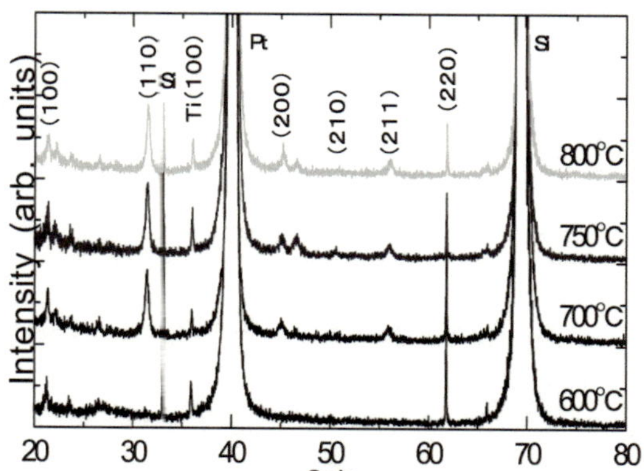

Fig. 4. XRD patterns of the BTS thin films annealed at different annealing temperatures

In Figure 4, XRD patterns of the films annealed at temperature ranging from 600⁰C to 800⁰C are showed. The films annealed at 600⁰C are still amorphous but for films annealed at 700⁰C and higher, crystal structure has been detected. The films have strong (110) peaks suggesting that the crystalline BTS films have a preferential orientation along (110) direction. The other peaks, assignable to a cubic perovskite type structure, are also present but their intensities are much smaller than the intensity of (110) peak. The preferred orientation and intensity ratios among the peaks revealed little distinct differences among these films as a function of annealing temperature. The average grain size was estimated from the half-width of the x-ray diffraction peak using Scherrer's formula to be in the 33.3 - 50 nm range.

For films fabricated at annealing temperatures of 700, 750 and 800^{0}C, leakage currents, C-V and temperature dependence of capacity (and through it, the permittivity dependence) were measured and analyzed. Except the temperature dependence of capacity, the other electrical measurements were performed at room temperature, well above the temperature of maximum permittivity.

The leakage current measurements showed that the films annealed at 750^{0}C have a higher leakage current than films annealed at 700^{0}C and 800^{0}C (Figure 5). The reason for this behavior is still not clearly understood. Because films with small leakage currents are desired the films annealed at 750^{0}C cannot be considered suitable for DB-mode infrared sensing applications. For this reason the attention was focused on the films annealed at 700^{0}C and 800^{0}C.

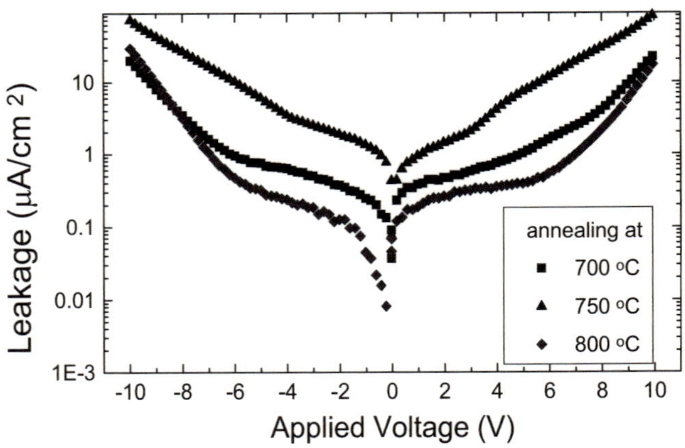

Fig. 5. Leakage current for BTS films annealed at different temperatures

The investigations of the temperature influence on the dielectric loss (Figure 6) revealed that the dielectric loss increases with increase in annealing temperature. Moreover, the dielectric loss for films annealed at 800^{0}C shows large temperature dependence compared with films annealed at 700 and 750^{0}C. On the other hand, the films annealed at 700^{0}C have the dielectric loss very little affected by the increase in the annealing temperature.

In Figure 7, temperature dependence of capacitance for films annealed at 700^{0}C and 800^{0}C has been plotted. The variation of capacitance for BTS samples annealed at 700^{0}C is more pronounced than for the samples annealed at 800^{0}C.

Reviewing the results obtained after physical and electrical properties in becomes clear that annealing at 700^{0}C is more suitable in obtaining BTS thin films with good properties for DB-mode of infrared sensor applications.

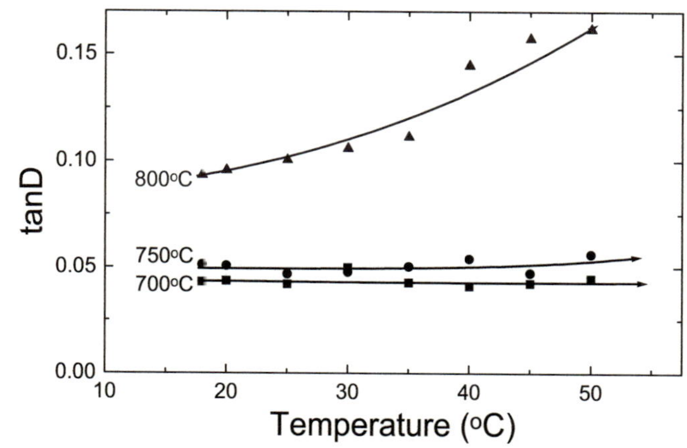

Fig. 6. Dielectric loss vs. sample temperature for BTS films annealed at different temperatures

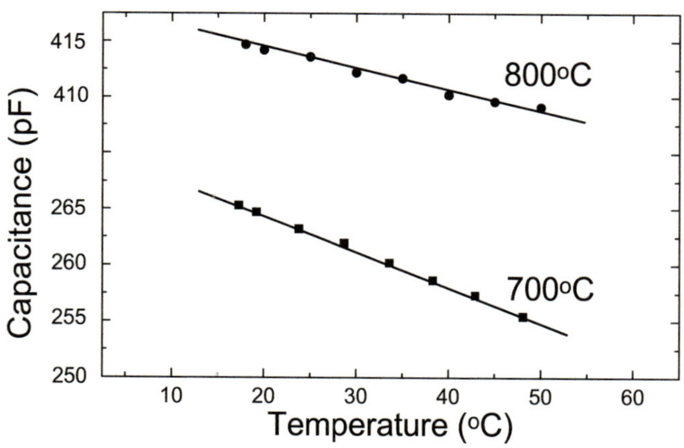

Fig. 7. Capacitance vs. sample temperature for BTS films annealed at 700°C and 800°C

The effect of postannealing temperatures on physical and electrical properties of BTS thin films was investigated keeping in mind that the films should be suitable for DB-mode of infrared sensor. The annealing temperature has been set to 700°C as a result of annealing temperature effect investigations performed earlier. After the top-electrode deposition, a postannealing treatment has been performed at temperatures of 200, 300 and 350°C in air and at 300°C, in vacuum for 60 minutes. The results of the investigations made on BTS samples are summarized in Table 2.

postannealing	Leakage density	TCD (25°C)	polarization	Grain size	Dielectric loss
not	200μA/cm^2	0.72-1 %/K	↓	Not affected	0.04
200°C	1μA/cm^2	1.3 %/K			0.04
300°C	100μA/cm^2	5.6 %/K			0.06
350°C	100μA/cm^2	1-1.1 %/K			0.04

Table 2. Postannealing effect on physical and electrical characteristics of BTS thin films. Leakage densities are given for an applied voltage of 10V

Only some electrical properties are affected by the treatment. Polarization in P-E hysteresis loops is increasing with the increase in postannealing temperature (not shown here). This can be explained considering the fact that a postannealing treatment is improving the metal-ferroelectric interface. The effect of oxygen diffusion during postannealing treatment should not be neglected while considering improvement in polarization. However, as we will show below, increase in polarization due only to improvement in film surface due to reduction in oxygen vacancies by oxygen diffusion from air cannot fully explain the tendency.
Analyzing the results summarized in Table 2 it can be seen that the current leakage of the BTS samples is the most affected by postannealing temperature being smaller for films postannealed at 200^{0}C. It can be observed that increase in postannealing temperature will not further improve the leakage currents of the samples. Y. Fukuda et al. (Fukuda et al., 1997) reported that, by increasing the postannealing temperature in the case of (Ba,Sr)TiO_3 thin films deposited on Pt/SiO_2/Si or $SrTiO_3$ substrates, the diffusion of the oxygen from the postannealing atmosphere is decreasing. Our results suggest the same effect by increasing the postannealing temperature because the leakage current, even if it is better than that for as-deposited samples, is increasing by increasing the annealing temperature.
Figure 8 is showing the I-E$^{1/2}$ characteristics of the leakage current for BTS samples postannealed at 200^{0}C and 350^{0}C along leakage current for samples that were not postannealed. The leakage behavior for samples postannealed at 300^{0}C is not shown to avoid overlay in the graphic because it shows almost the same behavior as samples annealed at 350^{0}C. It can be seen in the figure that postannealing treatment decreases Schottky leakage currents. The Schottky currents can be described by (Sze, 1981; Fukuda et al., 1998):

$$J = A^{*}T^{2}\exp\left(-q\Phi_{B} / kT\right)\exp\left(q\alpha E_{\text{int}}^{1/2} / kT\right) \tag{2}$$

$$\alpha = \left(q / 4\pi\varepsilon_{i}\right)^{1/2} \tag{3}$$

where A^{*} and Φ_B are the effective Richardson constant for electrode emission and the Schottky barrier height between cathode and the ferroelectric thin film, T is the temperature in Kelvin, E_{int} is the electric field at the interface and ε_i is the dynamic dielectric constant of the ferroelectric media.

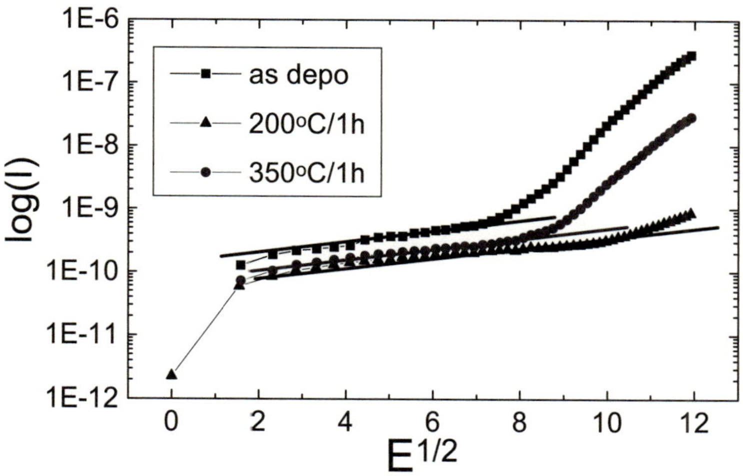

Fig. 8. J-$E^{1/2}$ characteristics of the leakage current for BTS samples, as-deposited and postannealed at 200^{0}C and 350^{0}C (Schottky currents)

In figure 8 it can be observed that the first part of the I-$E^{1/2}$ characteristics can be plotted with a straight line, suggesting that the leakage is mainly due to Schottky currents. Moreover, the plotted lines seem to be almost parallel to each other. Similar result has been obtained by Fukuda while investigating the effects of postannealing in oxygen ambient on leakage properties of (Ba,Sr)TiO_3 thin film capacitors (Fukuda et al., 1998). Because the plotted lines are parallel, all parameters except Φ_B in the equation (2) are almost equal in all cases. Increase in Schottky barrier and, thus, decrease in the leakage can be explained considering the oxygen that was diffused during postannealing treatment. Since, a correlation between the oxygen vacancy in the film and the diffusion coefficient can be made (Fukuda et al., 1997; Fukuda et al., 1998), an increase in the Schottky barrier can be explained by decrease in the oxygen vacancy concentration at the metal-ferroelectric film interface.

Focusing the attention back to table 2, it can be seen that TCD is highest for samples postannealed at 300^{0}C reaching 5.6% at 25^{0}C. Even if the leakage behavior for samples postannealed at 300^{0}C and 350^{0}C is almost similar, we expect a difference in oxygen vacancy concentration due to different oxygen diffusion coefficients.

In order to understand how postannealing at 300^{0}C is improving the value of TCD, the postannealing treatment has been performed in air as well as in high vacuum conditions. In this way the effect of presence of oxygen in the postannealing atmosphere can be better understood. Physical and electrical properties (especially leakage current and TCD versus sample temperature) were again investigated but this time the attention has been focused into noticing any particular differences among samples.

Post-annealing after electrode deposition in air or vacuum was found to have little effect on the BTS XRD peaks, indicating that the crystalline structure is not changed after the post-annealing. AFM observation (not shown here) revealed a root-mean-square (RMS) roughness of 1 to 3 nm.

The chemical change induced by the postannealing in films was obtained after XPS investigations (Figure 9). The attention was focused upon the chemical shifts that were clearly visible in the samples. The peaks were carefully calibrated using the Pt peaks and viewing the carbon peaks for confirmation.

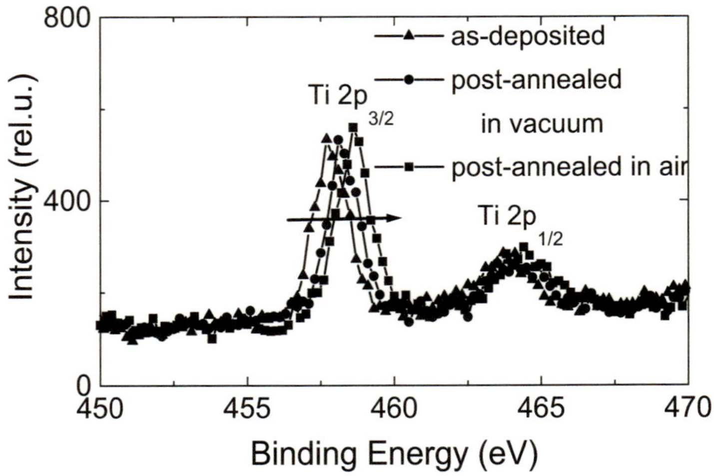

Fig. 9. XPS spectra for BTS thin films as-deposited and postannealed at 300°C for 1 hour in vacuum and air.

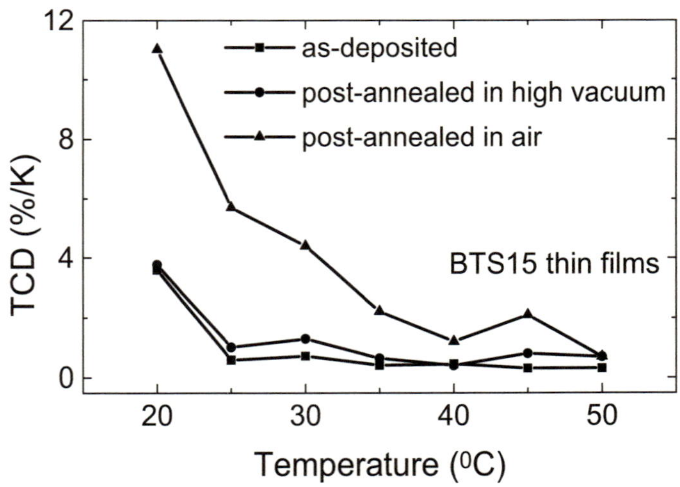

Fig. 10. TCD vs.film temperature for BTS thin films as-deposited and postannealed at 300°C for 1 hour in vacuum and air

The exact binding energy of an electron depends upon the formal oxidation state of the atom from which it was extracted and local chemical and physical environment. The postannealing after electrode deposition was performed in air as well in vacuum to study the influence of diffused oxygen to the chemical properties of the near-surface layer of the BTS thin films. For postannealed films in vacuum or air, the Ti peaks are shifting towards higher binding energies than Ti peaks for as-deposited films. The presence of O_2 in the air can explain why the Ti peaks for the sample postannealed in air are shifting more than the Ti peaks for the sample postannealed in vacuum. Chemically speaking, the presence of O_2 in postannealing atmosphere causes oxygen diffusion into the BTS thin films that will be responsible for the reduction in concentration of the oxygen vacancies near the surface, increasing the oxidation state of the Ti, causing the shift of the Ti peaks position towards higher binding energies in XPS investigations.

An important electrical measurement is the investigation of the temperature dependence of the capacitance (i.e. dielectric constant). Figure 10 shows the TCD behavior for BTS thin films as-deposited and postannealed in air and vacuum. The films post-annealed at 300°C in air have TCD values reaching more than 5.4 %/K at 25^{0}C and 11 %/K at 20^{0}C, which is very high compared with similar reported values for TCD. The improvement in TCD values makes the BTS thin film very promising for realizing highly sensitive dielectric-bolometer mode of infrared sensor.

2.3 DB-mode of infrared sensor using BTS thin films as active materials

Because of the principle of operation, a dielectric-bolometer mode is expected to offer high sensitivity compared with other detectors (Noda et al., 1999; Balcerak, 1999; Radford et al., 1999; Noda et al., 1999). This aspect, along with other advantages offered, such as chopper free device and low operation voltages are good reasons to consider the DB-mode a good choice in fabricating an infrared sensor.

Following the results obtained for ferroelectric BTS thin films, integration into a simple infrared sensing structure will confirm that the BTS can be considered a good candidate for DB-mode of infrared sensing applications.

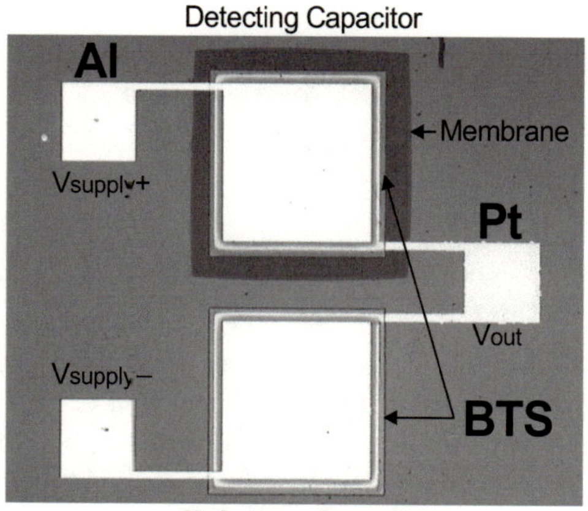

Fig. 11. Picture view of the infrared sensor cell

In order to investigate what are the sensor capabilities of a structure containing BTS thin film as detecting layer, a simple structure was made, containing a simple capacitance ratio sensor that will sense any capacitance difference between detector and reference capacitors. A picture view of the fabricated structure is shown in Figure11.
Fabrication of the structure on silicon was made with the use of silicon micro machining process. The fabrication steps are shown in Figure 12. Only the detector-capacitor is constructed on a membrane, the reference capacitor will stay on $SiO_2/Si_3N_4/SiO_2/Si$ substrate.

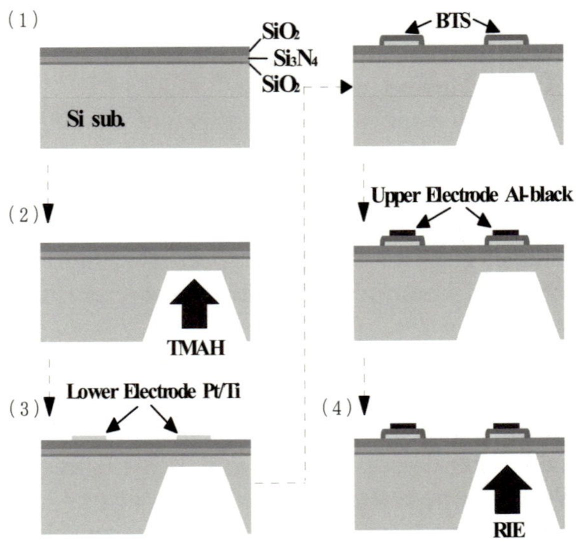

Fig. 12. Process of infrared sensor fabrication

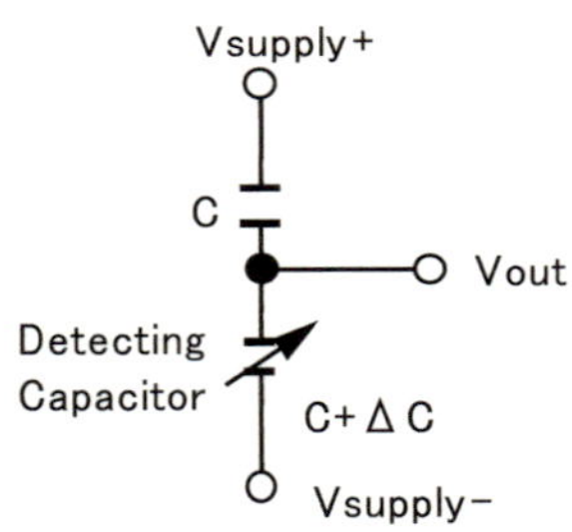

Fig. 13. Detection circuit for infrared measurement

Schematically, an infrared sensing cell can be represented as in Figure 13. A sensing cell is composed of serially connected capacitors. This sensor cell is operating on the principle of sensing the change in the capacitance of the detector-capacitance relative to reference capacitance. Because of the construction, when the sensing cell is exposed to infrared radiation, the temperature at the ferroelectric BTS material site for the detecting capacitor is higher compared with the one for the reference capacitor. Different temperatures are responsible for different dielectric constant values at the detector and reference capacitors that translate into different capacitance values. The variation of the capacitance of the detector-capacitor relative to the value of the capacitance in reference capacitor is detected as a voltage change. Because this voltage signal is very small, amplification is required for the detection.

The infrared response evaluation system is showed schematically in Figure 14. In infrared response evaluation, the temperature of a black body radiator (600^{0}C to room temperature range) is used as source of infrared rays. The infrared rays were focused with germanium lens so that the radiation will fall mainly on the single element sensor. A function generator was used to apply sinusoidal waves with voltage amplitude of 3V, offset of 1.5V and frequency of 1kHz to both capacitors. An almost 180 degree reversal of the phase was used in the capacitors in order to minimize the output signal. When infrared radiation will fall on the detecting capacitor, heating will cause a change in the value of capacitance. This change will affect the "equilibrium" state in the circuit and a V_{out} signal will be detected. The output voltage is then amplified through the band-pass filter of 1 kHz for which lock-in amplifier was substituted and observed as an output waveform with an oscilloscope. Furthermore, using the high-speed Fourier transform (FFT) function built in the oscilloscope, the output signal is extracted.

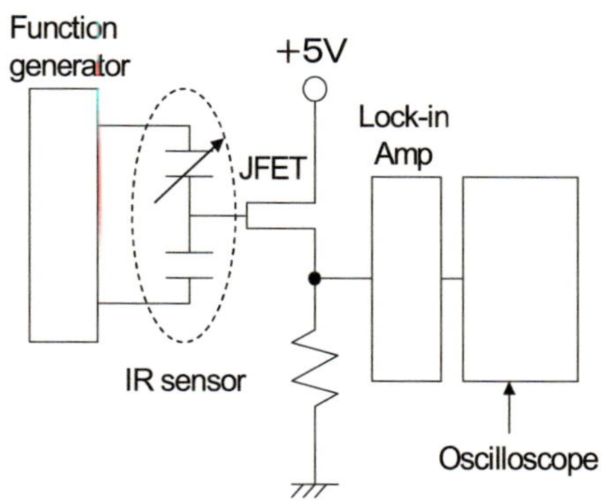

Fig. 14. Infrared response evaluation system

The optimization of the DB operation conditions has to be made before making any comment about the sensing properties of the ferroelectric BTS thin films. Running a set of experiments such as DB output voltage behavior at different applied voltages considering the low leakage behavior of the films at low applied voltages or DB output voltage behavior at different applied frequencies are essential in increasing device sensitivity.

Blackbody temperature dependence of DB output as a parameter of the operation amplitude of supply voltage is showed in Figure 15. For the same sensing cell structure and the same applied frequency, DB output signal is increased by increase in applied voltage amplitude.

Blackbody temperature dependence of DB output as a parameter of the applied frequency of supply voltage is shown in Figure 16.

It can be seen that the DB output level increases with decreasing the frequency of the supplied voltage. The reason for this behavior is considered to be the fact that not the entire voltage amplitude is applied to the series capacitor structure while the frequency is increased.

It can be concluded now that the optimal DB operation conditions are:

1. Larger amplitude of supply voltage. The amplitude should be, however, small enough to ensure small leakage currents through the BTS thin film. 3 to 5 V amplitude for the applied voltage is considered here;
2. Low operation frequency for the applied voltage. 10 or 100Hz is considered in this experiment.

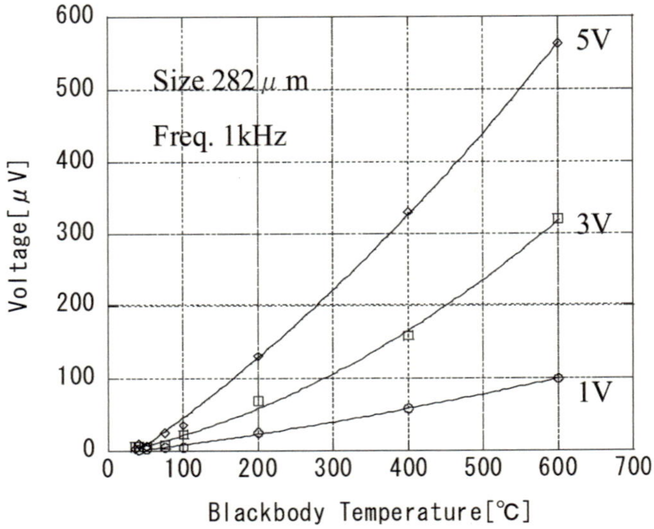

Fig. 15. DB output as a parameter of the operation amplitude of supply voltage

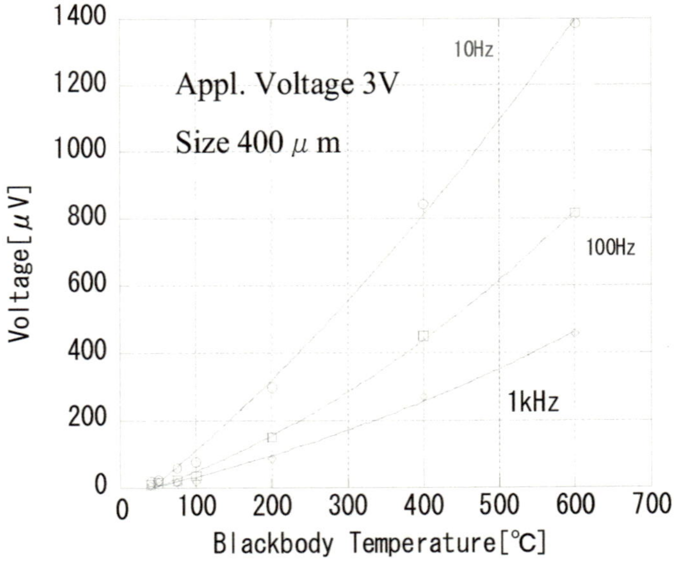

Fig. 16. DB output as a parameter of the operation frequency of supply voltage

As a result of optimization, sensing properties of the BTS ferroelectric thin film can be investigated. The output voltage for infrared sensing cells containing BTS thin films as deposited and postannealed at 300⁰C in air for 60 minutes is shown in Figure 17.

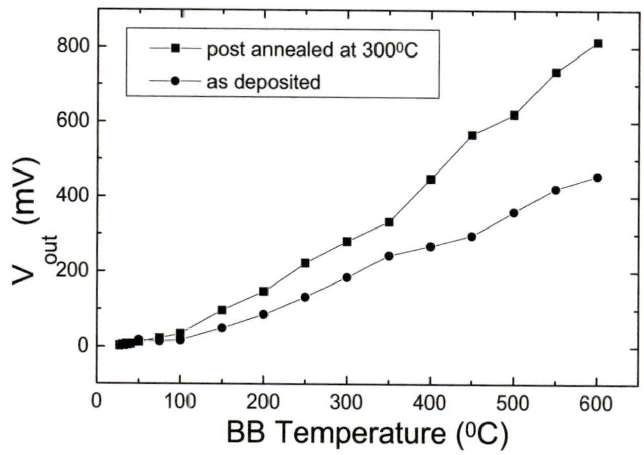

Fig. 17. Output voltage vs. blackbody temperature for sensing cells containing BTS thin films as deposited and postannealed at 300⁰C in air

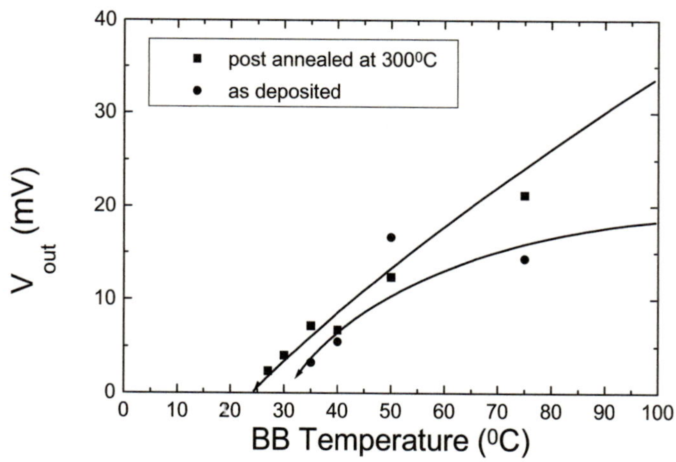

Fig. 18. Output voltage for blackbody temperatures below 100⁰C for sensing cells containing BTS thin films as deposited and postannealed at 300⁰C in air

The importance of postannealing can be clearly seen from this figure. Moreover, a closer look at the response while exposing to the infrared radiations emitted by the black body when its temperature was below 100⁰C (Figure 18) revealed that a stable detection of infrared radiation emitted by a blackbody when its temperature was 27⁰C was successfully obtained. However the output signal is less than expected because 75% of the incident radiation is reflected by the top electrode; only a maximum of 25% of incident radiation will cause the heating in the sensing cell. Voltage responsivity (*Rv*) and specific detectivity (D^*) were calculated to be 0.1 KV/W and $3x10^8$ $cmHz^{1/2}W$, respectively, being in the same range as thin metal oxide film bolometers.

BTS obtained by metal-organic decomposition process can be successfully used as active material in fabrication of DB-mode of infrared sensor. As demonstrated above, temperatures lower than the temperature of a human body can be successfully detected by this type of infrared sensor cell using BTS deposited by metal-organic decomposition process as active material.

3. Preparation and characterization of $(Ba_{0.6},Sr_{0.4})TiO_3$ thick films for application to embedded multilayered capacitor structures

The demand of miniaturization and increased functionality in electronic devices trigered the need of finding ways to increase densification of components on electronic boards and 3D packing. There is a need to improve current technologies or develop new ones in order to cope with the problems that arise with miniaturization and 3D packing. The use of high temperatures during fabrication are not desirable since can trigger unwanted chemical reactions, interdiffusion, shrinkage and/or alteration of electrical properties for the component already present on the circuit board. A relatively new deposition method called the Aerosol Deposition (AD) technique based on room temperature impact consolidation (RTIC) phenomena can be a good alternative in film formation at room temperature (Akedo & Lebedev, 1999; Akedo et al., 1999; Akedo & Lebedev, 2001; Akedo, 2004; Akedo, 2006). In this way, the problems linked with relatively high temperatures needed for film formation using the current (more popular) technologies can be avoided and embedding of dense ceramics into low temperature substrate becomes possible.

As mentioned earlier, barium strontium titanate is an extensively investigated ferroelectric material due to its good electrical properties in bulk and thin film form being a leading candidate for applications in many electronic devices. Barium strontium titanate is currently considered as an attractive material in sensing, memory, capacitor and RF and microwave applications (Kirchoefer et al., 2002; Acikel et al., 2002; Hwang et al., 1995; Zhu et al., 2004; Tissot, 2003). But many important issues, such as improving dielectric constant values, dielectric loss and leakage, still need further attention in order to improve film quality and device performance. Regarding the AD-deposited $(Ba_{0.6},Sr_{0.4})TiO_3$ (BST) films, there are few reports regarding the film particularities. The logical ways to improve film properties are by tempering with film chemical composition, deposition conditions and post-film-formation treatments, and metallization. However, preliminary results have shown that post-film formation annealing is not helpful to improve the properties of the AD-fabricated BST films (Popovici et al., 2009). The substrate is also playing an important role in improving the AD-fabricated BST film properties since a soft substrate is suitable in ensuring that the films are less stressed (Popovici et al., 2008).

For AD process, powder condition is one of the most important factors since humidity, physical characteristics of the particle and particle aggregation are affecting the deposition rate and film properties.

Below, only the results on the investigation regarding the quality of commercially available (raw) powder used in the AD-deposition and improvement by heat treatments to allow the fabrication of $(Ba_{0.6},Sr_{0.4})TiO_3$ (BST60) layers with higher dielectric constants will be discussed due to space constrains in writing this article.

3.1 Aerosol Deposition method as alternative technique in BST thick film deposition

BST60 thick films were grown by the AD technique on Cu substrates using raw and thermally treated powders. To investigate the effect of powder thermal treatment on AD-fabricated BST60 thick films properties, powder from the same lot has been thermally treated for 1 h at 800 or 900^{0}C in O_2 atmosphere.

The AD system used in film fabrication is represented schematically in figure 19. Powder aerosols are formed by oxygen flowing in the vacuum powder chamber at a rate of 4 l/min and transported through connecting tubes to the vacuum deposition chamber where the particle are ejected through the nozzle toward a moving substrate for deposition. A schematic representation of the consolidation process by AD is shown in figure 20. During AD deposition, the particles will suffer a plastic like deformation and fracture upon impact with the substrate. This plastic like deformation and the fracture of the impacting particles are essential to ensure the formation of very dense AD films.

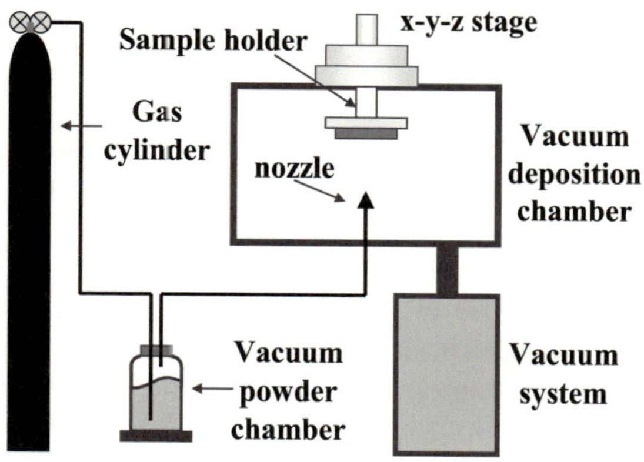

Fig. 19. Schematic representation of AD system

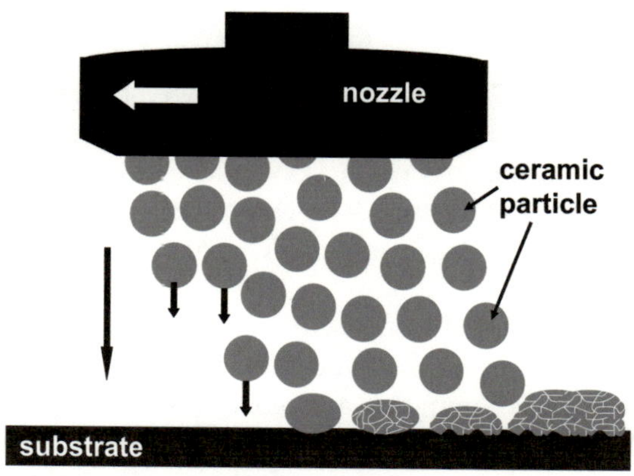

Fig. 20. Schematics of consolidation process by AD

Before being used in AD, the powder is optimized by ball milling to allow fabrication of high quality films at high deposition rates and with minimum consumption of powder. The average particle size after ball milling should fall within the optimum range (considered to be 0.7-1.4 μm for the BST powder). The thickness of the fabricated BST60 films was around 3 μm and their density was estimated to be in the 92-93% range from the theoretical density, being independent of the powder condition. To examine the electrical properties of the films, Pt/Ti electrodes were deposited by RF sputtering on BST60 thick films to form capacitor structures.

3.2 AD-fabricated BST thick film properties

In BST powder synthesis, there are a number of reports that suggest that the presence of $BaCO_3$ in BT and BST powders is difficult to prevent whatever the fabrication route is used (Henningh & Mayr, 1978; Coutures et al., 1992; Hennings & Schreinemacher, 1992; Stockenhuber et al., 1993; Lemoine et al., 1994; Ries et al.,2003). Moreover, there are reports that suggest that BT is thermodynamically unstable in H_2O having a pH below 12 (Lencka & Riman, 1993; Abicht et al., 1997; Voltzke et al.,1999). The BST is expected to show a similar problem since it is a BT-based material. The most probable chemical reaction with water is shown below:

$$BaTiO_3(s)+9H_2O \rightarrow Ba(OH)_2+8H_2O+TiO_2(s). \quad (4)$$

The formed solid TiO_2 (amorphous) will remain in the outer shell of the initial BT (or BST) particle and will act as a barrier in the further removal of Ba by water (Voltzke et al., 1999). Upon air exposure, $Ba(OH)_2$ will react as follows:

$$Ba(OH)_2+CO_2(air) \rightarrow BaCO_3(s)+H_2O \quad (5)$$

The instability of strontium titanate (STO) material in water is not confirmed, therefore, only, only the instability of Ba^{2+} ions in H_2O is considered here.

Whatever the reasons for the presence of $BaCO_3$ as a secondary phase in BST and BT powders, $BaCO_3$ formation must be controlled to ensure the fabrication of BST or BT films with the desired properties since AD is a room temperature process and post-film-formation thermal treatments at elevated temperatures are not reccomended.

X-ray photoelectron spectroscopy (XPS) has been used to clarify the presence of the secondary phase in the powder and films and the effect of powder annealing. The powder specimens were prepared on Al plate using a commercial double-sided adhesive tape on which the powder adhered. The tape was well covered with powder to avoid the occurrence of tape-related peaks in the XPS spectra. For calibration purpose and to avoid charging due to electron photoemission, a very thin layer of Au was deposited on the surface of the samples by RF sputtering. Six elements were detected on the surface of the investigated samples: C, Au, O, Ba, Sr and Ti. The Au peaks were used to calibrate the XPS profiles.

The XPS profiles of the C 1s peaks of raw powder, the AD-fabricated film obtained from this powder, and powder recovered from the deposition chamber are shown in figure 21. The C 1s peak located near 284.8 eV is commonly attributed to C-C and C-H bonds. The C 1s peak located near 288.45 eV is assumed to be correlated with the C state in CO_3^{2-} of $BaCO_3$ (Viviani et al, 1999) since the other possible chemical states for carbon, C-O and CO_2, should reveal peaks located at binding energies that are higher with approximately 2 eV (Viviani et

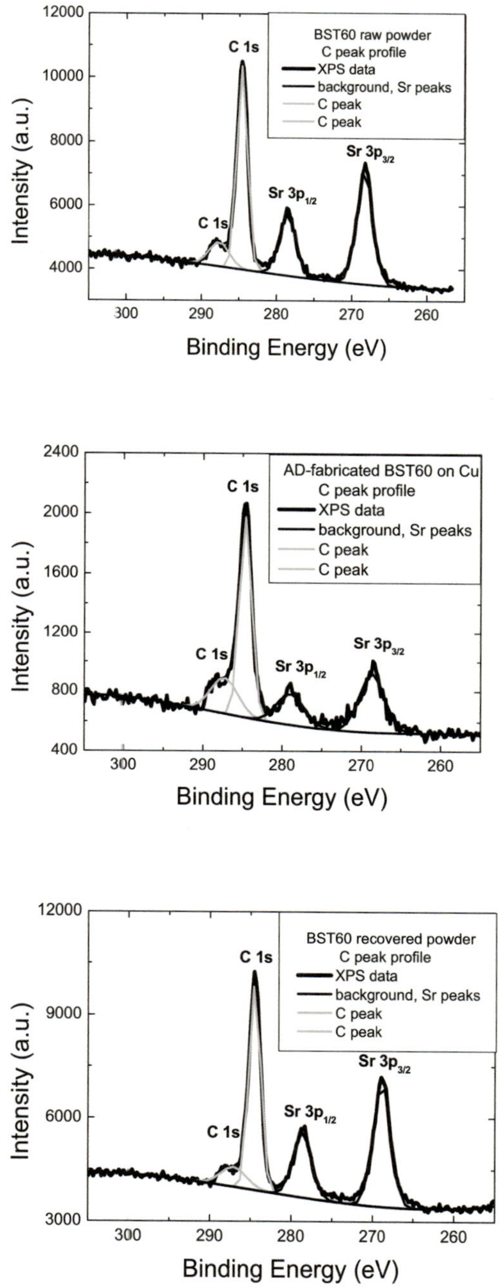

Fig. 21. XPS C1s peak profile for raw and annealed powders

al, 1999) and 7 eV (Wagner et al.,1979), respectively, than those of the reference C 1s peak. Comparing the relative intensities of the CO_3^{2-}-related C 1s peak (relative to Sr $3p_{3/2}$ peak) for the raw powder, AD-fabricated BST60 film, and powder recovered from the deposition chamber it can be concluded that the relative intensity of the carbonate phase is higher in the AD-fabricated film than in the powder.

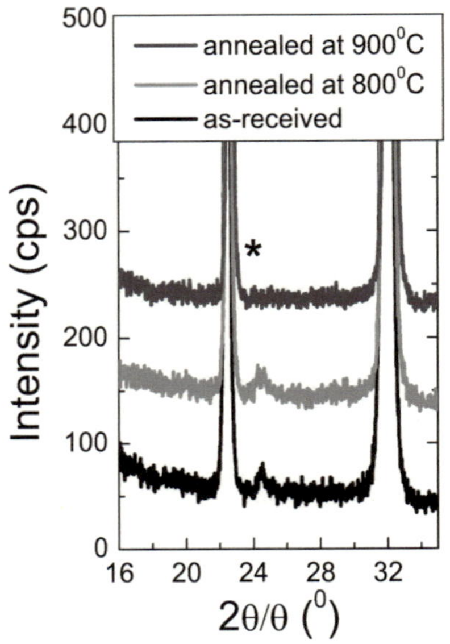

Fig. 22. XRD profiles of BST raw and thermally treated powders

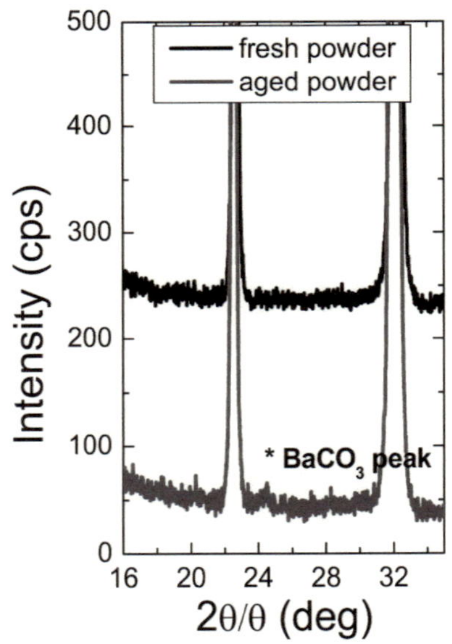

Fig. 23. XRD profiles of 900⁰C treated powder immediately after treatment and after one year

The XRD profiles of the as-received and 800 and 900⁰C thermally treated powders show that the crystallinity is retained in all films. However, a closer observation of the XRD pattern near 2theta=24⁰ revealed the presence of an additional peak that can be assigned to the orthorhombic $BaCO_3$ phase (figure 22). Since the concentraton of $BaCO_3$ second phase is high enough to surpass the sensitivity limit of the X-ray diffraction system, it can be also assumed that a peak related to this phase will also appear in XPS results. On this ground, it has been assumed that the $BaCO_3$ second phase will be responsible for the appearance of a relatively high intensity peak in the C 1s XPS peak profile of the BST60 powder, peak that should be located near 289 eV.

This is another reason for linking the peak located at 288.45 eV in C 1s XPS profile to the C state in CO_3^{2-} of the $BaCO_3$ second phase. It should be notted that for powders thermally treated at 900^{0}C for 1 h, peaks generated by the presence of the second carbonate phase were not observed in the XRD profile, suggesting that this temperature is suitable for the removal of the secondary phase in the BST60 powders. To further test the BST powder instability against humidity and CO_2 in air some treated powder was intentionally placed in atmospheric conditions for 1 year. In figure 23, the XRD patterns near 2theta=24^0 are shown for freshly treated powder and powder aged in air for 1 year away from dust. The carbonate peak becomes visible again suggesting that humidity and CO_2 from air were sufficient to trigger Ba^{2+} ion removal from BST60 powders and formation of $BaCO_3$ in the outershell of BST particles.

The effect of powder thermal treatment on the physical and electrical properties of the AD-fabricated BST60 films was also analyzed. As shown in figure 24, the dielectric constant of AD-fabricated films using 900^{0}C treated powders is highest among the investigated films being close to 200 for a wide frequency range. The dielectric loss in all samples was below 0.06 and no marked changes in this parameter were observed. Since the grain size is similar in all the films, the difference in dielectric constant is not due to its dependence on grain size.

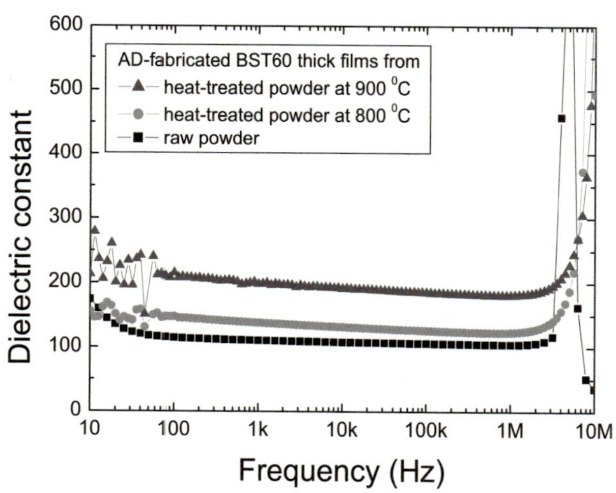

Fig. 24. Frequency dependence of dielectric constant for AD BST thick films deposited from raw and thermally treated powders

Due to the unique way deposition of films take place in AD, the material in the outershell of the crystalline particles participating in the consolidation process will always be found to form grain boundaries in the as-deposited AD films. The increase in dielectric constant can be correlated with the improvement of the grain boundary regions since, by minimizing the ammount of the secondary phase, the low dielectric constant carbonate phase will be less present at the grain boundary. Moreover, the leakage in AD films is improved by annealing the powder at 900⁰C (figure 24).

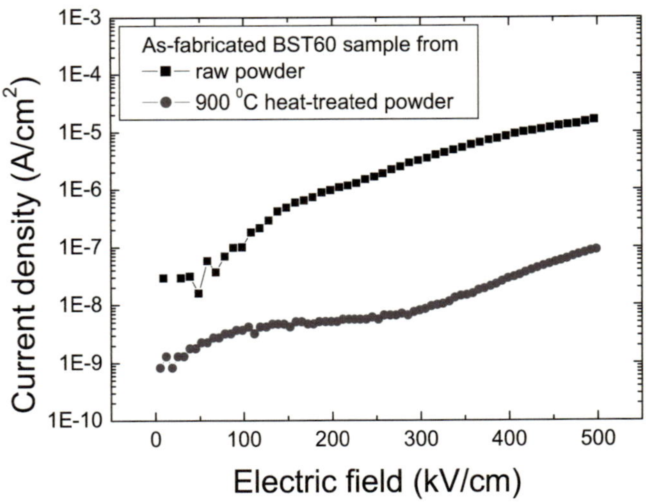

Fig. 25. Leakage of BST films deposited from raw and 900⁰C thermally treated powders

For the AD-fabricated BST60 thick films obtained from 900⁰C treated powders, the leakage currents stay below 10^{-7} A/cm^2 for applied electric fields up to 500 kV/cm. A reduction in leakage current of one or two orders of magnitude is observed for the AD-fabricated films obtained from 900⁰C treated powders as compared with the films fabricated from raw powders. This is another indication that the grain boundaries of the BST60 films were modified following the thermal treatment of the powders at 900⁰C prior to deposition. The removal of the secondary phase from the grain boundary regions by thermally treating the powder at 900⁰C increases the resistivity of the grain boundary regions, reducing the leakage currents through the grain boundary.

Thermal treatment of commercially available BST powder at 900⁰C is a good approach to increase the overall performance of the AD-fabricated BST thick films deposited at room temperature and to make them more attractive for embedded multilayer capacitor applications.

4. Conclusions

BT-based materials represent a class of materials with a wide range of applications. Here, we showed how substitution in A or B site can make these materials suitable for different applications. Two examples of how BT-based materials can show potential in specific applications are discussed.

$Ba(Ti_{0.85},Sn_{0.15})$ (BTS) ferroelectric thin films have been prepared by metal-organic decomposition (MOD) technique. Annealing and postannealing temperatures were optimized to obtain films with suitable electrical properties for application in DB-mode of infrared sensing. Annealing at 700ºC has as the result the minimization of the leakage current and the dielectric loss. Also, the capacitance of a capacitor containing BTS thin film crystallized at this temperature will decrease more rapidly with an increase in film temperature than on a similar capacitor made with BTS thin films annealed at 800ºC. Considering the results, annealing at 700ºCis suitable to fabricate with good properties for application to DB-mode of infrared sensing. Applying a postannealing treatment to the capacitors after top-electrode deposition can improve further the electrical properties of the BTS thin films. It has been found that, postannealing at 300ºC in air for 60 minutes, even if the leakage is higher than in the case of postannealing at 200ºC, will increase the value of the temperature coefficient of dielectric constant (TCD) from 1 %/K to 5.6 %/K at 25ºC comparing with only 1.3 %/K for films postannealed at 200ºC. A closer look on the leakage current behavior on BTS films postannealed in air reveals that an increase in postannealing temperature will reduce the oxygen diffusion from the air into the films that translates as higher leakage currents for samples postannealed at temperatures higher than 200ºC than on the samples postannealed at 200ºC. However these values are still smaller than that of as-deposited BTS thin films. TCD values are higher for samples postannealed at 300ºC suggesting that some degree of oxygen deficiency in the film is needed in order to obtain satisfactory values for TCD. A close investigation regarding the importance of postannealing at 300ºC revealed the important role played by the oxygen vacancies to the value of TCD. Postannealing in air-free environment will not do much improvement to the electrical properties of the film except a relatively small increase in polarization observed in P-E hysteresis loops. On the other hand, postannealing in air will promote oxygen diffusion into the film and, as a result, a change in electrical properties of the dead-layer and a change in the lattice parameters of the crystalline BTS thin films. It can be observed that postannealing at 300ºC for 60 minutes is an important condition in order to fabricate BTS thin films suitable for DB-mode of infrared sensing. The results obtained after BTS film investigations were used in the fabrication of a simple-structures infrared sensing cell. The cell consist in a series of two capacitors, one used as reference capacitor and the other, fabricated on a membrane to reduce the thermal loss, used as detector-capacitor.After optimization of the BD operation mode (application of sinusoidal waves with a voltage amplitude of 3 to 5V and a frequency between 10Hz and 100Hz) sensing properties of the films were revealed. A stable infrared detection was possible even for objects (in this case a black body) heated at temperatures of 27ºC. Good figures-of-merit such as voltage responsivity (Rv) of 0.1 KV/W and specific detectivity (D^*) of $3x10^8$ $cmHz^{1/2}W$ were also calculated making BTS material a strong candidate for application in DB-mode of infrared sensing.

$(Ba_{0.6},Sr_{0.4})TiO_3$ (BST60) thick films were fabricated on Cu substrates by Aerosol Deposition (AD) method. The quality of the raw powder has been checked and optimized in order to increase the dielectric constant of the fabricated films without the need of post-film-formation annealing procedure. Carbonate phase has been observed in the raw powders and it was successfully reduced by thermally treating the powder at 900ºC. The AD-fabricated films obtained from the 900ºC treated powder show a dielectric constant of 200 being much higher that the dielectric constant of the AD-films obtained from the as-received powders. The leakage currents in the films fabricated from 900ºC treated powders stay below 10^{-7} A/cm^2 when the applied electric filed is less than 500 kV/cm and it is at least one order of magnitude smaller than for films obtained from as-received powders. The above results indicate that thermally treating the powder at 900ºC is a good way to improve the AD-fabricated BST60 thick films electrical properties. This results represent a step forward in our goal of ceramic fabrication at room temperature aiming integration into embedded multilayered ceramic capacitor structures in electronic devices.

5. Acknowledgment

The first author acknowledge the support of Japanese Government Scholarship (Monbukagakusho) program in making possible his research and study at Osaka University. This research has also been partially supported by the NEDO project on "The next generation MEMS (Fine MEMS) project" in Japan.

6. References

Mueller, V.; Beige, H. & Abicht, H-P. (2004). Non-Debye dielectric dispersion of barium titanate stannate in the relaxor and diffuse phase-transition state. *Applied Physics Letters*, Vol. 84, Issue 8, (2004), pp. 1341-1343, ISSN 1077-3118

Lu, S.G.; Xu, Z.K. & Chen, H. (2004). Tunability and relaxor properties of ferroelectric barium stannate titanate ceramics. *Applied Physics Letters*, Vol. 85, Issue 22, (2004), pp. 5319-5321, ISSN 1077-3118

Yasuda, N.; Ohwa H. & Asano S. (1996). Dielectric Properties and Phase Transitions of $Ba(Ti_{1-x}Sn_x)O_3$ Solid Solution. *Japanese Journal of Applied Physics*, Vol. 35, (1996), pp. 5099-5103, ISSN 1347-4065

Xiaoyong, W.; Yujun, F. & Xi, Y. (2003). Dielectric relaxation behavior in barium stannate titanate ferroelectric ceramics with diffused phase transition. *Applied Physics Letters*, Vol. 83, Issue 10, (2003), pp.2031-2033, ISSN 1077-3118

Jiwei, Z.; Bo, S.; Xi, Y. & Liangying, Z. (2004). Dielectric and ferroelectric properties of $Ba(Sn_{0.15}Ti_{0.85})O_3$ thin films grown by a sol-gel process, *Materials Research Bulletin*, Vol. 39 (September 2004), pp. 1599-1606, ISSN 0025-5408

Fukuda, Y.; Haneda, H.; Sakaguchi, I.; Numata, K.; Aoki, K. & Nishimura, A. (1997). Dielectric Properties of $(Ba,Sr)TiO_3$ Thin Films and their Correlation with Oxygen Vacancy Density. *Japanese Journal of Applied Physics*, Vol 36 (1997), pp. L1514-L1516, ISSN 1347-4065

Sze, S.M. (1981). *Physics of semicondictor devices*, Wiley-Interscience, New-York, USA, ISBN 0-471-05661-8

Fukuda, Y.; Numata, K.; Aoki, K. ; Nishimura, A.; Fujihashi, G.; Okamura, S.; Ando, S. & Tsukamoto, T. (1998). Effects of Postannealing in Oxygen Ambient on Leakage Properties of $(Ba,Sr)TiO_3$ Thin-Film Capacitors. *Japanese Journal of Applied Physics*, Vol. 37 (1998), pp. L453- L455, ISSN 1347-4065

Noda, M.; Hashimoto, K.; Kubo, R.; Tanaka, H.; Mukaigawa, T.; Xu, H. & Okuyama, M. (1999), A new type of dielectric bolometer mode of detector pixel using ferroelectric thin film capacitors for infrared image sensor. *Sensors and Actuators A*, Vol. 77, (Septemper 1999), pp. 39-44, ISSN 0924-4247

Balcerak, R.S. (1999). Uncooled IR imaging: technology for the next generation. *Proceedings of the 25th SPIE Conference on Infrared Technology and Applications*, Vol. 3698, pp. 110-118, ISBN 9780819431721, Orlando, FL, USA, April 5-9, 1999

Radford, W.; Murphy, D.; Finch, A.; Hay, K.; Kennedy, A.; Ray, M.; Sayed, A.; Wyles, J.; Wyles, R. & Varesi, J. (1999). Sensitivity improvements in uncooled microbolometer FPAs. *Proceedings of the 25th SPIE Conference on Infrared Technology and Applications*, Vol. 3698, pp. 119-130, ISBN 9780819431721, Orlando, FL, USA, April 5-9, 1999

Noda, M.; Mukaigawa, T.; Hashimoto, K.; Kiyomoto, T.; Xu, H.; Kubo, R.; Tanaka, H.; Usuki, T. & Okuyama, M. (1999). Simple detector pixel of dielectric bolometer mode and its device structure for an uncooled IR image sensor. *Proceedings of the 25th SPIE Conference on Infrared Technology and Applications*, Vol. 3698, pp. 565-573, ISBN 9780819431721, Orlando, FL, USA, April 5-9, 1999

Noda, M. ; Inoue, K. ; Ogura, M. ; Xu, H. ; Murakami, S. ; Kishihara, H. & Okuyama, M. (2002). An uncooled infrared sensor of dielectric bolometer mode using a new detection technique of operation bias voltage. *Sensors and Actuators A*, Vol. 97-98, (April 2002), pp. 329-336, ISSN 0924-4247

Akedo, J. & Lebedev, M. (1999). Microstructure and Electrical Properties of Lead Zirconate Titanate ($Pb(Zr_{52}/Ti_{48})O_3$)Thick Films Deposited by Aerosol Deposition Method. *Japanese Journal of Applied Physics*, Vol. 38, (1999), pp. 5397-5401, ISSN 1347-4065

Akedo, J.; Minami, N.; Fukuda, K.; Ichiki, M. & Maeda, R. (1999). Electrical properties of direct deposited piesoelectric thick film formed by gas deposition method annealing effect of the deposited films. *Ferroelectrics*, Vol. 231 (1999), pp. 285-292, ISSN 1563-5112

Akedo, J. & Lebedev, M. (2001). Influence of Carrier Gas Conditions on Electrical and Optical Properties of $Pb(Zr,Ti)O_3$ Thin Films Prepared by Aerosol Deposition Method. *Japanese Journal of Appied. Physics*, Vol. 40, (2001), pp. 5528-5532, ISSN 1347-4065

Akedo, J. (2004). Aerosol Deposition Method for Fabrication of Nano Crystal Ceramic Layer, *Materials Science Forum*, Vol. 449-452, (2004), pp. 43-48, ISSN 1662-9752

Akedo, J. (2006). Aerosol Deposition of Ceramic Thick Films at Room Temperature: Densification Mechanism of Ceramic Layers. *Journal of the American Ceramic Society*, Vol. 89, Issue 6, (June 2006), pp. 1834-1839, ISSN 1551-2916

Kirchoefer, S.W. ; Cukauskas, E.J.; Barker, N.S.; Newman, H.S. & Chang, W. (2002). Barium-strontium-titanate thin films for application in radio-frequency-microelectromechanical capacitive switches. *Applied Physics Letters*, Vol. 80, Issue 7, (2002), pp. 1255-1257, ISSN 1077-3118

Acikel, B.; Taylor, T.R.; Hansen, P.J.; Speck, J.S. & York, R.A. (2002). A new high performance phase shifter using $Ba_xSr_{1-x}TiO_3$ thin films. *IEEE Microwave and Wireless Components Letters*, Vol. 12, Issue 7, (2002), pp. 237-239, ISSN 1531-1309

Hwang, C.S; Park, S.O.; Cho, H.J.; Kang, C.S.; Lee S.I.; & Lee, M.Y. (1995). Deposition of extremely thin $(Ba,Sr)TiO_3$ thin films for ultra-large-scale integrated dynamic random access memory application. *Applied Physics Letters*, Vol. 67, Issue 19, (1995), pp. 2819-2821, ISSN 1077-3118

Zhu, H.; Miao, J.; Noda, M. & Okuyama, M. (2004). Preparation of BST ferroelectric thin film by metal organic decomposition for infrared sensor. *Sensors and Actuators A*, Vol. 110, Issues 1-3, (February 2004), pp. 371-377, ISSN 0924-4247

Tissot, J.L. (2003). Uncooled focal plane infrared detectors: the state of the art (in French). *Comptes Rendus Physique*, Vol. 4, Issue 10, (December 2003), pp. 1083-1088, ISSN 1631-0705

Popovici, D.; Tsuda, H. & Akedo, J. (2009). Postdeposition annealing effect on (Ba0.6,Sr0.4)TiO3 thick films deposited by aerosol deposition method. *Journal of Applied Physics*, Vol. 105, Issue 6, (March 2009), pp. 061638-1-061638-5, ISSN 1089-7550

Popovici, D.; Tsuda, H. & Akedo, J. (2008). Fabrication of $(Ba_{0.6},Sr_{0.4})TiO_3$ Thick Films by Aerosol Deposition Method for Application to Embedded Multilayered Capacitor Structures. *Japanese Journal of Applied Physics*, Vol. 47, (September 2008), pp. 7490-7493, ISSN 1347-4065

Hennings, D. & Mayr, W. (1978), Thermal decomposition of (BaTi) citrates into barium titanate. *Journal of Solid State Chemistry*, Vol. 26, Issue 4, (December 1978), pp. 329-338, ISSN 0022-4596

Coutures, J.P.; Odier, P. & Proust, C. (1992). Barium titanate formation by organic resisns formed with mixed citrate. *Journal of Materials Science*, Vol. 27, No. 7, (1992), pp. 1849-1856, ISSN 1573-4803

Hennings, D. & Schreinemacher, S. (1992). Characterization of hydrothermal barium titanate. *Journal of European Ceramic Society*, Vol. 9, Issue 1, (1992), pp. 41-46, ISSN 0955-2219

Stockenhuber, M.; Mayer, H. &. Lercher, J.A (1993). Preparation of Barium Titanates from Oxalates. *Journal of American Ceramic Society*, Vol. 76, Issue 5, (May 1993), pp. 1185-1190, ISSN 1551-2916

Lemoine, C.; Gilbert, B.; Michaux, B.; Pirard, J.P. & Lecloux, A.J. (1994). Synthesis of barium titanate by the sol-gel process. *Journal of Non-Crystalline Solids*, Vol. 175, Issue 1, (September 1994), pp. 1-13, ISSN 0022-3093

Ries, A.; Simoes, A.Z.; Cilense, M.; Zaghete, M.A. & Varela, J.A. (2003). Barium strontium titanate powder optained by polymeric precursor method. *Materials Characterization*, Vol. 50, Issues2-3, (March 2003), pp. 217-221, ISSN 1044-5803

Lencka, M.M. & Riman, R.E. (1993). Thermodynamic modeling of hydrothermal synthesis of ceramic powders. *Chemistry of Materials*, Vol 5, Issue 1, (January 1993), pp. 61-70, ISSN 1520-5002

Abicht, H.-P.; Voltzke, D.; Roder, A.; Schneider, R. & Voltersdorf, J. (1997). The influence of the milling liquid on the properties of barium titanate powders and ceramics. *Journal of Materials Chemistry*, Vol. 7, Issue 3 (1997), pp. 487-492, ISSN 1364-5501

Voltzke, D.; Gablenz, S.; Abicht, H.-P.; Schneider, R.; Pippel, E. & Woltersdorf, J. (1999). Surface modification of barium titanate powder particles. *Materials Chemistry and Physics*, Vol. 61, Issue 2, (October 1999), pp. 110-116, ISSN 0254-0584.

M. Viviani, M.; Buscaglia, M.T.;. Nanni, P; Parodi, R.; Gemme, G. & Dacca, A. (1999). XPS investigation of surface properties of $Ba_{(1-x)}Sr_xTiO_3$ powders prepared by low temperature aqueous synthesis. *Journal of the European Ceramic Society*, Vol.19, Issues 6-7, (June 1999), pp. 1047-1051, ISSN 0955-2219

Wagner, C.D.; Riggs, W.M.; Davis, L.E. & J.F. Moulder, J.F. (1979). *Handbook of X-ray photoelectron spectroscopy*, ed. G.E. Muilenberg (Perkin-Elmer Corporation, Physical Electronics Division 1979), p.38

15

Lead-Free Ferroelectric Ceramics with Perovskite Structure

Rigoberto López-Juárez[1], Federico González[2] and María-Elena Villafuerte-Castrejón[1]
[1]*Instituto de Investigaciones en Materiales, Universidad Nacional Autónoma de México*
[2]*Departamento de Ingeniería de Procesos e Hidráulica, Universidad Autónoma Metropolitana-Iztapalapa México*

1. Introduction

Ferroelectric ceramics were discovered in the 1940s in polycrystalline barium titanate (von Hippel et al., 1946; Wul & Goldman, 1945), since then, there has been a continuous succession of new materials and technology developments that have led to a significant number of industrial and commercial applications.

Structurally speaking there are four types of ferroelectric ceramics: (1) perovskites, (2) the tungsten-bronze group, (3) pyrochlores and (4) the bismuth layer-structure group. Of these, the perovskites (ABO_3) are by far the most important category. The families with composition $BaTiO_3$, $PbZr_{1-x}Ti_xO_3$ (PZT), PZT:La (PLZT), $PbTiO_3$ (PT), $Pb(Mg_{1/3}Nb_{2/3})O_3$ (PMN) and $(K_{0.5}Na_{0.5})NbO_3$ (KNN) represents most of the ferroelectric ceramics manufactured in the world (Haertling, 1999).

In this chapter the structure of calcium titanium oxide ($CaTiO_3$), the ferroelectrics ceramics $BaTiO_3$, $Na_{0.5}Bi_{0.5}TiO_3$ (NBT), $K_{0.5}Bi_{0.5}TiO_3$ (KBT) are described as well the concept of hysteresis loop, ferroelectric domains and why lead free materials are now in the top of the interest in ferroelectric and piezoelectric materials. The aim of this chapter is to present results of the synthesis, characterization and piezoelectric properties of two lead free piezoelectric compounds: $K_{0.5}Na_{0.5}NbO_3$ and $(K_{0.48}Na_{0.52})_{0.96}Li_{0.04}Nb_{0.85}Ta_{0.15}O_3$.

1.1 Perovskite structure

The mineral perovskite is calcium titanate, with chemical formula $CaTiO_3$, its ideal structure has space group *Pm-3m*. Most of the commercially important ferroelectric materials have perovskite related crystal structure. The family of the perovskite oxides has generic composition ABO_3, where A is 12 fold coordinated with respect to oxygen (Fig. 1c) and B is octahedrally coordinated by oxygen (Fig. 1a and 1b). The A site is at the corner of the cube, the B site is at the center, and there is an oxygen at the middle of each face. Alternatively, the structure could be represented with the B site at the corner, the A site at the center and O ions at the midpoint of each edge, respectively.

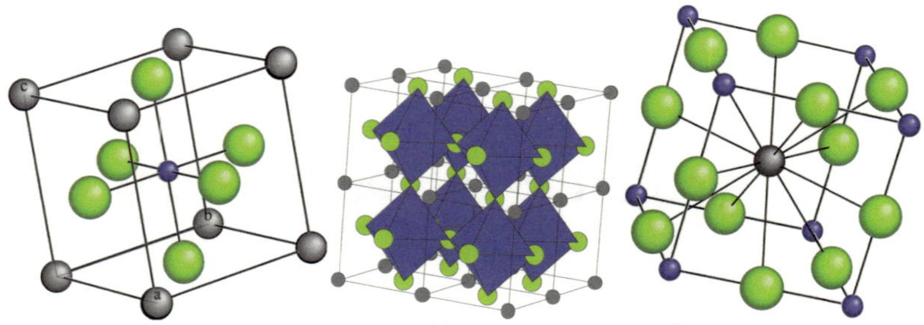

a) b) c)

Fig. 1. The unit cell of the ABO_3 ideal cubic perovskite.

The perovskite type structure is enormously tolerant to variations in composition and distortions due to its ability to adapt a mismatch between the equilibrium A-O and B-O bond lengths, allowing the existence of a large number and variety of stoichiometric compounds. Those distortions, for instance tetragonal (Fig. 2), orthorhombic, rhombohedral and monoclinic, give rise to changes in the crystal symmetry, and one or more cations shift from high-symmetry positions in the lattice, producing ferroelectric or antiferroelectric behavior. In other words, the center of positive and negative charge within the unit cell is no longer coincident, which is the origin of the spontaneous polarization. However, in a ferroelectric material the spontaneous polarization is necessary but not sufficient, since it also requires the reorientation of the polarization by an electric field.

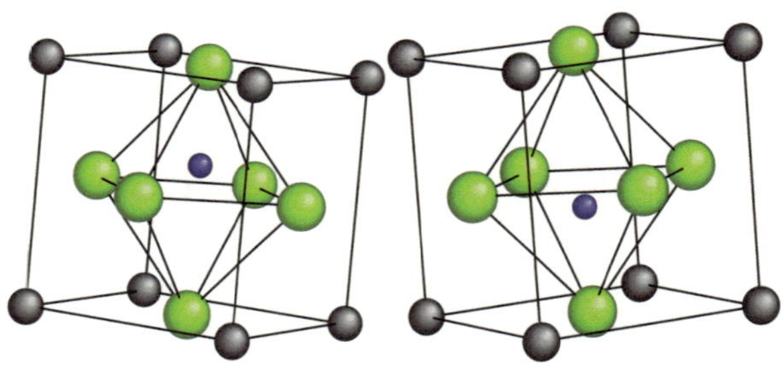

Fig. 2. Tetragonal ferroelectric distortion of the perovskite structure, illustrating two polarization states.

2. Some characteristics of ferroelectric materials

2.1 Hysteresis loop: the fingerprint of ferroelectricity

As mentioned-above, a distinctive feature of ferroelectricity is the reorientation of the polarization by an electric field. Thus, the observation of some evidence of switching is fundamental to establish the ferroelectricity. The experimental evidence is given by the electric hysteresis loop; actually, the term ferroelectric was coined in analogy with the similar magnetic loop *M-H* (magnetization versus magnetic field) obtained from a ferromagnetic material, with the obvious exception that iron is not necessarily present in a ferroelectric. In its standard form, the *P-E* (polarization versus electric field) hysteresis loop is symmetric and the remnant polarization and coercive field are straightforwardly determined. The remnant polarization is the saturation polarization at zero field, and the coercive field, if the complete loop is determined, is the field value at zero polarization. It is crucial to be aware of the potential artifacts associated with the measurement of *P-E* loops (Scott, 2008). These loops must show saturation and have a concave region in *P* versus *E* for being considered satisfactory.

2.2 Ferroelectric domains

The volume regions of the material with the same polarization orientation are referred to as ferroelectric domains. When the sample is under zero field and strain-free conditions, all the domain states have the same energy; but if an electric field is applied, the free energy of the system is lowered by aligning the polarization along the field. Thus, large applied electric fields can permanently reorient the polarization between the allowed domain states, which are restricted by crystallography. As a result, even ceramics, constituted by polycrystals randomly oriented can be electrically poled to produce net piezoelectric coefficients. Much of the importance of ferroelectric materials is due to their properties, leading to a wide range of applications. Among these applications are high dielectric constant capacitors, piezoelectric sonar, ultrasonic transducers, ultrasonic motors, actuators and pyroelectric detectors. Special mention is reserved for the ferroelectric memories, field effect and cooling devices.

2.3 A way to improve the electromechanical properties of ferroelectric ceramics: morphotropic phase boundary (MPB) and polymorphic phase transition (PPT)

In analogy to the characteristics of the PZT ($PbZr_{1-x}Ti_xO_3$) phase diagram, which presents a MPB between tetragonal and trigonal phases (Jaffe, 1971) (which means literally *the boundary between two forms*), where the electromechanical properties exhibit an outstanding behavior, a lot of work has been conducted in different ferroelectric ceramic systems in order to form MPBs. The renaissance of the issue was initiated with the finding of Noheda of a monoclinic phase which acts as a bridge between the trigonal and tetragonal phases in the PZT system (Noheda et al., 1999). Generally speaking, the enhancement of electromechanical properties is due to the larger number of possible polarizable directions in the monoclinic phase. Furthermore, enhancement of electromechanical properties has been observed in PPTs, they are temperature-dependent phase transitions, in contrast to the MPB which is composition-dependent and almost vertical. At the PPTs the electromechanical properties are improved. In general, PPTs are above room temperature; therefore, some research has the aim to modify the materials by the addition of dopants in order to shift PPT´s to room temperature. At the PPT´s, the increased polarizability associated with the transition leads to increased dielectric and piezoelectric properties (Guo et al., 2004).

2.4 The environmental issue: lead-free based materials

The most widely used ferroelectric ceramics are those based on the $PbTiO_3$-$PbZrO_3$ solid solution, generically called PZT. The PZT is composed of about 60 wt.% of lead, which rises ecological concerns; thus, some countries have legislated to replace this material by lead-free ceramics (European commission, 2008) since lead is a toxic element that affects the human health and the environment. Consequently, in recent years diverse systems are being investigated, among them, barium titanate (Yoon et al., 2007), bismuth-alkaline metal titanates and niobates (Hao et al., 2005; Jing et al. 2003; Ma et al., 2006), especially the $K_{0.5}Na_{0.5}NbO_3$ solid solution abbreviated KNN (Du et al., 2006; Saito et al., 2004).

3. Important lead-free ferroelectric ceramics with perovskite structure

3.1 $BaTiO_3$

The first oxide with perovskite-type structure exhibiting ferroelectric behavior was $BaTiO_3$ (BTO) (von Hippel et al., 1946; Wul & Goldman, 1945). It played a major role in demonstrating that ferroelectric ceramics had piezoelectric response through the poling process. At these days, the prevailing opinion was that ceramics could not be piezoelectrically active, because the randomly oriented dipoles would, on the whole, cancel out each other. This was proved not to be true for ferroelectrics ceramics, in which the dipoles could be permanently aligned or reoriented with an electric field. One of the fundamental issues in the understanding of ferroelectricity and piezoelectricity in ceramics was the discovery of the unusually high dielectric constant of BTO (Jaffe, 1958). Although BTO does not exhibit high piezoelectric constants, it has high relative permittivity. For this reason, BTO is the most widely used material in capacitors. Billions of BTO condensers are still made annually, at a cost of less than one cent per capacitor (Scott, 2007). However, BTO has an important drawback, its relatively low Curie temperature (~120 °C) (Merz, 1949). While advances in order to improve the piezoelectric properties and to increase the Curie temperature are concurrently underway, they have had little success. The observation of large and colossal permittivity (10^4-10^6) (West, 2006, Yu et al., 2004) in the BTO, has consolidate it as a material for capacitors. For instance, $(Ba_{0.92}Ca_{0.08})(Ti_{0.95}Zr_{0.05})O_3$ has high piezoelectric coefficient d_{33} = 365 pC/N and high planar electromechanical factor k_p = 48.5%; nevertheless, the Curie temperature diminishes to ~ 110°C. On the other hand, solid solutions of BTO with ferroelectrics of higher Curie temperature have been studied in order to increase the T_C of the system; unfortunately, although the T_C increases, the effects on the dielectric properties are undesirable. In the solid solution 0.80BTO-0.20$(K_{0.5}Bi_{0.5})TiO_3$ the T_C reaches a value around 240°C, but the relative permittivity at room temperature and at the T_C, has lower values than the pure BTO (Haertling, 1999; Takenaka, 2008). The colossal permittivity observed in BTO, is attributed to an interfacial polarization and is achieved in nanomaterials by the activation of a high number of carriers and their trapping at the interfaces (Guillemet-Fritsch et al., 2008).

3.2 $Na_{0.5}Bi_{0.5}TiO_3$ and $K_{0.5}Bi_{0.5}TiO_3$

Bismuth sodium titanate $Na_{0.5}Bi_{0.5}TiO_3$ (BNT), was discovered 50 years ago (Smolenskii et al., 1961), it shows strong ferroelectric properties with a significantly remnant polarization of 38 $\mu C/cm^2$, and a Curie temperature of 320°C. However, this ceramic has disadvantages such as high conductivity and large coercive field (~73 kV/cm), which cause problems in the poling process. Data on exact piezoelectric properties of the BNT ceramic are insufficient due to the as-mentioned difficulties at the poling process. On the other hand, the BNT ceramic needs a high sintering temperature (>1200°C) to obtain dense samples. It is thought that the vaporization of Bi^{+3} ions occurred during the sintering process at temperatures higher than 1200°C, resulting in the poor poling treatments because of the high conductivity. As in the case of BTO, there have been efforts to improve the piezoelectric response of NBT by the substitution of one or more of its ions. Different authors have studied solid solutions of NBT with BTO, $K_{0.5}Bi_{0.5}TiO_3$ (Takenaka et al., 2008) and KNN (Nagata et al., 2003; Yao et al., 2009; Zhang 2008). All these attempts try to exploit the morphotropic or polymorphic phase boundaries, where it is known that an improvement of dielectric and piezoelectric properties exist. In addition, some rare earths such as La, Y, Ce and some transition metals such as Co, Nb and Mn (Li et al., 2004; Nagata & Takenaka, 2001; Takenaka et al., 1990; Zhou et al., 2009) have been used. Some results are promising, but still more work is needed to improve the dielectric and piezoelectric properties simultaneously.

Just as the NBT, the KBT was discovered 50 years ago (Smolenskii et al., 1961). KBT has tetragonal symmetry at room temperature and a relatively high T_C of 380°C (Buhrer, 1962). KBT has a better dielectric response and similar piezoelectric response than NBT (Lin et al., 2006). In view of the fact that low density materials are difficult to pole, one of the main challenges of KNT is to obtain enough dense ceramics.

3.3 $K_{0.5}N_{0.5}NbO_3$ (KNN)

The pioneering work on $KNbO_3$-$NaNbO_3$ solid solution was carried out in the mid-50s of last century (Shirane et al., 1954). KNN is a specific composition on a complete solid solution of antiferroelectric $NaNbO_3$ and ferroelectric $KNbO_3$, namely, 50:50. This composition is close to the MPB between two orthorhombic phases, resembling the PZT system. Undoubtedly, the KNN and the derived compounds are the most promising lead-free ferroelectric materials demonstrated by the results published some years ago (Saito et al., 2004). The major contribution of this work was to show the modification of the PPT, present in KNN (Shirane et al., 1954), by the addition of Li^{+1} and Ta^{+5}. Since then, this system has been caused a lot of interest and many studies have been done on this field. In fact, our research deals with this material which is presented in section 6. The main obstacles in the processing of KNN are the synthesis and sintering steps that will be treated in next two sections. These difficulties occur, since the alkaline elements undergo sublimation at the high temperature required to achieve the adequate densification, which changes the initial stoichiometry considerably. This problem has been addressed through different methods, one of these involves densification improvement by the addition of some oxides such as CuO, MnO_2, CeO_2 (Gao et al., 2009; Yang et al., 2010; Yin et al., 2010). According to these researches, it is believed that these compounds form a liquid phase at low temperature, thus promoting densification. Another approach involves addition of A and B elements into the ABO_3 structure of the KNN solid solution. In the A site, several cations can be added, e. g. Li^+, Ba^{2+}, La^{3+}, Bi^{3+}, whereas for the B site it is possible to introduce Ti^{4+}, Sb^{5+} or Ta^{5+} (Ahn et al., 2009; Hagh et al., 2009; Jiang et al., 2009). The ion substitution can induce phase transformation and consequently a better performance of materials. A third way to improve densification is by reducing the particle size of the synthesized powders; however, since the conventional ceramic method does not achieve considerable reduction of particle size, then, the sol-gel, Pechini and hydrothermal methods have been used. Furthermore, the chemical homogeneity of the KNN compound with Li^{+1} and Ta^{+5} dopants synthesized by the conventional solid state reaction route has revealed an inhomogeneous distribution of Nb^{5+}, Ta^{5+}, K^+ and Na^+ cations, which leads to a considerable detriment of the piezoelectric properties, being one reason for the discrepancy among the data reported by several authors for the same or similar composition (Y. Wang et al., 2007). All these issues are addressed in the subsequent sections, which are the central part of our contribution.

4. Synthesis of KNN and co-doped KNN

This section will be dedicated to briefly review some methods used for the synthesis of KNN and related compositions. The ceramic method is discussed first, and then the chemically methods used in an effort to obtain chemical homogeneous powders. These include the sol-gel, Pechini and hydrothermal methods. They have produced some interesting results, but there are still some issues that must have the attention of the researchers.

4.1 Conventional ceramic method

For the synthesis of KNN lead-free ferroelectrics, the initial point is the ceramic method (CM), this is the simplest method for the production of ceramic materials. The conventional method is well-known and extensively used, and was the first method reported for the synthesis of KNN (Egerton & Dillon, 1959; Jaeger & Egerton, 1962; Shirane et al., 1954), since it is simple and low cost. Basically, it consists of mixing carbonates and oxide powders of the desired elements. The process is carried out in a conventional ball mill, or in mills that supply more energy as the attrition or planetary ball mills (high energy mills), with the purpose to obtain a homogeneous mixture of the powders. The process is performed in liquid media for a better mixing; the most popular liquids are absolute ethanol and acetone, the former being cheaper and with low toxicity. During grinding, the powders undergo grain size reduction, and become amorphous if high energy milling is used. Once the mixture is ready, this is calcined at an adequate temperature, which depends on composition. In the case of the lead-free ceramics based in KNN, these temperatures are between 800 and 950° C. The heat treatment should be carried out for several hours. Finally, the crystalline powders are grinded again to reduce the particle size for their subsequent pressing and sintering. The advantages of this method are the inexpensive equipment and low cost of reagents. On the other hand, high temperature calcinations and long time of the heat treatments usually results in considerable loss of alkaline elements; furthermore, two steps of grinding are also needed.

4.2 The sol-gel method

Taking into account the characteristics of the powders obtained by means of the conventional method, the so-called chemical routes have been investigated for the synthesis of lead-free ferroelectric ceramic powders. Among them, the sol-gel method (Shiratori et al., 2005; Chowdhury et al., 2010) has been reported to produce KNN nanometric powders. The technique consists of mixing metal-organic compounds (mainly alkoxides) in an organic solvent, the subsequent addition of water generates two reactions, hydrolysis and polymerization, producing the gel which is dried and calcined for obtaining crystalline ceramics. The method has some advantages, such as the nanometric and chemical homogeneity of the powders and the low crystallization temperature (Shiratori et al., 2005). The disadvantages of this procedure are the utilization of metal-organic chemicals, which are expensive. Besides, they need of a strict control of the conditions for the reaction since they generally possess a different hydrolysis rate and must be handled under free moisture atmosphere for avoiding the rapid decomposition of alkoxides. The addition of organic compounds is necessary to improve the dispersion and to obtain fine powders.

4.3 Pechini method

One of the chemical methods that have attracted attention in the synthesis of ceramic materials is the Pechini method. The process implies the formation of a polymeric resin between an organic acid and an alcohol (generally ethylene-glycol). The precursor solution should be heated to evaporate the solvent and to promote the formation of the resin. Once the resin is obtained, it is crushed and calcined at different temperatures to observe the crystallization evolution. As in the case of sol-gel, the Pechini route also uses niobium moisture sensitive reagents, so that the problems are similar in both methods. Despite these drawbacks, the very fine powders obtained are promising to produce dense ceramics, but there are not reports on the piezoelectric properties of ceramics synthesized by this method, only the synthesis of KNN powders is reported (Chowdhury et al., 2009). In this study the authors used an ammonium niobate oxalate hydrate instead the alkoxide. With this approach nanometric powders were synthesized.

4.4 Hydrothermal method

With the aim to obtain KNN ceramic powders at low temperature and to avoid the loss of sodium and potassium, the hydrothermal method have been used recently (Sun et al., 2007; Maeda et al., 2010; N. Liu, et al., 2009). This method involves placing the reagents into a pressurized reactor or autoclave, the reaction is carried out at low temperature (< 300°C) where the pressure generated depends on the temperature at which the reactor is heated. The studies reported until now suggest a processing time of 6-24 hours at the desired temperature. Nevertheless, these studies also indicate that the resultant products are composed of two phases, a sodium rich phase and another with greater quantity of potassium. The reagents that have been used in these experiments are potassium and sodium hydroxides, whit a KOH/NaOH molar ratio between 3/1 and 4/1, and the total concentration around 6 M of hydroxides. Alternatively, the synthesis of KNN has also been reported by means of the microwave-hydrothermal method at 160°C for 7 hours with an alkalinity of 6 M (Zhou et al., 2010) the authors underline that improved piezoelectric constant d_{33} was obtained (126 pC/N), compared with other reports (80 and 90 pC/N), but important parameters like k_p and $tan\ \delta$ were not reported.

As a final comment for this synthesis section, it is important to mention that the powder characteristics obtained by any synthesis method may aid the sintering stage, therefore the powders should be chemically pure i.e. without secondary phases, the calcination temperature (except in hydrothermal synthesis) must be as low as possible to avoid the considerable loss of alkaline compounds, and the nanometric powders are more suitable since these contribute to an additional driving force for sintering.

5. Sintering of KNN and related compositions

Just like the synthesis stage, the sintering process in the KNN lead-free ferroelectric ceramics is a crucial step to produce materials with high electromechanical properties. It has been found that a narrow sintering range exists (Y. Wang et al., 2007) where the materials experience considerable changes in the grain size, density, appearance of secondary phases, liquid phase, and then the piezoelectric and ferroelectric properties change as well. In the text below, are discussed some of the sintering methods used for the conformation of KNN ceramics. First, the conventional sintering (CS), then the hot pressing (HP) and finally the spark plasma sintering (SPS) are going to be described.

5.1 Conventional sintering

The method consists of pressing the powders in a uniaxial press or through cold isostatic pressing. Then, the green pellets are heat treated in a high temperature furnace. The sintering temperature depends upon the composition for pure KNN samples the temperature is set between 1020 and 1120°C. The method is simple and economic comparing with HP or SPS which will be described in the next sections. Most studies about KNN and related compositions use conventional sintering (Chang et al., 2007; Egerton & Dillon, 1959; Hao et al., 2009; Park et al., 2007; Saito & Takao, 2006; Y. Wang et al., 2007; Zuo et al., 2007), and just some papers report lead-free piezoceramics sintered by HP or SPS. In conventional sintering two steps are commonly used during the treatment, first the binder burn out at 400-500°C, and then the sintering at high temperature proceeds. This high temperature stage is performed from 1 to ~ 12 hours. For instance sintering a Li doped KNN composition gave optimal results when the time was set at 8 h (Wang et al., 2010), but it is common to use 2 h. It has been observed the influence of the heating rate over the properties, these rates are close to 4-5°C/min (Du et al., 2006). The fundamental objective to investigate these issues is to determine the effects on the grain growth and hence on the ferroelectric and piezoelectric properties. Most of the investigations try to search for sintering conditions that avoid or reduce at least, the loss of alkaline elements. Combining the ceramic method for the synthesis and the conventional sintering results in low density materials. For this reason, the HP and the SPS methods are being explored, mainly the later, for the improvement of density and the correspondingly enhancement in the electromechanical performance of ceramics.

5.2 Hot pressing

This method has the advantage that pressure and temperature are simultaneously applied, being able to obtain a better densification. Nevertheless, the sintering temperatures are as high as in the conventional sintering. Furthermore, few data on electromechanical properties have been reported by means of this technique (Jaeger & Egerton, 1962). The piezoelectric properties have been improved considerably using this method, compared as those sintered conventionally.

5.3 Spark plasma sintering

The SPS technique is not new in the field of sintering, but its use was not exploited for sintering lead-free piezoelectric ceramics. Very recently it was applied for sintering KNN (K. Wang et al., 2008), and related compositions (Abe et al., 2007; Shen et al., 2010). The advantage of the SPS over CS or HP is that it requires lower temperatures and shorter times for producing ceramics with densities close to the theoretical values. Commonly, heating rates are around 100°C/min, so in few minutes the sintering temperature is achieved; as a result the sintering time is reduced by several hours. This is possible due to the heating mechanism. In this method, a very high electric current is passed through the sample and pressure is applied simultaneously, and liquid phase is generated rapidly which assist the densification, but for more details the reader is encouraged to revise some specialized publications on the subject (Hungría et al., 2009; Tokita, 1993). This sintering method allows reducing the loss of alkaline elements because of the low sintering temperature and short holding time; nevertheless, additional heat treatment is required to eliminate oxygen vacancies (Abe et al., 2007; Wang et al., 2007). In Fig. 3 the SEM images of KNLNT sintered samples by CS and SPS are shown (López-Juárez et al., 2011b), it is clearly observed the difference in densification (porosity). The difference in densification level affects directly the piezoelectric and dielectric properties.

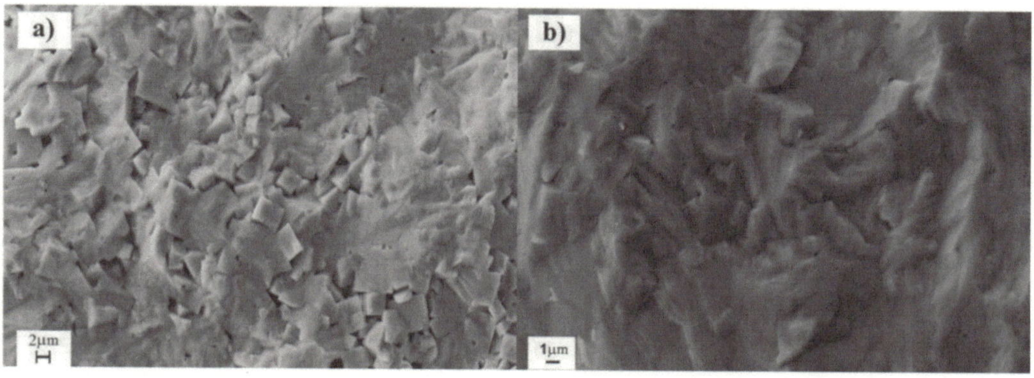

Fig. 3. SEM images of fractured samples sintered by: a) CS at 1200 °C and b) SPS at 900 °C.

6. Synthesis of $K_{0.5}Na_{0.5}NbO_3$ and $(K_{0.48}Na_{0.52})_{0.96}Li_{0.04}Nb_{0.85}Ta_{0.15}O_3$ by spray drying

As already mentioned, the key problem with the synthesis of KNN is there are no stable niobium chemical reagents to use in sol-gel, Pechini or whatever the method employed. The only stable niobium compound is the oxide (Nb_2O_5). Then, the synthesis of KNN based ceramics has been reviewed in previous sections, emphasizing the chemical methods used until now. In this work a new approach is described as is reported elsewhere (López et al., 2010, 2011b). The spray drying method was employed to synthesize chemically homogeneous powders. For this purpose the chelation of niobium and/or tantalum is necessary. In our preparation method it was possible to synthesize lead-free ferroelectric ceramics stabilizing niobium with an organic acid, by previously dissolving Nb_2O_5 and precipitating the corresponding hydrated oxide (López et al., 2010), this is also applicable to Ta_2O_5 because it behaves in a similar manner. Actually, tantalum is introduced into the KNN solid solution structure. The $K_{0.5}Na_{0.5}NbO_3$ and $(K_{0.48}Na_{0.52})_{0.96}Li_{0.04}Nb_{0.85}Ta_{0.15}O_3$ compositions were synthesized. It was probed that the crystallization can be set at 800°C with a heating time of 1 hour. Finally, the microwave-hydrothermal method was tested for KNN synthesis, and interesting results are going to be released.

6.1 Characterization of the synthesized powders

In Fig. 4 the X-Ray diffraction patterns of the two compositions are shown. The most interesting feature is that the powders are chemically pure when calcined at 800°C for 1 h, irrespective of the composition. It is observed that the as sprayed powders are amorphous in both compositions. For the KNN powders, the subsequent heat treatment at 600°C generates the formation of two phases; the $K_6Nb_{10}O_{30}$ phase (JCPDS 70-5051) with tetragonal structure and the KNN perovskite phase with orthorhombic lattice.

When powders were calcined at 700°C the amount of tetragonal phase diminishes considerably, this fact is noticed by the reduction in the Bragg reflections corresponding to the tetragonal phase, and at 750°C only perovskite phase is observed. The calcination temperature and time are lowered compared with those required in the synthesis by the ceramic method. For the KNLNT composition (Fig. 4b), once the powders were thermally treated at 600°C several Bragg reflections appeared, corresponding to the tetragonal

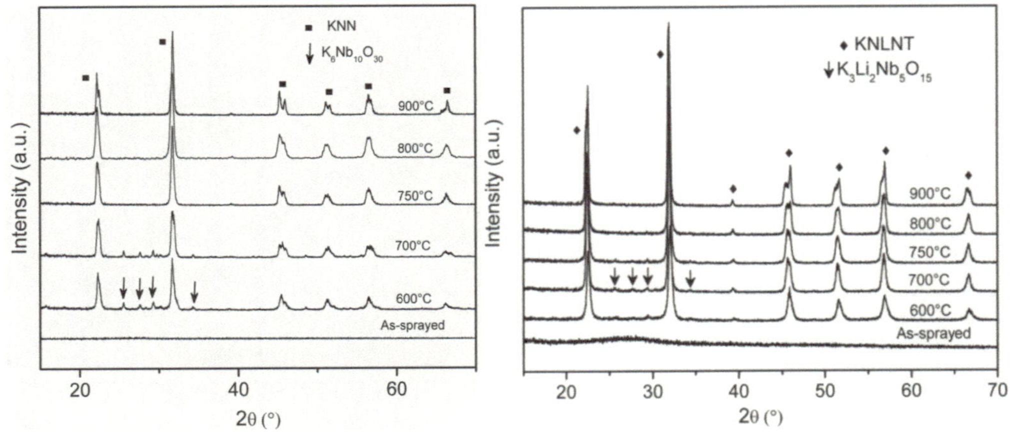

Fig. 4. a) KNN powders calcined at different temperatures (Left), b) KNLNT powders calcined at different temperatures (Right) (López et al., 2010).

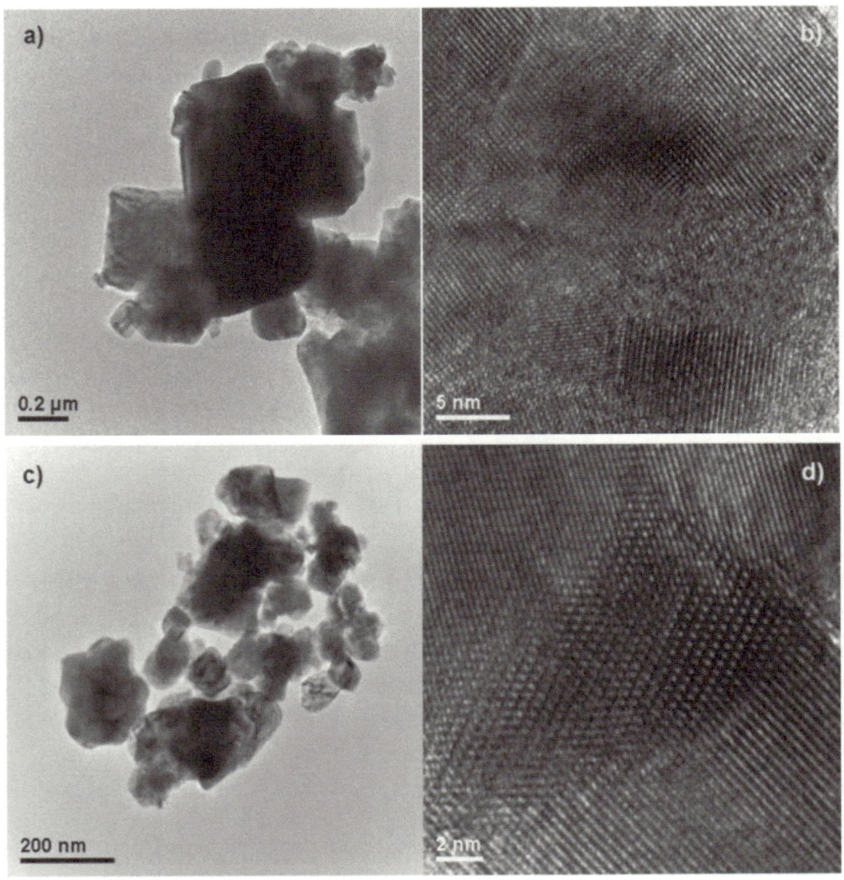

Fig. 5. a) BF and b) HR images of KNN, c) BF and d) HR pictures of KNLNT powders calcined at 800°C for 1 h.

secondary phase $K_3Li_2Nb_5O_{15}$ (JCPDS-ICDD 52-0157) and those expected for KNLNT phase were observed. The secondary phase diminish for the samples calcined at 750°C and at 800°C only reflections of the KNLNT phase remains. The TEM images of the calcined powders reveal the fine particle size with average grains < 0.3 μm (Fig. 5). In Fig. 5a the bright field and 5b the high resolution images of KNN powders are shown. The typical cubic shape of KNN is clearly seen (Jenko et al., 2005). The KNLNT crystalline powders are shown in Figs. 5c and 5d, a bright field (BF) image and the corresponding high resolution (HR) picture are observed, where it is shown that the addition of tantalum has inhibited the grain growth, as reported before (Saito & Takao, 2006) compared with KNN. In the high resolution image the coalescence between two nanocrystals is depicted with the crystalline planes well developed. The average grain size for KNN powders was 281 nm and 100 nm for KNLNT, the measurements were done over several bright field TEM images using the ImageJ software. The particle size achieved by the spray drying route is comparable with those results of sol-gel and Pechini method. The small grains contain high surface energy that is one of the driving forces for sintering. It is known that in addition to the surface energy, the pressure and chemical reactions would aid the sintering (Rahaman, 2006), the later being uncommon for this purpose.

The combination of pressure and heating at the same time in hot-pressing or spark plasma sintering, does not require necessarily very small particles in order to obtain high density materials, but for conventional sintering it is desirable to synthesize finer particles to get high densities in bulk ceramics.

6.2 Sintering, piezoelectric and ferroelectric properties of lead-free ceramics

The synthesis is only the first step in the processing of ferroelectric ceramics. Pressing and sintering are another two important issues for completing the whole process. Then, in the following paragraphs the sintering and the properties of the prepared ceramics are presented.

6.2.1 Sintering of KNN powders

The consolidation and sintering of ferroelectrics ceramics with KNN composition is described in our previous publication (López et al., 2011a). For a sort of clarity, here we are giving some more details. The calcined powders at 800°C for 1 h were pressed in a uniaxial press at 443 MPa. Then the pressed samples were placed into a high temperature furnace and sintered for 2 hours. The heating rate was set at 7°C/min, the sintering temperature was established at 1060-1120°C, that is the ideal sintering treatment for which the highest piezoelectric properties were measured. The density was measured by the Archimedes method in distilled water.

In Fig. 6 the images of KNN sintered samples are shown. Evidently, with increasing of the sintering temperature the grain size increases. But, first the density increases when passing from 1060°C to 1080°C, and then decreases at 1100 and 1120°C, this coincides with the SEM images, where considerable pores are seen in the sample sintered at 1060°C and diminished at 1080°C (Fig. 7b). The higher density was that of the sample sintered at 1080°C, 4.33 g/cm^3 (96% of theoretical value, 4.51g/cm^3 being the reference value). The low density of the samples sintered at 1100 and 1120°C are due to the formation of liquid phase and the considerable volatilization of alkaline elements (Jenko et al., 2005; López et al., 2011a; K. Wang et al., 2010).

Fig. 6. SEM images of KNN sintered samples at: a) 1060, b) 1080, c) 1100 and d) 1120°C for 2 h (López et al., 2011a).

6.2.2 Piezoelectric, dielectric and ferroelectric properties of KNN

For the piezoelectric properties evaluation, the measurements were done on the poled samples, also the dielectric constant and losses were acquired, the d_{31} and k_p parameters were calculated with modeling the impedance profile as reported elsewhere (Alemany et al., 1995; Pardo et al., 2010). The d_{33} piezoelectric constant was measured with a d_{33}-meter, this parameter was measured for the sample with better k_p and d_{31}. The ferroelectric loops were obtained in a Radiant workstation at room temperature. The dielectric constant and dielectric losses are shown in Fig. 7, the $T_{O\text{-}T}$ and T_C are clearly observed, the $T_{O\text{-}T}$ is close to 200°C as reported in several works (Du et al., 2006; Egerton & Dillon, 1959). The Curie temperature also agrees well with that reported previously which is near to 420°C (Ringgaard & Wurlitzer, 2005; Singh et al., 2001). The dielectric constant and dielectric losses are improved when the sintering temperature is 1060 and 1080°C, but at 1100 and 1120°C the dielectric constant diminishes and *tan* δ increases, this is directly related with density that depends on the liquid phase formation and vacancies generated when potassium and sodium are lost.

The piezoelectric properties are also related with the remnant polarization (P_r) and coercive field (E_C). These are extracted from the hysteresis loops shown in Fig. 8. The P_r and E_C are improved for the sample sintered at 1080°C and are degraded for the sample sintered at 1120°C where the ferroelectric loop is rounded; this behavior is typical of a conduction process related to high concentration of vacancies (Chen et al., 2007; Kizaki et al., 2007). If the phase diagram is invoked, 1120°C is close to the melting point of the KNN composition (~1140°C) then it is obvious that high volatilization of alkaline elements takes place, generating also oxygen vacancies for electro-neutrality within the crystals. The ferroelectric properties are also summarized in table 1. The existence of vacancies is common in KNN lead-free ferroelectric ceramics after being sintered at high temperature. The measurement of leaking current has been used for the indirect determination of vacancies. The higher the electrical conduction the higher the concentration of vacancies (Kizaki et al., 2007). According to the authors knowledge, it has never been reported the observation of vacancies by HR-TEM in KNN lead-free ferroelectrics. In Fig. 9 the bright field and high resolution images of the KNN sintered sample at 1080°C are shown, this sample was mechanically polished with SiC paper following with alumina powder with 50 nm grain size and finally ion-milled.

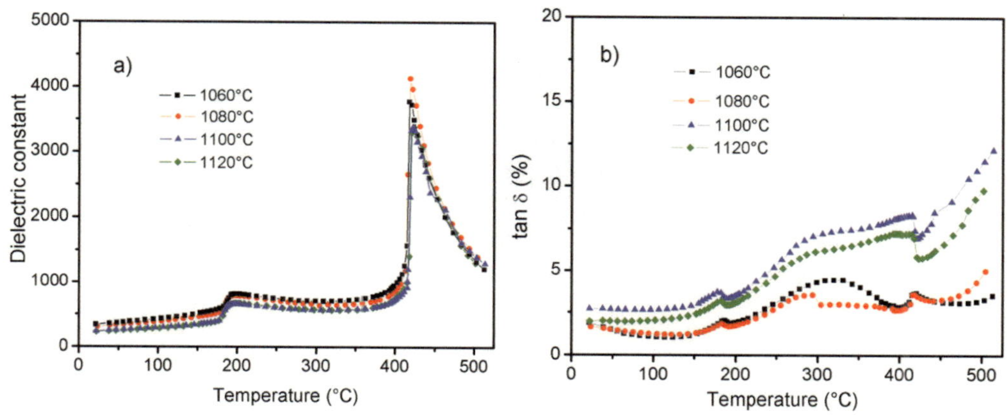

Fig. 7. a) Dielectric constant and b) *tan* δ for KNN sintered pellets at 100 kHz (López et al., 2011a).

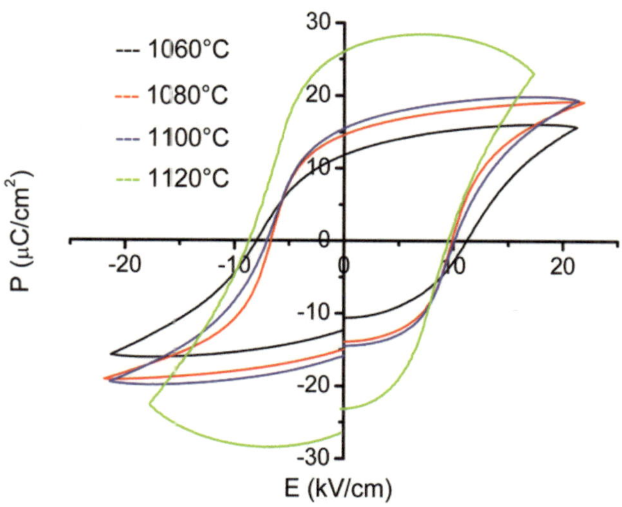

Fig. 8. Ferroelectric loops of KNN ceramics.

In the high resolution image the atomic columns are observed, and some vacancies are highlighted with arrows that are seen as unfilled gaps. The image was taken in the [111] direction of the orthorhombic lattice, the typical hexagonal geometry of atomic columns is clearly distinguished.

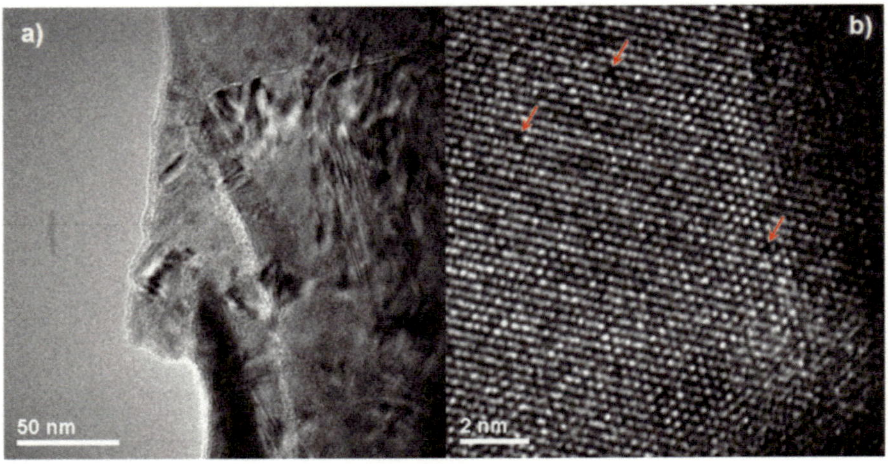

Fig. 9. TEM images of KNN sintered at 1080°C, a) BF and b) HR.

The piezoelectric properties obtained for the sintered samples are summarized in table 1. As was stated above, the best properties are those for the sintered pellet at 1080°C. In table 2, the properties of KNN reported for several authors are shown. It can be seen that the properties of the ceramics processed by spray drying and conventionally sintered are comparable with those previously reported.

Parameter	Sintering temperature (°C)			
	1060	1080	1100	1120
k_p	0.33	0.36	0.23	0.20
$-d_{31}$ (pC/N)	29.6	30	20.9	19.1
tan δ (%) (100 kHz)	1.9	1.6	2.7	2
T_C (°C)	418	419	423	423
$2P_r$ (μC/cm^2)	23.2	29	30.57	51
$2E_C$ (kV/cm)	19.1	16.5	17.23	18.1
S_{11}^E (10^{-12} m^2 N^{-1})	11.082	9.131	9.949	12.04
S_{12}^E (10^{-12} m^2 N^{-1})	- 2.807	-2.171	-3.008	-3.058
ρ (gr/cm^3)	4.27	4.33	4.28	4.17

Table 1. Piezoelectric and ferroelectric properties of KNN.

k_p	d_{33} (pC/N)	$-d_{31}$ (pC/N)	ε_r	ρ (g/cm³)	tan δ (%)	T_C (°C)	Reference
0.36	80	32	290 (100 kHz)	4.24	2 (100 kHz)	---	Egerton & Dillon, 1959
0.45	160	49	420 (100 kHz)	---	1.4 (100 kHz)	---	Jaeger & Egerton, 1962
0.32	107	---	264 (1 kHz)	4.09	---	---	Maeda et al., 2010
0.40	120	---	500 (1 kHz)	4.4	---	400	Du et al., 2006
0.39	70 - 90	45	400 (1 kHz)	4.28	2.5 (1 kHz)	390	Ringgaard & Wurlitzer, 2005
---	---	---	----	---	1.3 (10 kHz)	395	Singh et al., 2001
0.34	---	46.2	453 (1 kHz)	---	35 (1 kHz)	402	L. Liu et al., 2009
0.36	117[1]	30	309 (100 kHz)	4.33	1.6 (100 kHz)	419	López et al., 2011a

Table 2. Piezoelectric and dielectric properties of $K_{0.5}Na_{0.5}NbO_3$ ceramics reported by several authors.

It is obvious the wide range of values of the piezoelectric properties, even when most authors use the conventional ceramic method for the production of ferroelectric ceramics. This remarks the sensitiveness of these materials to any small processing variation, the moisture sensitivity of alkaline carbonates, the calcination temperature, the heating time and, finally, the temperature at the sintering stage. All these steps considered together influence the poling process and then and final performance of the ceramics.

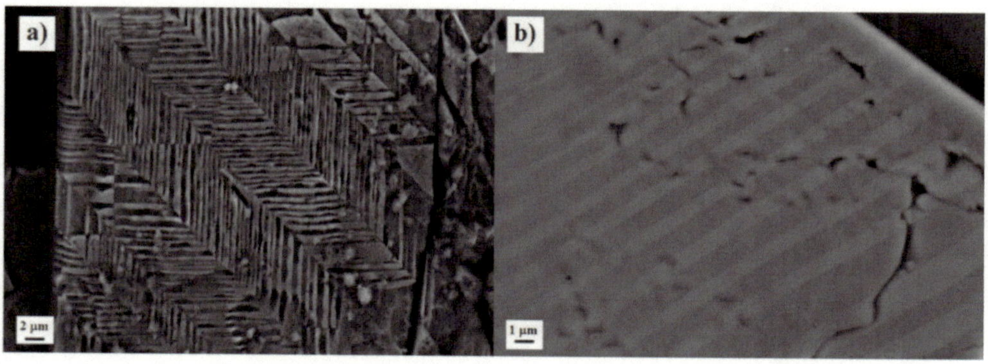

Fig. 10. Ferroelectric domains in KNN synthesized by microwave-hydrothermal method and sintered at 1080°C.

To conclude this section it should be mentioned the efforts to achieve the low temperature synthesis of KNN powders as was underlined at the hydrothermal synthesis section. This method requires heat treatment for a long time at low temperature (<300°C) if the conventional hydrothermal method is used. But recently, the microwave-hydrothermal technique it is being explored for the synthesis of some inorganic materials. Although, the influence of microwaves on the reaction system it is not well understood until now. In the case of potassium-sodium niobate, the synthesis was proved to proceed faster with the aid of microwaves. The sintered powders experienced extremely grain growth with grain size

[1] This value was not reported in (López et al., 2011a).

average > 60 μm where the ferroelectric domains were reveled easier. In Fig. 10 the ferroelectric domains are shown in the ceramics sintered at 1080°C for 4 hours in air. As it is observed, the ferroelectric domains are large enough to be revealed by contrast using backscattered electrons (at 1 pA and 20kV), and in etched samples with hydrofluoric acid. The domain structure was found to be constituted mostly by 90° and 180° ferroelectric domains, for more details refer to the work by López et al. (López et al., 2011c).

6.2.3 Sintering of KNLNT

The synthesized powders by spray drying with $(K_{0.48}Na_{0.52})_{0.96}Li_{0.04}Nb_{0.85}Ta_{0.15}O_3$ composition were pressed and sintered at 1100-1150°C in air for 2 hours. The SEM pictures of these samples are shown in Fig. 11. The grain size was smaller comparing with the KNN composition. It is reported that the addition of Ta^{+5} into the KNN structure inhibits grain growth (Saito & Takao, 2006). When the sintering temperature was set at 1100°C, the grains grow inhomogeneously, but increasing the temperature to 1120 and 1130°C, these are bigger with the characteristic cubic shape, and for 1150°C, the grains grew even more because the faster mass transport rate. The densities and other piezoelectric parameters are shown on table 3. As it was observed for KNN ceramics, here the behavior of density and piezoelectric properties first increase and then fall for samples sintered at higher temperature. The factors which affect the densification of KNN pellets are also present for KNLNT composition. Evidently, the high temperature used for the sintering is because of the Ta^{+5} content, and the low concentration of Li^{+1} on the system, despite that, it is well known that lithium aids densification at lower temperatures as compared with KNN.

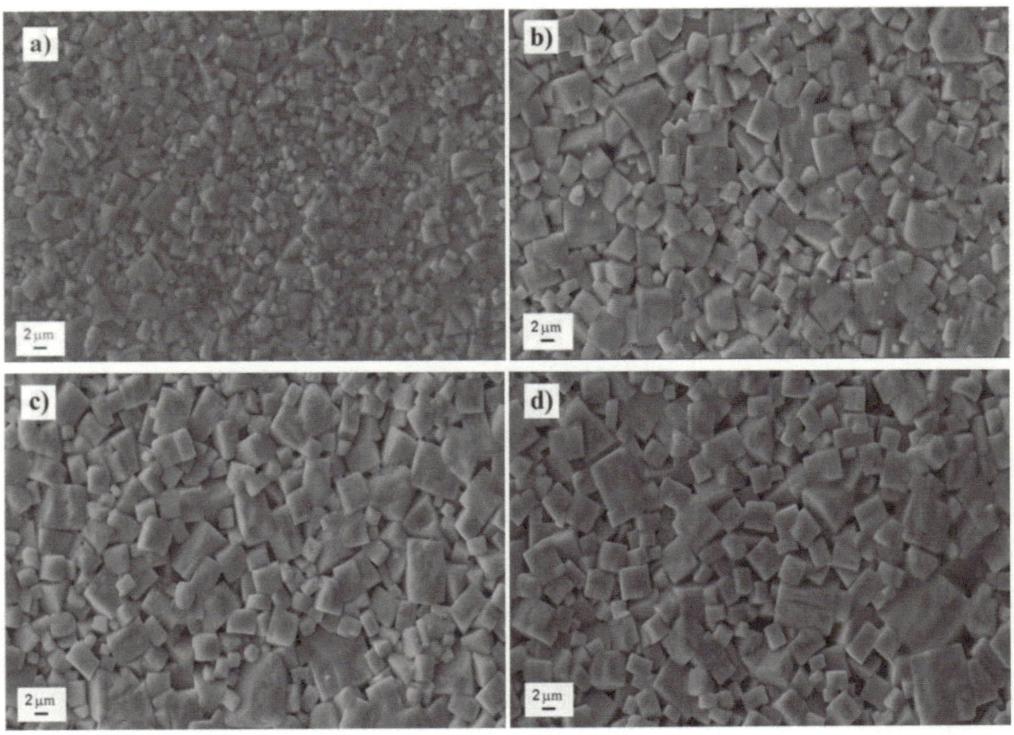

Fig. 11. SEM images of KNLNT sintered pellets, a) 1100, b) 1120 and c) 1130 (López et al., 2010), 1150°C for 2 h.

6.2.4 Piezoelectric, dielectric and ferroelectric properties of KNLNT

The KNLNT samples were processed and poled as reported before (López et al., 2010). In table 3, the properties obtained for the samples sintered at 1100 and 1150°C are also included. The values of piezoelectric properties increase with increasing sintering temperature, reaching maximum values for 1120°C, and then decrease for 1130 and 1150°C. In these materials is common to reach equilibrium between sintering temperature and piezoelectric properties. The temperature must be high enough to guarantee the diffusion of matter, eliminate pores and to promote grain growth, but not too high for avoiding the volatilization of alkaline elements and the formation of secondary phases, which degrades the piezoelectric properties. The presence of secondary phases, liquid phase or a high concentration of vacancies would result in incomplete poling because the leakage current leads to poor performance. Then the optimum sintering temperature of KNLNT ceramics was established at 1120°C. At this temperature the density, k_p and other parameters were improved. In Figs. 12 and 13 the dielectric constant and *tan* δ are shown for samples sintered at 1120 and 1130°C. The ε_r curves for both samples are similar, except that the magnitude of ε_r is higher for 1120°C. Furthermore, only one phase transition is observed at 354°C, from tetragonal to cubic (T_C). In addition, the dielectric constant has no dispersion with frequency and temperature.

Parameter	Sintering temperature (°C)			
	1100	1120	1130	1150
k_p	0.30	0.41	0.40	0.29
$-d_{31}$ (pC/N)	49.01	55.86	50.55	37.12
$tan\delta$(%)(100kHz)	0.7	1	1	1.2
T_C (°C)	360	354	354	350
$2P_r$ (μC/cm^2)	24.00	25.84	21.43	12.24
$2E_C$ (kV/cm)	20.92	25.70	26.77	19.82
S^E_{11} (10^{-12} m^2 N^{-1})	12.33	11.80	12.24	13.12
S^E_{12} (10^{-12} m^2 N^{-1})	-4.15	−2.786	-3.64	-4.271
ρ (gr/cm^3)	4.49	4.57	4.56	4.37

Table 3. Piezoelectric properties of KNLNT sintered ceramics.

It has been demonstrated that the magnitude of the dielectric constant is affected by extrinsic variables as the porosity (Fang et al., 1993), i.e. the air trapped within the samples has a low ε that is averaged in the measured signal. The grain size also influences the dielectric response (Fang et al., 1993; Hoshina et al., 2008), but there are the intrinsic effects due to the introduction of cations in the A and B sites of the ABO_3 structure. More precisely, when donor-type doping of KNN with La^{+3} and Ta^{+5} is performed (Hao et al., 2009) and in $BaTiO_3$ (Morrison et al., 1999), the dielectric constant at the T_C decreases and the transition became broad, this behavior it is not well understood but it is suggested that the many different transitions collapse at once resulting in the broad pick at T_C. In the KNLNT it seems that the Ta^{+5} doping induces that the ε_r was lower at the transition compared with pure KNN, the contribution of grain size was discarded. The porosity of the sintered samples is of the same order and considering that well defined grain boundaries exist within them, the diminishment of ε_r is attributed to Ta^{+5} doping. The phenomena need more careful revision taking in mind the different ions introduced and the interactions among them as well as the type of bonding within the matrix atoms. Another important feature is the low dielectric losses in KNLNT (Fig. 13), in KNN it was found that high losses are related with high concentration of oxygen vacancies (Chen et al., 2007; Kizaki et al., 2007), then the high concentration of vacancies and the low activation energy needed to move them (Kizaki et al., 2007; L. Liu et al., 2009) increase the dielectric losses. This means that the KNLNT samples are thermally stable and have low conductivity due to vacancies. Fig. 14 shows the ferroelectric loops of the sintered samples. The four samples show saturated loops, the only difference between them are the remnant polarization and coercive field. The highest P_r is encountered for the sample sintered at 1120°C, and is reported in table 3. The other samples have lower P_r, finding the lowest value for the heat treatment at 1150°C.

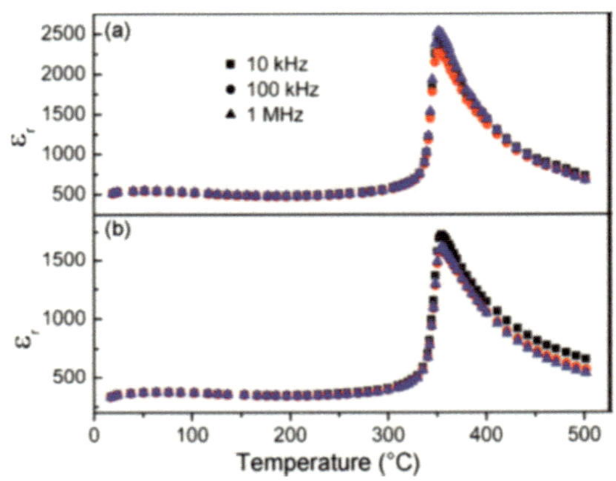

Fig. 12. Dielectric constant for KNLNT sintered samples, a) 1120 and b) 1130°C for 2 h.

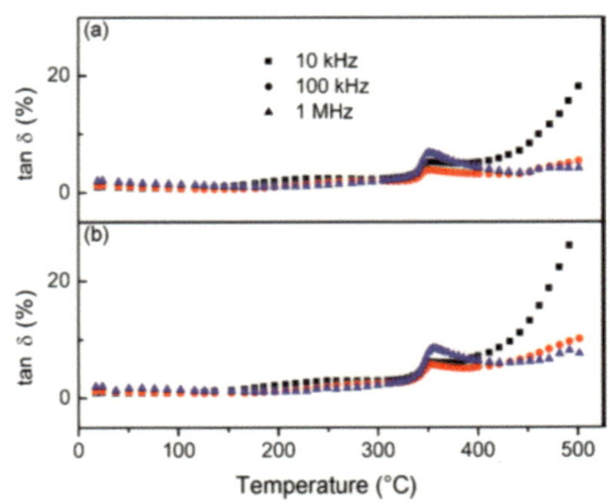

Fig. 13. Dielectric losses for KNLNT sintered samples, a) 1120 and b) 1130°C for 2h (López et al., 2010).

This is completely in agreement with the piezoelectric properties of the sintered ceramics. Probably, the most important issue is the crystal structure of the ceramics, if it is assumed that for KNLNT the polymorphic phase transition (PPT) is close to room temperature, and invoking the assumption that orthorhombic and tetragonal phases coexist at the PPT, then the polarization will be enhanced and the piezoelectric properties improved as well. This is evident when compared the properties found in KNN and KNLNT.

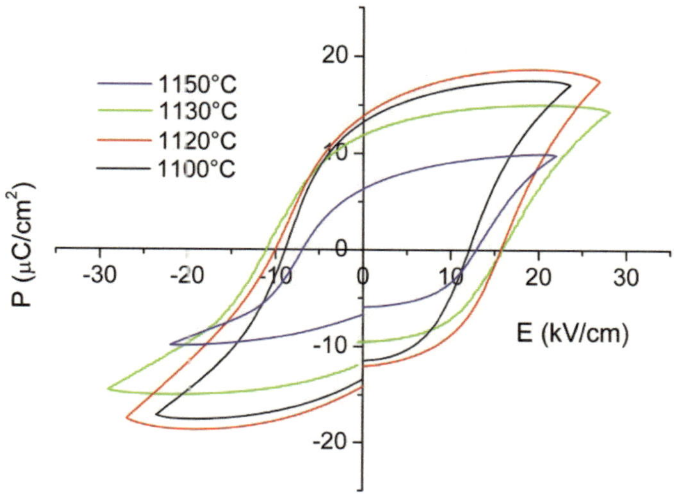

Fig. 14. Ferroelectric loops of KNLNT sintered ceramics.

In table 4, the comparison of several compositions is presented. In the same way as the results for pure KNN, the properties cover a wide range of values also for KNLNT and related compositions. It is observed that in compositions with $BaTiO_3$ k_p and d_{33} are reduced. On the other hand antimony increases these properties in general, but the T_C is shifted to lower values.

Composition	k_p	d_{33} (pC/N)	ε_r	*tan* δ (%)	T_C (°C)	Reference
$0.95(Na_{0.5}K_{0.5})NbO_3$-$0.05CaTiO_3$	0.41	241	1316(10 kHz)	9.0(10 kHz)	306	Park et al., 2007
$(Na_{0.5}K_{0.5})_{0.97}La_{0.01}Nb_{0.95}Ta_{0.05}O_3$	0.37	119	815(10 kHz)	4.5(10 kHz)	340	Hao et al., 2009
$(K_{0.44}Na_{0.52}Li_{0.04})(Nb_{0.84}Ta_{0.10}Sb_{0.06})O_3$	0.48	299	1865(1 kHz)	2.1(1 kHz)	265	Hagh et al., 2007
$(K_{0.5}Na_{0.5})_{0.096}Li_{0.04}Nb_{0.775}Ta_{0.225}O_3$	0.48	208	1146	< 3	320	Lin et al., 2007
$(Na_{0.52}K_{0.4375})(Nb_{0.9175}Sb_{0.04})O_3$–$0.0425LiTaO_3$	0.48	310	1644(10 kHz)	---	337	Fu et al., 2008
$(K_{0.38}Na_{0.52}Li_{0.04})(Nb_{0.86}Ta_{0.10}Sb_{0.04})\ O_{2.97}$	0.37	195	1060(1 MHz)	2.8(1 MHz)	276	Rubio-Marcos et al., 2007
$(Li_{0.04}K_{0.44}Na_{0.52})(Nb_{0.85}Ta_{0.15})O_3$	0.38	200	---	---	---	Li et al., 2008
$0.97Li_{0.06}(Na_{0.5}K_{0.5})_{0.94}NbO_3$–$0.03BaTiO_3$	0.32	128	800(1 kHz)	---	370	Kakimoto et al., 2010
$0.995(K_{0.5}Na_{0.5})_{0.95}(LiSb)_{0.05}Nb_{0.95}O_3$-$0.005BaTiO_3$	0.42	209	1100(10 kHz)	2.6(10 kHz)	344	Zang et al., 2010
$0.92(K_{0.5}Na_{0.5})NbO_3$–$0.08AgTaO_3$	0.41	183	683(10 kHz)	3.3(10 kHz)	356	Wang et al., 2010
$(K_{0.48}Na_{0.52})_{0.96}Li_{0.04}Nb_{0.85}Ta_{0.15}O_3$	0.41	186	465(10 kHz)	1.0(10 kHz)	354	López et al., 2010

Table 4. Piezoelectric properties of KNLNT and related compositions.

7. Conclusion

In the last few years several researches were performed in the lead-free ferroelectric materials for piezoelectric applications. These works were focused in the improvement of the piezoelectric properties by doping potassium-sodium niobate ceramics. The dopants are those of the A and B occupancy within the perovskite structure. Most of them shift the T_C and the polymorphic phase boundary ($T_{O\text{-}T}$) to room temperature, thus improving the piezoelectric performance. It is believed that this is due to the coexistence of the orthorhombic-tetragonal phases at room temperature, and this allows a better poling process because of the existence of more crystallographic directions for the polarization. Despite the improvement in properties, there are troubles in the synthesis, processing and poling of the sintered ceramics. The high volatilization of potassium, sodium and lithium is one of the most notable drawbacks of lead-free ferroelectric ceramics based in the KNN solid solution. For this reason, there is also a great search for low temperature synthesis and sintering of these materials. Then, the microwave-hydrothermal, sol-gel and Pechini synthesis methods are being proved, which in combination with the spark plasma sintering are some alternatives for processing KNN and related ceramics; nevertheless, these need more investigation to find appropriate conditions.

In this work, the spray drying synthesis method was used to synthesize KNN and KNLNT ceramics. The obtained results pointed out that this method is a promising option for avoiding inhomogeneous distribution of cations within the ceramics, and the losses of alkaline elements. This was possible because the temperature and time at the calcination stage were reduced. The sintered ceramics shown good piezoelectric and ferroelectric properties, i.e. high k_p, low *tan* δ and ε, and moderate E_C and P_r. In the case of KNLNT the ideal sintering temperature, depends greatly upon the doping elements. In the KNLNT the improvement of the piezoelectric properties was attributed to the polymorphic phase transition close to room temperature.

8. Acknowledgment

The authors want to thank to Dr. Fernando Rubio and Prof. José F. Fernández (ICV-CSIC) for the evaluation of the d_{33} constant. Special thanks to Omar Novelo-Peralta (Instituto de Investigaciones en Materiales-Universidad Nacional Autónoma de México [IIM-UNAM]) and Ariosto Medina (Instituto de Investigaciones Metalúrgicas-Universidad Michoacana de San Nicolás de Hidalgo) for the SEM and TEM images, respectively. The authors are indebted to Dr. M.P. Cruz (CNyN-UNAM) for the access to the Radiant workstation for ferroelectric loops measurements. Finally, R. López-Juárez wants to acknowledge to CONACyT-México and to IIM-UNAM for the PhD scholarships.

9. References

Abe, J.; Kobune, M.; Kitada, K. & Yazawa, T. (2007). Effects of Spark-Plasma Sintering on the Piezoelectric Properties of High-Density (1-x)($Na_{0.5}K_{0.5}$)NbO_3-x$LiTaO_3$ Ceramics. *J. Korean Phys. Soc.*, Vol. 51, No. 2, (Aug 2007), pp. 810-814, ISSN 03744884

Ahn, C.W.; Park, C.S.; Choi, C.H.; Nahm, S.; Yoo, M.J.; Lee, H.G. & Priya, S. (2009). Sintering behavior of lead-free (K,Na)NbO_3-based piezoelectric ceramics *J. Am. Ceram. Soc.*, Vol. 92, No. 9, (Sept 2009), pp. 2033–2038, ISSN 0002-7820

Alemany, C.; Gónzalez, A.M.; Pardo, L.; Jiménez, B.; Carmona, F. & Mendiola, J. (1995). Automatic determination of complex constants of piezoelectric lossy materials in the radial mode. *J. Phys. D: Appl. Phys.*, Vol. 28, No. 5, (May 1995), pp. 945-956, ISSN 0022-3727

Buhrer, C. F. (1962). Some properties of bismuth perovskites: *J. Chem. Phys.*, Vol. 36, No. 3, (Feb 1962), pp. 798-803, ISSN 0021-9606

Chang, R.C.; Chu, S.Y.; Lin, Y.F.; Hong, C.S. & Wong, Y.P. (2007). An investigation of $(Na_{0.5}K_{0.5})NbO_3$–$CaTiO_3$ based lead-free ceramics and surface acoustic wave devices. *J. Europ. Ceram. Soc.*, Vol. 27, No. 16, (2007), pp. 4453–4460, ISSN 0955-2219

Chen, K.; Xu, G.; Yang, D. & Wang, X. (2007). Dielectric and Piezoelectric Properties of Lead-Free $0.95(K_{0.5}Na_{0.5})NbO_3$–$0.05LiNbO_3$ Crystals Grown by the Bridgman Method. *J. Appl. Phys.*, Vol. 101, No. 4, (2007), pp. 044103(1-4), ISSN 0021-8979

Chowdhury, A.; Bould, J.; Zhang, Y.; James, C. & Milne, S.J. (2010). Nano-powders of $Na_{0.5}K_{0.5}NbO_3$ made by a sol–gel method. *J. Nanopart. Res.*, Vol. 12, No. 1, (Jan 2010), pp. 209-215, ISSN 1388-0764

Chowdhury, A.; O'Callaghan, S.; Skidmore, T.A.; James, C. & Milne, S.J. (2009). Nanopowders of $Na_{0.5}K_{0.5}NbO_3$ Prepared by the Pechini Method. *J. Am. Ceram. Soc.*, Vol. 92, No. 3, (Mar 2009), pp. 758–761, Online ISSN 1551-2916

Du, H.; Li, Z.; Tang, F.; Qu, S.; Pei, Z. & Zhou, W. (2006). Preparation and piezoelectric properties of $(K_{0.5}Na_{0.5})NbO_3$ lead-free piezoelectric ceramics with pressure-less sintering. *Mater. Sci. Eng. B*, Vol. 131, No. 1-3, (July 2006), pp. 83–87, ISSN 0921-5107

Egerton, L. & Dillon, D.M. (1959). Piezoelectric and Dielectric Properties of Ceramics in the System Potassium Sodium Niobate. *J. Am. Ceram. Soc.*, Vol. 42, No. 9, (Sept 1959), pp. 438–442, Online ISSN 1551-2916

European commission. (2008). Recast of the WEEE and RoHS Derectives proposed, In: *European commission: Environment*, last access 10-03-2011, Available from: http://ec.europa.eu/environment/waste/weee/index_en.htm

Fang, T.T.; Hsieh, H.L. & Shiau F.S. (1993). Effects of Pore Morphology and Grain Size on the Dielectric Properties and Tetragonal-Cubic Phase Transition of High-Purity Barium Titanate. *J. Am. Ceram. Soc.*, Vol. 76, No. 5, (1993), pp. 1205–1211, Online ISSN 1551-2916

Fu, J.; Zuo, R.; Wu, Y.; Xu, Z. & Li, L. (2008). Phase Transition and Electrical Properties of Li- and Ta-Substituted $(Na_{0.52}K_{0.48})(Nb_{0.96}Sb_{0.04})O_3$ Piezoelectric Ceramics. *J. Am. Ceram. Soc.*, Vol. 91, No. 11, (Nov 2008), pp. 83771–3773, Online ISSN 1551-2916

Gao, D.; Kwok, K.W.; Lin, D. & Chan, H.L.W. (2009). Microstructure, electrical properties of CeO_2-doped $(K_{0.5}Na_{0.5})NbO_3$ lead-free piezoelectric ceramics. *J. Mater. Sci.*, Vol. 44, No. 10 (May 2009), pp. 2466–2470, ISSN 0022-2461

Guillemet-Fritsch, S.; Valdez-Nava, Z.; Tenailleau, C.; Lebey, T.; Durand, B. & Chane-Ching, J.Y. (2008). Colossal permittivity in ultrafine grain size BaTiO3-x and $Ba_{0.95}La_{0.05}TiO_{3-x}$ materials. *Adv. Mater.* Vol. 20, No. 3, (Feb 2008), pp. 551–555, ISSN 0935-9648

Guo, Y.; Kakimoto, K. & Ohsato, H. (2004). Phase transitional behavior and piezoelectric properties of $(Na_{0.5}K_{0.5})NbO_3$-$LiNbO_3$ ceramics. *Appl. Phys. Lett.*, Vol. 85, No. 18, (Nov 2004), pp. 4121-4123, ISSN 0003-6951

Haertling, G. H. (1999). Ferroelectric ceramics: History and technology. *J. Am. Ceram. Soc.*, Vol. 82, No. 4, (Apr 1999), pp. 797–818, ISSN 0002-7820

Hagh, N.M.; Jadidian, B. & Safari, A. (2007). Property-processing relationship in lead-free (K, Na, Li)NbO_3-solid solution system. *J. Electroceram.*, Vol. 18, No. 3-4, (Aug 2007), pp. 339–346, ISSN 1385-3449

Hagh, N.M.; Kerman, K.; Jadidian, B. & Safari, A. (2009). Dielectric and piezoelectric properties of Cu^{2+}-doped alkali niobates. *J. Eur. Ceram. Soc.*, Vol. 29, No. 11, (Aug 2009), pp. 2325–2332, ISSN 0955-2219

Hao, J.; Xu, Z.; Chu, R.; Li, W.; Li, G. & Yin, Q. (2009). Relaxor behavior and dielectric properties of (La, Ta)-modified $(K_{0.5}Na_{0.5})NbO_3$ lead-free ceramics. *J. Alloy compd.*, Vol. 484, No. 1-2, (Sept 2009), pp. 233–238, ISSN 0925-8388

Hao, J.; Wang, X.; Chen, R. & Li, L. (2005). Synthesis of $(Bi_{0.5}Na_{0.5})TiO_3$ nanocrystalline powders by stearic acid gel method. *Mater. Chem. Phys.*, Vol. 90, Nos. 2-3, (Apr 2005), pp. 282–285, ISSN 0254-0584

Hoshina, T.; Takizawa, K.; Li, J.; Kasama, T.; Kakemoto, H. & Tsurumi, T. (2008). Domain Size Effect on Dielectric Properties of Barium Titanate Ceramics. *J. Journal of Appl. Phys.*,Vol. 47, No. 9, (2008), pp. 7607–7611, ISSN 0021-4922

Hungría, T.; Galy, J. & Castro, A. (2009). Spark Plasma Sintering as a Useful Technique to the Nanostructuration of Piezo-Ferroelectric Materials. *Adv. Eng. Mater.*, Vol. 11, No. 8, (Aug 2009), pp. 615-631, Online ISSN 1527-2648

Jaeger, R.E.; & Egerton, L. (1962). Hot Pressing of Potassium–Sodium Niobates. *J. Am. Ceram. Soc.*, Vol. 45, No. 5, (May 1962) pp. 209–213, Online ISSN 1551-2916

Jaffe, B.; Cook, W.R. & Jaffe H. (1971). *Piezoelectric Ceramics*, Academic Press, ISBN 0-12-379550-8, London

Jaffe, H. (1958). Piezoelectric Ceramics. *J. Am. Ceram. Soc.*, Vol. 41, No. 11, (Nov 1958), pp. 494–98, ISSN 0002-7820

Jenko, D.; Benčan, A.; Malič, B.; Holc, J. & Kosec, M. (2005). Electron Microscopy Studies of Potassium Sodium Niobate Ceramics. *Microsc. Microanal.*, Vo. 11, No. 06, (Nov 2005), pp. 572–580, ISSN 1431-9276

Jiang, M.; Liu, X.; Chen, G. & Zhou, C. (2009). Dielectric and piezoelectric properties of $LiSbO_3$ doped 0.995 $K_{0.5}Na_{0.5}NbO_3$-0.005$BiFeO_3$ piezoelectric ceramics. *Mater. Lett.*, Vol. 63, No. 15, (June 2009), pp. 1262–1265, ISSN 0167-577X

Jing, X.; Li, Y. & Yin, Q. (2003). Hydrothermal synthesis of $Na_{0.5}Bi_{0.5}TiO_3$ fine powders. *Mater. Sci. Eng. B*, Vol. 99, Nos. 1-3, (May 2003), pp. 506–510, ISSN 0921-5107

Kakimoto, K.; Ando, K. & Ohsato, H. (2010). Grain size control of lead-free $Li_{0.06}(Na_{0.5}K_{0.5})_{0.94}NbO_3$ piezoelectric ceramics by Ba and Ti doping. *J. Eur. Ceram. Soc.*, Vol. 30, No. 2, (Jan 2010), pp. 295–299, ISSN 0955-2219

Kizaki, Y.; Noguchi, Y. & Miyayama, M. (2007). Defect Control for superior Properties in Single Crystals. *Key Eng. Mater.*, Vol. 350, (Oct 2007), pp. 85-88, ISSN 1662-9795

Li, H.D.; Feng, C.D. & Yao, W.L. (2004). Some effects of different additives on dielectric and piezoelectric properties of $(Bi_{1/2}Na_{1/2})TiO_3$-$BaTiO_3$ morphotropic phase boundary composition. *Mater. Lett.*, Vol. 58, Nos. 7-8, (Mar 2004), pp. 1194-1198, ISSN 0167-577X

Li, J.F.; Zhen, Y.; Zhang, B.P.; Zhang, L.M. & Wang, K. (2008). Normal sintering of (K, Na)NbO_3-based lead-free piezoelectric ceramics. *Ceram. Int.*, Vol. 34, No. 4, (May 2008), pp. 783–786, ISSN 0272-8842

Lin, D.; Kwok, K.W. & Chan, H.L.W. (2007). Microstructure, phase transition, and electrical properties of $(K_{0.5}Na_{0.5})_{1-x}Li_x(Nb_{1-y}Ta_y)O_3$ lead-free piezoelectric ceramics. *J. Appl. Phys.*, Vol. 102, No. 3, (Aug 2007), pp. 034102(1-7), ISSN 0021-8979

Lin, D.M.; Xiao, D.Q.; Zhu, J.G. & Yu, P. (2006). Piezoelectric and ferroelectric properties of $[Bi_{0.5}(Na_{1-x-y}K_xLi_y)_{(0.5)}]TiO_3$ lead-free piezoelectric ceramics. *Appl. Phys. Lett.*, Vol. 88, No. 6, (Feb 2006), p. 062901, ISSN 0003-6951

Liu, L.; Fan, H.; Fang, L.; Chen, X.; Dammak, H. & Thi, M.P. (2009). Effects of Na/K evaporation on electrical properties and intrinsic defects in $Na_{0.5}K_{0.5}NbO_3$ ceramics. *Mater. Chem. Phys.*, Vol. 117, No. 1, (Sept 2009), pp. 138–141, ISSN 0254-0584

Liu, N.; Wang, K.; Li, J.F. & Liu, Z. (2009). Hydrothermal Synthesis and Spark Plasma Sintering of (K,Na)NbO_3 Lead-Free Piezoceramics. *J. Am. Ceram. Soc.*, Vol. 92, No. 8, (Aug 2009), pp. 1884–1887, Online ISSN 1551-2916

López, R.; González, F. & Villafuerte-Castrejón, M.E. (2010). Structural and electrical characterization of $(K_{0.48}Na_{0.52})_{0.96}Li_{0.04}Nb_{0.85}Ta_{0.15}O_3$ synthesized by spray drying. *J. Europ. Ceram. Soc.*, Vol. 30, No. 6, (Apr 2010), pp. 1549–1553, ISSN 0955-2219

López, R.; González, F.; Cruz, M.P. & Villafuerte-Castrejon, M.E. (2011a). Piezoelectric and ferroelectric properties of $K_{0.5}Na_{0.5}NbO_3$ ceramics synthesized by spray drying method. *Mater. Res. Bull.*, Vol. 46, No. 1, (Jan 2011), pp. 70–74, ISSN 0025-5408

López-Juárez, R.; González-García, F.; Zárate-Medina, J.; Escalona-González, R.; Díaz de la Torre, S. & Villafuerte-Castrejón, M.E. (2011b). Piezoelectric properties of Li–Ta co-doped potassium–sodium niobate ceramics prepared by spark plasma and conventional sintering. *J. Alloy compd.*, Vol. 509, No. 9, (Mar 2011), pp. 3837–3842, ISSN 0925-8388

López-Juárez, R.; Novelo-Peralta, O.; González-García, F.; Rubio-Marcos, F. & Villafuerte-Castrejón, M.E. Ferroelectric domain structure of lead-free potassium-sodium niobate ceramics. *J. Europ. Ceram. Soc.*, (In press, 2011c), ISSN 0955-2219

Ma, Y.J.; Cho, J.H.; Lee, Y.H. & Kim, B.I. (2006). Hydrothermal synthesis of $(Bi_{1/2}Na_{1/2})TiO_3$ piezoelectric ceramics. *Mater. Chem. Phys.*, Vol. 98, No. 1, (July 2006), pp. 5–8, ISSN 0254-0584

Maeda, T.; Takiguchi, N.; Ishikawa, M.; Hemsel, T. & Morita, T. (2010). (K,Na)NbO_3 lead-free piezoelectric ceramics synthesized from hydrothermal powders. *Mater. Lett.*, Vol. 64, No. 2, (Jan 2010), pp. 125–128, ISSN 0167-577X

Merz, W.J. (1949). The electric and optical behavior of $BaTiO_3$ single-domain crystals. *Phys. Rev.*, Vol. 76, No. 8, (Oct 1949), pp. 1221-1225, ISSN 0031-899X

Morrison, F.D.; Sinclair, D.C. & West, A.R. (1999). Electrical and structural characteristics of lanthanum-doped barium titanate ceramics. *J. Appl. Phys.*, Vol. 86, No. 11, (Dec 1999), pp. 6355-6366, ISSN 0021-8979

Nagata, H. & Takenaka, T. (2001). Additive effects on electrical properties of $(Bi_{1/2}Na_{1/2})TiO_3$ ferroelectric ceramics. *J. Eur. Ceram. Soc.*, Vol. 21, Nos. 10-11, pp. 1299-1302, ISSN 0955-2219

Nagata, H.; Yoshida, M.; Makiuchi, Y. & Takenaka, T. (2003). Large piezoelectric constant and high curie temperature of lead-free piezoelectric ceramic ternary system based on bismuth sodium titanate-bismuth potassium titanate-barium titanate near the morphotropic phase boundary. *Jpn. J. Appl. Phys. Pt. 1*, Vol. 42, No. 12, (Dec 2003), pp. 7401-7403, ISSN 0021-4922

Noheda, B.; Cox, D.E.; Shirane, G.; Gonzalo, J.A.; Cross, L.E. & Park, S.E. (1999). A monoclinic ferroelectric phase in the $Pb(Zr_{1-x}Ti_x)O_3$ solid solution. *Appl. Phys. Lett.*, Vol. 74, No. 14, (Apr 1999), pp. 2059-2061, ISSN 0003-6951

Park, H.Y.; Cho, K.H.; Paik, D.S. & Nahm, S. (2007). Microstructure and piezoelectric properties of lead-free $(1-x)(Na_{0.5}K_{0.5})NbO_3$-$x$$CaTiO_3$ ceramics. *J. Appl. Phys.*, Vol. 102, No. 11, (Dec 2007), pp. 124101(1-5), ISSN 0021-8979

Pardo, L.; García, A.; Brebøl, K.; Curecheriu, L.P.; Mitoseriu, L.; Mercadelli, E. & Galassi, C. (2010). Piezoelectric characterization of lead-free ferroelectric ceramics. *Processing and Application of Ceramics*, Vol. 4, No. 3, (2010), pp. 199–207

Rahaman, M.N. (2006). *Ceramic processing and sintering* (second edition), Marcel Dekker, ISBN 0824709888, New York, USA

Ringgaard, E., & Wurlitzer, T. (2005). Lead-free piezoceramics based on alkali niobates. *J. Eur. Ceram. Soc.*, Vol. 25, No. 12, (2005), pp. 2701–20706, ISSN 0955-2219

Rojac, T.; Bencan, A.; Ursic, H.; Malic, B. & Kosec, M. (2008). Synthesis of a Li- and Ta-modified $(K,Na)NbO_3$ solid solution by mechanochemical activation. *J. Am. Ceram. Soc.*, Vol. 91, No. 11, (Nov 2008), pp. 3789–3791, ISSN 0002-7820

Rubio-Marcos, F.; Ochoa, P. & Fernandez, J.F. (2007). Sintering and properties of lead-free $(K,Na,Li)(Nb,Ta,Sb)O_3$ ceramics. *J. Eur. Ceram. Soc.*, Vol. 27, No. 13-15, (2007), pp. 4125–4129, ISSN 0955-2219

Saito, Y., & Takao, H. High Performance Lead-free Piezoelectric Ceramics in the $(K,Na)NbO_3$-$LiTaO_3$ Solid Solution System. *Ferroelectrics*, Vol. 338, No. 1, (2006), pp. 17–32, ISSN 0015-0193

Saito, Y.; Takao, H.; Tani, T.; Nonoyama, T.; Takatori, K.; Homma, T.; Nagaya, T. & Nakamura, M. (2004). Lead-free piezoceramics. *Nature*, Vol. 432, No. 7013, (Nov 2004), pp. 84–87, ISSN 0028-0836

Scott, F.J. (2007). Applications of modern ferroelectrics. *Science*, Vol.315, No. 5814, (Feb 2007), pp. 954-959, ISSN 0036-8075

Scott, F.J. (2008) Ferroelectrics go bananas. *J. Phys.: Condens. Matter.*, Vol. 20, No. 2, (Jan 2008), p. 021001, ISSN 0953-8984

Shen, Z.Y.; Li, J.F.; Wang, K.; Xu, S.; Jiang, W. & Deng, Q. (2010). Electrical and Mechanical Properties of Fine-Grained Li/Ta-Modified $(Na,K)NbO_3$-Based Piezoceramics Prepared by Spark Plasma Sintering. *J. Am. Ceram. Soc.*, Vol. 93, No. 5, (May 2010), pp. 1378–1383, Online ISSN 1551-2916

Shirane, G.; Newnham, R. & Pepinsky, R. Dielectric properties and phase transitions of $NaNbO_3$ and $(Na,K)NbO_3$. *Phys. Rev.*, Vol.96, No.3, (Nov 1954), pp. 581-588, ISSN 0031-899X

Shiratori, Y.; Magrez, A. & Pithan, C. (2005). Particle size effect on the crystal structure symmetry of $K_{0.5}Na_{0.5}NbO_3$. *J. Europ. Ceram. Soc.*, Vol. 25, No. 12, (July 2005), pp. 2075–2079, ISSN 0955-2219

Singh, K.; Lingwal, V.; Bhatt, S.C.; Panwar, N.S. & Semwal, B.S. (2001). Dielectric properties of potassium sodium niobate mixed system. *Mater. Res. Bull.*, Vol. 36, No. 13-14, (Nov 2001), pp. 2365–2374, ISSN 0025-5408

Smolenskii, G.A.; Isupov, V.A.; Agranovskaya, A.I. & Krainik, N.N. (1961). New ferroelectrics of complex composition .4. *Sov. Phys. Solid State*, Vol. 2, No. 11, (1961), pp. 2651-2654, ISSN 0038-5654

Sun, C.; Xing, X.; Chen, J.; Deng, J.; Li, L.; Yu, R.; Qiao, L. & Liu, G. (2007). Hydrothermal Synthesis of Single Crystalline (K,Na)NbO_3 Powders. *Eur. J. Inorg. Chem.*, Vol. 2007, No. 13, (May 2007), pp. 1884–1888, Online ISSN 1099-0682

Takenaka, T. ; Sakata, K. & Toda, K. (1990). Piezoelectric properties of $(Bi_{1/2}Na_{1/2})TiO_3$-based ceramics. *Ferroelectrics*, Vol. 106, pp. 375-380, ISSN 0015-0193

Takenaka, T.; Nagata, H. & Hiruma, Y. (2008). Current Developments and Prospective of Lead-Free Piezoelectric Ceramics. *Jap. J. Appl. Phys.*, Vol. 47, No. 5, (May 2008), pp. 3787-3801, ISSN 0021-4922

Tokita, M. (1993). Trends in advanced SPS Spark Plasma Sintering Systems and Technology. *J. Soc. Powder Technol. Jpn.*, Vol. 30, No. 11, (Nov 1993), pp. 790–804, ISSN 0386-6157

von Hippel, A.; Breckenridge, R. G.; Chesley, F. G. & Tisza, L. (1946). High Dielectric Constant Ceramics. *Ind. Eng. Chem.*, Vol. 38, No. 11, (Nov 1946), pp. 1097–109, ISSN 0019-7866

Wang, K. & Li, J.F. (2010). Low-Temperature Sintering of Li-Modified (K, Na)NbO_3 Lead-Free Ceramics: Sintering Behavior, Microstructure, and Electrical Properties. *J. Am. Ceram. Soc.*, Vol. 93, No. 4, (Apr 2010), pp. 1101–1107, Online ISSN 1551-2916

Wang, K.; Zhang, B.P.; Li, J.F. & Zhang, L.M. (2008). Lead-free $Na_{0.5}K_{0.5}NbO_3$ piezoelectric ceramics fabricated by spark plasma sintering: Annealing effect on electrical properties. *J. Electroceram.*, Vol. 21, No. 1-4, (Dec 2008), pp. 251-254, ISSN 1385-3449

Wang, Y.; Damjanovic, D.; Klein, N.; Hollenstein, E. & Setter, N. (2007). Compositional Inhomogeneity in Li- and Ta-Modified (K, Na)NbO_3 Ceramics. *J. Am. Ceram. Soc.*, Vol. 90, No. 11, (Nov 2007), pp. 3485–3489, Online ISSN 1551-2916

Wang, Y.; Qibin, L. & Zhao, F. (2010). Phase transition behavior and electrical properties of $[(K_{0.50}Na_{0.50})_{1-x}Ag_x](Nb_{1-x}Ta_x)O_3$ lead-free ceramics. *J. Alloy compd.*, Vol. 489, No. 1, (Jan 2010), pp. 175–178, ISSN 0925-8388

Wang, Y.; Wu, J.; Xiao, D.; Zhu, J.; Yu, P.; Wu, L. & Li, X. (2008). Piezoelectric properties of (Li, Ag) modified $(Na_{0.5}K_{0.50})NbO_3$ lead-free ceramics with high Curie temperature. *J. Alloy compd.*, Vol. 459, No. 1-2, (July 2008), pp. 414–417, ISSN 0925-8388

West, A. R. (2006). Inorganic functional materials: Optimization of properties by structural and compositional control. *Chem. Rec.*, Vol. 6, No. 4, (Aug 2006), pp. 206-216, ISSN 1527-8999

Wul, B. M. & Goldman, I. M. (1945). Dielectric Constants of Titanates of Metals of the Second Group. *Dokl. Akad. Nauk SSSR*, Vol. 46, pp. 154–57 (1945)

Yang, W.; Jin, D.; Wang, T. & Cheng, J. (2010). Effect of oxide dopants on the structure and electrical properties of $(Na_{0.5}K_{0.5})NbO_3$-$LiSbO_3$ lead-free piezoelectric ceramics. *Physica B: Cond. Matter*, Vol. 405, No. 7, (Apr 2010), pp. 1918–1921, ISSN 0921-4526

Yao, Z.H.; Liu, H.X.; Chen, L. & Cao, M.H. (2009). Morphotropic phase boundary and piezoelectric properties of $(Bi_{0.5}Na_{0.5})(1-x)(Bi_{0.5}K_{0.5})(x)TiO_3$-$0.03(Na_{0.5}K_{0.5})NbO_3$

ferroelectric ceramics. *Mater. Lett.*, Vol. 63, No. 5, (Feb 2009), pp. 547-550, ISSN 0167-577X

Yin, Q.; Yuan, S.; Dong, Q. & Tian, C. (2010). Effect of CuO and MnO_2 doping on electrical properties of $0.92(K_{0.48}Na_{0.54})NbO_3$-$0.08LiNbO_3$ under low-temperature sintering. *J. Alloy Comp.*, Vol. 491, Nos. 1-2, (Feb 2010), pp. 340–343, ISSN 0925-8388

Yoon, S.; Baik, S.; Kim, M.G.; Shin, N. & Kim, I. (2007). Synthesis of tetragonal barium titanate nanoparticles via alkoxide-hydroxide sol-precipitation: Effect of water addition. *J. Am. Ceram. Soc.*, Vol. 90, No. 1, (Jan 2007), pp. 311–314, ISSN 0002-7820

Yu, J.; Paradis, P.F.; Ishikawa, T.; Yoda, S.; Saita, Y.; Itoh, M. & Kano, F. (2004). Giant Dielectric Constant of Hexagonal $BaTiO_3$ Crystal Grown by Containerless Processing. *Chem. Mater.*, Vol. 16, No. 21, (Oct 2004), pp. 3973-3975, ISSN 0897-4756

Zang, G.Z.; Yi, X.J.; Xu, Z.J.; Pu, X.P.; Fu, P. & Zhao, L.M. (2010). Lead-free $(K_{0.5}Na_{0.5})_{0.95}(LiSb)_{0.05}Nb_{0.95}O_3$-$BaTiO_3$ Piezoceramics. *J. Electroceram.*, Vol. 25, No. 1, (Aug 2010), pp. 85-88, ISSN 1385-3449

Zhang, S.T.; Kounga, A.B.; Aulbach, E.; Granzow, T.; Jo, W.; Kleebe, H.J. & Rodel, J. (2008). Lead-free piezoceramics with giant strain in the system $Bi_{0.5}Na_{0.5}TiO_3$-$BaTiO_3$-$K_{0.5}Na_{0.5}NbO_3$. I. Structure and room temperature properties. *J. Appl. Phys.*, Vol. 103, No. 3, (Feb 2008), p. 034107, ISSN 0021-8979

Zhao, W.; Zhou, H.P.; Yan, Y.K.; Liu, D. & Liu, S.S. (2007). Influence of different dopants on the piezoelectric properties of the $Na_{1/2}Bi_{1/2}TiO_3$-$BaTiO_3$- lead-free ceramics, *Key Eng. Mater.*, Vols. 336–338, pp. 105-108, ISSN 1013-9826,

Zhou, C.; Liu, X.; Li, W. & Yuan, C. (2009). Dielectric and piezoelectric properties of Y_2O_3 doped $(Bi_{0.5}Na_{0.5})_{(0.94)}Ba_{0.06}TiO_3$ lead-free piezoelectric ceramics. *Materials Res. Bull.*, Vol. 44, No. 4, (Apr 2009), pp. 724–727, ISSN 0025-5408

Zhou, Y.; Yu, J.; Guo, M. & Zhang, M. (2010). Microwave Hydrothermal Synthesis and Piezoelectric Properties Investigation of $K_{0.5}Na_{0.5}NbO_3$ Lead-Free Ceramics. *Ferroelectrics*, Vol. 404, No. 1, (2010), pp. 69–75, ISSN 0015-0193

Zuo, R.; Ye, C. & Fang, X. (2007). Dielectric and Piezoelectric Properties of Lead Free $Na_{0.5}K_{0.5}NbO_3$–$BiScO_3$ Ceramics. *J. Journal of Appl. Phys.*, Vol. 46, No. 10A, (2007), pp. 6733–6736, ISSN 0021-4922

16

Synthesis of PZT Ceramics by Sol-Gel Method and Mixed Oxides with Mechanical Activation Using Different Oxides as a Source of Pb

J. M. Yáñez-Limón[1], G. Rivera-Ruedas[1], F. Sánchez De: Jesús[2], A. M. Bolarín-Miró[2], R. Jiménez Riobóo[3] and J. Muñoz-Saldaña[1]

[1]*Centro de Investigación y Estudios Avanzados del IPN. Unidad Querétaro. Libramiento Norponiente No. 2000 Fracc. Real de Juriquilla, Querétaro, Qro*

[2]*Área Académica de Ciencias de la Tierra y Materiales, Universidad Autónoma del Estado de Hidalgo, CU, Carr. Pachuca-Tulancingo, Pachuca, Hgo,*

[3]*Instituto de Ciencia de Materiales de Madrid Consejo Superior de Investigaciones de Científicas Cantoblanco, Madrid,*

[1,2]*México*

[3]*Spain*

1. Introduction

Lead titanate zirconate is a ferroelectric material that presents excellent piezoelectric and pyroelectric properties, applied to actuators transductors and generators. Despite of the study of new lead-free ferroelectric materials that are being studied nowadays, few of these can compete with the properties of PZT (Aman et al., 2010; Ky et al., 2010; Sawawuchi,1952; Shrout & Zhang, 2007; Wei et al., 2010; Zhang et al., 2010; Zhou et al.,2004.

It is well known that within the different compositions of PZT (Zr/Ti) the compositions closer to the morphotropic line (composition 53/47) show the best ferroelectric, piezoelectric and pyroelectric properties (Jaffe et al., 1954). Which has been attributed to the coexistence of a mixture of rhombohedral and tetragonal phases, as well as to the existence of a monoclinic phase (Noheda et al., 2000). The width of the morphotropic phase boundary varies and depends on the homogeneity of the composition, the synthesis method and the conditions of the mixture processing (Shirane & Takeda,1952a; 1952b; Sooksaen et al., 2008).

Despite the existence of different well established methods for the preparation of ferroelectric ceramics, the necessity of controlling stoichiometry, doping, grain size, porosity, the homogeneity of the obtained phases, the control of the particle size in submicrometric powders, the obtention of thin films through different techniques, as well as lowering costs of processing and utilization of low-toxicity reactives, make important the exploration of variants in synthesis methods (Charles et al., 1992, Gringber & Rappe, 2007, Hammer & Hoffman, 1998, Heywang etal., 2008; Pontes et al., 2004).

There are several routes reported in the literature for synthesis of ferroelectric PZT ceramics (Guarany et al., 2007, Legrand et al., 2007) with the subsequent densification by pressureless sintering in oxygen rich atmosphere. The mechanochemical process is based on mechanical energy (associated e.g. to high energy ball milling) instead of thermal energy to provide the activation energy for solid-state reaction and has several advantages over both conventional solid-state reaction and wet-chemical processes, including the use of low-cost raw materials, simplicity of the process, and the ability to obtain fine particles (Hurtado-Macias et al.,2008). For instance, some reports show that PZT powders were successfully prepared by mechanochemical synthesis in a significantly shorter time in comparison with literature data, using a planetary ball mill (Brankoviĉ et al., 2003a; Brankoviĉ et al., 2003b; Legrand et al, 2007; Schwartz et al., 2004). However, the mechanosynthesis process, using different kind of high energy ball milling, is a complex process and hence involves optimization of a number of variables that are sensitive to the type of milling process to achieve product phases with Perovskite structures.

In the case of the mixed oxides and posterior calcination and sintering, in this work a mechanoactivation stage is added in the oxides stoichiometric mixture without introducing excess of Pb, through high energy milling, exploring milling times in rough conditions during 4, 8 and 12 hrs in a spex-8000.D mill. Additionally, in this work, despite the traditional method where PbO (litharge) is utilized as source of Pb, the process of obtention of PZT with two additional Pb oxides, that in general are not so frequently used in the literature, PbO_2 (plattnerite) and Pb_3O_4 (minium) are studied. Kinetic studies are performed by means of x-ray diffraction, Rietveld analysis during milling stages (Kong et al., 2008), and thermal treatments at 300°C (at this temperature the changes obtained in the milling time are manifested), 500, 700, and 900°C, in compositions near the morphotropic line, 55/45, 53/47, and 52/49 for the three different sources of lead. According with the phase diagram of PZT, these compositions show rhombohedral phases, a mixture of tetragonal/rhombohedral and tetragonal phases respectively. In this case it t is important to find the best combination of processing parameters to synthesize pure PZT, and compare the results of the mechanochemical synthesis of $Pb(Zr_{53}Ti_{47})O_3$, using different Pb oxides as precursors by carrying out a quantification of the phase contents.

In the case of sol-gel processing and with the purpose of comparing the materials obtained, the same compositions of the oxides mixture combined with mechanicoactivation were sintered. The 2-metoxyethanol route (Zhang et al.,2001) was utilized again without the introduction of lead excess in the stoichiometry of the initial solution. The sol-gel process has been used since it has the advantages of low temperature processing, high purity and very good composition control (Coffman et al., 1994). Sol-gel precursors can be used for the preparation of PZT thin films by spin or dip coating and also for the synthesis of PZT powders, that can be used also for the preparation of bulk samples [11-13]. In this work the 2-metoxyethanol polymeric sol-gel route was used following the method described by Coffman et al (Coffman & Dey, 1994, Coffman et al., 1996) to prepare solutions of PZT precursor, inducing thereafter a gelation stage by adding ethylenglycol and water. Using these precursors it was possible to synthesize crystalline PZT powders at low calcination temperatures. The chosen compositions of PZT were those of Zr/Ti ratios close to the morphotropic phase boundary in both sides of the MPB the tetragonal and romboedric phases (55/45, 53/47, 51/49).

Morphological, structural, dielectric permittivity, hysteresis cycles and Pyroelectric response were conducted, which then are compared with the corresponding results obtained by the oxide mixture and mechanicoactivation for the three different sources of Pb.

Comparing both routes of synthesis regarding costs, security and speed, the mechanicoactivation route is the most favored, nevertheless because of the purity of the powders obtained and the control of the Phases. The sol-gel method is also appropriate, with the problem of the use of the toxic reactive 2-metoxiethanol, which must be handled very carefully, additionally the cost of the precursors utilized is high. However, in this work ceramics with similar characteristics and ferroelectric behaviors in both synthesis routes were obtained.

2. Synthesis of materials

2.1 Mixture of powders oxides with mechanical activation

PZT powders were synthesized by a combination of mechanosynthesis and calcination treatments. Commercial powders of lead oxide (PbO, Fisher Sigma Aldrich 99.9%), lead oxide IV (PbO_2 Sigma Aldrich 97%), lead oxide II-IV (Pb_3O_4, Fermont 99.8%) were used as Pb sources, zirconium dioxide (ZrO_2, Sigma Aldrich 99.9%) and titanium dioxide (TiO_2, Fermont 99.99%) were mixed with the corresponding Pb oxide following stoichiometric ratios to obtain PZT samples with a composition of Zr/Ti=53/47, that corresponds to a tetragonal phase close to the morphotropic phase boundary (MPB). The stoichiometric balance for each lead oxide was done according to Eq. 1.

$$\begin{aligned}
&A) PbO + 0.53ZrO_2 + 0.47TiO_2 \rightarrow Pb(Zr_{0.53}Ti_{0.47})O_3 \\
&B) PbO_2 + 0.53ZrO_2 + 0.47TiO_2 \rightarrow Pb(Zr_{0.53}Ti_{0.47})O_3 \\
&C) \frac{1}{3}Pb_3O_4 + 0.53ZrO_2 + 0.47TiO_2 \rightarrow Pb(Zr_{0.53}Ti_{0.47})O_3 + \frac{1}{6}O_2
\end{aligned} \qquad (1)$$

The samples and mixture powders will be identified with the corresponding source of lead oxide A(PbO), B(PbO_2) and C(Pb_3O_4). Powder mixtures in stoichiometric ratio (53/47 composition) were milled at 4, 8 and 12 h with zirconia balls of $1.27x10^{-2}$ m in diameter in a nylamid cylindrical vial ($60x10^{-6}$ m^3) at room temperature in air atmosphere using a high energy mixer/mill. The charge ratio (CR) was 10:1. Thereafter, the milled powders were uniaxially pressed at 588 MPa to obtain green samples with a diameter of $1.27x10^{-2}$ m. Calcination treatments were done during 4 h at 300, 500, 700 and 900 °C with a heating and cooling rates of 2 and 5 °C/min, in air atmosphere, respectively.

2.2 Sol-gel method

The synthesis of PZT powders was carried out starting with the precursor solution of PZT prepared by the sol-gel process using the 2-metoxyethanol route following the method described by Coffman et al., (Coffman & Dey, 1994, Corrman et al., 1996). Precursors were prepared using lead acetate, Ti-isopropoxide and Zr-n-propoxide (99.9, 97 and 70% purity, respectively, Sigma-Aldrich), as starting materials. Precursors reaction was activated with 2-metoxyethanol leading to the corresponding metal-methoxyethoxides (lead acetate previously dehydrated) and thereafter the three metal-methoxyethoxides (Pb, Zr and Ti) were mixed and kept in the gelation process adding ethylenglycol and the required amount of water. The gel was dried at 100°C and heat treated in two stages. The first heat treatment was carried out at 400°C for two hours with a heating rate of 9°C/min. The obtained powder agglomerates were manually milled in a mortar and then calcined at 850°C for 4h with a scan rate of 9°K/min to promote the formation of the Perovskite structure of PZT. After this second heat treatment, the carbon compounds were eliminated leading to nanometric PZT powders with high purity.

3. Experimental results

3.1 Kinetics aspects in the PZT formation during mechanical milling and thermal treatment

For comparison purposes, the structure of commercial powders used as precursors was characterized by XRD and is shown in Figure 1. Previous to the milling, a mixture of orthorhombic and tetragonal phases was identified in the PbO powder, whereas in PbO_2 and Pb_3O_4, platnerite and minium were identified, both having tetragonal phase. Further on, TiO_2 and ZrO_2 powders showed anatase (tetragonal) and monoclinic phases, respectively. These results are important to identify phase transformations after milling and heat treatments. Figure 1 also shows the structural evolution due to milling at times from 4 to 12 h of all powder mixtures. Increasing the milling time leads to a clear diminution of peaks intensity, which causes difficulties in the quantification of phases. Nevertheless, few comments of clear effects regarding phase transformations are given in this section.

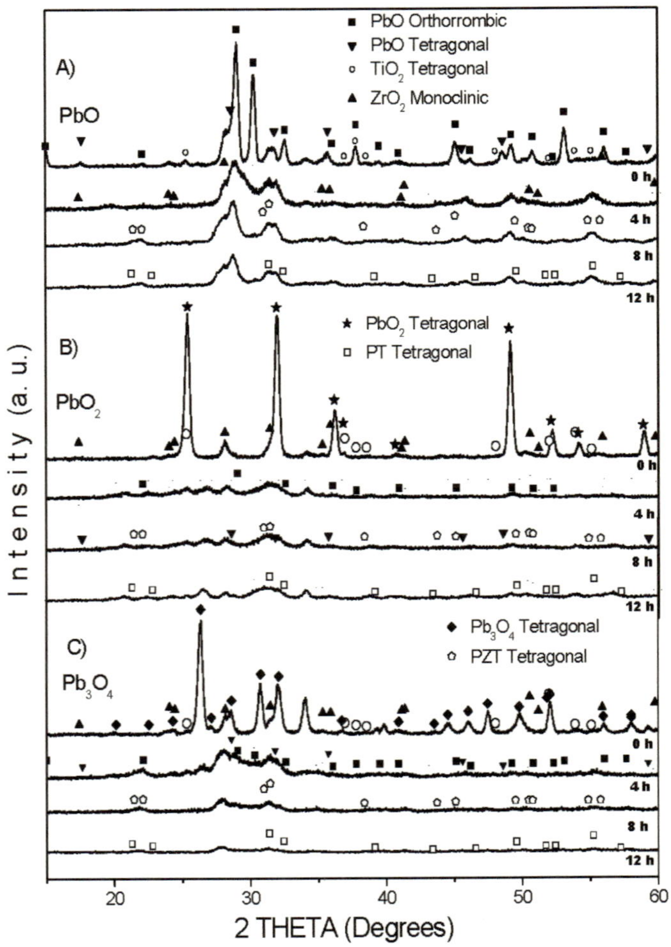

Fig. 1. Evolution of the XRD patterns with milling time of the mixture starting oxides, of three studied set of samples, A) (ZrO_2, TiO_2,PbO), B) (ZrO_2, TiO_2,PbO_2) and C) (ZrO_2, TiO_2,Pb_3O_4)

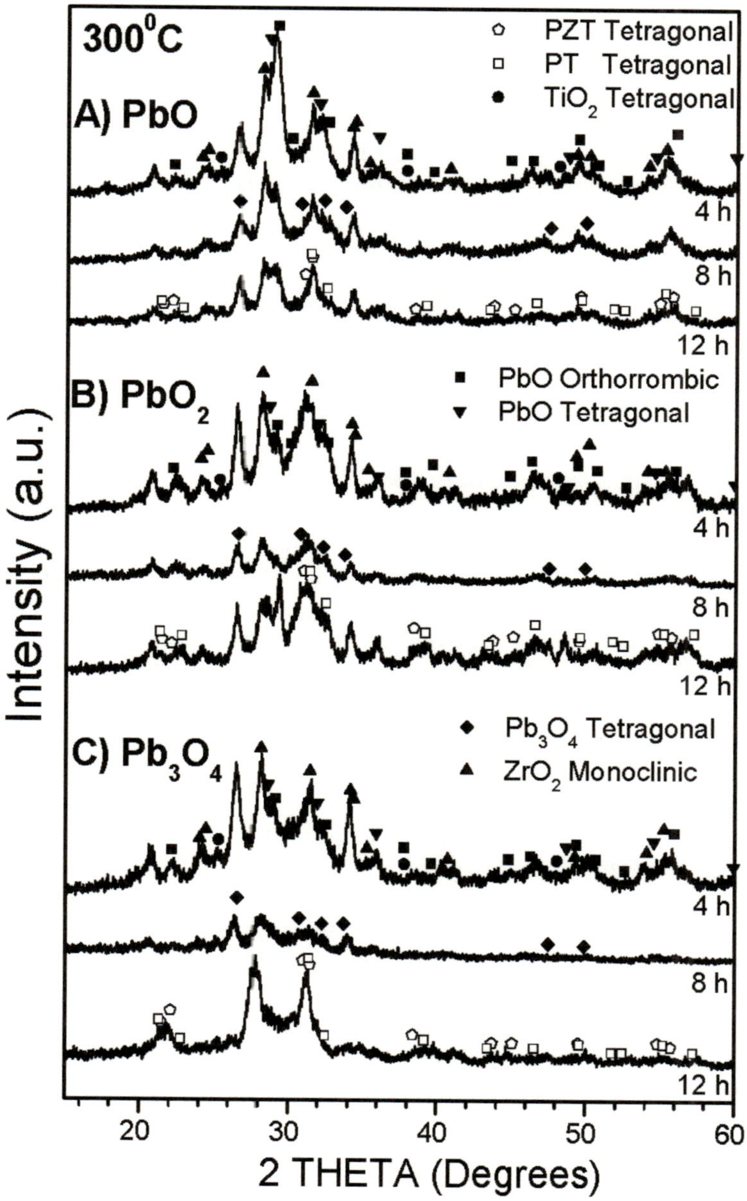

Fig. 2. XRD-patterns evolution of set samples: A (PbO), B (PbO2) and C (Pb3O4). Thermal treatment at 300°C after milling time at 4, 8 and 12 h

In the case of mixtures with PbO, 4 h of milling are enough to cause almost a full phase transformation from orthorhombic to tetragonal of the PbO powder (PbO-t), which originally showed two phases. For this particular case, at 4h of milling the ZrO_2 powder kept the monoclinic phase and additional peaks can be found around 20, 22, 31-32, 46 and 55-56°

in 2 theta scale. The new peaks are identified with a mixture of PZT and $PbTiO_3$ (PT), both having tetragonal phase. Increasing the milling time causes further transformations, where the correspondent XRD patterns are constituted principally by PbO, ZrO_2 PZT and PT. The powder mixtures prepared with B(PbO_2) before milling show the Platnerite phase (tetragonal phase). After 4 h of milling, the intensity of PbO_2 diffraction peaks are drastically reduced. The XRD pattern has the PbO_2 and ZrO_2 visible as well as additional peaks appearing around 21, 22, 27, 38.8, and 56.5° of 2-theta, which were identified as PZT and PT both in tetragonal phases. Increasing the milling time to 8 and 12 h doesn´t cause further changes in the phase constitution. For the mixture prepared with Pb_3O_4, the XRD peaks show the expected phases corresponding to the starting oxides, Pb_3O_4 (minium, tetragonal phase), ZrO_2 (monoclinic) and TiO_2 (tetragonal). After 4 h of milling the XRD pattern also shows a dramatic phase transformation observed with the diminution intensity of the Pb_3O_4 tetragonal peaks. This effect was completed increasing the milling time to 8 and 12 h. Additional peaks at 22 and 55°, which correspond to the tetragonal phase of PZT already, appear at 4h of milling. At 8 and 12 h of milling the powder mixture is conformed by the PZT, PT and ZrO_2 phases.

Summarizing, in all mixtures, varying the Pb source (PbO, PbO_2 and Pb_3O_4), the mechanochemical activation produces an effect of particle size diminution of the original oxides that causes diminution in intensity of the diffraction peaks, and gives place to small quantities of $PbTiO_3$ and PZT. Heat treatments carried out at low temperatures (300 °C) will be used to further analysis of the influence of milling time on the phase contents in the different powder mixtures, which will be discussed in the next paragraphs.

Figures 2 to 4 show the structural evolution due to thermal annealing (T_a = 300, 500, 700°C and 900°C) for each set of samples after they were milled at (t_m= 4, 8 and 12 hours). As mentioned before, heat treatments at the lowest temperature helps to quantify the effect of milling time, since no dramatic phase transformations are expected at 300°C (Babushkin et al., 1996). For instance, the phase transformation from orthorhombic to tetragonal phases of PbO, which happens by increasing the milling time from 4 to 12 h, already discussed in Figure 1 is herewith confirmed.

Fig. 2 shows the XRD patterns of the samples annealed at 300°C, some peaks are well defined and permit us to identify the phase transformation due to milling effect. For example the Pb_3O_4, ZrO_2, and PbO (orthorhombic) phases are observed in the three set of samples, additionally to PZT and PT presents in the samples without thermal annealing. In the C samples the peaks (100), (110) and (112) at 22, 32 and 56° respectively are best resolved corresponding to the perovskite PZT phase. For t_m=12 h the XRD patterns of samples A and B are constituted by PT, PZT, PbO, Pb_3O_4 and ZrO_2, at difference the C samples show the PT, PZT, ZrO_2 and PbO phases, the Pb_3O_4 phase is not observed.The Rietveld refinement analysis of these patterns clearly revealed the formation of PZT and PT, where the former increases from 7 to 12 wt.% as a function of milling time. The PT content remains constant around 12 wt.% irrespective of milling time. Further on, the PbO-t content diminished (from 15 to 10 wt.%) increasing the milling time, giving place to PT and PZT

Fig.3 shows the XRD patterns for the set of samples annealed at 500°C. At this temperature the formation of PZT phase independently of the previous milling time is clearly observed. Although significant differences among these set of samples are observed. The A samples, besides the PZT phase, show ZrO_2 and PbO phases. The B sample also shows the ZrO_2 and PT phases as well. And the C samples also show the PbO phase, but their peaks show lower intensity than samples A and B at 8 and 12 h.

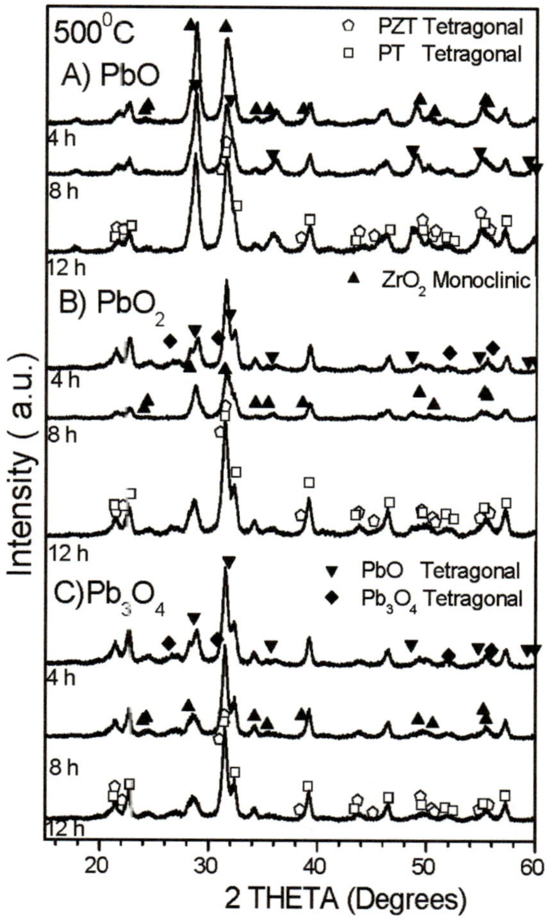

Fig. 3. XRD-patterns evolution of set samples A(PbO), B(PbO_2) and C(Pb_3O_4). Thermal treatment at 500°C after milling time at 4, 8 and 12 h.

Fig. 4a shows the effect of thermal annealing at T_a=700°C. The A samples at 4 and 8 h of milling show the PZT phase (91 and 90 %wt respectively) and a small peak associated with PbO (around 10% wt). At 12 h of milling aside from the PZT phase, the pyrochlore phase (10%wt) is also present, in these sense, long periods of milling in the PbO case give place to the undesirable pyrochlore phase. In the B samples with 4 h of milling, the PZT (97%wt) phase and pyrochlore (3%wt) phase are shown. Nevertheless, in this case there is a reversed effect because at 8 and 12 h of milling only the PZT (94% wt) and PbO (6%wt) phases are present. In the C samples the PZT phase (around 97%wt) at all milling times in its tetragonal phase and a small peak associated with PbO (around 3%wt) are observed. Fig. 4b shows the XRD patterns of the samples annealed at 900°C during 4 hrs. In the A samples at 4 h of milling, only the PZT phase is observed. Additionally At 8 h of milling, the PbO (4%wt) and pyrochlore phases (3%wt) are present. Finally at 12 h of milling, besides the PZT, the PbO phase is also observed. For the set of samples B and C at all milling times the perovskite in its tetragonal phase was obtained.

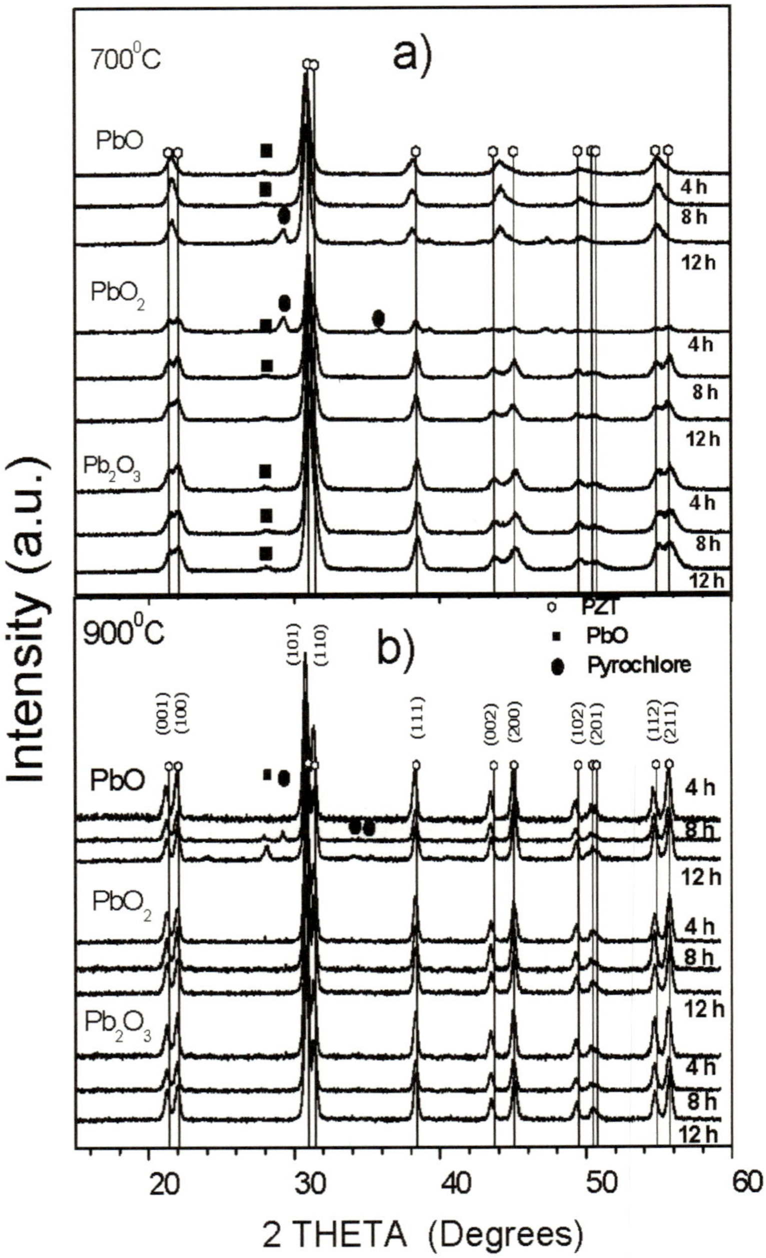

Fig. 4. Evolution of XRD-patterns from the set of samples A(PbO), B(PbO_2) and C(Pb_3O_4). Thermal treatment at a) 700°C and b) 900°C, after milling time at 4, 8 and 12 h.

Fig. 5 shows the representative results of quantitative analysis (obtained by Rietveld analysis) of the evolution of crystalline phases observed during the mechanical milling (4,8 and 12 h) with a corresponding heat treatment at 300 , 500, 700 and 900 ° C, for the sample A (PbO), similar results were obtained in samples B and C. On the other hand Figure 6 shows the comparative analysis of the concentrations evolution of ZrO_2 and PZT for the three sets of samples prepared with different lead oxides as a function of thermal treatment after they were submitted to a milling process during 4h.

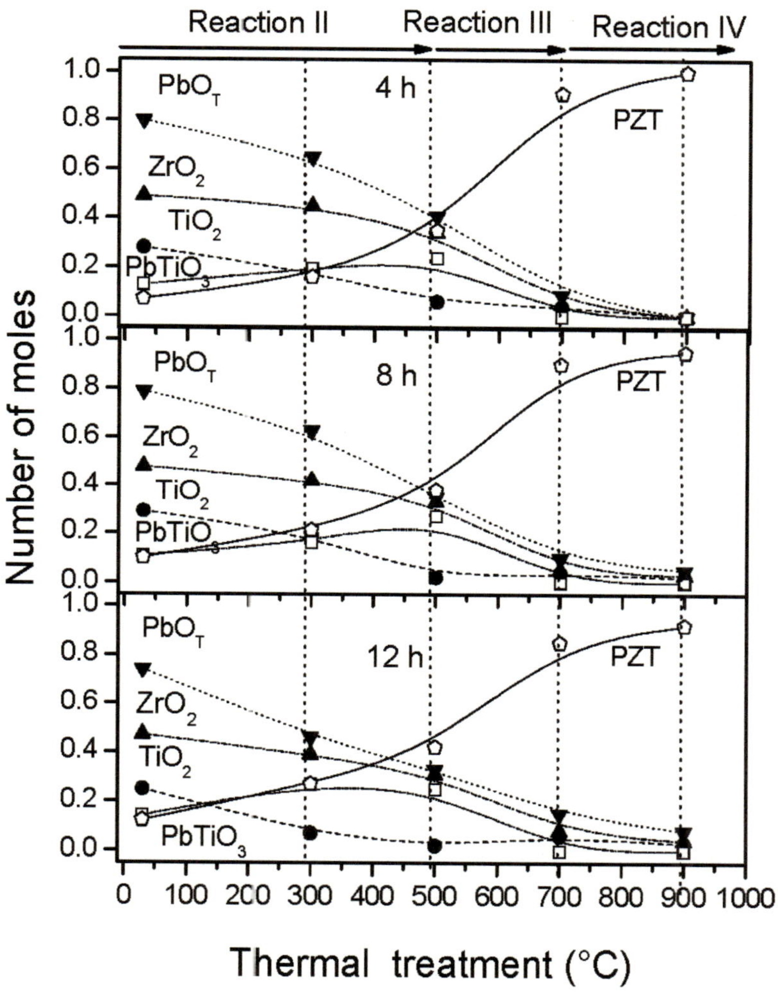

Fig. 5. Evolution concentration of the mixture oxides after 4 h of milling as a function of the thermal treatment in the obtention of PZT using PbO precursor.

The concentration of oxides shown in Figures 5 and 6 at 30 ° C, correspond to the molar amount quantified by the Rietveld method of powder mixture subjected to a grinding 4, 8 and 12 hours without heat treatment, which started from a stoichiometric ratio, depending on the type of oxide used for the composition 53/47 of PZT.

In the work of (Babushkin & Lindbach, 1996) related to the kinetics of formation of PZT obtained by the traditional method of mixing oxides, four regions of transformation are established, which may be susceptible to particle size, impurities and morphology of the starting powders. These regions are defined by the temperatures of treatment as follows:

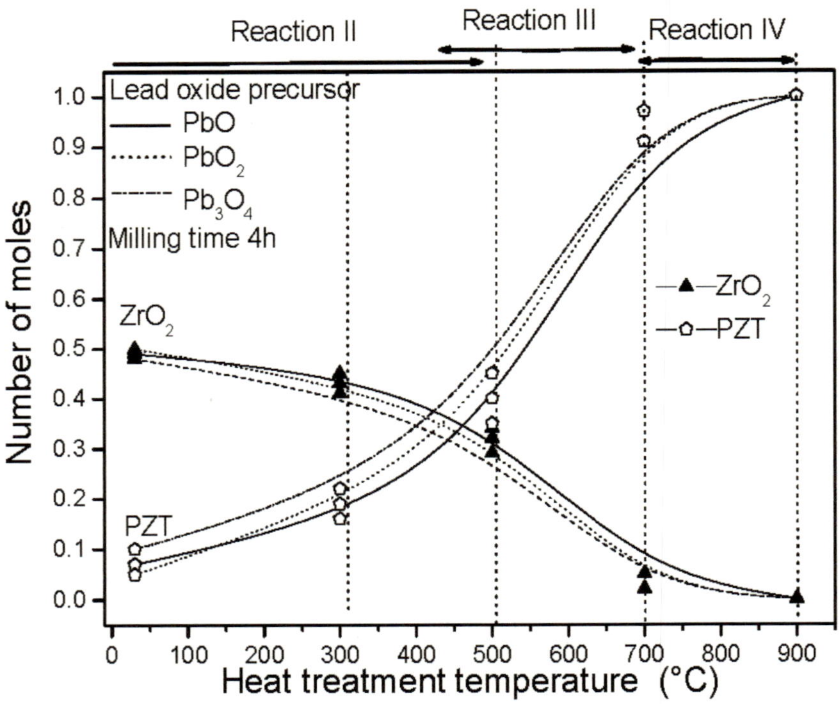

Fig. 6. Evolution concentration of ZrO_2 and PZT as a function of thermal treatment after a milling time of 4 h, for mixture powders with A (PbO), B(PbO_2) and (Pb_3O_4) as source of Pb.

i. (T<350°C) no reaction is present.
ii. $PbO + TiO_2 \rightarrow PbTiO_3$, from 350 to approx. 630°C, formation of PT.
iii. $PbTiO_3 + PbO + ZrO_2 \rightarrow Pb(Zr_xTi_{1-x})O_3$ from 650 to approx. 950 °C, reaction of ZrO_2 and formation of PZT.
iv. $Pb(Zr_xTi_{1-x})O_3 + PbTiO_3 \rightarrow Pb(Zr_xTi_{1-x})O_3$; among 700 and 950 °C, complete reaction of PZT.

The synthesis of $Pb(Zr_x,Ti_{1-x})O_3$ by solid state reaction has been reported to be mainly attributed to Pb^{2+} ion diffusion, which is necessarily enhanced by the starting powders ranging from submicrometric to nanometric sizes. The mechanism of reaction starts with the formation of tetragonal $PbTiO_3$, which reaches a maximum at temperatures close to 680°C, with a subsequent reaction of the remaining ZrO_2 and TiO_2 leading to the complete solid state reaction of mixtures of $PbTiO_3$ and PZT at temperatures above 800°C.

In the case of oxide mixtures with mechanical activation, the reactions that occur with the milling and heat treatments can be summarized into four stages, following the scheme of (Babushkin & Lindbach) . The first stage is developed during the milling process, and the rest of the reactions of the stage are shown in Figure 5 and 6. The effect of the high energy milling is to lower the temperatures of the reaction in the formation of PZT. The reaction regions observed are the following:

i. Mechanical activation by high energy milling, which leads to phase transformations in the oxide precursors and the formation of PT and PZT with concentrations between 7 and 12% (30 ° C).
ii. Increased concentration of PT (17 to 33%) and PZT (16 to 42%), with the reduction of ZrO2 (48 to 35%) in a temperature range from 300 to 500 ° C.
iii. $PbTiO_3 + PbO + ZrO_2 \rightarrow Pb(Zr_xTi_{1-x})O_3$ (increase of the molar concentration of PZT up to 83%) in a temperature range from 400 to 700 ° C.
iv. $Pb(Zr_xTi_{1-x})O_3$ $PbTiO_3$ + $\rightarrow Pb(Zr_{x'}Ti_{1-x})O_3$ PZT complete transformation. (between 700 and 900 ° C)

Now, we attempt to compare our obtained results with the the mechanochemical activation by high energy ball milling of the powders respect to the known kinetic process.
After 4 h of milling, a mixture of phases of the starting powders with partially reacted PT and PZT can already be observed. Thus, such milling conditions allow us to have a premature mixture of reactions II and III, which increases with the milling time and happens before any heat treatment. After heat treatments at 700ºC, high PZT concentrations are obtained between 85 to 97% and the full reaction of PZT is already completed at 900ºC. Again compared with the typical kinetic reaction kinetic of PZT, which is typically completed at temperatures higher than 900ºC, the mechanochemical activation allows to lower the calcination temperatures and high concentration of PZT is obtained at 700ºC.
One of the main differences of the results obtained in this study with those reported by (Babushkin, & Lindbach), is that the activation by mechanical milling allows the transformation of phases at temperatures below 350 ° C, including the formation of PT and PZT which appear during milling (at concentrations of 7-12%) and are increased with heat treatments. Typically, the reaction process of ZrO_2 initiates at 650ºC, but the milling step allow (Figures 5 and 6) that in this case starts his reaction from 300 ° C. The formation of PZT and consumption of ZrO_2 with heat treatment is very similar for all three types of samples studied obtained with different types of oxides. From the beginning, the PbO starts to decrease, contributing as zirconia to the formation of PZT. At temperatures between 300 and 500 ° C the highest concentration of PT is shown, which like the PbO and ZrO_2, after 500 ° C contribute to the formation of PZT, at this temperature there is an appreciable increase in PZT concentration.

3.2 Electrical properties

3.2.1 Hysteresis cycles

Figure 7 shows the curves of hysteresis loops of PZT samples with 53/47 composition obtained by sol-gel method and mixed oxide (PbO and Pb_3O_4 as sources of Pb). Fig 7A) shows that the samples obtained by mixing oxides have a higher remanent polarization than that obtained by sol-gel, and within those obtained by mixture oxides, the sample obtained using Pb_3O_4 has a remanent polarization higher than the sample prepared with PbO.
The coercive fields have very similar values for the samples obtained by sol-gel and that obtained with Pb_3O_4, however the sample with PbO has a higher coercive field. In fig. 7B) it can be seen that the variation of the ratio of the remanent polarization to maximum polarization (Pr / Pm) as a function of applied bias field, has values very similar to the sol-gel samples, and to that obtained with PbO (about 88% for bias fields of 45 kV / cm) . On the other hand the sample with Pb_3O_4 presents values close to 98%, indicating that it virtually retains its polarization value after removing the bias field.

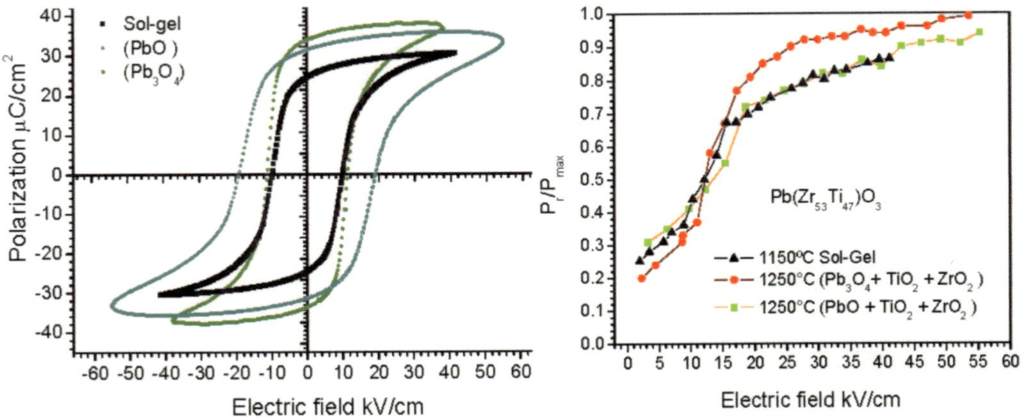

Fig. 7. A) Histeresys cycles for PZT (53/47) samples obtained by sol-gel and mixture oxides using PbO, and Pb_3O_4. B) Evolution of the ratio (Pr/Ps) as a function of the maximum electric field applied in the same samples as in the case of A) .

Table 1 shows the comparative parameters among the PZT samples obtained with oxide mixtures and by the sol-gel method, like the density, remnant polarization, coercive field, Curie temperature. In general, it can be observed that the values of the densification are higher than 93% of the theoretical value; the samples obtained with Pb_3O_4 show higher remnant polarization and lower values of coercive fields and their Curie temperature values are between 388 and 400°C. From this comparative values it is possible to establish that those samples obtained with Pb_3O_4 showed the best ferroelectric values.

Zr/Ti	*Precursor*	$T_{sint.}$ (°C)	E_{max} (kV/cm)	P_r (μC/cm²)	E_c (kV/cm)	P_r/P_{max}	ρ(g/cm³	T_c (°C))
55/45	Pb_3O_4	1250	27.08	25.55	10.814	0.90	7.52	392
	PbO	1250	27.23	19.13	11.765	0.80	7.48	388
	Sol-gel	1150	34.99	13.18	8.77	0.77	7.83	393
53/47	Pb_3O_4	1250	53.62	34.02	11.301	0.99	7.47	396
	PbO	1250	55.34	31.35	18.045	0.94	7.54	394
	Sol-gel	1150	41.18	34.65	9.89	.87	7.87	396
51/49	Pb_3O_4	1200	33	29.73	10.874	0.89	7.88	400
	PbO	1200	42.11	23.45	10.181	0.82	7.7	397
	Sol-gel	1150	45.4	33.38	11.54	0.86	7.78	400

Table 1. Comparative ferroelectric values of PZT samples obtained with mixture oxides and by the sol-gel method.

3.2.2 Dielectric function

Figure 8 shows the dielectric permittivity and dielectric loss as a function of temperature for the composition 53/47, obtained atA) 10 kHz and B) 1 MHz. The maximum of the dielectric permittivity is used to estimate the Curie temperature (data showed in Table1), where the samples suffer a phase transition from ferroelectric to paraelectric state. In general, the dielectric permittivity shows a strong dependence on temperature and varies from 1000 at 200°C to 20000 close to the Curie temperature. The samples obtained by sol-gel and Pb_3O_4 show similar values for 10 khz and 1 Mhz, nevertheless, the sample obtained with PbO shows minor values at 1 Mhz.

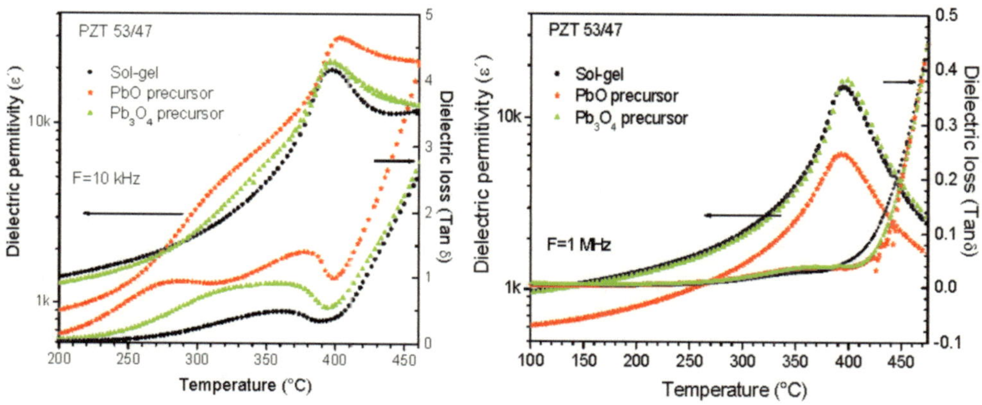

Fig. 8. Dielectric permittivity and dielectric loss of the samples obtained by mixture oxides and sol-gel, composition 53/47 as a function of temperature. A) Curves obtained at 10 khz and B) Curves obtained at 1 Mhz.

It is important to point out here that the samples obtained with PbO_2 although the corresponding structural phase of all compositions of PZT reported here were obtained, the corresponding electrical characterization was not measured, because it shows a high conductivity, due to the high vacancies concentration.

3.2.3 Photopyroelectric response

Figure 9 shows the photopyroelectric signal as a function of the modulation frequency, using a pohotopyroelectric system, (Mandelis & Zver, 1985, Marinelli et al., 1990, Balderas-López etal., 2007) of samples obtained by the sol-gel method and by mixture oxides with mechanical activation. For both set of samples the composition 53/47 shows the higher signal, and the samples obtained by the sol-gel method show a slightly higher signal than the samples obtained by the mixture oxides. For purpose of using these samples as photopyroelectric detectors they have a similar behaviour, nevertheless, the samples obtained by the sol-gel method show best response and it is inferred that they have a higher pyroelectric coefficient.

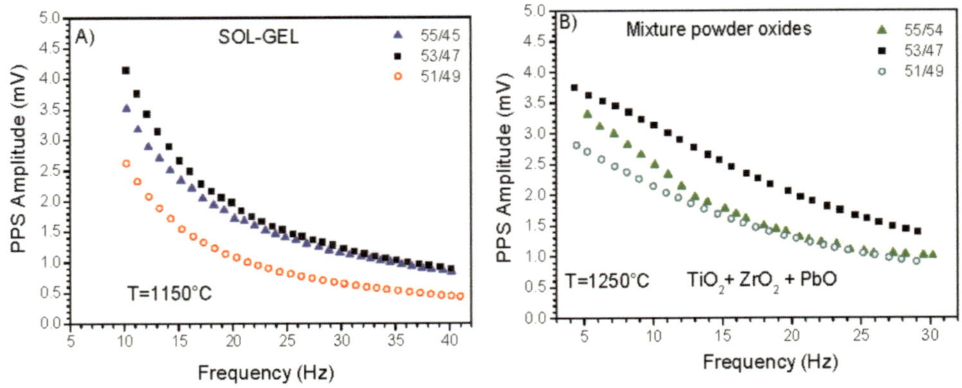

Fig. 9. Pyroelectric response in PZT samples with compositions 55/45, 53/47 and 51/49 obtained by A) Sol-gel method and B) mixture oxides with mechanical activation using PbO as Pb source

4. Conclusion

The mechanical activation stage in the oxide mixtures process is a critical step, since it allowed to obtain PZT ceramics using the common Pb oxides (PbO, PbO_2 and Pb_3O_4) combined with a further thermal treatment. The mechanical activation process produces particle size reduction, promotes the transformation of PbO to its tetragonal phase and the formation of $PbTiO_3$ and PZT, thus decreasing the synthesis temperature of PZT powders. These ceramic powders are homogeneous and with submicrometric size, and therefore highly reactive, this favours the reactivity of ZrO_2, leading to the early formation of PZT (350°C) compared to synthesis temperature of traditional methods. This result is important, since it allows to avoid lead oxide evaporation during the heat treatment for the reaction to form the Perovsquite phases at 900°C.

The mechanism of phase transformation of the mixtures by milling seems to be the compatible with the crystalline structure of the raw materials to the perovsquite structure. PbO in its orthorhombic phase transforms to tetragonal phase during milling, and then the perovskite phase of $PbTiO_3$ and PZT is formed. Increasing its concentration for the thermal treatment from 300°C, 500°C and 700°C. The samples A, B and C at 4h of milling and 700°C of thermal treatment reach concentrations around 91, 97 and 97 % of PZT respectively. A milling time of four hours is the best condition to promote the early formation of PZT in the three set of samples with different Pb oxides.

Comparing both routes of synthesis regarding costs, security and speed, the mechanicoactivation route is the most favoured. Nevertheless because of the purity of the powders obtained, and the control of the phases, the sol-gel method is also appropriate, with the problem of the use of the toxic reactive 2-metoxiethanol, which must be handled very carefully. Additionally the cost of the precursors utilized is high. In this work however ceramics with similar characteristics and ferroelectric behaviours from both synthesis routes were obtained.

5. Acknowledgment

Financial support from Conacyt-México, through the project 82843, is acknowledged. M. G. Rivera-Ruedas is grateful with Conacyt by the scholarship, also the technical assistance of M. en T. Rivelino Flores Farias is acknowledged.

6. References

Aman, U., Chang, A. , Ali, H. Ill, K. (2010). The effects of sintering temperatures on dielectric, ferroelectric and electric field-induced strain of lead-free Bi0.5(Na0.78K0.22)0.5TiO3 piezoelectric ceramics synthesized by the sol-gel technique, Current Applied Physics, Vol. 10, No. Issue 6, pp. 1367e1371, (November 2010)

Babushkin, T. Lindbach, J. Luc, J. Leblais, Kinetic Aspects of the Formation of Lead Zirconium Titanate. (1996). Journal of the European Ceramic. Society. Vol. 16, pp. 1293-1298.

Balderas-López J. A., Mandelis A. and. García J. A. (2001). Measurements of the thermal diffusivity of liquids with a thermal-wave resonator cavity. Analytical Sciences, Vol. 17, pp.s519-s522

Brankoviĉ Z, Brankoviĉ G,. Jovalekiĉ Ĉ, and Varela J. A.. (2003). PZT ceramics obtained from mechanochemically synthesized powders. J. of Materials Science, Materials in electronics, Vol. 14, pp37-41.

Brankoviĉ Z., Brankoviĉ G., Jovalekiĉ Ĉ,. Maniette Y, Cilense M. and Varela J. A., (2003) Materials Science and Engineering A, Vol. 345,pp243-248.

Charles, D. Lekeman E. Payne A. (1992). Processing effects in the Sol-Gel preparation of PZT dried gels powders, and ferroelectric thin layers, Journal of the American Ceramic Society, vol. 75, No Issue 11, pp. 3091-3096 (November 1992).

Coffman P.R. and . Dey S. K. (1994). J. of Sol-gel Science and Technology. Vol 1, pp. 251-265.

Coffman P.R., Barlingay C. K, Gupta A., and Dey S.K. (1996). J. of Sol-gel Sci. and Technol. Vol. 6, pp. 83-106.

Coffman, P. Dey, S. (1994), Structure evolution in the PbO-ZrO2-TiO2 sol-gel system: Part I, Characterization of prehydrolyzed precursors, Journal of Sol-Gel Science and Technology, Vol. 1, No. Issue pp. 251-265, (1994).

Grinberg, I. Rappe, M. (2007), Nonmonotonic TC Trends in Bi-Based Ferroelectric Perovskite Solid Solutions, Physical Review Letters , vol. 98,No. issue (January 2007),pp. 037603-1-4,

Guarany C. A., Araújo B. E., Silva P. R. J and Saitovitch H. (2007). Physica B. Condensed Matter, Vol.389 pp.130-134.

Hammer M., M Hoffmann. (1998), Detailed X-ray Diffraction Analyses and Correlation of Microstructural an Electromechanical Properties of La-doped PZT Ceramics, Journal of Electroceramics, Vol. 2:No. Issue 2, pp. 75-84, (August 1998).

Heywang, W., Lubitz, K. Wersing, W. (2008), Piezoelectricity, Evolution and future of a technology,(2008) Editorial Springer Series in Materials Science, ISSN 0933-033x Springer , Verlag, Heidelberg, Berlín.

Hurtado-Macías A., Muñoz-Saldaña J., Espinoza Beltrán F. J., Swain T., M. V.and. Schneider G. A. (2008). Journal of Physics D: Applied Physics Vol.4, 035407.

Jaffe, B., Roth R. Marzullo, S. (1954), Piezoelectric Properties of Lead Zirconate Lead Titanate Solid Solution Ceramics, Journal of Applied Physics, vol. 52, No. Issue 6, pp. 809-810.

Ky, P., Ali, H., Chang, A. , Il, K., Soon, J. Jae, L. (2010) Giant strain in Nb-doped Bi0.5(Na0.82K0.18)0.5TiO3 lead-free electromechanical Ceramics. Materials Letters , Vol. 64, No. Issue 22 pp. 2219–2222.

Legrand C., Da Costa A., Desfeux R., Soyer C., Rèmiens D. (2007). Piezoelectric evaluation of ion beam etched Pb(Zt,Ti)O3 thin films by piezoresponse force microscopy. Applied Surface Science, Vol. 253, pp.4942–4946.

Mandelis A. Zver M. M. (1985). Theory of photopyroelectric spectroscopy of solids. J. Appl. Phys. Vol. 57, pp.4421-4430.

Marineli M., Murtas F., Mecozzi M. G, Zammit U., Pizzoferrato R, Scudieri F., S. Martellucci, and Marinelli M. (1990). Simultaneous Determination of Specific Heat, Thermal Conductivity and Thermal Diffusivity at Low Temperature Via the Photopyroelectric Technique. Appl. Phys. A. Vol. 51, 387-393.

Noheda B., Cox D., and Shirane G., (2000), Stability of the monoclinic phase in the ferroelectric perovskite PbZr1ÀxTixO3, Physical Review B, Vol. 63, pp. 014103(1-9) (December 2000)

Pontes F. M., Leite E. R., Nunes M. S. J., Pontes D.S.L, Longo E, Magnani R., Pizani P. S,. Varela J. A. (2004). Preparation of Pb(Zr,Ti)O_3 thin films by soft chemical route. *J. of the European Ceramic Society*. Vol, 24, pp 2969-2976.

Sawawuchi E. (1953). Ferroelectricity versus Antiferroelectricity in Solid Solutions of PbZrO3 and PbTiO3Journal of the Physical Society of Japan, Vol. 5 No issue 5, pp 615-629, (September 1952)

Schwartz R. W. (1997). Chemical Solution Deposition of Perovskite Thin Films. *Chem Matter*. Vol. 9, pp. 2325-2340.

Shirane G., Takeda A., (1952). Phase Transitions in Solid Solutions of PbZrO3 and PbTiO3 (I) Small Concentrations of PbTiO3. Journal of the Physical Society of Japan, Vol. 7 No issue 1, pp 5-1, (February 1952)

Shirane G., Takeda A., (1952). Phase Transitions in Solid Solutions of Lead Zirconate and Lead Titanate:II". Journal of the Physical Society of Japan, Vol. 7, No issue 1, pp 5-1. (February 1952).

Shrout, R. & Zhang, J. (2007), Lead-free piezoelectric ceramics: Alternatives for PZT?", Journal of Electroceramics, Vol. 19, No. 1, pp. 111–124, (February 2007)

Sooksaen, P., Hongart J., Tippawan A. Utumporn M., (2008), Crystallization and analysis of perovskite crystals inferroelectric-based glasses, Chiang Mai Journal of Science, Vol. 35 No. Issue 3, pp. 427-436, ISSN 0125 - 2526 (September 2008).

Wei, L, Zhijun, X., Ruiqing, Ch., Peng, F. Guozhong, Z. (2010), High piezoelectric d33 coefficient in (Ba1−xCax)(Ti0.98Zr0.02)O3 lead-free ceramics with relative high Curie temperature, Materials Letters, Vol. 64, pp. 2325–2327, (July 2010)

Zhang Q. and Whatmore R. (2001). Sol-gel PZT and Mn-doped PZT thinfilms for pyroelectric applications. *J. Phys. D: Appl. Phys.*, Vol. 34, pp. 2296-2301.

Zhou, H., Hoatson, L. Vold, L. (2004). Local structure in perovskite relaxor ferroelectrics: high-resolution 93Nb 3QMAS NMR. *Journal of Magnetic Resonance*, Vol. 167, No. issue 2, pp. 42–252,

17

Flexible Ferroelectric $BaTiO_3$ – PVDF Nanocomposites

V. Corral-Flores and D. Bueno-Baqués
Research Center for Applied Chemistry, Mexico

1. Introduction

Ferroelectric materials are considered as smart materials, since they can be configured to store, release or interconvert electrical and mechanical energy in a well-controlled manner. Their exceptionally large piezoelectric compliances, pyroelectric coefficients, dielectric susceptibilities and electro-optic properties make them very attractive for nanotechnology-related applications such as high energy density capacitors, pyroelectric thermal imaging devices, gate insulators in transistors, electro-optic light valves, thin-film memory elements, multiferroic transducers, energy harvesters, etc. (Alpay et al., 2009, Nonnenmann & Spanier, 2009; Scott, 2007; Leionen et al., 2009)

The most common ferroelectric materials in commercial applications are ceramics, such as lead zirconate-titanate (PZT), barium titanate (BTO), calcium-copper titanate $CaCu_3Ti_4O_{12}$ (CCTO), sodium niobate ($NaNbO_3$), among others, which present a high dielectric constant, high dipole moment and high electromechanical coupling coefficient. Ferroelectric ceramics have been recently synthesized by solvothermal (Wada et al., 2009), coprecipitation (Hu, et al., 2000), sol-gel (Kobayashi et al., 2004), and template assisted methods (Rorvik et al., 2009), in order to obtain nanostructured materials. Considering the toxicity of lead and its compounds, there is a general awareness for the development of environmental friendly lead-free materials (Panda, 2009; Jia et al., 2009). In the development of this work, we have chosen BTO for its excellent ferro-, piezo-, and di-electric properties.

Barium titanate presents the perovskite crystal structure, which has the general: formula $A^{2+}B^{4+}O_3^{2-}$, where A represents a divalent metal ion (barium) and B represents tetravalent metal ions (titanium in this case). Above the Curie temperature (T_C), the crystal has a cubic symmetry, a centrosymmetric microstructure where the positive and negative charges coincide. Below T_C, crystals have a tetragonal symmetry. This form has no center of symmetry, in each unit cell exhibits an electric dipole that can be reoriented by an applied electric field. The material is then called ferroelectric.

Ceramics, however, are brittle and require high temperature processing. By the other side, ferroelectric polymers present good mechanical properties, can be formed in complex shapes at low temperature, are flexible and have high dielectric strengths; although the ferroelectric properties and dielectric constant are lower than ceramics. Poly(vinylidene fluoride) (PVDF) is an electroactive polymer that exhibits polymorphism. Its most common crystalline phases are: α, β, γ and δ phases; also known as form II, I, III and IV respectively.

Each form has its own characteristic unit cell due to chain conformation. α phase crystallizes in an orthorhombic cell, where two chains are opposite packed canceling the individual dipole moments. The chain conformation consists of alternating trans and gauche sequences. In β phase, two chains in all-trans planar zigzag conformation are packed into an orthorhombic unit cell. The fluorine atoms are positioned on one-side of the unit cell resulting in a net dipole moment of 2.1 debye, the highest among all phases. In γ phase, two opposite chains conform a monoclinic crystal lattice, where only a fraction of dipole moments are cancelled. δ phase is formed when α phase is electrically poled, and one of the chains align parallel to the other, resulting in a weak net dipole moment. The crystal lattice parameters are identical to α phase (Schwartz, 2002).

A hybrid ceramic-polymeric composite is a convenient solution to tune both mechanical and electrical properties. In this respect, several systems have been already developed, such as CCTO - poly(vinylidene fluoride – trifluoroethylene) [P(VDF-TrFE)] (Arbatti et al., 2005), BTO – PVDF (Chanmal & Jog, 2008), MWCNTs - BTO - PVDF (Dang et al., 2003), Sm/Mn doped $PbTiO_3$ - epoxy (Li et al., 2003), and PZT - Rubber (Qi et al., 2010). Composites are complex, heterogeneous and usually anisotropic systems. Its properties are affected by many variables, including constituent material properties, geometry, volumetric fraction, interface properties, coupling properties between the phases, porosity, etc. Connectivities between the phases play a very important role in the ultimate properties of the composites. The connectivity has great importance in a multiphase material because it heavily influences the mechanical, electrical and thermal fluxes between the phases. From matrix-loaded composites to highly sophisticated arrangements, composites can be designed to tailor the acoustic impedance, coupling constant and mechanical quality factor, as compared to bulk ceramics.

In nanotechnology applications, ferroelectric ceramics have to overcome some size scaling challenges, since their main properties can be dramatically affected when the grain size decreases to a certain limit, where the material suffers either changes in Tc, phase transition or variations in its polarization state (Eliseev & Morozovska, 2009). In a similar manner, ferroelectric polymers have to be processed in a way that enhances its crystalinity and favours the growing of the polar phase. These two issues must be carefully addressed when processing hybrid ceramic-polymeric composites.

Electrospinning is a versatile technique widely used to produce either polymer (An et al., 2006; Koombhongse et al., 2001) or ceramic nanofibers (Lu et al., 2006; Azad, 2006). Even nanocomposites have been produced by this technique (Saeed et al., 2006; Wang et al., 2004). The major components are a high voltage power supply, a container with a metallic tip to feed the polymer solution and a grounded collector. Electrospinning occurs when the electrical forces at the surface of a charged polymer solution droplet overcome the surface tension. The solution is ejected as an electrically charged jet towards the oppositely charged electrode, while the solvent evaporates, leading to the formation of dry nanofibers. When the jet flow away from the droplet to the target, it undergoes a series of electrically driven bending instabilities, following a complex path that gives rise to a series of looping and spiraling motions. The jet elongates, and this stretching significantly reduces its diameter (Reneker et al., 2000). Since this technique involves high electric fields, it is then expected to enhance the formation of polar phases in polymorph polymers, such as PVDF (Ramakrishna et al., 2010).

Template-assisted synthesis is a simple method to produce one-dimensional nanostructures and nanotube arrays. The templates, such as porous anodic alumina, have pores in which a solution containing the desired components can be incorporated, forming the nanotubes after solvent evaporation (Rorvik et al., 2009).

This chapter covers several configurations and connectivities of hybrid barium titanate - poly(vinylidene fluoride) (BTO - PVDF) nanocomposites and nanostructures:

i. BTO nanoparticles embedded in a PVDF matrix.
ii. BTO nanoparticles embedded in PVDF nanofibers.
iii. BTO nanofibrous membranes in a PVDF matrix.
iv. BTO - PVDF nanotube arrays.

Dielectric properties and polarization hysteresis are presented and discussed. This study is expected to further expand the understanding and range of applicability of these functional nanostructured materials.

2. Experimental

The ceramic nanoparticles were obtained by microwave assisted hydrothermal method, while the nanofibrous membranes were synthesized by electrospinning technique. PVDF (Kynar 761 kindly supplied by Arkema) was processed by spin-coating and electrospinning to obtain films and nanofibers, respectively. BTO - PVDF nanotube arrays were prepared by template assisted synthesis, combining sol-gel and sol-humectation in a porous membrane. All chemicals used in this study were reagent grade purchased from Sigma-Aldrich.

2.1 BTO nanoparticles embedded in a PVDF matrix

BTO nanoparticles were synthesized in two steps procedure. First, TiO_2 nanoparticles were obtained by direct precipitation from a $TiCl_4$ solution in ice-cold water after seven days of reaction. Second, TiO_2 nanoparticles were subjected to microwave-assisted hydrothermal conditions in a CEM oven model MARS 5. A barium hydroxide aqueous solution under a Ti:Ba molar ratio of 1:1.8 was used as the reaction media; the hydrothermal reaction took place at 150°C for 15 min. After washing, the nanoparticles were capped with 3-aminopropyl triethoxysilane at acidic media (acetic acid was added until a pH of 4 was reached) under microwave-assisted hydrothermal conditions at 150°C for 60 min. Samples at this stage were named as BTO-MWHT.

BTO-PVDF films were obtained by spin-coating PVDF-N,N-dimethylformamide (DMF) solutions containing silane-capped BTO nanoparticles at the following BTO:PVDF weight ratios: 1:10, 1:20 and 1:100. PVDF concentration in DMF was kept at 12 wt.% for the suspensions with high content of BTO, and 14 wt.% for the 1:100 ratio. Samples after this procedure were named as BTO-MWHT-PVDF-SC.

2.2 BTO nanoparticles embedded in PVDF nanofibers

BTO-PVDF suspensions (as described in 2.1) were electrospun at 15 kV and 20μA in a horizontal set-up. The feeding rate was 0.5 ml/hr and the tip-to-collector distance was varied from 10 to 15 cm. Electrospun samples were named as BTO-PVDF-ES.

2.3 BTO nanofibrous membranes embedded in a PVDF matrix

BTO nanofibrous membranes were obtained by electrospinning a 1 M precursor solution containing the corresponding metal ions. Titanium butoxide and barium acetate at a 1:1 molar ratio were dissolved in methoxyethanol and acetic acid. Poly(vinyl pirrolidone) was added to facilitate the electrospinning process. The following conditions were used: feed rate of 0.5 ml/hr, 15 kV and tip-to-collector distance in the range of 10 to 15 cm. The electrospun nanofibrous membranes were sintered at 800°C for 2 hours to obtain the ceramic phase. Then the BTO nanofibrous membranes (named as BTO-ES) were embedded in a PVDF matrix by spin-coating a 15 wt.% PVDF solution over the membranes in a conductive Pt-Si substrate. After deposition, samples were heat-treated at 60°C for 1 hour to crystallize the beta phase of the polymer. The nomenclature for these samples was set as BTO-ES-PVDF-SC.

2.4 BTO – PVDF nanotube arrays

These arrays were obtained in two steps: synthesis of BTO nanotubes and wetting of the nanotubes in a PVDF solution. To synthesize the BTO nanotubes, alumina templates (Whatman Anodisc 13) were immersed in the precursor solution described in 2.3, and then the excess solution was wiped off. In order to crystallize the BTO ceramic phase, the samples were heat treated at 400°C for 1 hr and then at 700°C for 2 hours under a heating and cooling rate of 2°C/min. To obtain the PVDF nanotubes, the templates containing BTO nanotubes were immersed in a 5wt.% PVDF-DMF solution, then the solvent was evaporated at room temperature. This set of samples was named as BTO-PVDF-NT.

3. Results

Crystalline phases were determined by X-ray diffraction (XRD, Siemmens D-5000), morphology was studied by scanning and transmission electron microscopy (SEM, Jeol JSM-740IF and TEM, Fei Titan 80), and infrared spectroscopy (FTIR, IR-Nexus 470) -using a micro-ATR (attenuated total reflectance) accessory- was used to determine the fraction of beta phase in the polymer.

Polarization hysteresis loops were measured with a 300 Hz driving signal amplified by a TEGAM HV. Sample response was collected with virtual ground charge/current amplifier. Field and current signals were digitized simultaneously and numerically processed to obtain the electric displacement and polarization. The dielectric properties were measured using a TEGAM 3550 impedance analyzer in a range from 100 Hz to 5 MHz.

3.1 Structural characterization

X-ray diffraction patterns are shown in Fig. 1. The tetragonal doublet corresponding to (002) and (200) in BTO was not clear at the test conditions, probably due to the peak widening, or a mixture of cubic and tetragonal phases. However, a pattern refinement (using the program PowderCell 2.4) revealed that the best fit is obtained from the tetragonal reference. The Miller indexes for tetragonal barium titanate are presented in Fig. 1. The crystallite size and cell parameters obtained from the refinement are presented in Table 1. According to these results, all samples presented small crystallite size in the range of 20 - 26 nm.

BTO surface functionalization with silane at hydrothermal conditions altered both the crystallite size and the degree of tetragonality, possibly due to the acidic conditions and the extended reaction time. Additionally, the small diffraction peaks present in this sample (Fig. 1) were identified as TiO_2, both rutile and anatase phases, which presumably were formed by the dissolution of BTO and subsequent leaching of barium during the functionalization reaction. Nevertheless, the surface functionalization of the nanoparticles was crucial for its proper dispersion in the PVDF matrix. A micrograph showing the BTO-MWHT-Silane nanoparticles is presented in Fig. 2. Average particle size was determined as 31.6 nm, slightly higher than the crystallite size estimated from XRD results. This difference can be attributed to the silane layer, and/or the possible presence of an amorphous (distorted) phase at the surface of the nanoparticles.

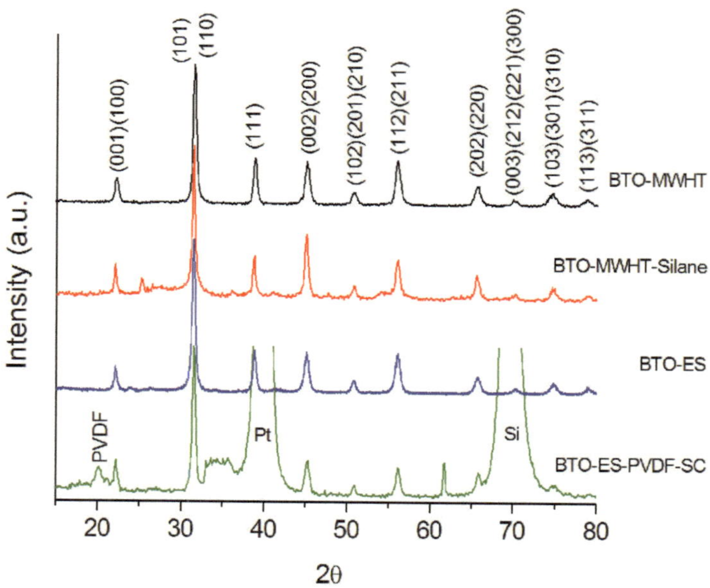

Fig. 1. X-ray diffraction of BTO samples obtained from different techniques.

Sample	d(nm)	a (Å)	c(Å)	DoT	Rp(%)
BTO-MWHT	20	4.0176	4.0458	1.00702	9.72
BTO-MWHT-Silane	24	4.0182	4.0192	1.00025	13.45
BTO-ES	23	4.0039	4.0317	1.00694	10.90

Table 1. Refinement results: crystallite size (d), cell parameters (a and c), profile R factor (Rp) and the degree of tetragonality (DoT) for BTO tetragonal phase.

Electrospun BTO nanofibers showed a single phase XRD pattern, very similar to that of the BTO nanoparticles (Fig. 1), however, its morphology is clearly different. Figure 3 shows BTO nanofibers obtained at 15 kV and tip-to-collector distance of 15 cm, before and after heat treatment. As-spun fibers are long and smooth, with average diameter of 110.4 ±48.2, ranging from 27 to 362 nm. Sintered nanofibers preserved its morphology while presenting a rough surface, revealing the presence of grains. These nanofibers presented a reduction in size, with an average diameter of 105.5 ±16.5 nm, ranging from 51 to 225 nm. Apparently, the thinnest nanofibers are fractured and lost during heat treatment.

When BTO nanofibrous membranes were immersed in PVDF to obtain hybrid ceramic-polimeric composites, the nanofibers were not modified in terms of morphology or crystallograpy. XRD (Fig. 1) showed the presence of silicon and platinum from the substrate, together with some other unknown reflections that are attributed also to the substrate. PVDF was confirmed from the peak at around 20.1 degrees in 2θ. According to several reports, α phase can be identified by diffraction peaks present at 17.83, 18.52, 20.1 and 25.88°, β phase at 20.44° and γ phase at 26.74° (Nasir et al., 2007; Esterly & Love, 2004; Gao et al., 2006). However, it was difficult to distinguish between alfa and beta phases. For this purpose, infrared spectroscopy was used.

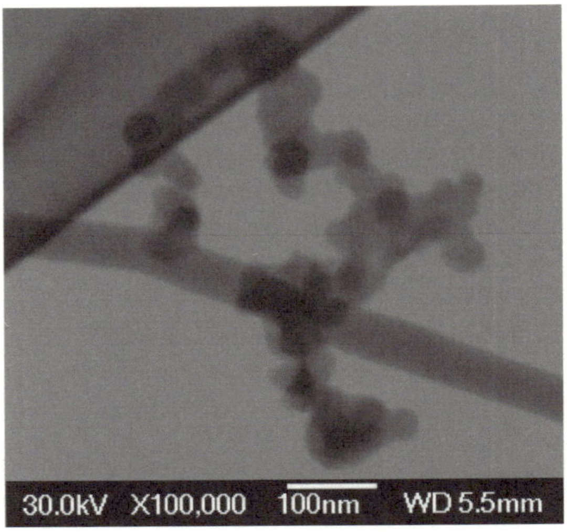

Fig. 2. STEM Micrograph of the BTO nanoparticles obtained by microwave-assisted hydrothermal method and capped with silane (sample BTO-MWHT-Silane).

Fig. 3. SEM micrographs of BTO nanofibers, collected at 15 kV and 15 cm between electrodes, (a) before heat treatment and (b) sintered at 800°C.

FTIR spectra are shown in Figure 4. The absorption bands corresponding to the crystalline phases are shown by dotted lines for clarity. Absorption bands at 762, 795 and 974 cm^{-1} correspond to α phase, bands at 839 and 1276 cm^{-1} are assigned to β phase, while γ phase is identified by the band at 1235 cm^{-1} (Yee et al., 2007). PVDF raw material showed predominantly alfa phase. Samples obtained either from electrospinning or spin-coating showed a substantial reduction on the content of this non-polar phase, while the presence of beta phase was enhanced in some cases.

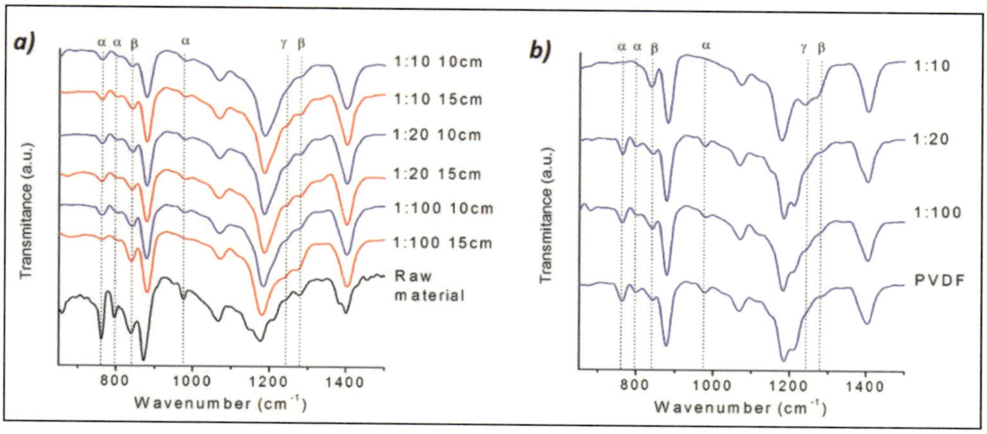

Fig. 4. FTIR-ATR spectra of BTO-PVDF samples obtained from (a) electrospinning technique and (b) spin-coating.

The fraction of beta phase was estimated from the equation proposed by Gregorio & Cestari (Gregorio & Cestari, 1994) and later used by several authors (Salimi & Youse, 2004; Jiang et al., 2007; Sobhani et al., 2007; Andrew & Clarke, 2008):

$$F(\beta) = \frac{A_\beta}{1.26A_\alpha + A_\beta} \tag{1}$$

Where A_α and A_β are the absorption band intensity for α and β phases, respectively. The fraction of beta phase of the samples containing PVDF in the present study is shown in Table 2. PVDF raw material presented a fraction of beta phase of 0.12. Samples obtained from electrospinning technique showed higher beta content than that of those from spin-coating. This could be attributed to differences in the methods, such as the solvent evaporation rate and the presence of the electric field during electrospinning.

BTO:PVDF ratio	Technique	Distance (cm)	Beta fraction
0:1	ES	10	0.69
0:1	ES	15	0.66
0:1	SC	-	0.38
1:10	ES	10	0.60
1:10	ES	15	0.65
1:10	SC	-	1.00
1:20	ES	10	0.63
1:20	ES	15	0.64
1:20	SC	-	0.46
1:100	ES	10	0.60
1:100	ES	15	0.86
1:100	SC	-	0.47

Table 2. Beta fraction of electrospun (ES) and spin-coated (SC) samples. Distance stands for the tip-to-collector distance in electrospinning set up.

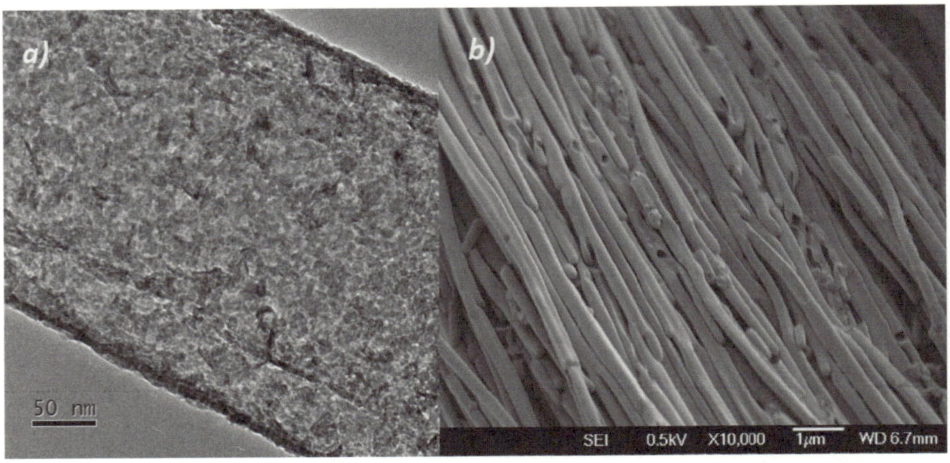

Fig. 5. Micrographs of (a) a BTO nanotube and (b) PVDF nanotubes, both obtained in an alumina template with porous diameter of 200 nm.

In order to release the nanotubes from the alumina template nanotube arrays, the template was dissolved in a 5M NaOH solution. TEM revealed the formation of ceramic BTO nanotubes with average wall thickness of 11 nm (as shown in Fig. 5a). The ceramic nanotubes were crystalline in nature and presented small crystallites in the order of 4 to 5 nm. PVDF nanotubes, observed by SEM (Fig. 5b), reproduced the inner morphology of the alumina templates, even small defects. The average diameter was consistent with that of the templates, and the length was expected to reach the template thickness (60 microns).

3.2 Electric behavior

Samples were clamped between highly smooth parallel contact plates cut from an Ir-Pt coated silicon wafer for electric polarization measurements, while Au-Pd was deposited as top electrode before acquiring the dielectric properties. The electric behaviour is discussed below.

3.2.1 Electric polarization

Electric polarization of electrospun samples (BTO-PVDF-ES) revealed a linear behaviour in most of the cases, due to the porosity inherent to the fibrous array. An estimate of the effective area in contact with the top electrode is 60%, being the rest essentially air. Samples containing a BTO:PVDF ratio of 1:10 presented a non-linear behaviour, especially when electrospun at a distance of 15 cm (Fig. 6). These results characterize a lossy ferroelectric material (Scott et al., 1998), and are in agreement with the fraction of beta phase calculated from FTIR spectra (Table 2) which is higher for the sample obtained at 15 cm. It is known that the electrospinning conditions such as the tip-to-collector distance and applied voltage play an important role in the morphology and size of the fibers (Azad, 2006), and additionally, can affect the crystallization of polar phases in the case of ferroelectric polymers (Yee et al., 2007).

A similar response was obtained from electrospun BTO fibers (sample BTO-ES), as shown in Fig. 7, due to the low density of the sample and the contribution of air in the measurement. However, the concave region in the graph is indicative of the ferroelectric properties of BTO. Addition of PVDF to the BTO fibers (sample BTO-ES-PVDF-SC) did not change this behaviour, note that the later sample was measured at a lower applied voltage.

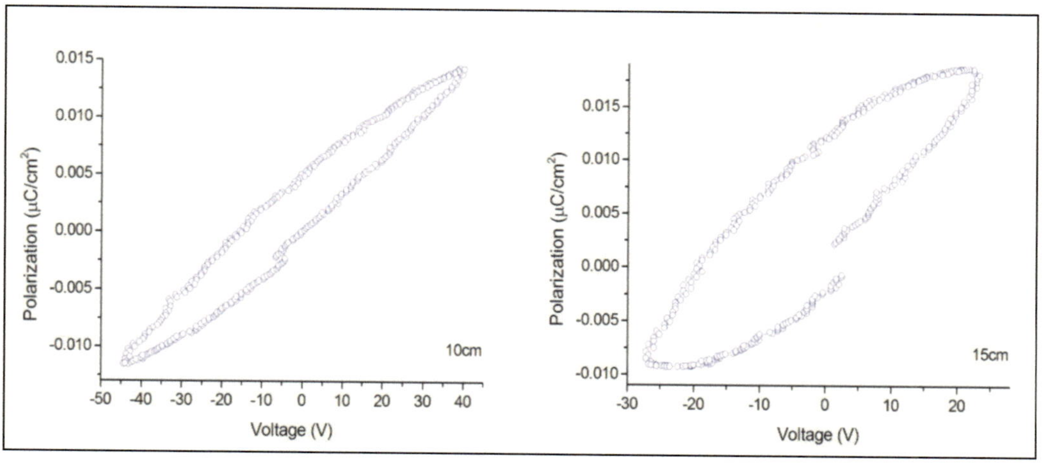

Fig. 6. Electric polarization of the electrospun samples with a BTO:PVDF ratio of 1:10. Tip-to-collector distance was 10 and 15 cm.

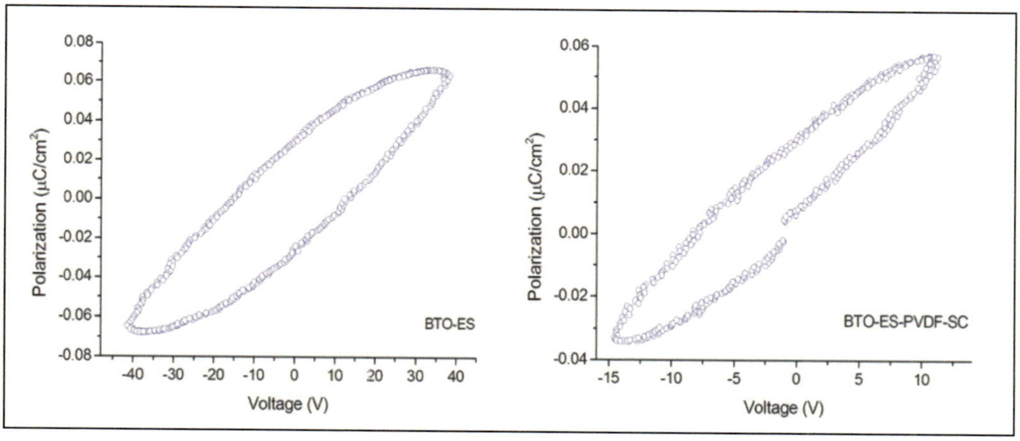

Fig. 7. Electric polarization of electrospun BTO and electrospun BTO with PVDF deposited by spin-coating.

Films of PVDF with BTO nanoparticles (samples BTO-MWHT-PVDF-SC), prepared by spin-coating, exhibited a ferroelectric response based on the amount of nanoparticles present. As shown in Fig. 8, the sample with the highest content of BTO nanoparticles (1:10 ratio) showed a nice ferroelectric hysteresis loop that reflects that in fact, the composition, geometry and connectivity of this sample are optimum to enhance the ferroelectricity of hybrid ceramic-polymeric composites.

In an opposite behaviour to the system BTO-MWHT-PVDF-SC, the nanotube arrays showed a better ferroelectric response when only the polymer was present. As shown in Fig. 9, the PVDF nanotube array (PVDF-NT) is ferroelectric, despite that the alumina template is a dielectric. The electric polarization was calculated taking into account the total area of the alumina templates (13 mm), however, the effective area in contact with the top electrode is much lower, estimated from SEM micrographs as 66% of the total template area.

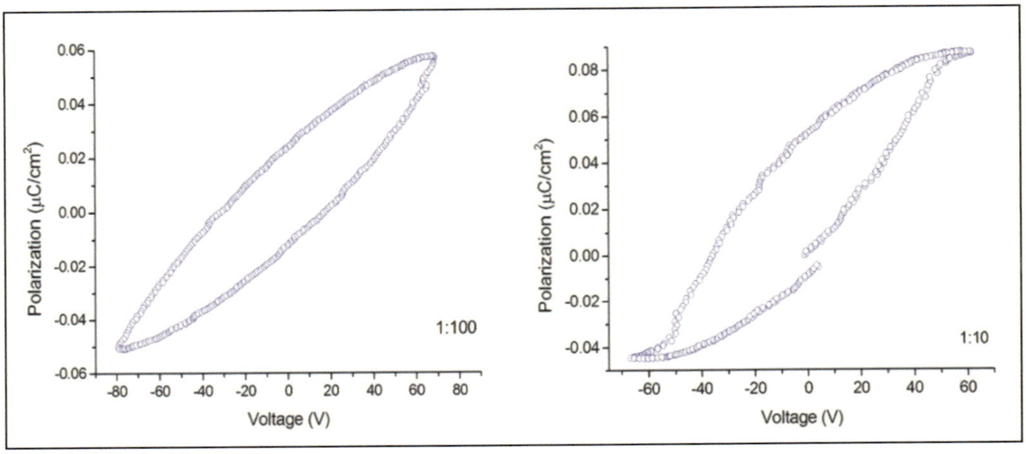

Fig. 8. Electric polarization of BTO-PVDF films deposited by spin-coating. BTO:PVDF weight ratio is presented.

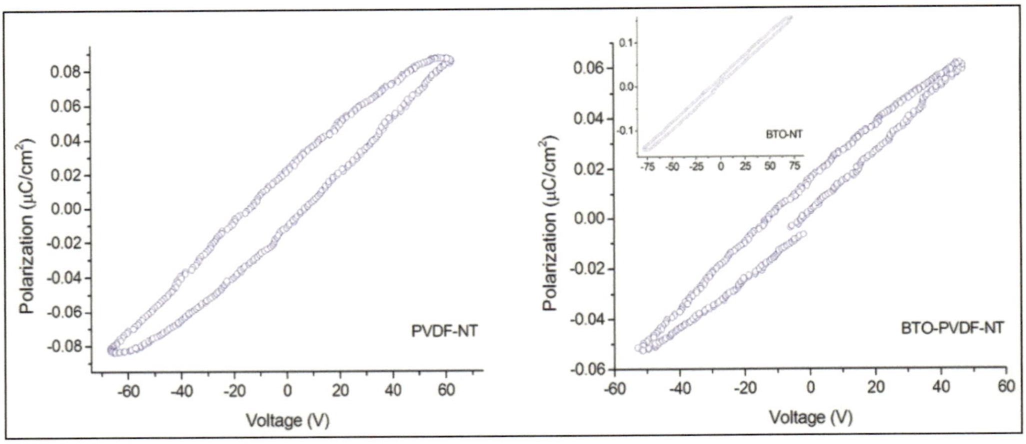

Fig. 9. Electric polarization of BTO-PVDF nanotube arrays. The response of the pure BTO nanotubes is presented as inset.

3.2.2 Dielectric spectroscopy and impedance analysis

Figure 10 shows the dielectric permittivity and loss of single phase and composite nanotube arrays measured at room temperature from 100 Hz to 3 MHz.

As observed, the dielectric permittivity decreases with the addition of PDVF layer on top of the BTO nanotube structure formed in the template. This could be related to the complex geometry of the composite structure, in which PVDF layer is encapsulating the BTO and alumina template. It is known (Tsangaris 1999 as cited in Patsidis & Psarras 2008) that Maxwell-Wagner-Sillas effect heavily influences the electrical behaviour of systems with complex electrical heterogeneities, where large accumulation of charges can take place at the interface of the system. In present composites, based on the fact that dielectric permittivity of BTO is considerable higher than alumina, large unbounded charges can arise from the ceramics interface. The arising limitation of field induced dipoles to follow the alternation of the electric field will shift the associated relaxation process to lower frequencies, as suggested by the behaviour of the dielectric loss in the case of the BTO-NT sample. With the addition of PVDF layer, competing phenomena could arise from trapped charges at the BTO - PVDF interface, leading to self balance stable state with a consequent change in the relaxation behaviour.

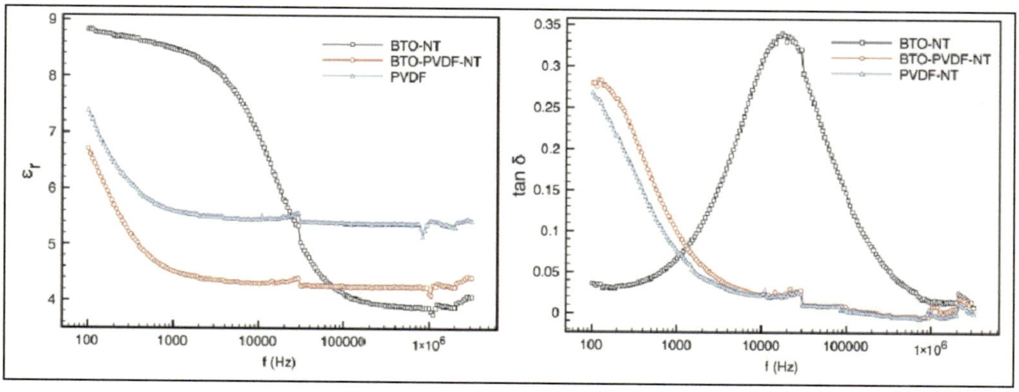

Fig. 10. Dielectric permittivity and dielectric loss of the nanotube arrays.

The real and imaginary impedance components increase with the inclusion of PDVF nanotubes at low frequencies, as shown in figure 11. This could indicate a better insulation at low frequencies introduced by the PVDF layer. At high frequencies the AC conductivity increases.

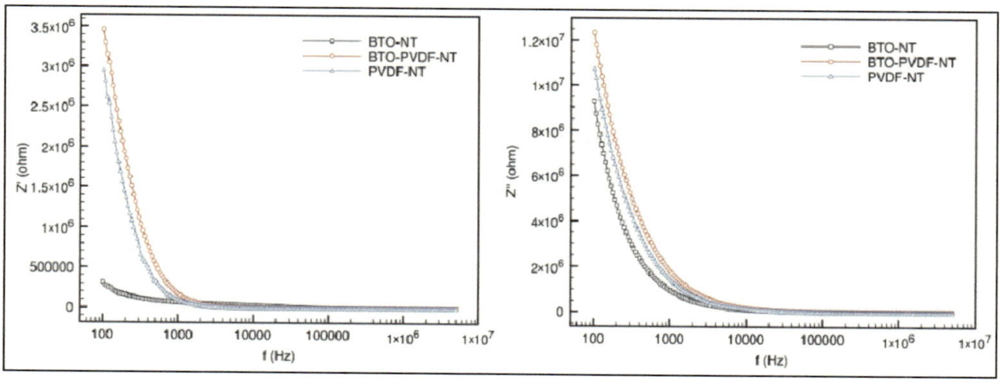

Fig. 11. Real and imaginary impedance components of the nanotube arrays.

Figure 12 shows the dielectric permittivity and dielectric loss of the BTO-MWHT-PVDF-SC film composites. As shown, dielectric permittivity increases with the increase in the ceramic loading and could be attributed to an enhanced polarization from dipole-dipole interactions in more closely aggregated nanocrystallites of the ceramic phase. Also, dielectric permittivity decreases for higher frequencies values, more rapidly for frequencies above 100 kHz. Several unset are observed for both compositions, which can be related to the dipole orientation polarization of the $BaTiO_3$ nanoparticles. Dielectric loss shows one noticeable relaxation at frequency close to 3 Hz for both composite samples. Due to the almost similar frequency values and shape it is suggested that this relaxation should be related to the polymer phase. This can be associated to cooperative motions in the main polymer chains related to the non-polar α phase in the PVDF. Several authors have discussed and proposed different mechanisms for this phenomenon (Kochervinskii, 2007 as cited in Chanmal & Jog, 2008).

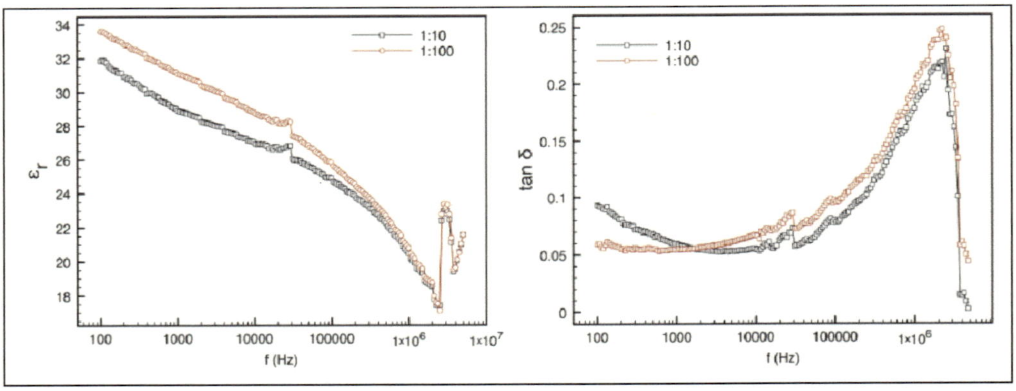

Fig. 12. Dielectric permittivity and dielectric loss for BTO-MWHT-PVDF-SC film composites

Real and imaginary parts of the impedance for the BTO-MWHT-PVDF-SC film composites are shown in figure 13. As can be observed in the figures, both impedance components are respectively very similar to that for the films with different BTO content. Also, real (Z′) and imaginary (Z′′) components decrease with the increase in frequency. The behaviour of Z′ indicates an increase in the AC conductivity of the material, while Z′′ suggest a Debye type relaxation process in the samples. The fact that both curves merge above 10 kHz could imply that dipole polarization in the BTO particles and trapped charges on their surfaces predominates over the space charge effects.

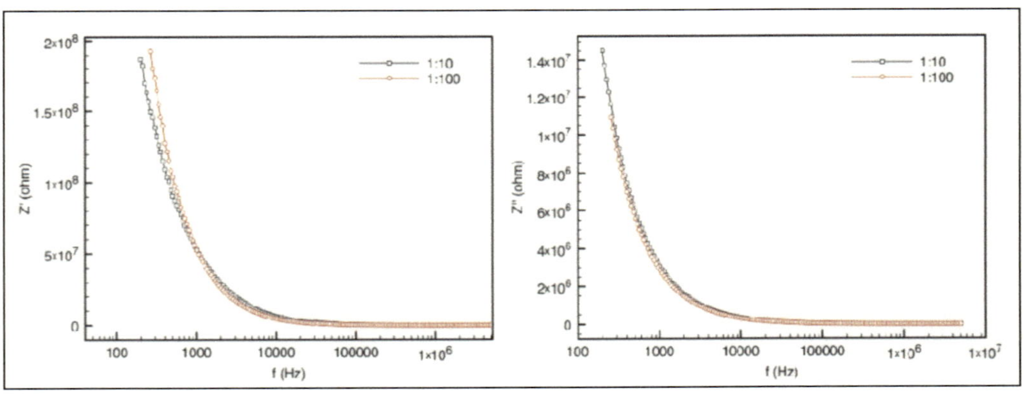

Fig. 13. Real and imaginary parts of the impedance for BTO-MWHT-PVDF-SC film composites as a function of frequency.

4. Conclusion

Barium titanate - Poly(vinylidene fluoride) hybrid composites with different morphologies were synthesized by combining several appropriate techniques: microwave-assisted hydrothermal, electrospinning, spin-coating, precipitation and sol-humectation. Different connectivities were obtained by the use of BTO nanoparticles and nanofibers in PVDF films and nanofibers. We demonstrated the suitability of such techniques to synthesize hybrid ceramic-polymeric ferroelectric nanocomposites. BTO crystallite size was 20 and 24 nm, when synthesized by microwave-assited hydrothermal method and electrospinning technique, respectively. Single tetragonal phase was obtained in both cases, with a degree of tetragonality or around 1.007. A silane functionalization by MWHT resulted in a slight growth of BTO nanoparticles accompanied by a partial dissolution, leading to the formation of secondary phases. Additionally, a decrease in the degree of tetragonality was detected. The fraction of beta phase present in the polymer was higher than the raw material, as revealed by FTIR analysis. Samples obtained by electrospinning technique presented, in general, higher content of beta phase than those obtained by spin-coating.

Electrospun samples of the system BTO-PVDF-ES presented an electric behaviour from linear dielectric to lossy ferroelectric when the amount of BTO nanoparticles increased. Sample with a BTO:PVDF weight ratio of 1:10 showed a good ferroelectric-like response when electrospun at a tip-to-collector distance of 15 cm. Similarly, electrospun BTO ceramic nanofibers showed a ferroelectric response, as well as the sample BTO-ES-PVDF-SC.

Films of the system BTO-MWHT-PVDF-SC showed the best ferroelectric results of all series. The sample with BTO:PVDF weight ratio of 1:10 showed a hysteresis loop indicative of ferroelectric properties. In contrast, nanotube arrays presented an opposite behaviour, since PVDF nanotube array presented a better response than hybrid PVDF-BTO and even BTO-NT.

Dielectric spectroscopy reveals a complex polarization and relaxation behaviour in the composites. Depending on the geometry and connectivity of the phases competing phenomena related to the charge accumulation at the interfaces and trapped charges at the surface of the ceramic phase takes place. However it is suggested that with the increase of the polymer content, the later mechanism prevails.

5. Acknowledgment

Authors gratefully acknowledge the support from L'Oréal, United Nations Educational, Scientific and Cultural Organization (UNESCO) and the Mexican Academy of Sciences through the scholarship "For Women in Science". This work was also supported by the Mexican council for Science and Techology (CONACyT) through the grants number 133815 and 83813.

6. References

Alpay, S. P., Nagarajan, V. & Rossetti, G. A. Jr. (2009) Recent developments in ferroelectric nanostructures and multilayers, J. Mater. Sci. Vol. 44, pp. (5021–5024)

An, H., Shin C. & Chase, G.G (2006). Ion exchanger using electrospun polystyrene nanofibers. *J. of Membrane Science,* Vol. 283, (2006), pp. (84-87)

Andrew, S. & Clarke, D.R. Effect of electrospinning on the ferroelectric phase content of polyvinylidene difluoride fibers, *Langmuir,* Vol. 24 (2008) pp. (670-672)

Arbatti, M., Shan, X. & Cheng, Z. Y. (2005) New High-Dielectric-Constant Polymer-Ceramic Composites, *Mater. Res. Soc. Symp. Proc.,* Vol. 847

Azad, A.M. Fabrication of yttria-stabilized zirconia nanofibers by electrospinning. *Mat. Lett.*, Vol. 60, (2006), pp. (67-72)

Chanmal, C. V. & Jog, J. P. (2008). Dielectric Relaxations in PVDF/$BaTiO_3$ Nanocomposites. *eXPRESS Polymer Letters*, Vol. 2, No. 4, (2008) pp. (294–301)

Dang, Z. M., Fan, L. Z., Shen, Y., & Nan, C. W. (2003) Dielectric behavior of Novel Three-Phase MWNTs/BaTiO3/PVDF Composites, *Mat. Sci. Eng.* Vol. B103, pp. (140-144)

Eliseev, E. A. & Morozovska, A. N. (2009) General approach for the description of size effects in ferroelectric nanosystems, *J. Mater. Sci.* Vol. 44, pp. (5149–5160)

Esterly, D.M. & Love, B.J. Phase transformation to β-poly(vinylidene fluoride) by milling. *J. Polym. Sci. B: Polym. Phys.*, Vol. 42 (2004) pp. (91-97)

Gao, K., Hu, X., Dai, C. & Yi, T. Crystal structures of electrospun PVDF membranes and its separator application for rechargeable lithium metal cells. *Mat. Sci. and Engineering B*, Vol. 131 (2006) pp. (100-105)

Gregorio, R.Jr. & Cestari, M. Effect of crystallization temperature on the crystalline phase content and morphology of poly(vinylidene fluoride), *J. Polym. Sci. B: Pol. Phys.*, Vol. 32, No. 5 (1994) pp. (859-870)

Hu, M. Z. C., Miller, G. A., Payzant, E. A. & Rawn, C. J. (2000) Homogeneous (co)precipitation of inorganic salts for synthesis of monodispersed barium titanate particles. *J. Mat. Sci.* Vol. 35, pp. (2927-2936)

Jia, Q., Shen, B., Hao, X. Song, S. & Zhai, J. (2009) Anomalous dielectric properties of $Ba_{1-x}Ca_xTiO_3$ thin films near the solubility limit, *Materials Letters*, Vol. 63, pp. (464–466)

Jiang, Y., Ye, Y., Yu, J., Wu, Z., Li, W., Xu, J. & Xie, G. Study of thermally poled and corona charged poly(vinylidene fluoride) films. *Polym. Eng. and Sci.* DOI 10.1002/pen 20817 (2007) pp. (1344-1350)

Kobayashi, Y., Nishikata, A., Tanase, T. & Konno, M. (2004) Size Effect on Crystal Structures of Barium Titanate Nanoparticles Prepared by a Sol-Gel Method. *Journal of Sol-Gel Science and Technology*, Vol. 29, pp. (49-55)

Koombhongse, S., Liu, W. & Reneker, D.H. (2001). Flat polymer ribbons and other shapes by electrospinning. *J. Polym. Sci. B: Polym. Phys.*, Vol. 39, (2001), pp. (2598-2606)

Leionen, M., Palosaari, J., Hannu, J. Juuti, J. & Jantunen, H. (2009) Piezoelectric energy harvesting for powering low power electronics. In: Paukkeri, A., Yla-Mella, J. and Pongracz, E. (eds.) Energy Research at the University of Oulu. Proceedings of the EnePro Conference, June 3rd, 2009, University of Oulu, Finland. Kalevaprint, Oulu, ISBN 978-951-42-9154-8, pp. (105-109)

Li, K., Chan, H. L. W., & Choy, C. L. (2003). Samarium and Manganese-Doped Lead Titanate Ceramic Fiber/Epoxy 1:3 Composite for High-Frequency Transducer Application. *IEEE Trans. On Ultrasonics, Ferroel. Freq. Control*, Vol. 50, No. 10, (October 2003)

Lu, X., Liu X., Zhang W., Wang, C. & Wei Y. (2006). Large-scale synthesis of tungsten oxide nanofibers by electrospinning. *J. Colloid and Interface Sci.*, Vol. 298, (2006), pp. (996-999)

Nasir, M., Matsumoto, H., Minagawa, M., Tanioka, A., Danno, T. & Horibe, H. Formation of β-phase crystalline structure of PVDF nanofibers by electrospray deposition:

additive effect of ionic fluorinated surfactant. *Polymer Journal,* Vol. 39, No. 7 (2007) pp. (670-674)

Nonnenmann, S. S. & Spanier, J. E. (2009) Ferroelectricity in chemical nanostructures: proximal probe characterization and the surface chemical environment, J. Mater. Sci. Vol 44, pp.(5205–5213)

Panda, P. K. (2009) Review: environmental friendly lead-free piezoelectric materials, *J. Mater. Sci.* Vol. 44, pp. (5049–5062)

Qi, Y., Jafferis, N. T., Lyons, K. Jr., Lee, C. M., Ahmad, H. & McAlpine, M. C. (2010) Piezoelectric Ribbons Printed onto Rubber for Flexible Energy Conversion. *Nanoletters,* Vol. 10, No. 2, pp. (524–528)

Ramakrishna, S., Jose, R., Archana, P. S., Nair, A. S., Balamurugan, R., Venugopal, J. & Teo, W. E. (2010). Science and engineering of electrospun nanofibers for advances in clean energy, water filtration, and regenerative medicine. *J. Mater. Sci.,* Vol. 45, pp. (6283–6312)

Reneker, D.H., Yarin, A.L., Fong, H. % Koombhongse, S. Bending instability of electrically charged liquid jets of polymer solutions in electrospinning. *J. Appl. Phys.,* Vol. 87, (2000) pp. (4531-4547)

Rorvik, P. M., Tadanaga, K., Tatsumiago, M., Grande, T. & Einarsrud, M. A. (2009). Template-assisted synthesis of $PbTiO_3$ nanotubes. *J. European Ceram. Soc.* Vol. 29, pp. (2575-2579)

Saeed, K., Park, S.Y., Lee, H.J., Back, J.B. & Huh, W.S. Preparation of electrospun nanofibers of carbon nanotube / polycaprolactone nanocomposite. *Polymer* Vol. 47, (2006) pp. (8019-8025)

Salimi, A. & Youse., A.A. Conformational changes and phase transformation mechanisms in PVDF solution-cast films. *J. Polym. Sci. B: Polym. Phys.,* Vol. 42 (2004) pp. (3487-3495)

Schwartz, M. (Ed.). (2002). *Encyclopedia of Smart Materials,* John Wiler and Sons, Inc., ISBN 0-471-17780-6, United States.

Scott, J.F. (2007) Applications of Modern Ferroelectrics. *Science,* Vol. 315 (16 February 2007), pp. (954-959)

Scott, J.F., Alexe, M., Zakharov, N.D., Pignolet, A., Curran, C., & Hesse, D. Nano-phase SBT-family ferroelectric memories. *Integrated Ferroelectrics,* Vol. 21 (1998) pp. (1-14)

Sobhani, H., Razavi-Nouri, M., Youse, A.A. Effect of flow history on poly(vinylidene fluoride) crystalline phase transformation. *J. Appl. Polym. Sci.* Vol. 104 (2007) pp. (89-94)

Wada, S., Nozawa, A., Ohno, M., Kakemoto, H., Tsurumi, T., Kameshima, Y. & Ohba, Y. (2009) Preparation of barium titanate nanocube particles by solvothermal method and their characterization, *J. Mater. Sci.* Vol. 44 pp. (5161–5166)

Wang, M., Singh, H., Hatton, T.A. & Rutledge, G.C. Field-responsive superparamagnetic composite nanofibers by electrospinning. *Polymer* Vol. 45 (2004) pp. (5505-5514)

Yee, W.A, Kotaki, M., Liu, Y. & Lu, X. Morphology, polymorphism behavior and molecular orientation of electrospun poly(vinylidene fluoride) fibers. *Polymer,* Vol. 48 (2007) pp. (512-521)

Patsidis, A. & Psarras, G.C. Dielectric behaviour and functionality of polymer matrix – ceramic $BaTiO_3$ composites. *eXPRESS Polymer Letters*, Vol. 2, No. 10, (2008) pp. (718–726)

18

Epitaxial Integration of Ferroelectric $BaTiO_3$ with Semiconductor Si: From a Structure-Property Correlation Point of View

Liang Qiao and Xiaofang Bi
Key Laboratory of Aerospace Materials and Performance (Ministry of Education), School of Materials Science and Engineering, Beijing University of Aeronautics and Astronautics, Beijing, People's Republic of China

1. Introduction

Ferroelectric thin films have attracted significant attention recently, due to their great potential for practical use in microelectronic and optoelectronic applications, such as non-volatile ferroelectric random access memories (NvFRAM), high-density capacitors, micro-electromechanical system (MEMS) and electro-optic devices etc. [1,2]. Among all the ferroelectric materials, $BaTiO_3$ (BTO) is most widely investigated not only for its simple chemical composition and remarkable properties (high dielectric constant and non-linear optical properties) but also because of their lead-free and environmental-friendly characteristic, which has become increasingly important from a long term point of view [3]. To date, a number of preparation techniques such as vacuum evaporation [4], sputtering [5], pulsed laser deposition (PLD) [6], molecular beam epitaxy (MBE) [7], metalorganic chemical vapor deposition (MOCVD) [8], sol-gel [9], and hydrothermal method [10] have been reported to fabricate BTO thin films.

On the other hand, newer device concepts may become realities if ferroelectricity can be coupled with conventional Si semiconductor technology. For example, ferroelectric field effect transistors, in which no power is required to maintain logic states, enter the realm of the possible if high-quality ferroelectric oxides can be grown epitaxially on Si with the right kinds of interfaces [11,12]. The criterion for successful integration with semiconductors is the minimum critical thickness for ferroelectricity, which must be of the order of a few nm in order to be compatible with very large scale integration. When in intimate contact with a semiconductor such that the polarization vector is normal to the interface, a ferroelectric oxide tends to develop a depolarizing field that suppresses the ferroelectric ground state. This depolarizing field can be eliminated if the ferroelectric oxide is sandwiched by metal electrodes which perfectly screen the polarization field. However, if the ferroelectric is in contact with a semiconductor and the interface state density is zero, there is very little screening of the polarization field, and the ferroelectric critical thickness becomes unacceptably large.

Meanwhile, the current request of downscaling needs for microelectronic devices has highlighted the importance of size effects as well as strain and dislocation influence on the properties of ferroelectric thin films [13-16]. Change of film strain field, either tensile or compressive, will elongate or shorten the corresponding lattice parameter. As a result, film tetragonality, polarization direction, domain configuration as well as dielectric constant along different axis will be subsequently engineered by that change of strain. However, due to large difference of lattice parameters and thermal expansion coefficients (TECs) between BTO and Si, most of the films reported to date are in an in-plane tensile strain state [12]. Although tensile BTO films with in-plane polarization are suitable for electric-optical applications [17], there are many applications in which a compressive strain leading to out-of-plane polarization is desired, e.g. the ferroelectric memory devices and quantum computing architecture [18,19]. An effective approach to control the strain state of BTO films is to choose an appropriate buffer layer by taking lattice constants and TECs into account. For example, Vaithyanathan *et al.* [20] has recently demonstrated that a $Ba_{0.7}Sr_{0.3}TiO_3$ layer of 300 Å enables a BTO film of 100 Å on Si to be in a compressive strain state, which is characterized by a complete *c*-axis orientation.

It is known that the well-developed $LaNiO_3$ (LNO) thin film serves not only as a crystallographic seed layer but also as a conductive layer. What is more, the lattice parameter of LNO has a good match with that of BTO (along [100] direction) and Si (along [110] direction) and the TEC of LNO (8.2×10^{-6} K^{-1}) is in between the values of BTO (10.4×10^{-6} K^{-1}) and Si (1.4×10^{-6} K^{-1}) [21,22]. Therefore, the strain state for the LNO buffer layer is expected to have a great influence on the strain state in its adjacent BTO layer, and a compressive strain in BTO on Si substrate could be induced by controlling the strain state in LNO. In this chapter, we will show that by insertion of a LNO buffer layer, high quality of BTO films can be integrated with Si(001) substrates and the structure-property correlations of BTO is largely dependent on the LNO buffer layer. A transformation from tensile to compressive strain state for BTO can be tuned by thickness of LNO. The corresponding microstructure and phase transition features of the BTO films will also be discussed.

2. Materials and experimental procedure

Both LNO and BTO films were fabricated on (001) Si substrates by multi-targets radio frequency (rf) co-sputtering method. Because of the insulating nature of the ferroelectric oxides, like BTO used here, the basic dc-sputtering doesn't work. Charges will build up at the surface of the insulating target and will eventually cancel out the external electric field. One solution to this problem is to apply an alternating electric field instead of a dc electric field. However, if the introduced alternating electric field is of low frequency, both electrons and ions can respond to the applied oscillating field, which will result in the damage of the film. For example, during one half of the cycle, sputtering of the cathode targets happens and during the other half cycle, sputtering of the anode substrates happens, which is known as re-sputtering and is detrimental to film growth. Radio frequency field is a good way to overcome this shortcoming, since the heavy Ar^+ ions cannot respond to a rf electric field while the electrons still can. Therefore, sputtering of the insulators is achievable without re-sputtering the deposited film [23-25]. Here, we utilize electromagnetic fields with a frequency of 13.56 MHz to generate the plasma. As shown in Fig. 1(a), the target is connected to the powered electrode as a "cathode", while the rest of the vacuum chamber, including substrates, is normally grounded as the "anode". The rf electric field generates plasma, but only the electrons are light enough to respond to the alternating filed at this frequency. The heavy Ar^+ ions "see" only the average electric field. The smaller area of the "cathode" results in a higher electron concentration during each half cycle compared to the electron concentration at the "anode", which will result in a net negative dc bias between the target and the substrate in a few cycles. It is this self-generated negative dc bias at the target that will drive and accelerate the Ar^+ ions towers the target, causing the sputtering of the target surface atoms.

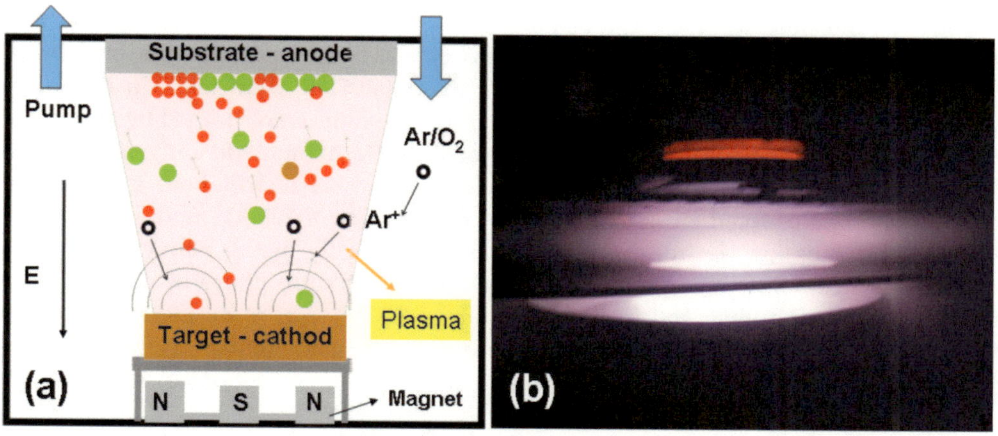

Fig. 1. (a) Schematic diagram of rf sputtering process. (b) The actual arcing and deposition process of our rf sputtering system

Si (100) wafers were first cleaned ultrasonically in acetone and methanol for 10 min each. During the whole deposition process base vacuum was below 3×10^{-4} Pa and substrate temperature was maintained at 600 °C. The deposition was generally performed either in vacuum or in a mixed atmosphere of oxygen (1.0 Pa) and argon (3.0 Pa) to prevent oxygen loss. After deposition, the as-deposited films were cooled down to room temperature (RT) in the mixed sputtering atmosphere and no additional annealing was applied. Typical puttering parameters are summarized in the following Table 1.

	LNO film	BTO film
Sputtering gas	$Ar+O_2$	$Ar+O_2$
Gas pressure	$Ar:O_2 = 3:1$	$Ar:O_2 = 3:1$
Input power (W)	71	67
Substract T (°C)	600	600
Target diameter (mm)	60	60
Deposition rate (Å/s)	1.0	0.8

Table 1. Typical sputtering conditions for LNO and BTO films

Specifically, LNO target was made by precursors of lanthanum nitrate hexahydrate and nickel acetate in the following method. First, lanthanum nitrate and nickel acetate powders were washed in distilled water, respectively. After drying, both of them with the precise stoichiometric ratio 1:1 were dissolved by acetic acid in a magnetic beater at a constant

temperature of 70 °C to give a homogeneous solution until it became steadily clear and transparent. Then, the homogeneous solution formed was dried at 200 °C with 4 h for the elimination of water and acetate, 600 °C with 4 h for the elimination of nitrate. The sintering process was carried out at 800 °C in an open atmosphere for 10 h. This process was repeated twice with intermittent grinding in an agate mortar. Finally the mixture was pelletized to the target shape, added with ethylene glycol (3% v/v) (in order to increase the viscosity of the powder), and sintered at 800 °C for 24 h.
Microstructure and crystallographic orientation of the films were characterized by θ-2θ scans of a 30 kV X-ray diffractometer (XRD, D/max2200PC, Rigaku) with Cu K_α radiation and a Ni filter. Grain size and surface morphology of the films were studied by atomic force microscope (AFM, SPI 3800N, SEIKO), performed at the tapping mode with sharp tips (BS-ElectricMulti75, resonance frequency of 75 Hz). Grain size distribution, interfaces and ferroelectric domains for the BTO/PLNO were observed by a 200 kV field emission high-resolution transmission electron microscope (HRTEM, 2100F, JEOL). Plan-view and cross sectional TEM samples were prepared through a standard procedure of cutting, gluing, slicing, grinding, dimpling, and finally ion milling, see Fig. 2(a) ~ (c). Electrical transport properties of the LNO films were evaluated by a domestic made four-probe testing system. Temperature dependent dielectric permittivity

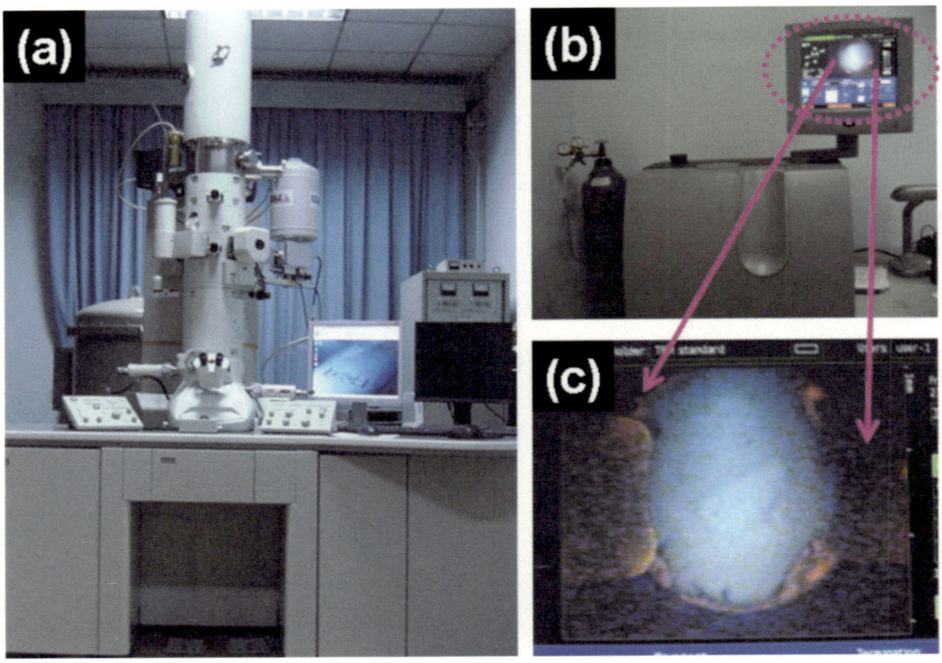

Fig. 2. (a) JOEL-2100F TEM system, (b) RS101 multi-functional ion milling system, and (c) the TEM sample was in-situ milling by the sputtering gun.

and dielectric loss for various BTO films were measured using an impedance analyzer (HP4194A, Hewlett-Packard Ltd.), at a bias voltage of 0.1 V, in the frequency range of 100 Hz–120MHz. Hysteresis loop was measured under an alternative electric field using a ferroelectric test system (TF 2000 analyzer, axiACCT, Germany). The electric properties were measured in a typical plate-capacitor setup. The top electrode layer was LNO, which was fabricated with the same deposition parameters as the bottom LNO and then patterned with a stainless iron mask of 200 mm in thickness and 1mm in diameter.

3. Growth mechanism and crystal structure

Fig. 3(a) shows typical XRD pattern for LNO films deposited on Si (100) substrate. As can beseen LNO film possesses a pseudocubic crystal structure with a (100) preferred orientation. In fact, the lattice parameters for cubic silicon are $a = b = c = 5.43$ Å. But for pseudocubic LNO, they are $a = b = c = 3.84$ Å. Although there is no direct lattice match between (100) silicon and (100) LNO, the diagonal length for the pseudocubic LNO equals to $\sqrt{2}\,a = 5.43$ Å, indicating that the atomic arrangement along the (110) direction for LNO matches well with that along the (100) direction for silicon. As a result, it can be inferred that the LNO films have grown epitaxially, when deposited on (100) Si wafers, via a 45° of azimuth angle rotation in *a-b* plane of LNO, leading to the formation of the highly (100)-oriented LNO films on the (100) Si substrate. Fig. 3(b) shows the schematic map of orientation and interatomic spacing relationship between Si (100) and LNO (100) lattices, it can be obtained that the epitaxial relationship is (001) LNO | | (001) Si and [110] LNO | | [100] Si. From AFM observation, it can be seen that the LNO films exhibit a uniform and dense microstructure with grain size about 30 nm, as shown in the insert of Fig. 3(a). The root-mean-square (rms) roughness measured from the figure is 8.75 nm over the area of 1μm×1μm, indicating that the films has a very smooth surface.

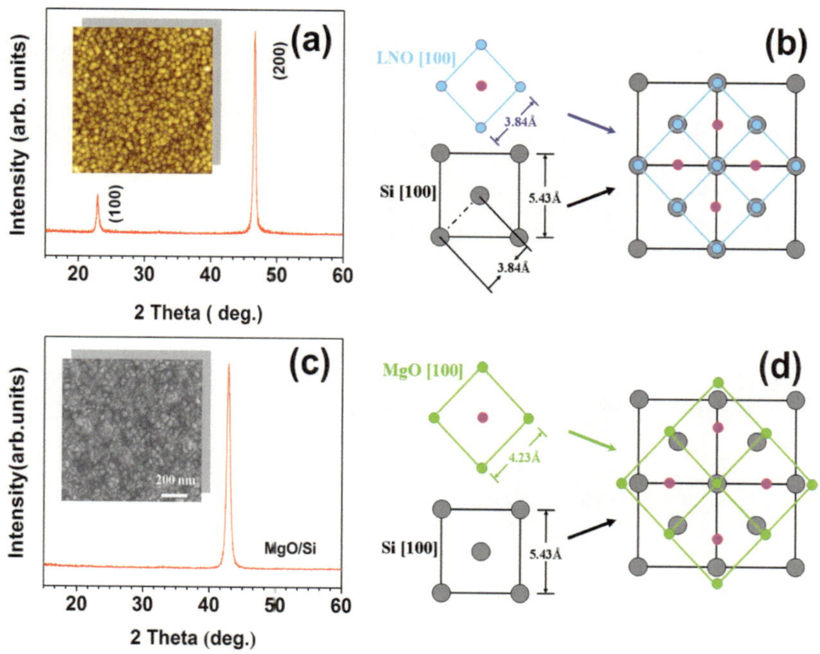

Fig. 3. XRD patterns of LNO (a) and MgO (c) films deposited on Si(001) substrate. (c) and (d) are schematic maps of orientation and interatomic-spacing relationship of LNO/Si (100) and MgO/Si (100), respectively. Insets in (a) and (c) are AFM and SEM images for LNO and MgO films, respectively.

Since magnesium oxide (MgO) is also a common buffer layer for the growth of thin film ferroelectric oxide materials and it exhibits superior stable chemical property, we also grow MgO films on Si substrate. Fig. 3(c) presents the XRD pattern of MgO(400nm)/Si. Obviously, MgO also exhibit seemingly high (100)-orientation and a good crystallinity (see inserted SEM image in Fig.3(c)). This can also be understood in term of the lattice matching between MgO and Si. Cubic MgO has a lattice constant of 4.23 Å, the lattice mismatch between MgO and Si in [100] direction is thus ~ 22.1%. This value is still too large to maintain a film with good crystallographic quality. On the other hand, as illustrated in Fig. 3(d), the lattice mismatch between the two materials in [110] direction is decreased to 10.2%, which is an acceptable value for the growth of highly oriented MgO film. However, it should be noted the lattice mismatch of MgO/Si is still significantly larger than that of LNO/Si, which will affect the structural quality of the upper ferroelectric BTO films.

Fig. 4(a) show XRD patterns of BTO films grown on bare Si(001) and MgO/Si(001) substrates, respectively. As can be seen, the BTO film grown directly on Si shows a typical random perovskite phase with no (100) orientation due to the large lattice mismatch between BTO (3.99 Å) and Si (5.43 Å). Insertion of a MgO intermediate layer between BTO and Si improves the (100) texture orientation of BTO film hugely and the BTO film also exhibit a good crystalline structure (Fig. 4(b)). However, a tiny BTO (110) peak still presents in the XRD pattern, indicating that MgO is not an ideal buffer layer material for highly (100) orientation growth of BTO because of its large lattice mismatch with Si substrate, as mentioned above. In contrast, BTO film grown on LNO/Si exhibits a

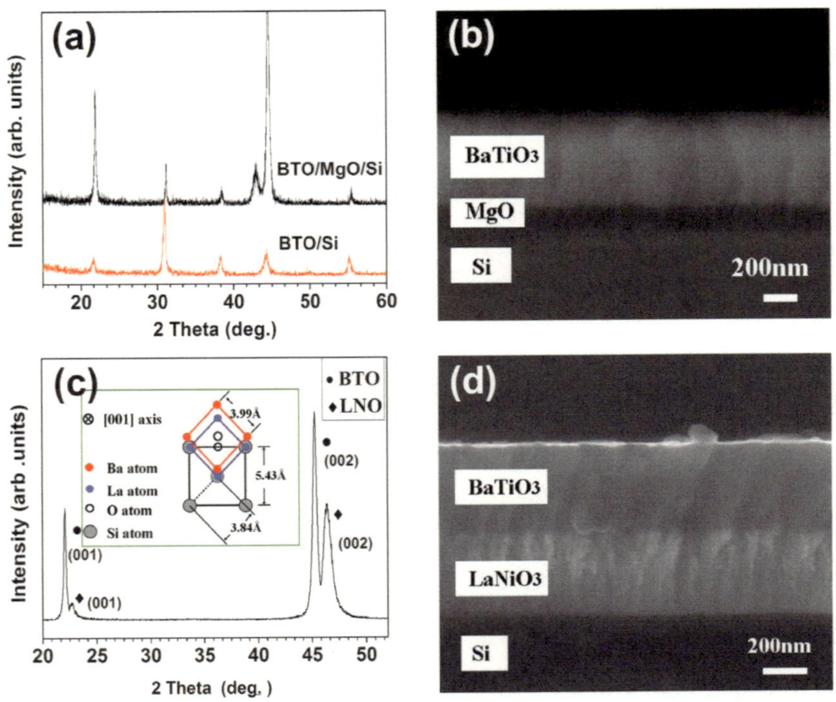

Fig. 4. (a) XRD patterns of BTO films grown on bare Si(001) and MgO/Si(001) substrates. (b) Cross-sectional SEM image for BTO/MgO/Si. (c) XRD pattern of BTO/LNO/Si(001). (D) (b) Cross-sectional SEM image for BTO/LNO/Si.

pure (001)-orientation structure, as indicated in Fig. 4(c). Since lattice mismatch between BTO and LNO is only ~ 3.76%, the pre-deposited LNO film will act as a seed layer for the (001) orientation growth of the BTO layer. An atomic arrangement relationship among the three kinds of lattices (Si substrate, conductive LNO and ferroelectric BTO) is also described in Fig. 4(c). In the film growth process, this little lattice mismatch is very important as it can induce a proper internal stress energy, which will not only maintain the (100) orientation growth but also induce an additional tetragonal distortion and change the ratio of c/a for the ferroelectric film. Fig. 4(d) shows the typical cross-section image of BTO/LNO/Si structure, in which both the BTO and LNO layers display dense microstructure and columnar grains extending over the entire film. The interface between LNO layer and BTO layer is distinct, smooth and of good quality.

The structural quality of BTO films on LNO/Si substrates is further demonstrated by in-plane XRD measurement and HRTEM cross-sectional interface observation. Fig. 5(a) is a typical in-plane {110} φ scan of the BTO film on LNO/Si. The presence of four accurately uniformly spaced (110) peaks reveals the fourfold symmetry of the film, confirming the good in-plane alignment between BTO film and LNO/Si substrate. The cross-sectional HRTEM image of the interface between [001] - LNO and [001] - BTO is presented in Fig. 5(b), along with the low magnification image for the bi-layer (inset). The consecutive lattice image shown in the figure proves the epitaxial growth relation between the BTO thin film and the LNO/Si substrate within a "cube-on-cube" fashion. An epitaxial growth relationship is as follows: (001) [001] BTO | | (001) [001] LNO | | (001) [001] Si and [110] BTO | | [110] LNO | | [100] Si.

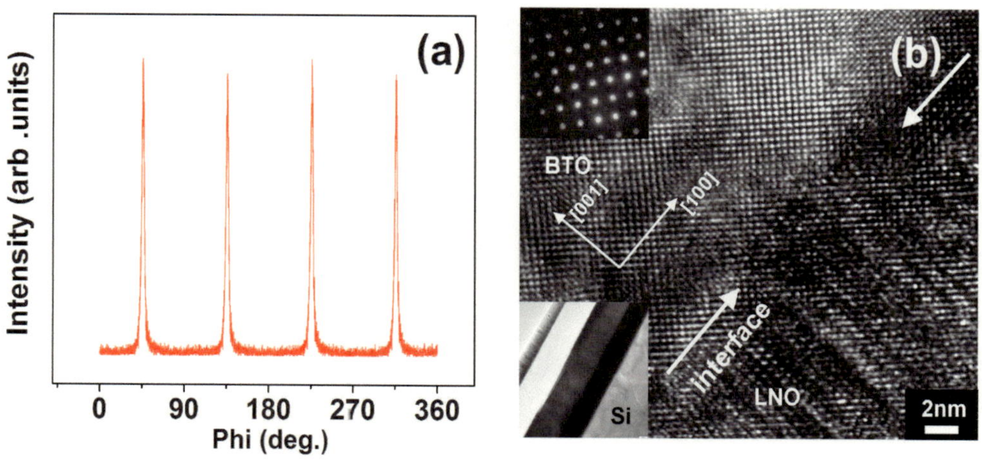

Fig. 5. (a) Typical in-plane {110} φ scan of the BTO film on LNO/Si. (b) HRTEM cross-sectional image of the BTO/LNO interface on Si.

Based on above discussions, the LNO film is a more suitable buffer layer for the growth of high quality BTO films. Since LNO films are also used as electrodes, we have investigated their electric properties. When the film thickness is 400 nm, the resistivity for the highly (100)-oriented LNO film at room temperature is 1.2×10^{-3} Ω·cm. Previous reports have shown that the resistivity for those highly (100)-oriented LNO thin films varies the range of $4.5 \times 10^{-4} \sim 1.0 \times 10^{-2}$ Ω•cm, depending on fabrication processes [26,27]. The variations in the resistivity could be attributed to composition deviation from typical ABO_3 chemical stoichiometry especially the oxygen loss, which were influenced by the deposition method and temperature. The lower resistivity of around 1.2 × 10^{-3} Ω·cm obtained in this work indicates that the highly (100)-oriented LNO films prepared by sputtering could be used as both buffer layer and bottom electrode, which is of significance for fabricating highly-oriented ferroelectric thin films.

4. Grain size effect

It is known that both microstructure and properties of ferroelectric films are dependent greatly on fabrication processes [28-31]. The extrinsic parameters are also assumed to be responsible for variations in microstructure dependence of ferroelectricty, and might be the reason for a broad dispersion in some data such as critical size below which ferroelectricity is eradicated. For example, in the last two decades, there had been a variety of different experimental critical thickness for epitaxial ferroelectric thin films [32-37]. However, recent theoretical and experimental studies have implied that there is no intrinsic thickness limit for ferroelectricity in thin films with thickness down to even several unit cells [38,39]. While for ferroelectric polycrystalline films, nano-particles, or nano-ceramics, there is still no unambiguous conclusion that if there exists a critical grain size responsible for the disappearance of the macroscopic ferroelectricity. To investigate the effect of grain size on the ferroelectric properties of BTO films, it is important to control the experiment so that the only different parameter for different BTO films is variable size of grains. The ferroelectric BTO thin films of 200 nm were first deposited onto the LNO buffer layers at room temperature. The as-deposited films were then annealed in an open-air atmosphere with different temperatures ranging from 400 ~ 800 °C for 2 hours in order to obtain different grain size.

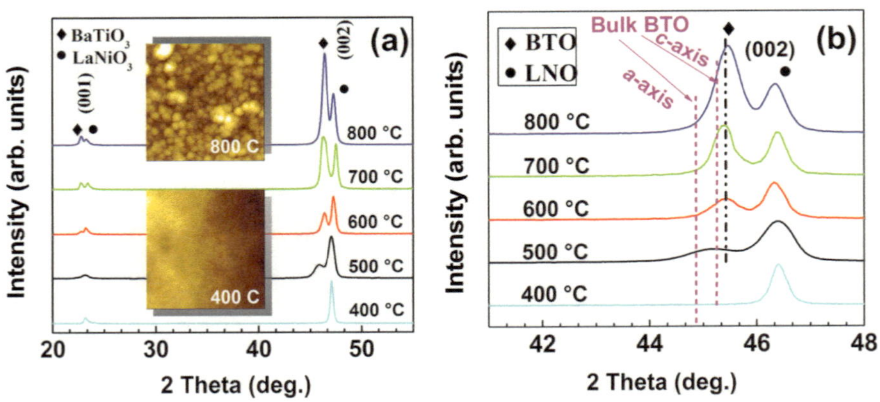

Fig. 6. (a) XRD patterns of the BTO films annealed at different temperatures ranging from 400°C to 800°C, (b) magnified view of XRD patterns in Fig. 6(a)

Fig. 6 (a) shows the XRD patterns of the BTO films annealed at several different temperatures. When the annealing temperature is 400 °C, only (100) and (200) peaks for LNO can be observed and no any other peaks related with BTO can be identified, indicating that the BTO film remain an amorphous structure at the temperature. When the annealing temperature is increased to 500 °C, a broadened (200) peak ascribed to the BTO start to appear. With further increasing the annealing temperature, the peaks become much sharper and are gradually intensified. From the results, it can be obtained that the BTO films start to crystallize at 500 °C from the amorphous phase and grow into a (100) preferred orientation on the LNO (100) buffered Si substrates as the annealing temperatures increase. The surface morphologies of BTO films annealed at different temperatures had also been analyzed by AFM observations, as shown in inset of Fig. 6(a). The 400 °C -annealed BTO film exhibits an amorphous structure without the formation of any distinct grains on the surface and increasing annealing temperature will increase the BTO grain size, which agrees well with the XRD results.

From the magnified view of XRD patterns in Fig. 6(b), the lattice constants as well as strain states for both BTO and LNO layers can also be extracted. The (002) peaks for all the LNO layers are positioned at the same diffraction angel, indicating the LNO films possess same lattice constant, which is calculated to be 3.91 Å. Compared with that of bulk LNO (3.84 Å), the larger value reveals that LNO films are under tensile strain. On the other hand, compared with the 2θ peak position of bulk BTO (purple dashed lines), these BTO (002) 2θ diffraction peaks are shifted to higher angles, indicating a decease in the out-of-plane lattice constants and an increase in the in-plane lattice constants in the BTO films. Since the LNO layer is very thin compared with Si, so the shift is suggested to be the result of the thermal strain induced during the cooling process caused by the difference of thermal expansion coefficients between BTO layer and Si substrate. This corresponds well with the experimental strain states for BTO when incorporated with Si substrate [40,41]. The strain state and its effect on the structure and properties of ferroelectric BTO films will be further discussed in the part 5. Nevertheless, as the (002) peaks for all the BTO layers are almost at the same position (black dashed line), so the BTO layers are under same strain state in spite of different annealing temperature.

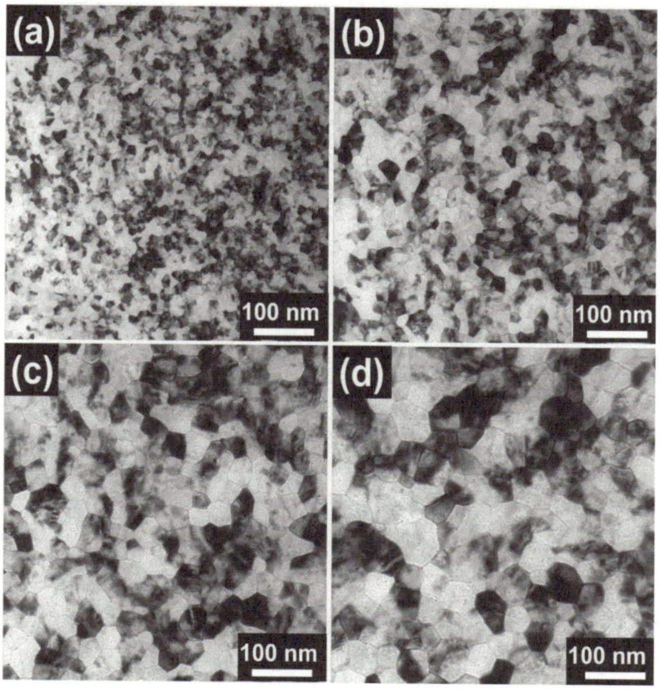

Fig. 7. Plan-View TEM images of the BTO film annealed at different temperatures: (a) 500°C, (b) 600°C, (c) 700°C and (d) 800°C. The inset is the electron diffraction arc for the 800°C - annealed film.

Microstructures for the annealed films were characterized by plan-view TEM observations, as shown in Fig. 7(a) ~ (d). It is clearly seen that the films annealed at temperatures ≥500 °C are crystallized and featured as uniform and cracks-free. The grain size is increased with increasing the annealing temperatures, changing from 14 to 55 nm in diameter. It is noted that the BTO films in this work exhibit much smaller grain sizes as compared to other reported BTO films [42,43]. The difference is considered to be the result of the influence of microstructure of the LNO buffer layer. It is known that grain size for a newly formed crystal is dependent on the nucleation rate and the growth rate, respectively. In this work, the LNO buffer layer with very fine grain size of 20 ~ 30 nm was used as seed layers for the ferroelectric BTO films. The grain boundaries in the LNO buffer layers will act as nucleation sites for the crystallization of the BTO films during the annealing processes. The smaller grain size of the LNO film leads to more nucleation sites for the crystallization of BTO films and, as a result, the BTO films grow into a microstructure characterized by fine and uniform grains.

Room temperature ferroelectric hysteresis loops of the BTO films annealed at different temperatures are displayed in Fig. 8(a). For BTO films annealed at 600 °C and above, obvious hysteretic shape of polarization vs electric field (*P-E*) curves are obtained. With increasing the annealing temperature, the *P-E* hysteresis loop starts to become much more erect and saturated, showing a typical ferroelectric characterization. The obtained remnant polarization (P_r) for 800 °C -annealed BTO is 2.0 μC/cm², similar to the values of other polycrystalline BTO films, e.g. 2.0 μC/cm² on $Pt/Ti/TiO_x/Si$ by Thomas *et al.* [44] and 1.0 μC/cm² on $Pt/SiO_2/Si$ by Huang *et al.* [45]. However, compared with the value for BTO single crystal (24 μC/cm²) or other epitaxial BTO films [46,47], the P_r is still much lower, probably due to fine grain size and the formation of in-plane tensile

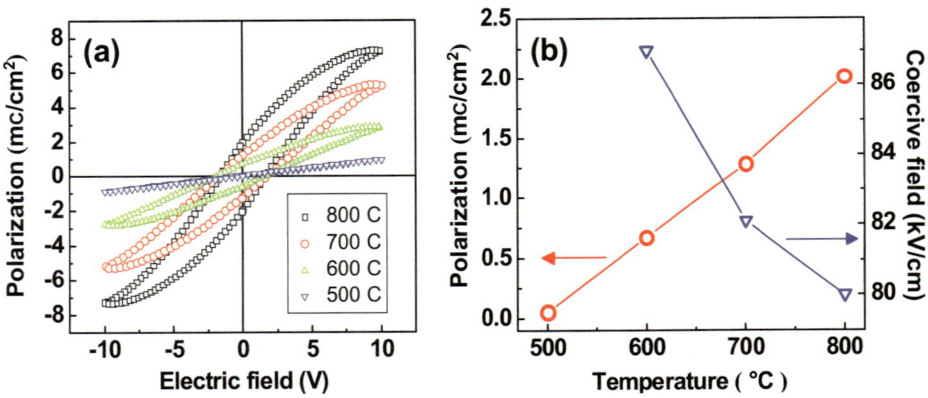

Fig. 8. (a) Hysteresis loops of the BTO films after annealing at different temperatures. (b) Annealing temperature dependence of the remnant polarization and coercive field for the BTO films.

strain state on the Si substrate, as had been obtained by the XRD analysis. On the other hand, for BTO film with the finest grain size of 14 nm (annealed at 500°C), it still exhibits some hysteresis characteristics with P_r ~ 0.08 μC/cm², although it is not obvious and shows almost linear dependence of *P(E)*. This corresponds well with the recently reported experimental results that the ferroelectricity does exist in nanocrystalline BTO ceramics with ultra fine grain size of 30 nm [48], 22 nm [49], and even 8 nm [50]. However, the significantly reduced P_r is indicative of a strong suppression of macroscopic ferroelectric character in the 14 nm BTO film, which may arises from either the frozen domain structure under an external field by grain boundary effects, such as the clamping of the domain walls and the hindrance of polarization switching, or the depolarization field originated by the low permittivity nonferroelectric grain boundaries [49]. Besides, the low crystallinity of 500 °C - annealed BTO is also a possible reason for the obtained low P_r. On the other hand, it should be noted that from XRD patterns for the BTO films, the separation of diffraction peaks (200) and (002) was not observed for all the annealed films. Meanwhile, the ferroelectricity is known to be attributed to the formation of tetragonal structure in materials. Thus, the different properties for the BTO thin films annealed at different temperatures suggest that a pseudocubic structure be formed in the BTO films having a larger grain size. The idea of the pseudocubic phase is based on a core-shell grain model in which individual grains consist of a cubic shell and a tetragonal grain interior [51]. Internal strains in tetragonal structure caused by the formation of nano-scale grains are believed to be responsible for its change to the so-called pseudocubic phase [52,53]. It can be considered that the obvious ferroelectricity observed in this work for the films after annealing at or above 600 °C is attributed to the formation of the pseudocubic phase, while the strong restrained ferroelectricity for the film annealed at 500 °C is due to the suppression of the tetragonal core caused by the smaller grain size.

To further study the change of the ferroelectricity with different grain size, Fig. 8(b) plots the P_r and coercive field (E_c) of the BTO films as a function of annealing temperatures. As the annealing temperature increases, the remnant polarization increases while the coercive field decreases. Since all the films have the same thickness and the same in-plane tensile strain state, their grain size is then responsible solely for their different ferroelectric behaviors. Theoretical calculations have demonstrated that the density of 90° domain walls is inversely proportional to the square root of the grain size [54]. It means that the density of domain walls is increased with the decrease of annealing temperature, and consequently a distance dependent repulsive force between neighboring domain walls is enhanced. This leads to a reduction of the mobility for domain walls and more difficulty in domain orientation, resulting in the reduction of P_r and the increases of E_c.

5. Strain engineering and phase transition

The microstructure of the film has a great influence on the corresponding physical properties. Recently, the investigation of strain effect has become increasingly important due to its great influence on the ferroelectric phase transition, domain formation, and polarization magnitude for ferroelectric thin films. Haeni *et al.* [55] have demonstrated that a proper epitaxial strain from a $DyScO_3$ substrate can increase ferroelectric Curie temperature by hundreds of degrees and produce room temperature ferroelectricity in $SrTiO_3$ which does not originally exhibit ferroelectric property at any temperature. Enormously high remnant polarization (P_r) was also reported by Choi *et al.* [47] in relatively thick epitaxial BTO films grown on a newly developed $DyScO_3$ substrate. The P_r observed is about 70 $\mu C/cm^2$, which is almost 2.7 times higher than the spontaneous polarization of bulk BTO (P_s = 26 $\mu C/cm^2$). In this section, we show that, the strain state and the dielectric, ferroelectric as well as domain configurations of ferroelectric BTO layer on LNO/Si is critically dependent their respective film thickness, and this can actually be utilized to engineer the strain and the physical properties of BTO.

5.1 Strain modeling

For ferroelectric thin films, internal strains are mainly induced by lattice distortion due to the different lattice parameters [56] and the incompatible thermal expansion coefficients (TECs) between the film and substrate (or buffer layers) [57], to the self-induced strain of phase transition during the cooling process [58], and to the inhomogeneous defect-related strains such as impurities or dislocations [41]. However, the contribution from the later two factors can be avoided by selecting suitable materials and exploring advanced film growth techniques.

Schematic Fig. 9 illustrates the formation and evolution of the strain in a typical epitaxy film growth process. At the film growth temperature, when atoms arrive at the surface of the substrate, they will initially adopt the substrate's in-plane lattice constant to form an epitaxial film [Fig. 9(a)]. As long as the film thickness (t) is smaller than the critical thickness (h_c) of the film/substrate system, the film will keep its coherence with substrate and maintain a fully strained layer [Fig. 9(b)]. When $t > h_c$, dislocations will appear at the interface or near interface region and the whole film relaxes. However, the relaxation is a dynamic controlled process, if the film thickness is not large enough than h_c, the relaxation may only occur partially [Fig. 9(c)]. Finally, during the cooling process,

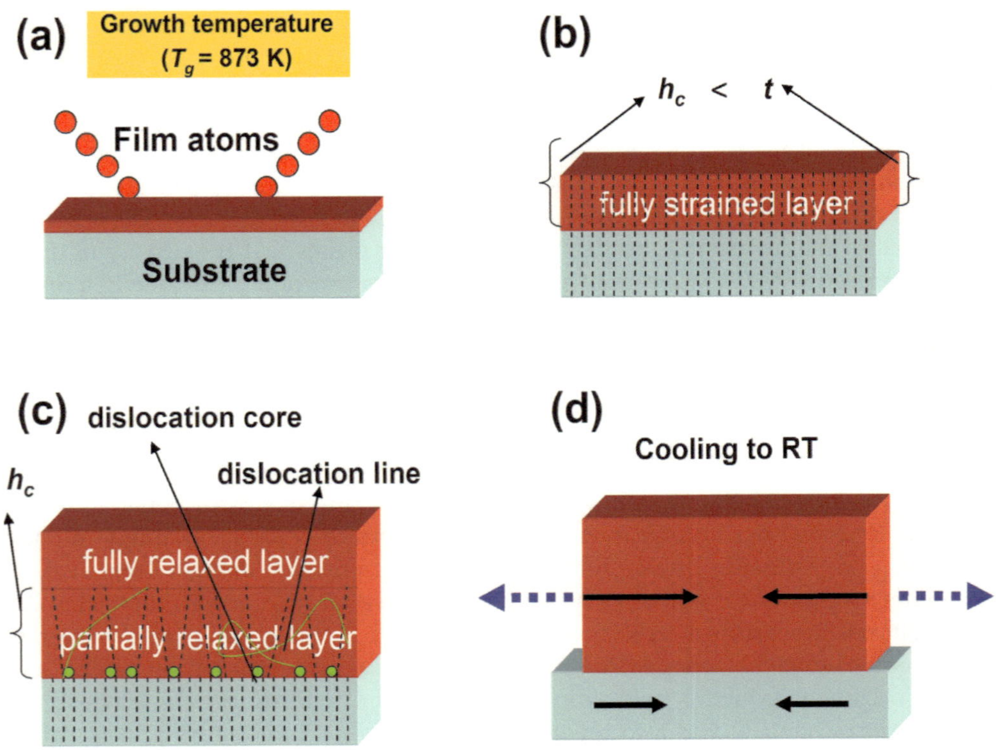

Fig. 9. An illustration of the strain formation and evolution in a typical epitaxy film growth process.

additional thermal strain may also be exerted on film due to the difference of the TECs between the film and substrate [Fig. 9(d)]. Therefore, the temperature dependent misfit strain in a thin film can be modeled simply by taking into account the combined contribution of the temperature dependent lattice strain [$S_m(T_g)$] and the thermal strain [$S_{therm}(T)$] [59], which can be approximated by the linear relation,

$$S_m(T) = S_m(T_g) + S_{therm}(T) \tag{1}$$

$$S_{therm}(T) = (\alpha_s - \alpha_f)(T - T_g) \tag{2}$$

where, T_g = 873 K, is the growth temperature, $S_{therm}(T)$ is the thermal strain, α_s and α_f are linear thermal expansion coefficients (TECs) of the substrate and prototypic cubic phase of the film. $S_m(T_g) = [a_s^*(T_g) - a_f(T_g)] / a_s^*(T_g)$ is the effective misfit strain of the film and substrate at T_g, $a_s^* = a_s(1 - \rho)$ is the effective lattice parameter of the substrate [60] and ρ is the dislocation density [61], which reflects the effect of strain relaxation induced by the appearance of misfit dislocations at the film/substrate interface at T_g.

For the convenience of understanding, we define an original misfit lattice strain $S_m^0(T_g)$, which means the actual original misfit strain between the as-grown film and the supporting substrate if the film does not relax at all at the growth conditions, as follows,

$$S_m^0(T_g) = [a_s(T_g) - a_f(T_g)] / a_s(T_g) \tag{3}$$

Taking into account the thermal expansion, the lattice constant of the film and substrate at T_g can be approximated by $a_f(T_g) = a_f(RT)[1 + \alpha_f(T_g - RT)]$ and $a_s(T_g) = a_s(RT)[1 + \alpha_s(T_g - RT)]$, respectively. As a matter of fact, the $S_m^0(T_g)$ does not really exist, because the film growth and relaxation occur simultaneously. However, we assume the film growth process and the strain relaxation process can occur in the following two successive steps. First, the film doesn't relax during the whole growth procedure (holding a $S_m^0(T_g)$) and then, when growth is done the relaxation process dominates and the as-grown film begins to relax only when the accumulated $S_m^0(T_g)$ exceeds the critical relaxation requirements. In this picture, the $S_m(T_g)$ in equation (1) can be thus equivalently and much more schematically divided into the combination of an original lattice strain $S_m^0(T_g)$ at T_g and a strain variation due to the formation of misfit dislocations [$S_{dis}(\rho, T_g)$] during relaxation,

$$S_m(T_g) = S_m^0(T_g) + S_{dis}(\rho, T_g) \tag{4}$$

In addition, structural factors such as growth defects, crystallinity, and oxygen vacancies may also contribute to the $S_m(T)$ [41], which is denoted by S_{other} in the following expression.

$$S_m(T) = S_m^C(T_g) + S_{dis}(\rho, T_g) + S_{therm}(T) + S_{other} \tag{5}$$

By analyzing the first three terms on the right side of equation (5), we can roughly estimate the final strain in the obtained film.

We start from the LNO buffer layer. Fig. 10 (a) shows the XRD patterns for various LNO films with different thickness. It is obvious the LNO (200) peak shifts toward high angles with increasing the film thicknesses, indicating a decrease in the lattice constant. Fig. 10 (b) shows the LNO thickness dependent lattice constant ($a = 2d_{002}$) and misfit strain ($S_m = (a - a_0)/a_0$, where a_0 is the lattice constant for freestanding bulk LNO) obtained from the XRD result at RT. As can be seen, the lattice parameters decrease with increasing the LNO thickness and become close to the bulk value (3.84 Å) for 600 nm LNO film.

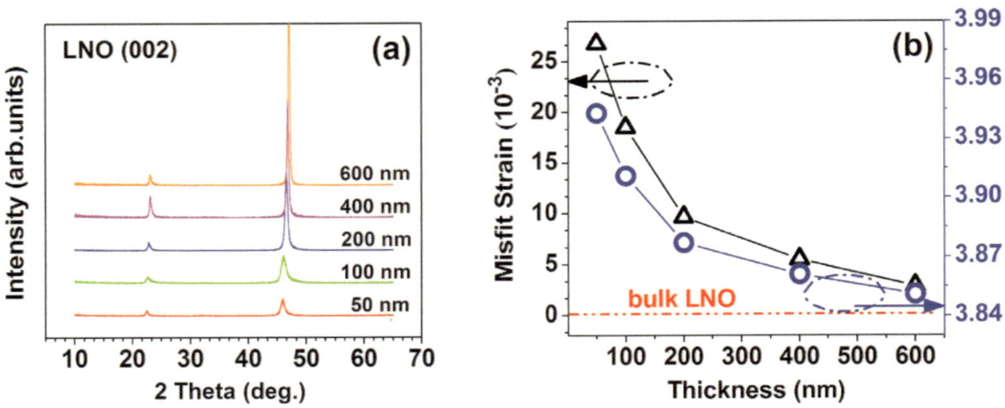

Fig. 10. (a) XRD patterns for LNO films with different thicknesses. (b) Calculated LNO thickness dependence of misfit strain and lattice constant, along with the lattice constant for bulk LNO.

For the LNO film directly grown on a Si substrate, using equations (3), we can calculate the origin misfit lattice strain and $S_m{}^0\ (T_g) \sim -3.68\times10^{-3}$. Based on elastic theory, the $S_m{}^0\ (T_g)$ will be fully relaxed by the formation of misfit dislocations at the film/substrate interface when the thickness of the film (h) is larger enough than the critical thickness (h_c) [62],

$$h_c = \left(\frac{b}{f}\right)\left[\frac{1}{4\pi(1+\upsilon)}\right]\left[\ln\left(\frac{h_c}{b}\right)+1\right] \tag{6}$$

where υ is the Poisson's ratio, f the relative misfit, and b the Burger's vector of misfit dislocations. Due to lack of v value for LNO, here we simply assume $\upsilon = 0.3$, a typical value for perovskite oxides [63], and h_c is estimated to be on the order of 23 nm for a 0.5%-misfit film. Considering that the film thickness $t >> h_c$, so the $S_m{}^0\ (T_g)$ will be fully relaxed by $S_{dis}\ (\rho, T_g)$, making $S_m{}^0\ (T_g)$ and $S_{dis}\ (\rho, T_g)$ negligible. The $S_m(T)$ in equation (5) is therefore attributed mainly to the thermal strain $S_{therm}(T)$ and S_{other}. Generally, due to large difference in TECs between LNO and Si, the induced thermal strain will make the LNO film under a tensile strain state with an enlarged lattice constant at room temperature, which is consistent with the former XRD results. Using equation (2) the thermal strain $S_{therm}(T)$ at RT for the LNO is estimated to be ~ 3.91×10^{-3}, while the XRD analysis shows that $S_m(RT)$ for the LNO films is decreased from 26.82×10^{-3} to 2.865×10^{-3}, as shown in the inset, when the thickness varies from 50 nm to 600 nm. The result also indicates that a strain in the LNO films induced by the Si substrate can be fully relaxed by increasing their thicknesses to a certain extent. Note that the difference between $S_m(RT)$ values and the thermal strain also confirms the contribution of structural parameters (S_{other}), as represented in equation (5).

5.2 Tensile strained BTO

Fig. 11(a) shows the XRD patterns for 200 nm BTO films grown on the 100 nm LNO buffered Si. In order to determine the in-plane lattice alignment and in-plane constant of BTO, samples were placed on a tilted holder with a set azimuth angle of $\psi = 45^o$, so that the (101) and (202) crystal planes are parallel to the detected surface of the films. As a result, the reflections for (101) and (202) planes in the film will become much easier to satisfy the

Prague's Law, $2d\sin\theta = \lambda$ (d is the lattice spacing, θ the diffraction angle and λ the x-ray wave length) [64], in the x-ray detecting process and obvious diffraction of (101) and (202) planes will occur at their own characteristic diffraction angle. The 45 ° tilted XRD θ - 2θ scans for BTO/LNO bi-layers are shown in Fig. 11(b). It is seen that only (101) and (202) reflections for LNO and BTO films are detected, implying the in-plane lattice alignment between [110] LNO and [110] BTO. Using lattice spacing d_{002} and d_{202} obtained from the Prague's Law ($d = \lambda/2\sin\theta$), the out-of-plane lattice constants ($a_{\perp}$) and in-plane lattice constants ($a_{||}$) for BTO can be calculated by the following equations [65],

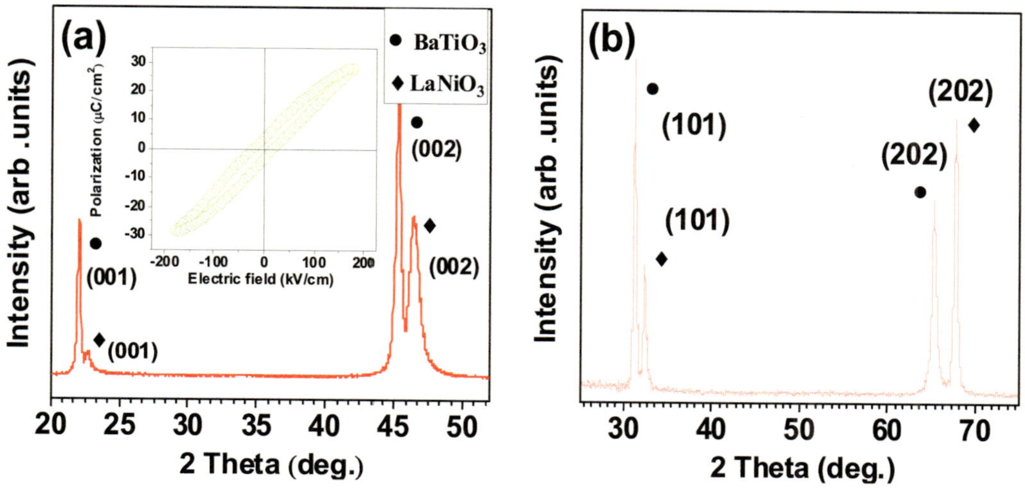

Fig. 11. (a) XRD patterns for 200 nm BTO thin film deposited on 100 nm LNO-buffered Si substrate. Inset shows the room temperature ferroelectric hysteresis loop for this BTO film. (b) 45° tilted in-plane scan for the BTO/LNO bilayer films.

$$a_{\perp} = 2d_{002} \tag{7}$$

$$a_{||} = \frac{2}{\sqrt{\frac{1}{d_{202}^2} - \frac{1}{d_{002}^2}}} \tag{8}$$

The obtained $a_{\perp}$ and $a_{||}$ for 200 nm BTO are 4.001 and 4.077 Å, respectively. Compared with bulk BTO (a = 3.992 Å and c = 4.036 Å), the BTO films are elongated along a-axis and compressed along c-axis. Besides, as out-of-plane lattice constants are always smaller than the in-plane lattice constants for both BTO films, thus it can be inferred that the BTO films are under an in-plane tensile strain state. Inset of Fig. 11(a) shows room temperature polarization and capacitance with electric field at 1 kHz. The small remnant P_r indicates that the film is nearly in an in-plane polarization state, that is, the polarization vectors mainly parallel to the film surface.

The temperature dependent dielectric permittivity and dielectric loss for the bilayer films were shown in Fig. 12(a). Over the temperature region, two broad but obvious peaks for the dielectric permittivity and dielectric loss are detected at 30 °C and 170 °C, respectively. This indicates that two phase transitions have occurred. The dielectric response can be explained by the misfit strain-temperature phase diagrams theory [66-71] for an epitaxial polydomain ferroelectric film grown on a "tensile" substrate. As shown in Fig. 12(b), the polydomain ferroelectric films have different phase states and domain configurations compared to epitaxial single-domain film or bulk materials. This results in the contribution of an extrinsic response (domain-wall movements) together with the intrinsic response (substrate induced strain) to the dielectric response in a small signal dielectric measurement in the plate-capacitor setup. The temperature dependent misfit strain can be approximated by equation (1). Since BTO film is pretty thick, the contribution of lattice strain can be neglected, and the total strain is subjected solely to the thermal strain. Thus, the misfit strain (S_m) at the ferroelectric phase transition temperature (443 K) is estimated to be $(\alpha_s\text{-}\alpha_f)(T\text{-}T_g) \sim 3.87 \times 10^{-3}$, which just lies in the predicated $a_1/a_2/a_1/a_2$ polydomain region [66]. It can be obtained that, when the film is cooled down from the deposition temperature to Curie temperature, a second order phase transition from cubic parelectric to pseudo-tetragonal $a_1/a_2/a_1/a_2$ ferroelectric phase occurs, leading to the appearance of the broad dielectric peak in the temperature-dependent dielectric curves. On the other hand, the second permittivity peak at 30 °C is suggested to be the result of the structural phase transition between the $a_1/a_2/a_1/a_2$ and $ca_1/ca_2/ca_1/ca_2$ polydomain states that is accompanied by the appearance of the out-of-plane polarization. This is also consistent with the observation of the small P_r at room temperature.

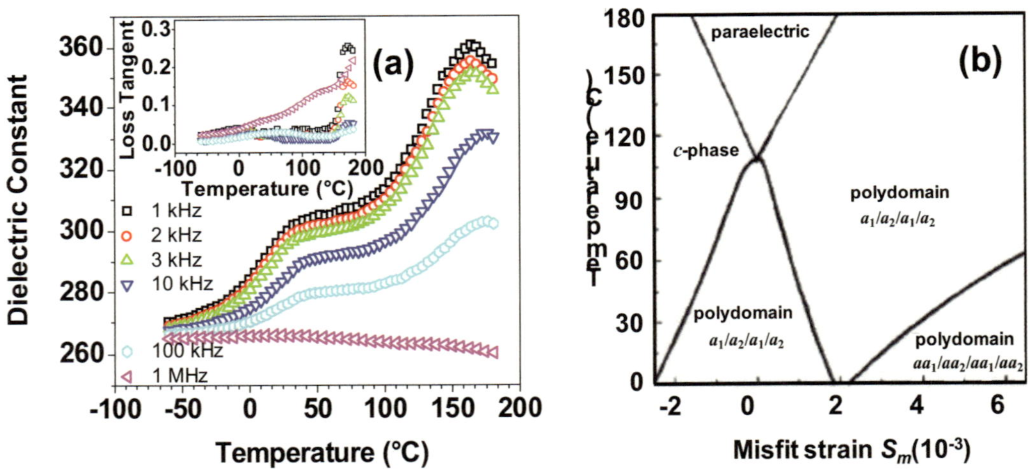

Fig. 12. (a) Temperature dependent dielectric permittivity and dielectric loss (inset) for the tensile-strained BTO film. (b) Schematic illustration of the misfit strain-temperature for BTO thin film.

Fig. 13(a) shows the plan-view HRTEM image of elastic domain pattern for the BTO film. The adjacent elastic domain walls form a coherent twin boundary lying along the surface of {110} twin planes for the minimization of in-plane elastic strain energy. Fig. 13(b) shows the cross-sectional TEM image of elastic domains. It can be clearly seen that the domain walls exhibit a blunt fringe contrast, because the polarization vectors in adjacent domains form an angle and they, as a result, are not in the same height with respect to the observation direction [72].

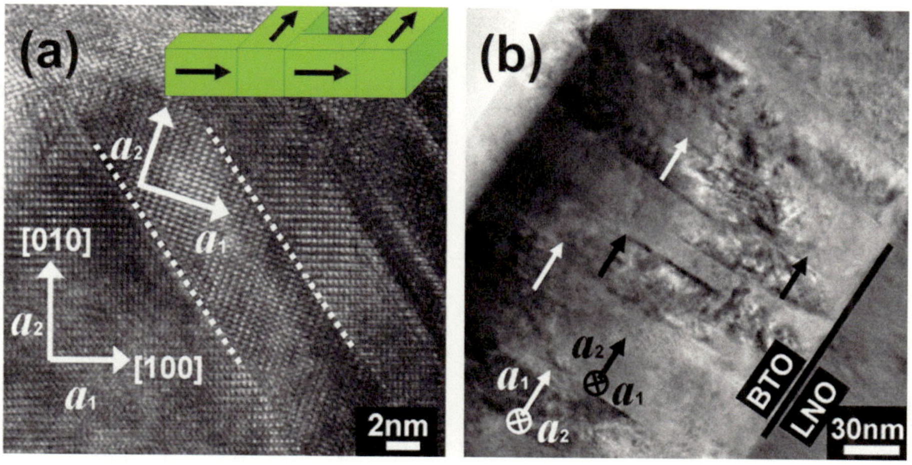

Fig. 13. (a) HRTEM plan-view image of elastic domain configurations, (b) cross-sectional image of elastic domains.

5.3 Compressive strained BTO

Fig. 14(*a*) and 14(*b*) show the XRD patterns of normal and 45°tilted θ-2θ scans of BTO(100 nm) on LNO(600 nm)/Si. Using above mentioned method, the in-plane and out-of-plane lattice constants for the BTO film are calculated to be a = 3.955 Å and c = 4.056 Å, respectively. Then the tetragonal distortion c/a is 1.025. Compared to bulk BTO (a = 3.992 Å and c = 4.036 Å) and other tensile strained BTO films on Si substrates (e.g. c = 3.975 Å by Meier *et al.* [40]), the BTO film is elongated along c-axis and compressed along a-axis, and corresponds well with the results obtained by Petraru *et al.* in BTO (56 nm)/STO (a = 3.925 Å and c = 4.125 Å) [73]. The unit cell volume can be estimated as $V_{film} = a \times a \times c \sim 63.444$ Å^3, which is smaller than that of the bulk ($V_{teg} \sim 64.318$ Å^3 and $V_{cubic} \sim 64.722$ Å^3) [74]. Therefore, the BTO film is under a compressive strain state.

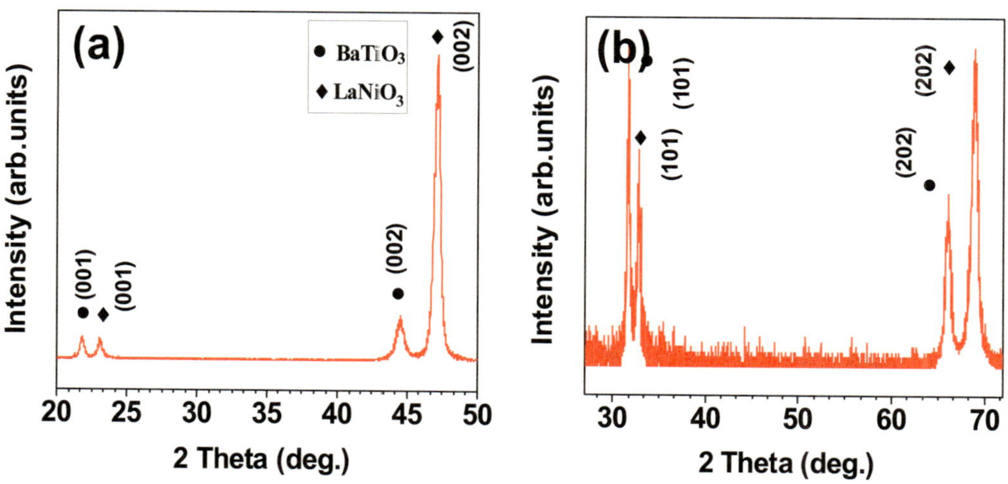

Fig. 14. XRD patterns of regular (a) and 45°tilted (b) θ-2θ scans of BTO(100nm)/LNO(600nm)/Si.

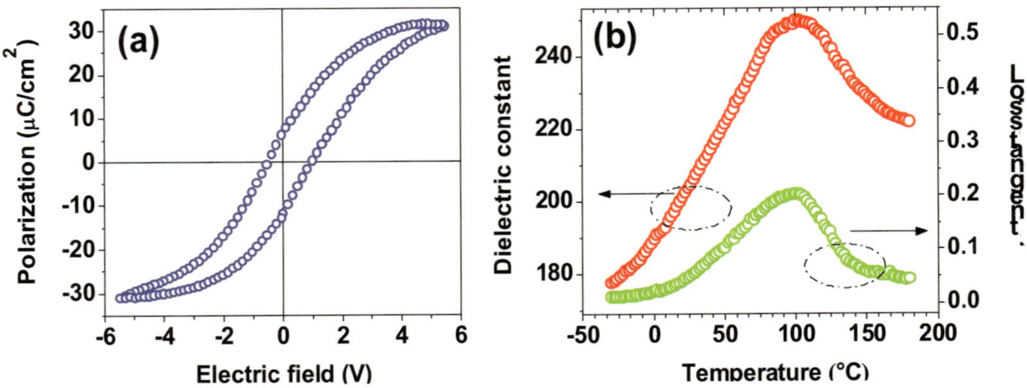

Fig. 15. Room temperature hysteresis loop (a) and temperature dependent dielectric response (b) for compressive strained BTO film.

Electrical properties of compressive strained BTO film have been investigated by ferroelectric and dielectric measurements. Hysteresis loop for the compressive BTO, as shown in Fig. 15(a), exhibits a well-defined shape, which is significantly different from those of tensile BTO films. The P_r is 10.2 μC/cm^2, much larger than 0.7 μC/cm^2 and 2.0 μC/cm^2 observed in tensile BTO films on Si substrate [41,44], which is apparently due to the compressive strain state induced by thick LNO layer. However, it should be noted that the obtained P_r is still smaller compared with the giant P_r values for other fully strained BTO films with purely *c*-domain structure on compressive oxide substrates, such as $SrTiO_3$ [46], $GdScO_3$ and $DyScO_3$ [47]. Temperature dependent dielectric permittivity and loss tangent curves exhibit a broad peak near 100 °C, showing a slight decrease in the ferroelectric to parelectric phase transition temperature (T_c) with respect to its bulk counterparts [75]. The strain state dependent T_c for BTO film had been extensively investigated, and it is very dependent on the film or buffer layer thickness [76,77], substrate chosen [78,79] as well as the microstructure and crystallinity [80,81] of the fabricated BTO films. For example, Huang *et al.* [76] had fabricated BTO films with wide range of thickness (35 ~ 1000 nm) on 400 nm LNO buffered Si substrates using Ar/O_2 mixed sputtering gas and found that all the films were tensile strained and the T_c was greatly reduced with decreasing the BTO film thickness. However, their BTO films were significantly (110)-oriented instead of (001)-oriented. On the other hand, based on the misfit strain-temperature phase diagrams theory for epitaxial polydomain ferroelectric thin films, both tensile and compressive epitaxial strain will substantially enhance the T_c for ideal homogeneous ferroelectric epitaxial films. However, it has recently been demonstrated that in thin films the inhomogeneous strain field resulted by the strain gradients in the growth direction of the film should also be considered, which, combined with the homogeneous strain field, will both influence the polarization and ferroelectric phase transition character of ferroelectric thin films [41,82,83]. In addition, Kato *et al.* [80] observed a marked decrease of T_c for 20 °C in polycrystalline BTO films on LNO(200nm)/Pt(400nm)/Si and Chen *et al.* [81] also reported a reduced T_c in polycrystalline multiferroic NiFe/BTO/Si.

In fact, the reduction of T_c for the ferroelectric crystals and films are commonly observed in a system under an external compressive stress [74,81]. Based on the soft mode theory, the phase transition for displacive ferroelectrics can be attributed to the frozen of soft mode in the center of Brillouin zone. The frequency of the soft mode (ω_T) is determined by the interaction between local restoring "short range" repulsions (R_0'), which prefers the undistorted paraelectric cubic structure, and "long range" Coulomb force, which stablizes the ferroelectric distortions [84],

$$\mu\omega_T^2 = R_0' - 4\pi(\varepsilon+2)(Z'e)^2/9V \quad (9)$$

where, μ is the reduced mass of the ions, $Z'e$ the effective ionic charge, V the volume of the unit cell, and ε the high frequency dielectric constant. The decreased lattice volume in the compressive BTO film ($V_{film} < V_{teg} < V_{cubic}$) leads to the decrease of average ion distance (r), which in turn increases the short range force and the Coulomb force as well. Since the short range force is proportional to r^{-n} (n = 10~11) while the Coulomb force to r^{-3}, the increase of the former with decreasing r is much faster than the latter [85,86]. The result leads to the stiffening of the soft mode, resulting in a lower ferroelectric transition temperature from a macroscopic point of view.

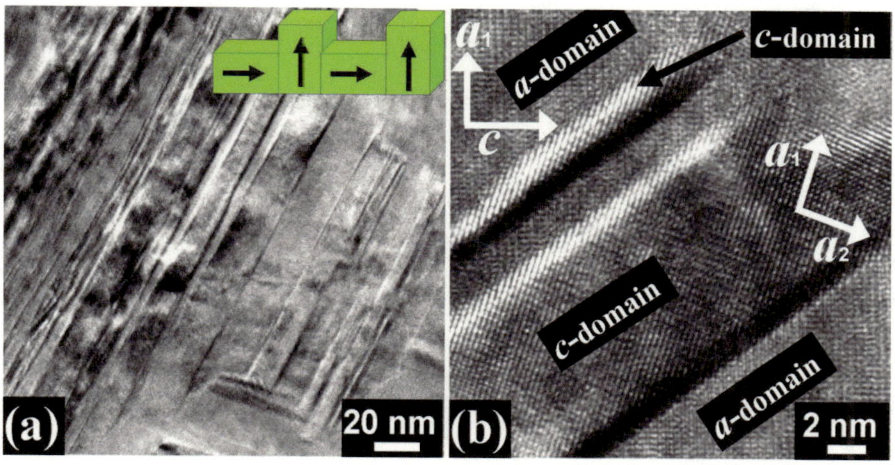

Fig. 16. (a) Plan-view TEM image of domain configurations and (b) HRTEM image of elastic domains for the compressive BTO film.

The compressive BTO exhibits very different domain configurations as compared with a tensile BTO, in which twining $a_1/a_2/a_1/a_2$ domain structure was observed. Fig. 16(a) shows plan-view TEM image of domains for the compressive BTO film, in which lamellar domain patterns are clearly observed. Further HRTEM observation, as shown in Fig. 16(b), reveals a $c/a/c/a$ domain pattern, in which c-domains have equal in-plane lattice parameters of a_1=a_2 with polarization vectors parallel to c-axis and a-domains have non-equal in-plane lattice parameters with polarization parallel to a-axis. These observations correspond well with the typical $c/a/c/a$ polydomain configurations in compressive ferroelectric films observed by Lee et al. [72] and Alpay et al. [87].

5.4 Phase transition

Fig. 17(a) shows the normal XRD pattern for a 300 nm BTO thin film grown on the 600nm LNO-buffered Si substrate. The lattice constants for BTO film are a = 3.982 and c = 4.053 Å, thus it can be inferred that the sputtered BTO film is under an in-plane compressive strain state. Fig. 17(b) and (c) demonstrate the HRTEM images of typical ferroelectric domains for the BTO film. It is seen that a BTO grain is distinctively split by the appearance of laminar domain configurations in order to minimize the in-plane elastic strain energy [88]. Similarly, for this compressive strained BTO, the observed domain wall between adjacent domains exhibits a blunt fringe contrast, indicating a *c*/*a*/*c*/*a* domain configuration.

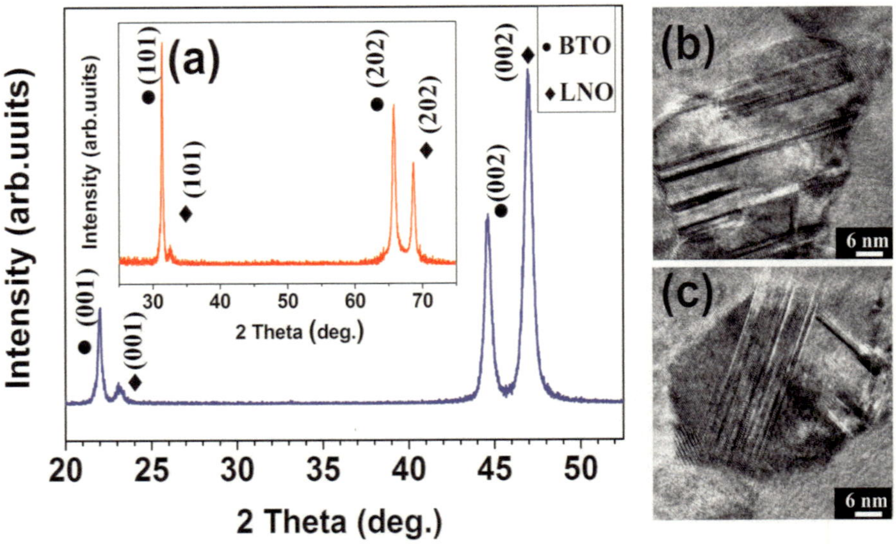

Fig. 17. (a) XRD θ - 2θ scan for 300 nm BTO on LNO(600nm)/Si. Inset is the 45° tilted XRD θ - 2θ scan for the same film. (b) and (c) HRTEM lattice image of typical ferroelectric domains inside a single BTO grain.

Fig. 18(a) and (b) show the temperature dependent dielectric constant (ε') and dielectric loss (tanδ) at different frequency of 1 - 500 kHz for the BTO film. It is observed that the Curie temperature (T_c), characterizing the ferroelectric to parelectric phase transition, is around 108 °C, which is lower than the value of typical T_c for BTO bulk or single crystal. On the other hand, in addition to the reduction of T_c, several other feathers are also evidenced in Fig. 18(a) and (b): (1) A broadened maximum in the dielectric constant appears at a wide temperature ranging from 80 °C to 120 °C, (b) the magnitude of the dielectric constant decreases, while T_c increases with increasing frequency, (c) the peak in dielectric loss is also frequency dependent and it shifts to higher temperatures with increasing frequency. The above observed strongly frequency dependent dielectric properties resemble the typical diffusive ferroelectric phase transition in ferroelectric relaxors rather than a normal ferroelectric phase transition, which shows a sharp anomaly at the T_c [89].
According to Smolensky and Uchino *et al.* [90,91], the diffuseness of the phase transition can be investigated by a modified Curie-Weiss (CW) law,

$$1/\varepsilon' - 1/\varepsilon'_m = (T - T_m)^\gamma / C \qquad (10)$$

where ε' is the dielectric constant at temperature T, ε'_m is the dielectric constant at T_m, γ is the critical exponent, and C is the Curie constant. A value of $\gamma = 1$ indicates a normal transition with ideal CW behavior, whereas $\gamma = 2$ indicates a diffusive transition behavior. The plot of $\log(1/\varepsilon' - 1/\varepsilon'_m)$ as a function of $\log(T - T_m)$ at 1 kHz is shown in the Fig. 19(a). By fitting the modified CW law, the exponent γ, determining the degree of the diffuseness of the phase transition, can be extracted from the slope of $\log(1/\varepsilon'-1/\varepsilon'_m)$ - $\log(T-T_m)$ plot. The relatively high γ value of 1.624 also indicates a relaxor behavior, which seems to be inconsistent with the predominant concept that BTO is a typical displacive ferroelectric material and should exhibit sharp dielectric transition [92].

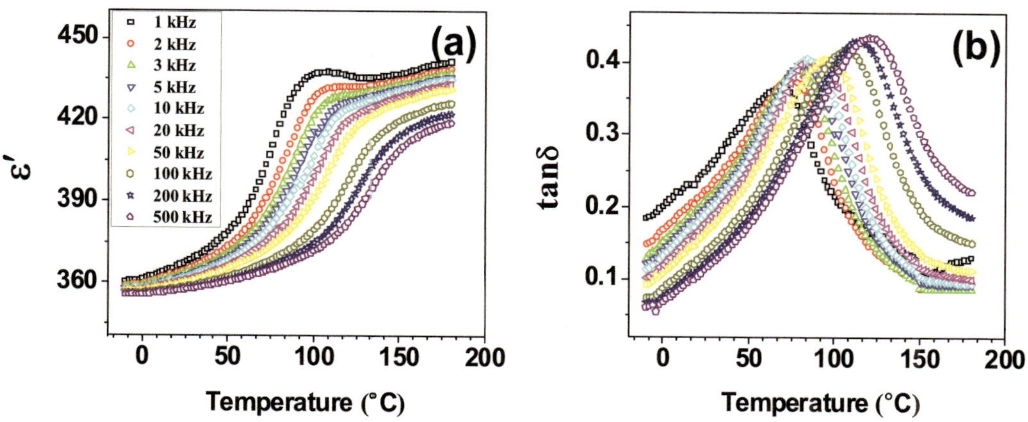

Fig. 18. Temperature dependent (a) dielectric constant and (b) loss tangent for the BTO film at frequency range of 1 kHz ~ 500 kHz.

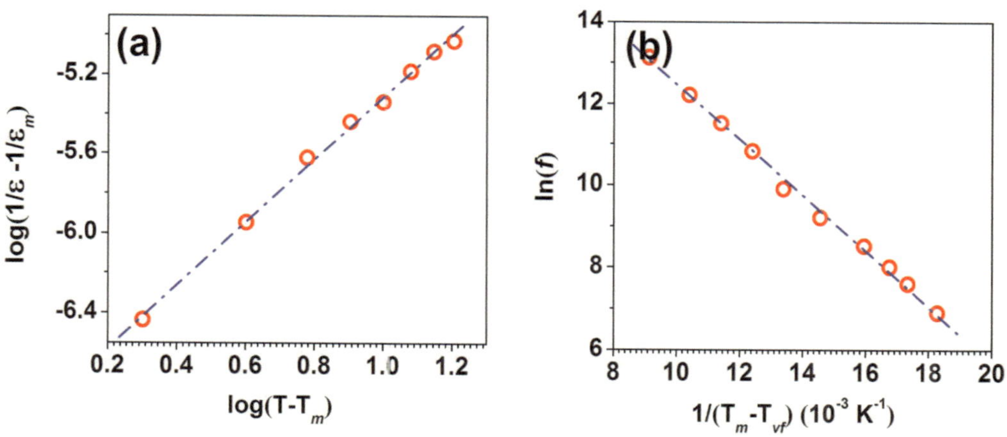

Fig. 19. (a) $\log(1/\varepsilon'-1/\varepsilon'_m)$ - $\log(T-T_m)$ plot for the BTO film at 1 kHz. (b) $\ln(f)$ – $1/(T_m-T_{vf})$ plot for the BTO film at 1 kHz. Symbol represents experimental data and solid dot line shows the fitting result.

However, recent nuclear magnetic resonance and Raman scattering studies had both evidenced the coexistence of the displacive character of transverse optical soft mode with the order-disorder character of Ti ions [93], especially in the BTO thin films. As the sputtering is proceed in an oxygen deficient atmosphere, thus the oxygen vacancies induced structural disorders and compositional fluctuations in the film may be responsible for the observed relaxor behavior. Similar diffusive transition had also been observed in BTO films on MgO and Pt-coated Si substrates [94,95].

The relaxor nature of the frequency dependent dielectric response of BTO film can also be examined by the Vogel-Fulcher (VF) relation [96],

$$f = f_0 \exp[-E_a/k(T_m - T_{vf})] \quad (11)$$

where f is the measuring frequency, f_0 is the characteristic relaxation frequency, E_a is the activation energy, T_m is the phase transition temperature at f, and T_{vf} is the freezing temperature of polarization-fluctuation. The $\ln(f) - 1/(T_m - T_{vf})$ plot with best fittings for the film is displayed in Fig. 19(b). The validity of VF relationship further demonstrates the relaxor behavior. From the slop of the fittings, the corresponding parameters can be obtained, $f_0 \sim 3.12108$ Hz, $T_{vf} \sim 327.3$ K and $E_a \sim 0.097$ eV.

6. Conclusions

High quality ferroelectric BTO thin films with (100)-preferred orientation have been grown on LNO buffered Si substrate by rf sputtering and the corresponding structure-property correlations have been discussed. Using combination of XRD and HRTEM, it is revealed that highly-oriented BTO film could be achieved on the lattice-mismatched Si in a "cube-on-cube" fashion with LNO as both buffer layer and conductive electrode layer. Polarization-switching measurement points out that while obvious ferroelectricity is obtained for BTO films with grain size larger than 22 nm, a weak ferroelectricity is still observed in BTO film of 14 nm grains, indicating that if a critical grain size exists for ferroelectricity it is less than 14 nm for BTO/LNO/Si system. We also demonstrate that due to their unique feature of gradient lattice constant and thermal expansion coefficient values for ferroelectric BTO, conductive LNO, and substrate Si, the BTO/LNO/Si system exhibits very interesting strain states. By choosing appropriate thicknesses for BTO and LNO, strain in ferroelectric BTO layer could be evolved from tensile strain to compressive strain state. The internal strain has a significant influence on the polarization, dielectric phase transition, and domain configuration for BTO film on Si and this can be used as a tool to engineer the properties of BTO films. The present work may have important implications on the future ferroelectric semiconductor devices.

7. Acknowledgements

This work is supported by the innovation Foundation of BUAA for PhD Graduates and program for New Century Excellent Talents in university (NCET-04-0160) and Innovative Research Team in University (IRT0512).

8. References

[1] Y. Yano, K. Iijima, Y. Daitoh, a T. Terashim, Y. Bando, Y. Watanabe, H. Kasatani and H. Terauchi, *J. Appl. Phys.* 76, 7833 (1994).
[2] S. Kim and S. Hishita, *Thin Solid Films* 281-282, 449 (1996).
[3] L. Qiao and X. F. Bi, *Thin Solid Films* 517, 3784 (2009).
[4] R. E. Avila, J. V. Caballero, V. M. Fuenzalida and I. Eisele, *Thin Solid Films* 348 44 (1999).
[5] T. Pencheva and M. Nenkov, *Vacuum* 48, 43 (1997).
[6] D. Y. Kim, S. G. Lee, Y. K. Park and S. J. Park, *Mater. Lett.* 40, 146 (1999).

[7] X. H. Wei, Y. R. Li, J. Zhu, Z. Liang, Y. Zhang, W. Huang and S. W. Jiang, *Appl. Surf. Sci.* 252, 1442 (2005).
[8] T. W. Kim, M. Jung, Y. S. Yoon, W. N. Kang, H. S. Shin, S. S. Yom and J. Y. Lee, 1993 *Solid State Commun.* 86, 565 (1993).
[9] K. Yao and W. G. Zhu, *Thin Solid Films* 408, 11 (2002).
[10] W. Xu, L. Zheng, H. Xin, C. Lin and O. Masanori, *J. Electrochem. Soc.* 143, 1133 (1996).
[11] S. A. Chambers, *Adv. Mater.* 22, 219 (2010).
[12] J. W. Reiner, A. M. Kolpak, Y. Segal, K. F. Garrity, S. I. Beigi, C. A. Ahn, and F. J. Walker, *Adv. Mater.* 22, 2929 (2010).
[13] M. P. Warusawithana, C. Cen, C. R. Sleasman, J. C. Woicik, Y. L. Li, L. F. Kourkoutis, J. A. Klug, H. Li, P. Ryan, L. P. Wang, M. Bedzyk, D. A. Muller, L. Q. Chen, J. Levy, and D. G. Schlom, *Science* 324, 367 (2009).
[14] J. Schwarzkopf and R. Fornari, *Prog. Crystal Growth Character. Mater.* 52, 159 (2006).
[15] A. K. Tagantsev, N. A. Pertsev, P. Muralt, and N. Setter, *Phys. Rev. B* 65, 012104 (2001).
[16] W. Y. Park, K. H. Ahn, and C. S. Hwanga. *Appl. Phys. Lett.* 83, 4387 (2003).
[17] S. B. Mi, C. L. Jia, T. Heeg, O. Trithaveesak, J. Schubert, and K. Urban, J. Cryst. Growth 283, 425 (2005).
[18] O. Auciello, J. F. Scott, and R. Ramesh, Phys. Today 51(7), 22 1998).
[19] J. Levy, Phys. Rev. A 64, 052306 (2001).
[20] V. Vaithyanathan, J. Lettieri, W. Tian, A. Sharan, A. Vasudevarao, Y. L. Li, A. Kochhar, H. Ma, J. Levy, P. Zschack, J. C. Woicik, L. Q. Chen, V. Gopalan, and D. G. Schlom, J. Appl. Phys. 100, 024108 (2006).
[21] Y. S. Touloukian, R. K. Kirby, R. E. Taylor, and T. Y. R. Lee, Thermal Expansion, Nonmetallic Solids, Thermophysical Properties of Matter (Plenum, New York, 1977), Vol. 13.
[22] L. Qiao and X. F. Bi, J. Cryst. Growth 310, 5327 (2008).
[23] L. W. Martin, Y. H. Chu, R. Ramesh, *Mater. Sci. Eng. Rep.* 68, 111 (2010).
[24] A. B. Posadas, M. Lippmaa, F. J. Walker, M. Dawber, C. H. Ahn, and J. M. Triscone, *Topics. Appl. Phys.* 105, 219 (2007).
[25] E. Kawamura, V. Vahedi, M. A. Lieberman, and C. K. Birdsall, *Plasma Sources Sci Technol* R45, 240 (1999).
[26] B. G. Chae, Y. S. Yang, S. H. Lee, M. S. Jang, S. J. Lee, S. H. Kim, W. S. Baek, S. C. Kwon, *Thin Solid Films* 410, 107 (2002).
[27] N. Wakiya, T. Azuma, K. Shinozaki, N. Mizutani, *Thin Solid Films* 410, 114 (2002).
[28] D. H. Bao, N. Mizutani, X. Yao and L. Y. Zhang, *Appl. Phys. Lett.* 77, 1041 (2000).
[29] Q. Zou, H. E. Ruda and B. G. Yacobi, *Appl. Phys. Lett.* 78, 1282 (2001).
[30] D. H. Bao, N. Wakiya, K. Shinozaki, N. Mizutani and X. Yao, *Appl. Phys. Lett.* 78, 3286 (2001).
[31] J. R. Cheng, L. He, S. W. Yu and Z. Y. Meng, *Appl. Phys. Lett.* 88, 152906 (2006).
[32] S. Schlag and H. F. Eicke, *Solid State Commun.* 91, 883 (1994).
[33] W. Zhong, B. Jiang, P. Zhang, J. Ma, H. Chen, Z. Yang and L. Li, *J. Phys.: Condens. Matter* 5, 2619 (1993).
[34] S. Chattopanhuay, P. Ayyub, V. R. Palkar and M. Multani, *Phys. Rev. B* 52, 13177 (1995).

[35] S. Li, J. A. Eastman, J. M. Vetrone, C. M. Foster, R. E. Newnham and L. E. Cross, *Jpn. J. Appl. Phys., Part I* 36, 5169 (1997).
[36] T. Maruyama, M. Saitoh, I. Sakay and T. Hidaka, *Appl. Phys. Lett.* 73, 3524 (1998).
[37] Y. S. Kim, D. H. Kim, J. D. Kim, Y. J. Chang, T. W. Noh, J. H. Kong, K. Char, Y. D. Park, S. D. Bu, J.-G. Yoon and J.-S. Chung, *Appl. Phys. Lett.* 86, 102907 (2005).
[38] J. Junquera and P. Ghosez, *Nature* 422, 506 (2003).
[39] D. D. Fong, G. B. Stephenson, S. K. Streiffer, J. A. Eastman, O. Auciello, P. H. Fuoss and C. Thompson, *Science* 304, 1650 (2004).
[40] A. R. Meier, F. Niu and B. W. Wessels, *J. Crystal Growth*, 294, 401 (2006).
[41] B. Dkhil, E. Defay and J. Guillan, *Appl. Phys. Lett.* 90, 022908 (2007).
[42] H. Huang, X. Yao, M. Q. Wang and X. Q. Wu, *J. Crystal Growth* 263, 406 (2004).
[43] Y. P. Guo, K. Suzuki, K. Nishizawa, T. Miki and K. Kato, *J. Crystal Growth* 284, 190 (2005).
[44] R. Thomas, V. K. Varadan, S. Komarneni and D. C. Dube, *J. Appl. Phys.* 90, 1480 (2001).
[45] L. M. Huang, Z. Y. Chen, J. D. Wilson, S. Banerjee, R. D. Robinson, I. P. Herman, R. Laibowitz and S. O'Brien, *J. Appl. Phys.* 100, 034316 (2006).
[46] Y. S. Kim, J. Y. Jo, D. J. Kim, Y. J. Chang, J. H. Lee, T. W. Noh, T. K. Song, J.-G. Yoon, J.-S. Chung, S. I. Baik, Y.-W. Kim and C. U. Jung, *Appl. Phys. Lett.* 88, 072909 (2006).
[47] K. J. Choi, M. Biegalski, Y. L. Li, A. Sharan, J. Schubert, R. Uecker, P. Reiche, Y. B. Chen, X. Q. Pan, V. Gopalan, L. Q. Chen, D. G. Schlom and C. B. Eom, *Science* 306, 1005 (2004).
[48] M. T. Buscaglia, M. Viviani, V. Buscaglia, L. Mitoseriu, A. Testino, P. Nanni, Z. Zhao, M. Nygren, C. Harnagea, D. Piazza andi C. Galass, *Phys. Rev. B.* 73, 064114 (2006).
[49] X. Y. Deng, X. H. Wang, H. Wen, L. L. Chen, L. Chen and L. T. Li, *Appl. Phys. Lett.* 88, 252905 (2006).
[50] X. H. Wang, X. Y. Deng, H. Wen and L. T. Li, *Appl. Phys. Lett.* 89, 162902 (2006).
[51] G. Liu, X. H. Wang, Y. Lin, L. T. Li and C. W. Nan, *J. Appl. Phys.* 98, 044105 (2005).
[52] Y. Park and H-G. Kim, *J. Am. Ceram. Soc.* 80(1), 106 (1997).
[53] T. Takeuchi, M. Tabuchi, H. Kageyama and Y. Suyama, *J. Am. Ceram. Soc.* 82(4), 939 (1999).
[54] G. Arlt, D. Hennings and G. de With, *J. Appl. Phys.* 58, 1619 (1985).
[55] J. H. Haenl, P. Irvin, W. Chang, R. Uecker, P. Reiche, Y. L. Li, S. Choudhury, W. Tian, M. E. Hawley, B. craigo, A. K. Tagantsev, X. Q. Pan, S. K. Streiffer, L. Q. Chen, S. W. Kirchoefer, J. Levy, and D. G. Schlom, *Nature* 430, 758 (2004).
[56] J. Q. He, E. Vasco, R. Dittmann, and R. H. Wang, *Phys. Rev. B* 73, 125413 (2006).
[57] H. D. Kang, W. H. Song, S. H. Sohn, H. J. Jin, S. E. Lee, and Y. K. Chung, *Appl. Phys. Lett.* 88, 172905 (2006).
[58] M. Jimi, T. Ohnishi, K. Terai, M. Kawasaki, M. Lippmaa, *Thin Solid Films* 486, 158 (2005).
[59] N. A. Pertsev, A. G. Zembilgotov, S. Hoffmann, R. Waser, and A. K. Tagantsev, J. Appl. Phys. 85, 1698 (1999).
[60] K. S. Lee and S. Baik, J. Appl. Phys. 87, 8035 (2000).
[61] R. Dittmann, R. Plonka, E. Vasco, N. A. Pertsev, J. Q. He, C. L. Jia, S. Hoffmann, and R. Waser, Appl. Phys. Lett. 83, 5011 (2003).
[62] R. People and J. C. Bean, Appl. Phys. Lett. 47, 322 (1985).

[63] J. M. Gere and S. P. Timoshenko, *Mechanics of Materials*, 4th ed. (PWS, Boston, 1997), p. 889.
[64] M. S. Rafique and N. Tahir, *Vacuum* 81, 1062 (2007).
[65] D. Y. Wang, Y. Wang, X. Y. Zhou, H. L. W. Chan and C. L. Choy, *Appl. Phys. Lett.* 86, 212904 (2005).
[66] N. A. Pertsev, V. G. Koukhar, R. Waser, and S. Hoffmann, *Integrated Ferroelectrics* 32, 235 (2001)
[67] N. A. Pertsev, A. G. Zembilgotov and A. K. Tagantsev, *Phys. Rev. Lett.* 80, 1988 (1998).
[68] N. A. Pertsev, A. G. Zembilgotov and A. K. Tagantsev, *Ferroelectrics* 223, 79 (1999).
[69] N. A. Pertsev and V. G. Koukhar, *Phys. Rev. Lett.* 84, 3722 (2000).
[70] V. G. Koukhar, N. A. Pertsev, and R. Waser, *Phys. Rev. B* 64, 214103 (2001).
[71] Y. L. Li and L. Q. Chen, *Appl. Phys. Lett.* 88, 072905 (2006).
[72] K. S. Lee, J. H. Choi, J. Y. Lee, and S. Baik, *J. Appl. Phys.* 90, 4095 (2001).
[73] A. Petraru, N. A. Pertsev, H. Kohlstedt, U. Poppe, R. Waser, A. Solbach, and U. Klemradt, J. Appl. Phys. 101, 114106 (2007).
[74] Z. H. Dai, Z. Xu, and X. Yao, Appl. Phys. Lett. 92, 072904 (2008).
[75] D. A. Tenne, X. X. Xi, Y. L. Li, L. Q. Chen, A. Soukiassian, M. H. Zhu, A. R. James, J. Lettieri, D. G. Schlom, W. Tian and X. Q. Pan, *Phys. Rev. B* 69, 174101 (2004)
[76] G. F. Huang and S. Berger, *J. Appl. Phys.* 93, 2855 (2003).
[77] L. Qiao and X. F. Bi, *J. Phys. D: Appl. Phys.* 41, 195407 (2008).
[78] K. M. Ring and K. L. Kavanagh, *J. Appl. Phys.* 94, 5982 (2003).
[79] M. E. Marssi, F. L. Marrec, I. A. Lukyanchuk and M. G. Karkut, *J. Appl. Phys.* 94, 3307 (2003).
[80] K. Kato, K. Tanaka, K. Suzuki and S. Kayukawa, *Appl. Phys. Lett.* 91, 172907 (2007).
[81] Y. C. Chen, T. H. Hong, Z. X. Jiang and Q. R. Lin, *J. Appl. Phys.* 103, 07E305 (2008).
[82] G. Catalan, B. Noheda, J. McAneney, L. J. Sinnamon, and J. M. Gregg, Phys. Rev. B 72, 020102R (2005).
[83] G. Catalan, L. J. Sinnamon, and J. M. Gregg, J. Phys.: Condens. Matter 16, 2253 (2004).
[84] W. Cochran, Phys. Rev. Lett. 3, 412 (1959).
[85] G. A. Samara, T. Sakudo, and K. Yoshimitsu, Phys. Rev. Lett. 35, 1767 (1975)
[86] R. E. Cohen, Nature 358, 136 (1992).
[87] S. P. Alpay, V. Nagarajan, L. A. Bendersky, M. D. Vaudin, S. Aggarwal, R. Ramesh, and A. L. Roytburd, *J. Appl. Phys.* 85, 3271 (1999).
[88] I. T. Kim, J. W. Jang, H. J. Youn, C. H. Kim and K. S. Hong, *Appl. Phys. Lett.* 72, 308 (1998).
[89] B. D. Qu, M. Evstigneev, D. J. Johnson and R. H. Prince, *Appl. Phys. Lett.* 72, 1394 (1998).
[90] G. A. Smolensky, *J. Phys. Soc. Jpn.* 28, 26 (1970).
[91] K. Uchino and S. Nomura, *Ferroelectr. Lett. Sect.* 44, 55 (1982).
[92] M. M. Kumar, K. Srinivas and S. V. Suryanarayana, *Appl. Phys. Lett.* 76, 1330 (2000).
[93] M. Tyunina and J. Levoska, *Phys. Rev. B* 70, 132105 (2004).

[94] S. Chattopadhyay, A. R. Teren, J. H. Hwang, T. O. Mason and B. W. Wessels, J. Mater. Res. 17, 669 (2002).

[95] R. Thimas, V. K. Varadan, S. Komarneni and D. C. Dube, *J. Appl. Phys.* 90, 1480 (2001).

[96] J. Xu and Y. Akishige, *Appl. Phys. Lett.* 92, 052902 (2008).

19

Nanostructured $LiTaO_3$ and $KNbO_3$ Ferroelectric Transparent Glass-Ceramics for Applications in Optoelectronics

Anal Tarafder and Basudeb Karmakar
Glass Science and Technology Section, Glass Division,
Central Glass and Ceramic Research Institute,
Council of Scientific and Industrial Research (CSIR, India),
India

1. Introduction

Ferroelectric bulk crystals are widely used in optoelectronic devices because of their well combination of extraordinary optical and electronic properties. Their crystal structure is non-centrosymmetric and due to this structural anisotropy they exhibit many nonlinear optical properties, for example, the electro-optic effect (change in optical index with electric field), harmonic generation (changing frequency of light), and photorefraction (index change in response to light), to name a few. However, preparation of their defect-free optical quality transparent single crystal is very difficult, lengthy process, and requires sophisticated costly equipment. In recent past, to triumph over these difficulties, much attention has been paid for development of transparent ferroelectric glass-ceramics by the high speed glass technology route because of its low cost of fabrication, tailoring of properties and flexibility to give desired shapes. Lithium tantalate ($LiTaO_3$, LT) and potassium niobate ($KNbO_3$, KN) single crystals are the most important lead-free ferroelectric materials with the $A^{1+}B^{5+}O_3$ type perovskite structure concerning the environmental friendliness. LT has the rhombohedral crystal structure with crystal symmetry class 3m (unit cell dimensions: a = 5.1530 Å and c = 13.755 Å), large nonlinear constant (d_{33} = 13.6 pm/V at 1064 nm), second harmonic generation (SHG) coefficient (d_{33}^{2w} = 40.0 with respect to KDP at 1060 nm) (Risk et al., 2003, JCPDS No. 29-0836, Moses, 1978) and Curie temperature (660°C). In contrast, KN has the orthorhombic crystal structure with crystal symmetry class mm2 (unit cell dimensions: a = 5.6896 Å, b = 3.9693 Å and c = 5.7256 Å), large nonlinear coefficient (d_{33} = 27.4 pm/V at 1064 nm) [Moses, 1978] and Curie temperature (435°C). Thus, they exhibit unique electro-optic, piezoelectric, acousto-optic, and nonlinear optical (NLO) properties when doped with rare-earth (RE) [$4f^{1-13}$] elements combined with good mechanical and chemical stability (Abedin et al., 1997, Zhu et al., 1995, Mizuuchi et al., 1995, Zgonik et al., 1993, Xue et al., 1998). Very recently, potassium niobate ceramics were investigated with an aim to develop environmental friendly lead-free piezoelectric and nonlinear materials (Ringgaard & Wurlitzer, 2005).

The electronic structure of each trivalent RE element consists of partially filled 4f subshell, and outer $5s^2$ and $5p^6$ subshell. With increasing nuclear charge electrons enter into the underlying 4f subshell rather than the external 5d subshell. Since the filled $5s^2$ and $5p^6$

subshells screen the 4f electrons, the RE elements have very similar chemical properties. The screening of the partially filled 4f subshells, by the outer closed $5s^2$ and $5p^6$ subshell, also gives rise to sharp emission spectra independent of the host materials. The intra-subshell transitions of 4f electrons lead to narrow absorption peaks in the ultra-violet, visible, and near-infrared regions.

In this chapter, we report synthesis, structure, properties and application of transparent ferroelectric $LiTaO_3$ (LT) and $KNbO_3$ (KN) nanostructured glass-ceramics. They were prepared by controlled volume (bulk) crystallization of their precursor glasses with and without RE dopant. The crystallization processes were studied by differential thermal analysis (DTA), X-ray diffraction (XRD), field emission scanning electron microscopy (FESEM), transmission electron microscopy (TEM), Fourier transform infrared reflection spectra (FT-IRRS), fluorescence and excited state lifetime analyses and dielectric constant measurement. The X-ray diffraction (XRD) patterns, selected area electron diffraction (SAED) and transmission electron microscopic (TEM) images confirm crystallization of $LiTaO_3$ and $KNbO_3$ nanocrystals in the transparent glass-ceramics.

2. Experimental procedure

2.1 Precursor glass and glass-ceramics preparation

The LT precursor glasses having molar composition $25.53Li_2O$-$21.53Ta_2O_5$-$35.29SiO_2$-$17.65Al_2O_3$ (LTSA) doped with RE ions (0.5 wt% oxides of Eu^{3+} and Nd^{3+} in excess) or undoped were prepared by the melt-quench technique. The melting of thoroughly-mixed batches was done at 1600°C. The quenched glass blocks were annealed at 600°C for 4 h to remove the internal stresses of the glass and then slowly cooled down (@ 1°C/min) to room temperature. The annealed glass blocks were cut into desired dimensions and optically polished for ceramization and to perform different measurements. The crystallization was carried out at 680°C in between 0-100 h duration.

The KN precursor glasses having composition (mol%) $25K_2O$-$25Nb_2O_5$-$50SiO_2$ (KNS) doped with Er_2O_3 (0.5 wt% in excess) or undoped were prepared similarly as mentioned above by the melt-quench technique. The well-mixed raw materials were melted in a platinum crucible in an electric furnace at 1550°C and the quenched glasses were annealed at 600°C to remove the internal stresses of these precursor glasses. They were transformed into nanostructured transparent glass-ceramics by heat-treatment at 800°C in between 0-200 h duration.

2.2 Characterization

The density of precursor glasses was measured using Archimedes principle using water as buoyancy liquid. The refractive indices of precursor glass and representative glass-ceramics (d) were measured either on a Pulfrich refractometer (Model PR2, CARL ZEISS, Jena, Germany) at wavelength (λ_e = 546.1 nm) or on a Metricon 2010/M Prism Coupler at different wavelength (λ = 473, 532, 633, 1064 and 1552 nm). Differential thermal analysis (DTA) of precursor glass powder was carried out up to 1000°C at the rate of 10°C/min with a SETARAM TG/DTA 92 or with a Netzsch STA 409 C/CD instrument from room temperature to 900°C at a heating rate of 10°C/min. to ascertain the glass transition temperature (T_g) and the crystallization peak temperature (T_P). XRD data were recorded

using a PANalytical X'Pert-PRO MPD diffractometer operating with CuK_α = 1.5406 Å radiation to identify the developed crystalline phases. The data were collected in the 2θ range from 10° to 80° with a step size of 0.05°.
A high resolution FE-SEM (Model: Gemini Zeiss Supra™ 35 VP, Carl Zeiss) was used to observe the microstructure of freshly fractured surfaces of the heat-treated nano glass-ceramics after etching in 1% HF solution for 2 minutes and coated with a thin carbon film. The TEM images and selected area electron diffraction (SAED) of powdered glass-ceramic sample were obtained from FEI (Model: Tecnai G^2 30ST, FEI Company) instrument. The FTIR reflectance spectra of all the glasses and glass-ceramics were recorded using a FTIR spectrometer (Model: 1615, Perkin-Elmer) in the wavenumber range 400-2000 cm^{-1} with a spectral resolution of ±2 cm^{-1} and at 15° angle of incidence. Optical absorption spectra were recorded on UV-Vis-NIR spectrophotometer (Model: Lambda 20, Perkin-Elmer) at room temperature. The UV-Vis fluorescence emission and excitation spectra of Eu^{3+} doped precursor glass and nano glass-ceramics were measured on a fluorimeter (Model:Fluorolog-II, SPEX) with 150 W Xe lamp as a source of excitation. The fluorescence decay curves were recorded on the same instrument attached with SPEX 1934D phosphorimeter using pulsed Xe lamp. On the other hand, the fluorescence emission and excitation spectra of rest of samples were measured on continuous bench top modular spectrofluorimeter (QuantaMaster, Photon Technology International) attached with gated Hamamatsu NIR PMT (P1.7R) as detector and Xe arc lamp as excitation source. The excited state lifetime was measured with the same instrument using a Xe flash lamp of 75 W. The dielectric constants of precursor glass and nano glass-ceramics were measured at room temperature using a Hioki LCR meter (Model: 3532-50 Hitester, Hioki) at 1 MHz frequency after coating the surfaces with a conductive silver paint followed by drying at 140°C for 1h. Second harmonic generations (SHG) at 532 nm in the undoped glass-ceramics have been realized under fundamental beam of Nd^{3+}:YAG laser source (1064 nm). The input energy of Nd^{3+}: YAG laser was fixed at 17 mJ. The input energy of laser was divided in two directions (50% energy in each direction) using reflecting neutral density filter. In one direction KDP was put for reference. The reference SHG signal was measured using photodiode. Second beam was passed through visible filter (which blocks all visible wavelengths but pass 1064 nm) and focused onto the test samples. The SHG generated from the sample was focused onto a second harmonic separator, which reflects 532 nm at 45° and transmit 1064 nm. The SHG signal reflected from SHG separator passed through IR filter was finally measured using PMT. The reference signals from photodiode and from PMT were measured simultaneously using Lecroy oscilloscope (bandwidth 1GHz).

3. Nanostructured $LiTaO_3$ ferroelectric glass-ceramics

3.1 Background

Lithium tantalate ($LiTaO_3$, LT) single crystal is one of the most important lead-free ferroelectric materials in the $A^{1+}B^{5+}O_3$ type perovskite family. The correlation of property alteration of LT single crystals, powders, thin films, glass-ceramics, etc. with processing parameters is an important area of exploration. In recent times researchers have demonstrated the property monitoring based on preparation of $LiTaO_3$ powders (Zheng et al., 2009) and thin films (Cheng et al., 2005, Youssef et al., 2008) by different methods. Luminescence properties of Ho^{3+}, Eu^{3+}, Tb^{3+} etc. doped $LiTaO_3$ crystals, an another important area of exploration, which have also been investigated by various researchers (Sokólska, 2002, Sokólska et al., 2001, Gasparotto et al., 2008, Gruber et. al., 2006). Rare-earth (RE) doped transparent $LiTaO_3$ nanocrystallite containing glass ceramics, in which RE ions selectively incorporated into the $LiTaO_3$ nanocrystals embedded in an oxide glassy matrix, can offer excellent luminescent properties due to the low phonon energy environment of $LiTaO_3$ nanocrystallites for luminescent ions, and good mechanical and chemical properties of oxide glassy matrix. This ability, combined with inherent nonlinear optical (NLO) properties of ferroelectric crystals, could offer a possibility to design self frequency doubling laser sources. Hence, this new material has attracted great attention in the continuous research for the development of novel optoelectronic devices (Jain, 2004, Romanowski et al., 2000, Hase et al., 1996). Mukherjee and Varma have reported the crystallization and physical properties of $LiTaO_3$ in a $LiBO_2$-Ta_2O_5 reactive glass matrix, however, they have not explored RE doped $LiTaO_3$ containing glass-ceramics (Mukherjee & Varma, 2004). As such, work performed on nanocrystalline $LiTaO_3$ containing aluminosilicate glass-matrix materials is very rare due to the difficulties in preparation of transparent precursor glass in general and glass–ceramics in particular which involves high temperature (about 1600°C) for its precursor glass melting (Ito et al., 1978). For this reason, the structure, dielectric and fluorescence properties of Eu^{3+}, Nd^{3+} and Er^{3+} ion doped transparent precursor glass and glass-ceramic composites of $LiTaO_3$ with heat-treatment time have been studied and reported elaborately by Tarafder et al., 2009 & 2010, Tarafder et al., DOI:10.1111/j.1744-7402.2010.02494.x. Second harmonic generation (SHG) from bulk $LiTaO_3$ glass-ceramics has also been studied (Tarafder et al., 2011). For better understanding, the structure, dielectric and fluorescence properties of Eu^{3+} and Nd^{3+} ion doped transparent precursor glass and glass-ceramic composites of $LiTaO_3$ with heat-treatment time have been reported elaborately along with the second harmonic generation (SHG) from bulk $LiTaO_3$ glass-ceramics.

4. Results and discussion

4.1 Differential thermal analysis (DTA)

The DTA curve of the Eu^{3+} doped precursor glass is shown in Fig. 1. This exhibits the inflection in the temperature range 680-715°C followed by the intense exothermic peak at 821°C (T_p) corresponding to the $LiTaO_3$ crystallization. The glass transition temperature (T_g) has been estimated to be 696°C from the point of intersection of the tangents drawn at the slope change as is marked in Fig. 1.

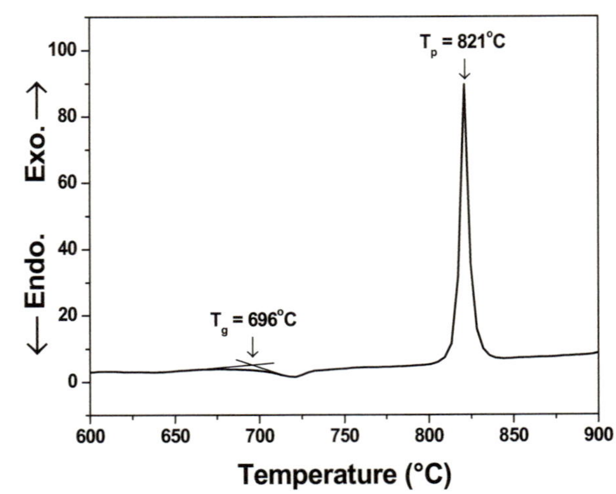

Fig. 1. DTA curve of Eu^{3+} doped precursor LTSA powdered glass.

4.2 Refractive index

The Eu^{3+} doped precursor LTSA glass samples were heat treated at 680°C near glass transition temperature for various heat-treatment durations (0, 1, 3, 5, 7, and 10 h) after nucleating at 650°C for 2 h. Similarly, the Nd^{3+} doped precursor LTSA glass samples were heat treated at 680°C for 0, 3, 5, 10, 20, 50 and 100 h and were labeled as a, b, c, d, e, f and g. The Nd^{3+} doped precursor glass and nano glass-ceramics are presented in Fig. 2. From the measured glass density (ρ) and refractive index (n_e) at wavelength λ_e = 546.1 nm, other related optical properties of Eu^{3+} doped precursor glass have been determined using relevant expressions and the same is presented in Table 1. Fig. 3 present Cauchy fitting based on measured refractive indices at five different wavelengths (see experimental procedure) and shows the dependences of the refractive index on the wavelength for Nd^{3+} doped precursor glass (a) and representative heat-treated glass-ceramics samples. In general, refractive index decreases with increasing wavelength due to dispersion. In addition to this, the refractive index of the glass-ceramics samples has increased in comparison with precursor glass (a) that can be seen in Fig. 3. The refractive indices n_F, n_D and n_C have been estimated at three standard wavelengths (λ_F = 486.1 nm, λ_D = 589.2 nm and λ_C = 656.3 nm respectively) from the dispersion curve (Figs. 3, curve a). Similarly, from the measured glass density (ρ) and refractive index (n_D) at wavelength λ_D = 589.2 nm, other related optical properties of Nd^{3+} doped precursor glass have also been determined and the results are presented in Table 1. From Table 1, it is clear that the LTSA glass under study has high values of refractive index and density. The large refractive indices of this glass are due to high ionic refraction (23.4) of Ta^{5+} ions (Volf, 1984) having an empty or unfilled d-orbital (outer electronic configuration: $5d^06s^0$) which contributes strongly to the linear and nonlinear polarizability (Yamane & Asahara, 2000). The high density of the glass has originated from the large packing effect of Ta in the glass matrix (Hirayama & Berg, 1963). For the same reason, this glass possesses a high value of molar refractivity and electronic polarizability. Due to formation of high refractive index $LiTaO_3$ (RI = 2.1834 at 600 nm (Lynch, 1975)), the heat-treated sample exhibit higher refractive indices as shown in Fig. 3, curve-d.

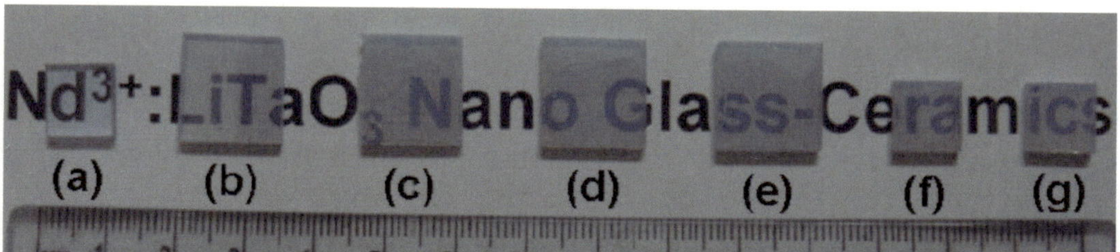

Fig. 2. (Color online) Photographs of Nd^{3+} doped precursor LTSA glass and LT nano glass-ceramics (thickness: 2 mm) laid over the writing to show their transparency respectively.

4.3 X-ray diffraction analysis

The X-ray diffractogram of Eu^{3+} doped precursor LTSA glass and cerammed glass-ceramics are shown in Fig. 4. The XRD pattern of the precursor glass exhibits broad humps characterizing its amorphous structure. With progression of heat-treatment, several diffraction peaks have been appeared in the glass-ceramics. From the analysis of these peaks it has been concluded that these peaks are attributed to rhombohedral $LiTaO_3$ (JCPDS Card

Properties	Corresponding value	
	Eu^{3+}:$LiTaO_3$ doped precursor glass	Nd^{3+}:$LiTaO_3$ doped precursor glass
Average molecular weight, M_{av}	142.47	142.37
Density, ρ (g.cm^{-3})	4.54	4.50
Refractive index	n_e (at 546.1 nm) → 1.7852	n_F (at 486.1 nm) → 1.8053 n_D (at 589.2 nm) → 1.7894 n_C (at 656.3 nm) → 1.7821
Molar refractivity, R_M (cm^3)	13.23	13.39
Electronic polarizability, α (cm^3)	1.84×10^{-21}	1.79×10^{-21}
RE^{3+} ion concentration, $N_{RE}3+$ (ions/cm^3)	5.71×10^{19}	5.66×10^{19}
Glass transition temperature, T_g (°C)	696	702
Crystallization peak, T_p (°C)	821	820

Table 1. Some measured and calculated properties of RE^{3+}:$LiTaO_3$ precursor glass

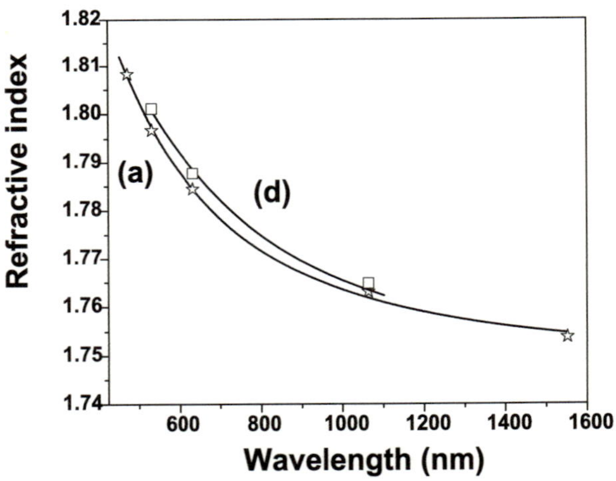

Fig. 3. Variation of refractive indices (Cauchy fitted) of Nd^{3+} doped (a) precursor LTSA glass and (d) 10 h heat-treated LT nano glass-ceramic as a function of wavelength.

File No. 29-0836) except a few diffraction peak around 2θ = 23.0°, 25.5°, 44.5° and 47.0° which are due to the formation of β-spodumene ($LiAlSi_2O_6$) crystal phase (JCPDS Card File No. 35-0797) in minor quantity. It is clearly evidenced from the XRD analysis that the peak of $LiAlSi_2O_6$ (2θ = 25.5°) is more prominent in sample of 5 h heat-treatment and it got diminished with respect to $LiTaO_3$ phase in longer heat-treated glass-ceramics, indicating the stabilization of $LiTaO_3$ nanocrystallites with increase in heat-treatment duration. From the full width at half maximum (FWHM) of the most intense diffraction peak (012) of

$LiTaO_3$, the average crystallite size (diameter, *d*) is calculated by using the Scherrer's formula (Cullity, 1978)

$$d = 0.9\lambda / \beta \cos\theta \tag{1}$$

where λ is the wavelength of X-ray radiation (CuK_α = 1.5406Å), β is the full width at half maximum (FWHM) of the peak at 2θ. The average crystallite size of each RE doped heat-treated glass-ceramics found to increase with heat-treatment duration.

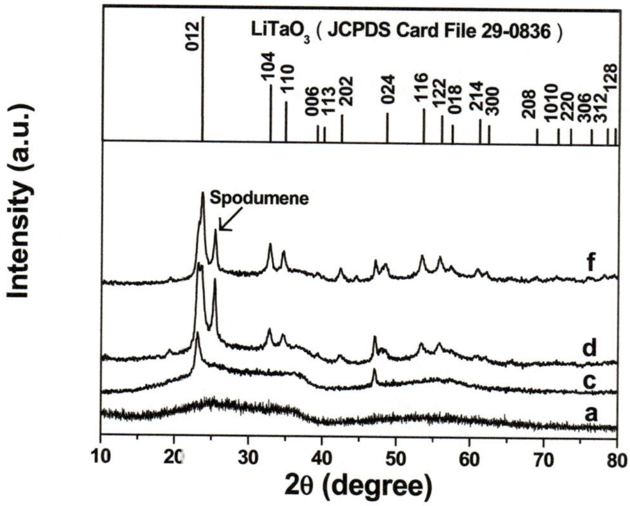

Fig. 4. XRD pattern of Eu^{3+} doped precursor powdered LTSA glass and LT nano glass-ceramics.

4.4 FESEM and TEM image analyses

The morphology and $LiTaO_3$ crystallite size of Eu^{3+} and Nd^{3+} doped nano glass-ceramics have been examined by FESEM and TEM image analyses. FESEM images of the fractured surface of Nd^{3+} doped nano glass-ceramics have been presented in Figs. 5(a)-(b). The Nd^{3+} doped glass-ceramics 5(a) is obtained by heat-treating the precursor glasses at 680°C for 5 h. Similarly, the glass-ceramics 5(b) is obtained by heat-treating the precursor glasses at 680°C for 20 h. From the FESEM micrographs, it is clearly observed that the glassy matrix of the heat-treated samples initially phase separated on nanometric scale followed by incipient precipitation of defined crystallites within the Li–Ta rich phase regions with increase in heat-treatment time. The droplets have irregular shapes and dispersed uniformly thought out the bulk glass matrix. The size of the droplets varies in the range 20-60 nm. TEM image of the Eu^{3+} doped 10 h heat-treated glass-ceramics (f) has been presented in Fig. 6(a). The SAED pattern of the observed crystalline phase is presented in Fig. 6(b). From this image, it is observed that many spheroidal $LiTaO_3$ crystallites precipitated homogeneously from the glass matrix and remained homogeneously dispersed in the residual glass matrix. The crystallite size from this TEM image of sample f found to be around 18 nm. The presence of fine spherical rings around the central bright region in SAED pattern discloses the existence of $LiTaO_3$ nanocrystallites in the glassy matrix.

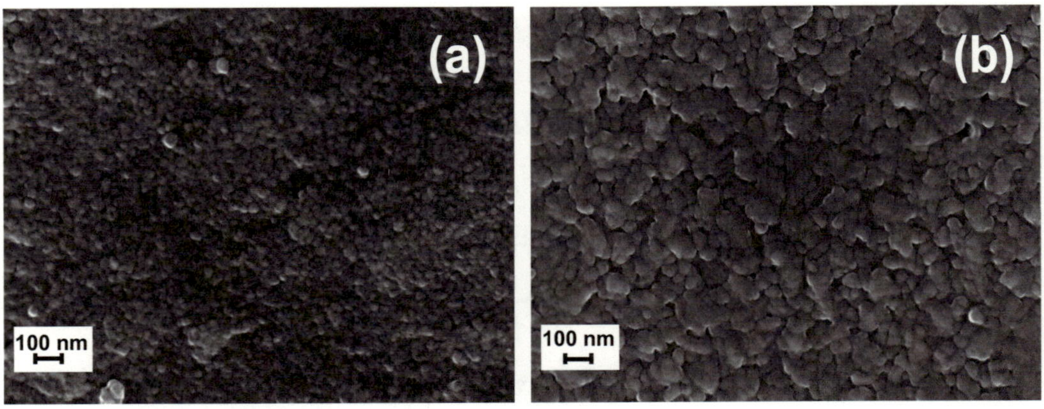

Fig. 5. FESEM image of Nd^{3+} doped samples (a) c and (b) e.

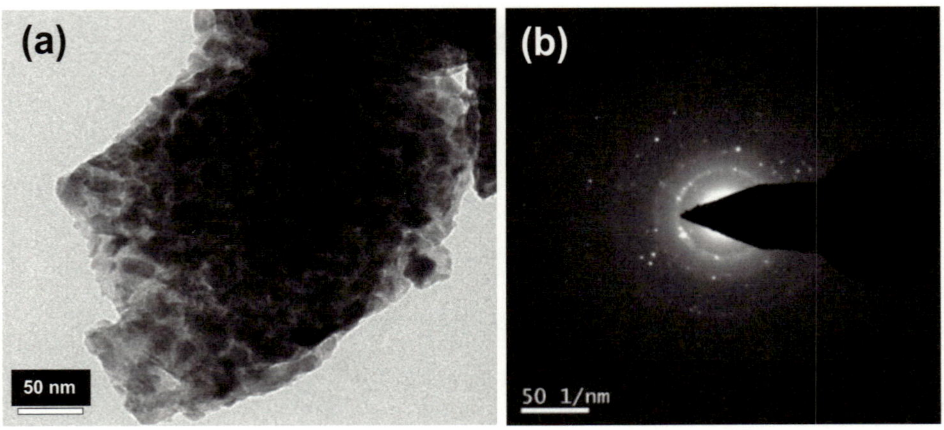

Fig. 6. (a) TEM image and (b) SAED pattern of Eu^{3+} doped glass-ceramics sample f.

4.5 Fourier transform infrared reflectance spectroscopy (FTIRRS)

The FTIR reflectance spectra of the Nd^{3+} doped precursor LTSA glass and heat-treated glass-ceramic samples in the wavenumber range 400-2000 cm^{-1} is shown in Fig. 7. It is seen from this figure that the precursor glass (curve-a) exhibits two broad reflection bands centered around 960 and 610 cm^{-1} as a result of wider distribution of silicon and tantalate structural units respectively. As alumina is one of the glass constituents, it prefers to enter into the silica rich phase and somewhere replace the Si^{4+} and the charge is compensated by Li^{+} ion. But, in order to maintain neutral charge condition, the later phase contains a higher amount of Li^{+} ions as the TaO_6 octahedra are negatively charged (Fukumi & Sakka, 1988, Samuneva et al., 1991). Hence, from the rearrangement of the glassy matrix it can be indicated that both the phase separated compositions begin to crystallize producing a nanostructure with the prolonged heat-treatment time. The appearance of a low intensity band at 735 cm^{-1} upon heat-treatment related to the stretching mode of Al-O bond of AlO_4 tetrahedra of β-spodumene (Burdick & Day, 1967). The prominent band occurred at 600 cm^{-1} corresponds to the stretching mode of O-Ta bond of TaO_6 octahedral units of lithium tantalate (Ono et al., 2001, Zhang et al., 1999). The reflection band centered at 600 cm^{-1} is assigned as $LiTaO_3$ crystal formation and the reflection band centered at 1000 cm^{-1} is assigned to Si-O stretching vibration of residual glass and β-spodumene crystal. The variation of Si-O (998 cm^{-1}) and Ta-O (602 cm^{-1}) stretching vibration bands intensities (here reflectivity) of Nd^{3+} doped samples with heat-treatment time is also recorded. It is seen that with progression of heat-treatment the band intensities increase rapidly initially and then become almost saturated after a certain time of heat-treatment (10 h). The gradual increase of relative intensity of band at 600 cm^{-1} clearly indicates formation of $LiTaO_3$ crystal with the increase of heat treatment time. The results of the FT-IRRS are in good agreement with that of XRD, FE-SEM and TEM studies. A similar observation has also been reported by Ito et al., 1978.

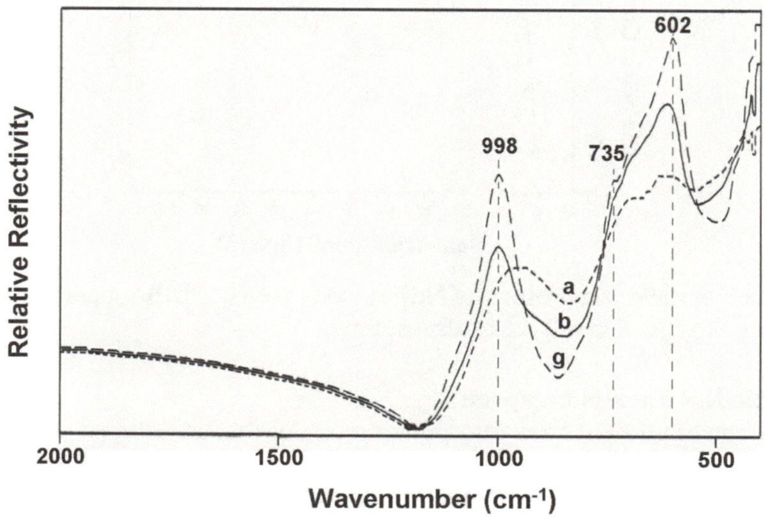

Fig. 7. FTIR-RS spectra of Nd^{3+} doped precursor LTSA glass and LT nano glass-ceramics.

4.6 Dielectric constant (ε_r)

The as prepared Eu^{3+} and Nd^{3+} doped LTSA precursor glasses exhibit relatively higher value (~20.0) of dielectric constant (ε_r) than the common vitreous silica (3.8) or soda-lime silicate (7.2) or borosilicate glasses (4.1-4.9) (Blech, 1986) due to high ionic refraction of Ta^{5+} ions (23.4) (Volf, 1984). This is due to its empty or unfilled d-orbital which contributes very strongly to its high polarizability (Yamane & Asahara, 2000, Risk et al., 2003). Its magnitudes show a sharp increase with increase in heat-treatment duration up to 5 h and thereafter it maintained saturation with a small decrease for any further heat treatment time as shown in Fig. 8. This suggests that, at the initial stages of heat treatment (1-3 h), separation of silica rich phase and Li–Ta enriched phases takes place and with the further heat-treatment, incipient precipitation of $LiTaO_3$ crystalline phase of high dielectric constant ($\varepsilon_r = 52$) (Moses, 1978) and spontaneous polarization (P_s = 0.50 C/m^2) (Risk et al., 2003) occurs gradually which becomes well defined at 5 h and attains the maximum volume fraction of the crystalline phase. Thus accumulation of Li^+ ions in the phase-separated glass matrix initially could cause a slight increase of dielectric constant and with further heat treatment time due to formation of stable $LiTaO_3$ ferroelectric crystals remarkably increase the dielectric constant reaching the highest value for 5 h heat treated sample and then maintain almost same on further course of heat-treatment. The variation in the dielectric constant (ε_r) values among the heat-treated nano glass-ceramics are mostly due to volume fraction of crystal phases contained and also the distribution of the $LiTaO_3$ phase in the microstructure (Vernacotola, 1994).

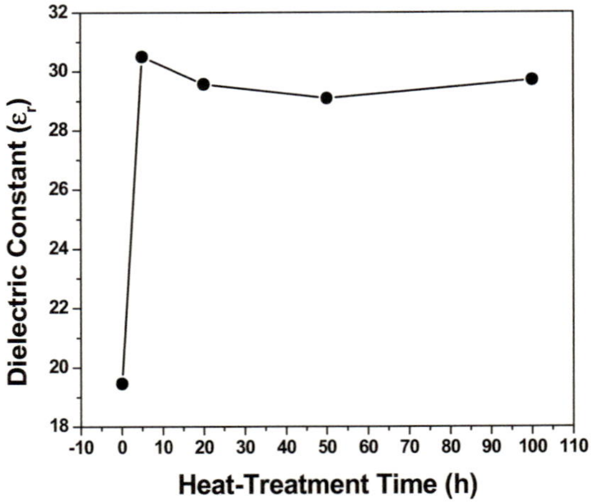

Fig. 8. Variation of dielectric constant of Nd^{3+} doped precursor LTSA glass and LT nano glass-ceramics as a function of heat-treatment time.

4.7 UV-Visible-NIR absorption spectra

The room temperature measured absorption spectra of the Nd^{3+} doped precursor glass (a) and 100 h heat-treated glass-ceramic samples (g) in the visible-NIR range have been presented in Fig. 9. The spectra reveal absorption peaks due to the $4f^3$-$4f^3$ forced electric dipole transitions from the ground $^4I_{9/2}$ state to different excited states of Nd^{3+} ion in $4f^3$ configuration. All the peaks $^4I_{9/2} \rightarrow {}^4G_{9/2}$ (512 nm), $^2K_{13/2}+{}^4G_{7/2}+{}^4G_{9/2}$ (526 nm), $^4G_{5/2}+{}^2G_{7/2}$ (583 nm), $^2H_{11/2}$ (626 nm), $^4F_{9/2}$ (679 nm), $^4F_{7/2}+{}^4S_{3/2}$ (739 nm), $^4F_{5/2}+{}^2H_{9/2}$ (806 nm) and $^4F_{3/2}$

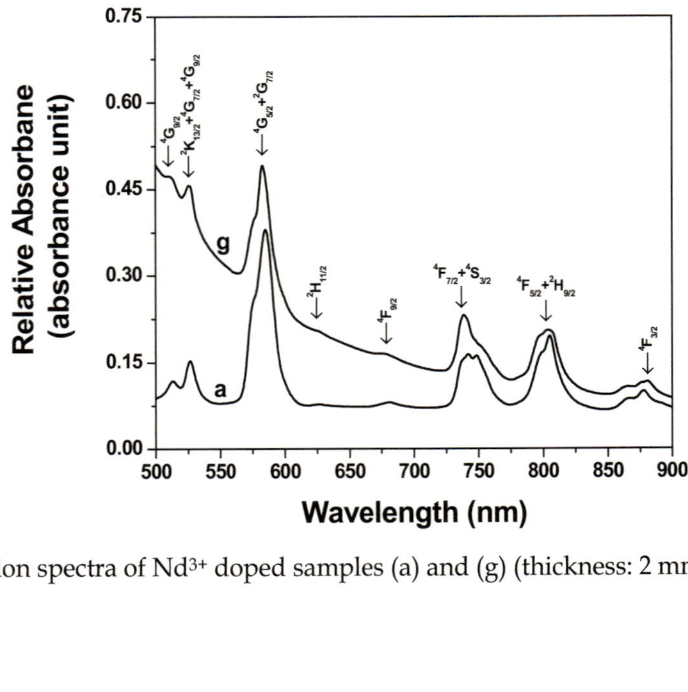

Fig. 9. Absorption spectra of Nd^{3+} doped samples (a) and (g) (thickness: 2 mm).

(880 nm) are assigned in accordance with Carnall's convention (Carnall et al., 1968, Chen et al., 2005). From this figure it is noticed that the base line of absorption spectra of heat-treated sample (g) has been elevated significantly with the diminished intensity of the absorption peak. This uplifting can be attributed to scattering of short wavelength light by the crystals (Beall & Duke, 1983 & 1969] or may be due to the difference in refractive index of crystalline phase (RI of $LiTaO_3$ is 2.1834 at 600 nm (Lynch, 1975)) with that of residual glassy matrix.

4.8 Photoluminescence spectra

The photoluminescence emission spectra of precursor glass (a) and Eu^{3+}:$LiTaO_3$ containing nano glass-ceramics (c and f) are recorded with an excitation at 392 nm and depicted in Fig. 10(A). All the spectra exhibit emissions from $^5D_0$ excited level to the ground state multiplets $^7F_{0,\ 1,\ 2,\ 3,\ 4}$ levels of Eu^{3+} ions with overall dominance of electric dipole (ED) transition $^5D_0 \rightarrow {}^7F_2$. The emission peak around 532 nm has been assigned to $^5D_1 \rightarrow {}^7F_2$ transition. In the perovskite type $LiTaO_3$ crystals, Li^+ and Ta^{5+} occupy octahedral sites with C_3 or nearly C_{3v} point symmetry. The Eu^{3+} ion is entering into the crystal ($LiTaO_3$) and it prefers to replace Li^+ over Ta^{5+} site forming $[REO_6]^{9-}$ octahedron (Wu & Zheng, 2002, Wu & Dong, 2005, Chang et al., 1993) due to the closeness of their ionic radii (Eu^{3+} = 0.95 Å, Li^+ =0.74 Å and Ta^{5+} = 0.64 Å). The local field asymmetry defining factor such as relative intensity ratio of I_{ED} to I_{MD} of Eu^{3+} doped glass and nano glass-ceramics has been estimated from their photoluminescence spectra and it was found that the asymmetric ratio of all samples was greater than unity, which implies that the Eu^{3+} ions take non-centrosymmetric sites.

The infrared photoluminescence spectra (λ_{ex} = 809 nm) of the Nd^{3+} doped samples (a, c and g) around 1069 nm are shown in Fig. 10(B). The emission band intensity around 1069 nm decreases with progression of heat-treatment. This decrease in emission intensity is due to the clustering of Nd^{3+} ions which is extremely sensitive to concentration quenching (Riello et al., 2006). Dejneka, 1998, has demonstrated in fluoride glasses that clustering thereby quenching occurs when the Eu^{3+}-Eu^{3+} ionic separation is around 40 Å. In the present case, the Nd^{3+}-Nd^{3+} ionic separation (R_i) in the precursor glass is found to be about 26 Å which was calculated using the relation (Pátek, 1970):

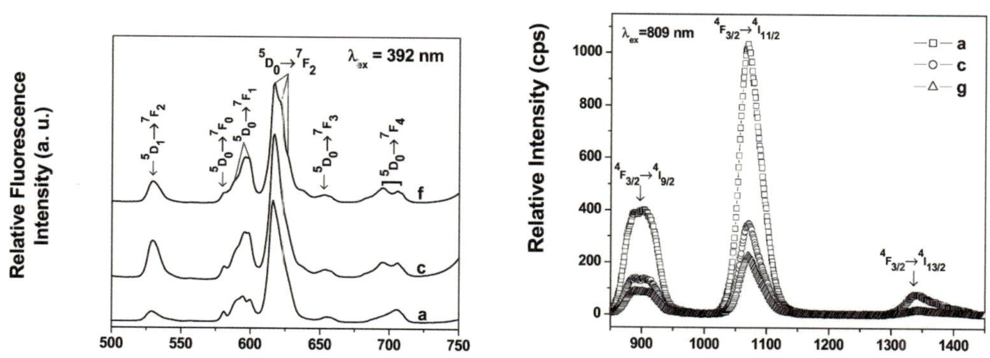

Fig. 10. Photoluminescence spectra of (A) Eu^{3+} and (B) Nd^{3+} doped precursor LTSA glass and LT nano glass-ceramics (thickness: 2 mm) respectively.

$$R_i(\overset{o}{A}) = (1 / N_{Nd^{3+}})^{1/3} \tag{2}$$

where $N_{Nd^{3+}}$ is the Nd^{3+} ion concentration. It is, therefore, seen that the Nd^{3+}- Nd^{3+} ionic separations (R_i) are in the quenching region. Theoretically, the rate of relaxation due to concentration quenching varies as R_i^{-6} (Campbell & Suratwala, 2000, Kang et al., 2001, Zgonik et al., 1993). With the progression of heat-treatment, the $LiTaO_3$ crystal phase has been formed and the Nd^{3+} ions partitioned into the residual glassy phase by reducing the inter-ionic separation less than 40 Å. This fact results in reduction in fluorescence intensity due to concentration quenching. The emission bands become sharper and take shapes as in crystalline host with progress of heat-treatment duration. All these observations indicate that the Nd^{3+} ions enter into the $LiTaO_3$ crystalline phase and therefore, environment around Nd^{3+} ions is changed with progression of heat-treatment.

4.9 Second harmonic generation (SHG)

The variation of SHG output power (nJ) with time period in bulk LTSA glass and LT glass-ceramics has been shown in Figs. 11(a) and (b). It is seen from the Fig. 11(a) that the precursor LTSA glass does not exhibit any SHG output. This is due to inversion symmetry of the precursor glass. It is also observed that for a constant heat-treatment time (10 h), the SHG output power increase from 1.08 to 1.875 nJ when the temperature is increased from 680°C to 850°C. This is attributed due to the formation of ferroelectric (non-centrosymmetric) $LiTaO_3$ crystals in the glassy matrix. Moreover, the increase in SHG output power with subsequent increase of heat-treatment temperature is due to the increase of $LiTaO_3$ crystallites content and their sizes in the glassy matrix.

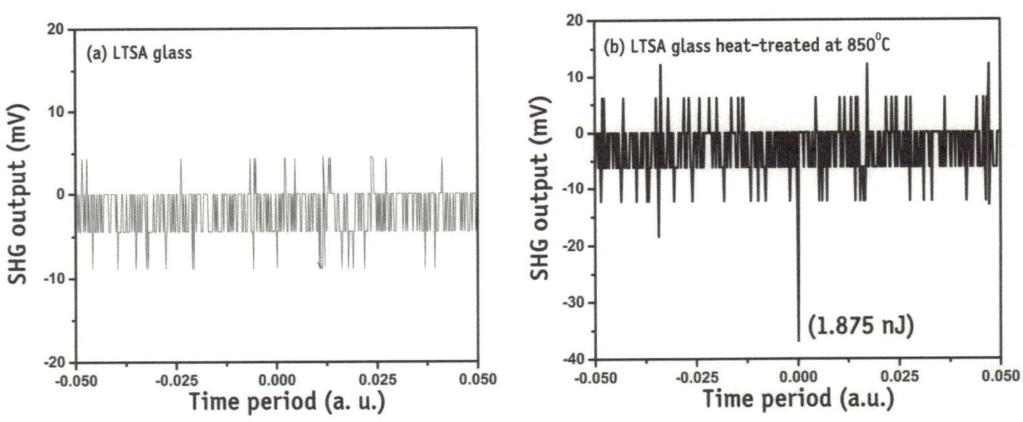

Fig. 11. Variation of SHG output power (nJ) with time period of precursor (a) LTSA glass and (b) 850°C heat-treated for 10 h LT glass-ceramic sample.

5. Nanostructured $KNbO_3$ ferroelectric glass-ceramics

5.1 Background

Ferroelectric potassium niobate ($KNbO_3$, KN) has the $A^{1+}B^{5+}O_3$ perovskite-type (orthorhombic) crystal structure (crystal symmetry class *mm2*; unit cell dimensions: a = 5.6896 Å, b = 3.9693 Å and c = 5.7256 Å) and having large nonlinear coefficient (d_{33} = 27.4

pm/V at 1064 nm). It is widely used in frequency doubling, tunable wave guiding, active laser host, holographic storage and surface acoustic wave (Xue & Zhang, 1998, Zgonik et al., 1993, Risk et al., 2003). Consequently, it becomes a subject of intense study. Very recently, potassium niobate ceramics were revisited in the interest of a search for environmental friendly lead-free piezoelectric and nonlinear materials (Ringgaard & Wurlitzer, 2005). Due to the low cost and high speed fabrication process of glass technology in comparison to single crystal preparation, with the flexibility of tailored properties to produce transparent nanostructures by controlled crystallization, ferroelectric $KNbO_3$ containing transparent glass-ceramics have generated increasing academic and technological interests. A large electro-optic effect (r_{42} = 380 pm/V for $KNbO_3$ (Ringgaard & Wurlitzer, 2005)) has been observed in a number of transparent glass-ceramic materials containing a ferroelectric crystalline phase. Several isothermal or non-isothermal studies have been carried out with a view to generate $KNbO_3$, $KNbSi_2O_7$, $K_3Nb_3Si_2O_{13}$ etc. ferroelectric crystal phases in various compositions of the K_2O-Nb_2O_5-SiO_2 (KNS) glass system (Pernice et al., 1999 & 2000, Golubkov et al., 2001, Tanaka et al., 2003, Aronne et al., 2004) and noticed strong second harmonic generation (SHG). It is observed that these glass-ceramics doped with rare earth (RE) ions, become very good luminescent media which are able to generate and amplify light. This application in combination with inherent nonlinear optical (NLO) properties of ferroelectric crystals, could offer a possibility to design self frequency doubling laser sources (Ringgaard, & Wurlitzer, 2005). For these reasons, the structure, dielectric and fluorescence properties of Eu^{3+}, Nd^{3+} and Er^{3-} ion doped transparent precursor glass and glass-ceramic composites of $KNbO_3$ with heat-treatment time have been studied and reported elaborately by Chaliha et al., 2009 & 2010. Second harmonic generation (SHG) from bulk $KNbO_3$ glass-ceramics has also been studied by Tarafder et al., 2011. For better understanding, the structure, dielectric and fluorescence properties of Er^{3+} doped transparent precursor glass and glass-ceramic composites of $KNbO_3$ with heat-treatment time have been reported elaborately along with the second harmonic generation (SHG) from bulk $KNbO_3$ glass-ceramics.

In comparison to the other rare earth ions, Er^{3+} has been extensively used as the most suitable active ion in several hosts by normal fluorescence for optical amplification at 1.5 µm. Trivalent erbium ions characterized by the $^4I_{13/2} \rightarrow {}^4I_{15/2}$ intra-4f transition play the key role in near infrared (NIR) emission (Dai et al., 2005). Besides, Er^{3+} doped glasses are chosen for getting better lasing property that can act as an eye-safe laser sources near 1550 nm. The photoluminescent emission intensity of erbium doped glasses is mainly dependent on the chemical environment of Er^{3+} ion because in the excited state Er^{3+} ion can de-excite by either photon emission at 1.53 µm, or non-radiative relaxation through coupling to a quenching site. Also, if the symmetry of the local crystal field around the erbium ion in the host glass is not distorted, the emission of erbium ion is forbidden. Therefore, the erbium ions must be incorporated in a non-centrosymmetric host material for strong optical emission (Winburn, 1985, Hui et al., 2007). Similarly, under favorable conditions strong green emission at 551 nm is possible upon normal high energy excitation.

6. Results and discussion

6.1 Differential thermal analysis (DTA)

The DTA curve of the Er^{3+} doped precursor glass is shown in Fig. 12. It exhibit the inflection in the temperature range 647-689°C followed by the intense exothermic peak at 759°C (T_p) corresponding to the phase crystallization. The glass transition temperature (T_g) has been estimated to be 681°C from the point of intersection of the tangents drawn at the slope change as is marked in Fig. 12.

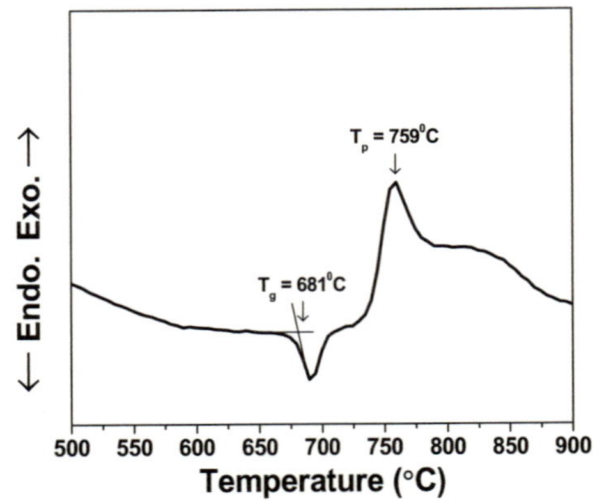

Fig. 12. DTA curve of Er^{3+} doped precursor powdered KNS glass.

6.2 Refractive index

The precursor glass was heat-treated at 800°C near glass transition temperature for various heat-treatment durations after nucleating at 750°C for 2 h. The prepared glass was transparent with body color of pink. From the measured glass density (ρ) and refractive index (n_e) at wavelength λ_e = 546.1 nm, other related optical properties of Er^{3+} doped precursor glass have been determined using relevant expressions and the same is presented in Table 2. The refractive indices of glass and heat-treated samples at five different wavelengths (473, 532, 633, 1064 and 1552 nm) are shown in Fig. 13. It is seen that the refractive index of 100 h heat-treated sample is higher as compared to the precursor glass. This is due to the formation of $KNbO_3$ crystals having high refractive index (2.2912 at 600 nm (Palik, 1998)). The large refractive index of this glass is due to the presence of highly

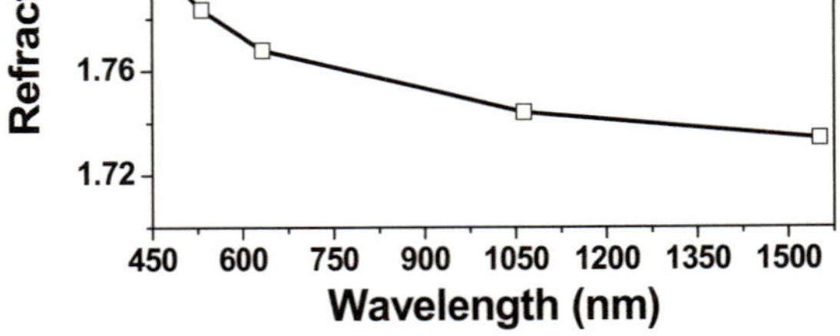

Fig. 13. Variation of refractive indices of precursor KNS glass and KN glass-ceramics obtained after heat-treatment for 100 h as a function of wavelength.

polarizable Nb^{5+} ions with high ionic refraction, 24.5 (Volf, 1984). It is observed that ions with an empty or unfilled d-orbital such as Nb^{5+} ion (outer electronic configuration: $4d^05s^0$) contributes very strongly to the linear and nonlinear polarizabilities (Yamane & Asahara, 2000). For the same reason, this glass is also possessing a high value of molar refractivity (R_M = 14.95 cm^3) and electronic polarizability (α = 5.592 x 10^{-24} cm^3) (Vernacotola & Shelby, 1994).

Properties	Corresponding value
	Er^{3+}:$KNbO_3$ doped precursor glass
Average molecular weight, M_{av}	120.05
Density, ρ (g.cm^{-3})	3.37
Refractive index	n_e (at 546.1 nm) → 1.78076 n_F' (at 480.0 nm) → 1.79612 n_C' (at 643.8 nm) → 1.76688
Molar refractivity, R_M (cm^3)	14.95
Electronic polarizability, α (cm^3)	5.592×10^{-24}
RE^{3+} ion concentration, $N_{RE}3+$ (ions/cm^3)	5.41×10^{19}
Glass transition temperature, T_g (°C)	681
Crystallization peak, T_p (°C)	759

Table 2. Some measured and calculated properties of Er^{3+}:$KNbO_3$ precursor glass

6.3 X-ray diffraction analysis

Fig. 14 shows the X-ray diffractogram of precursor glass along with the glass-ceramic samples. The amorphous nature of the as-prepared glass is indicated by the XRD pattern consisting of only a broad and halo band at around 29° diffraction angle. The structuring of this halo band takes place in the XRD pattern of the heat-treated glass-ceramic samples of 1-100 h along with the appearance of other well defined peaks around 25°, 28.5°, 30°, 32.8° and

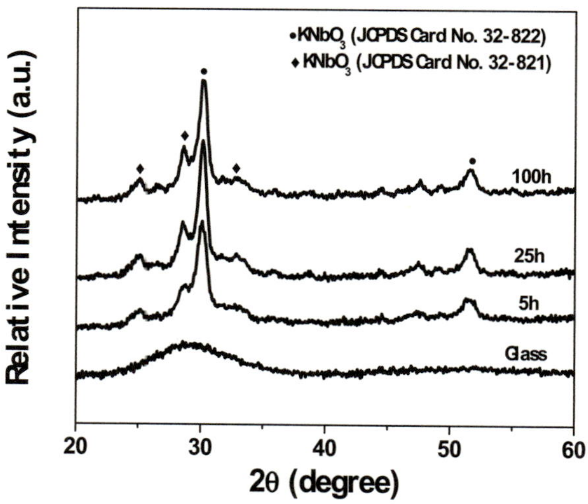

Fig. 14. XRD pattern of precursor KNS glass and glasses heat-treated at 800°C for different duration.

51.5° diffraction angles, which confirms the precipitation crystalline phase in the amorphous matrix. The crystalline phase resembles the JCPDS cards 32-821 and 32-822 of known potassium niobate. The calculated average crystallite sizes lie in the range 5-12 nm.

6.4 FESEM and TEM image analyses

The FESEM photomicrographs of the sample heat-treated at 800°C for 3 and 50 h duration are taken and from the FESEM micrographs, it is clearly observed that the glassy matrix of the heat-treated samples initially phase separated on nanometric scale followed by incipient precipitation of defined crystallites within the Nb-K rich phase regions on prolonged heat-treatments. The TEM image and SAED pattern of the sample heat treated for 50 h at 800°C have been presented in Figs. 15(a) and 15(b) respectively. The crystallite size from this TEM image of 50 h heat-treated sample found to be around 20 nm. The presence of fine spherical rings around the central bright region in SAED pattern discloses the existence of $KNbO_3$ nanocrystallites in the glassy matrix.

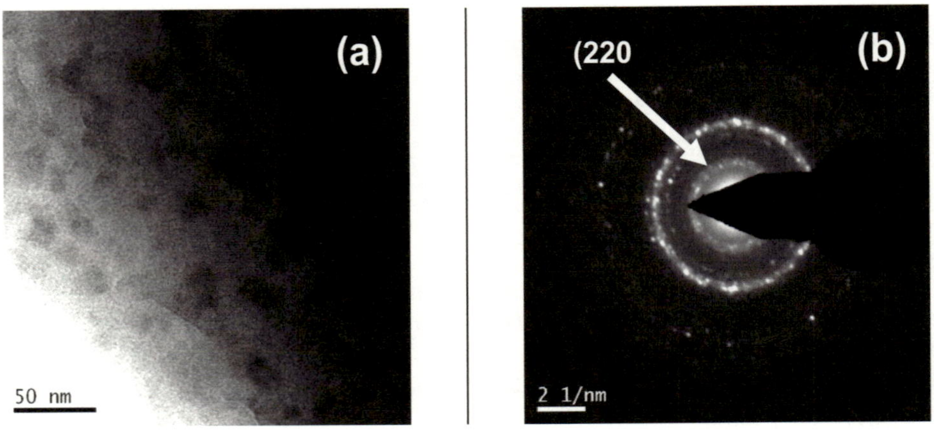

Fig. 15. (a) TEM image and (b) SAED pattern of Eu^{3+} doped 50 h heat-treated sample.

6.5 Fourier transform infrared reflectance spectroscopy (FTIRRS)

Fig. 16 shows the comparative FTIR reflectance spectra (FTIRRS) of the precursor glass and samples heat-treated at 800°C for 2 and 100 h duration in the wavenumber range 500-2000 cm^{-1}. Its inset shows the reflectivity at 930 and 749 cm^{-1} of precursor glass and heat-treated glasses as a function of heat-treatment time. It is seen that the FTIRR spectrum of the precursor glass exhibits a broad reflection band centered at 930 cm^{-1} as a result of wider distribution of SiO_4 structural units. The FTIR reflectance spectra of 2 and 100 h reveal narrowing of the main reflection band with additional features arising at 1128, 749 and 598 cm^{-1} in comparison to the precursor glass (Pernice et al., 1999). In the FTIRR spectra, the stretching modes of the Si-O-Si bonds of the SiO_4 tetrahedra with nonbridging oxygen (NBO) atoms are active in 900-1000 cm^{-1} range and the stretching modes of the Nb-O bonds in the NbO_6 octahedra occur in the 700-800 cm^{-1} range (Samuneva et al., 1991). The reflection band at 1128 cm^{-1} and 930 cm^{-1} wavenumber can be related to the asymmetric and symmetric stretching vibration modes of Si-O bonds in SiO_4 tetrahedra respectively, while the band at 749 cm^{-1} is due to the Nb-O stretching modes of distorted NbO_6 octahedra (de Andrade et al., 2000, Silva et al., 2006). The band observed around 598 cm^{-1} is assigned as ν_2 bending vibrational modes of the Si-O bonds in the SiO_4 tetrahedra.

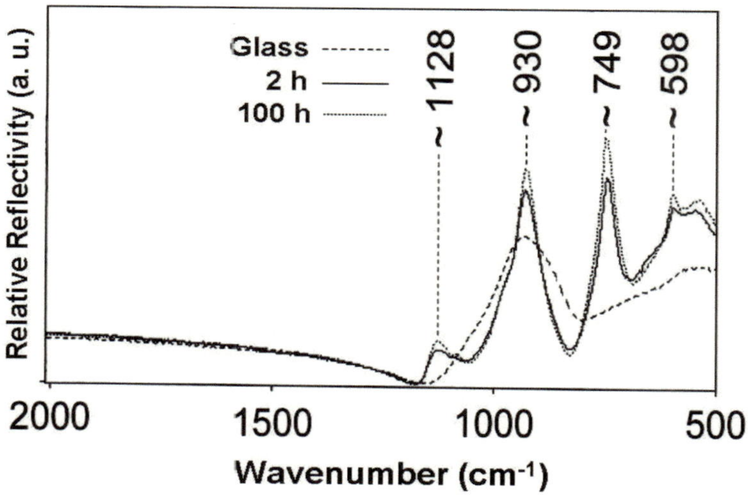

Fig. 16. FTIR-RS spectra of Er^{3+} doped precursor KNS glass and KN nano glass-ceramics.

6.6 Dielectric constant (ε_r)

Fig. 17 shows the magnitude of dielectric constant of precursor glass and heat-treated glass-ceramics. From this Fig., it is observed that the dielectric constant increase steeply from ε_r =17 (for precursor glass) to ε_r =31 (for 100 h heat-treated glass-ceramics) and thereafter it maintains almost saturation for any further heat treatment time. This suggests that on heat-treatment, at the initial stages, separation of silica rich phase and K-Nb enriched phases takes place and with the further prolonged heat-treatments incipient precipitation of $KNbO_3$ having high dielectric constant (ε_r =137) (Simões et al., 2004) and spontaneous polarization, P_s = 0.41 C/m^2 (Risk et al., 2003). The variation in crystallite size distributions and also the distribution of the $KNbO_3$ phase in the microstructure are the causes for the differences in the dielectric constant values amongst the heat-treated samples (Vernacotola, 1994).

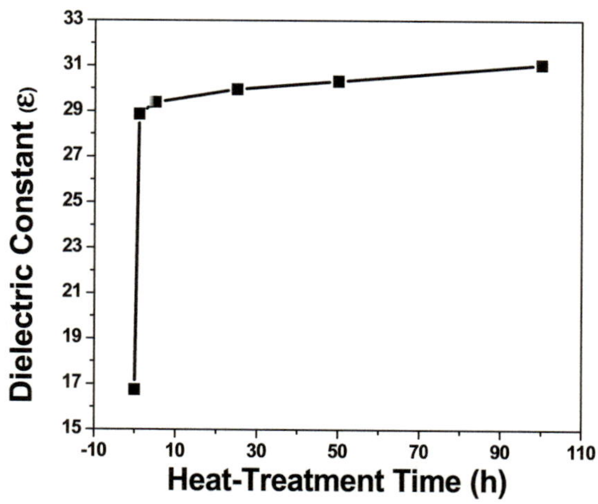

Fig. 17. Variation of dielectric constant of Er^{3+} doped precursor KNS glass and KN nano glass-ceramics as a function of heat-treatment time.

6.7 UV-visible-NIR absorption spectra

Fig. 18 shows the representative UV-Vis-NIR absorption spectra of Er^{3+} doped precursor glass and heat-treated samples for 100 h duration. The absorption occurs due to the *4f-4f* electric dipole transitions from the ground $^4I_{15/2}$ state to different excited state of Er^{3+} ions. The absorption spectra have very strong absorption edges below 350 nm and exhibit ten numbers of distinct absorption peaks which are similar to those appeared in silicate, tellurite and chloro sulphide glasses (Mandal et al., 2004, Bhaktha et al., 2006, Qian et al., 2008, Nayak et al., 2007, Lin et al., 2004, Schweizer et al., 1997). All the peaks were assigned in accordance with Carnall's convention as $^4I_{15/2} \rightarrow {}^2G_{9/2}$ (365 nm), $^4G_{11/2}$ (377 nm), $^2H_{9/2}$ (406 nm), $^4F_{5/2} + {}^4F_{3/2}$ (450 nm), $^4F_{7/2}$ (488 nm), $^2H_{11/2}$ (521 nm), $^4S_{3/2}$ (544 nm), $^4F_{9/2}$ (651 nm), $^4I_{9/2}$ (799 nm) and $^4I_{11/2}$ (978 nm) (Carnall et al., 1968). From this figure it is noticed that the base lines of the absorption spectra of heat-treated samples have been elevated significantly with diminishing intensities of the absorption peaks. The uplifting of the base line for glass-ceramic sample due to scattering imparted by the nanocrystallite phase is discussed as follows.

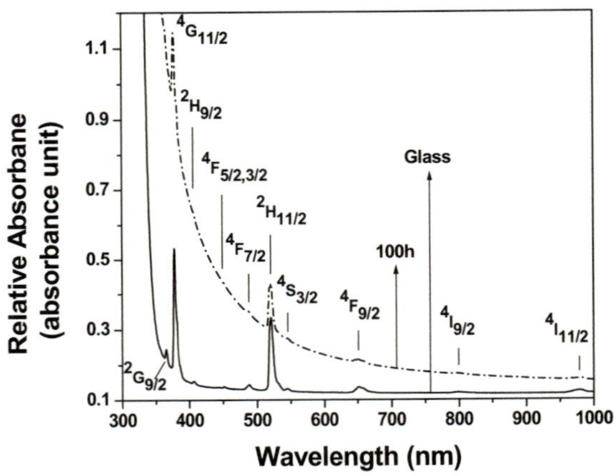

Fig. 18. Absorption spectra of the precursor KNS glass and glass heat-treated at 800°C for 100 h.

6.8 Photoluminescence spectra

Fig. 19 depicts the infrared emission spectra of precursor KNS glass and samples heat-treated for 2 and 100 h durations. The directly excited 980 nm emission spectra of the as prepared glass and the heat-treated samples exhibits emission from $^4I_{13/2}$ excited level to the $^4I_{15/2}$ ground level with Stark splitting at 1537 and 1566 nm. With increase in heat-treatment time the peak at 1537 nm for glass shifted to 1540 nm for 100 h heat-treated sample. The peak intensity ratio at 1566 nm to 1540 nm for precursor glass is ~0.93 which decreases down to ~0.79 for the heat-treated samples. It is generally seen that the glass-ceramics samples show more intense photoluminescence than the precursor glass. In this case it is found that the fluorescence intensity first decreases for the glass-ceramics heat-treated for shortest duration (2h) and then increases but with low intense than precursor glass. The reason behind this may be the fact that in the shorter durations of heat-treatment the samples are phase separated and stable $KNbO_3$ phase grows with longer heat-treatment duration.

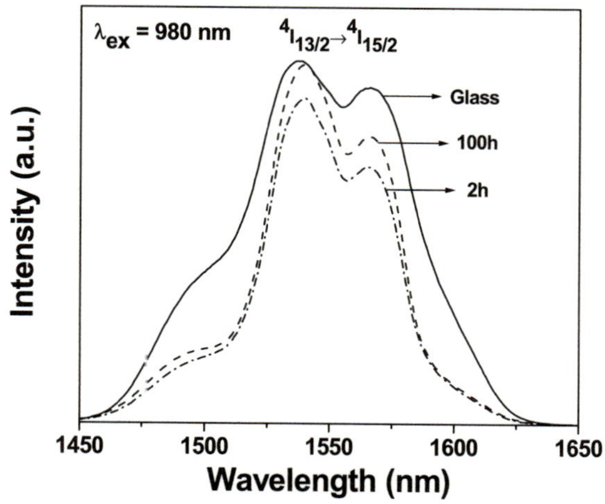

Fig. 19. (a) Near infrared emission spectra (λ_{ex} = 980 nm) of the precursor KNS glass and glasses heat-treated at 800°C for 2 and 100 h (a. u. = arbitrary unit).

6.9 Second harmonic generation (SHG)

The SHG output power (nJ) of KNS glass and KN glass-ceramics heat-treated at 800°C for 10 h is shown in Figs. 20(a) and (b). It is seen from the Fig. 20(a) that the precursor glass does not exhibit any SHG output. This phenomenon once again explained on the basis of inversion symmetry of the precursor glass. With heat-treatment of the precursor glass at 800°C for 10 h duration, the SHG output power increase to 39.74 nJ. This is attributed due to the formation of ferroelectric (non-centrosymmetric) $KNbO_3$ crystals in the glassy matrix.

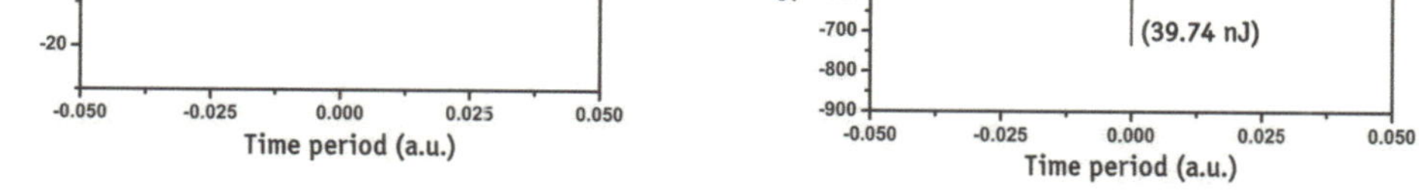

Fig. 20. Variation of SHG output power (nJ) with time period of precursor (a) KNS glass and (b) 800°C heat-treated for 10 h KN glass-ceramic sample.

7. Conclusions

The precursor glasses having molar composition $25.53Li_2O$-$21.53Ta_2O_5$-$35.29SiO_2$-$17.65Al_2O_3$ doped with RE oxides (0.5 wt% of Eu_2O_3 and Nd_2O_3 in excess) and $25K_2O$-$25Nb_2O_5$-$50SiO_2$ (KNS) doped with Er_2O_3 (0.5 wt% in excess) were prepared by melt-quench technique. The precursor glasses and the resulting LT and KN nano glass-ceramics were characterized by studying their thermal, structural, optical, dielectric properties. The results of XRD, FESEM, TEM and FT-IRRS confirmed the formation of nanocrystalline LT phases in the LTSA glass matrices and KN phase in the KNS glass matrix. The nanocrystallite size of LT and KN evaluated from TEM images found to vary in the range 14-36 nm and 5-12 nm respectively. The dielectric constants found to increase with heat-treatment time due to ferroelectric LT and KN formation. The photolminescence studies indicate the incorporation of RE ions into LT and KN crystal lattice. The derived LT and KN nano glass-ceramics also exhibit considerable SHG output which is very important for exploitation in self frequency doubling laser devices.

8. References

Abedin, K. S.; Tsuritani, T.; Sato, M. & Ito, H. (1997). Integrated intracavity quasi-phase-matched second harmonic generation based on periodically polled Nd:$LiTaO_3$. *Appl. Phys. Lett,* 70, 1, 10-12, ISSN 0003-6951.

Aronne, A. ; Sigaev, V. N. ; Pernice, P. ; Fanelli, E. & Usmanova, L. Z. (2004). Non-isothermal crystallization and nanostructuring in potassium niobium silicate glasses. *J. Non-Cryst. Solids,* 337, 2, 121-129, ISSN 0022-3093.

Beall, G.H. & Duke, D. A. (1969). Transparent Glass-Ceramics. *J. Mater. Sci.* 4, 4, 340-352, ISSN 0022-2461.

Beall, G. H. & Duke, D. A. (1983). *Glass-Ceramic Technology in Glass Science and Technology, Vol. 1: Glass-Forming Systems,* D.R. Uhlmann, N.J. Kreidl (Eds.), Academic Press, p. 403, ISBN 978-0127067018, New York,.

Bhaktha, S. N. B.; Boulard, B. S.; Chaussedent, Chiappini, A.; Chiasera, A.; Duval, E.; Duverger, C.; Etienne, S.; Ferrari, M.; Jestin, Y.; Mattarelli, M.; Montagna, M.; Monteil, A.; Moser, E.; Portales, H. & Vishunubhatla, K. C. (2006). Erbium-activated modified silica glasses with high $^4I_{13/2}$ luminescence quantum yield. *Opt. Mater,* 28, 11, 1325–1328, ISSN 0925-3467.

Blech, I. A. (1986). *Properties of materials in electronics engineering handbook,* 2nd Edition, D.G. Fink, D. Christiansen Ed., McGraw-Hill, New York, p. 6-30.

Burdick, V. L. & Day, D. E. (1967). Coordination of aluminium ions in tricalcium aluminate. *J. Am. Ceram. Soc.* 50, 2, 97-101, ISSN 1551-2916.

Campbell, J. H. & Suratwala, T. I. (2000). Nd-doped phosphate glasses for high-energy/high-peak-power lasers. *J. Non-Cryst. Solids,* 263-264, 318-341, ISSN 0022-3093.

Carnall, W. T.; Fields, P. R. & Rajnak, K. (1968). Electronic energy levels in the trivalent lanthanide aquo ions. I. Pr^{3+}, Nd^{3+}, Pm^{3+}, Sm^{3+}, Dy^{3+}, Ho^{3+}, Er^{3+}, and Tm^{3+}. *J. Chem. Phys.* 49, 10, 4424-4442.

Chaliha, R. S.; Annapurna, K.; Tarafder, A.; Tiwari, V. S.; Gupta, P. K. & Karmakar, B. (2009). Luminescence and dielectric properties of nano-structured Eu^{3+}:K_2O-Nb_2O_5-SiO_2 glass-ceramics. *Solid State Sci.,* 11, 8, 1325–32, ISSN 1293-2558.

Chaliha, R. S.; Annapurna, K.; Tarafder, A.; Tiwari, V. S.; Gupta, P. K. & Karmakar, B. (2010). Structure, dielectric and optical properties of transparent Nd^{3+}: $KNbO_3$ nanocrystalline glass-ceramics. *Opt. Mater.* 32, 9, 1202–1209, ISSN 0925-3467.

Chaliha, R. S.; Annapurna, K.; Tarafder, A.; Tiwari, V. S.; Gupta, P. K. & Karmakar, B. (2010). Optical and dielectric properties of isothermally crystallized nano $KNbO_3$ in Er^{3+}-

doped K_2O-Nb_2O_5-SiO_2 glasses. *Spectrochim. Acta A Mol. Biomol. Spectrosc.* 75, 1, 243–250, ISSN 1386-1425.

Chang, Y. M.; Yeom, T. H.; Yeung, Y. Y. & Rudowicz, C. (1993). Superposition model and crystal- field analysis of $^4A_2$ and $^2{}_aE$ states of Cr^{3+} ions at C_3 sites in $LiNbO_3$. *J. Phys. C: Conds. Mater.* 5, 34, 6221-6230, ISSN 0953-8984.

Chen, Y.; Huang, Y.; Huang, M.; Chen, R. & Luo, Z. (2005). Effect of Nd^{3+} on the spectroscopic properties of bismuth borate glasses. *J. Am. Ceram. Soc.* 88, 1, 19-23, ISSN 0002-7820.

Cheng, Z. X.; Kimura, H.; Ozawa, K.; Miyazaki, A. & Kanna, C. V. (2005). Ferroelectric lithium tantalate thin film derived from peroxide. *J. Alloys Compd.* 402, 1-2, 208-212, ISSN 0925-8388.

Cullity, B. D. (1978). *Elements of X-ray diffraction*, 2nd Edition, Addison-Wesley Publishing Co., p. 101, ISBN-978-0201011746, London.

Dai, S.; Wu, J.; Zhang, J.; Wang, G.; & Jiang, Z. (2005). The spectroscopic properties of Er^{3+}-doped TeO_2–Nb_2O_5 glasses with high mechanical strength performance. *Spectrochim. Acta A Mol. Biomol. Spectrosc.* 62, 1-3, 431-437, ISSN 1386-1425.

Dejneka, M. (1998). The luminescence and structure of novel transparent oxyfluoride glass-ceramics. *J. J. Non-Cryst. Solids,* 239, 1-3, 149-155, ISSN 0022-3093.

de Andrade, J. S.; Pinheiro, A. G.; Vasconcelos, I. F.; deArujo, M. A. B.; Valente, M. A. & Sombra, A. S. B. (2000). Structural studies of $KNbO_3$ in niobate glass-ceramics. *J. Phys. Chem. Solids,* 61, 6, 899-906, ISSN 0022-3697.

Fukumi, K. & Sakka, S. (1988). Coordination state of Nb^{5+} ions in silicate and gallate glasses as studied by raman spectroscopy. *J. Mater. Sci.* 23, 8, 2819-2823, ISSN 0022-2461.

Gasparotto, G.; Lima, S. A. M.; Davolos, M. R.; Varela, J. A.; Longo, E. & Zaghete, M. A. (2008). Luminescence properties of Eu^{3+}- and Mg^{2+}-doped $LiTaO_3$ obtained via the polymeric precursor method. *J. Lumin.* 128, 10, 1606-1610, ISSN 0022-3093.

Golubkov, V. V.; Dymshits, O. S.; Zhilin, A. A.; Redin, A. V. & Shepilov, M. P. (2001). Crystallization of Glasses in the K_2O–Nb_2O_5–SiO_2 System. *Glass Phys. Chem.* 27, 6, 504-511, ISSN 1087-6596.

Gruber, J. B.; Allik, T. H.; Sardar, D. K.; Yow, R. M.; Scripsick, M. & Wechsler, B. (2006). Crystal growth and spectroscopic characterization of Yb^{3+}:$LiTaO_3$. *J. Lumin.* 117, 2, 233-238, ISSN 0022-2313.

Hase, H.; Nasu, H.; Mito, A.; Hashimoto, T.; Matsuoka, J. & Kamiya, K. (1996). Second harmonic generation from surface crystallized Li_2O-Ta_2O_5-SiO_2 glass. *Jpn. J. Appl. Phys.* 35, 10, 5355-5356, ISSN 0021-4922.

Hirayama, C. & Berg, D. (1963). Dielectric properties of glasses in the system Nb_2O_5-Na_2O-SiO_2. *J. Am. Ceram. Soc.* 46, 2, 85-88, ISSN 1551-2916.

Hui, Y. Y. ; Shih, P-H. ; Sun, K-J. & Lin, C-F. (2007). Enhancement of 1.5 μm emission in erbium-doped spin-on glass by furnace annealing. *Thin Solid Films.* 515, 17, 6754-6757, ISSN 0040-6090.

Ito, S.; Kokubo, T. & Tashiro, M. (1978). Transparency of $LiTaO_3$-SiO_2-Al_2O_3 glass-ceramics in relation to their microstructure. *J. Mater. Sci.* 13, 5, 930-938, ISSN 0022-2461.

Jain, H. (2004). Transparent ferroelectric glass-ceramics. *Ferroelectrics,* 306, 111-127, ISSN 0015-0193.

Kang, U.; Zhilin, A. A.; Logvinov, D. P.; Onushchenko, A. A.; Savost'yanov, V. A.; Chuvaeva, T. I. & Shashkin, A. V. (2001). Transparent Nd^{3+}-activated glass-ceramics in the Li_2O-Al_2O_3-SiO_2 system: physicochemical aspects of their preparation and optical characteristics. *Glass Phys. Chem.* 27, 4, 344-352, ISSN 1087-6596.

Lin, H.; Liu, K.; Pun, E. Y. B.; Ma, T. C.; Peng, X.; An, Q. D.; Yu, J. Y. & Jiang, S. B. (2004). Infrared and visible fluorescence in Er^{3+}-doped gallium tellurite glasses. *Chem. Phys. Lett.* 398, 1-3, 146-150, ISSN 0009-2614.

Lynch, C. T. (1975). *CRC Handbook of Materials Science,* Vol. III, CRC Press, p. 170, Cleveland, Ohio.

Mandal, D.; Banerjee, H. D.; Goswami, M. L. N. & Acharya, H. N. (2004). Synthesis of Er^{3+} and Er^{3+}:Yb^{3+} doped sol-gel derived silica glass and studies on their optical properties. *Bull. Mater. Sci.* 27, 4, 367-372, ISSN 0250-4707.

Mizuuchi, K. & Yamamoto, K. (1995). Harmonic blue light generation in bulk periodically poled $LiTaO_3$. *Appl. Phys. Lett.* 66, 22, 2943-2945, ISSN 0003-6951.

Moses, A. J. (1978). *The practicing scientist's handbook.* Van Nostrand Reinhold Company, p. 558, ISBN 0442 25584 5 New York.

Mukherjee, P. & Varma, K. B. R. (2004). Crystallization and physical properties of $LiTaO_3$ in a reactive glass matrix ($LiBO_2$-Ta_2O_5). *Ferroelectrics,* 306, 1, 129-143, ISSN 0015-0193.

Naranjo, B.; Gimzewski, J. K. & Putterman, S. (2005). Observation of nuclear fusion driven by a pyroelectric crystal. *Nature,* 434, 1115-1117, ISSN 0028-0836.

Nayak, A.; Kundu, P. & Debnath, R. (2007). Strong green emission from Er^{3+} in a fluorine containing (Pb, La)–tellurite glass without using up-conversion route. *J. Non. Cryst. Solids,* 353, 13-15, 1414-1417, ISSN 0022-3093.

Ono, H.; Hosokawa, Y.; Shinoda, K.; Koyanagi, K. & Yamaguchi, H. (2001). Ta-O phonon peaks in tantalum oxide films on Si. *Thin Solid Films,* 381, 1, 57-61, ISSN 0040-6090.

Palik, E. D. (1997) *Handbook of optical constants of solids III,* Academic Press, p. 834, ISBN 0125444230, San Diego.

Pátek, K. (1970). *Glass Lasers,* Butterworth & Co (Publishers) Ltd., p.95, ISBN 978-0592027784, London.

Pernice, P.; Aronne, A.; Sigaev, V. N.; Sarkisov, P. D.; Molev, V. I. & Stefanovich, S. Y. (1999). Crystallization behavior of potassium niobium silicate glasses. *J. Am. Ceram. Soc.* 82, 12, 3447-3452, ISSN 0002-7820.

Pernice, P.; Aronne, A.; Sigaev, V. N. & Kupriyanova, M. (2000). Crystallization of the $K_2O \cdot Nb_2O_5 \cdot 2SiO_2$ glass: evidences for existence of bulk nanocrystalline structure. *J. Non-Cryst. Solids,* 275, 3, 216-224, ISSN 0022-3093.

Qian, Q.; Wang, Y.; Zhang, Q. Y.; G. F.; Yang, Yang, Z. M. & Jiang, Z. H. (2008). Spectroscopic properties of Er^{3+}-doped Na_2O–Sb_2O_3–B_2O_3–SiO_2 glasses. *J. Non. Cryst. Solids,* 354, 18, 1981-1985, ISSN 0022-3093.

Riello, P.; Bucella, S.; Zamengo, L.; Anselmi-Tamburini, U.; Francini, R.; Pietrantoni, S.; & Munir, Z. A. (2006). Erbium-doped LAS glass ceramics prepared by spark plasma sintering (SPS). *J. Eur. Ceram. Soc.* 26, 15, 3301-3306, ISSN 0955-2219.

Ringgaard, E. & Wurlitzer, T. (2005). Lead-free piezoceramics based on alkali niobates. *J. Eur. Ceram. Soc.* 25, 12, 2701-2706, ISSN 0955-2219.

Risk, W. P.; Gosnell, T. R. & Nurmikko, A. V. (2003). *Compact blue-green lasers,* Cambridge University Press, ISBN 9780521623186, Cambridge.

Romanowski, W. R.; Sokólska, I.; Dzik, G. D. & Gołab, S. (2000). Investigation of $LiXO_3$ (X=Nb, Ta) crystals doped with luminescent ions: Recent results. *J. Alloys Compd.* 300-301, 152-157, ISSN 0925-8388.

Samuneva, B.; Kralchev, St. & Dimitrov, V. (1991). Structure and optical properties of niobium silicate glasses. *J. Non-Crys. Solids*, 129, 1-3, 54-63, ISSN 0022-3093.

Schweizer, T.; Brady, D. J. & Hewak, D. W. (1997). Fabrication and spectroscopy of erbium doped gallium lanthanum sulphide glass fibres for mid-infrared laser applications. *Opt. Exp.* 1, 4, 102-107, ISSN 1094-4087.

Silva, E. N.; Ayala, A. P.; Guedes, I.; C. Paschoal, W. A.; Moreira, R. L.; C. -K. Loong, & Boatner, L. A. (2006). Vibrational spectra of monazite-type rare-earth orthophosphates. *Opt. Mater.* 29, 2-3, 224-230, ISSN 0925-3467.

Simões, A. Z.; Ries, A.; Riccardi, C. S.; Gonzalez, A. H.; Zaghete, M. A.; Stojanovic, Cilense, B. D. M. & Varela, J. A. (2004). Potassium niobate thin films prepared through polymeric precursor method. *Mat. Lett.* 58, 20, 2537-2540, ISSN 0167-577X.

Sokólska, I. (2002). Infrared to visible conversion of radiation in some Ho^{3+}-doped oxide and fluoride crystals. *J. Alloys Compd.* 341, 1-2, 288-293, ISSN 0925-8388.

Sokólska, I.; Kück, S.; Dzik, G. D. & Baba, M. (2001). The up-conversion processes in Ho^{3+} doped $LiTaO_3$. *J. Alloys Compd.* 323-324, 273-278, ISSN 0925-8388.

Tanaka, H.; Yamamoto, M.; Takahashi, Y.; Benino, Y.; Fujiwara, T. & Komatsu, T. (2003). Crystalline phases and second harmonic intensities in potassium niobium silicate crystallized glasses. *Opt. Mater.* 22, 1, 71-79, ISSN 0925-3467.

Tarafder, A.; Annapurna, K.; Chaliha, R. S.; Tiwari, V. S.; Gupta, P. K. & Karmakar, B. (2009). Processing and properties of Eu^{3+}:$LiTaO_3$ transparent glass–ceramic nanocomposites. *J. Am. Ceram. Soc.* 92, 9, 1934-1939, ISSN 0002-7820.

Tarafder, A.; Annapurna, K.; Chaliha, R. S.; Tiwari, V. S.; Gupta, P. K. & Karmakar, B. (2010). Structure, dielectric and optical properties of Nd^{3+} doped $LiTaO_3$ transparent ferroelectric glass-ceramic nanocomposites. *J. Alloys Compd.* 489, 1, 281-288, ISSN 0925-8388.

Tarafder, A.; Annapurna, K.; Chaliha, R. S.; Tiwari, V. S.; Gupta, P. K. & Karmakar, B. (DOI:10.1111/j.1744-7402.2010.02494.x.). Effects of nano $LiTaO_3$ crystallization on dielectric and optical properties in Er^{3+}-doped Li_2O-Ta_2O_5-SiO_2-Al_2O_3 glasses. *Int. J. Appl. Ceram. Technol.*

Tarafder, A.; Annapurna, K.; Chaliha, R. S.; Satapathy, S.; V. Tiwari, S.; Gupta, P. K. & Karmakar, B. (2011). Second harmonic generation in ferroelectric $LiTaO_3$ and $KNbO_3$ containing bulk nano glass-ceramics. *J. Nonlinear Opt. Phys. Mater.* 20, 1, 49-61, ISSN 0218-8635.

Vernacotola, D. E. (1994). Alkali Niobium and Tantalum Silicate Glasses and Ferroelectric Glass-Ceramics. *Key Engg. Mater.* 94-95, 379-408.

Vernacotola, D. E. & Shelby, J. E. (1994). Potassium niobium silicate glasses. *Phys. Chem. Glasses*, 35, 4, 153-159, ISSN 0031-9090.

Volf, M. B. (1984). *Chemical approach to glass*, Elsevier, p.125, ISBN 0444996354, Amsterdam.

Wu, S. Y. & Zheng, W. C. (2002). EPR parameters and defect structures for two trigonal Er^{3+} centers in $LiNbO_3$ and MgO or ZnO codoped $LiNbO_3$ crystals. *Phys. Rev. B*, 65, 22, 224107-1-6, ISSN 0163-1829.

Winburn, D. C. (1985). *Practical Laser Safety*, Dekker, ISBN 0824773489, New York.

Wu, S. Y. & Dong, H. N. (2005). Studies of the EPR g factors and the local structure for the trigonal Nd^{3+} center in $LiNbO_3$:Nd^{3+} crystal. *J. Alloy. Comp.*, 386, 1-2, 52-56, ISSN 0925-8388.

Xue, D. & Zhang, S. (1998). Linear and nonlinear optical properties of $KNbO_3$. *Chem. Phys. Lett.* 291, 3-4, 401–406 ISSN 0009-2614.

Yamane, M. & Asahara, Y. (2000). *Glasses for photonics*, Cambridge University Press, p.173, ISBN 0 521 58053 6, Cambridge, UK.

Youssef, S.; Asmar, R. Al., Podlecki, J.; Sorli, B. & Foucaran, A. (2008). Structural and optical characterization of oriented $LiTaO_3$ thin films deposited by sol-gel technique. *Eur. Phys. J. Appl. Phys.* 43, 1, 65-71, ISSN 1286-0042.

Zgonik, M.; Schlesser, R.; Biaggio, I.; Voit, E.; Tscherry, J. & Giinter, P. (1993). Materials constants of $KNbO_3$ relevant for electro- and acousto-optics. *J. Appl. Phys.* 74, 2, 1287–1297, ISSN 0021-8979.

Zhang, J. Y.; Boyd, I. W.; Dusastre, V. & Williams, D. E. (1999). Ultraviolet annealing of tantalum oxide films grown by photo-induced chemical vapor deposition. *J. Phys. D: Appl. Phys.* 32, 19, L91-95, ISSN 0022-3727.

Zheng, F.; Liu, H.; Liu, D.; Yao, S.; Yan, T. & Wang, J. (2009). Hydrothermal and wet-chemical synthesis of pure $LiTaO_3$ powders by using commercial tantalum hydroxide as starting material. *J. Alloys Compd.* 477, 1-2, 688-691, ISSN 0925-8388.

Zhu, S.; Zhu, Y.; Yang, Z.; Wang, H.; Zhang, Z.; Hong, J.; Ge, C. & Ming, N. (1995). Second-harmonic generation of blue light in bulk periodically poled $LiTaO_3$. *Appl. Phys. Lett*, 67, 3, 320-322, ISSN 0003-6951.

20

Ferroelectricity in Silver Perovskite Oxides

Desheng Fu[1] and Mitsuru Itoh[2]
[1]*Shizuoka University,*
[2]*Tokyo Institute of Technology*
Japan

1. Introduction

Many ferroelectric oxides possess the ABO_3 perovskite structure (Mitchell, 2002), in which the A-site cations are typically larger than the B-site cations and similar in size to the oxygen anion. Figure 1 shows a schematic drawing for this structure, where the A cations are surrounded by 12-anions in the cubo-octahedral coordination and the B cations are surrounded by 6-anions in the octahedral coordination. An ideal perovskite exhibits a cubic space group *Pm3m*. This structure is centrosymmetric and cannot allow the occurrence of ferroelectricity that is the presence of a switchable spontaneous electric polarization arising from the off-center atomic displacement in the crystal (Jaffe et al., 1971; Lines & Glass, 1977). The instability of ferroelectricity in the perovskite oxides is generally discussed with the Goldschimidt tolerance factor (t) (Goldschmidt, 1926 and Fig. 2),

$$t = (r_A + r_O) / \sqrt{2}(r_B + r_O),$$

where r_O, r_A, and r_B are the ionic radii of the O, A, and B ions. For a critical value t=1, the cubic paraelectric phase is stable. This unique case can be found in $SrTiO_3$, which has an ideal cubic perovskite structure at room temperature and doesn't show ferroelectricity down to the absolute 0 K (Müller & Burkard, 1979). However, ferroelectricity can be induced by the substitution of O^{18} in this quantum paraelectric system at $T<T_c$~23 K (Itoh et al., 1999). When t>1, since the B-site ion is too small for its site, it can shift off-centeringly, leading to the occurrence of displacive-type ferroelectricity in the crystal. Examples of such materials are $BaTiO_3$ and $KNbO_3$ (Shiozaki et al., 2001). On the other hand, for t<1, the perovskite oxides are in general not ferroelectrics because of different tilts and rotations of BO_6 octahedra, which preserve the inversion symmetry. But exceptions may be found in the Bi-based materials, in which large A-site displacement is observed. This large A-site displacement is essentially attributed to the strong hybridization of Bi with oxygen (Baettig et al., 2005). Similar cases are observed in Pb-based materials, which commonly have large Pb displacement in the A-site (Egami et al., 1998) and strong covalent nature due to the unique stereochemistry of Pb (Cohen, 1992; Kuroiwa et al., 2001).
Although $BaTiO_3$- and $PbTiO_3$-based ceramics materials have been widely used in electronic industry (Uchino, 1997; Scott, 2000), there remain some importance issues to be solved. One of such challenges is to seek novel compounds to replace the Pb-based materials, which have a large Pb-content and raises concerns about the environmental pollution (Saito et al. ,2004; Rodel et al.,2009).

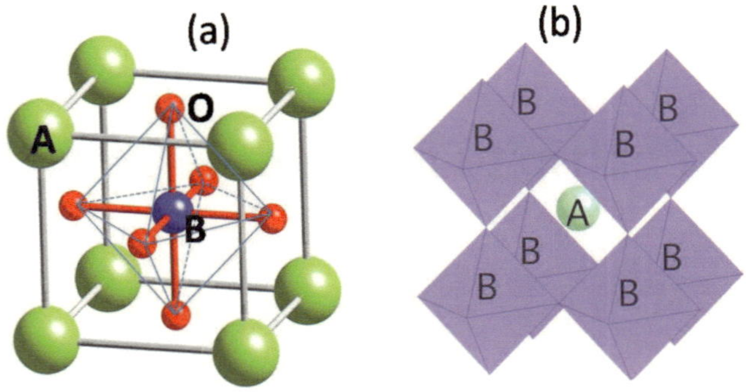

Fig. 1. The structure of an ABO_3 perovskite with the origin centered at (a) the B-site ion and (b) the A-site ion.

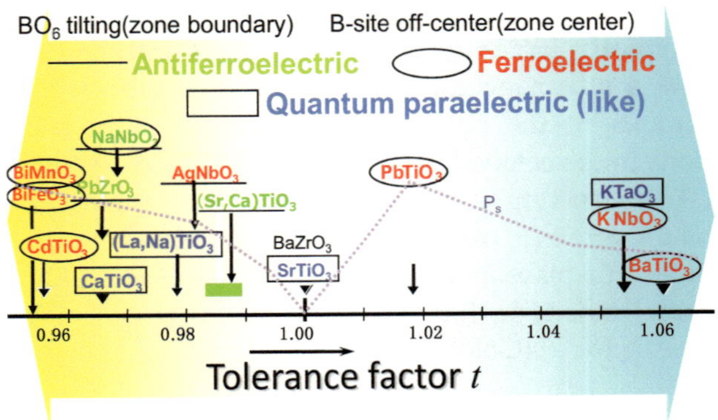

Fig. 2. Tolerance factor of typical dielectric oxides.

The discovery of extremely large polarization (52 μC/cm²) under high electric field in the $AgNbO_3$ ceramics (Fu et al., 2007) indicates that Ag may be a key element in the designs of lead-free ferroelectric perovskite oxides (Fu et al., 2011a). With the advance of first-principles calculations (Cohen, 1992) and modern techniques of synchrotron radiation (Kuroiwa et al., 2001), we now know that the chemical bonding in the perovskite oxides is not purely ionic as we have though, but also possesses covalent character that plays a crucial role in the occurrence of ferroelectricity in the perovskite oxides (Cohen, 1992; Kuroiwa et al., 2001). It is now accepted that it is the strong covalency of Pb with O that allows its large off-center in the A-site. Although Ag doesn't have lone-pair electrons like Pb, theoretical investigations suggest that there is hybridization between Ag and O in $AgNbO_3$(Kato et al., 2002; Grinberg & Rappe, 2003,2004), resulting in a large off-center of Ag in the A-site of perovskite $AgNbO_3$ (Grinberg & Rappe, 2003,2004). This prediction is supported by the results from X-ray photoelectron spectroscopy, which suggest some covalent characters of the chemical bonds between Ag and O as well as the bonds between Nb and O (Kruczek et al., 2006). Moreover, bond-length analysis also supports such a theoretical prediction. Some of the bond-lengths (~2.43 Å) in the structure (Sciau et al., 2004; Yashima et al. 2011) are significantly less than the sum of Ag^+ (1.28 Å) and O^{2-} (1.40 Å) ionic radii (Shannon, 1967). All these facts make us believe that $AgNbO_3$ may be used as a base compound to develop novel ferroelectric materials. Along such a direction, some interesting results have been obtained. It was found that ferroelectricity can be induced through the chemical modification of the $AgNbO_3$ structure by substitution of Li (Fu et al., 2008, 2011a), Na(Arioka, 2009; Fu et al., 2011b), and K (Fu et al., 2009a) for Ag. Large spontaneous polarization and high temperature of para-ferroelectric phase transition were observed in these solid solutions. In the following sections, we review the synthesis, structure, and dielectric, piezoelectric and ferroelectric properties of these solid solutions together with another silver perovskite $AgTaO_3$ (Soon et al.,2009, 2010), whose solid solutions with $AgNbO_3$ are promising for the applications in microwaves devices due to high dielectric constant and low loss (Volkov et al. 1995; Fortin et al., 1996; Petzelt et al., 1999; Valant et al., 2007a).

2. $AgNbO_3$

2.1 Synthesis

Both single crystal and ceramics of $AgNbO_3$ are available. Single crystal can be grown by a molten salt method using NaCl or V_2O_5 as a flux (Łukaszewski et al., 1980; Kania, 1989). Ceramics samples can be prepared through a solid state reaction between Nb_2O_5 and silver source (Francombe & Lewis, 1958; Reisman & Holtzberg, 1958). Among the silver sources of metallic silver, Ag_2O and Ag_2CO_3, Ag_2O is mostly proper to obtain single phase of $AgNbO_3$(Valant et al., 2007b). For silver source of Ag_2O, thermogravimetric analysis indicates that phase formation can be reach at a firing temperature range of 1073-1397 K (Fig. 3). The issue frequently encountered in the synthesis of $AgNbO_3$ is the decomposition of metallic silver, which can be easily justified from the color of the formed compounds. Pure powder is yellowish, while grey color of the powder generally indicates the presence of some metallic silver. It has been shown that the most important parameter that influences the phase formation is oxygen diffusion (Valant et al., 2007b). In our experiments to prepare the $AgNbO_3$ ceramics, we first calcined the mixture of Ag_2O and Nb_2O_5 at 1253 K for 6 hours in O_2 atmosphere and then sintered the pellet samples for electric measurements at 1323 K for 6 hours in O_2 atmosphere (Fu et al., 2007). Insulation of these samples is very excellent, which allows us to apply extremely high electric field to the sample (Breakdown field >220 kV/cm. For comparison, $BaTiO_3$ ceramics has a value of ~50 kV/cm.)

2.2 Electric-field induced ferroelectric phase

Previous measurements on *D-E* hysteresis loop by Kania *et al.* (Kania et al., 1984) indicate that there is small spontaneous polarization P_s in the ceramics sample of $AgNbO_3$. P_s was estimated to be ~0.04 μC/cm² for an electric field with an amplitude of E=17 kV/cm and a frequency of 60 Hz. Our results obtained at weak field have confirmed Kania's reports (Fig. 4 and Fu et al., 2007). The presence of spontaneous polarization indicates that $AgNbO_3$ must be ferroelectric at room temperature. The good insulation of our samples allows us to reveal a novel ferroelectric state at higher electric field. As shown in Fig.4, double hysteresis loop is distinguishable under the application of E~120 kV/cm, indicating the appearance of new ferroelectric phase. When E>150 kV/cm, phase transformation is nearly completed and very large polarization was observed. At an electric field of E=220 kV/cm, we obtained a value of 52 μC/cm² for the ceramics sample. Associating with such structural change, there is very large electromechanical coupling in the crystal. The induced strain was estimated to be 0.16% for the ceramics sample (Fig.5). The *D-E* loop results unambiguously indicate that the atomic displacements are ordered in a ***ferri***-electric way rather than an ***anti-ferro***electric way in the crystal at room temperature.

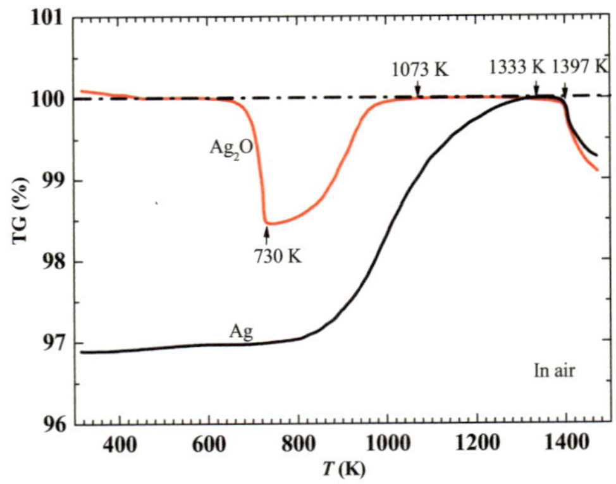

Fig. 3. Thermogravimetric curves of the $AgNbO_3$ formation in air using Ag_2O or metallic Ag powder as the silver source (Valant et al., 2007b). The curves are normalized to a weight of single-phase $AgNbO_3$. Temperatures of phase formation completed and decomposition are also indicated. For case of Ag_2O, decomposition of Ag_2O into Ag and oxygen occurs at temperature of ~730 K.

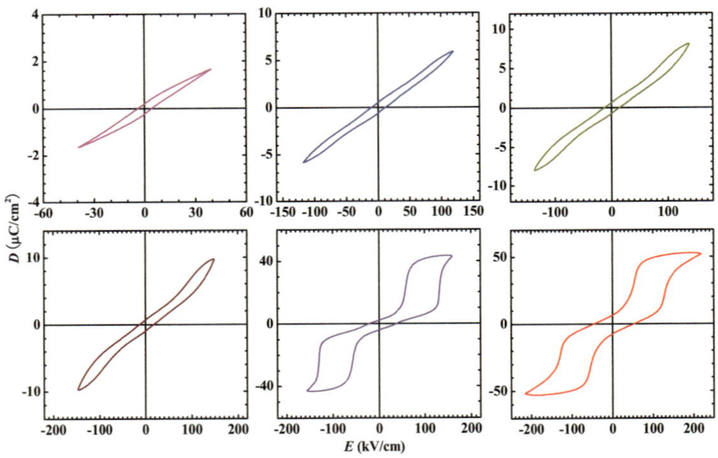

Fig. 4. D-E hysteresis loops for poly-crystalline $AgNbO_3$ at room temperature.

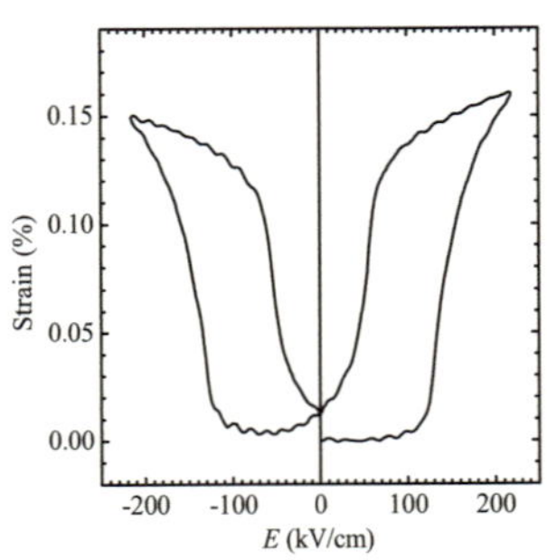

Fig. 5. Strain vs electric field for poly-crystalline $AgNbO_3$ at room temperature.

2.3 Room-temperature structure

There are many works attempting to determine the room-temperature structure of $AgNbO_3$ (Francombe & Lewis, 1958; Verwerft et al., 1989; Fabry et al., 2000; Sciau et al., 2004; Levin et al., 2009). However, none of these previous works can provide a *non*-centrosymmetric structure to reasonably explain the observed spontaneous polarization. Very recently, this longstanding issue has been addressed by R. Sano *et al.* (Sano et al., 2010). The space group of $AgNbO_3$ has been unambiguously determined to be $Pmc2_1$ (No. 26) by the convergent-beam electron diffraction (CBED) technique, which is *non*-centrosymmetric and allows the appearance of ferroelectricity in the crystal (Fig.6).

$Pmc2_1$ (T=298K)				
Site	x	y	z	U (Å^2)
Ag1 4c	0.7499(3)	0.7468(3)	0.2601(5)	0.0114(2)
Ag2 2b	1/2	0.7466(6)	0.2379(5)	0.0114(2)
Ag3 2a	0	0.7424(4)	0.2759(6)	0.0114(2)
Nb1 4c	0.6252(2)	0.7525(5)	0.7332(2)	0.00389(18)
Nb2 4c	0.1253(2)	0.24159	0.27981	0.00389(18)
O1 4c	0.7521(9)	0.7035(12)	0.7832(24)	0.0057(5)
O2 2b	1/2	0.804(3)	0.796(3)	0.0057(5)
O3 4c	0.6057(7)	0.5191(18)	0.4943(18)	0.0057(5)
O4 4c	0.6423(7)	0.0164(18)	0.539(2)	0.0057(5)
O5 2a	0	0.191(3)	0.256(3)	0.0057(5)
O6 4c	0.1339(9)	0.0410(17)	0.980(2)	0.0057(5)
O7 4c	0.1154(8)	0.4573(17)	0.5514(19)	0.0057(5)

Table 1. Structural parameters of $AgNbO_3$ at T=298K (Yashima et al., 2011). Number of formula units of $AgNbO_3$ in a unit cell: Z=8. Unit-cell parameters: a = 15.64773(3) Å, b = 5.55199(1) Å, c = 5.60908(1) Å, $\alpha = \beta = \gamma$ = 90 deg., Unit-cell volume: V = 487.2940(17) Å^3. U (Å^2)=Isotropic atomic displacement parameter.

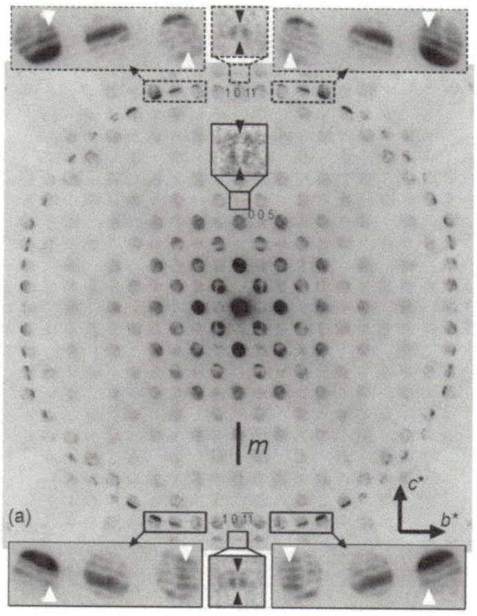

Fig. 6. Convergent-beam electron diffraction (CBED) pattern of $AgNbO_3$ taken at the [100] incidence. In contrast to a mirror symmetry perpendicular to the b*-axis, breaking of mirror symmetry perpendicular to the c*-axis is seen, indicating that spontaneous polarization is along the c-direction (Taken by R. Sano & K. Tsuda (Sano et al., 2010)).

On the basis of this space group, M. Yashima (Yashima et al., 2011) exactly determined the atom positions (Table 1) in the structure using the neutron and synchrotron powder diffraction techniques. The atomic displacements are schematically shown in Fig.7. In contrast to the reported centrosymmetric *Pbcm* (Fabry et al. 2000; Sciau et al. 2004; Levin et al. 2009), in which the Ag and Nb atoms exhibit antiparallel displacements along the *b*-axis, the $Pmc2_1$ structure shows a ***ferri***-electric ordering of Ag and Nb displacements (Yashima et al., 2011) along the *c*-axis of $Pmc2_1$ orthorhombic structure. This polar structure provides a reasonable interpretation for the observed polarization in $AgNbO_3$.

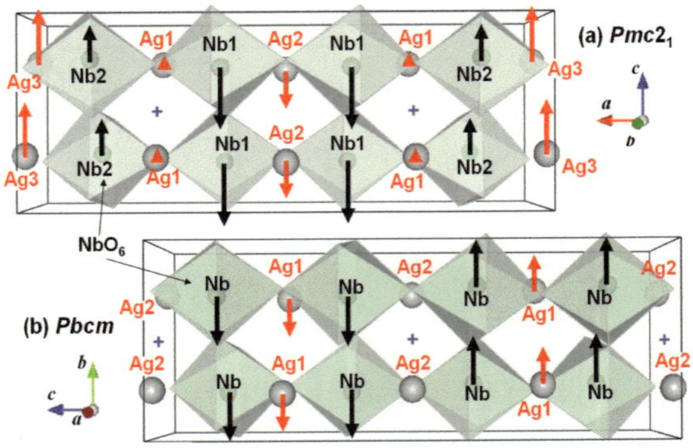

Fig. 7. (a) *Ferri*electric crystal structure of $AgNbO_3$ ($Pmc2_1$) at room temperature. The atomic displacements along the *c*-axis lead to the spontaneous polarization in the crystal. (b) For comparison, the patterns for the previously reported *Pbcm* (Sciau et al., 2004) are also given. A cross (+) stands for the center of symmetry in the *Pbcm* structure. (by M. Yashima (Yashima et al., 2011)).

2.4 Dielectric behaviours and phase transitions

Initial works on the phase transitions of $AgNbO_3$ and their influence on the dielectric behaviors were reported by Francombe and Lewis (Francombe & Lewis, 1958) in the late 1950s, which trigger latter intensive interests in this system (Łukaszewski et al., 1983; Kania, 1983, 1998; Kania et al., 1984, 1986; Pisarski & Dmytrow, 1987; Paweczyk, 1987; Hafid et al., 1992; Petzelt et al., 1999; Ratuszna et al., 2003; Sciau et al., 2004). The phase transitions of $AgNbO_3$ were associated with two mechanisms of *displacive* phase transition: tilting of oxygen octahedra and displacements of particular ions (Sciau et al., 2004). Due to these two mechanisms, a series of structural phase transitions are observed in $AgNbO_3$. The results on dielectric behaviors together with the reported phase transitions are summarized in Fig.8. Briefly speaking, the structures of the room-temperature (Yashima et al., 2011) and the high temperature phases ($T > T^{O1\text{-}O2}$=634 K) (Sciau et al., 2004) are exactly determined, in contrast, those of low-temperature (*T*<room temperature) and intermediate phases (T_C^{FE}=345 K $< T < T^{O1\text{-}O2}$) remain to be clarified. In the dielectric curve, we can see a shoulder around 40 K. It is unknown whether this anomaly is related to a phase transition or not. It should be noticed that Shigemi *et al.* predicted a ground state of *R3c* rhombohedra phase similar to that of $NaNbO_3$ for $AgNbO_3$ (Shigemi & Wada, 2008). Upon heating, there is an anomaly at T_C^{FE}=345 K, above which spontaneous polarization was reported to disappear (Fig.8 (c) and Kania et al., 1984). At the same temperature, anomaly of lattice distortion was observed (Fig.8 (c) and Levin et al., 2009). The dielectric anomaly at T_C^{FE}=345 K was attributed to be a ferroelectric phase transition. Upon further heating, there is a small peak at *T*=453 K, which

		AgNbO$_3$			
Atom		573 K	645 K	733 K	903 K
Nb	*x*	0.2460(15)	1/4	0	0
	y	0.2422(10)	1/4	0	0
	z	0.1256(6)	0	0	0
	U (Å^2)	1.12(6)	1.31(4)	1.38(4)	1.63(2)
Ag(1)	*x*	-0.2507(35)	0	0	1/2
	y	1/4	-0.001(2)	1/2	1/2
	z	0	1/4	1/2	1/2
	U (Å^2)	2.78(7)	1.98(6)	2.80(3)	3.47(4)
Ag(2)	*x*	-0.2556(26)	0		
	y	0.2428(18)	0.494(3)		
	z	1/4	1/4		
	U (Å^2)	1.73(6)	3.54(8)		
O(1)	*x*	0.3022	0.2827(7)	0	0
	y	1/4	1/4	0	0
	z	0	0	1/2	1/2
	U (Å^2)	1.31(6)	2.27(5)	3.14(5)	3.60(4)
O(2)	*x*	-0.0292(14)	0	0.2782(3)	
	y	0.0303(14)	0.2259(8)	0.2218	
	z	0.1140(6)	0.0205(8)	0.0358(4)	
	U (Å^2)	1.71(7)	1.94(4)	2.08(4)	
O(3)	*x*	0.5262(13)	0.2710(7)		
	y	0.4778(14)	0.2456(11)		
	z	0.1375(5)	1/4		
	U (Å^2)	1.17(6)	2.56(6)		
O(4)	*x*	0.2155(23)			
	y	0.2622(24)			
	z	1/4			
	U (Å^2)	1.67(7)			
cell	*a* (Å)	5.5579(5)	7.883(1)	5.5815(3)	3.9598(3)
	b (Å)	5.5917(6)	7.890(1)		
	c (Å)	15.771(2)	7.906(1)	3.9595(3)	

Table 2. Structural parameters for high temperature phases at 573 K(*Pbcm*), 645 K(*Cmcm*), 733 K (*P4/mbm*), and 903 K(*Pm3m*) (Sciau et al., 2004).

is not so visible. However, it can be easily ascertained in the differential curve or in the cooling curve. This anomaly is nearly unnoticed in the literatures (Łukaszewski et al., 1983; Kania, 1983, 1998; Kania et al., 1986; Pisarski & Dmytrow, 1987; Paweczyk, 1987; Hafid et al., 1992; Ratuszna et al., 2003). The detailed examination of the temperature dependence of the 220_o *d*-spacing (reflection was indexed with orthorhombic structure) determined by Levin *et al.* (Levin et al., 2009 and Fig.8(c)) reveals anomaly that can be associated with this dielectric peak. These facts suggest that a phase transition possibly occurs at this temperature. Around 540 K, there is a broad and frequency-dependent peak of dielectric constant, which is also associated with an anomaly of 220_o *d*-spacing. However, current structural investigations using x-ray and neutron diffraction do not find any symmetric changes associated with the dielectric anomalies at 456 K and 540 K, and the structure within this intermediate temperature range was assumed to be orthorhombic *Pbcm* (Sciau et al., 2004). At $T=T_C^{AFE}=631$ K, there is a sharp jump of dielectric constant due to an antiferroelectric phase transition (Pisarski et al., 1987; Sciau et al., 2004). The atomic displacement patterns in the antiferroelectric phase (*Pbcm*) are shown in Fig.7 (b) using the structural parameters determined by Ph Sciau *et al.* (Sciau et al., 2004). For $T>T_C^{AFE}$, there are still three phase transitions that are essentially derived by the tilting of oxygen octahedral and have only slight influence on the dielectric constant. For conveniences, the tilting of octahedral (Sciau et al., 2004) described in Glazer's notation is given in Fig.8 and the structure parameters (Sciau et al., 2004) are relisted in Table 2.

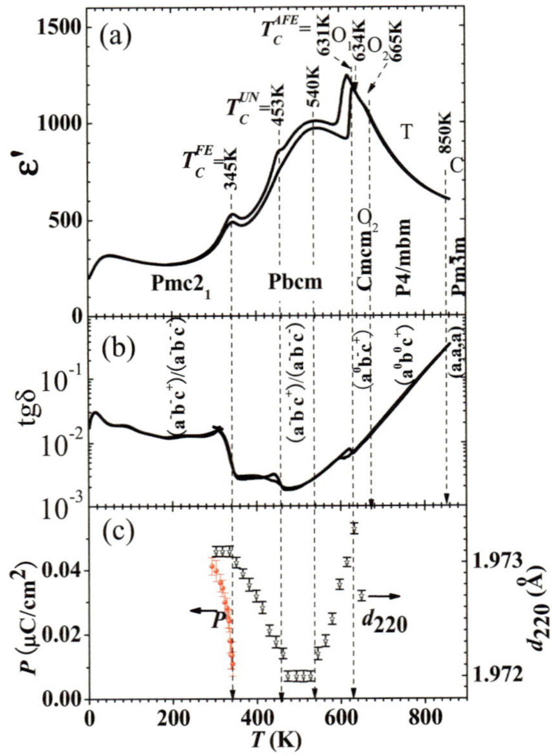

Fig. 8. Temperature dependence of (a) dielectric constant, (b) dielectric loss, and (c) polarization (Kania et al., 1984) and 220_O *d*-spacing of the lattice (Levin et al., 2009).

3. $(Ag_{1-x}Li_x)NbO_3$ solid solution

Li can be incorporated into the Ag-site of $AgNbO_3$. However, due to the large difference of ionic radius of Li^+(0.92Å), and Ag^+(1.28Å)(Shannon, 1976), the solid solution is very limited. Nalbandyan *et al.*(Nalbandyan et al., 1980), systematically studied the stable and metastable phase equilibrias and showed that solid solution limit is narrow (x~0.02) for the stable phase, but is relatively wide (x~0.12) for the metastable phase(Sakabe et al., 2001; Takeda et al., 2003;

Fu et al., 2008, 2011a). With a small substitution of Li for Ag ($x>x_c$=0.05~0.06), a ferroelectric rhombohedra phase is evolved in the solid solution (Nalbandyan et al., 1980; Sakabe et al., 2001; Takeda et al., 2003;Fu et al., 2008, 2011a). In this solid solution, the strong off-center of small Li (Bilc & Singh, 2006) plays an important role in triggering the ferroelectric state with large spontaneous polarization (Fu et al., 2008, 2011a).

3.1 Synthesis

Single crystals of $(Ag_{1-x}Li_x)NbO_3$ can be obtained by the melt growth process (Fu et al., 2008). Stoichiometric compositions of Ag_2O, Li_2CO_3, and Nb_2O_5 were mixed and calcined at 1253 K for 6 h in an oxygen atmosphere. The calcined powder was milled, put into an alumina container, and melted at 1423 K for 4 h in an oxygen atmosphere. The melt was cooled to 1323 K to form the crystal at a rate of 4 K/h, followed by furnace cooling down to room temperature. Using this process, single crystals with size of 1–3 cm can be obtained for the $(100)_p$ (Hereafter, subscript p indicates pseudocubic structure) growth face. Due to the volatility of lithium at high temperature, the exact chemical composition of the crystal is generally deviated from the starting composition and is required to be determined with methods like inductively coupled plasma spectrometry.
Ceramics samples can be prepared by a solid state reaction approach. Mixtures of Ag_2O, Nb_2O_5, and Li_2CO_3 were calcined at 1253 K for 6 h in O_2 atmosphere, followed by removal of the powder from the furnace to allow a rapid cooling to prevent phase separation. The calcined powder was milled again and pressed to form pellets that were sintered at 1323 K for 6 h in O_2 atmosphere, followed by a rapid cooling.

3.2 Structure

The structural refinements using the powder X-ray diffraction data suggest that $(Ag_{1-x}Li_x)NbO_3$ solid solution with $x>x_c$ has the space group of $R3c$ (Fu et al., 2011a). Table 3 lists the structural parameters of this model for composition x=0.1. Figure 9 shows a schematic drawing for this structure. In this rhombohedral $R3c$ phase, the spontaneous polarization is essentially due to the atomic displacements of the Ag/Li, Nb, and O atoms along the pseudocubic [111] direction.

$Ag_{0.9}Li_{0.1}NbO_3$ (R3c, No.161, T=room temperature)					
a(Å)	5.520		α	90	
b(Å)	5.520		β	90	
c(Å)	13.79		γ	120	
V(Å^3)	364.0				
Atom	Site	x	y	z	U (Å^2)
Ag/Li	6a	0	0	0.2545(8)	0.5
Nb	6a	0	0	0.0097(8)	0.5
O	18b	0.5533	1	0.2599(9)	0.5

Table 3. Structural parameters for rhombohedra structure of (Ag,Li)NbO_3 solid solution.

3.3 Ferroelectric and piezoelectric properties

Evolution of the polarization state in $Ag_{1-x}Li_xNbO_3$ solid solutions is shown in Fig.10. Basically, when $x<x_c$, the solid solutions have the *ferri*electric state of pure $AgNbO_3$ with a small spontaneous polarization at zero electric field. In contrast, when $x>x_c$, a normal

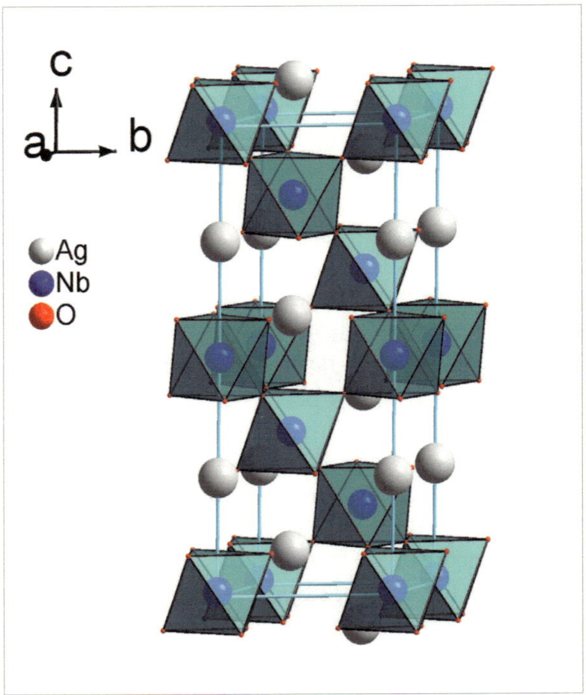

Fig. 9. Schematic structure of $Ag_{1-x}Li_xNbO_3$ (x = 0.1) with symmetry $R3c$ (No.161).

ferroelectric state with large value of remanent polarization (P_r) is observed. All ceramics samples show P_r value comparable to P_S of $BaTiO_3$ single crystal (26 μC/cm²) (Shiozaki et al., 2001). Moreover, the polarization in $Ag_{1-x}Li_xNbO_3$ solid solution is very stable after switching. Large P_r value and ideal bistable polarization state of $Ag_{1-x}Li_xNbO_3$ ceramics may be interesting for non-volatile ferroelectric memory applications. Measurements on single crystal samples (Fig.11) indicate that saturation polarization along the $<111>_p$ rhombohedra direction ($P_s^{<111>}$ ~40 μC/cm²) is greatly larger than that along the $<001>_p$ tetragonal direction ($P_s^{<001>}$ ~24 μC/cm²) and the ratio between them is $\sqrt{3}$ (Fu et al., 2008), which is in good agreement with results of structural refinements. This suggests that the polar axis is the $<111>_p$ direction of pseudo-cubic structure.

The strain-E loops indicate that there are good electromechanical coupling effects in $Ag_{1-x}Li_xNbO_3$ crystals. Although the spontaneous polarization is along the $<111>_p$ axis, the $<001>_p$-cut crystal shows larger strain and less hysteresis than the $<111>_p$-cut one (Fig.11 (b) and (c)). These phenomena are very similar to those reported for the relaxor-ferroelectric crystals (Wada et al., 1998). The most significant result exhibited from $Ag_{1-x}Li_xNbO_3$ single crystal is its excellent g_{33} value that determines the voltage output of the piezoelectric device under the application of an external stress (Fu et al., 2008). The g_{33} value together the d_{33} value and dielectric constants for the $<001>_p$-cut single crystals are shown in Fig. 12. The high g_{33} value is a direct result from the large d_{33} constant and the low dielectric constant of the single crystal.

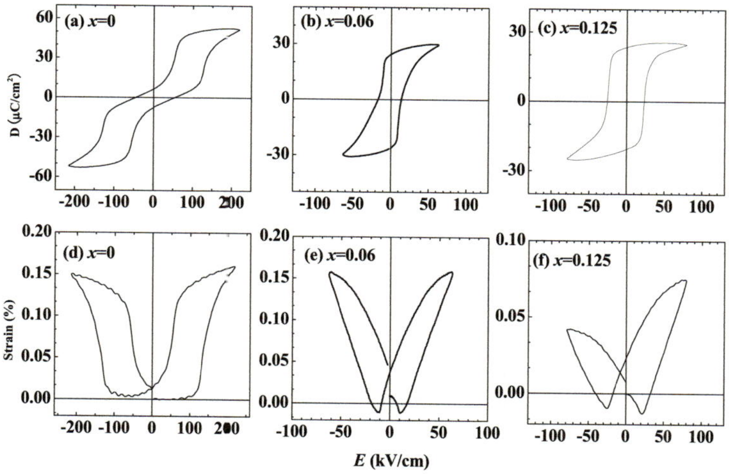

Fig. 10. Typical *D-E* and strain-*E* loops for the $Ag_{1-x}Li_xNbO_3$ ceramics samples.

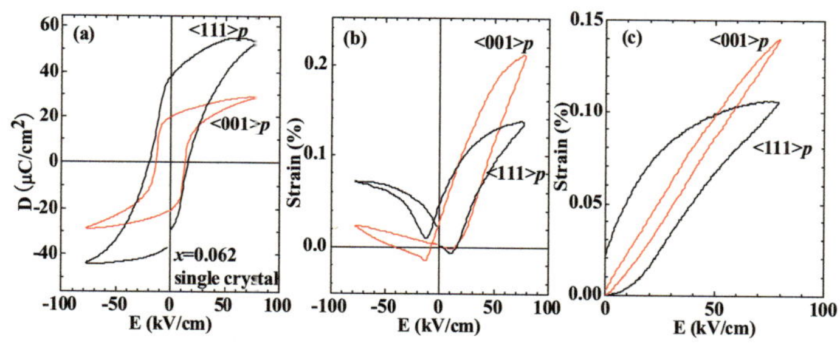

Fig. 11. (a) *D-E* loops, (b) strain vs *E* for bipolar electric field, and (c) strain vs *E* for unipolar field for $Ag_{1-x}Li_xNbO_3$ single crystal with x=0.062.

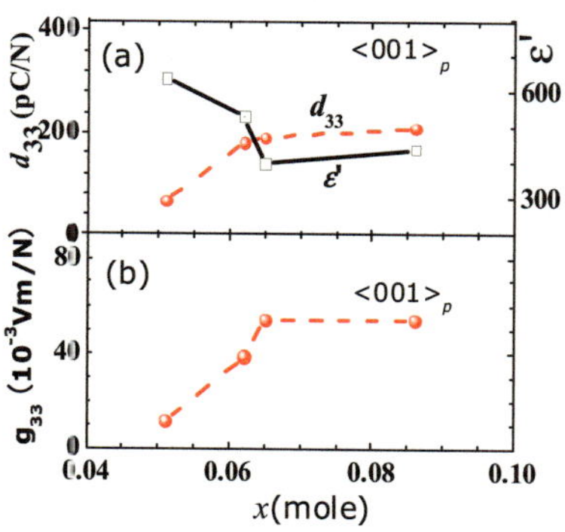

Fig. 12. Composition dependence of dielectric constant ε, d_{33} and g_{33} for the <001>$_p$-cut $Ag_{1-x}Li_xNbO_3$ single crystals.

3.4 Dielectric behaviours and proposed phase diagram

Figure 13 shows the dielectric behaviours of the ferroelectric $Ag_{1-x}Li_xNbO_3$ solid solutions. For comparison, the temperature dependence of dielectric constant of $AgNbO_3$ is also shown. It can be seen that solid solution with $x > x_c$ shows different temperature evolutions of the dielectric constant as compared with $AgNbO_3$. In sharp contrast to the complex successive phase transitions in $AgNbO_3$, ferroelectric $Ag_{1-x}Li_xNbO_3$ solid solutions ($x>x_c$) show only two phase transitions in the temperature range of 0-820 K. Polarization measurements suggest that the high temperature phase ($T > T_C^{FE}$) is nonpolar, thus it seems

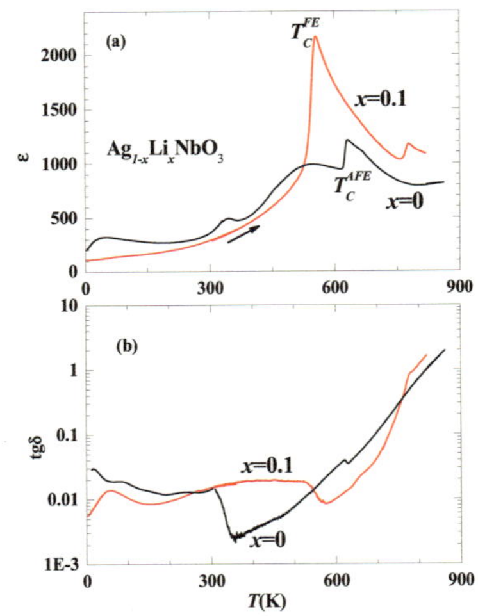

Fig. 13. Typical dielectric behaviours of $Ag_{1-x}Li_xNbO_3$ ceramics in comparison with that of pure $AgNbO_3$.

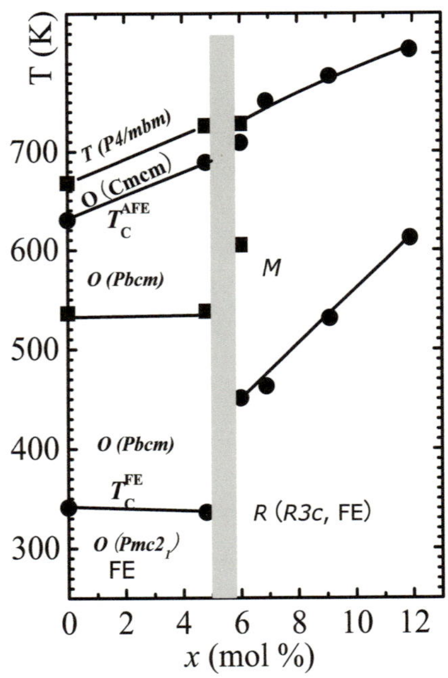

Fig. 14. Phase diagram proposed for Ag_{1-x} Li_xNbO_3 solid solution. The gray area indicates the phase boundary (Fu et al., 2011a).

that the higher temperature phase transition is not related to a ferroelectric phase transition. On the basis of the dielectric measurements, the phase diagram of $Ag_{1-x}Li_xNbO_3$ solution is summarized in Fig.14, where *T*, *O*, *R* and *M* represent tetragonal, orthorhombic, rhombohedra, and monoclinic symmetries, respectively. At room temperature, structure transformation from *O* to *R* phase at x_c dramatically changes the polar nature of $Ag_{1-x}Li_xNbO_3$. It is a ***ferri***electric with small spontaneous polarization in the *O* phase, but becomes a ferroelectric with large polarization in the *R* phase.

4. $(Ag_{1-x}Na_x)NbO_3$ solid solution

The ionic radius of Na^+ (1.18 Å) is comparable to that of Ag^+ (1.28 Å) (Shannon, 1976), allowing to prepare the $Ag_{1-x}Na_xNbO_3$ solid solution within the whole range of composition (x=0-1) (Kania & Kwapulinski, 1999). Kania *et al.* previously performed investigation on the dielectric behaviors and the differential thermal analysis for the $Ag_{1-x}Na_xNbO_3$ solid solutions, and stated that the solid solution evolves from disordered antiferroelectric $AgNbO_3$ to normal antiferroelectric $NaNbO_3$. As described in section §2.3, we now know that $AgNbO_3$ is not *antiferroelectric* but rather is *ferrielectric* at room temperature (Yashima et al., 2011). Moreover, recent reexamination on the polarization behaviors of stoichiometric and non-stoichiometric $NaNbO_3$ polycrystallines indicates that the reported clamping hysteresis loop of $NaNbO_3$ can be interpreted by pining effects while stoichiometric $NaNbO_3$ is intrinsically *ferroelectric* (Arioka et al., 2010). Therefore, reexamination on this solid solution is necessary. Evolution of the polarization with composition clearly indicates that the solid solution evolves from *ferrielectric* $AgNbO_3$ to *ferroelectric* $NaNbO_3$ (Fu et al., 2011b).

4.1 Synthesis

$Ag_{1-x}Na_xNbO_3$ solid solution was prepared by a solid state reaction approach. Mixtures of Ag_2O(99%), Nb_2O_5 (99.99%), and Na_2CO_3 (99.99%) were calcined at 1173 K for 4 h in O_2 atmosphere. The calcined powder was ground, pressed into pellet with a diameter of 10 mm at thickness of 2 mm, and sintered with the conditions listed in Table 4.

Composition	Temperature	Time	atmosphere
x=0,0.1,0.2	1273K	5h	O_2
x=0.4, 0.5, 0.6	1323K	5h	O_2
x= 0.8,0.9,1	1373K	5h	O_2

Table 4. Sintering conditions for $Ag_{1-x}Na_xNbO_3$.

4.2 Polarization

Figure 15 shows the change in polarization with composition in the $Ag_{1-x}Na_xNbO_3$ solid solutions. For a wide range of composition x<0.8, the solid solution possesses the characteristic polarization behaviours of pure $AgNbO_3$: small spontaneous polarization at E=0 but large polarization at E> a critical field. This fact suggests that the solid solution is *ferri*electric within this composition range. This is also supported by the temperature dependence of dielectric constant (Fig.16). On the other hand, for the Na-rich composition, particularly, x>0.8, we observed large remanent polarization with value close to the saturation polarization at high field. This result indicates that a normal ferroelectric phase is stable in these compositions. Therefore, polarization measurements show that the $Ag_{1-x}Na_xNbO_3$ solid solution evolves from *ferrielectric* $AgNbO_3$ to *ferroelectric* $NaNbO_3$ (Fu et al., 2011b).

4.3 Dielectric properties

The dielectric properties of the $Ag_{1-x}Na_xNbO_3$ solid solutions are summarized in Fig.16. The change in dielectric behaviours with composition is very similar to that observed in polarization measurements (Fig.15). For $x \leq 0.8$, the solid solution shows successive phase transitions similar to pure $AgNbO_3$. In contrast, it has the characteristic phase transition of pure $NaNbO_3$ for $x>0.8$. The composition dependence of transition temperature derived from the dielectric measurements is shown in Fig.17. Two noticed features may be seen: (a) the thermal hysteresis is extremely large for antiferroelectric phase transition and reaches a value greater than 100 K at $x \sim 0.5$. Such large thermal hysteresis is rarely observed in normal polar phase transition. (b) It seems that there is a phase boundary at x=xc~0.8 (Fu et al., 2011b), around which structural transformation between *ferri-* and *ferro*-electric phases occurs.

5. $(Ag_{1-x}K_x)NbO_3$ solid solution

$(Ag_{1-x}K_x)NbO_3$ solid solutions are available only for very limited composition (Weirauch & Tennery, 1967; Łukaszewski, 1983; Kania, 2001). Weirauch *et al.* reported that solid solution of $AgNbO_3$ in $KNbO_3$ was limited to slightly less than 6 mole % and solid solution of $KNbO_3$ in $AgNbO_3$ was limited to less than 0.5 mole % (Weirauch & Tennery, 1967). However, our process indicates that $KNbO_3$ and $AgNbO_3$ can be alloyed with each other within 20 mole % (Fu et al., 2009a). Apparently, the reported solid solution limit is dependent on the process. In our samples, we found that a ferroelectric phase with large spontaneous polarization can be induced by substitution of K for Ag for $x>x_{c1}=0.07$. This ferroelectric phase shows nearly composition-independent ferroelectric phase transition. On the other hand, the K-rich solid solution ($x>0.8$) possesses the ferroelectric phase transition sequence of pure $KNbO_3$ and the transition temperature is dependent on the composition.

5.1 Synthesis

$(Ag_{1-x}K_x)NbO_3$ solid solutions were prepared by a solid state reaction method. Mixture of Ag_2O, Nb_2O_5, and K_2CO_3 were calcined at 1173 K for 6 h in O_2 atmosphere with a slow heating rate of 1 K/min. The calcined powder was milled again, pressed in a 6-mm steel die with a pressure of 10 MPa to form the pellets, which were then preheated at 773 K for 2 h, followed by a sintering at temperature of 1253–1323 K (1323 K for x=0–0.1 and 1.00, 1273 K for x=0.15 and 0.90, and 1253 K for x=0.17 and 0.80, respectively) for 3 h in O_2 atmosphere. The atmosphere and heating rate are found to have significant influences on the phase stability of the solid solution.

5.2 Structural change with composition

Figure 18 shows the change in lattice parameters with composition in the $Ag_{1-x}K_xNbO_3$ solid solutions. When the amount of substitution is small, the solid solution possesses the orthorhombic structure of pure $AgNbO_3$. In the phase boundary $x_{c1}=0.07$, structural transformation occurs and the phase changes into a new orthorhombic structure. In this new ferroelectric phase, the lattice constants show linear increase with composition. Interestingly, the orthorhombic distortion angle β is nearly independent with the composition. In the K-rich region ($x>x_{c3}=0.8$), the solid solution has the orthorhombic structure of pure $KNbO_3$ at room-temperature, which is also ferroelectric. In contrast to nearly unchanged orthorhombic angle β in the Ag-rich orthorhombic phase, β shows monotonous decreases with x in the K-rich phase.

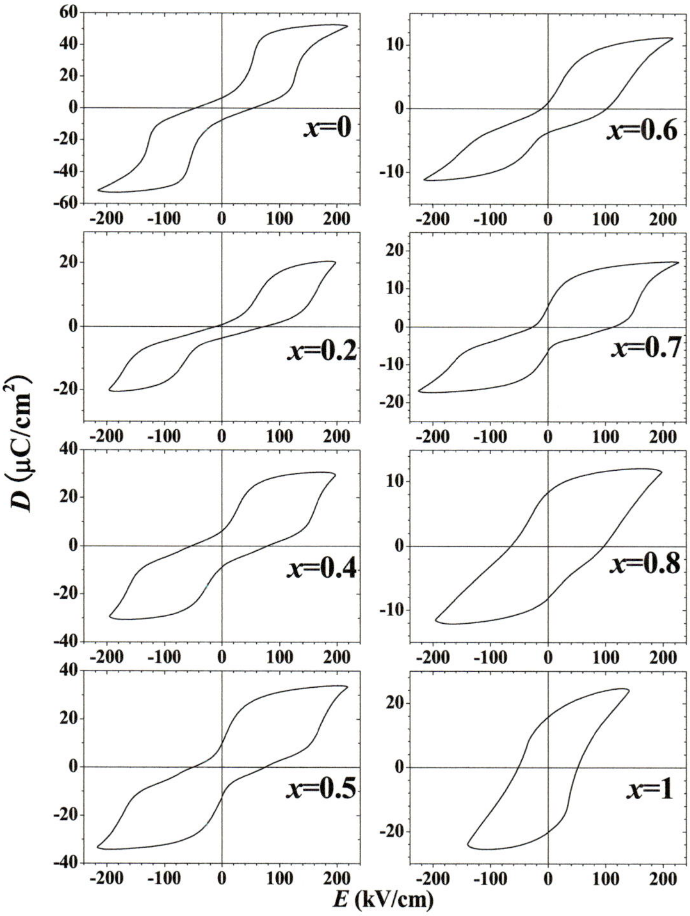

Fig. 15. Composition dependence of *D-E* loops obtained at room temperature for $Ag_{1-x}Na_xNbO_3$ solid solutions

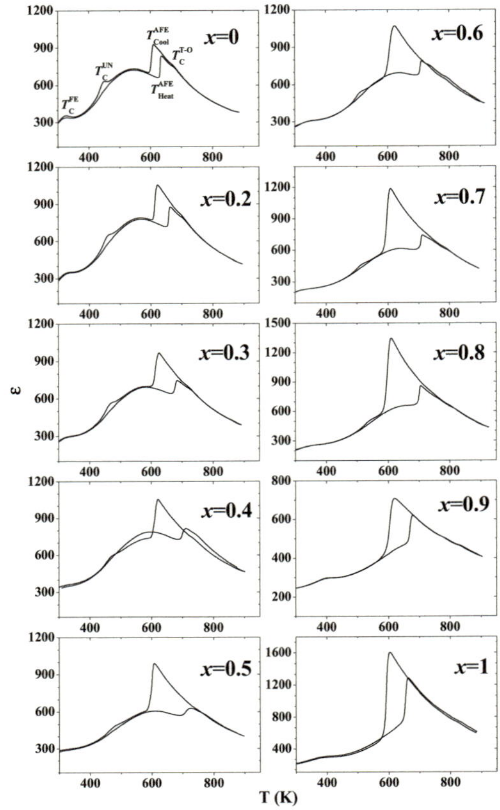

Fig. 16. Temperature dependence of dielectric constants for $Ag_{1-x}Na_xNbO_3$ solid solutions.

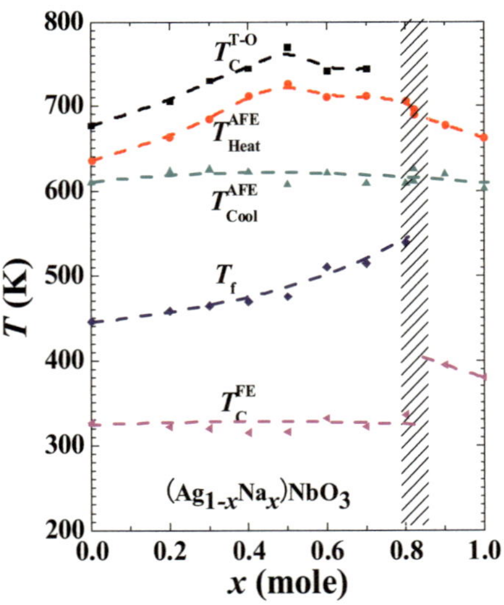

Fig. 17. Composition dependence of phase transition temperatures detected from dielectric measurements for $Ag_{1-x}Na_xNbO_3$ solid solutions.

5. 3 Ferroelectric and piezoelectric properties

Figure 19 shows typical results of polarization and strain behaviors observed at room temperature for the $Ag_{1-x}K_xNbO_3$ solid solutions. Similar to pure $AgNbO_3$, merely a small spontaneous polarization was observed in samples with $x<x_{c1}$. However, when $x>x_{c1}$, a normal *D-E* loop with large value of remanent polarization P_r was observed. A value of P_r=20.5 μC/cm² was obtained for a ceramics sample with x =0.10, which is greatly larger than that observed for $BaTiO_3$ ceramics (Fu et al., 2010). These results show that Ag-rich orthorhombic phase ($x_{c1}<x<x_{c2}$) is really under a ferroelectric state with large polarization. For K-rich region $x>x_{c3}$, normal *D-E* loops were also obtained. Associating with the evolution into the ferroelectric phase, butterfly strain curve were observed. The piezoelectric constants determined from the piezo-d_{33} meter generally have values of 46–64 pC/N for these ferroelectric samples (Fu et al., 2009a).

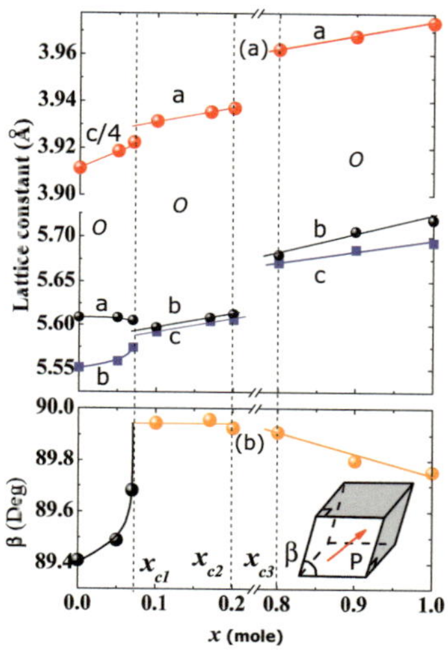

Fig. 18. Lattice parameters change with composition in $Ag_{1-x}K_xNbO_3$ solid solution. β is the orthorhombic distortion angle. The inset shows the orthorhombic distortion due to the ferroelectric displacements along the <110>$_p$ direction of the pseudocubic structure.

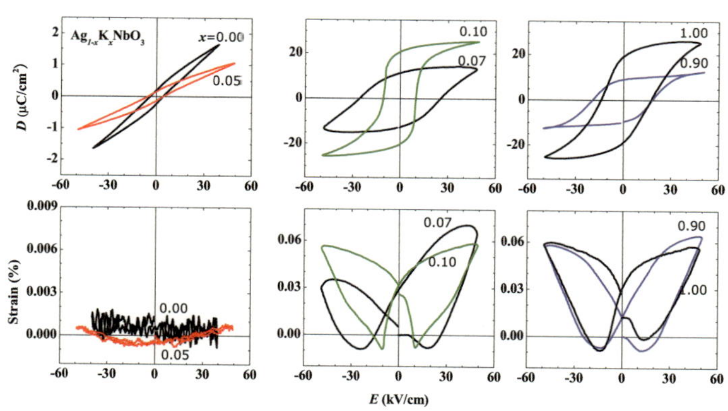

Fig. 19. Changes of polarization and strain under the application of electric field with composition at room temperature.

5.4 Dielectric behaviours and proposed phase diagram

Associating with the change in structure, dielectric behaviours of the $Ag_{1-x}K_xNbO_3$ solid solution also change with composition. As shown in Fig.20, the temperature dependence of dielectric constant can be sorted by three types: (1) $AgNbO_3$-type for $x<x_{c1}=0.07$, (2) $KNbO_3$-type for K-rich region $x>x_{c3}=0.8$, and (3) a new type for the intermediate composition $x_{c1}<x<x_{c2}$. In this intermediate composition, dielectric measurements indicate that there are two phase transitions within the temperature range of 0-750 K. One transition locates at T_{c2}~420 K with thermal hysteresis and shows small change in dielectric constant, which seems to be due to a ferro-to-ferro-electric phase transition. Another phase transition occurs at T_{c1}~525 K. Around this transition, the dielectric constant changes sharply and follows exactly the Curie-Weiss law. The Curie-Weiss constant was estimated to be 1.47 $*10^5$ K for x=0.1 sample, which is a typical value for the displacive ferroelectric transition, suggesting that this is a displacive type ferroelectric transition. A phase diagram is proposed in Fig.21, in which PE, FE, and AFE represent the paraelectric, ferroelectric, and antiferroelectric phases, respectively. When carefully comparing the phase transition temperature with the orthorhombic angle β(Fig.18) due to the ferroelectric distortion (For $x_{c1}<x<x_{c2}$ and $x>x_{c3}$. In contrast, for $x< x_{c1,}$ β change is basically due to the oxygen-octahedral tilting.) , one might find that there is a correlation between β and the temperature of the ferroelectric phase transition.

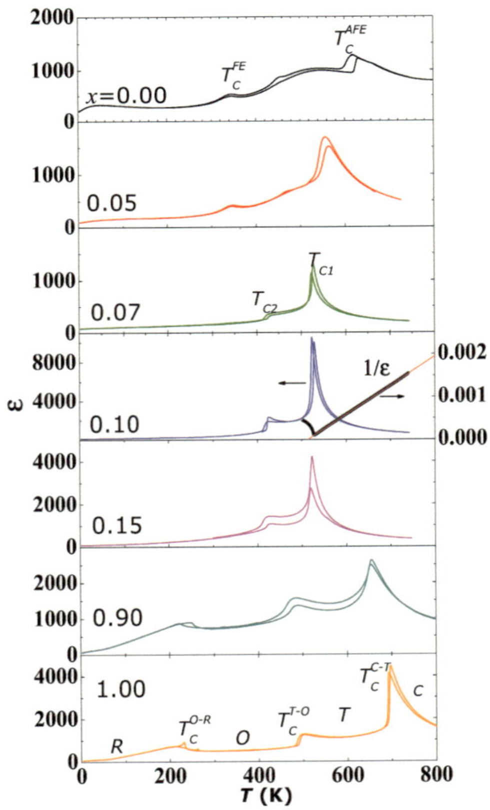

Fig. 20. Temperature dependence of dielectric constant for $Ag_{1-x}K_xNbO_3$ solid solution.

6. $AgTaO_3$

$AgTaO_3$ is another oxide of the two discovered silver perovskites (Francombe & Lewis, 1958). It is generally accepted that $AgTaO_3$ undergoes a series of phase transitions from rhombohedral phase ($T \leq 685$ K) to monoclinic phase (650 K$\leq T \leq$703 K) and then to tetragonal phase (685 K$<T\leq$780 K), and finally to cubic phase at $T_{T\text{-}C}$= 780 K upon heating (Francombe & Lewis, 1958; Kania, 1983; Paweczyk, 1987; Kugel et al.,1987; Hafid et al., 1992). However, due to the coexistence regions between rhombohedral and monoclinic, and monoclinic and tetragonal, the actual transition temperatures from rhombohedral to monoclinic $T_{R\text{-}M}$ as well as that from monoclinic to tetragonal $T_{M\text{-}T}$ still remain uncertain. Furthermore, the ground state and the origins that trigger these phase transitions still remain to be addressed (Wołcyrz & Łukaszewski,1986; Kugel et al.,1987;Soon et al., 2010).

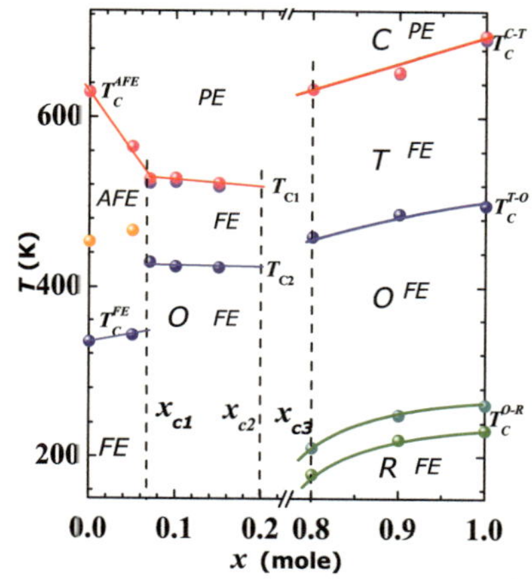

Fig. 21. Phase diagram proposed for the $Ag_{1-x}K_xNbO_3$ solid solution. Phase boundaries are indicated by the dashed lines. *C, T, O,* and *R* indicate the cubic, tetragonal, orthorhombic, and rhombohedral symmetries, respectively. PE, FE and AFE represent paraelectric, ferroelectric, and antiferroelectric phases, respectively.

6.1 Synthesis

Although single crystal of $AgTaO_3$ is available through a flux method (Łukaszewski et al., 1980; Kania, 1989), it is extremely difficult to prepare its dense ceramics sample (Francombe & Lewis, 1958;Kania,1983) for electrical measurements. Since decomposition of $AgTaO_3$ occurs at 1443±3 K in atmosphere (Valant et al., 2007b), sintering cannot be performed at higher temperatures to obtain dense ceramics. However, this long-standing synthesis difficulty now can be solved by a processing route involving the conventional solid-state reaction and sintering in environment with a high oxygen pressure at ~13 atm (Soon et al., 2010). In this synthesizing route, Ag_2O and Ta_2O_5, first underwent a grind mixing and was calcined at 1273 K for 6 hours. The calcined powder was then pressed into a pellet in 6 mm in diameter. Sintering was carried out by placing the powder compact into a sealed zirconia tube that was connected to a pressure control valve (Fig.22). Prior to the sintering, oxygen gas at ~6.25 atm was filled into the sealed zirconia tube after the evacuation. Upon heating, the pressure of sealed oxygen gas increased and reached ~13 atm when the powder compact was sintered at 1573K for 2 hours. This eventually led to formation of dense polycrystalline $AgTaO_3$.

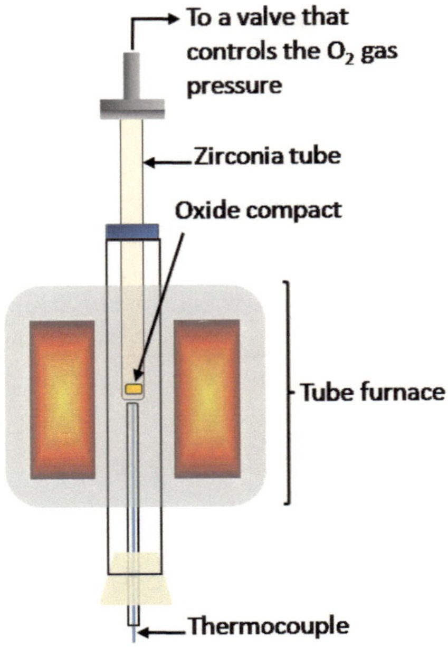

Fig. 22. Schematic diagram of the self-customized furnace employed for the sintering at high oxygen pressure.

6.2 Phase formation and dielectric behaviors

X-ray diffraction analyses (Fig.23) suggest the $AgTaO_3$ is rhombohedral with $R\bar{3}c$ symmetry at room temperature (Wołcyrz & Łukaszewski, 1986). Diffraction patterns obtained at 68.4 K remain unchanged, indicating that such nonpolar phase persists down to low-temperatures. This is also supported by the measurements on the dielectric constants (Fig.24) and heat capacity (Fig.25), in which no anomaly was probed in the low temperatures.

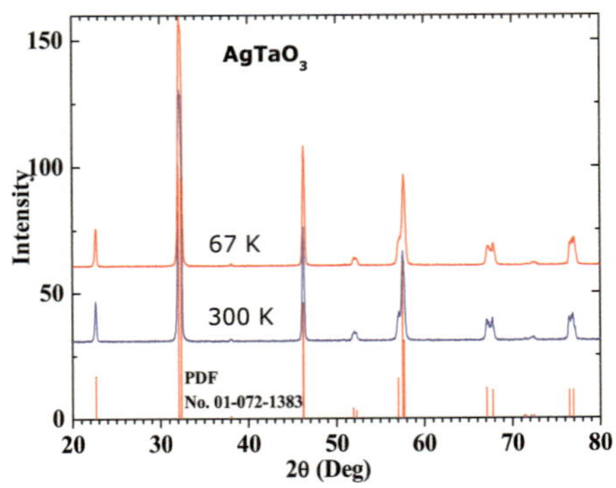

Fig. 23. XRD traces for $AgTaO_3$ obtained at 300 K and 68.4 K together with the standard pattern given by the powder diffraction file (PDF) No. 01-072-1383.

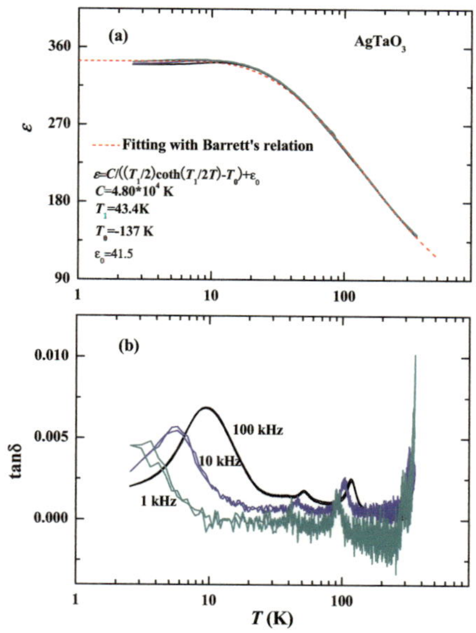

Fig. 24. Temperature dependence of (a) ε together with the fitting to the Barrett's relation indicated by the dashed line and (b) tanδ for $AgTaO_3$ ceramics samples.

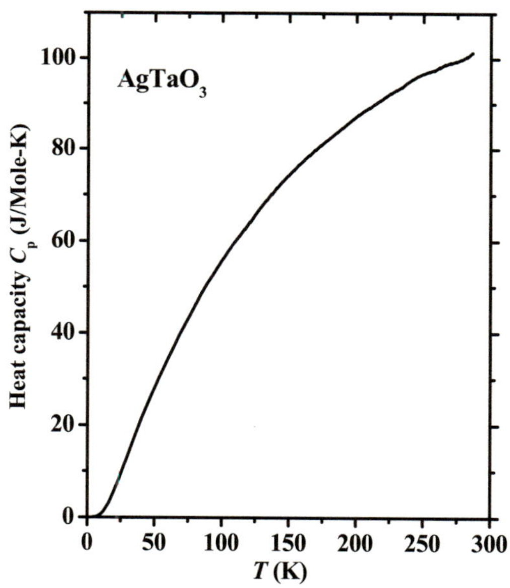

Fig. 25. Heat capacity for $AgTaO_3$.

Although several frequency-dependent peaks are seen in the dielectric loss (Fig.24(b)), it can be reasonably attributed to polarization relaxations due to defects (Soon et al., 2010). Interestingly, within the low-temperature region, the dielectric behavior follows the Barrett's relation (Barrett, 1952) that is characteristic for the quantum paraelectric system (Abel, 1971; Höchli & Boatner,1979; Itoh et al., 1999), suggesting that $AgTaO_3$ may be a quantum paraelectric. On the other hand, two step-like dielectric anomalies corresponding to the phase transitions from monoclinic to tetragonal and tetragonal to cubic were observed at 694 and 780 K, respectively, upon heating the samples (Fig.26). This observation is in agreement with the previous reports (Kania,1983;Kugel et al., 1987). Furthermore, the temperature dependence of $1/\varepsilon$ for $AgTaO_3$ shows non-linear behavior, which is similar to that of $KTaO_3$, obeying the modified form of the Curie-Weiss law $\varepsilon=\varepsilon_L+C/(T-T_0)$ (Rupprecht & Bell, 1964).

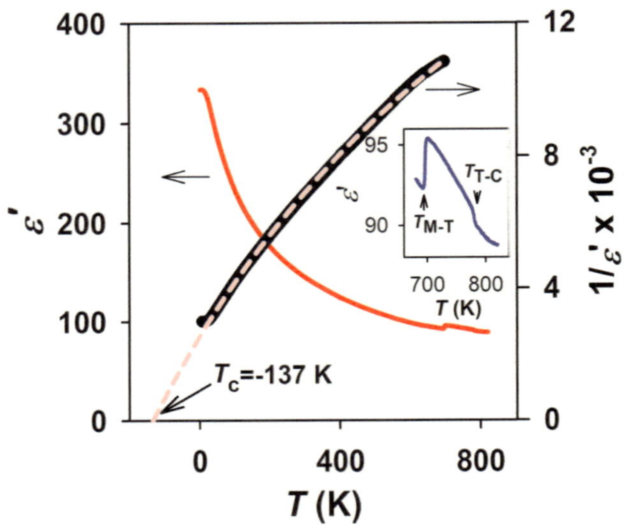

Fig. 26. Temperature dependences of ε' and $1/\varepsilon'$ fitted to the modified Curie-Weiss law $\varepsilon=\varepsilon_L+C/(T-T_0)$ (dashed line, ε_L= 38.2, C=4.7*10^4 K, T_0= -137 K) for $AgTaO_3$ measured at 1 MHz upon heating. The inset further shows the step-like phase transitions from monoclinic to tetragonal and then to cubic at $T_{M\text{-}T}$= 694 K and $T_{T\text{-}C}$ =780 K, respectively.

7. $(Ag_{1-x}Li_x)TaO_3$ solid solution

Similar to the case of $AgNbO_3$, ~12 mole% of Li can be incorporated into the Ag-site of $AgTaO_3$ to form $(Ag_{1-x}Li_x)TaO_3$ (Soon et al., 2009). Although the transition temperature is lower than room temperature, ferroelectricity can be induced in this solid solution due to the strong off-centering nature of the small Li ions.

7.1 Synthesis

$(Ag_{1-x}Li_x)TaO_3$ was prepared by the conventional solid-state reaction with Ag_2O, Ta_2O_5 and Li_2CO_3, where the powder mixture was calcined at 1273 K for 6 hours. The calcined powder was then pressed into a pellet with 6 mm in diameter. Sintering was carried out by the same high-pressure process used for pure $AgTaO_3$, which eventually led to formation of dense ceramics samples of $(Ag_{1-x}Li_x)TaO_3$

7.2 Dielectric behaviours and confirmation of ferroelectric phase

Figures 27 & 28 plot the temperature dependence of dielectric constant ε and loss tanδ for $(Ag_{1-x}Li_x)TaO_3$ with $x\leq$ 0.12 obtained at frequencies ranging from 1 kHz to 1 MHz, respectively. It can be seen that a dielectric peak was gradually induced by Li^+ substitution in $AgTaO_3$. In contrast to the single peak of the dielectric constant, there are two to four peaks of the dielectric loss within the same temperature window. Since the additional loss peaks do not associate with a remarkable change in the dielectric constant, it is very likely due to the defect effects dependent with sample processing. It can be seen that a well-defined peaks has occurred in the ~MHz frequency regions for the substitution at extremely low level, for example, at x=0.02. This indicates the existence of local polarization in the doped crystal (Vugmeister & Glinchuk, 1990; Samara, 2003). This fact again suggests that $AgTaO_3$ is actually under a critical state of the quantum paraelectric. Any slight modification will lead to the appearance of observable polarization in the system. For small substitution, the location of the dielectric anomaly depends on the observed frequency. Figure 29 gives an evaluation on this frequency dependence. In the figure, the temperature-axis is scaled with the peak position of 1 MHz, making it easy to see the change with composition. For x=0.008, the frequency dispersion is very strong, peak position of the dielectric constant shifts about 50% with respect to that of 1 MHz for f= 1 kHz. However, for x=0.035, such change is less than 2%, meaning that ferroelectric phase transition temperature T_c is well defined in the sample. Thus, we can infer that a macroscopic ferroelectric phase is evolved below T_c in this composition. This was also confirmed by the results shown in Fig.30, in which ferroelectric loop was obtained for $T<T_c$ for a sample with x=0.12 that has the same dielectric behaviors to that of x=0.035.

On the basis of the above results, a phase diagram is proposed in Fig.31, in which PE and FE represents the paraelectric and ferroelectric phases, respectively. As mentioned above, since the peak of the dielectric constant is strongly dependent with frequency for $0<x<0.035$, the gray zone in the phase diagram may be attributed to dipole-glass phase(Vugmeister & Glinchuk, 1990;Pirc & Blinc, 1999;Samara, 2003) or a phase with nanosized ferroelectric domains (Fisch, 2003; Fu et al., 2009b), which remains to be addressed by further investigations.

8. Concluding remarks

Our recent works reveal that silver perovskites are of great interests from either the view-point of fundamental research or that of application research in the fields of ferroelectric or piezoelectric. Promising ferroelectric and piezoelectric properties have been demonstrated in some compounds such as $(Ag,Li)NbO_3$ and $(Ag,K)NbO_3$ alloys, but further works are required to improve the material performance, to understand the basic physics of the ferroelectricity/piezoelectricity of the materials, and to seek novel promising compounds among the discovered solutions or alloys with other ferroelectric systems. Moreover, integration techniques of thin films are also a direction for the future works when considering the practical applications.

9. Acknowledgment

Part of this work was supported by the Collaborative Research Project of Materials and Structures Laboratory of Tokyo Institute of Technology, and Grant-in-Aid for Scientific Research, MEXT, Japan.

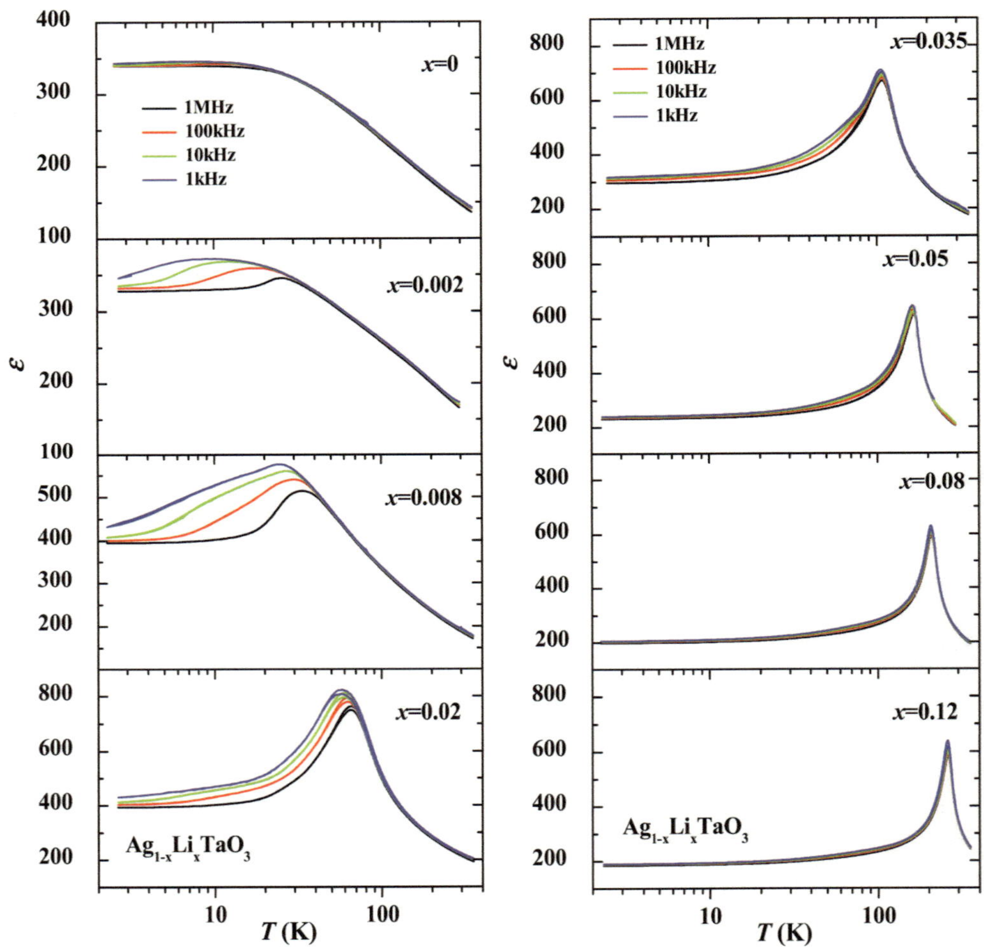

Fig. 27. Dielectric constant $\varepsilon'(T)$ for $Ag_{1-x}Li_xTaO_3$ with $0 \le x \le 0.12$.

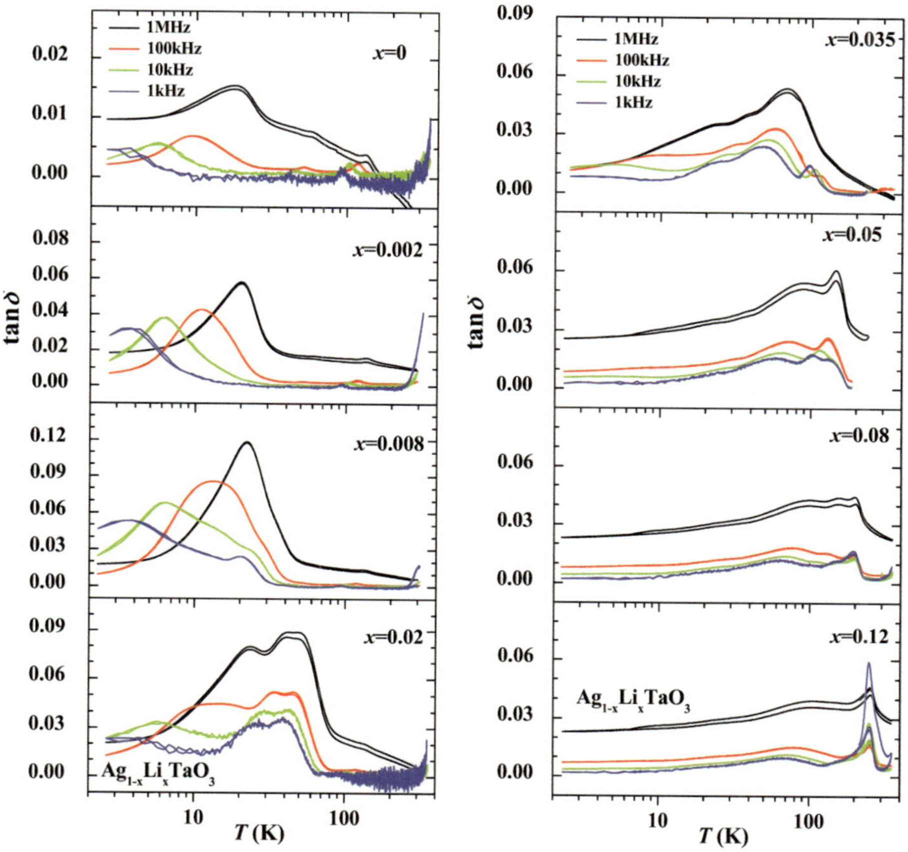

Fig. 28. Dielectric loss for $Ag_{1-x}Li_xTaO_3$ with $0 \leq x \leq 0.12$.

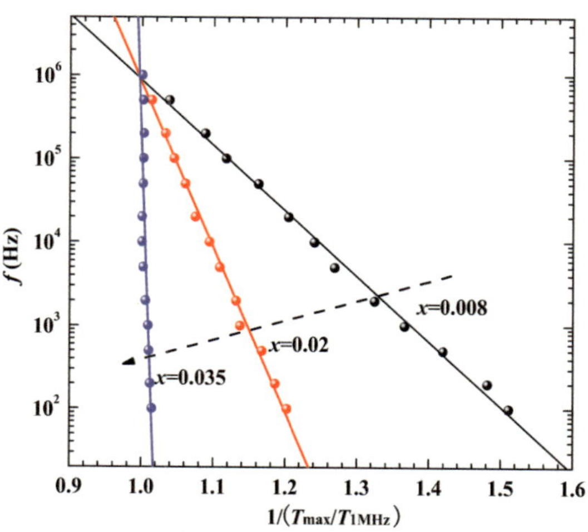

Fig. 29. Relationship between frequency and *T*max in $Ag_{1-x}Li_xTaO_3$. *T*max was normalized by the *T*max of 1 MHz.

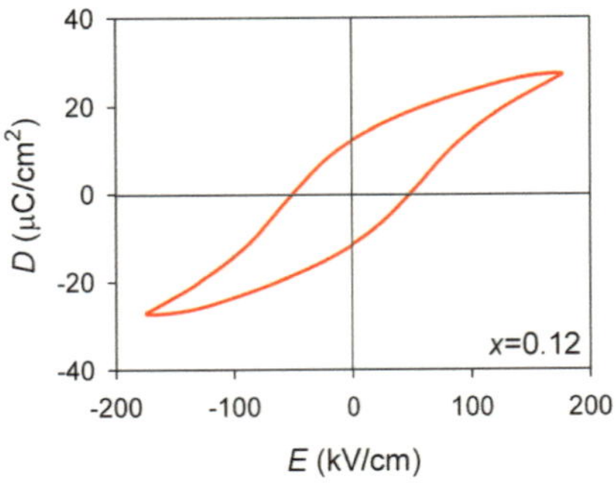

Fig. 30. Hysteretic *D-E* loop for $Ag_{1-x}Li_xTaO_3$ with x=0.12 obtained at 0.1 Hz and 77 K, indicating the ferroelectric state at $T< T_c$ (=258 K).

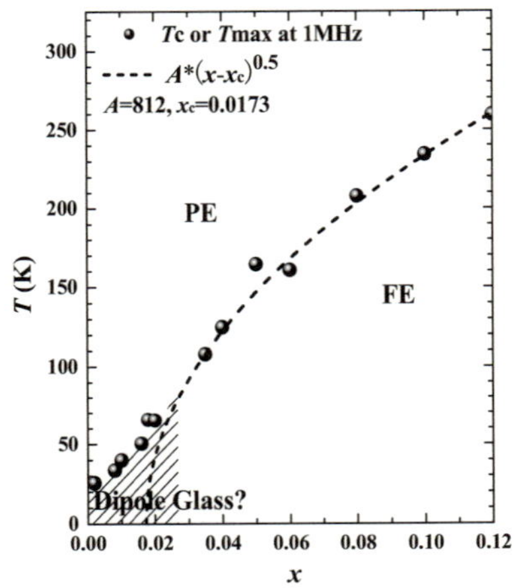

Fig. 31. Proposed phase diagram for $Ag_{1-x}Li_xTaO_3$. The dashed-dotted line shows the fitting to Morf-Schneider's relationship that is proposed for a quantum phase transition (Morf et al., 1977). PE and FE denote paraelectric and ferroelectric states, respectively. The gray zone may be a dipolar-glass state (Pirc & Blinc, 1999) or a state with nanosized ferroelectric domains (Fisch, 2003; Fu et al., 2009b).

10. References

Abel, W. R. (1971). Effect of pressure on the static dielectric constant of $KTaO_3$, *Phys. Rev. B* 4, pp. 2696-2701

Arioka, T. (2009) *Undergraduate thesis of Tokyo Institute of Technology*, Tokyo

Arioka, T.; Taniguchi, H.; Itoh, M.; Oka, K.; Wang, R. & Fu, D. (2010). Ferroelectricity in $NaNbO_3$: Revisited, *Ferroelectrics* 401, pp. 51-55

Baettig, P.; Ederer, C. & Spaldin, N. A. (2005). First principles study of the multiferroics $BiFeO_3$, Bi_2FeCrO_6, and $BiCrO_3$: Structure, polarization, and magnetic ordering temperature, *Phys. Rev. B* 72, pp. 214105

Barrett, J. H. (1952). Dielectric Constant in perovskite type crystals, *Phys. Rev.* 86, pp. 118-120

Bilc, D. I. & Singh, D. J. (2006). Frustration of tilts and A-site driven ferroelectricity in $KNbO_3$-$LiNbO_3$ alloys, *Phys. Rev. Lett.* 96, pp. 147602

Cohen, R. E. (1992). Origin of ferroelectricity in perovskite oxides, *Nature* 358, pp. 136–138

Eglitis, R. I.; Postnikov, A. V.& Borstel, G. (1997). Semiempirical Hartree-Fock calculations for pure and Li-doped $KTaO_3$, *Phys. Rev. B* 55, pp. 12976-12981

Egami, T.; Dmowski, W.; Akbas M. & Davies, P. K. (1998). Local structure and polarization in Pb containing ferroelectric oxides in *First-Principles Calculations for Ferroelectrics – 5th Williamsburg Workshop* ed Cohen R., pp 1–10, AIP, New York

Fabry, J.; Zikmund, Z.; Kania, A. & Petricek, V. (2000). Silver niobium trioxide, $AgNbO_3$, *Acta Cryst.* C56, pp. 916-918

Fisch, R. (2003). Random-field models for relaxor ferroelectric behavior, *Phys. Rev. B* 67, pp. 094110

Fortin, W. & Kugel, G. E.; Grigas, J. & Kania, A. (1996). Manifestation of Nb dynamics in Raman, microwave, and infrared spectra of the $AgTaO_3$-$AgNbO_3$ mixed system, *J. Appl. Phys.* 79, pp. 4273-4282

Francombe, M. H. & Lewis, B. (1958). Structural and electrical properties of silver niobate and silver tantalate, *Acta Cryst.* 11, pp. 175-178

Fu, D.; Endo, M.; Taniguchi, H.; Taniyama, T. & Itoh, M. (2007). $AgNbO_3$: A lead-free material with large polarization and electromechanical response, *Appl. Phys. Lett.* 90, pp. 252907

Fu, D.; Endo, M.; Taniguchi, H.; Taniyama, T.; Koshihara, S. & Itoh, M. (2008) Piezoelectric properties of lithium modified silver niobate perovskite single crystals, *Appl. Phys. Lett.* 92, pp. 172905

Fu, D.; Itoh, M. & Koshihara, S. (2009a). Dielectric, ferroelectric, and piezoelectric behaviors of $AgNbO_3$–$KNbO_3$ solid solution, *J. Appl. Phys.* 106, pp. 104104

Fu, D.; Taniguchi, H.; Itoh, M.; Koshihara, S.; Yamamoto, N. & Mori, S. (2009b). Relaxor $Pb(Mg_{1/3}Nb_{2/3})O_3$: A ferroelectric with multiple inhomogeneities, *Phys. Rev. Lett.* 103, pp. 207601.

Fu, D.; Arioka, T.; Taniguchi, H.; Taniyama, T. & Itoh, M. (2011b). Ferroelectricity and electromechanical coupling in $(1 -x)AgNbO_3$-$xNaNbO_3$ solid solutions, *Appl. Phys. Lett.* 99, pp. 012904

Fu, D.; Itoh, M. & Koshihara, S. (2010). Invariant lattice strain and polarization in $BaTiO_3$–$CaTiO_3$ ferroelectric alloys, *J. Phys.: Condens. Matter* 22, pp. 052204

Fu, D.; Endo, M.; Taniguchi, H.; Taniyama, T.; Itoh, M. & Koshihara, S. (2011). Ferroelectricity of Li-doped silver niobate $(Ag,Li)NbO_3$, *J. Phys.: Condens. Matter* 23, pp. 075901

Goldschmidt, V.M. (1926). *Shrifter Norske Videnskaps-Akad. Oslo, I: Mat.-Naturv. Kl.* No. 2, pp. 8

Grinberg, I. &Rappe, A. M. (2003). Ab initio study of silver niobate in *Fundamental Physics of Ferroelectrics*, edited by Davies, P. K. & Singh, D. J., pp 130–138,American Institute of Physics, New York

Grinberg, I. & Rappe, A. M. (2004). Silver solid solution piezoelectrics, *Appl. Phys. Lett.* 85, pp. 1760-1762

Hafid, M.; Kugel, G. E.; Kania, A.; Roleder, K. & Fontana, M. D. (1992). Study of the phase-transition sequence of mixed silver tantalate niobate ($AgTa_{1-x}Nb_xO_3$) by inelastic light-scattering, *J. Phys. Conden. Matter.* 4, pp. 2333-2345

Höchli, U. T. & Boatner, L. A. (1979). Quantum ferroelectricity in $K_{1-x}Na_xTaO_3$ and $KTa_{1-y}Nb_yO_3$, *Phys. Rev B.* 20, pp. 266-275

Itoh, M.; Wang, R.; Inaguma, Y.; Yamaguchi, T.; Shan, Y-J. &Nakamura, T. (1999). Ferroelectricity induced by oxygen isotope exchange in strontium titanate perovskite, *Phys. Rev. Lett.* 82, pp. 3540-3543

Jaffe, B.; Cook, Jr., W. R. & Jaffe, H. (1971). *Piezoelectric Ceramics,* Academic, London

Kania, A. (1983). $AgNb_{1-x}Ta_xO_3$ solid solutions-dielectric properties and phase transitions, *Phase Transit.* 3, pp. 131-139

Kania, A.; Roleder, K. & Lukaszewski, M. (1984). The ferroelectric phase in $AgNbO_3$, Ferroelectrics 52, pp. 265-269

Kania, A.; Roleder, K.; Kugel, G. E. & Fontana, M. D. (1986). Raman scattering, central peak and phase transitions in $AgNbO_3$, *J. Phys. C: Solid State Phys.* 19, pp. 9-20

Kania, A. (1989). Flux growth of $AgTa_xNb_{1-x}O_3$ (ATN) solid solution single crystals, *J. Cryst. Growth* 96, pp. 703-704

Kania, A. (1998). An additional phase transition in silver niobate $AgNbO_3$, *Ferroelectrics* 205, pp. 19-28

Kania, A. & Kwapulinski, J. (1999). $Ag_{1-x}Na_xNbO_3$ (ANN) solid solutions: from disordered antiferroelectric $AgNbO_3$ to normal antiferroelectric $NaNbO_3$, *J. Phys.: Condens. Matter.*11, pp. 8933-8946

Kania, A. (2001). Dielectric properties of $Ag_{1-x}A_xNbO_3$ (A: K, Na and Li) and $AgNb_{1-x}Ta_xO_3$ solid solutions in the vicinity of diffuse phase transitions, *J. Phys. D Appl. Phys.*34, pp. 1447-1455

Kato, H.; Kobayashi, H. & Kudo, A. (2002). Role of Ag^+ in the band structures and photocatalytic properties of $AgMO_3$ (M: Ta and Nb) with the perovskite structure, *J. Phys. Chem. B* 106, pp. 12441-12447

Kruczek, M.; Talik, E. & Kania, A. (2006). Electronic structure of $AgNbO_3$ and $NaNbO_3$ studied by X-ray photoelectron spectroscopy, *Solid State Commun.* 137, pp. 469-473

Kugel, G. E.; Fontana, M. D.; Hafid, M.; Roleder, K.; Kania, A. & Pawelczyk, M. (1987). A Raman study of silver tantalate ($AgTaO_3$) and its structural phase transition sequence, *J. Phys. C: Solid State Phys.* 20, pp. 1217-1230

Kuroiwa, Y.; Aoyagi, S.; Sawada, A.; Harada, J.; Nishibori, E.; Takata, M. & Sakata, M. (2001). Evidence for Pb-O Covalency in Tetragonal $PbTiO_3$, *Phys. Rev. Lett.* 87, pp. 217601

Levin, I.; Krayzman, V.; Woicik, J. C.; Karapetrova, J.; Proffen, T.; Tucker, M. G. & Reaney, I. M. (2009). Structural changes underlying the diffuse dielectric response in $AgNbO_3$, *Phys. Rev.* B 79, pp. 104113

Lines, M. E. & Glass, A. M. (1977). *Principle and Application of Ferroelectrics and Related Materials,* Oxford, Clarendon

Łukaszewski, M.; Kania, A. & Ratuszna, A. (1980). Flux growth of single crystals of $AgNbO_3$ and $AgTaO_3$, *J. Cryst. Growth* 48, pp. 493-495

Łukaszewski, L.; Pawelczyk, M.; Handerek, J. & Kania, A. (1983). On the phase transitions in silver niobate $AgNbO_3$, *Phase Transit.* 3, pp. 247-258

Łukaszewski, L. (1983). Dielectric properties of $Ag_{1-x}K_xNbO_3$ solid solutions, *Ferroelectrics* 44, pp. 319-324

Mitchell, R. H. (2002). *Perovskites: modern and ancient,* Almaz press, Ontario

Morf, R.; Schneider, T. & Stoll, E. (1977). Nonuniversal critical behavior and its suppression by quantum fluctuations, *Phys. Rev. B* 16, pp. 462-469

Müller, K. A. & Burkard, H. (1979). $SrTiO_3$: An intrinsic quantum paraelectric below 4 K, *Phys. Rev. B* 19, pp. 3593-3602

Nalbandyan, V. B.; Medviediev, B. S. & Beliayev, I. N. (1980). Study of the silver niobate-lithium niobate systems, *Izv.Akad. Nauk SSSR, Neorg. Mater.* 16, pp. 1819-1823

Paweczyk, M. (1987). Phase transitions in $AgTa_xNb_{1-x}O_3$ solid solutions, *Phase Transit.* 8, pp. 273-292

Petzelt, J.; Kamba, S.; Buixaderas, E.; Bovtun, V.; Zikmund, Z.; Kania, A.; Koukal, V.; Pokorny, J.; Polivka, J.; Pashkov, V.; Komandin, G. & Volkov, A. (1999). Infrared and microwave dielectric response of the disordered antiferroelectric $Ag(Ta,Nb)O_3$ system, *Ferroelectrics* 223, pp. 235-246

Pirc, R. & Blinc, R. (1999). Spherical random-bond–random-field model of relaxor ferroelectrics, *Phys. Rev. B* 60, pp. 13470-13478

Pisarski, M. & Dmytrow, D. (1987). Phase transitions in ceramic $AgNbO_3$ investigated at high hydrostatic pressure, *Ferroelectrics* 74, pp. 87-93

Ratuszna, A.; Pawluk, J. &Kania, A. (2003). Temperature evolution of the crystal structure of $AgNbO_3$, *Phase Transit.* 76, pp. 611-620

Reisman, A. & Holtzberg, F. (1958). Heterogeneous equilibria in the systems Li_2O–, Ag_2O–Nb_2O_5 and oxide-models. *J. Am. Chem. Soc.* 80, pp. 6503-6507

Rodel, J. Jo, W.; Seifert, K. T. P.; Anton, E.; Granzow, T. & Damjanovic, D. (2009). Perspective on the development of lead-free piezoceramics, *J. Am. Ceram. Soc.* 92, pp. 1153 -1177

Rupprecht, G. & Bell, R. O. (1964). Dielectric constant in paraelectric perovskites, *Phys. Rev.* 135, pp. A748-A752

Saito, Y.; Takao, H.; Tani, T.; Nonoyama, T.; Takatori, K.; Homma, T.; Nagaya, T. & Nakamura, M. (2004). Lead-free piezoceramics, *Nature* 432, pp. 84-87

Sakabe, Y.; Takeda, T.; Ogiso, Y. & Wada, N. (2001). Ferroelectric Properties of (Ag, Li)(Nb,Ta)O_3 Ceramics, *Jpn. J. Appl. Phys. Part* 1, 42, pp. 5675-5678

Samara, G. A. (2003).The relaxational properties of compositionally disordered ABO_3 perovskite. *J. Phys.: Condens. Matter* 15, pp. R367-R411

Sano, R., Morikawa, D.; Tsuda, K.; Fu, D. & Itoh, M. (2010). Space group determination of $AgNbO_3$ at room temperature by CBED method, *Meeting abstracts of the physical society of Japan* 65 (issue 2, part 4), pp. 916

Sciau, Ph; Kania, A.; Dkhil, B.; Suard, E. & Ratuszna, A. (2004). Structural investigation of $AgNbO_3$ phases using x-ray and neutron diffraction, *J. Phys.: Condens. Matter* 16, pp. 2795-2810

Scott, J. F. (2000). *Ferroelectric Memories*, Springer , Berlin

Shannon, R. D. (1976). Revised effective ionic radii and systematic studies of interatomic distances in halides and chalcogenides, *Acta Crystallogr., Sect. A: Cryst. Phys., Diffr., Theor. Gen.Crystallogr.* 32, pp. 751-767

Shigemi, A.; Wada, T. (2008). Crystallographic phase stabilities and electronic structures in $AgNbO_3$ by first-principles calculation, *Molecular Simulation* 34, pp. 1105-1114

Shiozaki, Y.; Nakamura, E. & Mitsui, T. (2001). *Ferroelectrics and Related Substances*, Vol. 36, Pt. A1, Springer, Berlin

Soon, H. P.; Taniguchi, H. & Itoh, M. (2009). Ferroelectricity triggered in the quantum paraelectric $AgTaO_3$ by Li-substitution, *Appl. Phys. Lett.* 95, pp. 242904

Soon, H. P.; Taniguchi, H. &Itoh, M. (2010). Dielectric and soft-mode behaviors of $AgTaO_3$, *Phys. Rev. B* 81, pp. 104105

Takeda, T.; Takahashi, Y.; Wada, N. & Sakabe, Y. (2003). Effects of substitution of Na and K ions for Ag ion in $(Ag,Li)NbO_3$ Ceramics, *Jpn. J. Appl. Phys.* Part 1, 42, pp. 6023-6026

Uchino, K. (1997). *Piezoelectric Actuators and Ultrasonic Motors*, Kluwer Academic, Boston

Valant, M.; Axelsson, A.; Alford, N. (2007a). Review of $Ag(Nb,Ta)O_3$ as a functional material, *J. Euro. Ceram. Soc.* 27, pp. 2549-2560

Valant, M.; Axelsson, A.; Zou, B. & Alford, N. (2007b). Oxygen transport during formation and decomposition of $AgNbO_3$ and $AgTaO_3$, *J. Mater. Res.* 22, pp. 1650-1655

Verwerft, M.; Dyck, D. V.; Brabers, V. A. M.; Landuyt, J. V. & Amelinckx S. (1989). Electron microscopic study of the phase transformations in $AgNbO_3$, *Phys. Stat. Sol. (a)* 112, pp. 451-466

Volkov, A. A.; Gorshunov, B. P.; Komandin, G.; Fortin, W.; Kugel, G. E.; Kania, A. & Grigas, J. (1995). High-frequency dielectric spectra of $AgTaO_3$-$AgNbO_3$ mixed ceramics, *J. Phys.: Condens. Matter* 7, pp. 785-793

Vugmeister, B. E. & Glinchuk, M. D. (1990). Dipole glass and ferroelectricity in random-site electric dipole systems, *Rev. Mod. Phys.* 62, pp. 993-1026

Wada, S., Park, S.-E.; Cross, L. E. & Shrout, T. R. (1998). Domain configuration and ferroelectric related properties of relaxor based single crystals, *J. Korean Phys. Soc.* 32, pp. S1290-S1293

Weirauch, D. F. &Tennery, V. (1967). Electrical, X-Ray, and thermal expansion studies in the system $KNbO_3$-$AgNbO_3$, *J. Am. Ceram. Soc.* 50, pp. 671-673

Wołcyrz, M. & Łukaszewski, M. (1986). The crystal structure of the room-temperature phase of $AgTaO_3$. *Zeitschrift für Kristallographie* 177, pp. 53-58

Yashima, M.; Matsuyama, S.; Sano, R.; Itoh, M.; Tsuda, K. & Fu, D. (2011). Structure of ferroelectric silver niobate $AgNbO_3$, *Chem. Mater.*23, pp. 1643-1645

Part 4

Thin Films

21

Amino-Acid Ferroelectric Thin Films

Balashova E.V. and Krichevtsov B.B.
Ioffe Physico-technical Institute of RAS, St-Petersburg, Russia

1. Introduction

The family of amino-acid ferroelectrics involves large number of crystals the chemical composition of which is based on combinations of different amino-acids (betaine $(CH_3)_3N^+CH_2COO^-$), sarcosine (CH_3NHCH_2COOH), glycine (H_2NCH_2COOH)) and non-organic acids (H_3PO_3, H_3AsO_4, H_3PO_4, HCl, H_2SO_4) or salts. The most well known example of amino-acid ferroelectrics - triglycine sulphate (TGS) - was discovered in 1956 (Matthias et al., 1956). After that the amino-acid ferroelectric crystals were synthesized on basis of sarcosine and in 80-th of the last century on basis of betaine amino-acids (Albers et al., 1988). The interest to betaine amino-acid ferroelectrics is concerned with large variety of phases (ferroelectric, ferroelastic, antiferroelectric, antiferrodistortive, incommensurate, glasslike state and so on), phase transformations, and with ferroelectric properties observed in these crystals. For example, a record value of dielectric constant at the ferroelectric phase transition $\varepsilon \approx 10^6$ has been observed in betaine arsenate crystals. Experimental and theoretical investigations of amino-acid ferroelectric single crystals has been carried out in large number of works and main results of these studies were summarized in review papers (Albers, 1988; Schaack, 1990).

Recently it was found (Balashova et al., 2008; 2011a) that thin films of betaine phosphite (BPI) and deuterated betaine phosphite (DBPI) can be manufactured by evaporation method on different substrates. The BPI films consist of large single-crystalline blocks and show ferroelectric properties mainly analogous to the bulk BPI crystals. The differences in dielectric behavior of films and bulk samples are related to film-substrate interaction and specifics of domain structure.

At present large attention is paid to ferroelectric thin films because of their potential applications in information storage systems, sensors of different fields, elements of microelectronics and so on (Tagantsev et al., 2010; Dawber et al. 2005; Ducharme et al. 2002). Also, the increased interest to multuferroic materials and, in particular, to composition of ferroelectrics and ferromagnets stimulates the search of ferroelectric films which can be prepared on different substrates without using high growth temperature. For these reasons the development and investigation of amino acide ferroelectric films seems to be of interest.

In this chapter we present results of preparation and studies of BPI, DBPI and TGS films which were published or accepted for publication during last three years (Balashova et al., 2008; 2009a,b; 2010; 2011a,b).

The chapter is organized as follows: Section 2 is devoted to short description of structural and dielectric properties of some amino acid ferroelectric crystals which were used for preparation of films (TGS, BPI, DBPI); in Section 3 the growth method, preparation of substrates and geometry of obtained structures is described; in Section 4 the results of study of block and crystal structure of films are presented; Sections 5-9 are devoted to experimental investigations of low-signal and strong-signal dielectric response in BPI, DBPI, and TGS films grown on different substrates, calculations of dielectric permittivity of films, thermodynamic description of dielectric anomaly and modeling of dielectric hysteresis loops; Conclusions summarize main results of investigations.

2. Amino-acid ferroelectric single crystals

In this section we present short description of structural and some dielectric properties of bulk amino acid ferroelectric crystals (BPI, DBPI, TGS) which were used for preparation of films.

2.1 TGS

Triglycine sulfate (TGS) $(CH_2NH_2COOH)_3 \cdot H_2SO_4$, a ferroelectric discovered in 1956 (Matthias et al., 1956), displays a large pyroelectric coefficient and a high Volt/Watt sensitivity, and, thus, may be considered a unique material for pyroelectric uses (Lal & Batra, 1993; Neumann, 1993). TGS undergoes second order phase transition from paraelectric to ferroelectric state at T_c = 322 K which is followed by change of structural space group of symmetry from $P2_1/m$ to $P2_1$. TGS unit cell contains two formula units. Lattice parameters values of TGS at RT are a = 9.392 Å, b = 12.734 Å, c = 5.784 Å and monoclinic angle β = 109.45 ° (Fletcher, 1976). The phase transition results in (1) continuous reorientation of the NH^{+3} group of the glycine about the ac plane making it a statistically averaged mirror in the high-temperature paraelectric phase; (2) disordering of the proton that connects the glycine groups making the two glycine ions indistinguishable in the high-temperature paraelectric phase.

The phase transition of TGS in ferroelectric state is accompanied by appearance of spontaneous polarization P_s along polar b axis and strong dielectric anomaly.

2.2 Betaines (BPI, DBPI)

Betaine phosphite (BPI), $(CH_3)_3NCH_2COO \cdot H_3PO_3$, is a compound of betaine amino acid, $(CH_3)_3N^+CH_2COO^-$, and inorganic acid H_3PO_3. Ferroelectricity in BPI was discovered by Albers et al. (Albers et al., 1988a; Albers, 1988b). BPI undergoes two phase transitions: antiferrodistortive ($P2_1/m$ (Z=2) → $P2_1/c$ (Z=4)) at T_{c1}=355 K and ferroelectric phase transition ($P2_1/c$ (Z=4) → $P2_1$ (Z=4)) at T_{c2} ≈198-224 K (Albers et al., 1988a; Fehst et al., 1993). Unit cell parameters: a = 11.191(3) Å, b = 7.591(3) Å, *c* = 12.447(6) Å , β = 116.62 (2)° at RT. In BPI structure the inorganic tetrahedral HPO_3 groups are linked by hydrogen bonds forming zig-zag chains along monoclinic b - axis. The betaine molecules are arranged almost perpendicular to the chains along x directions and linked by one hydrogen bond to the inorganic group. The ordering of hydrogen ions in the hydrogen bonds in the chains results in ferroelectric phase transition. The spontaneous polarization below T_{c2} occurs along monoclinic b - axis. The transition temperature T_{c2} appears to be sensitive to a small percentage of impurities or to crystalline defects. Deuteration of the hydrogen bonds can increase the ferroelectric phase transition temperature T_{c2} up to 310 K (Bauch et al. 1995).

Fig.1 show the temperature dependence of dielectric constant ε_b in BPI at ferroelectric phase transition. Maximal values of $P_s \approx 1.7\ 10^{-2}$ Cm^{-2} in BPI are smaller than in TGS ($P_s \approx 4.5\ 10^{-2}$ Cm^{-2} Albers et al., 1988).
The antiferrodistortive phase transition in BPI at T_{c1} is not accompanied by appearance of polarization. Nevertheless the temperature dependence of dielectric constant exhibit small anomaly at $T = T_{c1}$ that indicates a connection between order parameters of antiferrodistortive and ferroelectric phase transition.

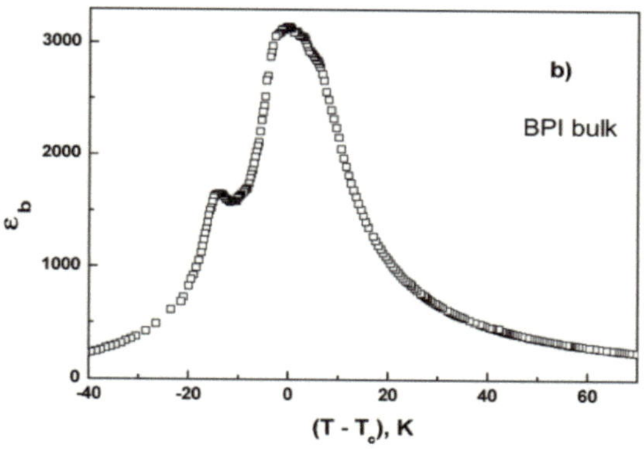

Fig. 1. Temperature dependence of dielectric constant along b axis in BPI (Balashova et al., 2002).

The dielectric and acoustic properties of BPI, and crystals of betaine phosphite with small admixture of antiferroelectric betaine phosphate, at the antiferrodistortive and ferroelectric phase transitions were explained using the thermodynamic approach based on Landau theory with account of $\xi\eta^2P^2$ ($\xi < 0$) term coupling the η nonpolar order parameter for high-temperature antiferrodistortive phase transition at T_{c1} and polarization P (Balashova & Lemanov, 2000, 2003b). The thermodynamic potential has a form:

$$F = \frac{1}{2}\alpha_1\eta^2 + \frac{1}{6}\gamma_1\eta^6 + \frac{1}{2\chi_0}P^2 + \frac{1}{4}\beta_2P^4 + \frac{1}{2}\xi\eta^2P^2 - PE \qquad (1)$$

where $\alpha_1 = \lambda_1(T - T_{c1})$, $\beta_1 = 0$ (the tricritical point), $\beta_2 > 0$, $\gamma_1 > 0$, $\xi < 0$, **E** is the macroscopic electric field; χ_0, the background dielectric susceptibility. Only one coefficient α_1 in this approach is temperature dependent. Since in the considered potential only one coefficient at η^2 term changes the sigh at T_{c1}, the ferroelectric phase transition at T_{c2} was called trigger phase transitions (Holakovsky, 1973). The thermodynamic potential (1) can be rewritten in a dimensionless form (Balashova et al.,2002; Balashova & Lemanov, 2003a)

$$f = \frac{1}{2}ta^2q^2 + \frac{1}{6}q^6 - ap^2 + \frac{1}{2}p^4 - q^2p^2 + 2ape \qquad (2)$$

where $t = (T - T_{c1})/\Delta T$ is the reduced temperature, $f = F \cdot \frac{8\beta_2^3\gamma_1^2}{\xi^6}$, $q^2 = \frac{2\beta_2\gamma_1}{\xi^2}\eta^2$, $p^2 = -\frac{2\beta_2^2\gamma_1}{\xi^3}P^2$, $e = \frac{\sqrt{2\gamma_1\beta_2\chi_0}}{(-\xi)^{3/2}}E$.

Parameter $\Delta T = \frac{\gamma_1}{\lambda_1 \chi_0^2 \xi^2}$ determines the temperature region of stability of the paraelectric antiferrodistortive phase (q ≠ 0, p = 0). The dimensionless parameter $a = \frac{2\beta_2\gamma_1}{\chi_0\xi^3} < 0$ defines the region of stability of the polar mixed phase (q ≠ 0, p ≠ 0) and the order (first, second or tricritical) of the phase transition into polar mixed phase. An important conclusion of these works is that the ferroelectric phase transition into the ($\eta \neq 0$, $P \neq 0$) state in BPI crystals is induced by the nonpolar order parameter η due to the η^2P^2 coupling and the temperature of the ferroelectric phase transition T_{c2} is determined by the coupling strength. This approach makes it possible adequately describe the nonlinear temperature dependences of the inverse dielectric constant in the antiferrodistortive phase of the $BPI_{1-x}BP_x$ (x = 0 – 0.1), including the phase transition region, the effect of the bias field on dielectric constant and the acoustic anomalies at the ferroelectric phase transition. In BPI the value of dimensionless parameter a is -2.5. Application of the model of coupled order parameters for betaine arsenat – deuterated betaine arsenate system was presented in ref. (Balashova et al., 1995).

3. Preparation of films

3.1 BPI films

Thin films of betaine phosphite (BPI) were grown on different substrates by evaporation method from the water solution of the BPI crystals at a temperature of 24°C. Single-crystalline quartz α-SiO_2 (Z–cut), lithium niobate $LiNbO_3$ (Y–cut) (Balashova et al., 2008; 2009a,b), α-Al_2O_3(110), $NdGaO_3$(001), and also fused quartz (Balashova et al.,2011) and glass were used as substrates. Before the film growth Al or Au interdigital structures (IDS) of electrodes were deposited on the substrates by the photolithographic method. Fig.2 shows schematically an arrangement of the IDS and the BPI film on substrate. The length, width and thickness of IDS electrodes were 4 mm×(25μm or 50 μm)×0.3 μm. The distance between electrodes was equal to the width of electrodes (25μm or 50 μm). The number N of pairs of electrodes in IDS was N = (35 or 40). Total aria of IDS was 35 mm^2. The thickness h of films measured by profilometer was h = (0.5-4)μm.

The aqueous solution of BPI crystals was deposited both in the IDS region and directly on the substrate surface. Thin layer of solution is practically invisible just after the rendering on the substrate but in some minutes the crystallization front moving from the border to the center of the substrate is observable when the substrate is oriented horizontally. If one of the borders of the substrate is higher than opposite, the crystallization front moves from the upper to the lower border. This shows that the crystallization process starts from the areas with the smallest thickness of solution layer.

The measurements of IDS resistance and dielectric response show that the crystallization process may be characterized by two stages. At the first stage, which takes several minutes, after the stop of the crystallization front, the resistance is about R ≈ (2-6) MOhm. Dielectric response shows considerable frequency dispersion of capacity and losses at T > 240 K. At this stage the block structure nevertheless is well observable in polarization microscope. In the second stage which takes several days the IDS resistance becomes higher than 20 MOhm. After this, the films exhibit low frequency dispersion of capacity and low value of dielectric losses at room temperature. Existence of two stages of crystallization is due to the fact that the BPI crystallization begins from the surface which is in direct contact with air.

Water evaporates from the surface, and the crystallization front gradually propagates down to the substrate. However, the crystallized part of the film hinders the evaporation from the layers near the film–substrate interface. For this reason the interface may contain non-crystallized regions for a fairly long time, which eventually crystallize. These regions have certain conductivity, which is responsible for the low frequency dispersion of the capacitance and losses during second stage of crystallization.

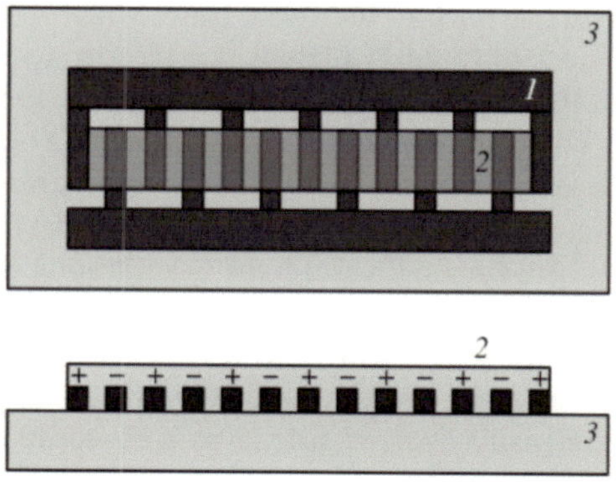

Fig. 2. Arrangement of (*1*) the IDS and (*2*) the BPI film on (*3*) substrates. The plus and minus signs identify the alternating charge distribution on the IDS electrodes when DC electric voltage is applied to IDS.

3.2 DBPI films

DBPI films with different degrees of deuteration were fabricated by evaporation from solutions of (i) BPI single crystals in heavy water D_2O, and (ii) DBPI single crystals, obtained by the recrystallization of BPI crystals in D_2O, and (iii) DBPI single crystals, grown by slow cooling from a solution of D_3PO_3 acid and betaine, in heavy water. The degree of deuteration was determined from the ferroelectric phase transition temperature, which, according to the data on DBPI single crystals, increases with an increase in the degree of deuteration (Bauch et al., 1995). $NdGaO_3(001)$, sapphire, and quartz α-$SiO_2(001)$ single crystals were used as substrates onto which interdigital gold structures were previously deposited by photolithography.

3.3 TGS films

In ref. (Wurfel & Barta, 1973; Wurfel et al., 1973) a polycrystalline ferroelectric TGS films with switching characteristics approaching those of a bulk crystal were prepared by sublimation in vacuum onto silicon substrates. Nevertheless, preparation of oriented (textured) films adaptable to present day planar technologies remains a topical problem. A study of the growth of TGS crystals from a saturated solution on single crystal silicon substrates and of the effect of various substrate surface treatments on the size and orientation, as well as the structure of crystallites, was reported in (Stekhanova et al., 2005). In this work TGS films were grown on substrates of fused quartz atop a layer of thermally deposited aluminum (Al/SiO_2), as well as on white sapphire (α-Al_2O_3) substrates with IDS of electrodes. The TGS films were prepared by evaporation of a saturated water solution of bulk crystals which was deposited on the substrate at room temperature. The thickness of TGS films was $h \approx 0.2$ μm.

4. Block and crystal structure of films

4.1 BPI films

Block structure of films can be visualized by means of polarizing microscope in reflection mode because BPI as well as other amino acid ferroelectrics belongs to low symmetry class and are characterized by strong birefringence that provides the possibility to observe different single crystalline areas in the film. Figure 3 presents images of a BPI film deposited on the Z-cut quartz surface in the IDS region, which were obtained in polarized light in reflection mode. The Z-cut quartz plate is not birefringent, and does not influence the contrast when rotated about an axis perpendicular to the surface, with the polarizers in the extinction position. The film deposited on the quartz surface induces birefringence. We readily see (Fig. 3) that when the crystal with the film is turned around the position of crossed polarizers (or when the crossed polarizers are turned relative to the crystal with the film), different areas of the film with different orientations of optical indicatrix main directions in the film plane become extinct. In each of these areas, extinction occurs after a turn through 90°. Thus, one may conclude that the BPI film is essentially a polycrystal with block dimensions much larger than the film thickness. The blocks dimensions may reach ~1 mm, which can be easily derived from Fig. 3 by comparing the blocks with the dimensions of the IDS electrodes and their separation, with the sum being 50 μm. Similar results were obtained for BPI films on lithium niobate $LiNbO_3$ and $NdGaO_3$ as well (Fig.4 and Fig.5).

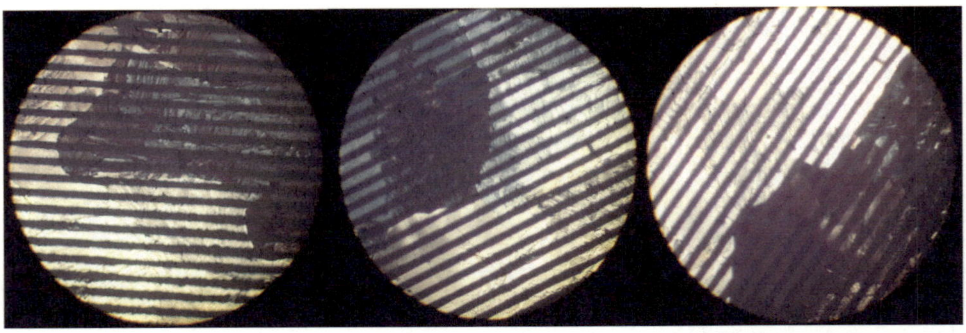

Fig. 3. Images of the BPI film grown on SiO_2(Z-cut) substrate obtained with a polarizing microscope operating in reflection for different orientations of the films relative to crossed polarizers (the IDS electrode separation is 25μm). Diameter of image is d = 1mm.

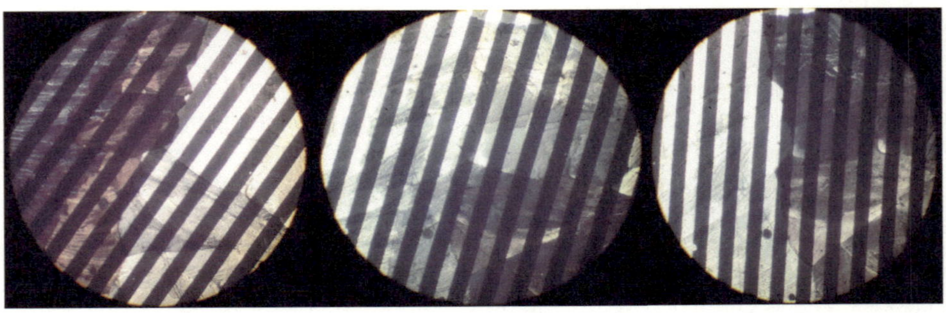

Fig. 4. Images of the BPI film grown on the $LiNbO_3$ substrate at different orientations of the films relative to crossed polarizers. The IDS electrode separation is 50 μm.

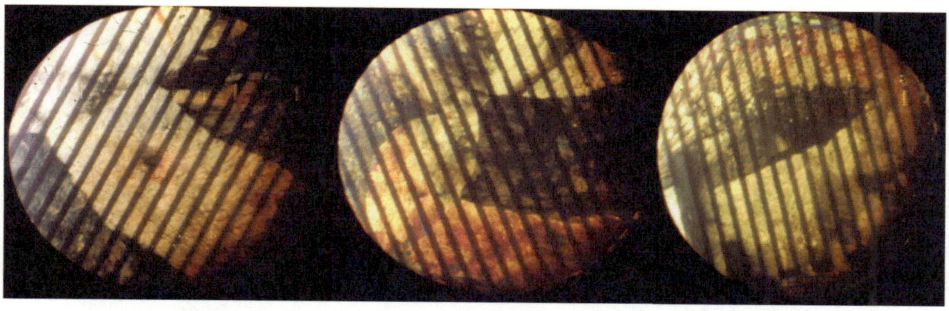

Fig. 5. Images of the BPI film grown on the $NdGaO_3(001)$ substrate at different orientations of the films relative to crossed polarizers.

4.2 DBPI films

The structure of single crystal blocks formed on the substrate surface during crystallization of DBPI is analogous to BPI films. Usually the number of blocks in the aria of the IDS structure usually does not exceed 5-10. Sometimes we managed to obtain the DBPI films with only two and even one single crystalline block per aria of the IDS. Fig. 6 demonstrates typical images of the DBPI film block structure in film with two blocks.

Fig. 6. Images of the DBPI film grown on the $NdGaO_3(001)$ substrate at different orientations of the films relative to crossed polarizers.

4.3 TGS films

Figure 7 and 8 presents images of TGS films grown on Al/SiO_2 and α-Al_2O_3 substrates obtained with a polarization microscope in reflection. The microscope field of view was 1 mm. Rotation of the films about crossed polarizers showed that the films are polycrystalline and consist of blocks measuring 0.1–0.3 mm for films on Al/SiO_2, and elongated blocks, 0.1 × 1 mm in size, on α-Al_2O_3. In blocks extinction was observed to occur each 90° of rotation of the film with respect to the crossed polarizers, thus evidencing the blocks to be single crystals.

4.4 X-Ray analysis

The orientation of crystallographic axes in blocks was determined by X-ray diffraction on a Dron 3 diffractometer (CuK_α radiation). Figure 9 shows a θ–2θ diffraction pattern for a DBPI film composed of two blocks (see Fig. 11). The presence of strong narrow lines in the diffraction patterns, which correspond to (200), (300), (400), (500), and (600) reflections, is indicative of a pronounced DBPI crystal structure, almost without foreign phases. The

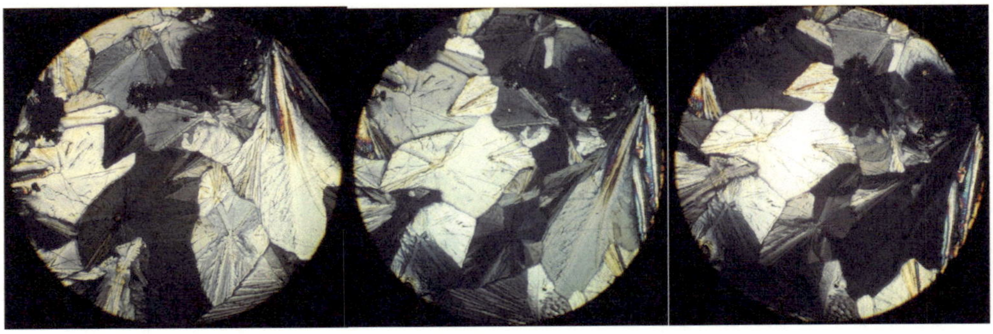

Fig. 7. Images of the TGS film grown on the Al/SiO_2 substrate at different orientations of the films relative to crossed polarizers.

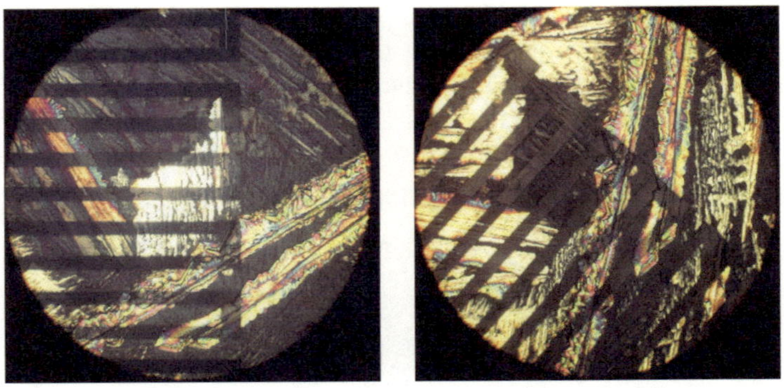

Fig. 8. Images of the TGS film grown on sapphire substrates with interdigital electrode structures at different orientations of the films relative to crossed polarizers.

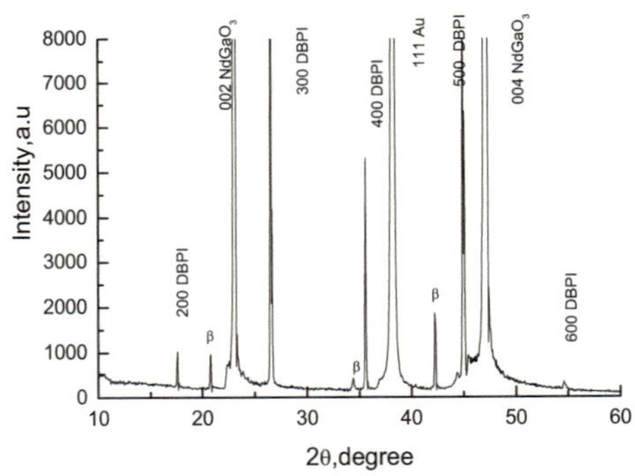

Fig. 9. θ–2θ diffraction patterns of the DBPI/$NdGaO_3$ structure, which is composed of two single crystal blocks (Fig. 6), with the identification of the peaks from the substrate, film, and gold interdigital electrodes. The bands denoted as β are due to the spurious CuK_{β} radiation.

absence of other reflections shows that the polar axis (monoclinic b axis) in both blocks is oriented in the substrate plane, and blocks differ by the orientation of b and c axes in the film plane. The (100) plane is parallel to the substrate surface in both blocks (correspondingly, the a^* axis is oriented perpendicularly to the film plane).

5. Small signal dielectric response in BPI films

The capacity and dielectric losses of the films were measured by a LCR meters MIT9216A at frequencies of f = 0.12, 1, 10, 100 kHz and by a E7-12 at f = 1 MHz with a drive voltage $U_{\sim}$ = 0.1 V in the temperature region T = (120–340) K. In the case of substrates with IDS the measured capacity of film/IDS/substrate structure is related basically to the in-plane orientation of electric field. The change of IDS/substrate capacity C_{sub} after the film growth reflects therefore the in-plane dielectric properties of the film.

5.1 Non-centrosymmetric substrates α-SiO_2, $LiNbO_3$

Figure 10 plots temperature dependences of the capacitance of the BPI/α-SiO_2 structure measured across the IDS electrodes at frequencies of 120 Hz, 1, 10, and 100 kHz. At room temperature, the IDS capacitance is increased by the presence of the BPI film by 13.7 pF to become 23 pF. As the temperature is lowered, the capacitance of the structure grows markedly and reaches a maximum at T ≈ 225 K, the temperature of the ferroelectric phase transition in a bulk crystal, after which it decreases with further lowering of temperature. There is practically no frequency dispersion, and the maxima in capacitance seen at different frequencies do not shift with temperature (Fig. 10). The variations of the permittivity of quartz in this temperature interval being small, all temperature-induced changes in the capacitance of the structure should be assigned to variation of the permittivity in the BPI film. Thus, the permittivity of the film at the maximum increases more than tenfold compared with the value at room temperature. Dielectric losses in the BPI/SiO_2 structure practically do not vary below room temperature and are less than 0.02 in the (0.12 –100) kHz frequency range.

Figures 11 and 12 present temperature dependences of the capacitance of the BPI/SiO_2 and BPI/$LiNbO_3$ structures which were obtained without and with a bias field U = 0, 9 and 18 V applied to the IDS. The maxima of capacitance in both structures in the absence of bias are seen to practically coincide in their temperature position (T ≈ 225 K).

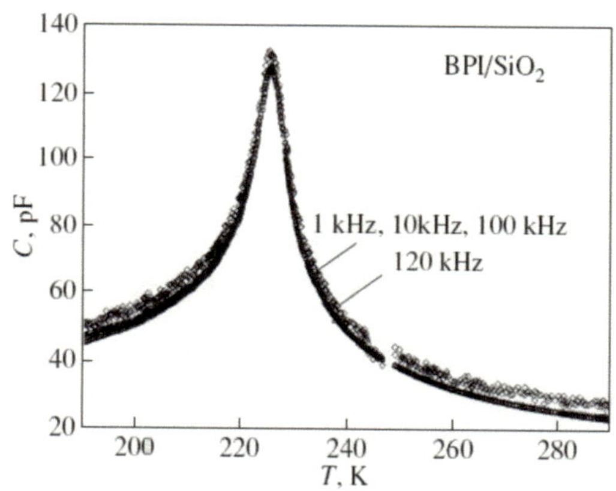

Fig. 10. Temperature dependences of the capacitance of the BPI/SiO_2 structure at frequencies of 120 Hz and 1, 10, and 100 kHz (Balashova et al., 2009a).

Application of a bias reduces the maximum capacitance, diffuses the maximum in temperature and shifts it toward higher temperatures, as is the case with bulk BPI crystals.

At room temperature, the capacitance of IDS on $LiNbO_3$ grows due to the film by about 3.8 pF. The film-induced capacitance at the maximum at 225 K increases almost by an order of magnitude, just as in the BPI/α-SiO_2 structure.

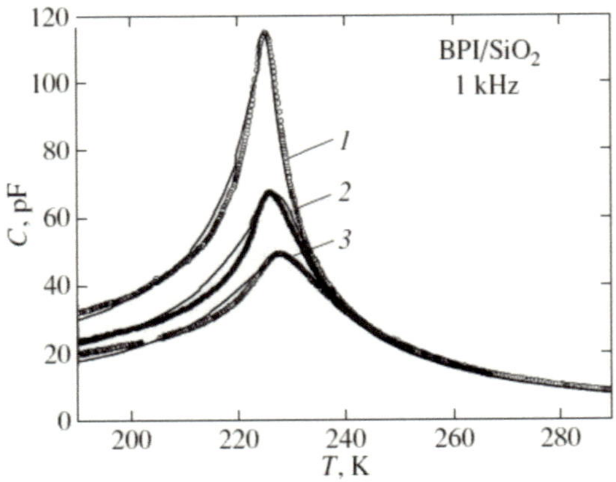

Fig. 11. Temperature dependences of the film capacitance in the BPI/SiO_2 structure measured without and with application of a bias field $U_{=}$ = (1) 0, (2) 9, and (3) 18 V. Solid lines plot the results of the calculation (a = -2.5 (Balashova et al., 2009a).

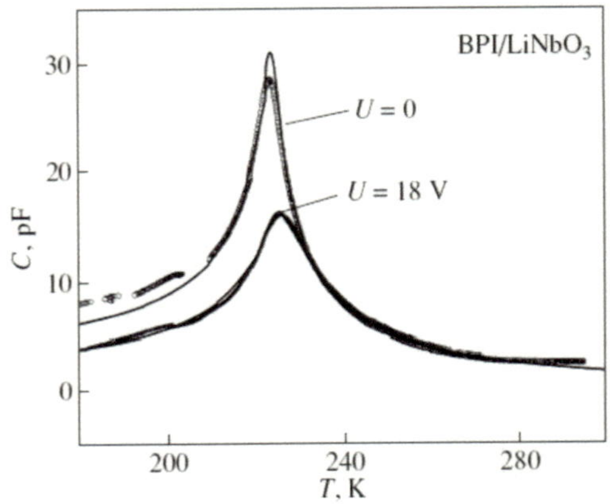

Fig. 12. Temperature dependences of the film capacitance in the BPI/$LiNbO_3$ structure measured without and with application of a bias field $U_{=}$ = 0 and 18 V to IDT. Solid lines plot the results of the calculation (a = -2.5 (Balashova et al., 2009a).

Figure 13 illustrates the behavior with temperature of the capacitance C and tanδ of the BPI/$LiNbO_3$ structure measured at frequencies of 1, 10, 100 kHz and 1 MHz. As in the case of the α-SiO_2 substrate, at the maximum the capacitance exhibits practically zero dispersion in frequency. The temperature at which the capacitance reaches maximum does not depend on frequency. As the temperature is increased above the room temperature, the permittivity reveals dispersion. The value of tanδ is very small in the region of the maximum capacitance and falls off with decreasing temperature at all frequencies, without exceeding 0.01. However, tanδ grows strongly above room temperature. As seen from Fig. 13, the temperature dependences of the capacitance and of tanδ measured at 100 kHz drops out of the general pattern of relations measured at four frequencies. In the (240–270) K interval, the capacitance at 100 kHz is larger than that at 1, 10 kHz and 1 MHz, while above 270 K it is smaller than at the other frequencies. In contrast to the other frequencies, at 100 kHz one observes a maximum of tanδ at approximately 270 K (Fig. 13b). Significantly, the room-temperature capacitance of IDS on lithium niobate without film, measured also at 100 kHz, was smaller than that at 0.12, 1, 10 kHz and 1 MHz. Thus, the unusual temperature dependences of the capacitance and tanδ of the BPI/$LiNbO_3$ structure at 100 kHz should be assigned not to the BPI film but rather to the properties of the substrate, the crystalline piezoelectric lithium niobate.

The anomalous behavior of the capacitance C and tanδ at temperatures above the temperature of the maximum in capacitance (Fig. 13a), which is observed in the BPI/$LiNbO_3$ structure at 100 kHz, suggests that at 100 kHz the structure falls into a region of resonance extending in temperature by about 100 K. For T > 270 K, the frequency turns out to be above the resonance, and for T < 270 K, below the resonance frequency. The low resonance frequency (~100 kHz) is typical of the bulk electromechanical resonance of the substrate, which is not connected in any way with the film properties. Indeed, at 100 kHz the wavelength of the shear acoustic mode in lithium niobate is approximately 3.6 cm, so that one half of the wavelength may be comparable with the substrate dimensions. The variation of the resonance frequency with temperature should be assigned to the acoustic wave velocity in lithium niobate increasing with decreasing temperature.

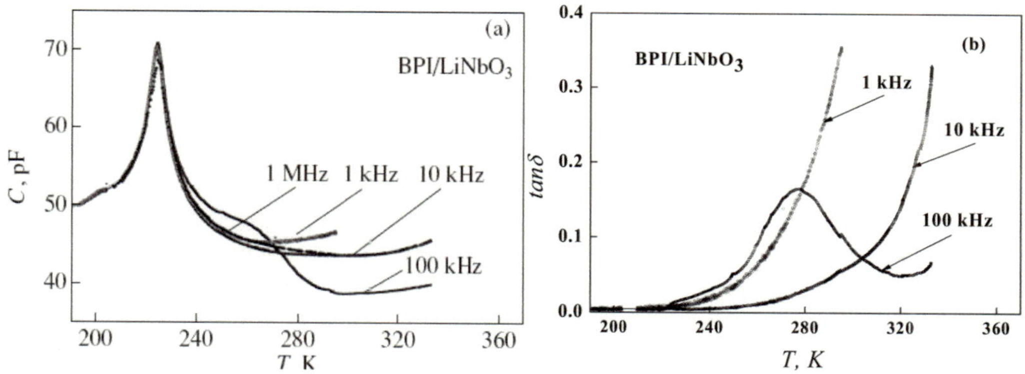

Fig. 13. Temperature dependences of (a) the capacitance and (b) tanδ of the BPI/$LiNbO_3$ structure at frequencies of 1, 10, and 100 kHz and 1 MHz (Balashova et al., 2009a).

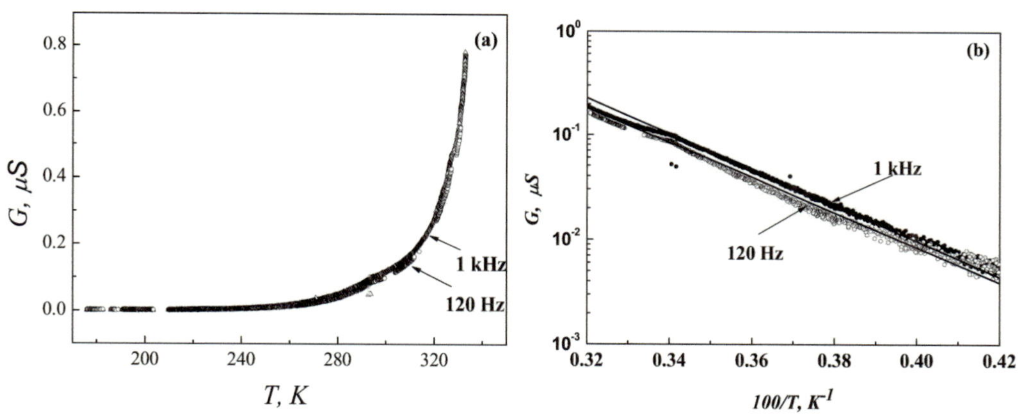

Fig. 14. (a) Temperature dependences of the conductivity G = Cωtanδ of the BPI/$LiNbO_3$ structure at frequencies of 120 Hz and 1 kHz on a linear scale. (b) Dependences of G on the inverse temperature on a semilogarithmic scale (Balashova et al., 2009a).

As the temperature is raised, tanδ grows strongly in the (294–340) K range to reach ~1 (Fig.13b). This increase at the frequencies of (0.12–1) kHz is inversely proportional to frequency. Figure 14 plots temperature dependences of the conductivity (G = Cω tanδ) derived from the behavior with temperature of C and tanδ at the frequencies of 120 Hz and 1 kHz for the BPI/$LiNbO_3$ structure. Examining Fig. 14, we see that the G(*T*) curves practically coincide at these frequencies. This suggests that the major contribution to these dependences is due to dc conductivity. In Fig. 14b, the G vs. T^{-1} relations are replotted in the log-linear coordinates. The graphs thus obtained can be fitted well with an exponential with an activation energy of about 10^3 K (~ 0.1 eV), a figure characteristic of thermally activated diffusion in films. Most probably, the conductivity here is mediated by diffusion of hydrogen atoms.

In bulk BPI crystals, a strong maximum in permittivity is observed only along the twofold axis (Y), along which spontaneous polarization appears at the ferroelectric phase transition. Perpendicular to the polar axis, permittivity does not reveal noticeable anomalies and is rather small. The maximum capacitance observed in BPI/SiO_2 and BPI/$LiNbO_3$ structures at T ≈ 225 K originates obviously from variation of permittivity in films, because the permittivity of substrates does not feature any noticeable changes in the temperature interval studied, and the temperature of the maximum coincides with that of the ferroelectric phase transition in bulk BPI crystals. In both structures, the permittivity of the film at its maximum exceeds by about an order of magnitude that at room temperature. As in single crystals, the maximum of capacitance does not shift in temperature with the frequency varied within a broad range, from 120 Hz to 1 MHz. Application of bias reduces the maxima in magnitude, diffuses them and shifts toward higher temperatures, just as the maxima in permittivity at the ferroelectric phase transition in a bulk BPI crystal.

Unlike a bulk BPI crystal, the phase transition in a film (Fig. 13b) is not accompanied by a pronounced increase of the loss tangent associated with domain wall motion. The absence of a domain contribution to dielectric losses should be apparently attributed to the fact that either domains in polycrystalline BPI films are pinned to defects near the interface with the substrate or the blocks with polarization oriented with the field in the film plane are single-domain because of a small depolarizing factor.

Thus, the temperature, field and frequency dependences of the capacitance of films are similar to those of bulk BPI crystals. At the same time, one could point out a few differences in the ferroelectric phase: (1) in zero bias, the capacitance of films decreases with temperature slower than it does in a bulk crystal; (2) in films one does not observe a domain contribution to dielectric losses; (3) in films, bias suppresses strongly the capacitance not only in the neighborhood of T_c but at lower temperatures as well; and (4) the maximum in the temperature dependence of permittivity in films is more narrow than that in a bulk crystal.

5.1.1 Dielectric permittivity of films

The permittivity of films can be calculated using the relations derived (Kino & Wagers, 1973) for the capacitance of an IDS located at the interface between the substrate and the film. In the case of a thin film, for $\gamma_0^I h < 1$ and the separation between the IDS electrodes equal to the electrode width, the expression for the capacitance of one pair of IDS electrodes reduces to a simple form

$$C_0 = \left(\frac{\pi^3 w \varepsilon_0}{32}\right)\{1 + \varepsilon_p^{II} + \varepsilon_p^{I}\tanh(\gamma_0^{I} h)\} \quad (3)$$

where w is the IDT electrode length, h is the film thickness, ε_0 is the permittivity of vacuum, $\varepsilon_p = [\varepsilon_{xx}\varepsilon_{zz} - \varepsilon_{xz}^2]^{1/2}$ (x is the direction in the substrate plane perpendicular to the IDS electrodes, z is the direction perpendicular to the substrate plane), and $\gamma_0^I = k\varepsilon_p\varepsilon_{zz}$ ($k = 2\pi/\lambda$, where λ is the spatial IDS field period). The upper indices I and II refer to the film and the substrate, respectively. At room temperature, the relative permittivity along and perpendicular to the polar axis in BPI crystals is $\varepsilon_b \approx 160$–200 ($\varepsilon_{xx}$) and $\varepsilon_a \approx \varepsilon_c \approx 10$ (ε_{zz}), respectively. For quartz, $\varepsilon_{xx} \approx \varepsilon_{zz} \approx 5$. The values of ε_p for the film and the substrate are $\approx$ 45 and $\approx$ 5, accordingly, and $\gamma_0{}^1 h \approx 0.28$ for the film thickness h = 1 μm and λ = 100 μm. For these values of the parameters, $\gamma_0{}^1 h << 1$, and Eq. (3) can be recast to the form

$$C_0 = (\frac{\pi^3 w \varepsilon_0}{32})\{1 + \varepsilon_p^{II} + \varepsilon_{xx}^{I} kh) \quad (4)$$

The total IDS capacitance (C_{IDS}) is a product of C_0 by the number of IDS electrode pairs. Calculations performed using Eqs. (3) and (4) show that in the absence of a BPI film the capacitance of the IDS (C_{sub}) on quartz and lithium niobate (h = 0) is 9.3 and 40 pF, respectively, in full agreement with the measurements. As follows from calculations, the presence of a film 1 μm thick with the relative permittivity $\varepsilon_{xx} = 200$ and $\varepsilon_{zz} = 10$ increases the IDS capacitance by 13.7 pF on the quartz substrate, and by 4 pF on lithium niobate, exactly what is observed experimentally (Table 1). Whence it follows that the permittivity of a BPI film on quartz at room temperature is close to that of a bulk BPI crystal along the monoclinic b axis. This suggests that the b axis in the single-crystal blocks of the film is oriented approximately perpendicular to the IDS electrodes.

Substrate	C_{sub} calc. (pF)	C_{sub} exp. (pF)	C_{IDS} calc. (pF)	C_{IDS} exp. (pF)
α-SiO_2	9.3	9.3	23.5 (**b**||**k**) 10.2 (**b**⊥**k**)	23.0
$LiNbO_3$	39.8	39.9	43.7 (**b**||**k**) 40.0 (**b**⊥**k**)	43.8

Table 1. Calculated and experimental values of C_{sub} and C_{IDS} for the IDS/substrate and the BPI/IDS/substrate structures at T=294 K.

Equation (4) is valid only for small values $\gamma_0{}^1 h < 0.5$, where the nonlinearity of the tanh($\gamma_0{}^1 h$) function is inessential. For large values of the argument, C is no longer linearly coupled with ε_{xx}. An increase in permittivity ε_{xx} in the region of $\gamma_0{}^1 h > 0.5$ will be accompanied by a slower growth of the IDS capacitance. Calculations show that for a film 1 μm thick on SiO_2 and λ = 100 μm, the lack of proportionality between C and ε_{xx} will produce a noticeable effect for $\varepsilon >$ 1500 (for $\varepsilon_{xx} = 2000$, Eq. (3) yields a capacitance 15% smaller than obtained from Eq. (4)). This should be borne in mind in describing temperature dependences of the permittivity of films, because in the region of the phase transition ε_{xx} can exceed this value. Note, however, that the largest values of permittivity are observed only within a narrow temperature interval in the neighborhood of T_c. At temperatures already a few degrees above or below T_c, the permittivity of BPI decreases considerably, and the condition $\gamma_0{}^1 h < 0.5$ is met. In the case of a BPI film on lithium niobate, with λ = 200 μm, the $\gamma_0{}^1 h < 0.5$ condition is certain to be met, thus validating the use of Eq. (4) in an analysis of the behavior with temperature of the IDS capacitance.

5.1.2 Thermodynamic description of dielectric anomaly

As shown in (Balashova&Lemanov,2000; Balashova et al.,2002), the ferroelectric phase transition at T = T_{c2} = 200–225 K in bulk BPI crystals is initiated by the order parameter η of the high-temperature antiferrodistortive phase transition (T_{c1} = 355 K), which is close to the tricritical point. To analyze the temperature dependences of the capacitance of the structures induced by the variation of permittivity in BPI films, we invoke the thermodynamic potential (1) used for describing the dielectric properties of bulk BPI crystals and include into it the elastic and striction energies:

$$F = \frac{1}{2}\alpha_1\eta^2 + \frac{1}{6}\gamma_1\eta^6 + \frac{1}{2\chi_0}P^2 + \frac{1}{4}\beta_2 P^4 + \frac{1}{2}\xi\eta^2P^2 - \frac{1}{2}s_0\sigma^2 - QP^2\sigma - PE \tag{5}$$

where $\alpha_1 = \lambda_1(T - Tc1)$, $\beta_1 = 0$, $\beta_2 > 0$, $\xi < 0$, **E** is the macroscopic field, χ_0 and s_0 are the background dielectric susceptibility and longitudinal elastic compliance, P is polarization, σ is stress, and Q is the electrostriction constant. The thermodynamic potential (3) can be reduced to a dimensionless form:

$$f = \frac{1}{2}ta^2q^2 + \frac{1}{6}q^6 - ap^2 + \frac{1}{2}p^4 - q^2p^2 - \frac{1}{2}\tilde{\omega}^2 - \Delta s^{1/2}p^2\tilde{\omega} + 2ape \tag{6}$$

where $t = (T - Tc)/\Delta T$ is reduced temperature, $\Delta T = \gamma_1/(\lambda_1\chi_0^2\xi^2)$ defines the temperature interval of stability of the paraelectric antiferrodistortive phase ($q \neq 0$, $p = 0$) (in BPI crystals, $\Delta T = T_{c1} - T_{c2}$, $f = F\cdot\frac{8\beta_2^3\gamma_1^2}{\xi^6}$, $q^2 = \frac{2\beta_2\gamma_1}{\xi^2}\eta^2$, $p^2 = -\frac{2\beta_2^2\gamma_1}{\xi^3}P^2$, $e = \frac{\sqrt{2\gamma_1\beta_2\chi_0}}{(-\xi)^{3/2}}E$, $\varpi^2 = \frac{8s_0\beta_2^3\gamma_1^2}{\xi^6}$,

and $\Delta s = \frac{2Q^2}{s_0\beta_2}$ is the relative value of the jump in the elastic compliance at T_{c2}. The dimensionless parameter $a = \frac{2\beta_2\gamma_1}{\chi_0\xi^3} < 0$ defines the region of stability of the mixed polar phase ($q \neq 0$, $p \neq 0$) and the order of the triggering ferroelectric phase transition. All information on the polarization response of the crystal and the closeness of the ferroelectric phase transition to the tricritical point is confined in the value of the dimensionless parameter a (in bulk BPI crystals, a = -2.5). Inclusion of the elastic and striction energies into the potentials (5) and (6) makes it possible to take into account the effect of substrate-induced film strains in the polarization response (Pertsev et al., 1988). Now the equations of state allowing for the film strains induced by the substrate take on the form

$$\begin{aligned} u_s &= \varpi + \Delta s^{1/2}p^2 \\ q\left(ta^2 + q^4 - 2p^2\right) &= 0 \\ p(-a + p^2 - q^2 - \Delta s^{1/2}\varpi) &= -ae \end{aligned} \tag{7}$$

where u_s is the longitudinal strain induced by the substrate in the film. The equations of state (7) lead one immediately to the following equation of state in the phase with $q \neq 0$, $p \neq 0$:

$$-ap + (1-\Delta s)p^3 - \Delta s^{1/2}u_s p - p\sqrt{2p^2 - ta^2} = -ae \tag{8}$$

The equation of state (8) yields a relation for susceptibility

$$\chi = \frac{x_o(-a + \Delta s^{1/2} p \frac{\partial u_s}{\partial e}) \cdot \sqrt{2p^2 - ta^2}}{[(-a - \delta u_s + 3(1 + \Delta s)p^2) \cdot \sqrt{2p^2 - ta^2} - 4p^2 + ta^2]} \tag{9}$$

Because BPI films evaporated on different substrates have about the same temperature of the maximum of permittivity which is close to the ferroelectric transition point in bulk BPI crystals, one may neglect the effect of static strain exerted by the substrate on the film and set $u_s \approx 0$ in the absence of bias. With a bias applied to the IDS, the ensuing piezoelectric effect gives rise to generation in the piezoelectric substrate of periodic strains, which act on the film with the IDS period. These strains, constant in time but periodic in space, can bring about a spatially periodic variation of the phase transition temperature in the film, which would depend on the sign and magnitude of strains and manifest itself in a broadening of the maximum of permittivity. No noticeable broadening of the maximum is, however, observed; on the contrary, the permittivity peak is more narrow than that in a bulk crystal.
The polarization response can be affected, in addition to static strains, by dynamic strains of the piezoelectric substrate, which vary in space and time under the action of applied field. As evident from Eq. (9), the effect of dynamic strains on the permittivity of a film is defined by the relation $\Delta s^{1/2}\, p\partial u_s/\partial e$. This relation is actually the product of the piezoelectric coefficients in the film and the substrate expressed in relative units. Expressed in dimensional units, it assumes the form QPd_s/s_0, where d_s is the piezoelectric coefficient of the substrate, and $QP/s_0 = h_f$ is the piezoelectric coefficient of the film which derives from linearized electrostriction. Significantly, in the frame of this approach the film is free with respect to static strains but clamped relative to the dynamic ones. The strains in the film originate only from dynamic strains in the substrate. The contribution of dynamic strains should depend on the domain state of the film. Consider the effect of dynamic strains for two variants of the domain state of a BPI film in the ferroelectric phase: (1) the film is single-domain; (2) the film has a periodic domain structure in accordance with the period of the bias applied to the IDS. The ac electric field generated by the IDS in the film can be approximated as

$$E(x,t) = E_{\sim}^{0} \cos(\omega t)\, M(kx) \tag{10}$$

where M(kx) is a meander-type function describing the spatial field distribution (M(kx) = 0 if the x coordinate is at the IDS electrodes, and M(kx) = ±1 if x is between the electrodes), $k = 2\pi/\lambda$, λ is the IDS field period, and E_0 is the electric field amplitude. A dc bias E_{dc} applied to the IDS should have the same spatial distribution as E, i.e., it will be described by the M(kx) function. Neglecting the scattering effects, one may assume that these fields in the film have only one component directed along the x axis.

1. The single-domain state of a film is defined by uniform spontaneous polarization $P = P_0$ in the film plane. The ac electric field applied to the IDS electrodes creates in the substrate and, hence, in the film dynamic strains $u_{sd} = d_s E_{\sim}^{0}\cos(\omega t)M(kx)$, where d_s is the piezoelectric coefficient of the substrate. The net variation of polarization in the film, mediated both by the electric field E and by the external dynamic strains generated by the piezoelectric effect in the film, can be written as

$$\Delta P(t, x) = (\chi + d_s e_f)E_{\sim}^{0}\cos(\omega t)\, M(kx) \tag{11}$$

where χ is the susceptibility and $e_f = \chi_0 h_f$ is the piezoelectric coefficient of the film originating from linearization of the electrostriction. As follows from Eq. (11), both contributions to polarization variation are synchronized both in time and in space with the ac electric field (10). This means that the contribution to the IDS capacitance due to the time-varying polarization $\Delta P(t)$, which is proportional to $M(kx)$, is determined in this case by the parameter $(\chi + e_f\, d_s)$.

2. The polydomain state which is created by a dc field applied to the IDS is characterized by a periodically varying direction of polarization $P = P_0M(kx)$. In this case, the contribution to polarization due to dynamic strains will alternate its sign in accordance with the variation of the direction of polarization. The total variation of polarization can be written as

$$\Delta P(t, x) = \chi E_{\sim}^{0}\cos(\omega t)\, M(kx) + e_f d_s\, E_{\sim}^{0}\cos(\omega t)\, M(kx)^2 =$$

$$(\chi + e_f d_s\, M(kx)) E_{\sim}^{0}\cos(\omega t)\, M(kx) \qquad (12)$$

As seen from Eq. (12), the contribution of dynamic strains to the variation of polarization is not synchronized in space with that of the electric field (10). The contribution of the film to the IDS capacitance will be determined by the magnitude of χ, and the alternating contribution of dynamic strains will disappear when summation is performed over x. Thus, the contribution of the substrate-induced dynamic strains to the IDS capacitance associated with the film can be significant only if the ferroelectric film is in the single-domain state. Formation of a periodic domain structure precludes appearance of the dynamic strain effect completely.
Equation (9) for the susceptibility corresponds to the case of a single-domain ferroelectric film where the contribution due to dynamic strains is described by the term $\Delta s^{1/2} p\partial u_s / \partial e$. In the case of a polydomain film with a period of the domain structure equal to that of IDS this term in Eq. (9) should be replaced with $\Delta s^{1/2} p(1 - 2\nu)\partial u_s / \partial e$, with a proper allowance for domains with the negative (ν) and positive ($1 - \nu$) direction of polarization.
The maxima in permittivity observed experimentally in BPI films turned out more narrow than the ones calculated from Eq. (9). This cannot be attributed to the presence or absence of the dynamic strain contribution, because it is small in the vicinity of the phase transition, where the spontaneous polarization is small. The narrowing of the dielectric anomalies may be associated with the presence of the depolarizing electric field which reduces the external field e by an amount (e – np) proportional to n, i.e., the effective depolarizing factor expressed in relative units ($n = 4\pi N\chi_0$). Now Eq. (9) for the susceptibility with allowance for dynamic strains and the depolarizing factor takes on the form

$$\chi = \frac{\chi_0(-a + \Delta s^{1/2}(1-2\nu)p)\dfrac{\partial u_s}{\partial e}) \cdot \sqrt{2p^2 - ta^2}}{[(-a(1+n) + 3(1+\Delta s)p^2) \cdot \sqrt{2p^2 - ta^2} - 4p^2 + ta^2]} \qquad (13)$$

where p(t) is calculated from the equation of state (8). In the absence of bias, the quantity $\chi(0)$ in the polar mixed phase ($q \neq 0$, $p \neq 0$) can be extracted from Eq. (13) using the values of p(t) calculated from Eq. (8) at $e = 0$, and that in the paraelectric antiferrodistorsive phase ($\eta \neq 0$, $p = 0$), from Eq. (13) with $p(t) = 0$. To obtain permittivity in the field $\chi(E)$ using Eq. (13), the values of p(t) are calculated numerically from the equation of state (8) for $e \neq 0$. Figures 11 and 12 plot the temperature dependences of the capacitance in different bias fields calculated using Eq. (13) for the BPI/SiO_2 and BPI/$LiNbO_3$ structures, respectively.

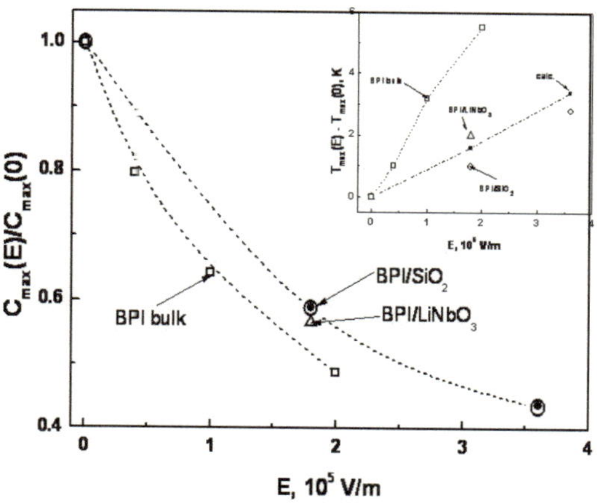

Fig. 15. Bias electric field dependences of the normalized capacity $C_{max}(E)/C_{max}(0)$ in the BPI/α-SiO_2 structure and bulk BPI crystal. The insert shows the electric field dependence of $T_{max}(E) - T_{max}(0)$. Empty circles are calculations (Balashova et al.,2009b).

Figure 15 presents comparison of experimental and calculated bias electric field dependences of the normalized capacity $C_{max}(E)/C_{max}(0)$ in the BPI/α-SiO_2, BPI/$LiNbO_3$ structures and bulk BPI crystal, and the electric field dependence of $T_{max}(E) - T_{max}(0)$ (T_{max} is a temperature of maximal capacitance).

Good agreement between the experimental and calculated dependences (Figs. 11, 12, 15) was reached by allowing for (1) the extent to which the film was in the singledomain state, (2) dynamic strains, and (3) depolarizing field. In the absence of bias, it was assumed that the film was single-domain ($\nu = 0$) and, accordingly, the contribution due to dynamic strains was fully taken into account. With the bias applied, the best fit to experimental curves was reached for $\nu \neq 0$. The value of ν defining the volume of domains of opposite sign in the maximum field was 25%. The contribution of dynamic strains was taken into account accordingly. The effective depolarizing factor in both structures turned out to be small, $N \approx 0.001$. The low depolarizing field could possibly be assigned to the appearance of uncompensated pinned charges at the free boundary of the films.

5.2 Centrosymmetric substrates: α-Al_2O_3, $NdGaO_3$

The temperature dependences of the capacitance and in an as grown BPI film (after first stage of crystallization) on sapphire α-Al_2O_3 are shown in Figs. 16a and 16b. The same dependences, but measured a month after film preparation (after second stage of crystallization), are presented in Figs. 16c and 16d. One can see that there is almost no change in the T_c temperature over time, and only the height of the capacitance peak at $T = T_c$ slightly decreases. The frequency dispersion of the capacitance and losses in the temperature range $T > 260$ K completely disappears a month after film preparation. In contrast, the dispersion of the dielectric response remains below T_c, which indicates that the motion of the domain walls contributes to the dielectric relaxation. This fact suggests that the BPI/α-Al_2O_3 films have a domain structure below T_c, and domain walls are more weakly bound with defects at the interface between the centrosymmetric substrate and film than in the case of substrates with piezoelectric properties (BPI/$LiNbO_3$ and BPI/α-SiO_2).

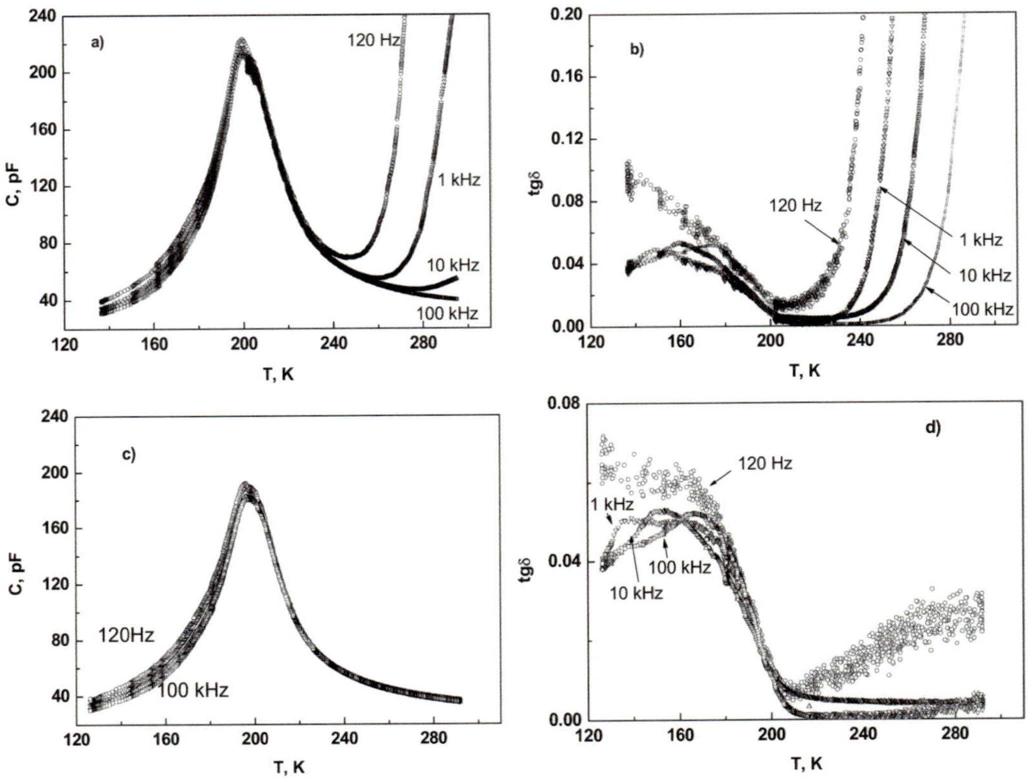

Fig. 16. Temperature dependences of the capacitance and tgδ for the BPI/α-Al_2O_3 structure, measured (a, b) immediately after preparation and (c, d) a month later (Balashova et al., 2011a).

Another difference in the properties of films grown on centrosymmetric and noncentrosymmetric substrates is the contribution that the interaction between the substrate and film makes to the structure capacitance. This interaction is due to the dynamic deformations, caused in a piezoelectric substrate by the electric field applied to the interdigital electrodes (reverse piezoelectric effect), which deform the film and lead to an additional contribution to the structure capacitance at $T < T_c$ (because of the direct piezoelectric effect). This interaction leads to a slower decrease in capacitance with a decrease in temperature below T_c than in the bulk crystal. Obviously, this effect should not occur in the case of centrosymmetric substrates, where dynamic deformations are absent. The dependences that the normalized capacitance C_n has on the difference $T - T_c$ for the BPI/α-SiO_2, BPI/$LiNbO_3$, and BPI/α-Al_2O_3 structures are compared in Fig. 17. The capacitance was normalized to make the dependence $C_n(T - T_c)$ for all structures coincide in the paraelectric phase. It can be seen that, for a structure with a centrosymmetric substrate (BPI/α-Al_2O_3), the capacitance decreases much more rapidly with a decrease in temperature below T_c. Even at $T - T_c = -40$ K, the capacitance of this structure almost coincides with the background, to which it tends at high temperatures ($(T - T_c) \approx 60$ K). In contrast, for the BPI/α-SiO_2 and BPI/$LiNbO_3$ structures, the capacitance at $T - T_c = -40$ K exceeds the background value by a factor of 2.

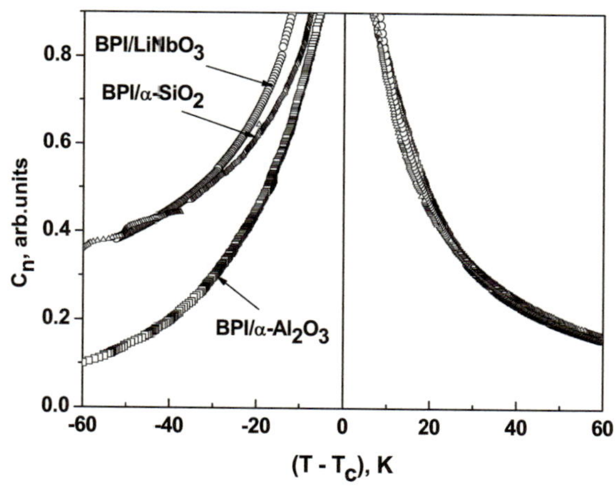

Fig. 17. Temperature dependences of the normalized capacitance of BPI based structures grown on noncentrosymmetric $LiNbO_3$ and α-SiO_2 substrates and on a centrosymmetric α-Al_2O_3 substrate (Balashova et al., 2011a).

6. Small signal dielectric response in DBPI films

The temperature dependences of the capacitance and dielectric losses in the as grown structure obtained from a solution of BPI single crystal in D_2O on an $NdGaO_3$ substrate are shown in Fig. 18. The phase transition temperature of the initial BPI single crystals used to grow the film was $T_c \approx 200$ K. The recrystallization of BPI from a solution in heavy water leads to an increase in T_c by 20 K, which corresponds to a degree of deuteration of about 20%. Analogously to the BPI/α-Al_2O_3 structure, there is the dispersion of capacitance and tanδ in the ferroelectric phase of the DBPI/$NdGaO_3$ structure, which is indicative of the contribution that domain walls make to the dielectric relaxation. A fast decrease in the capacitance with a decrease in temperature below T_c is observed for both DBPI/$NdGaO_3$ and BPI/α-Al_2O_3 structures.

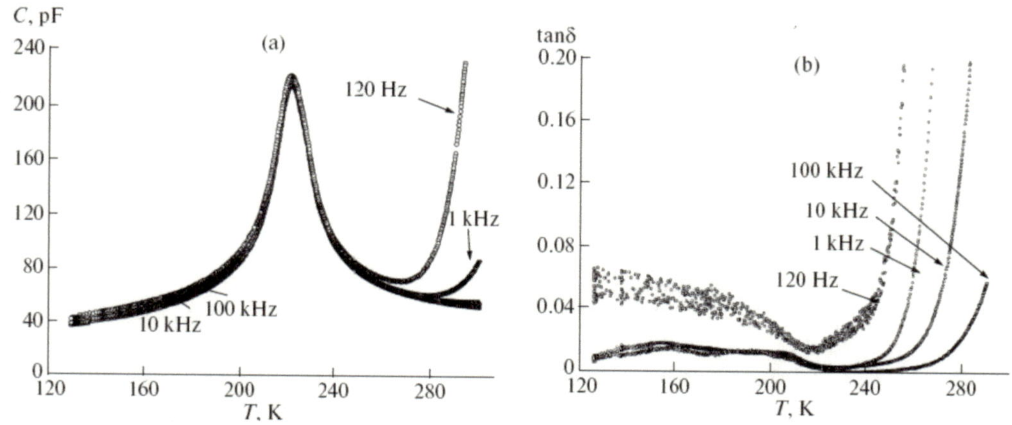

Fig. 18. Temperature dependences of the capacitance (a) and tanδ (b) for the DBPI/$NdGaO_3$ structure (Balashova et al., 2011a).

Figure 19 shows the temperature dependences of the capacitance of the structures grown from a solution of DBPI crystals (obtained by the recrystallization of BPI in heavy water) in D_2O. The permittivity peak for these structures is in the temperature range of 270–285 K, which corresponds to the degree of deuteration 80–90%. In comparison with the BPI-based samples, the capacitance anomaly is significantly broadened in these structures, which is especially pronounced for the film grown on an α-Al_2O_3 substrate. This broadening can be caused by the nonuniform distribution of the degree of deuteration in the film, which should diffuse the phase transition. The difference in the heights of capacitance peaks for the structures shown in Fig. 19 is due to the different orientations of the polar axes in blocks with respect to the normal to the direction of interdigital electrodes and with some differences in film thickness.

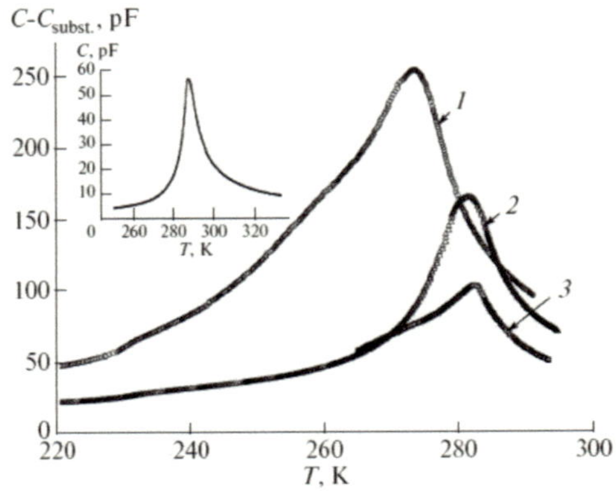

Fig. 19. Temperature dependences of the capacitance introduced into the structure by the DBPI film ($C - C_{substr}$), measured at f = 100 kHz, in DBPI-based structures on different substrates: (*1*) α-Al_2O_3, (*2*) SiO_2, and (*3*) $NdGaO_3$. The inset shows the temperature dependence of the capacitance in the initial DBPI crystal (Balashova et al., 2011a).

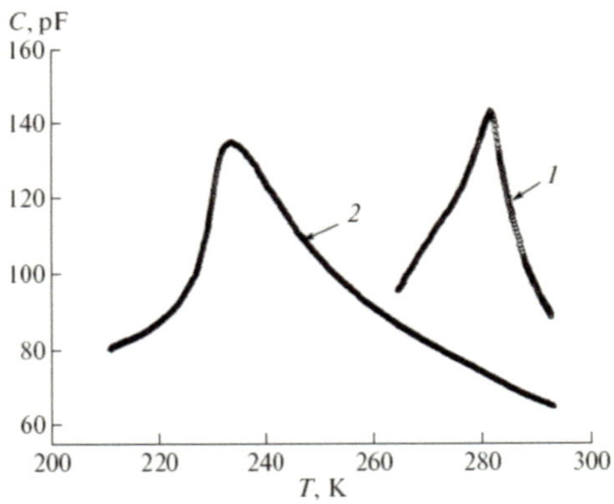

Fig. 20. Temperature dependences of the capacitance for the DBPI/$NdGaO_3$ structure measured (1) immediately after preparation and (2) after a month (Balashova et al., 2011a).

The structures based on highly deuterated DBPI are less stable than the BPI-based structures. This manifests itself in a slow decrease in the transition temperature of these structures with time. Figure 20 shows the temperature dependences of the capacitance of the DBPI/$NdGaO_3$ structure measured at f = 100 kHz immediately after the growth and a month later. The relatively narrow capacitance peak at T = 282 K observed in the as-grown structure is transformed into a diffuse anomaly with a maximum at T = 233 K. This indicates that very slow changes occur in the film structure which are absent in the bulk DBPI crystals. One suggests that this behavior is caused by a gradual decrease in the deuterium concentration in the film due to the diffusion induced replacement of deuterium with protons from the environment.

A study of variation of the dielectric anomaly in DBPI/$NdGaO_3$ structures with time has shown that the position of capacitance maximum gradually decreases in the temperature scale. This is followed by a variable broadening the dielectric anomaly. Figure 21 presents (a) the temperature T_{max} corresponding to maximum of capacitance, and (b) maximal value of capacitance C_{max} and width W of dielectric anomaly (W is a temperature interval in which $C(T)-C_{sub}$ exceeds $(C_{max}-C_{sub})/2$) versus time t (in days) after preparation of the structure.

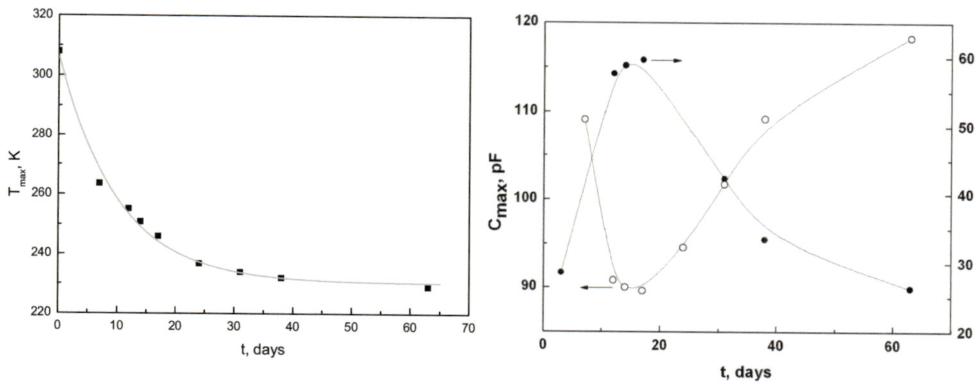

Fig. 21. (a) Time dependence of the temperature T_{max} corresponding to maximal capacitance C_{max} and (b) time dependence of C_{max} and width W of dielectric anomaly on DBPI/$NdGaO_3$ structure.

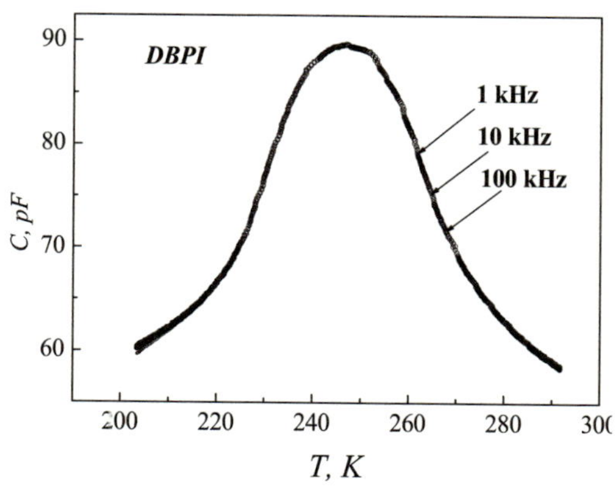

Fig. 22. Temperature dependence of capacity in DBPI/$NdGaO_3$ structure measured in 12 days after preparation at different frequencies.

Time dependence of T_{max} can be well approximated by exponential function $T_m = T_c(bulk) - \Delta T[1-\exp(-t/\tau)]$, where $T_c(bulk) = 308$ K is the phase transition temperature for bulk DBPI crystals used for preparation of film, $\Delta T = 77$ K, and $\tau \approx 10$ days. The width of dielectric anomaly W shows their maximal value after 12 days after preparation and much lower values for t = 7 and 63 days. In contrast to that C_{max} has their minimal value at t = 12 days. It is worth noting that the broadening of dielectric anomaly is not accompanied by dispersion of capacity and appearance of dielectric losses. Fig.22 shows the temperature dependence of capacity measured on different frequencies in t = 12 days after preparation when maximal broadening of dielectric anomaly has been observed. The absence of frequency dispersion proves that the broadening of dielectric anomaly is not related with appearance in films of "relaxors" as it takes place in the case of "diffused" phase transitions (Cross, 1987).

The gradual lowering of the transition temperature T_c and variable broadening of dielectric anomaly indicate that the crystal structure of film remains unchanged with time but the concentration and distribution of deuterium and hydrogen ions in the crystal structure is time dependent. This can happens due to substitution of deuterium ions D^+ by hydrogen ions H^+ from the environment on the surface of film and interdiffusion of H^+ and D^+ ions inside the film. Note that interdiffusion of D^+ and H^+ should go between crystallographic positions in which these ions link the tetrahedra HPO_3 by hydrogen bonds because ions in these positions are responsible for the temperature of phase transition. As a result the distribution of deuterium (hydrogen) ions in a film depends on time and is non-uniform along film thickness. The concentration of D^+ ions is changed with time from the case of totally deuterated film just after the crystallization to the case of DBPI film with small deuteration degree ($T_c \approx 230$ K, D ≈ 20 %). The speed of this process depends on parameters describing the substitution $H^+ \rightarrow D^+$ on the surface and interdiffusion $H^+ \leftrightarrow D^+$ inside of film.

The non-uniform distribution of D^+ ions results in distribution of phase transition temperature inside of film. For this reason different regions of the film differently contribute to total capacity C of the structure. Minimal value of W are observed for t = 63 days that is close to W value in BPI films. This indicates that after two months the DBPI film has uniform (or very close to uniform) distribution of deuterium ions (and also T_c). Maximal broadening of dielectric anomaly suggest the biggest dispersion of deuterinm concentration in the film. Comparing the maximal and minimal values of W we can estimate the temperature interval ΔT_1 in which phase transition temperatures T_c of different regions of film are distributed $\Delta T_1 = W_{max} - W_{min} \approx 30$ K. This corresponds to distribution of the deuteration degree in different regions of the film from 30% up to 60%.

7. Small signal dielectric response in TGS films

The dielectric properties of TGS films grown on Al/SiO_2 were studied in the direction normal to the film plane. For this purpose, a silver electrode was pasted onto the upper surface of the film. An aluminum layer served as the bottom electrode. The dielectric properties of TGS films on α-Al_2O_3 substrates were studied in the film plane. We measured temperature dependences of the capacity and loss tangent of the interdigitated structure following film deposition.

Figure 23 displays temperature dependences of the capacity and loss tangent in $TGS/Al/SiO_2$ (a) and TGS/α-Al_2O_3 (b) films measured on different frequencies within a range of 0.12–100 kHz. In both cases, the behavior of the capacity with temperature passes through maxima at the ferroelectric transition point in a bulk crystal T_c = 322 K. In films on Al/SiO_2, the capacity at its maximum measured at 100 kHz increases about threefold compared with that at room temperature. The capacity of the interdigital structure on α-Al_2O_3 substrates measured without a TGS film was about 19 pF at room temperature and increased with the film in place by about 3 pF. Because the permittivity of the substrate depends only weakly on temperature, the increase of the capacity of the structure should be assigned to the increase of the film permittivity. The capacity of TGS/α-Al_2O_3 films measured with the interdigital structure at its maximum at T = 322 K was found to exceed about sevenfold that at room temperature.

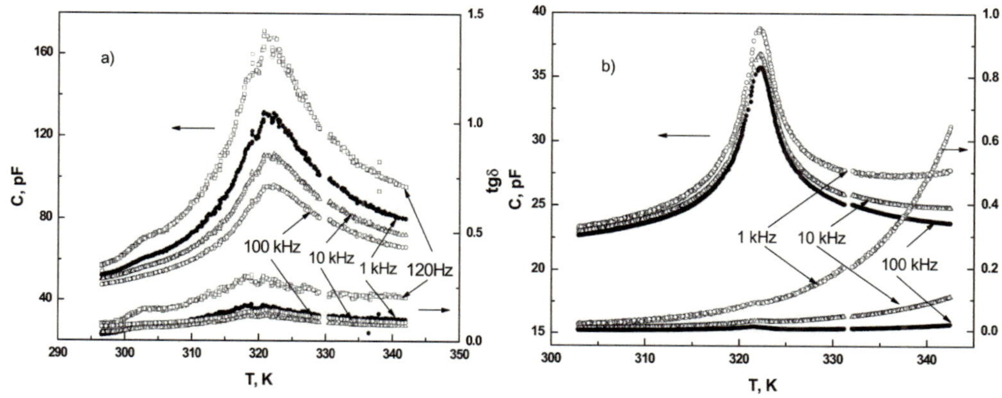

Fig. 23. Temperature dependences of the capacity and tanδ of TGS films grown on substrates of (a) Al/SiO_2 and (b) Al_2O_3 and studied at different frequencies (Balashova et al.,2010).

The maximum permittivity of films on an Al/SiO_2 substrate did not exceed $\varepsilon \approx 100$ whereas in bulk crystals used to grow the films the permittivity measured along the polar axis was $\varepsilon \approx 2000$. This suggests that the fraction of crystallites with the polar axes oriented perpendicular to the film plane is relatively small. The permittivity of TGS/α-Al_2O_3 films turned out to be substantially larger, $\varepsilon \approx 1000$. This suggests strongly that the polar axis orientation in TGS crystallites is close to the direction orthogonal to the interdigitated electrodes. The permittivity of the films was calculated with the use of the relation (3). The total capacity of the interdigitated structure is found as the product of C_0 by the number of electrode pairs N.

$$C = \left(\frac{\pi^3 wN\varepsilon_0}{32}\right)\left\{1 + 9.5 + (6.5\varepsilon_{xx})^{0.5}\tanh\left[\frac{2\pi h}{\lambda}\left(\frac{\varepsilon_{xx}}{6.5}\right)^{0.5}\right]\right\} \tag{14}$$

where ε_0 is the permittivity of vacuum, N = 35, w = $6 \cdot 10^{-3}$ m, $\varepsilon_p^I(T) \approx (\varepsilon_{xx}(T)\varepsilon_{zz})^{0.5}$, assuming $\varepsilon_{zz} \approx 6.5$ to be temperature independent. Calculations yield 19.3 pF for the capacity of an interdigitated structure on white sapphire without a TGS film (h = 0), in full agreement with experiment. To calculate the capacity of an interdigitated structure with the film in place, we substituted in Eq. (14) the experimental temperature dependences of permittivity along the polar axis of the bulk TGS crystals used in film preparation. The Curie–Weiss constant of the crystals was C_+ = 3500 K, and the relative permittivity at the maximum ε =

2000. In calculations using Eq. (14), the film thickness *h* served as a fitting parameter. The results of the calculations are confronted in Fig. 24 with the experimental relation for the capacity of an interdigital structure coated by a TGS film. The experimental temperature dependences of the capacity of the structure measured at 100 kHz were found to fit the calculated values for an average film thickness h_m = 0.2 μm. The excess of the experimental over calculated values of the capacity of the structure in ferroelectric phase can tentatively be assigned to the contributions of domain wall motion to the permittivity of a film and a bulk crystal being different.

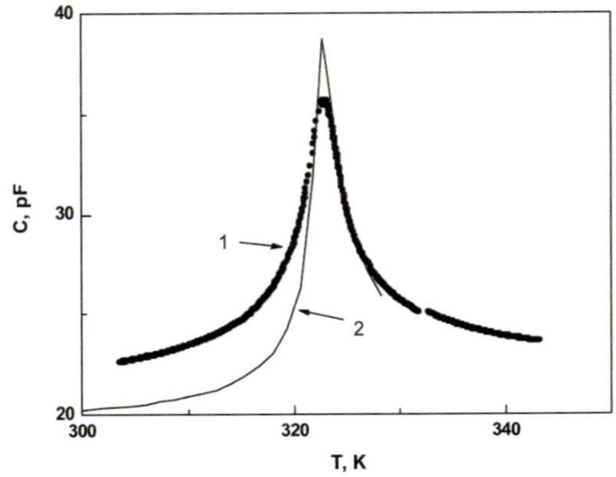

Fig. 24. (1) Experimental and (2) calculated temperature dependences of the capacity of the TGS/α-Al_2O_3 structure. The calculation was performed for the case of the film polar axis perpendicular to the electrodes of the interdigital structure (Balashova et al.,2010).

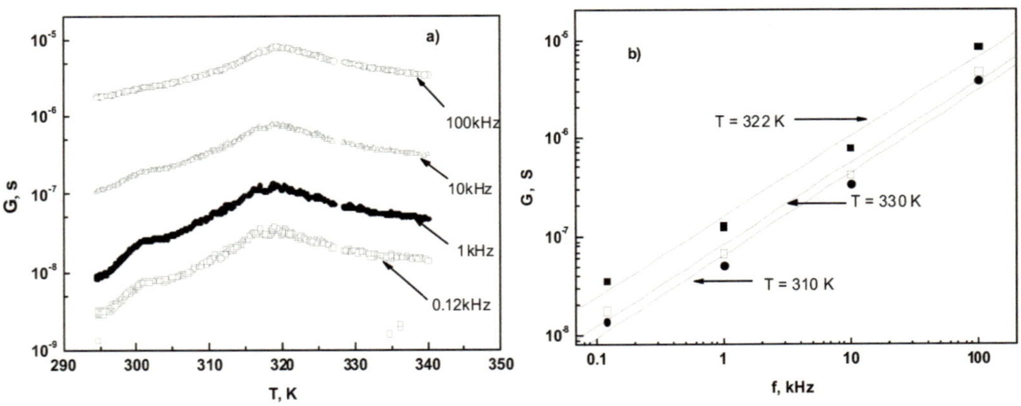

Fig. 25. (a) Temperature dependences of the conductivity on a semilogarithmic scale and (b) frequency dependences of the conductivity on a double logarithmic scale obtained at different temperatures for TGS/Al/SiO2 films (Balashova et al.,2010).

In TGS films on an Al/SiO_2 substrate, the maximum values of tanδ are observed in the region of the maximum in capacity. By contrast, in TGS/α-Al_2O_3 films one found in this region weak anomalies in tanδ which become manifest against a strong growth of losses with increasing temperature. A similar behavior of tanδ with temperature at different frequencies was observed also in films of betaine phosphite (BPI) grown on $LiNbO_3$

substrates (Fig.13b). The temperature dependences of permittivity and tanδ in $TGS/Al/SiO_2$ and $TGS/\alpha\text{-}Al_2O_3$ films reveal frequency dispersion in the frequency and temperature intervals studied (Fig. 23). To analyze the nature of this dispersion, calculations of the conductivity G =ωCtanδ, where ω is circular frequency, and C, the film capacity, were performed.

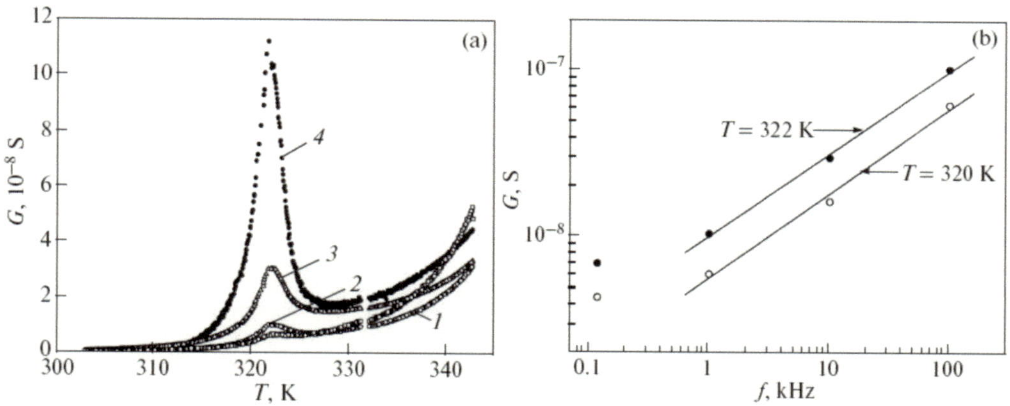

Fig. 26. (a) Temperature dependences of the conductivity in $TGS/\alpha\text{-}Al_2O_3$ films obtained at frequencies of (1) 0.12, (2) 1, (3) 10, and (4) 100 kHz. (b) frequency dependences of the conductivity on a double logarithmic scale obtained for two temperatures (Balashova et al.,2010).

Figure 25a plots temperature dependences of the conductivity G, and Fig. 25b, frequency dependences of the conductivity of $TGS/Al/SiO_2$ for three temperatures. In the low-frequency region (f < 1 MHz), the film conductivity can be mediated by dc and ac conductivity: $G(\omega) = G_{dc} + G_{ac}$, where G_{dc} is the frequency-independent dc conductivity, and $G_{ac} = A\omega^s$ ($s \leq 1$), the frequency dependent ac conductivity (Pike, 1972). The straight lines in Fig. 25b suggest that in $TGS/Al/SiO_2$ films the dc conductivity contributes very little, the main contribution being provided by the ac conductivity G_{ac}. The frequency dependences of the ac conductivity (Fig. 25b) permit determination of the parameter s $\approx$0.82. The deviation of the frequency dependence of the ac conductivity from a linear course (s = 1) signals a manifestation of specific conductivity mechanisms. The most plausible of them appears to be hopping charge transport between localized states separated by an energy barrier. The theoretical model of such a conductivity considered in (Pike, 1972) allows for carrier hopping between barrier-separated states distributed randomly in the bulk of the sample. This model was successfully used to describe the low-frequency conductivity in thin films of scandium oxide (Pike, 1972), as well as in betaine phosphate crystals with a 5% addition of BPI (Huttun et al. 1991). The value of the *s* parameter in the frequency dependence of ac conductivity permits one to estimate the energy difference between the ground state at the energy minimum and the free state in which a carrier can travel over the lattice, $W_m = 6kT/(1 - s)$. Calculations yield $W_m \sim 0.9$ eV for the $TGS/Al/SiO_2$ structure. The values of W_m and ε can now be used to determine the Bohr radius of a localized carrier, $a \approx e^2/2\varepsilon W_m \approx$ 10 A that is close to the TGS lattice parameter.

The temperature and frequency plots of conductivity in the $TGS/\alpha\text{-}Al_2O_3$ structures displayed in Fig. 26 differ noticeably from those for $TGS/Al/SiO_2$. In $TGS/\alpha\text{-}Al_2O_3$ films, at temperatures above (T >325 K) and below (T < 315 K) the phase transition point T_c = 322 K

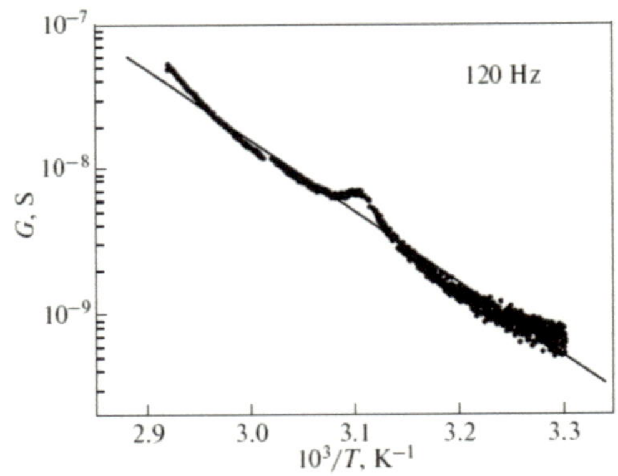

Fig. 27. Temperature dependence of the conductivity in TGS/α-Al_2O_3 films at f = 120 Hz (Balashova et al.,2010).

the frequency dependence of conductivity is fairly weak, so that the temperature dependences of conductivity drawn for different frequencies practically coincide (Fig. 26a). This suggests that within these temperature regions the dominant contribution comes from dc conductivity mediated by thermally activated diffusion. Around the phase transition point, 315 < T < 325 K, however, an additional contribution to conductivity appears that displays strong dispersion (Fig. 26a). The frequency difference of conductivity in the vicinity of the phase transition is shown graphically for two temperatures in Fig. 26b. Above f = 1 kHz, this relation is linear with a slope s ≈ 0.5. In the low-frequency domain, f < 1 kHz, the conductivity is seen to depend very weakly on frequency. This suggests that it is dominated by dc conductivity. Figure 27 plots the temperature dependence of conductivity for f =120 Hz drawn on a log-normal scale. The activation energy estimated from this slope is $E_a \approx 0.9$–1 eV. The same value of activation energy was obtained for DTGS crystals (Shilnikov et al., 2001). As follows from Fig. 27, near the phase transition conductivity passes through a small maximum. A similar maximum in conductivity at frequencies of 1–10 Hz was observed in DTGS single crystals (Shilnikov et al., 2001). Significantly, the activation energy $E_a \approx 0.9$–1 eV in TGS films exceeds by about an order of magnitude that observed in BPI films ($E_a \approx 0.1$ eV) grown by evaporation on $LiNbO_3$ (Y) and α-SiO_2 (Z) substrates. The manifestation of ac conductivity in the temperature interval 315 < T < 325 K suggests operation of an additional conductivity mechanism near the phase transition that was absent far from T_c. Viewed in the context of the hopping conduction model (Pike, 1972), this could signal formation of barriers and a decrease in the degree of carrier localization in the phase transition region. Note, however, that if the potentials are of the Coulomb type (V ~ 1/εr, where r is the distance from the local center to the carrier), the degree of carrier localization in this temperature interval should increase rather than decrease. Because close to T_c permittivity grows, the Coulomb potentials localizing the carriers become more narrow, thus reducing the potential overlap between neighboring centers. Whence it follows that the ac conductivity becoming manifest in the vicinity of T_c is not associated with hopping conduction and is rather determined by other mechanisms. Studies of TGS crystals revealed the existence within a narrow temperature interval close to T_c of a low-frequency relaxation contribution to permittivity whose magnitude is very sensitive to the presence of defects and to the crystal quality. The most probable mechanism of this contribution, as well as of the ac conductivity near T_c in TGS/α-Al_2O_3 structures, is the motion of domain walls or small-radius domain nuclei, which contribute noticeably to permittivity and mediate the low-frequency dielectric dispersion near the phase transition (Pawlaczyk, 1993).

8. Strong signal dielectric response in DBPI films

Strong signal dielectric response (dielectric hysteresis loops) was investigated by means of Sawyer-Tower method in the temperature range T =(120 – 340) K and frequency range f = (0.06 – 3) kHz. AC voltage with amplitude of $U_E \approx 100$ V (that corresponds to electric field of $E \approx 2*10^6$ V/m between electrodes), was applied to electrodes of IDS, and the voltage U_p proportional to polarization between electrodes was measured from the capacitance C = 0.3 μF included in series with the IDS.

Here we present the results of investigation of strong signal dielectric response in DBPI films with small deuteration degree ~20% ($T_c \approx 227$ K) consisting of one single crystal block (Balashova et al. 2011b). Dielectric loops in films with high degree of deuteration (~ 80%) will be published elsewhere. Fig. 28 shows the dependence of U_p on U_E at frequency f = 250 Hz for different temperatures. In paraelectric region at T > 240 K a dependence $U_P(U_E)$ is linear and the structure represents a linear dielectric. In vicinity of T_c the dependence $U_P(U_E)$ is nonlinear and at $T < T_c$ hysteresis loops appear.

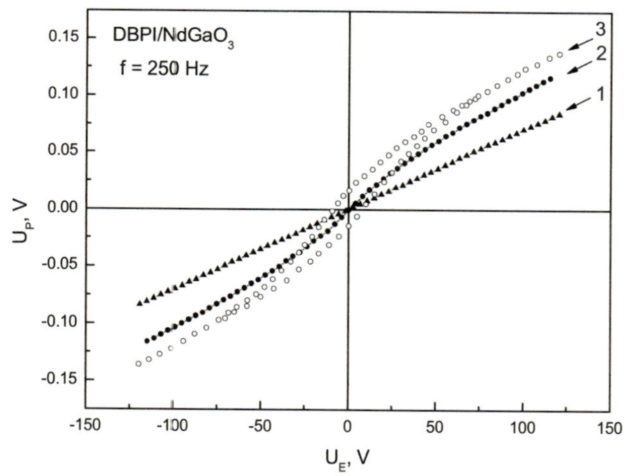

Fig. 28. $U_p(U_E)$ dependence in DBPI/$NdGaO_3$ structure at f = 250 Hz for three temperatures 1 – 258 K, 2- 232 K, 3 – 155 K (Balashova et al., 2011b).

Hysteresis loops, obtained after subtraction of linear dependence $U_P(U_E)$ which is observed for large values of U_E when the structure is in saturation, for different frequencies at T = 192 K and T =144 K (below T_c = 227K) are shown in Fig. 29. For low frequencies f ~ (0.06 – 0.1) kHz the hysteresis loops are very narrow. The width of loops considerably increases with frequency and for f ~ (2.5-3) kHz very wide loops are observed.

Fig. 30 shows the temperature dependence of magnitude of U_p voltage in saturated state of the film ($U_P{}^s$). The $U_P{}^s(T)$ does not depend on frequency. Non-zero values of $U_P{}^s$ in the region $T_c < T < 240$ K are related with dielectric non-linearity of ferroelectrics above the phase transition temperature. Below T_c the temperature dependence of $U_P{}^s$ is analogous to one of spontaneous polarization P_s.

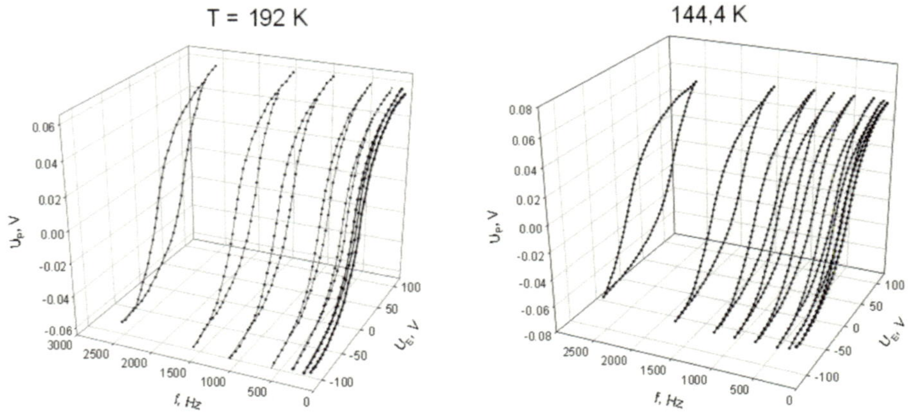

Fig. 29. Hysteresis loops in DBPI/$NdGaO_3$ structure for different frequencies at T = 182 K and 144.4 K (Balashova et al., 2011b).

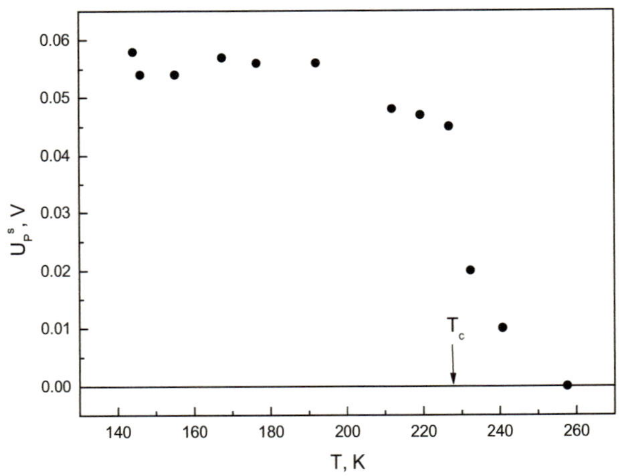

Fig. 30. Temperature dependence of U_P^s in DBPI/$NdGaO_3$ structure (Balashova et al., 2011b).

Fig.31 presents the temperature dependence of remnant voltage U_P^r ($U_P^r = U_P$ at U_E=0) for different frequencies (a) and frequency dependence of U_P^r for different temperatures (b). In contrast to U_P^s the remnant voltage exhibits strong frequency dispersion. The magnitude of U_P^r increases with frequency. Opening the loop takes place at a temperature T_{op} which depends on the frequency f. For f = 60 Hz and f = 120 Hz the non-zero values of U_P^r are observed for T < 180 K and T < 220 K correspondingly. At the further increase in frequency T_{op} increases up to T_{op} = 240 K for f = 3 kHz.

Strong signal dielectric response in DBPI/IDS/substrate structures is principally different from that observed in ferroelectric single crystals with use of plane parallel capacitance geometry. It is well known that in the last case the most wide dielectric loops are observed for the lowest frequencies. An increase of frequency results in gradual narrowing and disappearance of the loop (Tagantsev et al. 2010). We have observed this effect in bulk DBPI samples. In structures under study a situation is absolutely opposite – the loop aria is increased with frequency (Fig. 29). The difference in frequency behavior is related with difference of saturated states of uniaxial ferroelectric in the case of plane parallel capacitance and IDS structure. In the first case the saturated state is a single domain state which appears at $U_E = U_E^s$(Fig. 32a,b). This state remains at reduction of U_P to zero because the depolarizing fields are compensated by charges on electrodes. When U_E is reversed a polarization switching takes place. This process begins with appearance of the nucleus with reversed polarization and following movement of domain walls. The situation is quite different in the case of IDS. Let us consider an uniaxial ferroelectric film which which polar axis is orthogonal to the electrodes of IDS. In this case a saturated state at $U_E = U_E^s$ is multidomain state consisting of stripe domains type "head-to head" (Fig.32c). At $U_E = -U_E^s$ the saturated state also consists on stripe domains with opposite direction of polarization in stripes (Fig. 32d). Because the saturated state is multidomain the transition from one configuration of domains to another one can occur by movement of domain walls. Necessity of nucleation of domains with reversed polarization direction thus is absent.

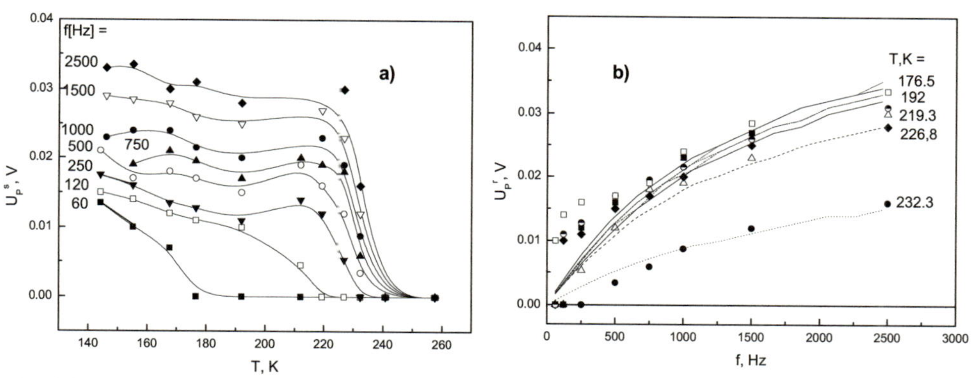

Fig. 31. (a) Temperature dependence of remnant voltage U_P^r for different frequencies and (b) frequency dependence of U_P^r for different temperatures (Balashova et al., 2011b).

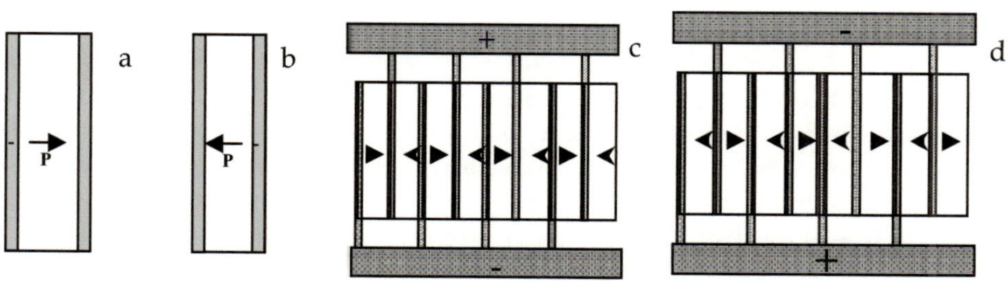

Fig. 32. Schematic image of saturated states of uniaxial ferroelectric in the scheme of plane parallel capacitance (a,b) and planar IDS (c, d) (Balashova et al., 2011b).

It is worth noting that an appearance in ferroelectrics of a system of "head-to head" domains is energetically non-profitable (Strukov &Levanyuk, 1998). Because in this case the domain walls are charged very strong depolarization fields appear. In IDS structure these fields can partially be compensated due to charges on IDS electrodes but full compensation doesn't arise because of a non-zero thickness of the film. As a result in IDS structure the system of stripe "head-to head" domains aspires to collapse and completely disappears at $U_E = 0$.

At low frequency of the driving field U_E the domain walls are in-time with change of applied field and dielectric hysteresis loops are absent or very narrow. At frequency increase, because the speed of movement of walls is limited, they are behind changes of the field and loops are open. As walls between "head-to head" domains are charged, their movement in electric field is accompanied by charge carrying. It is similar to occurrence in the structure of electric conductivity, however, in this case the conductivity is defined not by free but bonded charges which arise only in ferroelectric state.

Opening of dielectric hysteresis loop with frequency increase can be described if we know nonlinear non-hysteretic dependence $U_P(U_E)$ which is observed at low frequency and suppose that at step-like change of U_E ($U_E \rightarrow U_E + \Delta U_E$) at t = 0 the voltage U_P aspires to their equilibrium value $U_P(U_E+\Delta U_E)$ according to exponential law, i.e. $U_P(t) = U_P(U_E) + \Delta U_P[1 - \exp(-t/\tau)]$, where $\Delta U_P = U_P(U_E + \Delta U_E) - U_P(U_E)$, and τ is the relaxation time. In Fig.31b the results of calculation are shown by lines. In calculations the non-hysteretic $U_P(U_E)$ dependence was approximated by Langevin function which describes the nonlinear $U_P(U_E)$ dependence for different temperatures with good accuracy. Calculations show that the magnitude of τ increases with temperature decrease from $\tau \approx 25$ μs at $T \approx T_c$ up to $\tau \approx 45$ μs at T = 150 K. It should be noted that considered model though describes basically the behavior U_P^r (f) and U_P^r (T) nevertheless is not able to describe these dependences precisely, in particular at low frequencies. In particular, there is opened a question why on frequencies f = (60-120) Hz the loop opening occurs at temperatures essentially smaller T_c, unlike higher frequencies at which it occurs at $T \sim T_c$ (Fig. 31a). Probably, it is connected with change of character of movement of domain walls at frequency increase.

9. Conclusions

Main results of studies presented in this chapter can be summarized as follows. First of all the possibility of growth of amino acid ferroelectric film on different substrates is shown. In contrast to other methods of ferroelectric film preparation the evaporation method does not need use of high substrate temperature and is characterized by relative simplicity and low cost. Films prepared by this method are not epitaxial and for this reason the film growth is not strongly related with a mismatch between substrate and film lattice parameters but is governed mainly by wettability of substrate by water solution. The temperature of ferroelectric phase transition is BPI and TGS films coincides with bulk samples that means that a substrate has no influence on T_c. In spite of the fact that the films are not epitaxial they are strongly connected with substrate and clearly exhibit a film-substrate interaction.

Most of films presented in the chapter are polycrystalline. The only exception is the DBPI film grown on $NdGaO_3$ substrate which consists of one single crystal block . BPI and DBPI films consist of large single crystal blocks which dimensions are varied from 0.1 to several tens of mm^2. In contrast to that TGS films consist of blocks with dimensions smaller than 1 mm^2. The polar axis in both TGS and BPI (DBPI) films is oriented in-plane of substrate. For this reason a study of dielectric properties of films was carried out by means of IDS of electrodes deposited on substrate.

Main features of dielectric properties of films coincide with bulk samples. All films exhibit pronounced dielectric anomaly at ferroelectric phase transition analogous to bulk crystals, and similar to bulk the values of dielectric permittivity and the electric field effect on dielectric permittivity. At the same time they show also considerable differences. The most pronounced among them is a transformation with time of deuteration degree observed in

DBPI films. This phenomenon is absent in DBPI crystals. The dielectric anomaly in DBPI films with high degree of deuteration slowly changes with time reflecting changes in distribution of D^+ ions in the film. This process is caused by substitution of deuterium ions by hydrogen ions on the surface and interdiffusion of these ions inside of the film. We believe that this process can be stopped by covering the film surface by thin dielectric layer. If so it would be a method for preparation films with controlled width of dielectric anomaly. Also it is worth noting a specific temperature behavior of dielectric permittivity below T_c in BPI films grown on non-centrosymmetric substrates. In these films in contrast to bulk samples and films grown on centrosymmetric substrates the dielectric anomaly does not exhibit a contribution of domain wall movement.

The possibility to grow films of uniaxial ferroelectrics consisting of large single crystal blocks on IDS structure of electrodes gives rise to observation of specific dielectric hysteresis loops showing very unusual frequency behavior. To the best of our knowledge such very strong increase of remnant voltage and coercive field with frequency has not been observed in ferroelectrics up to now. At low frequency very narrow hysteresis loops with strong non-linear P-E behavior related with ferroelectric domain wall movement have been found.

It should be noted that the evaporation method has also some disadvantages as compared with epitaxial growth. Up to now it is not possible to control the orientation of film crystal axis in the plane of substrate. In most cases the polar axis in blocks is oriented not far from direction orthogonal to electrodes of IDS and therefore dielectric anomaly is large. Nevertheless in some films the anomaly is not strong showing that projection of polar axis on electrodes is small.

The further investigations of structures based on amino acid ferroelectric films could be related with increase of amino acid ferroelectrics used for preparation of films (glycine phosphite, betaine arsenate and so on), probing of different substrates, and study of acoustic and optical properties of obtained structures.

10. Acknowledgment

The authors thanks Prof. Dr. V.V. Lemanov for helpful discussions. This research was partly supported by RFBR (grant 10-02-00557).

11. References

Albers, J.; Klöpperpieper, A.; Rother, H.J. & Haussühl, S. (1988). Ferroelectricity in betaine phosphite, *Ferroelectrics*, Vol.81, pp. 27-30, ISSN 0015-0193.

Albers, J. (1988). Betaine compounds - a new family with ferroelectric and incommensurate phases, *Ferroelectrics*, Vol.78, pp. 3-10, ISSN 0015-0193

Balashova, E.V.; Lemanov, V.V.; Tagantsev, A.K.; Sherman, A.B. & Shomuradov, Sh.H. (1995). Betaine arsenat as a system with two instabilities, *Phys.Rev.B*, Vol. 51, No.14, pp. 8747-8753, ISSN 1098-0121

Balashova, E. V. & Lemanov, V. V. (2000). Acoustic and dielectric properties of betaine phosphite and a phenomenological model with coupled order parameters, *Ferroelectrics*, Vol.247(4), pp. 269-281, ISSN 0015-0193

Balashova, E. V.; Lemanov, V. V. & Klöpperpieper A. (2002). Effect of Electric Field on the Dielectric Permittivity of Betaine Phosphite Crystals in the Paraelectric Phase, *Phys.Sol.State*, Vol.44, No.8, pp. 1597-1603, ISSN 1063-7834

Balashova, E. V. & Lemanov, V. V. (2003). Dielectric Properties of Betaine Phosphite–Betaine Phosphate Solid-Solution Crystals in the Improper Ferroelastic Phase, *Phys.Sol.State*, Vol.45, No.7, pp. 1310-1314, ISSN 1063-7834 [*Fiz.Tverd.Tela*, Vol.45, No.7, pp. 1250-1255]

Balashova, E. V. & Lemanov, V. V. (2003). Dielectric and Acoustic properties of Some Betaine and Glycine Compounds, *Ferroelectrics*, Vol.285, pp. 1179-205, ISSN 0015-0193

Balashova, E.V.; Krichevtsov, B.B.& Lemanov, V.V. (2008). Ferroelectric betaine phosphite films: Growth, optical imaging, and dielectric properties, *J.Appl.Phys.* Vol.104, pp.126 104(1-3), ISSN 0021-8979

Balashova E.V.; Krichevtsov B.B. & Lemanov V.V. (2009). Dielectric properties of betaine phosphite films, *Fiz. Tverd. Tela* (St. Petersburg), Vol.51, No.3, pp. 525-532 [*Phys.Solid State* Vol.51, pp. 560-567, ISSN 1063-7834].

Balashova, E.V.; Krichevtsov; B.B. & Lemanov, V.V. (2009). Growth, optical imaging, and dielectric properties of ferroelectric betaine phosphite and triglycine sulphate films, *Integr. Ferroel.*, Vol.106, pp. 29-39, ISSN 1058-4587

Balashova, E.V.; Krichevtsov, B.B. & Lemanov, V.V. (2010). Permittivity and Conductivity of Triglycine Sulfate films on Al/SiO_2 and α-Al_2O_3 Substrates. *Fiz. Tverd. Tela* (St. Petersburg) Vol.52, No.1. pp.119-123 [*Phys.Solid State* Vol.51, No.1, pp.126-131, ISSN 1063-7834].

Balashova, E.V.; Krichevtsov, B.B.; Zaitseva, N.V.; Pankova, G.A.; Frederiks, I.D & Lemanov, V.V. (2011). Dielectric properties of betaine phosphite and deuterated betaine phosphite films, *Crystallography report*, Vol.56, No.1, pp. 38-43, ISSN 1063-7745 [Kristallografiya, Vol.56, No.1, pp. 42-47, ISSN 0023-4761]

Balashova, E.V.; Krichevtsov, B.B. & Lemanov, V.V. (2011). *Fiz. Tverd. Tela* (St. Petersburg) in press, [*Phys.Solid State,* in press].

Banys, J.; Sobiestianskas, R.; Völkel, G.; Klimm, C.; Klöpperpieper, A. (1996). Microwave dielectric dipersion in deuterated betaine phosphite, *Phys. Status Solidi* (a), Vol.155, No.2, pp. 541-545, ISSN 1862-6300

Bauch, H.; Banys, J.; Bötteher, R.; Pöple, A.; Völkel, G.; Klimm, C. & Klöpperpieper, A. (1995). Structural phase transitions in partially deuterated betaine phosphite studied by dielectric and electron paramagnetic resonance methods, *Ferroelectrics*, Vol. 163, No 1, pp. 59-68, ISSN 0015-0193

Cross, L.E. (1987). Relaxor ferroelectrics, *Ferroelectrics*, Vol. 76, No 1, pp. 241-267, ISSN 0015-0193

Dawber, M; Rabe, K.M. & Scott, J.F. (2005). Physics of thin-film ferroelectric oxides, *Rev. Mod. Phys.* Vol. 77, No. 4, pp. 1083 - 1130, ISSN 0034-6861

Ducharme, S.; Palto, S.P. & Fridkin, V.M. (2002). *Ferroelectrics and dielectric thin films,* In: Handbook of thin film materials, Vol.3, Chapter 11, Academic Press, San Diego

Fehst, I.; Paasch, M.; Hutton, S.L.; Braune, M.; Böhmer, R.; Loidl, A.; Dörffel, M.; Narz, Th.; Haussühl, S. & McIntyre, G.J. (1993). Paraelectric and ferroelectric phases of betaine phosphite: structural, thermodynamic, and dielectric properties. *Ferroelectrics,* Vol.138, pp.1-10, ISSN 0015-0193

Fletcher, S.R.; Keve, E.T. & Skapcki, A.C. (1976). Structural studies of triglycine sulphate part i: low radiation dose (structure a). *Ferroelectrics,* Vol.14, pp. 775-787, ISSN 0015-0193

Holakovsky, J. (1973). A new type of ferroelectric phase transition, *Phys.Stat.Sol. (b),* Vol.56, No.2, pp. 615-619, ISSN 0370-1972

Hutton, S.L.; Fehst, I.; Böhmer, R.; Braune M.; Mertz, B.; Lunkenheimer, P.; Loidl, A. (1991). Proton glass behavior and hopping conductivity in solid solutions of antiferroelectric betaine phosphate and ferroelectric betaine phosphite, *Phys.Rev.Lett.,* Vol. 66, No.15 pp. 1990-1993, ISSN 0031-9007

Kino G. S. & Wagers R. S. (1973). Theory of interdigital couplers on nonpiezoelectric substrates, *J. Appl. Phys.,* Vol.44, No.4, pp. 1480-1488, ISSN 0021-8979

Lal R.B. & Batra A.K. (1993). Growth and properties of triglycine sulfate (TGS) crystals: Review, *Ferroelectrics,* Vol.142, pp. 51-82, ISSN 0015-0193

Matthias, B.T.; Miller, C.E. & Remeika, J.P. (1956). Ferroelectricity of Glycine Sulfate, *Phys.Rev.,* Vol.104, pp.849-850.

Neumann N. (1993). Modified triglycine sulphate for pyroelectric infrared detectors, *Ferroelectrics,* Vol.142, pp. 83-92, ISSN 0015-0193

Pawlaczuk, Cz. (1993). Domain wall dielectric response in the phase transition region of some ferroelectrics, *Ferroelectrics,* Vol. 140, No 1, pp. 127-132, ISSN 0015-0193

Pertsev, N.A.; Zembilgotov & Tagantsev, A.K.(1998). Effect of Mechanical Boundary Conditions on Phase Diagrams of Epitaxial Ferroelectric Thin Films, *Phys.Rev.Lett.,* Vol. 80, No.9, pp. 1988-1991, ISSN 0031-9007

Pike, G.E. (1972). ac Conductivity of scandium oxide and a new hopping model for conductivity, *Phys.Rev.B,* Vol.6, No4, pp. 1572-1580, ISSN 1098-0121

Schaack G. (1990). Experimental results on phase transitions in betaine compounds, *Ferroelectrics,* Vol.104, pp. 147-158, ISSN 0015-0193

Shilnikov, A.P.; Pozdnyakov, A.P.; Nesterov, V.N.; Fedorikhin, V.A. & Shuvalov, L.A. (2001). The contributions from different mechanisms of motion of domain boundaries to the effective permittivity of triglycine sulfate crystals in moderate low-and infralow-frequency fields, *Phys. Sol State,* Vol.43, No 1, pp. 1576-1579, ISSN 1063-7834

Stekhanova. Z. D.; Yatsenko, O. B.; Milovidova, S. D.; Sidorkin, A.S. & Rogazinskaya, O. V. (2005). Properties of triglycine sulfate crystals grown from aqueous solutions, *Appl Chem.,* Vol.78, No.1, pp. 42-49, ISSN 1070-4272 [*Zh. Prikl. Khim.*(St. Petersburg), Vol.78 (1), pp. 45]

Strukov, B. A. & Levanyuk, A. P. (1998). *Ferroelectric Phenomena in Crystals,* Springer, ISBN 10354063121, Berlin.

Tagantsev, A. K., Cross, L. E. & Fousek, J. (2010). *Domains in Ferroic Crystals and Thin Films,* Springer. ISBN 978-1-4419-1416-3, NY.

Wood, E.A. & Holden, A.N. (1957). Monoclinic Glycine Sulfate: crystallographic data, *Acta Crystallogr.,* Vol.10, pp. 145-146, ISSN 0108-2701

Wurfel, P. & Batra, I.P. (1973). Depolarization-field-induced instability in thin ferroelectric films - experiment and theory, *Phys.Rev.B,* Vol.8, No.11, pp. 5126-5133, ISSN 1098-0121

Wurfel, P.; Batra, I.P. & Jacobs, J.T. (1973). Polarization instability in thin ferroelectric films, *Phys.Rev.Lett.,* Vol.30, No.24, pp. 1218-1221, ISSN 0031-9007

22

$BiFeO_3$ Thin Films Prepared by Chemical Solution Deposition with Approaches for Improvement of Ferroelectricity

Yoshitaka Nakamura, Seiji Nakashima and Masanori Okuyama
Graduate School of Engineering Science, Osaka University
Japan

1. Introduction

Bismuth ferrite ($BiFeO_3$) is one of the most promising lead-free ferroelectric materials. Both bulk and thin film forms have been investigated and found a remarkably large ferroelectric polarization as well as *G*-type antiferromagnetism, often called it as multiferroics having ferroelectricity, magnetism, and ferroelasticity. There are several thin film fabrication methods including pulsed laser deposition, metal organic chemical vapour deposition, sputtering, and chemical solution deposition for this material. In this chapter, we describe the basic property of $BiFeO_3$ and the general outline of chemical solution deposition.

1.1 Crystal structure of $BiFeO_3$

Bulk $BiFeO_3$ shows a rhombohedral symmetry (point group *R*3c) with lattice parameters of a = 0.396 nm and α = 89.6° at room temperature (RT). Its transition temperature was investigated by Roginskaya et al. and a structural change showed at around 1100 K (Roginskaya et al., 1963). A detailed crystalline structure above the transition temperature has been investigated by Palai et al.. They reported that the transition temperature from rhombohedral to orthorhombic symmetry was at around 1100 K as well as orthorhombic to cubic symmetry at around 1200 K (Palai et al., 2008).

1.2 Electrical property of $BiFeO_3$

Rhombohedral $BiFeO_3$ has a ferroelectric polarization along [111]. Polarization of the single crystal was measured and showed small polarizations of 6.1 μC/cm^2 along [111] and 3.5 μC/cm^2 along [100] at 80 K (Teague et al., 1970). These small polarizations are regarded as low sample-quality now. Recent reports showed saturated polarizations of ~60 μC/cm^2 with single crystals as well as ~40 μC/cm^2 with ceramics measured at RT (Lebeugle et al., 2007; Shvartsman et al., 2007).

Recent study has also been focused on $BiFeO_3$ thin films. A saturated polarization of 55 μC/cm^2 along [001] was reported from epitaxial films on $SrRuO_3/SrTiO_3$ (001) substrates measured at RT (Wang et al., 2003; Li et al., 2004). Films also showed ferroelectric polarizations of 80 μC/cm^2 on $SrTiO_3$ (110) and 100 μC/cm^2 on $SrTiO_3$ (111), having an equivalent relationship between $\sqrt{3}P_{(001)}$, $\sqrt{2}P_{(101)}$, and $P_{(111)}$ (Li et al., 2004). Moreover, a giant ferroelectric polarization beyond 150 μC/cm^2 has been reported from polycrystalline $BiFeO_3$

thin films with a tetragonal structure measured at 80 K (Yun et al., 2004). Recent theoretical calculation showed a polarization of about 100 μC/cm^2 in a rhombohedral structure as well as about 150 μC/cm^2 in a tetragonal structure. These values showed a good agreement with the experimental ones (Ederer et al., 2006; Ricinschi et al., 2006).

1.3 General outline of chemical solution deposition

Chemical solution deposition (CSD) is one of the thin film fabrication methods, and it includes spin-coating, drying and annealing processes. Precursor solution is deposited onto a substrate by a spin-coating process. After the spin-coating process, a film dying process is carried out to evaporate the solvent and decompose metal-organic compounds in the precursor. An amorphous film is obtained at this stage. These processes are repeated several times to obtain a desired film thickness. For the film crystallization, an annealing process is carried out. It is usually carried out by a rapid thermal annealing (RTA) equipment to crystallize and densify the film. Higher heating rate usually decomposes metal organic compounds quickly and then desired oxide films with a higher density can be obtained (Schwartz, 1997).

There are some advantages for CSD; (i) uniformity of the molecules in precursor solutions and thin films, (ii) control of the film thickness by changing the solution concentration or the coating speed, (iii) control of the composition ratio by mixing solutions, (iv) film fabrication in ambient pressure, (v) synthesis of a non-equilibrium phase by the low-temperature process. However, there are some disadvantages for this method; (i) possibility of cracks in a film fabrication process, (ii) contamination which results in a difficulty of the manufacturing process, (iii) films with low-coherency comparing with other thin film fabrication methods such as pulsed laser deposition, chemical vapor deposition, and molecular beam epitaxy.

1.4 Precursor solutions for $BiFeO_3$

Precursor solutions for the CSD method are distinctly important. They consist of metal organic compounds and solvent which determine process parameters such as drying and annealing temperatures, film thickness per one spin-coating process, and coating affinity to the substrates. In this chapter, $BiFeO_3$ thin films were prepared by CSD with precursor solutions using 2-ethylhexanoate bismuth [$Bi(OCO(CH)(C_2H_5)C_4H_9)_3$] and trisacetylacetonato iron [$Fe(C_5H_7O_2)_3$] as metal organic materials, and toluene as a solvent.

2. Ferroelectric property of $BiFeO_3$ thin films prepared by CSD with controlling Bi/Fe ratio in the precursor solution

In this section, we demonstrate the $BiFeO_3$ thin film growth with controlling Bi/Fe ratio of the precursor solutions. Composition ratio affects the crystal growth and the electric property of the films. We obtain both good crystallinity and ferroelectric polarization of 85 μC/cm^2 with films using 10 mol% Bi-excess solution (Nakamura et al., 2007; Nakamura et al., 2008).

2.1 Film preparation by CSD with controlling Bi/Fe ratio

$BiFeO_3$ thin films were deposited on a Pt (200 nm)/TiO_2 (40 nm)/SiO_2 (600 nm)/Si substrate by CSD using precursor solutions of different Bi/Fe ratios: 10 mol% Fe-excess (10%Fe-ex.), stoichiometric, 5 mol% Bi-excess (5%Bi-ex.), 10 mol% Bi-excess (10%Bi-ex.), and 20 mol% Bi-excess (20%Bi-ex.). The precursor solution was spin-coated at 3000 rpm for 30 s and dried at 250°C for 5 min in air. These processes were repeated 20 times to obtain a film thickness of 250 nm. Then the films were annealed at 450 °C for 15 min in nitrogen atmosphere using the RTA equipment. For electrical measurement, Pt top electrodes with a diameter of 190 μm and a thickness of 100 nm were formed on the films by rf sputtering at RT. We confirmed by inductively coupled plasma (ICP) analysis that the composition ratios of $BiFeO_3$ thin films were the same as the precursor solutions.

2.2 Crystal structure

Figure 1 shows θ-2θ scans of the XRD patterns of the $BiFeO_3$ thin films with different Bi/Fe ratios. All the films show polycrystalline perovskite phase mainly. However, the 10%Fe-ex. $BiFeO_3$ film has small amount of a $Bi_2Fe_4O_9$ phase and the 20%Bi-ex. film shows a Bi_2O_3 phase. This indicates that excessive Fe or Bi compounds in the precursor solution tend to form impurity phase. Comparing the peak intensities corresponding to the (010) and (110) planes, the 10%Bi-ex. and 20%Bi-ex. films show higher diffraction intensities, indicating that they are crystallized well. This result suggests that Bi compounds in the precursor solution contribute the promotion of the film crystallization but that 5 mol% excess Bi is insufficient for the crystallization of such films.

2.3 Surface texture and raman spectrum

Figures 2(a-e) show the atomic force microscope (AFM) images of $BiFeO_3$ films taken in a 20 × 20 μm^2 area. All the $BiFeO_3$ thin films show a rosette structure, which consists of circular regions with an uneven texture and outer regions with a flat surface. These structures were also reported in $PbZrO_3$ thin films prepared by the sol-gel method (Alkoy et al., 2005). Table I shows the percentages of circular regions, RMS roughnesses and total boundary lengths surrounding the circular regions evaluated from Figs. 2(a-e). As can be seen in Table I, the percentage of circular region and RMS roughness tend to increase with an increase in Bi/Fe ratio. On the other hand, the total boundary length is ~200 μm and seems to have no systematic dependence.

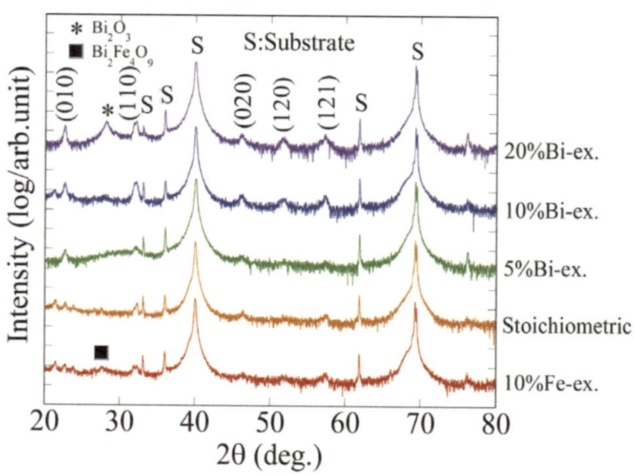

Fig. 1. XRD θ-2θ patterns of $BiFeO_3$ thin films prepared using 10 mol%Fe-excess (10%Fe-ex.), stoichiometric, 5 mol% Bi-excess (5%Bi-ex.), 10 mol% Bi-excess (10%Bi-ex.), and 20 mol% Bi-excess (20%Bi-ex.) precursor solutions.

Figure 3(a) shows an AFM image of the 10%Bi-ex. $BiFeO_3$ thin film with white circles marking the measurement location of Raman spectroscopy. A laser with a 0.7 μm spot size and an excitation wavelength of 515 nm was applied to the film surface labelled "Circular region" and "Outer region" in Fig. 3(a). Figure 3(b) shows the Raman spectra measured at RT in each measurement location shown in Fig. 3(a). The spectrum measured in $BiFeO_3$ ceramic is also shown as a reference. As shown in Fig. 3(b), the spectrum measured in the circular region is almost similar to that of $BiFeO_3$ ceramic consisting of polycrystalline grains. On the other hand, the spectrum measured in the outer region has a broad shape, found frequently in amorphous materials, and is different from that of $BiFeO_3$ ceramic.

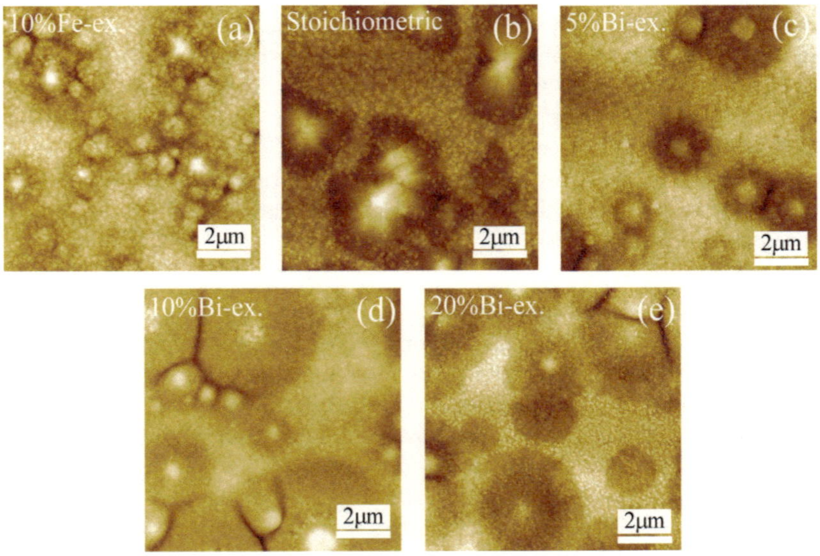

Fig. 2. 20 × 20 μm^2 surface AFM images for the $BiFeO_3$ films of (a) 10%Fe-ex., (b) stoichiometric, (c) 5%Bi-ex., (d) 10%Bi-ex., and (e) 20%Bi-ex.. Each film shows a rosette structure, which consists of circular regions and outer regions.

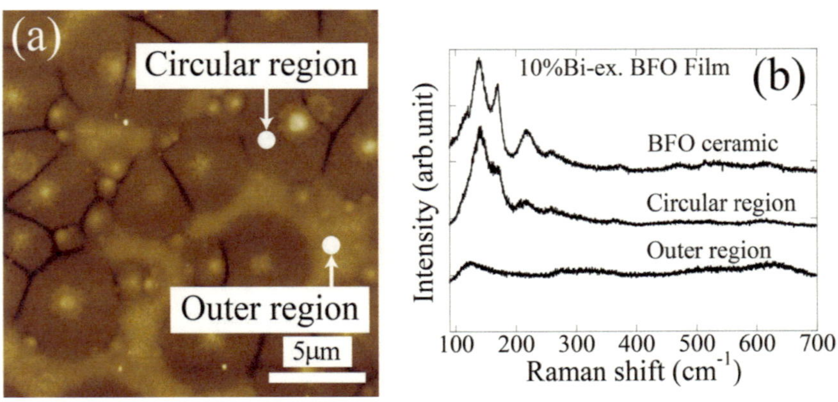

Fig. 3. (a) AFM image of 10%Bi-ex. $BiFeO_3$ thin film with white circle marking the measurement location of Raman spectroscopy. (b) Raman spectra measured at RT for each location shown in (a). Spectrum of $BiFeO_3$ ceramic is also shown as a reference.

Sample	Circular region area (%)	RMS roughness (nm)	Boundary length (mm)
10%Fe-ex.	46.0	3.6	238
Stoichiometric	48.1	5.1	225
5%Bi-ex.	41.1	5.6	196
10%Bi-ex.	85.9	4.8	165
20%Bi-ex.	53.7	6.8	208

Table 1. Circular region areas, RMS roughnesses, and boundary lengths for all samples.

The same results are also obtained in $BiFeO_3$ thin films prepared using the other precursor solutions with different Bi/Fe ratios. These results indicate that the circular regions have a $BiFeO_3$ crystalline phase, while the outer regions have an amorphous $BiFeO_3$ phase. Moreover, it can be considered that each phase exists from the top to the bottom of the film in a vertical direction because the excitation can sufficiently penetrate up to the bottom of the film. Consequently, the $BiFeO_3$ thin films of 10%Bi-ex. and 20%Bi-ex. have more circular regions and show good crystallinity, as shown in Fig. 1. This tendency is also observed in the $Pb(Zr,Ti)O_3$ (PZT) thin film prepared by the sol-gel method. Excessive Pb compounds in the precursor solution promote the formation of PZT and lead to show more circular regions (Alkoy et al., 2005). From Figs. 1 and 2, however, the 20%Bi-ex. film shows a Bi_2O_3 phase, and the area of circular regions does not seem to increase so much compared with that in the 10%Bi-ex. film. This result suggests that excessive Bi compounds in the precursor solution are more reactive, thereby they promote the formation of $BiFeO_3$. However, the major amount of Bi compounds tends to form Bi_2O_3 as well as $BiFeO_3$, therefore the circular region does not seem to increase so much.

2.4 Ferroelectric property

Figure 4(a) shows the leakage current density versus electric field (*J–E*) property of the $BiFeO_3$ thin films measured at RT. A comparatively larger leakage current is obtained in the films that contain more Bi. Figure 4(b) shows *P–E* hysteresis loops of the $BiFeO_3$ thin film measured at RT with a scanning frequency of 20 kHz. The 10%Bi-ex. $BiFeO_3$ thin film shows more squareness in hysteresis loop than the other films. The remanent polarizations (P_r) for the maximum applied electric field of 1.2 MV/cm are 30, 38, 28, 85, and 53 μC/cm² for the films of 10%Fe-ex., stoichiometric, 5%Bi-ex., 10%Bi-ex., and 20% Bi-ex., respectively.

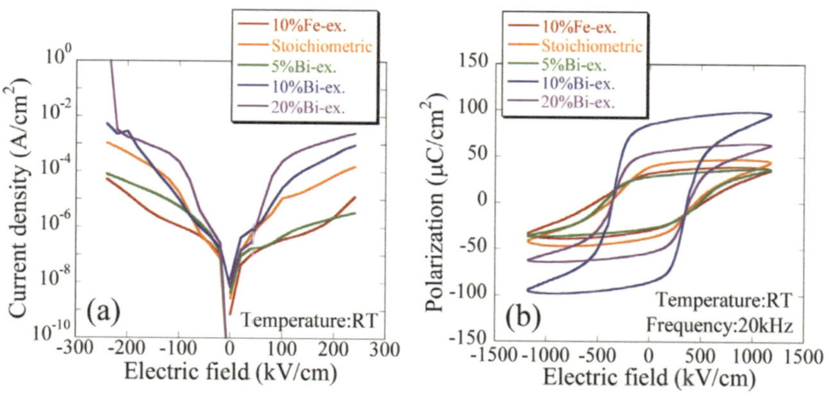

Fig. 4. (a) *J–E* characteristics of $BiFeO_3$ thin films measured at RT. (b) *P–E* hysteresis loops measured at RT.

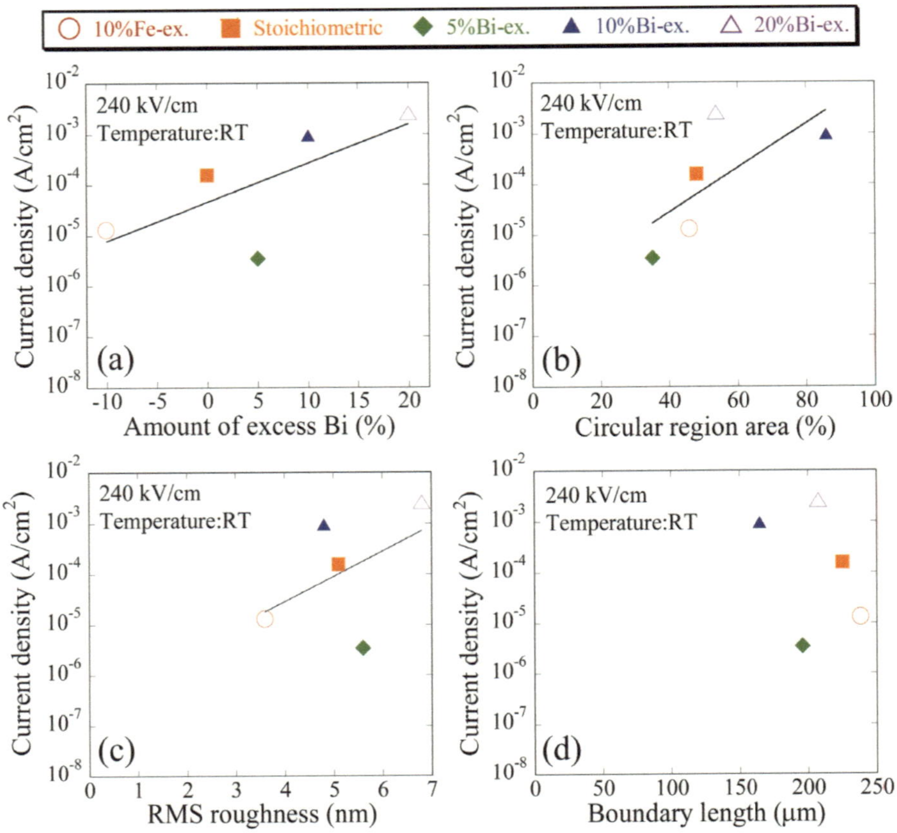

Fig. 5. Leakage current of $BiFeO_3$ thin films at 240 kV/cm as function of (a) amount of excess Bi, (b) percentage of circular region area, (c) RMS roughness, and (d) boundary length.

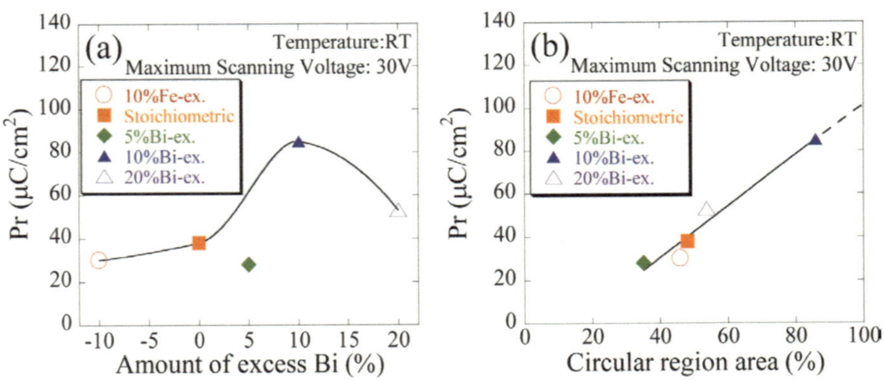

Fig. 6. Remanent polarization of $BiFeO_3$ thin films as function of (a) amount of excess Bi and (b) percentage of circular region area.

2.5 Relationship between surface texture and ferroelectricity

To investigate the influences of the surface texture and Bi/Fe ratio on the leakage current of $BiFeO_3$ thin films, we consider the amount of excess Bi, percentage of circular region area,

RMS roughness, and boundary length at the surface between crystal and amorphous phases, as shown in Table I. Figures 5(a-d) show the leakage current measured at 240 kV/cm versus (a) amount of excess Bi, (b) circular region area, (c) RMS roughness, and (d) boundary length. As shown in Figs. 5(a-c), leakage current tends to exponentially increase with an increase in the amount of excess Bi, circular region area, and RMS roughness although some scattering of the data is observed in Fig. 5(c). On the other hand, the length between the circular regions and the outer regions does not seem to affect the leakage current as shown in Fig. 5(d). These results suggest that the $BiFeO_3$ thin film prepared using the Bi excess precursor solution tends to have more circular regions that have $BiFeO_3$ crystals and to have a larger RMS roughness. From these leakage trends, leakage current mainly passes through circular regions consisting of crystalline $BiFeO_3$ rather than through outer amorphous region, and that current is increased by a rough surface. We further investigate the influences of the surface texture and Bi/Fe ratio on the ferroelectric polarization of $BiFeO_3$ thin films. We plot the amount of excess Bi and percentage of circular region area that has $BiFeO_3$ crystals, as shown in Fig. 3(b). Figures 6(a) and 6(b) show the remanent polarization measured at RT versus (a) amount of excess Bi and (b) circular region area. The remanent polarization increases with an increase in Bi ratio below the 10%Bi-ex. $BiFeO_3$ film. However, the 20%Bi-ex. film decreases its remanent polarization because of the mixed phase of $BiFeO_3$ and Bi_2O_3. As shown in Fig. 6(b), the remanent polarization linearly increases with an increase in the percentage of the circular region area. From the extrapolated line in Fig. 6(b), fully crystallized $BiFeO_3$ thin films are expected to show 100 $\mu C/cm^2$. According to leakage and polarization plots in Fig. 5 and Fig. 6, a 10 mol% Bi-excess solution gives $BiFeO_3$ thin films the best ferroelectric property with more circular regions.

3. Insertion effect of Bi-excess layer on $BiFeO_3$ thin films

In section 2, Bi-excess solution, or precursor solution with excessive Bi compounds, promotes film crystallization, leading to a good ferroelectricity. In this section, we demonstrate the insertion effect of Bi-excess layer to the stoichiometric $BiFeO_3$ thin films to improve the crystal growth and ferroelectricity of the films (Nakamura et al., 2007; Nakamura et al., 2008).

3.1 Insertion effect

Insertion effect, inserting Bi-excess $BiFeO_3$ layer to the film, is aiming to promote the crystal growth of the film and to obtain a good ferroelectricity. There are some reports that ferroelectric thin films prepared by CSD show a non-crystalline layer at the interface between the thin film and the electrode. Such a layer is reported as an interfacial layer which degrades the ferroelectric property of the film (Grossmann et al., 2002). These reports suggest that the low crystallinity part is concentrated at the interface between the film and the electrode. To improve the low crystallinity part, an insertion layer promoting crystal growth will be effective.

In our $BiFeO_3$ thin films, a thin film with stoichiometric solution shows low crystallinity with a small polarization, and a film with 10 mol% Bi-excess solution shows high crystallinity and a large polarization. Thus an insertion layer with 10 mol% Bi-excess solution is expected to be effective. To investigate the insertion effect of Bi-excess layers, three types of thin films were prepared on $Pt/TiO_2/SiO_2/Si$ substrates, as shown in Fig. 7: stoichiometric $BiFeO_3$ thin film with Bi-excess top layer (Bi-T), bottom layer (Bi-B), and top and bottom layer (Bi-TB). Then the films were annealed at 450 °C for 15 min in a nitrogen atmosphere using the RTA process. For the electrical measurement, Pt top electrodes with a diameter of 190 μm were formed by rf sputtering.

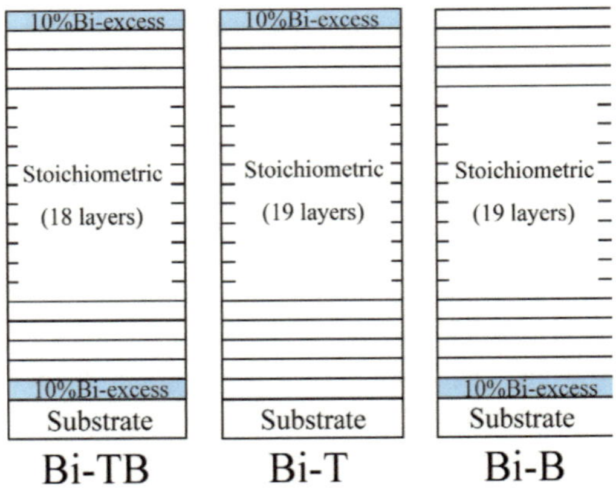

Fig. 7. Schematic models of $BiFeO_3$ thin films inserting Bi-excess top and bottom layer (Bi-TB), top layer (Bi-T), and bottom layer (Bi-B).

3.2 Crystal structure

Figure 8 shows the θ–2θ scans of XRD patterns of the $BiFeO_3$ thin films with Bi-excess top and bottom layer (Bi-TB), top layer (Bi-T), and bottom layer (Bi-B). These results show that all the films exhibit mainly polycrystalline perovskite single phase without nonperovskite phases such as $Bi_2Fe_4O_9$ and Bi_2O_3. Comparing peak intensities corresponding to (010) and (110) planes, the crystallinity of Bi-TB is the best. Bi-T is the second best, followed by Bi-B. This result indicates that the Bi-excess top layer improves the crystallization in the annealing process. Moreover, this tendency suggests that the crystallization is produced from the surface to the bottom using the RTA process. As for the difference between Bi-TB and Bi-T, crystallinity of $BiFeO_3$ film can be enhanced by inserting the Bi-excess layer on the top surface and the bottom.

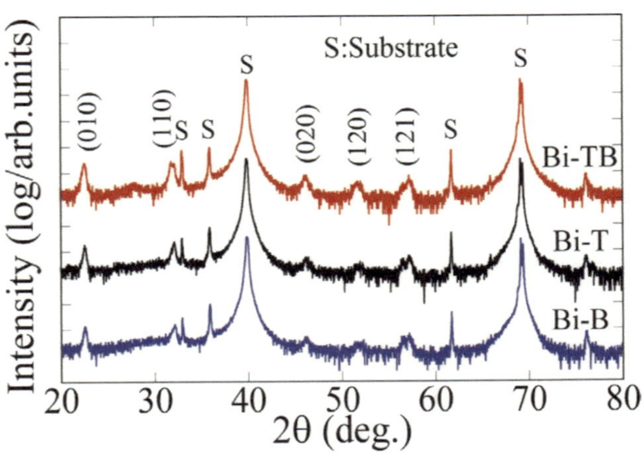

Fig. 8. XRD θ-2θ patterns of Bi-TB, Bi-T, and Bi-B $BiFeO_3$ thin films.

3.3 Surface texture and raman spectrum

Figures 9(a-c) show the AFM images of $BiFeO_3$ films taken over a 10×10 μm^2 area. As can be seen in Fig. 9(a), Bi-TB forms more grains than the others. On the other hand, Bi-T and Bi-B form finer grains as well as larger grains, as shown in Figs. 9(b) and 9(c). In addition, Bi-T seems to form larger grains than the film of Bi-B. The surface RMS roughness is estimated as 7.7, 6.7, and 5.0 nm for the films of Bi-TB, Bi-T, and Bi-B, respectively. The number of grains and the surface roughness increase with increasing crystallinity, comparing Fig. 9 with Fig. 8. To investigate the difference between finer and larger grains, Raman spectroscopy was carried out. A laser with a 0.7 μm spot size irradiated the points labelled A–C, which form large grains, and D–F, which form fine grains, as shown in Figs. 9(a-c). Figures 9(d) and 9(e) show Raman spectra measured at RT. These figures also include the spectrum measured in $BiFeO_3$ ceramic, as a reference. As shown in Fig. 9(d), the spectra measured in the areas A–C are almost the same as the spectrum of $BiFeO_3$ ceramic. On the other hand, the spectra measured in the areas D–F are different from the spectrum of $BiFeO_3$ ceramic, as shown in Fig. 9(e). These results indicate that the areas A–C have good $BiFeO_3$ crystals while the areas D–F seem to be amorphized. Moreover, the area at which the $BiFeO_3$ crystal spectrum was observed is the largest in Bi-TB. This result relates that Bi-TB crystallizes the best, comparing Figs. 9 and 8.

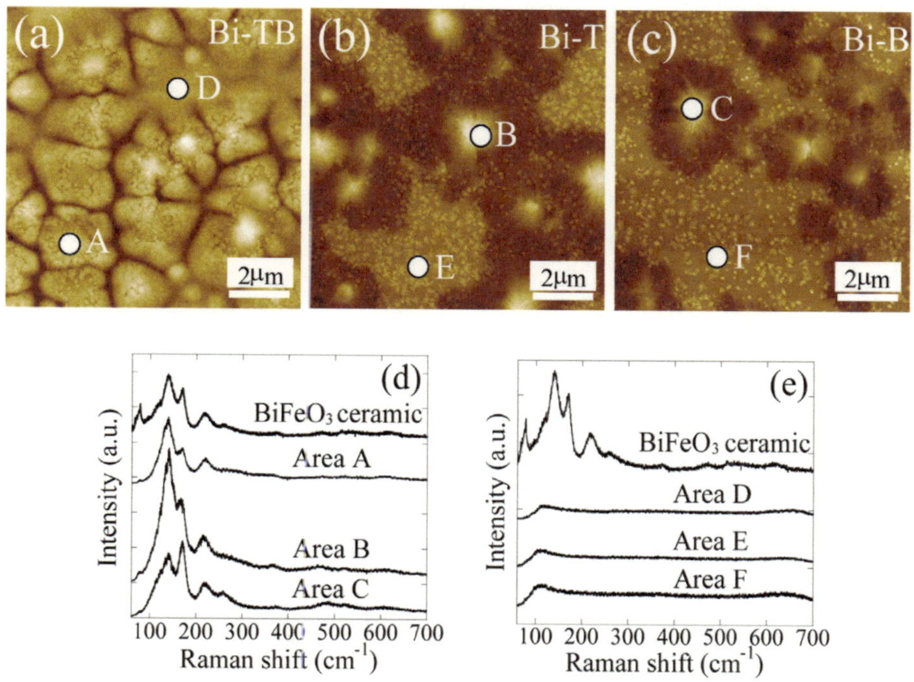

Fig. 9. 10×10 μm^2 surface AFM images with markings of the typical locations of Raman spectroscopy for the films of (a) Bi-TB, (b) Bi-T, and (c) Bi-B, respectively. Areas A–C form large grains, while areas D–F form fine grains. (d) and (e) Raman spectra measured at RT for the locations shown in Figs. 9(a)–9(c).

3.4 Ferroelectric property

Figure 10 shows the leakage current density versus electric field (*J–E*) of $BiFeO_3$ thin films measured at (a) RT and (b) 80 K. When the electric field is lower than 300 kV/cm, the leakage currents are almost unchanged among three types of films both at RT and 80 K. This suggests that the amorphous phase of the surface limits the conduction in the case of lower electric field, as mentioned in Abe et al. (Abe et al., 1993). On the other hand, when the electric field is higher than 300 kV/cm at 80 K, the leakage current becomes large for the film of Bi-TB. Therefore, it is suggested that the amorphous phase includes defects that limit the carrier emission at the interface, and the leakage current increases at higher electric fields in the Bi-TB film. In the case of Bi-T and Bi-B, the amorphous phase might suppress the leakage current at high electric field. Figure 11 shows ferroelectric polarization versus electric field (P–E) hysteresis loops of $BiFeO_3$ thin film at (a) RT and (b) 80 K, respectively. At RT, the remanent polarizations (P_r) for maximum applied electric field of 1.0 MV/cm are 55, 26, and 17 μC/cm² for the films of Bi-TB, Bi-T, and Bi-B, respectively. In addition, the coercive field of Bi-TB is 385 kV/cm, which is the lowest in the three types of films. At 80 K, the remanent polarizations for maximum applied electric field of 2.0 MV/cm are 65, 46, and 32 μC/cm² for the films of Bi-TB, Bi-T, and Bi-B, respectively. The remanent polarization of Bi-TB is about twice that of the film prepared by stoichiometric solution (28 μC/cm² at RT, and 38 μC/cm² at 80 K). These results show that $BiFeO_3$ thin film of Bi-TB gives the best ferroelectric property among the three types of films, which is attributed to the good crystallinity of the $BiFeO_3$ film, comparing Figs. 11 and 8.

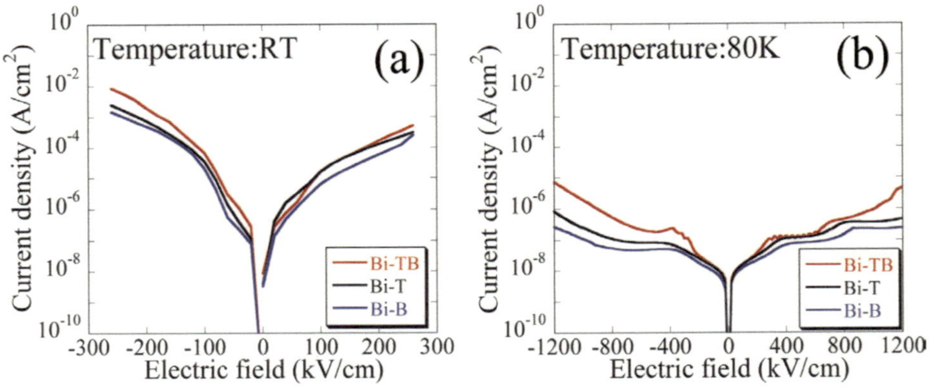

Fig. 10. Leakage current characteristics of $BiFeO_3$ thin films measured at (a) RT and (b) 80 K.

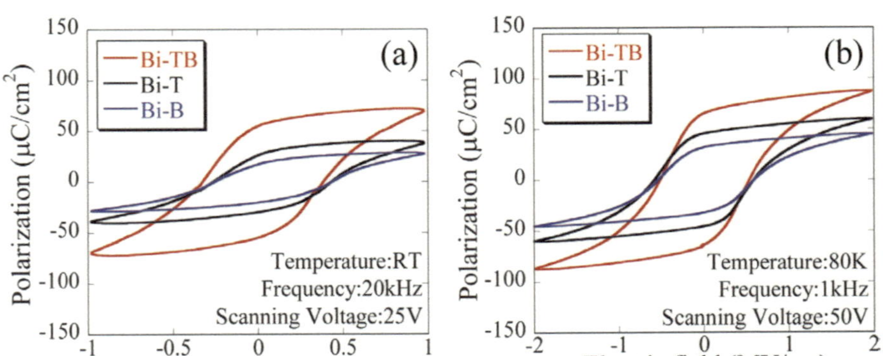

Fig. 11. P–E hysteresis loops of $BiFeO_3$ thin films measured at (a) RT and (b) 80 K.

4. Improvement of ferroelectricity of $BiFeO_3$ thin films by postmetallization annealing and electric field application

In this section, we describe the postmetallization annealing and electric field application by using 10 mol% Bi-excess $BiFeO_3$ thin film which shows good ferroelectricity in section 2. These are the ways to improve ferroelectricity of $BiFeO_3$ thin films. Postmetallizaton annealing is the electrode annealing process to reduce the leakage current which has already reported in several thin film materials such as $BaTiO_3$, (Ba,Sr)TiO_3, (Pb,Sr)TiO_3, and PZT after the deposition of top electrodes (Lee et al., 2004; Joo et al., 1997; Chung et al., 2001; Thakoor, 1994). Electric field application is to apply a high electric field to reverse its polarization reversal easily. It is typically carried out in bulk materials such as PZT (Kamel et al., 2007). These two approaches are expected to be effective to improve ferroelectric properties of $BiFeO_3$ thin films (Nakamura et al., 2009).

4.1 Film preparation methods

$BiFeO_3$ thin films were deposited on a Pt/TiO_2/SiO_2/Si substrate by CSD using 10mol % Bi-excess precursor solution. Spin-coating and drying processes were the same as in chapter 2. These processes were repeated 20 times to obtain a film thickness of 250 nm. Then, the films were treated by the RTA process at 450 °C for 20 min in nitrogen atmosphere. For the electrical measurement, Pt top electrodes were formed on the $BiFeO_3$ film by rf sputtering. After the deposition of Pt top electrodes, the sample was divided into three pieces and labelled as BFO, BFO-N, and BFO-O, respectively. Then the postmetallization annealing was carried out for 5 min at 300 °C in nitrogen atmosphere for BFO-N, and oxygen atmosphere for BFO-O by the RTA process. Finally, the following three films were obtained; BFO (as prepared film without postmetallization annealing), BFO-N (the film with the annealing in nitrogen atmosphere), and BFO-O (the film with the annealing in oxygen atmosphere).

4.2 Improvement of ferroelectric property of $BiFeO_3$ thin films by postmetallization annealing

Figure 12(a) shows the θ-2θ patterns of the XRD of the $BiFeO_3$ thin films with and without the postmetallization annealing. All the films consist mainly of polycrystalline perovskite phase, but a Bi_2O_3 phase is slightly observed. Evaluating the diffraction peak intensity of each film, it does not change among three films. This result indicates that the crystallinity of $BiFeO_3$ does not change by the postmetallization annealing. We note that the intensity of an observed Bi_2O_3 phase is much smaller than the $BiFeO_3$ phase and its intensity does not change after the postmetallization annealing. Therefore, it is enough to evaluate the dielectric property of $BiFeO_3$ and the annealing effect of the electrode. There may be a possibility of a peak shift due to a strain relaxation between the $BiFeO_3$ film and the Pt electrode by the postmetallization annealing, however, it is hard to observe the strain relaxation from such a small 190 μm diameter dot electrode because an incident X-ray beam width is about 2 mm, making it difficult to analyze the crystalline property of the small area. To confirm the strain relaxation, a 100 nm thick Pt film was deposited on the whole $BiFeO_3$ film surface and then the postmetallization annealing was carried out. Figure 12(b) shows the XRD patterns near (010) peak before and after the Pt deposition and the postmetallization annealing in nitrogen atmosphere. The peak intensity is decreased due to the deposition of the Pt film, but the peak shift is not observed. This result suggests that a clear strain relaxation does not occur near the interface between the $BiFeO_3$ film and the Pt electrode after the postmetallization annealing.

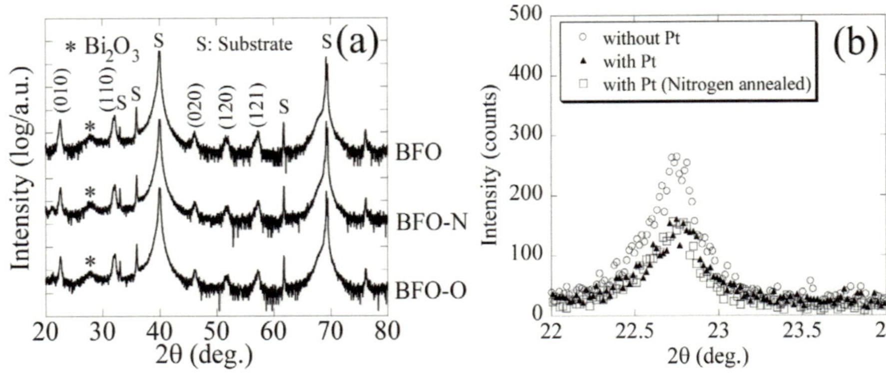

Fig. 12. (a) XRD θ-2θ scans of as-prepared (BFO), N_2 annealed (BFO-N), and O_2 annealed (BFO-O) $BiFeO_3$ thin films. (b) Expanded scans near the (010) peak before and after the Pt deposition and the postmetallization annealing. A 100 nm thick of Pt film was deposited on the whole surface of the $BiFeO_3$ film.

Figures 13(a) and 13(b) show the *J-E* characteristics of $BiFeO_3$ thin films measured at (a) 80 K and at (b) RT. The leakage current is suppressed in both the BFO-N and BFO-O films at 80 K. This suppression is also observed at RT. Joo et al. reported that the leakage current is suppressed by the postmetallization annealing in oxygen atmosphere due to the reduction of oxygen vacancies in a film (Joo et al., 1997). Contrary to their result, our result suggests that the postmetallization annealing improves the contact between the $BiFeO_3$ film and the Pt electrode or the reduction of defects near the interface between the $BiFeO_3$ film and the Pt electrode as suggested in Pt/PZT/Pt capacitors, rather than the compensation of oxygen vacancies in the $BiFeO_3$ film.

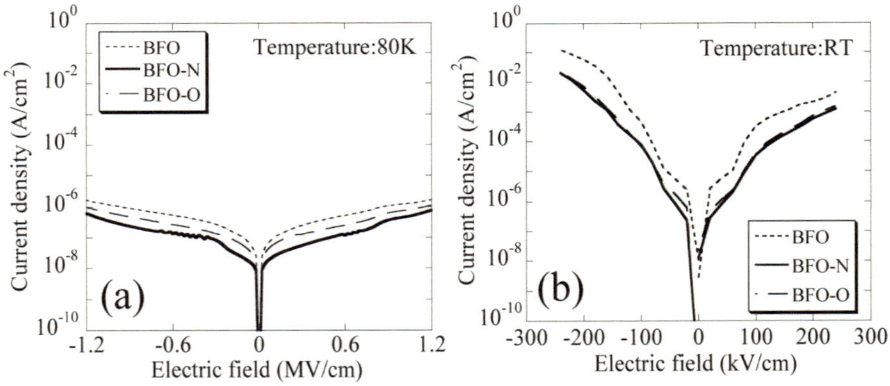

Fig. 13. Current density-electric field (*J-E*) characteristics of $BiFeO_3$ thin films measured at (a) 80 K and at (b) RT.

Figures 14(a) and 14(b) show frequency dependences of the dielectric constant and the dielectric loss tan δ measured at (a) 80 K and at (b) RT. The dielectric constant and the loss tangent of $BiFeO_3$ thin films measured at 80 K are found to be 185 and 0.061, 172 and 0.045, and 186 and 0.048 for the BFO, BFO-N, and BFO-O film with a measuring frequency of 1 MHz, respectively. In addition, the frequency variability of from 10^3 to 10^6 Hz is 20.5% BFO, 16.9% BFO-N, and 18.3% BFO-O. The reduction in the frequency variability and the dielectric loss may be due to the reduction in the leakage current as shown in Fig. 13(a). The same tendencies are also observed in the BiFeO$_3$ films measured at RT, as shown in Fig. 13(b). The frequency variability from 10^3 to 10^6 Hz is 30.6% BFO, 23.3% BFO-N, and 23.4% BFO-O measured at RT. These results indicate that the frequency variability of and the dielectric loss are successfully reduced by the postmetallization annealing.

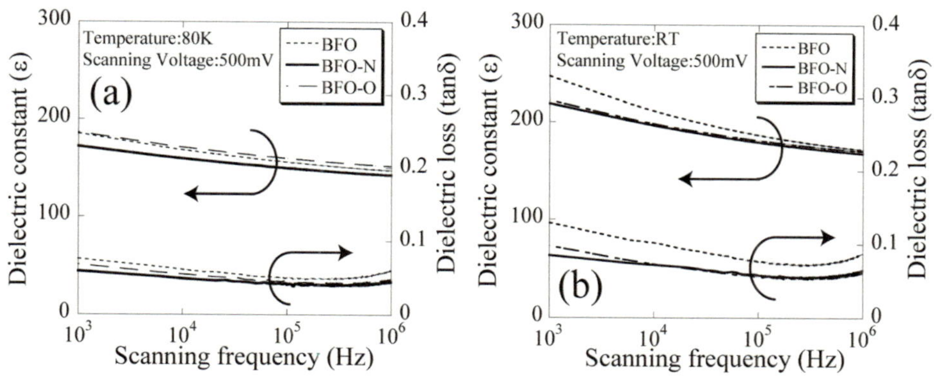

Fig. 14. Dielectric constant-frequency (ε-F) characteristics of BiFeO$_3$ thin films measured at (a) 80 K and at (b) RT.

Figures 15(a) and 15(b) show *P-E* hysteresis loops of BiFeO$_3$ thin films measured at (a) 80 K and at (b) RT with a scanning frequency of 20 kHz. The remanent polarizations (P_r) measured at 80 K under the maximum applied electric field of 1.2 MV/cm are 91, 87, and 89 μC/cm^2 for the films of BFO, BFO-N, and BFO-O, respectively. In addition, the double coercive field ($2E_c$) is reduced at about 90 kV/cm in the nitrogen-annealed film. The slight reduction of coercive field is also observed at RT, as shown in Fig. 15(b). This reduction of the coercive field may be due to the improvement of the contact between the Pt electrode and the BiFeO$_3$ film or the reduction in defects near the interface between BiFeO$_3$ film and the Pt electrode, as mentioned above.

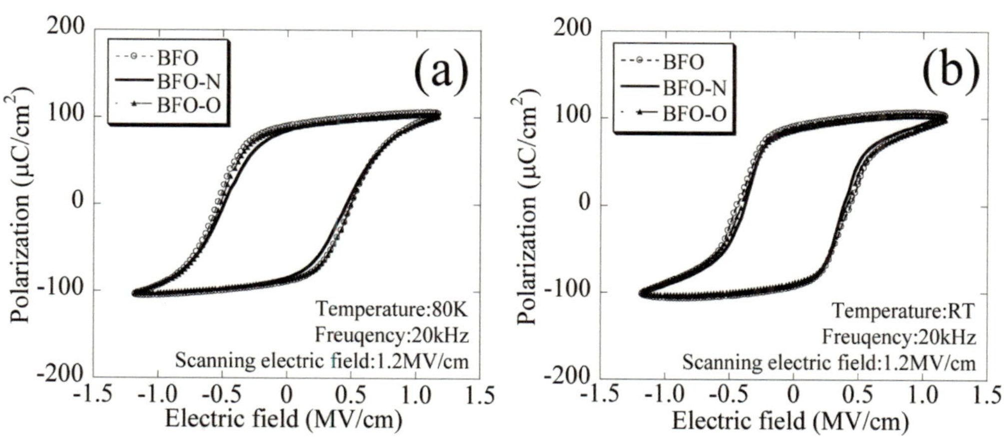

Fig. 15. *P-E* hysteresis loops measured at (a) 80 K and at (b) RT under 20 kHz triangular scanning voltage of as-prepared (BFO), N$_2$ annealed (BFO-N), and O$_2$ annealed (BFO-O) BiFeO$_3$ thin films.

4.3 Improvement of ferroelectric property of $BiFeO_3$ thin films by electric filed application

To evaluate the effect of the electric field application, *P-E* hysteresis loops were measured at 80 K in a following order; the first measurement was carried out by the applied voltage from 5 to 70 V, which corresponds to 0.2 MV/cm to 2.8 MV/cm, and then, the second measurement was carried out from 70 to 5 V with 1 kHz triangular wave. The temperature was set at 80 K to reduce the thermal effect and to apply high electric field. This measurement was carried out for the BFO-N film because it has the best insulation and ferroelectric characteristics among three films. Figures 16(a-d) show hysteresis changes in BFO-N film before (black lines) and after (red lines) applying the electric field of 0.4, 1.2, 2.0, and 2.8 MV/cm corresponding 10, 30, 50, and 70 V, respectively at 80 K. After applying the electric field of 2.8 MV/cm, which corresponds to 70 V, the shape of hysteresis loop is dramatically changed in the second measurement. In the sequence from Figs. 16(a-d), the remanent polarizations of the first measurement are 0.19, 29.9, 74.5, and 104 μC/cm^2 under the maximum field of 0.4, 1.2, 2.0, and 2.8 MV/cm, respectively. In addition, the remanent polarizations of the second measurement are 17.4, 84.4, 97.6, and 106.5 μC/cm^2 under the maximum field of 0.4, 1.2, 2.0, and 2.8 MV/cm, respectively.

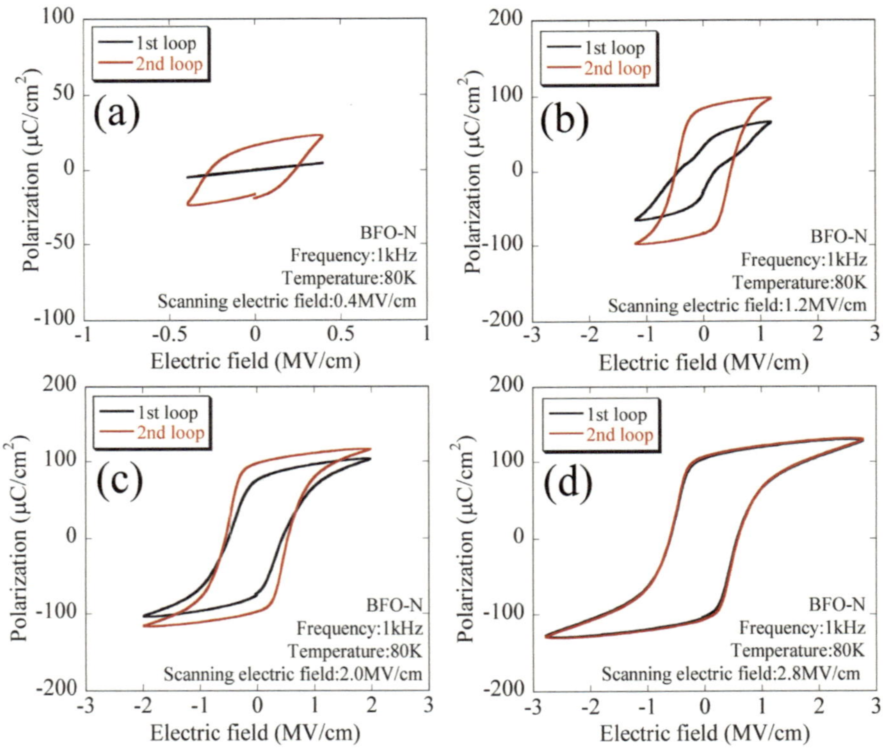

Fig. 16. *P-E* hysteresis of $BiFeO_3$ thin films measured at 80 K under the maximum field of (a) 0.4 MV/cm (10 V), (b) 1.2 MV/cm (30 V), (c) 2.0 MV/cm (50 V), and (d) 2.8 MV/cm (70 V), respectively. Hysteresis loops were measured from 5 to 70 V the first measurement and then measured from 70 to 5 V the second measurement.

Moreover, the leakage current is reduced at about 1 order of magnitude after the *P-E* measurement, as shown in Fig. 17. Dependences of P_r and E_c on the applied electric field are shown in Fig. 18. P_r and E_c obtained in the first measurement are gradually increased with increase in the electric field, and then, the second measurement keeps large P_r value even in lower electric field. The third measurement from 5 to 70 V was carried out and values of P_r and E_c are almost the same as the second data, as shown in Fig. 18. Kohli et al. reported that the hysteresis of the PZT thin film was changed by applying pulse electric field because of the removal of 90° domain pinning (Kohli et al., 1998). Okamura et al. reported that the wake-up phenomenon, which shows an increase in the remanent polarization by applying switching pulses and removing the locked polarizations in the $SrBi_2Ta_2O_9$ film (Okamura et al., 2000). The improvement of ferroelectricity is possibly due to the relaxation of pinned domains or locked polarizations in the $BiFeO_3$ film by applying high electric field.

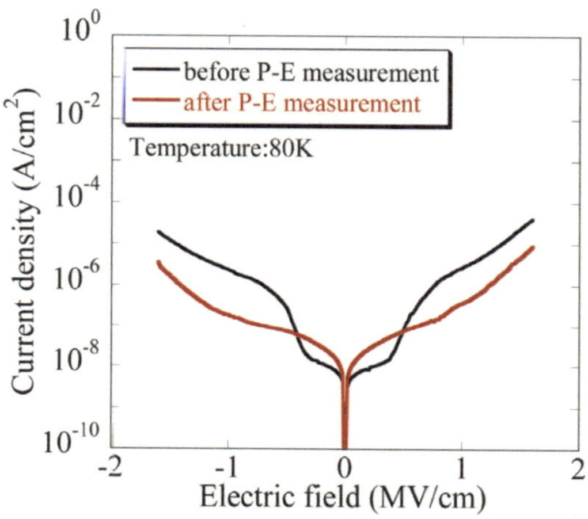

Fig. 17. Current density-electric field (*J-E*) characteristics of $BiFeO_3$ thin films measured at 80 K. The measurement was carried out before and after the *P-E* measurement shown in Fig. 5 (d).

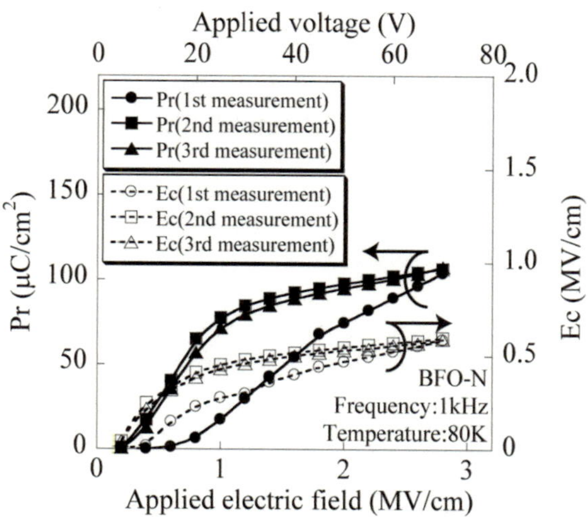

Fig. 18. Applied field dependences of remanent polarization and coercive field at 80 K for the first, the second, and the third measurement.

5. Conclusion

We describe $BiFeO_3$ thin films prepared by CSD with several approaches to improve its ferroelectricity. Controlling Bi/Fe ratio in the precursor solution contributes the promotion of the film crystallization and shows a large polarization of 85 $\mu C/cm^2$ with 10 mol% Bi-excess solution. Insertion of the 10 mol% Bi-excess layer to the stoichiometric $BiFeO_3$ films also promotes the film crystallization, leading to the improvement of the ferroelectricity. Ferroelectric property of films using 10 mol% Bi-excess solution can further improve by the postmetallization annealing as well as the electric field application. These are the effective methods to improve ferroelectricity of $BiFeO_3$.

6. Acknowledgments

The authors thank Takaaki Nakamura, Hideo Fukumura, and Professor Hiroshi Harima of Kyoto Institute of Technology for conducting Raman spectroscopy.

7. References

Roginskaya, Yu. E.; Venevtsev, Yu. N.; Fedulov, S.A. & Zhdanov, G.S. (1963). *Soviet Physics Crystallography*, 8, 610

Palai, R.; Katiyar, R. S.; Schmid, H.; Tissot, P.; Clark, S. J.; Robertson, J.; Redfern, S. A. T., Catalan, G. & Scott, J. F. (2008). β phase and γ-β metal-insulator transition in multiferroic $BiFeO_3$. *Physical Review B*, 77, 014110-1-014110-11

Teague, J. R.; Gerson, R. & James, W. J. (1970). DIELECTRIC HYSTERESIS IN SINGLE CRYSTAL $BiFeO_3$. *Solid State Communications*, 8, 1073-1074

Lebeugle, D.; Colson, D.; Forget, A. & Viret, M. (2007). Very large spontaneous electric polarization in $BiFeO_3$ single crystals at room temperature and its evolution under cycling fields. *Applied Physics Letters*, 91, 022907-1-022907-3

Shvartsman, V. V.; Kleemann, W.; Haumont, R. & Kreisel, J. (2007). Large bulk polarization and regular domain structure in ceramic $BiFeO_3$. *Applied Physics Letters*, 90, 172115-1- 172115-3

Wang, J.; Neaton, J. B.; Zheng, H.; Nagarajan, V.; Ogale, S. B.; Liu, B.; Viehland, D.; Vaithyanathan, V.; Schlom, D. G.; Waghmare, U. V.; Spaldin, N. A.; Rabe, K. M.; Wuttig, M. & Ramesh, R. (2003). Epitaxial $BiFeO_3$ Multiferroic Thin Film Heterostructures. *Science*, 299, 1719-1722

Li, J.; Wang, J.; Wuttig, M.; Ramesh, R.; Wang, N.; Ruette, B.; Pyatakov, A. P.; Zvezdin, A. K. & Viehland, D. (2004). Dramatically enhanced polarization in (001), (101), and (111) $BiFeO_3$ thinfilms due to epitiaxial-induced transitions. *Applied Physics Letters*, 84, 5261-5263

Yun, K. Y.; Ricinschi, D.; Kanashima, T.; Noda, M. & Okuyama, M. (2004). Giant Ferroelectric Polarization Beyond 150 $\mu C/cm^2$ in $BiFeO_3$ Thin Film. *Japanese Journal of Applied Physics*, 43, L647-L648

Ederer, C. & Spaldin, N. A. (2005). Effect of Epitaxial Strain on the Spontaneous Polarization of Thin Film Ferroelectrics. *Physical Review Letters*, 95, 257601-1-257601-4

Ricinschi, D.; Yun, K. Y. & Okuyama, M. (2006). A mechanism for the 150 $\mu C/cm^2$ polarization of BiFeO3 films based on first-principles calculations and new structural data. *Journal of Physics: Condensed Matter*, 18, L97-L105

Schwartz, W. R. (1997). Chemical Solution Deposition of Perovskite Thin Films. *Chemistry of Materials*, 9, 2325-2340

Nakamura, Y.; Yun, K. Y.; Nakashima, S. & Okuyama, M. (2007). Sol-Gel Preparation and Characterization of Multiferroic $BiFeO_3$ Thin Films with Various Bi/Fe Ratio. *Integrated Ferroelectrics*, 95, 226-233

Nakamura, Y.; Nakashima, S. & Okuyama, M. (2008). Influences of Surface Texture and Bi/Fe Ratio on Electric Properties of $BiFeO_3$ Thin Films Prepared by Chemical Solution Deposition. *Japanese Journal of Applied Physics*, 47, 7250-7253

Nakamura, Y.; Nakashima, S.; Ricinschi, D. & Okuyama, M. (2007). The Effect of Bi-excess Surface Layers on $BiFeO_3$ Thin Films Prepared by Chemical Solution Deposition. *2007 MRS Fall Proceedings*, 1034-K11-10

Nakamura, Y.; Nakashima, S.; Ricinschi, D. & Okuyama, M. (2008). THE INSERTION EFFECT OF Bi-EXCESS LAYERS ON STOICHIOMETRIC $BiFeO_3$ THIN FILMS PREPARED BY CHEMICAL SOLUTION DEPOSITION. *Functional Materials Letters*, 1, 19–24

Nakamura, Y.; Nakashima, S. & Okuyama, M. (2009). Improvement of ferroelectric properties of $BiFeO_3$ thin films by postmetallization annealing and electric field application. *Journal of Applied Physics*, 105, 061616-1-061616-4

Alkoy, E. M.; Alkoy, S. & Shiosaki, T. (2005). Effects of Ce, Cr and Er Doping and Annealing Conditions on the Microstructural Features and Electrical Properties of $PbZrO_3$ Thin Films Prepared by Sol–Gel Process. *Japanese Journal of Applied Physics*, 44, 6654-6660

Grossmann, M; Lohse, O; Bolten, D; Boettger, U.; Schneller, T. & Waser, R. (2002). The interface screening model as origin of imprint in $PbZr_xTi_{1-x}O_3$ thin films. I. Dopant, illumination, and bias dependence. *Journal of Applied Physics*, 92, 2680-2687

Abe, K. & Komatsu, S. (1993). Dielectric Constant and Leakage Current of Epitaxially Grown and Polycrystalline $SrTiO_3$ Thin Films. *Japanese Journal of Applied Physics*, 32, 4186-4189

Lee, E. J. H.; Pontes, F. M.; Leite, E. R.; Longo, E.; Magnani, R.; Pizani, P. S. & J. A. Varela (2004). Effects of post-annealing on the dielectric properties of Au/BaTiO3/Pt thin film capacitors. *Materials Letters*, 58, 1715–1721

Joo, J. H.; Jeon, Y. C.; Seon, J. M.; Oh, K. Y.; Roh, J. S. & Kim, J. J. (1997). Effects of Post-Annealing on the Conduction Properties of Pt/(Ba, Sr)TiO_3/Pt Capacitors for Dynamic Random Access Memory Applications. *Japanese Journal of Applied Physics*, 36, 4382-4385

Chung, H. J. ; Chung, S. J.; Kim, J. H. & Woo, S. I. (2001). The effect of post-annealing on the electrical properties of (Pb,Sr)TiO_3 thin films prepared by liquid source misted chemical deposition for ultra large-scale integration (ULSI) Dynamic random access memory (DRAM) capacitor. *Thin Solid Films*, 394, 213-218.

Thakoor, S. (1994). Enhanced fatigue and retention in ferroelectric thin film memory capacitors by post-top-electrode anneal treatment. *Journal of Applied Physics*, 75, 5409-5414

Kamel, T. M.; Kools, F. X. N. M. & With, G. (2007). *Journal of the European Ceramic Society*, 27, 2471–2479

Kohli, M.; Muralt, P. & Setter, N. (1998). Removal of 90° domain pinning in (100)$Pb(Zr_{0.15}Ti_{0.85})O_3$ thin films by pulsed operation. *Applied Physics Letters*, 72, 3217-3219

Okamura, S.; Takaoka, M.; Nishida, T. & Shiosaki, T. (2000). Increase in Switching Charge of Ferroelectric $SrBi_2Ta_2O_9$ Thin Films with Polarization Reversal. *Japanese Journal of Applied Physics*, 39, 5481-5484

23

Strontium Barium Niobate Thin Films for Dielectric and Electro-Optic Applications

Mireille Cuniot-Ponsard
*Laboratoire Charles Fabry de l'Institut d'Optique, C.N.R.S., Univ Paris Sud
France*

1. Introduction

The existence of strontium barium niobate crystals ($Sr_xBa_{1-x}Nb_2O_6$, noted SBN:100x) was first reported in 1960 (Francombe, 1960) and first large SBN single crystals were grown by Ballman and Brown over the range $0.25<x<0.75$ (Ballman & Brown, 1966). In 1970, the Bell Telephone Laboratories had published successive thorough investigations of the optical, electrical, and structural properties of SBN crystals. The high values of the electro-optic and pyroelectric coefficients oriented further work mainly towards holographic and pyroelectric applications.

The development of SBN films started at the USSR Academy of Science in Novosibirsk (Baginsky et al., 1978) using a RF sputtering technique. Different deposition techniques have been then investigated: mainly sol-gel process, metal-organic chemical vapor deposition (MOCVD), and pulsed laser deposition (PLD).

Thin SBN films are particularly attractive for their potential use as low voltage electro-optic (e-o) waveguides. Electro-optic light modulation is a key function in light-wave technologies, mostly realized by exploiting the linear e-o Pockels effect in ferroelectric bulk crystals like lithium niobate (LN) for primary example. Optimizing the performance of an e-o modulator involves minimizing the half-wave voltage-length product ($V_\pi L$) and the drive power (P). A considerable decrease in the required $V_\pi L$ and P values, by three orders of magnitude, is expected from the replacement of bulk crystals by thin film waveguides about 1μm thick. Beside LN and SBN, the ferroelectric materials which have been considered in the literature in view of preparing electro-optic thin films are mainly $BaTiO_3$ (BT), $(Ba,Sr)TiO_3$ (BST), and $(Pb,La)(Zr,Ti)O_3$ (PLZT).

The implementation of Pockels e-o effect in thin film waveguides also opens up the path to the realization of electrically-tunable photonic crystal (PC) devices. Through the engineering of photonic band gaps, PC structures enable developing the functionality and reducing radically the size of optical devices. Theoretically position and shape of a photonic band gap can be electro-optically controlled. This e-o control considerably broadens the scope of PC structures potential functionality. The future deployment of photonic technology largely rests on the tunability of PC characteristics.

The Pockels electro-optic effect is the expression of the dielectric non linear properties in the range of optical frequencies where ionic displacement is negligible and relative dielectric permittivity is reduced to its electronic component. The interest for dielectric non linear properties at lower frequencies, and particularly in the microwave region, has focused in the literature on another material (BST). This chapter devotes a significant part to the excellent dielectric non linear properties of SBN thin films. The great potential of SBN thin films is linked to our ability to control preparation and characterization. Through the successive sections, this chapter highlights some of the traps likely to deadlock research efforts.

2. Strontium barium niobate (SBN) crystals: main features

Strontium barium niobate ($Sr_xBa_{1-x}Nb_2O_6$, noted SBN : 100x) crystallizes in the region $0.25 < x < 0.75$ with the tetragonal tungsten bronze (TTB) structure represented in Figure 1 (Jamieson et al., 1968). The arrangement of NbO_6 octahedra in the form of five-members rings provides three types of interstitial sites: trigonal sites are vacant, tetragonal (A1) and pentagonal (A2) sites are partially occupied (5/6) by the divalent Sr and Ba atoms, and partially vacant (1/6) for reasons of electroneutrality. In this structure NbO_6 octahedra are not equivalent and two types must be distinguished. For both types the octahedral axes are not perfectly perpendicular to the (a, b) plane but slightly tilted from the polar c-axis (about 8°). Five formula units are necessary to form the unit cell depicted in the left part of Figure 1. Cell dimensions decrease with increasing the Sr/Ba ratio due to the smaller atomic radius of Sr from {a=b ≈ 12.48 Å, c ≈ 3.98 Å} when x≈25% to {a=b ≈12.43 Å, c ≈ 3.91 Å} when x≈75% at room temperature. This double variation in lattice parameters and chemical composition modifies significantly the Curie temperature Tc of the ferroelectric crystal: Tc decreases from about 220°C when x≈25% to about 60°C when x≈75% (Ballman & Brown, 1967).
Above Tc the displacement of metallic atoms from their mean oxygen planes along the c-axis becomes zero except for one of the two types of Nb atoms (80% of them), which are distributed above and below oxygen planes with equal probability. The symmetry point group of the crystal transforms from *4mm* to $\bar{4}2m$, which is a non polar but also a non-centrosymmetric class. Birefringence and second harmonic generation exist above Tc.
SBN is a disordered crystal since each interstitial site A1 or A2 may be either occupied or vacant, and, if occupied, either by a Sr or a Ba atom. Local composition may change from cell to cell. As a result SBN is a ferroelectric relaxor exhibiting a broad phase transition.

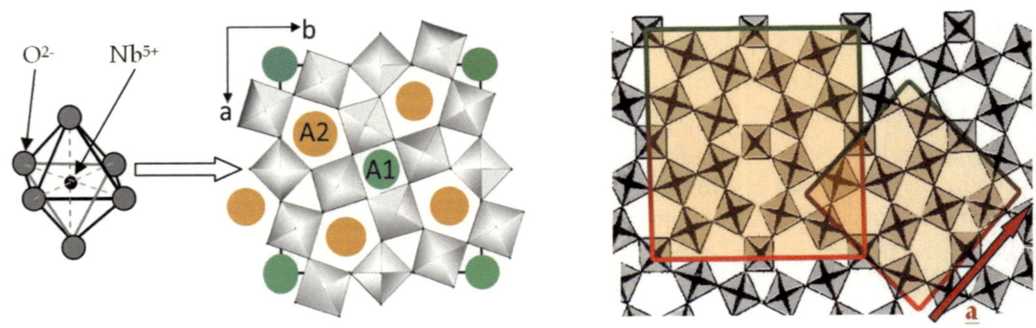

Fig. 1. View along the polar c-axis of the strontium barium niobate tetragonal tungsten bronze structure (After Jamieson et al., 1968). Rings made of five NbO_6 octahedra form three types of interstitial sites. The tetragonal (A1) and pentagonal (A2) positions are partially occupied by Sr and Ba atoms (5 /6) and partially vacant (1/6).

SBN crystals are optically uniaxial negative ($n_e<n_o$) at room temperature. Compared to the ordinary index n_o in the (a, b) plane, the extraordinary index n_e along the polar c-axis is much more sensitive to both Sr content and temperature (Venturini et al., 1968). At λ=633 nm and room temperature, (n_o, n_e) vary from (2.314, 2.259) to (2.312, 2.299) when x varies from 25 to 75%, respectively. The domain of transparency is 0.35µm - 6 µm.
Ferroelectric, dielectric and non linear optic properties of SBN crystals are very sensitive to the Sr/Ba ratio consistently with the impact of this ratio on Curie temperature (Glass, 1969; Lenzo, 1967). Increasing the Sr content reduces the interval between room and Curie temperatures, thus inducing a drastic enhancement of the dielectric permittivity, pyroelectric coefficient and non linear optic properties. Exceptionally large values of the linear electro-optic coefficient have been obtained (r_{33}=1340 pm/V at λ=633 nm) with a 75% Sr content (Lenzo et al., 1967).

3. SBN thin films: preparation

3.1 Stoichiometry

The ternary phase diagram shown in Figure 2 (Carruthers & Grasso, 1970) indicates the different phases that may crystallize when mixing the three basic oxides SrO, BaO, and Nb_2O_5. The coloured field denotes the existence region of TTB SBN. The ternary solubility was found to extend from about 4% excess (Ba+Sr) to about 1% excess Nb_2O_5.

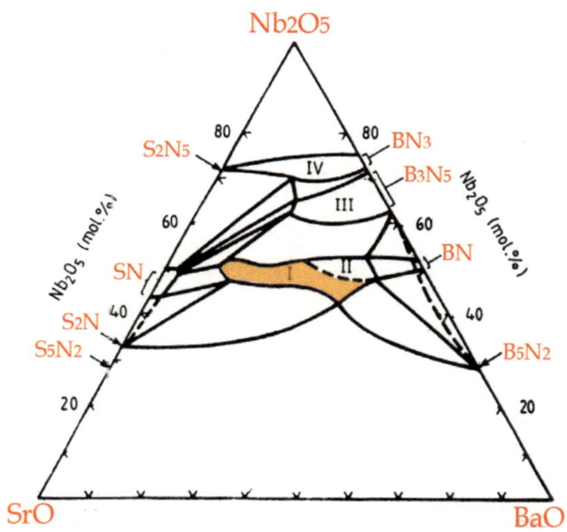

Fig. 2. Ternary phase relationships in the room temperature isotherm of the system BaO-SrO-Nb_2O_5, respectively denoted B-S-N in the chemical formula given along the axes. The coloured field is the existence region of the tetragonal tungsten bronze SBN phase (Carruthers & Grasso, 1970)

Stoichiometry of the deposit is a necessary but not sufficient condition for achieving single phase SBN thin films. Each deposition technique possesses its specific conditions for SBN stoichiometry, which must be established. In the case of the RF magnetron sputtering technique, the mechanisms which control the target-film composition transfer have been probed and their understanding exploited for stoichiometry control (Cuniot-Ponsard et al., 2003a). An illustration is given in Figure 3: two deposition parameters, the R.F. power and

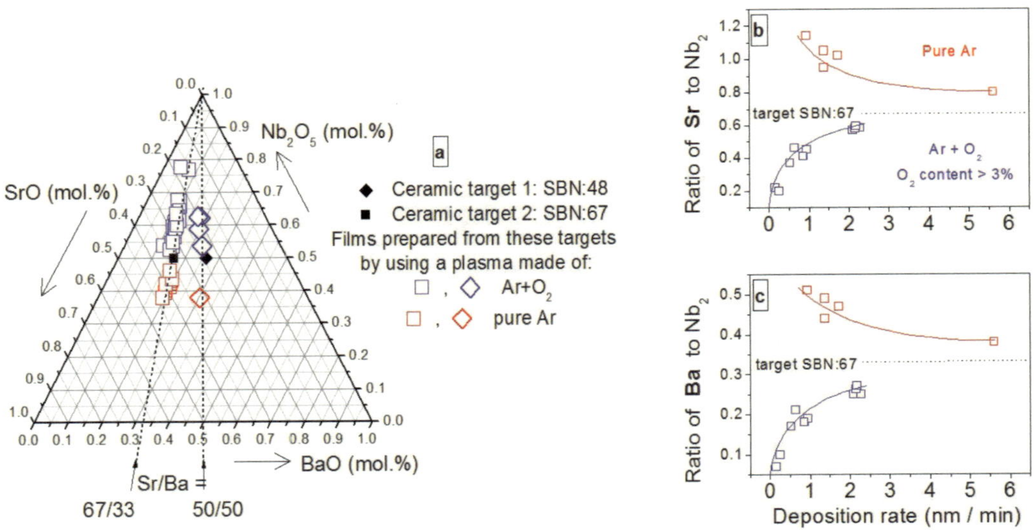

Fig. 3. (a) :Targets and sputtered films composition as determined from electron microprobe analysis. Differences in film composition have been obtained by varying R.F. power and oxygen percentage in the plasma. The other deposition parameters were fixed. (b) and (c) : Ratio of Sr and Ba to Nb_2, respectively, in films sputtered from a SBN:67 target as a function of deposition rate when this deposition rate is exclusively varied by modifying R.F. power. Oxygen percentage in the plasma is either zero (red symbols), or larger than 3% (3 to 15%) (blue symbols).

oxygen percentage in the plasma, have been varied and the resulting film composition is plotted in the ternary phase diagram. Targets and films composition has been determined by electron microprobe analysis using a single crystal of known composition (SBN: 60) as a standard. The film composition appears to be very sensitive to the varied parameters and large deviations from stoichiometry are observed. The presence of oxygen in the plasma changes the sign of the deviation: a lack of niobium becomes an excess niobium. This result indicates that the sputtering yield of niobium from the target is lower than that of the two other metallic atoms. On the other hand the Sr/Ba ratio is mainly determined by the target composition and independent of deposition parameters, which suggests that divalent Sr and Ba atoms have similar properties in terms of both sputtering yield and reactivity with oxygen.

In both cases of non zero and zero oxygen percentages in the plasma, the opposite deviations from the target stoichiometry depicted in Figures 3b and 3c unambiguously decrease with increasing R.F. power. As might be expected, increasing R.F. power reduces the difference between Nb and (Sr, Ba) sputtering yields but it may be hazardous for the target integrity. Another way towards stoichiometry stands out in these figures, which consists in adjusting the oxygen percentage between 0 and 3%. Then the deposition rate can be chosen independently as low as desired. Stoichiometry is necessary to attempt to crystallise a SBN single phase and it is not usually spontaneous. It is often inferred from the observation of an X-ray diffraction spectrum consistent with that expected from SBN. The next section discusses this point.

3.2 Crystallization: the X-ray diffraction traps

The implementation of the Pockels e-o effect in SBN films requires (001) oriented films. Compared to the X-ray θ - 2θ patterns of polycrystalline SBN (Fig. 4), a (001) oriented SBN film is expected to display only the two peaks arrowed in the figure at 2θ ≈ 22.5° and 45.9°. In Figure 4 the powder X-ray spectrum calculated from the atom distribution model of Jamieson et al. is given for reference (Fig. 4a). As highlighted by vertical dashed lines (Fig. 4b-4c), an increase of the Sr content results in a slight shift of the SBN peaks towards higher 2θ values due to the decrease in lattice constants. The difficulty of preparing a single SBN phase increases with increasing the Sr/Ba ratio close to the limit where the SBN phase becomes unstable (see Fig. 2). This explains the presence in the spectrum of the target SBN: 67 (Fig.4b) of parasitic peaks that issue from the phase $SrNb_2O_6$, noted SN in Fig. 2 (S for SrO and N for Nb_2O_5).

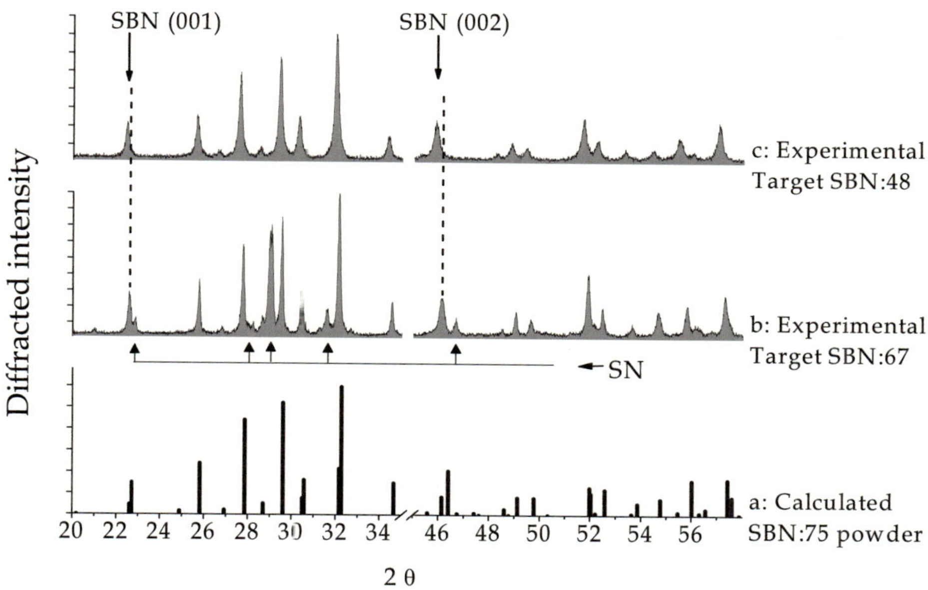

Fig. 4. X-ray θ - 2θ scan of a tungsten bronze SBN powder calculated from the atom distribution model for SBN : 75 (a) and experimental X-ray θ - 2θ scans measured from two polycrystalline targets SBN: 67 (b) and SBN: 48 (c). Target SBN: 67 is not single phase.

In the ternary phase diagram shown in Figure 2 some phases ($Sr_2Nb_{10}O_{27}$, $BaNb_2O_6$, $BaNb_6O_{16}$, $Ba_3Nb_{10}O_{28}$, respectively noted S_2N_5, BN, BN_3 and B_3N_5 in the diagram), have a c-cell parameter very close to that of the TTB SBN phase and consequently diffract their (001) and (002) peaks in very close angular locations as illustrated in Figure 5. Experimental data in the figure issue from the Joint Committee Powder Diffraction Standard - International Centre for Diffraction Data. The other cell parameters of these parasitic phases are also close to those (or to multiples) of the TTB SBN cell, so that the lattice-match and oriented (001) growth are similarly probable. Consequently, and taking into account the likely occurrence of a shift induced by epitaxial stress, the position of the (001) and (002) peaks in the X-ray pattern of a (001) oriented film cannot be considered as a reliable signature of the TTB SBN phase. On the other hand, the ratio of the (001) to the (002) peak intensities provides an unambiguous means of distinguishing SBN from three of these four phases: S_2N_5 , BN_3 and B_3N_5. Phi-scan measurements are capable to rule out the remaining ambiguity.

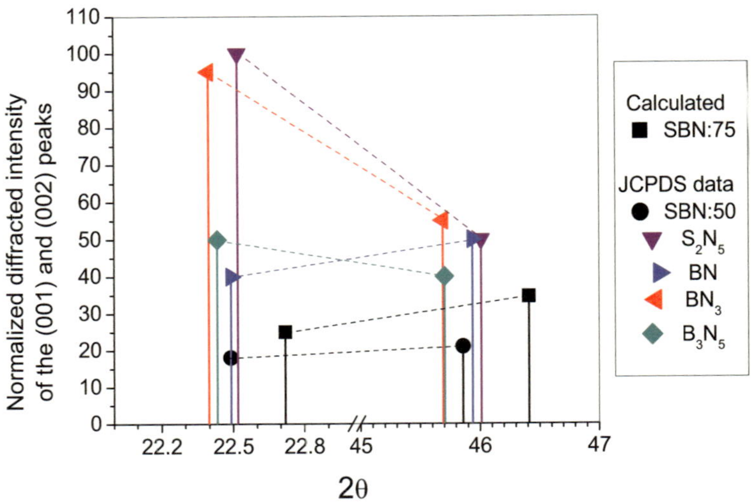

Fig. 5. Position and normalized diffracted intensity of the (001) and (002) peaks for the phases S_2N_5, BN, BN_3, B_3N_5, and SBN, from reported experimental data (Joint Committee Powder Diffraction Standard – JCPDS - International Centre for Diffraction Data), and calculated from the reported tetragonal tungsten bronze structure of SBN:75 (Jamieson et al., 1968)

3.3 Epitaxial crystallization on MgO and Pt covered MgO substrates

Magnesium oxide is the most common substrate used to grow epitaxial SBN thin films. Among the few substrate materials that can withstand temperatures higher than 700°C without reacting with the film, the reasons for choosing MgO are a small lattice mismatch (+1.2 to +1.6%, depending on Sr content, the positive sign implying a film tensile stress) and a large index contrast with SBN ($\Delta n = n_{SBN} - n_{MgO} \approx 0.5$ at $\lambda = 1.5$ μm). The oriented growth (001) SBN // (001) MgO has been reported from various deposition techniques including Metal Organic Chemical Vapor Deposition (MO-CVD) (Lu et al., 1994), Pulsed Laser Deposition (PLD) (Schwyn Thöny et al., 1994), Plasma Enhanced-CVD (Zhu et al., 1995), sol-gel synthesis (Sakamoto et al., 1996) and R.F. sputter deposition (Cuniot-Ponsard et al., 2003a). Figure 6 shows the X-ray diffraction spectrum of a (001) oriented SBN thin film prepared by RF magnetron sputtering of a ceramic target SBN: 60. The film was deposited amorphous then crystallized by using rapid thermal annealing. Rocking curves of the (001) and (002) peaks indicate a full width at half maximum of 0.65°.

When a bottom electrode is needed, a conductive crystalline substrate or a conductive epitaxial coating must be used. Among the possible conductive materials to be deposited on (001) MgO, platinum has been extensively studied for devices with ferroelectric, magneto-resistive, or magneto-optic applications, due to its chemical stability at high temperatures.

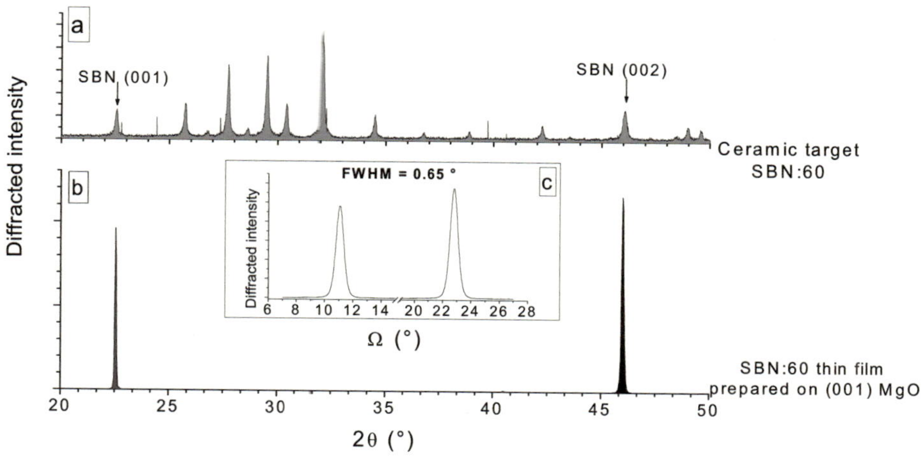

Fig. 6. (a) XRD θ - 2θ scan of a polycrystalline SBN:60 target as reference (b) XRD θ - 2θ scan of a film prepared from the previous target on a (001) MgO substrate by using the RF magnetron sputtering technique. (c) Rocking curves of the (001) and (002) peaks indicating a full width at half maximum of 0.65°.

Whatever deposition technique the authors used to prepare Pt thin films onto (001) MgO, the two orientations (111) and (001) were observed and found dominant at low and high deposition temperatures, respectively. However the reported temperature range in which the film orientation switches varies with the deposition technique: the lower the average energy of the depositing species, the higher the substrate temperature necessary to yield a dominant (001) Pt texture (Lairson et al., 1992; Narayan et al., 1994; McIntyre et al., 1995). This suggests that the incident kinetic and subsequent thermal energies of the depositing species are complementary regarding an energy threshold for the (001) Pt growth. The involvement of kinetic phenomena in the orientation development of the films has also been advanced. A decrease in the deposition rate has been reported (Ahn & Baik, 2002) to cause a drastic lowering of the orientation switching temperature (≈500°C→200°C). A similar lowering (≈500°C→350°C) can also be obtained with seeded substrates [Lairson et al., 1992). The epitaxial orientation relationship is of the type [110] Pt // [110] MgO for both (001)Pt and (111)Pt crystallites on (001)MgO, which means cube-on-cube orientation for (001) Pt. Calculations have been carried out (MIntyre, 1997), which assign minima of the Pt/(001)MgO interface energy to these two experimental in-plane alignments.

The (001) oriented growth of SBN onto Pt coated (001) MgO occurs exclusively on (001) oriented Pt crystallites (Cuniot-Ponsard et al., 2006). Figure 7 shows the X-ray symmetric patterns of two films simultaneously deposited and crystallized onto dominant (001) Pt and (111) Pt coatings, respectively. These spectra demonstrate a strong correlation between a (001) oriented SBN growth and the (001) orientation of underlying platinum. The integrated intensities of (001) SBN and (002) Pt reflections are found proportional: both are reduced to about 7% of their initial value from a spectrum (red) to the other (blue). Two other groups have published results about SBN thin films grown onto Pt coated MgO substrates [Sakamoto et al., 1996; Koo et al., 2000a]. Both prepared SBN by using a sol-gel process.

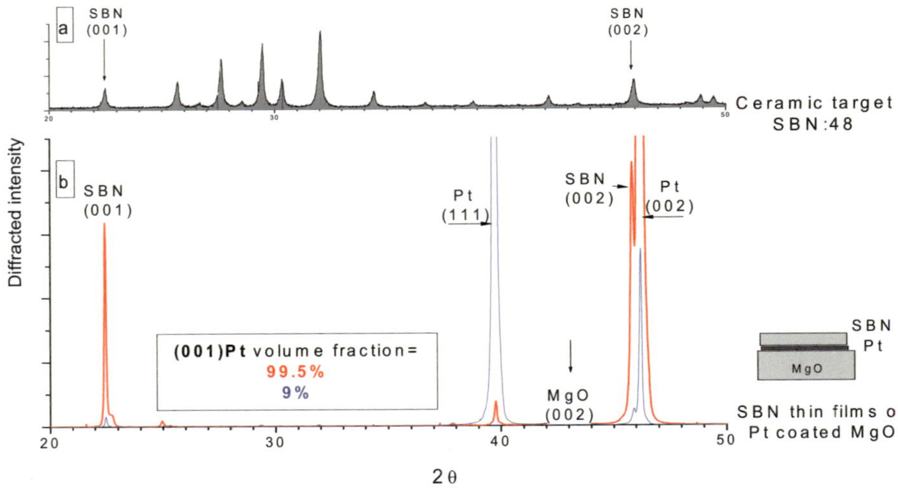

Fig. 7. Influence of the Pt crystallographic orientation on subsequent SBN growth: (a) XRD θ - 2θ scan of a polycrystalline SBN: 48 target as reference (b) XRD θ - 2θ scans of two films sputtered from the previous target, simultaneously deposited and crystallized on two Pt coated MgO (001) substrates only different in the (001) Pt volume fraction (99.5% and 9%).

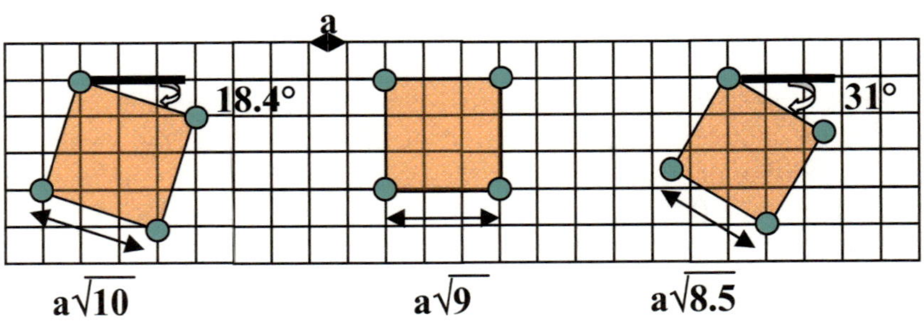

Fig. 8. Representation of the in-plane orientations experimentally observed for (001) SBN/(001) MgO. The large lattice constant of crystalline SBN is between a × $\sqrt{8.71}$ and a × $\sqrt{8.78}$ depending on Sr content if a is the lattice constant of cubic MgO (a= 0.421 nm).

Despite the small lattice mismatch between SBN and MgO cell parameters in the film plane, aligned orientations have been rarely observed and were never dominant. For (001) SBN/(001) MgO two couples of mirror symmetric in-plane orientations have been regularly reported in the literature since 1994 (Lu et al., 1994): ± 31° and ± 18.4° with respect to MgO cell axes (Fig. 8). The latter has been found present and dominant in all SBN thin films except in those we prepare by RF magnetron sputtering and in some samples prepared by RF plasma assisted Pulsed Laser Deposition (Scarisoreanu et al., 2008). This rotation by ± 18.4° implies a significant increase of the lattice misfit δ: $+1.2\%<\delta<+1.6\%$ (no rotation) becomes $+6.7\%<\delta<+7.1\%$ (for $25\%<x<75\%$). An explanation involving the role of electrostatic energy across the interface has been proposed (Schwynn Thöny et al., 1994) and rejected (Willmott et al., 2005) in favour of another explanation based on the existence of a SN thin layer at the interface. On the contrary, a rotation by ± 31° changes the sign of the lattice misfit but does not significantly modify its magnitude : +1.2%<δ<+1.6% (no rotation) becomes -1.6% <δ<-1.2% (for 25%<x<75%).

XRD phi-scans performed on SBN films prepared by sputtering demonstrate an epitaxial growth on both (001) MgO (Fig. 9a & b) and (001) Pt covered (001) MgO (Fig.9c & d). Two sets of planes, (211) and (311), have been selected on the basis of the high intensity they are expected to diffract at 2θ ≈ 27.7° and 32.1° respectively (see Fig.4). The only in-plane orientations which account for both (211) and (311) results are mirror symmetric (± α) to the MgO cell axes. The peaks location predicted from simulation for these two orientations is indicated in the figures with triangular symbols. As mentioned above, α is found equal to ± 31° for films deposited on MgO (Fig. 9a & 9b), which is consistent with a minimized lattice misfit. Among the parasitic phases likely to crystallize, SBN stoichiometry favours $SrNb_2O_6$ (SN), which grows epitaxially and (100) oriented on (001) MgO (Cuniot-Ponsard et al., 2003b). The presence of SN in the film prepared on MgO explains the parasitic peaks observed in the Φ scan. When the film is grown on (001) Pt / (001) MgO (Fig.9c & d), the SBN cell is found rotated by ± 18.4° with respect to the Pt and MgO cell axes. Contrary to the case discussed above of SBN/MgO, and due to the lower lattice constant of the cubic Pt cell (a=0.391 nm), the orientation ± 18.4° corresponds to the minimized lattice misfit in the case SBN/crystalline Pt (Fig. 8).

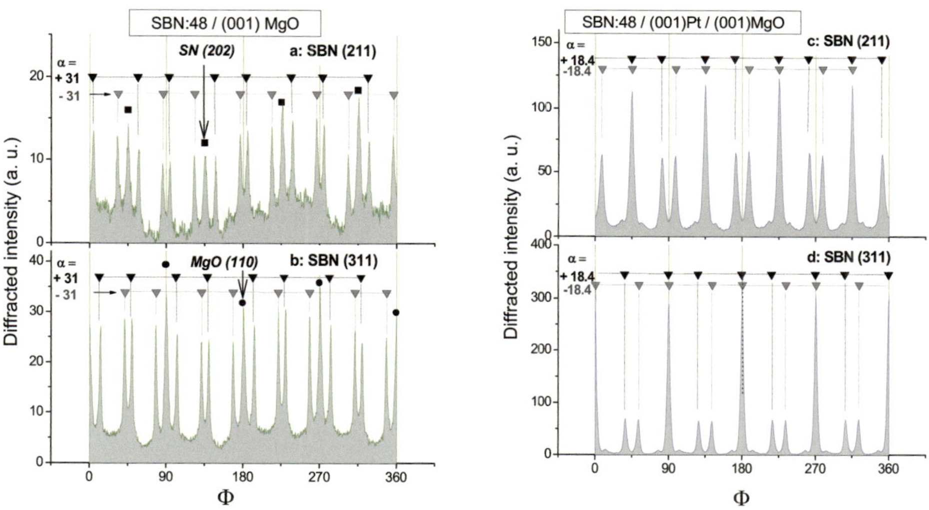

Fig. 9. Phi-scan measurements of the planes SBN(211), SBN(311) obtained from SBN: 48 thin films prepared by R.F. magnetron sputtering on (001) MgO (a,b) or (001) Pt / (001) MgO (c,d). The peaks location predicted from simulation for the two mirror symmetric in-plane orientations is denoted in the figures with triangular symbols. The presence of the parasitic phase $SrNb_2O_6$ (SN) in the film prepared on MgO is responsible for additional peaks assigned to SN (202) reflection.

A stoichiometric composition of the film is necessary but not sufficient to prevent the crystallisation of a mixture of parasitic phases like $SrNb_2O_6$ (SN) and $BaNb_2O_6$ (BN) of identical mean composition. The crystallization of a single phase SBN necessitates a minimum thermal energy whose amount increases when approaching the composition limits of SBN phase stability. An illustration is given in Figure 10 which compares the X-ray diffraction patterns of two stoichiometric films prepared simultaneously by RF magnetron sputtering of a SBN: 60 target, then crystallized by rapid thermal annealing in identical conditions except the annealing duration. When the minimum thermal energy needed for SBN crystallization is not supplied (blue spectrum), a mixture of oriented SN and BN phases is obtained instead of a single phase SBN. Besides sufficient oxygen pressure during crystallization is necessary to prevent oxygen vacancies in the film structure. These oxygen vacancies are responsible for a strong loss of transparency in the near-infrared region (Fig.11a & b) and for large dielectric loss at low frequencies (Fig.11 c).

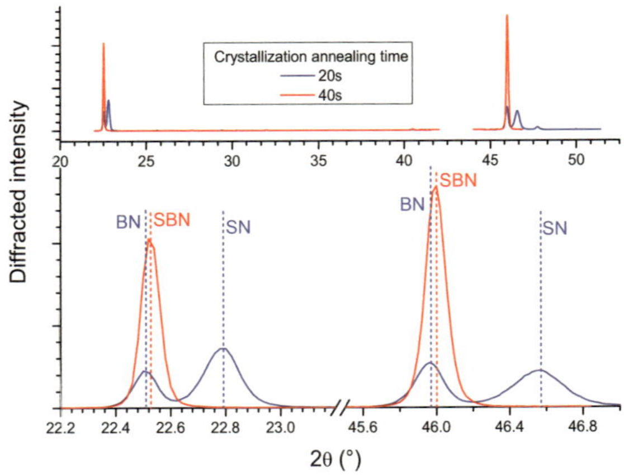

Fig. 10. X-ray diffraction spectra of two films prepared amorphous simultaneously by sputtering a SBN:60 ceramic target, then crystallized by rapid thermal annealing during either 20s or 40 s. A mixture of two parasitic phases $SrNb_2O_6$ (SN) and $BaNb_2O_6$ (BN) or a single phase SBN is obtained depending on the thermal energy supplied to the stoichiometric amorphous film.

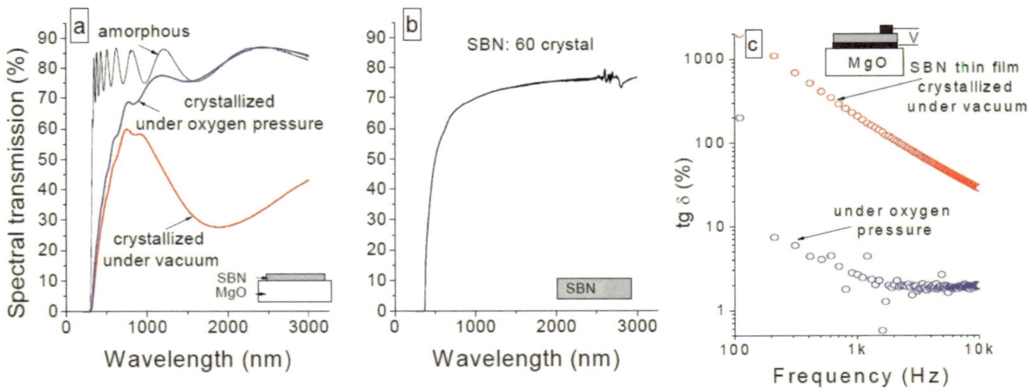

Fig. 11. Optical and electrical signatures of oxygen vacancies in SBN. (a) Optical transmission spectra of SBN films simultaneously prepared on MgO substrates and crystallized either under vacuum or oxygen pressure. The spectrum of the amorphous film before crystallization is also shown. (b) Optical transmission spectrum of a SBN: 60 crystal for reference. (c) Dielectric loss tangent versus frequency measured on Pt/SBN/Pt structures when SBN crystallization is achieved either under vacuum or oxygen pressure.

4. SBN thin films: dielectric properties

The dielectric properties of SBN crystals have been thoroughly investigated by Glass (Glass, 1969) as functions of temperature, frequency, dc electric field and Sr content. Concerning SBN thin films, the first detailed publication on ferroelectric properties (Antsigin, 1985) reported film properties similar to those of crystals. Such a complete similarity was not observed again until now. About fifteen papers have been published since, which report, often briefly, some dielectric properties of SBN films prepared by different techniques. The results vary strongly from a paper to another. Glass has shown that the dielectric permittivity deduced from the capacitance of a crystalline SBN disk sandwiched between two electrodes varied strongly with the nature of the electrodes, the specimen thickness and its degree of polarity. This influence of interfaces on the measured dielectric permittivity, necessarily enhanced in thin film measurements, is a first likely reason for the scattering of reported results. A second reason is to be found in the usual method used to record a Polarization-Electric field (P-E) hysteresis loop, which proceeds by integrating the current intensity induced by voltage variation in the circuit. The dielectric loss current is also integrated and may significantly contribute to form the observed hysteresis loop.

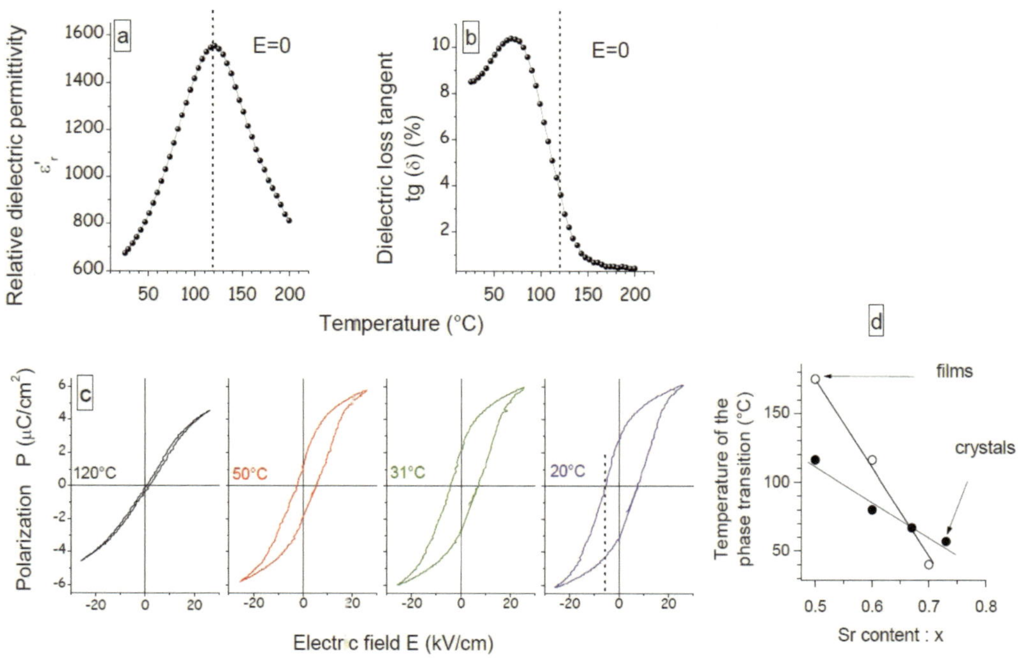

Fig. 12. Ferroelectric properties of a Pt / SBN / Pt structure prepared by sputtering from a SBN:0.60 target.(a-b) Zero field temperature dependence of the weak field relative dielectric permittivity and loss tangent at 10 kHz. (c) Polarization-electric field hysteresis loops recorded at a frequency of 33 Hz for different measurement temperatures. (d) Comparison of the phase transition temperature versus Sr content reported for crystals (Glass, 1969) and for epitaxial SBN films (Boulay, 2007).

Dielectric properties of epitaxial SBN thin films prepared by RF magnetron sputtering on (001) Pt covered (001) MgO substrates are presented in Figures 12, 13 & 14 (Boulay et al., 2007). Top platinum dots (0.5 and 0.3 mm in diameter) have been deposited by using R.F. magnetron sputtering. Capacitance and dielectric loss of the Pt / SBN / Pt sandwich structure are measured using a Hewlett Packard impedance analyzer. The weak field frequency is chosen equal to 10 kHz, which is high enough to eliminate the possible contribution of parasitic capacitances, and low enough to keep the effect of the measurement circuit inductance negligible. The polarization - field hysteresis loops are recorded at a frequency of 33 Hz using a Radiant Technologies RT66A analyzer.

A ferroelectric-paraelectric phase transition is observed in a temperature range depending on the Sr / Ba ratio (Fig. 12). The decrease of the transition temperature in the films with increasing Sr content is consistent with the evolution observed in crystals although it is of higher magnitude (Fig. 12d). As expected from a ferroelectric material, the shape of the polarization-electric field (P-E) loop becomes slimmer with increasing temperature and the hysteresis behaviour vanishes at the transition temperature(Fig.12c). The relative dielectric permittivity ε_r' and dielectric loss tangent ($tg\delta = \varepsilon''/\varepsilon'$, where ε' and ε'' are the real and imaginary parts of the dielectric permittivity) show a strong dependence on applied dc electric field (Fig.13). This field dependence varies with temperature similarly to the dielectric permittivity, and is maximum at the transition temperature. The electrical tunability of dielectric permittivity $\delta\varepsilon'/\varepsilon'\delta E$ is zero when electric field is equal to the coercive value, and maximum in the vicinity of this value. At room temperature the electrical tunability of SBN thin film dielectric permittivity is measured to be 1.8% cm/kV for SBN: 60 and 1.6% cm/kV for SBN: 67 (Fig. 14).

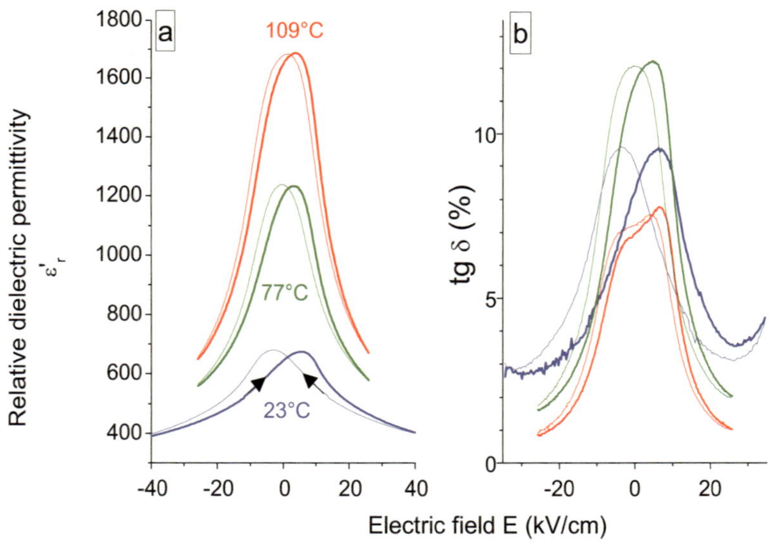

Fig. 13. Dielectric properties of a Pt / SBN / Pt structure prepared by sputtering from a SBN: 60 target. Electric field dependence of the weak field relative dielectric permittivity (a) and loss tangent (b) is measured at 10 kHz for different measurement temperatures.

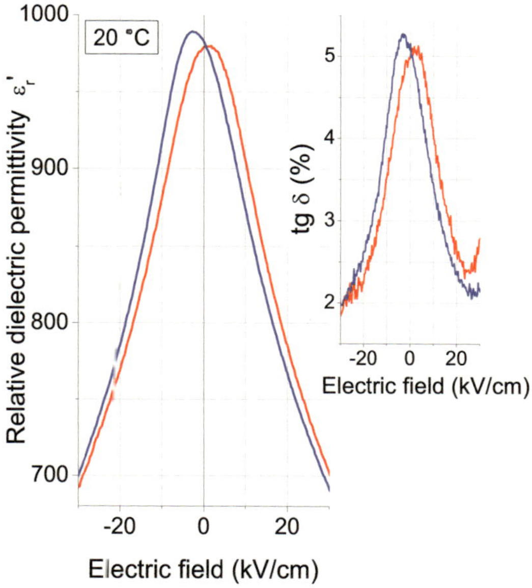

Fig. 14. Relative dielectric permittivity and loss tangent of a Pt / SBN / Pt structure prepared by sputtering from a SBN: 67 target, measured at 10 kHz and at room temperature. The electrical tunability of dielectric permittivity ($\delta\varepsilon'/\varepsilon'\delta E$) is 1.6% cm/kV (16% μm/V) at low dc electric field.

These values may be compared to that given by Glass for the electrical tunability of a SBN: 67 crystal (Glass, 1969) which is 4.4% cm/kV. They should also be compared to the electrical tunability of (Ba, Sr)TiO_3 (BST) thin films which have been extensively investigated in view of exploiting this property in microwave applications. A value of 0.25% cm/kV at low frequency (1MHz) was recently presented as an excellent result for BST thin films (Nozaka et al., 2008). The electrical tunability of SBN thin films prepared by PLD on MgO substrates has been measured by Moon et al. between 0.5 and 20 GHz and found to be 0.5% cm/kV (Moon et al., 2004). The same authors compared phase shifters based on (Ba, Sr)TiO_3 and SBN thin films between 0.5 and 20 GHz, and concluded that these two materials were competitors for microwave tunable devices (Moon et al., 2005).

Some of the dielectric properties reported here are comparable with those reported by Liu et al. (Liu et al., 1995), Zhao et al. (Zhao et al., 2004), and Guerra et al. (Guerra et al., 2008) for SBN thin films prepared by using other techniques.

The values measured for the transition temperature, the dielectric permittivity and the coercive field, are consistent with the values reported for crystals. On the contrary the values measured for the polarization and the electrical tunability are about 5 times and 3 times lower, respectively, than those reported for crystals of similar composition. The dielectric tunability, as well as the electro-optic effect, involves the derivative $\delta^2 P/\delta E^2$, where P and E are polarization and electric field. They are therefore expected to be (E,P)-dependent, and a correlated drop of polarization and electrical tunability is not surprising. Several possible reasons exist for a polarization lower in films than in crystals. They are presently under investigation.

5. SBN thin films: electro-optic properties

In this section, the dielectric permittivity is investigated at optical frequencies where ionic displacement is negligible and relative dielectric permittivity is reduced to its electronic component $\varepsilon_r = n^2$ (n is the complex index refractive of SBN).The electrical tunability of dielectric permittivity in this frequency range is the electro-optic (e-o) effect. It may be expressed in terms of an electric-field-induced distortion of the refractive index ellipsoid as:

$$\Delta(\frac{1}{n_{ij}^2}) = \sum_k r_{ijk} E_k \tag{1}$$

where *n*, *r*, and *E* represent, respectively, refractive index, electro-optic coefficient, and electric field. Only the first order terms (Pockels effect) have been kept in (1). The exceptionally high e-o coefficient in SBN is r_{33} (using reduced notation for r_{333}). A method enabling a reliable measurement of this coefficient in thin films is presented and applied to SBN films prepared by sputtering on (001)Pt/(001)MgO substrates (Cuniot-Ponsard et al., 2011).

5.1 Usual experimental methods and difficulties

Methods used to determine the electro-optic coefficients of a thin film may be classified in three types: interferometric, polarimetric, and prism-coupling methods. In interferometric methods, an interference pattern is created with light beams, at least one of which propagates inside the e-o film. The electric-field-induced response of the e-o film causes a variation in phase and amplitude of the wave emerging from the film, and results in a variation (ΔI) in pattern light intensity. The different e-o coefficients are determined independently from the measurement of ΔI, by varying the direction of the linear polarization. In polarimetric methods the electric field induces a change ($\Delta\varphi$) in the relative phase between the two orthogonal components of a linearly polarized light beam. In most cases $\Delta\varphi$ is converted into an output light intensity variation ΔI by using an analyzer. From the measurement of $\Delta\varphi$ or ΔI, an "effective" e-o coefficient is determined, which is a linear combination of two e-o coefficients. Polarimetric methods do not enable the value of each e-o coefficient to be determined separately. In the prism-coupling technique (also called ATR for Attenuated Total Reflection) the film is sandwiched between two electrodes and forms a waveguide in which light is coupled by using a prism. The electric-field-induced response of the film modifies coupling incident angles, thus causing a variation ΔI in the intensity of the light beam emerging from the prism.

In all cases, the determination of e-o coefficients is carried out from measurement of the variation ΔI induced in output light intensity. The electric-field-induced response being a priori a combination of electro-optic, converse piezoelectric and electro-absorptive effects, the induced output signal ΔI is a function of the variations in refractive index (Δn_{ij}), in dimensions (Δe_i), and in extinction coefficients (Δk_{ij}), which may be expressed as:

$$\Delta I = \sum_{i,j} \frac{\partial I}{\partial n_{ij}} \Delta n_{ij} + \sum_{i} \frac{\partial I}{\partial e_i} \Delta e_i + \sum_{i,j} \frac{\partial I}{\partial k_{ij}} \Delta k_{ij} \tag{2}$$

Two sources of potential error affect the determination of e-o coefficients. The first is the use of approximations in view of simplifying the above expression (2): they consist in neglecting optical effects like multiple reflections, and/or some of the electric-field-induced effects like converse-piezoelectricity and electro-absorption. The further analysis of SBN films electric-field-induced response demonstrates the significant errors that can be induced by such approximations. The second source of potential error is the extreme sensitivity of derivatives appearing in (2) to a variation in the values of physical parameters of the films interacting with light (active film and electrodes). An illustration of this sensitivity is given in Figure 15. The example of a Fabry-Perot interferometric set-up has been chosen. The output reflected intensity I and the derivative $\partial I/\partial n_o$ (n_o is the ordinary refractive index of the active film) have been rigorously calculated for the transverse electric polarization TE without using any simplifying approximation. Figure 15 shows the plots of $\partial I/\partial n_o$ versus angle of incidence θ calculated from two sets of parameters with a small change only in the active film thickness value (e = 0.686 or 0.690 μm). The comparison between the two plots shows that if the actual thickness value is 0.686 μm instead of a believed value of 0.690 μm, this relative error of about +0.6% in one parameter value can induce a dramatic error in the value of the derivative $\partial I/\partial n_o$. Indeed the true value of the derivative will be multiplied by (-1) if measurement is performed at $\theta \approx 31°$, and by (1/3) if measurement is performed at $\theta \approx 34°$. Values of Δn_o and of the corresponding e-o coefficient will consequently be inferred erroneous in the inverse proportions, that is, obtained with the wrong sign if measurement is performed at $\theta \approx 31°$, or overestimated by a factor 3 if measurement is performed at $\theta \approx 34°$. From this simple demonstration it appears that determination of e-o coefficients is liable to drastic errors and must be necessarily checked independent of angle of incidence.

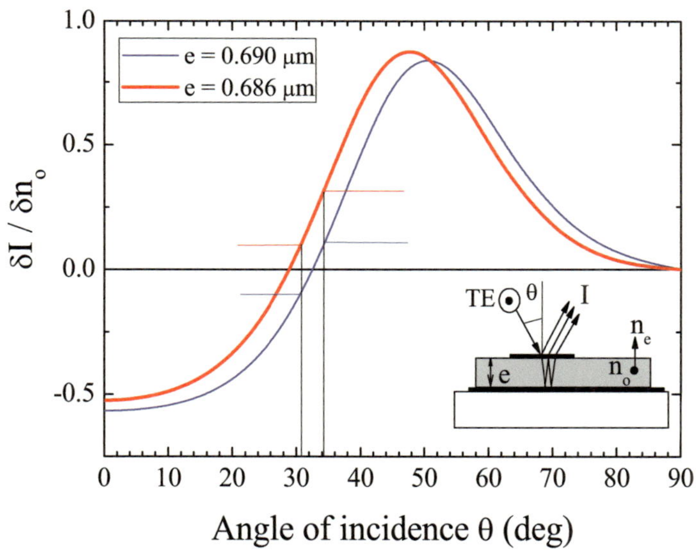

Fig. 15. Example of evolution of the derivative $\partial I/\partial n_o$ when slightly modifying one parameter, here the electro-optic film thickness value e, in a Fabry-Perot interferometric set-up. A variation in film thickness of only +0.6% (e = 0.686 →0.690 μm) can strongly modify the calculated derivative value: $\partial I/\partial n_o$ is multiplied by (-1) if measurement is performed at $\theta \approx 31°$, and by (1/3) if measurement is performed at $\theta \approx 34°$.

5.2 Principle of a reliable characterization method

The method exploits interferences obtained by reflection on the stack made of the e-o film sandwiched between two electrodes (Fig. 16). Reflectivity (R), and variation in reflectivity (ΔR) induced by an ac modulating voltage, are recorded versus incident angle θ, successively for TE and TM polarizations. According to the above expression (2):

$$\Delta R_{TE,\theta} = \left(\frac{\partial R}{\partial n_o}\right)_{TE,\theta} \Delta n_o + \left(\frac{\partial R}{\partial e}\right)_{TE,\theta} \Delta e + \left(\frac{\partial R}{\partial k_o}\right)_{TE,\theta} \Delta k_o \qquad (3)$$

where Δe, Δn_o, Δk_o are the electric-field-induced variations of, respectively, the active film thickness e, its ordinary refractive index n_o, its ordinary extinction coefficient k_o.

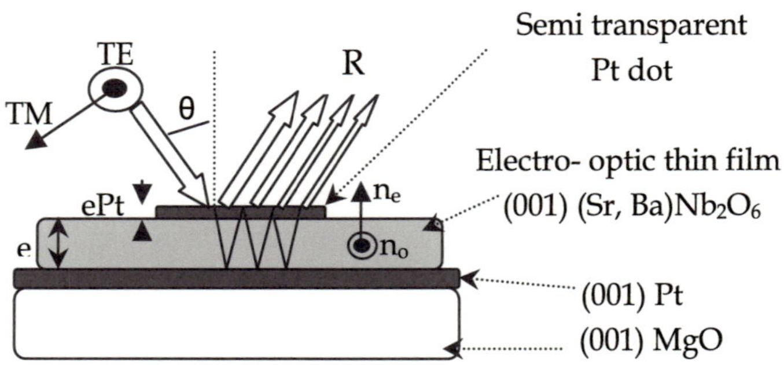

Fig. 16. Fabry-Perot interferometric configuration. Reflectivity (R), and variation in reflectivity (ΔR) induced by an ac modulating voltage, are recorded versus incident angle θ, successively for transverse electric TE and transverse magnetic TM polarizations

The calculation of reflectivity R is performed from Fresnel formulae by taking into account all the multireflection effects in the stack. The value of each derivative is numerically calculated. The determination of the three unknowns Δn_o, Δe, Δk_o appearing in (3) is then achieved by selecting three experimental data $[\Delta R_{TE}(\theta_1), \Delta R_{TE}(\theta_2), \Delta R_{TE}(\theta_3)]$ and solving the inferred system of three linear equations. At this step, the e-o coefficient r_{13} and the converse-piezoelectric coefficient d_{33} can be deduced:

$$r_{13} = \frac{-2\, e\, \Delta n_o}{n_o^3\, \Delta V} \qquad (4)$$

$$d_{33} = \frac{\Delta e}{\Delta V} \qquad (5)$$

where ΔV is the ac modulating voltage amplitude. The refractive index of the film seen by TM polarization (n_θ) and its variation under electric field (Δn_θ) are dependent on θ, which prevents the use of a similar procedure. If θ_F represents the refracting angle of the light wave in SBN, n_θ and Δn_θ verify:

$$\frac{1}{n_\theta^2} = \frac{\cos^2\theta_F}{n_o^2} + \frac{\sin^2\theta_F}{n_e^2} \qquad (6)$$

$$n_\theta = n_o \times \left(1 - \sin^2\theta \times \left(\frac{1}{n_e^2} - \frac{1}{n_o^2}\right)\right)^{1/2} \tag{7}$$

$$\Delta n_\theta = -\frac{n_\theta^3\,\Delta V}{2\,e} \times \left(r_{13}\cos^2\theta_F + r_{33}\sin^2\theta_F\right) \tag{8}$$

According to (2):

$$\Delta R_{TM,\theta} = \left(\frac{\partial R}{\partial n_\theta}\right)_{TM,\theta}\Delta n_\theta + \left(\frac{\partial R}{\partial e}\right)_{TM,\theta}\Delta e + \left(\frac{\partial R}{\partial k_\theta}\right)_{TM,\theta}\Delta k_\theta \tag{9}$$

Derivatives appearing in (9) are calculated numerically like they were calculated for TE polarization except that dependence of index on θ must be taken into account. A value of incident angle is selected for which $(\partial R/\partial k_\theta)_{TM,\theta} = 0$, and equation (9) becomes a linear equation with a single unknown (Δn_θ). The e-o coefficient r_{33} is deduced from the solution Δn_θ through above expression (8).
From values of r_{13}, d_{33}, Δk_o, the electric-field induced response ΔR_{TE} (θ) may be calculated versus incident angle and compared to the experimental data. For TM polarization, calculation of R_{TM} (θ) and ΔR_{TM} (θ) versus incident angle may be carried out by considering n_e as a parameter. Calculation must be checked to account for the whole of experimental data : R_{TE} (θ), R_{TM} (θ), ΔR_{TE} (θ) and ΔR_{TM} (θ).

5.3 Electric-field induced response of SBN thin films

The previous procedure was carried out using a He-Ne laser radiation (λ=633 nm) and applied to the determination of SBN: 60 film coefficients. It yields the results indicated below (10).

$$\begin{aligned} r_{13} &= +8.5 \pm 1.3\ pm/V \\ d_{33} &= \Delta e/\Delta V = +21 \pm 4\ pm/V \\ \Delta k_o/\Delta V &= (+9.8 \pm 0.6)\times 10^{-6} \\ r_{33} &= +38.9 \pm 0.5\ pm/V \\ r_{33}/r_{13} &= +4.5 \pm 0.6 \end{aligned} \tag{10}$$

Electro-optic, converse-piezoelectric and electro-absorptive components calculated from (10) are shown in Fig. 17 and Fig. 18 for TE an TM polarizations, respectively. As may be noticed, none of the three contributions is negligible and any attempt to account for experimental data without considering all of them would have been necessarily doomed to failure. Due to the opposite signs of Δn_o and Δe, phase shifts induced by electro-optic and converse-piezoelectric effects partially compensate one another so that the sum of these two contributions does not prevail over the electro-absorptive component in the final signal $\Delta R_{TE,\theta}$. A similar compensation is to be expected in a waveguide configuration if similarly low modulating frequencies are used. Measurements are therefore performed in unclamped conditions and the elasto-optic effect is likely responsible for a part of the measured coefficients. At high frequencies above piezo-electric resonance which are used in information optical processing, the converse-piezoelectric effect is clamped and cannot counteract the electro-optic performance by inducing phase shift compensation as in the present measurements.

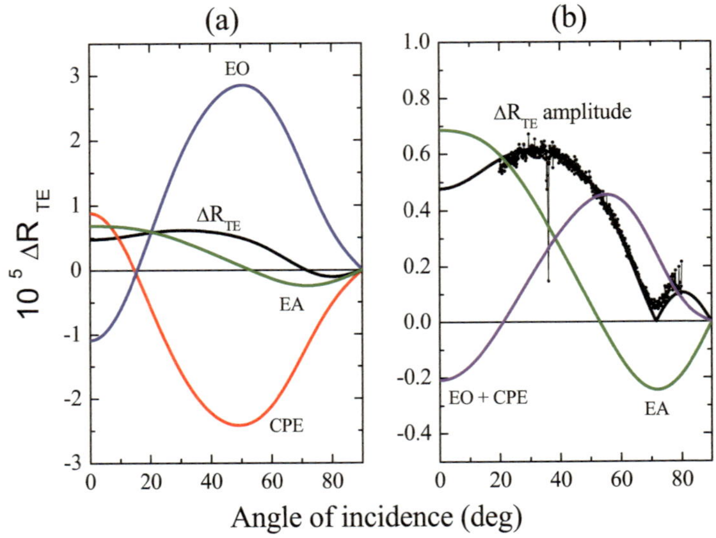

Fig. 17. (a) Electro-optic (EO), converse-piezoelectric (CPE) and electro-absorptive (EA) components of the electric-field induced variation in TE reflectivity (ΔR_{TE}) calculated from characterization procedure results. (b) Comparison between calculated (black continuous line) and experimental (black dots) ΔR_{TE} amplitudes. Amplitude of modulated voltage applied to SBN film is 1V. The plot of (EO+CPE) and (EA) components underlines the significant contribution of electro-absorptive effect to total TE response.

Figure 18 shows that the agreement between experimental and calculated data may be improved by varying the single adjustable parameter of this calculation: n_e. The resulting value of n_e ($n_e = n_o - 0.04$) is consistent with the birefringence value reported in the literature for crystalline SBN of similar composition at λ=633 nm ($n_e \approx n_o - 0.03$).

A few groups performed measurements of r_{33} in their SBN films. They reported r_{33} = 350 pm/V (Trivedi et al.,1996), r_{33} = 173.4 pm/V (Koo et al.,2000b), and r_{33} = 186 pm/V (Li et al.,2008) for SBN:60 films. Multiple reflections in the film, converse-piezoelectric and electro-absorptive effects were neglected in the quoted reports and the authors did not control that results were consistent when varying incident angle. The e-o coefficient values previously reported for SBN films in the literature may be suspected of error for various reasons and would need to be confirmed.

Concerning SBN crystals, very few reports exist in the literature on the separate measurements of r_{13} and r_{33} in SBN crystals. At λ=633 nm and for a composition SBN:60, e-o coefficients have been determined separately in one paper (Zhang et al., 1991) which reports: r_{13}=37 pm/V and r_{33}=237 pm/V from measurements that do not enable specifying signs. On the other hand, the converse-piezoelectric coefficient d_{33} in SBN crystals was reported to be about 95 pm/V. Compared to their crystalline counterpart, the three

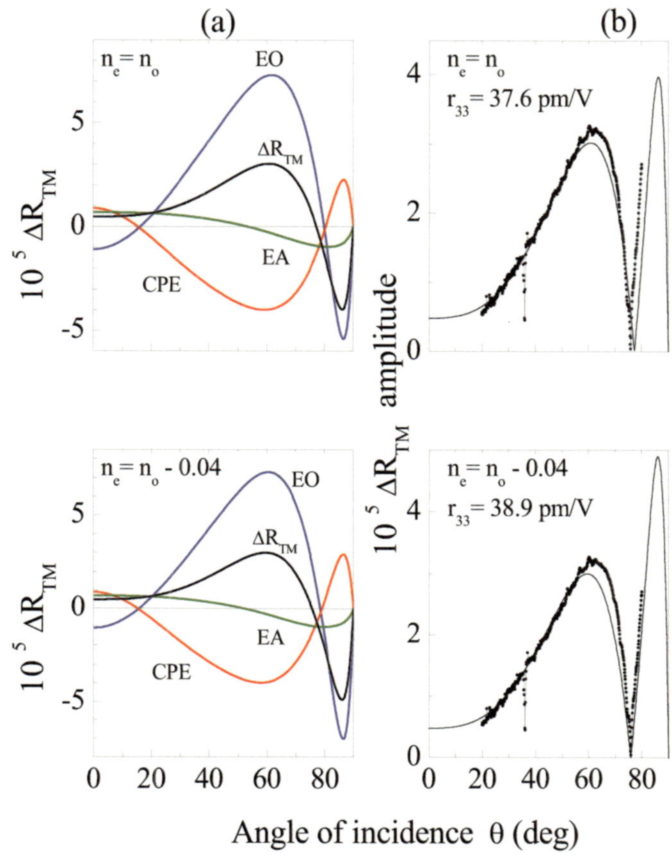

Fig. 18. Electric-field induced response for TM polarization (ΔR_{TM}). The three electro-optic (EO), converse-piezoelectric (CPE) and electro-absorptive (EA) contributions are detailed in (a) and comparison with experimental response is given in (b). Amplitude of modulated voltage applied to SBN film is 1V. Extraordinary refractive index n_e is the single adjustable parameter of characterization procedure; two cases are shown: $n_e = n_o$ (top) and $n_e = n_o - 0.04$ (bottom). The latter corresponds to the best calculation-experiment fit. The electro-optic coefficient r_{33} determined from characterization procedure and used for ΔR_{TM} calculation is equal to +37.6 (top) and +38.9 pm/V (bottom).

coefficients r_{13}, r_{33}, d_{33} of the film appear reduced in similar proportions by a factor about 5±1. The magnitude of polarization in our SBN films is also measured about 5 times lower than that reported for SBN:60 single domain crystals. As already mentioned above, a correlation between these results is not surprising.

Although lower than those of SBN crystals, the application relevant e-o coefficients measured on SBN: 60 thin films (r_{33} = +38.9 pm/V , $r_{eff} = r_{33} - (n_o/n_e)^3 r_{13}$ = +29.9 pm/V) are larger than those of a crystal of lithium niobate at the same wavelength λ=633 nm (r_{33} = +30.9 pm/V, r_{eff} = +20.1 pm/V). A further improvement of these SBN film coefficients is expected from the understanding of a lower polarization in films and from an increase in Sr content. The SBN thin film path is therefore proved to be competitive with regard to e-o modulation. Beside the e-o coefficient, the refractive index and the dielectric permittivity of the material are also involved in the e-o performance. The low dielectric permittivity of lithium niobate is an advantage to be taken into account when comparing thin film paths.

6. Conclusion

An overview of the questions relative to the preparation, dielectric and electro-optic properties of strontium barium niobate thin films has been proposed, with a special focus on epitaxial growth in view of electro-optic applications. Results obtained with films prepared by RF magnetron sputtering have been presented.

The polarization, dielectric tunability at low frequency and electro-optic coefficients of epitaxial SBN thin films prepared on (001)Pt/(001)MgO are found lower than those of crystals of the same composition in similar proportions. This correlation is not surprising and an understanding of the lower polarization in films should indicate the way towards a further improvement of the desired properties.

Even lower than those of their crystalline counterpart, the non linear properties of SBN epitaxial thin films are competitive with those of (Ba, Sr)TiO3 thin films for dielectric tunability and with those of crystalline $LiNbO_3$ for electro-optic properties.

7. References

Ahn, K.H. & Baih, S (2002). Change of growth orientation in Pt films epitaxially grown on MgO (001) substrates by sputtering. *Journal of Materials Research,* Vol. 17, No. 9 (September 2002), pp.2334-2338

Antsigin, V.D.; Egorov, V.M.; Kostsov, E.G.; Malinnovsky, V.K. & Sterelyukhina, L.N. (1985). Ferroelectric properties of thin strontium barium niobate films. *Ferroelectrics,* Vol. 63 (1985), pp. 235-242

Baginsky, I.L.; Gudaev, O.A.; Detinenko, V.A.;, Kostsov, E.G.; Malinovski, V.K.; Nesterikhin, Yu.E. & Pokrosky, L.D. (1978). Some peculiarities strontium barium niobate films and their electrophysical properties. *Ferroelectrics, Vol. 22 (1978, pp. 783-784.*

Boulay, N; Cuniot-Ponsard, M; Desvignes, J.M. & Bellemain (2007), A. Dielectric and ferroelectrics properties of $Sr_xBa_{1-x}Nb_2O_6$ (SBN: x) thin films. *Ferroelectrics,* Vol. 353 (2007), pp. 10-20.

Carruthers, J.R. & Grasso, M. (1970). Phase equilibria relations in the ternary system BaO-SrO-Nb_2O_5. *Journal of Electrochemical Society* Vol. 117, No. 11, (November 1970), pp. 1426-1430

Cuniot-Ponsard, M; Desvignes, J.M.; Ea-Kim, B & Leroy, E (2003a). Radio frequency magnetron sputtering deposition of hetero-epitaxial strontium barium niobate thin films ($Sr_xBa_{1-x}Nb_2O_6$). *Journal of Applied Physics,* Vol. 93, No.3, (February 2003), pp. 1718-1724.

Cuniot-Ponsard, M; Desvignes, J.M & Leroy, E (2003b). R. F. magnetron sputtering deposition of hetero-epitaxial $Sr_xBa_{1-x}Nb_2O_6$ thin films: the role of temperature. *Ferroelectrics,* Vol. 288, (2003), pp. 159-168.

Cuniot-Ponsard, M; Desvignes, J.M. & Bellemain,A (2006). Epitaxial growth of Sr_x Ba_{1-x} Nb_2 O_6 (SBN) thin films on Pt coated MgO substrates: the determining control of platinum crystallographic orientation. *Journal of Materials Science,* Vol. 41, (2006), pp. 5302-5309

Cuniot-Ponsard, M; Desvignes, J.M.; Bellemain, A & Bridou, F (2011). Simultaneous characterization of the electro-optic, converse-piezoelectric, and electro-absorptive effects in epitaxial (Sr, Ba)Nb_2O_6 (SBN) thin films.. *Journal of Applied Physics*, Vol. 109, 014107 (January 2011), pp. 1-11.

Francombe, M.H. (1960). . *Acta Cristallographica*, Vol. 13, (1960), pp. 131

Glass, A.M. (1969). Investigation of the electrical properties of $Sr_{1-x}Ba_xNb_2O_6$ with special reference to pyroelectric detection. *Journal of Applied Physics*, Vol. 40, No. 12, (November 1969), pp. 4699-4713

Guerra, J.de Los S.; Mendes, R.G. ; Eiras, J.A. ; Santos, I.A. & Araujo, E.B. (2008). Investigation of nonlinear dielectric properties in $Sr_{0.75}Ba_{0.25}Nb_2O_6$ relaxor ferroelectric thin films. *Journal of Applied Physics*, Vol.103 (2008), 014102, pp.1-8.

Jamieson; P.B.; Abrahams, S.C. & Bernstein, J.L. (1968). Ferroellectric tungsten bronze-type structures. I. Barium strontium niobate $Ba_{0.27}$ $Sr_{0.75}$ Nb_2 $O_{5.78}$. *The Journal of Chemical Physics*, Vol. 48, No. 11, (June 1968), pp. 5048-5057

Koo,J.; Jang, J.H.; Bae, B-S. (2000a). Optical and electrical properties of ferroelectric SBN thin films prepared by sol-gel process. *Journal of Sol-Gel Science and Technology*, Vol. 19, (2000), pp. 611-614

Koo,J.; Changho, L;Jang, J.H.; No, K. & Bae, B-S. (2000b). Measurement of the linear electro-optic coefficients of sol-gel derived strontium barium niobate thin films using a two beam polarization interferometer. *Applied Physics Letters*, Vol. 76, No. 19 (May 2000), pp. 2671-2673

Lairson, B.M.; Visokay, M.R; Sinclair, R; Hagstrom, S & Clemens, B.M. (1992). Epitaxial Pt(001), Pt(110), and Pt(111) films on MgO(001, MgO(110), MgO(111), and Al_2O_3(0001). *Applied Physics Letters* Vol. 61, No. 12 , (September 1992),pp. 1390-1392

Lenzo, P.V.; Spencer, E.G. & Ballman, A.A. (1967). Electro-optic coefficients of ferroelectric strontium barium niobate. *Applied Physics Letters*, Vol. 11, No. 1, (July 1967), pp. 23-24.

Li, X.T.; Du, P.Y.; Ye, H; Mak, C.L. & Wong, K.H. (2008). Electro-optic properties of epitaxial $Sr_{0.6}Ba_{0.4}Nb_2O_6$ filmsgrown on MgO substrates using $Li_xNi_{2-x}O$ buffer layer. *Applied Physics A*, Vol. 92 (2008), pp. 397-400.

Lin, W-J., Tseng, T-U.; Lin, S-P.; Tu, S-L.; Yang, S-J; Harn, J-J; Liu, K-S. & Lin, I-N. (1995). Growth of epitaxial-like $(Sr_{0.5}Ba_{0.5})Nb_2O_6$ ferroelectric films. *Japanese Journal of Applied Physics*,Vol. 34 (1995), pp. L625-L627.

Lu, Z; Feigelson, R.K.; Route, R.K. ; Hiskes, R. & Dicarolis, S.A. (1994). Growth of (001)-oriented SBN thin films by solid source MOCVD. *Materials Research Society Proceedings*, Vol. 335 (1994), pp. 59-64

McIntyre, P.C.;Maggiore, C.J. & Nastasi, M (1995). Orientation selection in thin platinum films on (001) MgO. *Journal of Applied Physics*, Vol. 77, No. 12 (June 1995), pp. 6201-6204

McIntyre, P.C.;Maggiore, C.J. & Nastasi, M (1997). Epitaxy of Pt thin films on (001) MgO-I. Interface energetic and misfit accommodation. *Acta Materials*, Vol. 45, No. 2 (1997), pp. 869-878.

Moon S.E.; Kwak,M.H.; Kim, Y-T.; Ryu, H-C.; Lee, S-J. & Kang, K-Y. (2004). Measurement of microwave dielectric properties of (Sr, Ba) Nb_2O_6 thin films. *Integrated Ferrolectrics*, Vol. 66 (2004), pp. 275-281

Moon S.E.; Kwak,M.H.; Kim, Y-T.; Ryu, H-C.; Lee, S-J. & Kang, K-Y. (2005). Comparison of microwave performance for ferroelectric phase shifters based on (001) oriented (Ba, Sr) TiO_3 and (Sr, Ba) Nb_2O_6 thin films. *Journal of the Korean Physical* Society, Vol. 46, No. 1 (January 2005), pp. 273-276

Narayan, J; Tiwari, P; Jagannadham, K & Holland, O.W. (1994). Formation of epitaxial and textured platinum films on ceramics-(100) MgO single crystals by pulsed laser deposition. *Applied Physics Letters,* Vol. 64, No. 16 (April 1994), pp. 2093-2095

Nozaka, T.; Mizutani, Y.; Gun, B.; Echizen, M.; Nishida, T., Takeda, H.; Uchiyama, K. & Shiosaki, T. (2008). Effect of interface structure on electrical properties of (Ba, Sr)TiO_3 thin films on glazed alumina substrate. *Japanese Journal of Applied Physics,* Vol. 47, No. 9 (2008), pp. 7494-7499.

Sakamoto, W.; Yogo, T; Kikuta, K.; Ogiso, K; Kawase, A. & Hirano, S. (1996). Synthesis of strontium barium niobate thin films through metal alkoxide. *Journal of American Ceramic Society,* Vol.79, No. 9 (1996), pp. 2283-2288

Schwyn Thöny, S.: Youde, K.E., Harris, J.S., Hesselink, Jr. & L. (1994). Growth of epitaxial strontium barium niobate thin films by pulsed laser deposition. *Applied Physics Letters* Vol. 65, No. 16, (October 1994), pp. 2018-2020

Trivedi,D; Tayebati, P. & Tabat, M. (1996). Measurement of large electro-optic coefficients in thin films of strontium barium niobate ($Sr_{0.6}Ba_{0.4}Nb_2O_6$). *Applied Physics Letters,* Vol.68, No. 23 (June 1996),pp.3227-3229

Venturini, E.L.; Spencer, E.G.; Lenzo, P.V. & Ballman, A.A. (1968). Refractive indices of strontium barium niobate. *Journal of Applied Physics* Vol. 39, (1968), pp. 343-344

Willmott, P.R., Herger, R., Patterson, B.D., Windiks, R. (2005). Experimental and theoretical study of the strong dependence of the microstructural properties of Sr_x Ba_{1-x} Nb_2 O_6 thin films as function of their composition. *Physical Review B,* Vol. 71 (April 2005), 144114.

Zhang, H.Y.; He, X.H. & Shih, Y.H. (1991). A new method for measuring the electro-optic coefficients with higher sensitivity anf higher accuracy. *Optics Communications,* Vol. 86 (1991), pp. 509-512

Zhao, J; Li, Y; Liu, X;Zhang,H & Wang, B. (2004). Preparation and electrical properties of SBN thin films derived from aqueous organic gels. *Materials Letters,* Vol. 58 (2004), pp. 1456-1460

Zhu, L.D.; Zhao, J.; Wang, F.; Norris, m95). Epitaxial electro-optical $Sr_xBa_{1-x}Nb_2O_6$ films by single source pasma-enhanced metalorganic chemical vapor deposition. *Applied Physics Letters,* Vol. 67, No. 13 (September 1995), pp. 1836-1838.